INTRODUCTION TO SYSTEM ANALYSIS

McGraw-Hill Series in Electrical Engineering

Consulting Editor
Stephen W. Director, Carnegie-Mellon University

Networks and Systems
Communications and Information Theory
Control Theory
Electronics and Electronic Circuits
Power and Energy
Electromagnetics
Computer Engineering
Introductory and Survey
Radio, Television, Radar, and Antennas

Previous Consulting Editors

Ronald M. Bracewell, Colin Cherry, James F. Gibbons, Willis W. Harman, Hubert Heffner, Edward W. Herold, John G. Linvill, Simon Ramo, Ronald A. Rohrer, Anthony E. Siegman, Charles Susskind, Frederick E. Terman, John G. Truxal, Ernst Weber, and John R. Whinnery

Networks and Systems

Consulting Editor
Stephen W. Director, Carnegie-Mellon University

Antoniou: *Digital Filters: Analysis and Design*
Belove and Drossman: *Systems and Circuits for Electrical Engineering Technology*
Belove, Schachter, and Schilling: *Digital and Analog Systems, Circuits, and Devices*
Bracewell: *The Fourier Transform and Its Applications*
Cannon: *Dynamics of Physical Systems*
Chirlian: *Basic Network Theory*
Desoer and Kuh: *Basic Circuit Theory*
Fitzgerald, Higginbotham, and Grabel: *Basic Electrical Engineering*
Glisson: *Introduction to System Analysis*
Hammond and Gehmlich: *Electrical Engineering*
Hayt and Kemmerly: *Engineering Circuit Analysis*
Hilburn and Johnson: *Manual of Active Filter Design*
Jong: *Methods of Discrete Signal and System Analysis*
Liu and Liu: *Linear Systems Analysis*
Papoulis: *The Fourier Integral and Its Application*
Peatman: *Design of Digital Systems*
Reid: *Linear System Fundamentals: Continuous and Discrete, Classic and Modern*
Sage: *Methodology for Large Scale Systems*
Schwartz and Friedland: *Linear Systems*
Temes and LaPatra: *Introduction to Circuit Synthesis*
Truxal: *Introductory System Engineering*

INTRODUCTION TO SYSTEM ANALYSIS

T. H. Glisson

Professor of Electrical and Computer Engineering
North Carolina State University

McGraw-Hill Book Company

New York St. Louis San Francisco Auckland Bogotá Hamburg
Johannesburg London Madrid Mexico Montreal New Delhi
Panama Paris São Paulo Singapore Sydney Tokyo Toronto

This book was set in Times Roman by Beacon Graphics Corporation.
The editors were Sanjeev Rao and Susan Hazlett;
the production supervisor was Phil Galea.
The drawings were done by ANCO/Boston.
The cover was designed by Rafael Hernandez.
Halliday Lithograph Corporation was printer and binder.

INTRODUCTION TO SYSTEM ANALYSIS

Copyright © 1985 by McGraw-Hill, Inc. All rights reserved. Printed in the United States of America. Except as permitted under the United States Copyright Act of 1976, no part of this publication may be reproduced or distributed in any form or by any means, or stored in a data base or retrieval system, without the prior written permission of the publisher.

234567890HALHAL898765

ISBN 0-07-023391-8

Library of Congress Cataloging in Publication Data

Glisson, T. H.
    Introduction to system analysis.

    (McGraw-Hill series in electrical engineering.
Networks and systems)
    Includes index.
    1. Electric network analysis.   2. System analysis.
I. Title.   II. Series.
TK454.2.G57   1985   621.319'2   84-10055
ISBN 0-07-023391-8

To Robin and Jack

CONTENTS

		Preface	xiii
Chapter 1		**Elements of System Analysis**	1
	1-1	Introduction	1
	1-2	Elementary Signals	4
	1-3	Elementary Systems	9
	1-4	Block Diagrams	20
	1-5	Fundamental Concepts	25
		Summary	35
		Problems	38
Chapter 2		**Linear Stationary Systems**	51
	2-1	Definition of a Linear Stationary System	51
	2-2	Convolution	57
	2-3	Interpretation of Impulse Response	69
	2-4	Systems Described by Differential Equations	76
		Summary	99
		Problems	100
Chapter 3		**Response to Sinusoidal Excitation**	109
	3-1	Linear Stationary Systems	109
	3-2	Nonlinear Static Systems	127
	3-3	Introduction to Frequency-Domain Analysis	138
		Summary	159
		Problems	162

Chapter 4 Fourier Series, Fourier Integral, and Fourier Transformation   175

- 4-1 Fourier Series   175
- 4-2 Fourier Integral   190
- 4-3 Fourier Transformation   200
- 4-4 Frequency-Domain System Analysis   222
- 4-5 Applications of Frequency-Domain Analysis   244
- Summary   271
- Problems   274

Chapter 5 Laplace Transformation   295

- 5-1 Definition and Properties of the Laplace Transformation   295
- 5-2 Application to Linear Stationary Systems   305
- 5-3 Interpretation of a System Function   320
- 5-4 One-Sided Laplace Transformation   359
- Summary   372
- Problems   379

Chapter 6 Digital Signals and Systems   392

- 6-1 Analog-to-Digital and Digital-to-Analog Conversion   394
- 6-2 Digital Signals and Systems   420
- 6-3 Fundamental Concepts   431
- Summary   440
- Problems   441

Chapter 7 Linear Shift-Invariant Digital Systems   449

- 7-1 Definition of a Linear Shift-Invariant System   449
- 7-2 Convolution   454
- 7-3 Interpretation of Delta Response   465
- 7-4 Systems Described by Difference Equations   470
- 7-5 Response to Sinusoidal Excitation   479
- 7-6 Digital Filters   490
- Summary   504
- Problems   506

Chapter 8 z Transformation   516

- 8-1 Definition and Fundamental Properties   517
- 8-2 Application to Linear Shift-Invariant Systems   528
- 8-3 Interpreting a System Function   536
- 8-4 Design of Recursive Digital Filters by Bilinear Substitution   557
- Summary   567
- Problems   568

Chapter 9 Computer-Aided Analysis and Design   576

- 9-1 Introduction to Simulation Using CSMP   576
- 9-2 Linear Stationary Systems   602

9-3	Computer-Aided Fourier Analysis	616
9-4	Interactive Computer-Aided Analysis	635
	Summary	655
	Problems	656

Appendixes 664

A	Mathematical Formulas	664
B	Fourier Series	670
C	Fourier Transformation	675
D	Laplace Transformation	677
E	z Transformation	679
F	Outline of CSMP	681
G	References	684

Index 685

PREFACE

This is a textbook for a first course in system analysis in electrical engineering and computer engineering curricula. Prerequisites assumed by the book are integral calculus, a first course in electric circuits, and a working knowledge of complex algebra. Neither prior nor concurrent study of differential equations and operational calculus is required. Chapters 1 to 5 provide all prerequisites for successful study of virtually all popular undergraduate textbooks on communication, control, power systems, instrumentation, and signal processing. These chapters also provide a meaningful introduction to system analysis for those who choose to pursue other areas of study.

The book began in 1975 as a set of notes for a required junior course in system analysis at North Carolina State University. The course bridges the gap between a sophomore course in circuits and senior electives in communication, control, instrumentation, power systems, and signal processing. This book has been used in various stages of revision in that course by more than 2500 students and by five different instructors. Explanations, examples, and problems given in the book have been tested thoroughly in the classroom.

The aim is to present concepts in a straightforward manner and in logical order and to illustrate definitions, principles, and procedures with an ample number of examples. The book contains almost 300 examples, more than 600 figures, and more than 450 problems. Approximately equal emphasis is placed on drill problems, multistep problems that require use of two or more principles, and substantive problems that require application of principles in realistic settings. A few problems require slight extensions of concepts developed in the body of the text.

The book stresses what I believe to be the three central ideas in system analysis: (1) methodology of system analysis is based on representing large systems as interconnections of simpler subsystems and on representing complicated signals as combinations of simpler signals; (2) performance of systems usually is described with reference to outputs for certain test inputs and in terms of a few fundamental properties, (e.g., realizability, stability, fidelity, and sensitivity to parameter variations); (3) the objective of system analysis is to determine whether a proposed system will perform as intended and, if not, to determine why, so that appropriate modifications can be made. These ideas are introduced in Chap. 1 and emphasized throughout the remainder of the book.

The first five chapters treat analog (continuous-time) systems. Chapter 1 describes elementary signals, elementary systems, block diagrams, and fundamental properties of systems. Chapter 2 treats time-domain methods (impulse response and convolution) for linear stationary systems, Chap. 3 treats sinusoidally excited systems, Chap. 4 treats frequency-domain methods (Fourier series and the Fourier integral), and Chap. 5 treats the Laplace transformation (both two-sided and one-sided).

Chapters 6 to 8 treat digital (discrete-time) systems. Chapter 6 describes analog-to-digital and digital-to-analog conversion, elementary digital systems, and fundamental properties of digital systems. Chapter 7 describes time- and frequency-domain methods for analysis of linear shift-invariant digital systems, and Chap. 8 treats the z transformation. Chapters 7 and 8 also introduce digital-filter design as a practical application of principles and techniques of digital system analysis.

Chapter 9, which provides an introduction to simulation and computer-aided analysis, can be studied in parallel with Chaps. 1 to 8. Section 9-1 provides an introduction to simulation using CSMP. Subsequent sections describe more advanced applications of simulation and also treat several other topics in computer-aided analysis, including numerical convolution, finding characteristic roots, finding partial-fraction coefficients, using fast Fourier transforms, and designing digital filters. Emphasis is on using existing software rather than developing algorithms.

I chose to treat analog and digital systems separately for two reasons: (1) In many curricula, as in ours, there is only one required course in systems analysis. I have found it impossible to carry a parallel treatment of analog and digital systems far enough in one semester to provide much of value to students who do not take a second course in system analysis. (2) I have found from my experience that pedagogical advantages of a parallel treatment are more apparent than real. Indeed, a parallel treatment fosters more confusion than insight; students become so concerned with parallels between analog and digital systems that they lose sight of more important relations between time-domain, frequency-domain, and complex-plane descriptions of signals and systems.

I chose to treat analog systems first for three reasons: (1) Most engineering applications of digital systems are found in analog systems (e.g., control systems, telephone networks, sonars, radars, and instrumentation systems). Intelligent design of digital systems for those applications requires some knowledge of analog systems. (2) Treating analog systems first makes use of and reinforces material just learned (in circuits and calculus) before it is forgotten. (3) Students who take only one course in systems and subsequently specialize in other areas benefit more from studying differential equations, convolution integrals, the Fourier integral, and the Laplace transformation than by exposure to analogous discrete-time topics.

A few other comments on content and organization of the book are in order. State-space methods are omitted entirely because most juniors in electrical engineering have neither the background, the inclination, nor the need for a meaningful treatment of those methods. Simple nonlinear systems are treated at an appropriate level. Some exposure to nonlinear systems appears essential because virtually every practical system contains at least one nonlinear element and because study of nonlinear systems promotes deeper understanding of linearity and of limitations on linear models. Finally, signals and system parameters are treated throughout as dimensioned quantities. This makes the subject less abstract, simplifies practical application of principles, promotes insight, and gives students a powerful method for checking correctness of relations and reasonableness of numerical results.

I am truly grateful for the quantity and quality of help and encouragement given by students, colleagues, staff, reviewers, friends, and family. My wife, Robin, and my son, Jack, have been understanding beyond belief. My students have given encouragement and much constructive criticism. The present and two past heads of our department, Nino Masnari, Larry Monteith, and George Hoadley, have provided the best environment imaginable for me and my project. Larry Monteith, Russell Pimmell, and Kenneth Williams taught from my notes and made many valuable suggestions. Kenneth Williams wrote many of the computer programs used to produce figures and verify examples. He also read the manuscript carefully and eliminated many errors. Sande Maxim and Nancy Tyson typed many revisions of the manuscript with perfection and cheerfulness. I appreciate the helpful comments of several competent and conscientious reviewers, including Don Childers, Steve Director, David Fisher, Syed Nasar, Ronald Rohrer, Lee Rosenthal, Andy Sage, Ron Schaffer, and Michael Silevitch. I took their suggestions to heart, and the book is better for it. I also wish to express my thanks to L. E. Schoonmaker, who was a friend indeed; to Charley Black and Bud Flood, who taught me a lot; and to Andy Sage, who was an inspiration when inspiration was scarce. Finally, I owe special thanks to Sy Matthews, whose careful reading and thoughtful criticism are responsible for much of what is good about the style and pedagogy of the text.

T. H. Glisson

INTRODUCTION TO SYSTEM ANALYSIS

CHAPTER
ONE

ELEMENTS OF SYSTEM ANALYSIS

The principal ideas introduced in this chapter are definitions of signal and system, definitions of five important signal models, the concept of a transfer characteristic, definitions of several important elementary systems, use of block diagrams for describing systems, and five important properties of systems.

In studying this chapter the reader may find it helpful to think of system analysis as a generalization (or analog) of circuit analysis. Circuit analysis deals with relations between voltages and currents. System analysis deals with relations between signals, which may be voltages, currents, temperatures, pressures, or other physical quantities that vary with time. Circuits are described by circuit diagrams, which are interconnections of idealized circuit elements (resistance, capacitance, inductance, and sources). Systems are described by block diagrams, which are interconnections of idealized elementary systems. Objectives of circuit analysis are to obtain and interpret relations between voltages and currents in an electric circuit. Objectives of system analysis are to obtain and interpret relations between signals in a system.

1-1 INTRODUCTION

We define signal, system, and system analysis; we discuss objectives of system analysis; and we describe some conventions regarding dimensions, units, and notation used in this book.

1-1A Signals and Systems

A *signal* is a physical (measurable) quantity that varies with time.† Examples are voltage across terminals of an electric circuit and temperature at a point in space. A

---
†In general, a signal may be a function of time and position, e.g., voltage on a long transmission line. Also, a signal may be a vector, e.g., electric field strength near an antenna. In this book a signal is a scalar function of time.

2 INTRODUCTION TO SYSTEM ANALYSIS

signal is represented as a function of time t, for example, $x(t)$. A *system* is a cause-and-effect relation between two or more signals. Signals identified as causes are called *inputs* or *excitations*. Signals identified as effects are called *outputs* or *responses*. For example, an audio amplifier can be regarded as a system whose input is voltage across the phono terminals and whose output is voltage across the speaker terminals.

Systems are represented by *block diagrams* (Fig. 1-1). The box represents a system, arrows entering the box represent inputs (excitations), and arrows leaving the box represent outputs (responses). A mathematical description of the system (of the relations between inputs and outputs) often is given in the box. For example, an audio amplifier can be represented as shown in Fig. 1-2. An input (a voltage across the phono terminals) is denoted by $x(t)$, the corresponding output (the voltage across the speaker terminals) is denoted by $y(t)$, and the relation between input and output is given in the box as $y(t) = Kx(t)$, where K denotes gain of the amplifier.

1-1B System Analysis

System analysis is the separation of systems into components for further study, which usually consists of examining the influence of one or more components on system performance. For example, an audio system might be separated into three components, as shown in Fig. 1-3, where $v(t)$ is motion of the phonograph stylus, $w(t)$ is voltage at the phono terminals, $x(t)$ is voltage at the speaker terminals, and $y(t)$ is pressure at a point in front of the speaker. Further study might show that the speaker has the poorest performance of the three components, suggesting that the speaker must be improved or replaced if performance of the audio system is to be improved.

System analysis plays an essential role in designing systems for communication, process control, data acquisition and processing, power generation and distribution, and other applications. Since construction of such systems is costly, it is economically necessary to have some assurance that a proposed system will perform as intended before construction is begun. *The central problem of system analysis is to determine whether a proposed system will perform as intended, and if not, why not, so the design can be corrected.*

Performance of a system usually is specified in terms of the output of the system for one or more test inputs. Consequently, the problem of system analysis as described above has two parts: (1) calculating the output of a system for one or more test inputs and (2), more important, *interpreting* the result of that calculation in terms of performance of the system. Test inputs used for specifying performance of systems are described in Sec. 1-2. Properties referred to in describing performance of systems are described in Sec. 1-5.

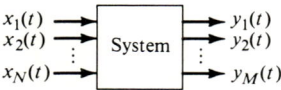

Figure 1-1 Block diagram for a system having inputs $x_1(t), x_2(t), \ldots, x_N(t)$ and outputs $y_1(t), y_2(t), \ldots, y_M(t)$.

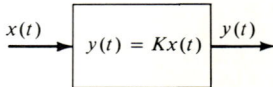

Figure 1-2 Block diagram for an audio amplifier.

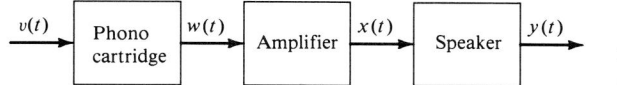

Figure 1-3 Block diagram for an audio system.

1-1C Dimensions, Units, and Notation

A signal is a measurable quantity. A signal has a dimension and is specified at any instant by a number *and a unit*. Otherwise, it may be impossible to use the signal in a meaningful calculation. For example, suppose we are told that the signal at the terminals of a loudspeaker is 5 cos ωt. What, exactly, have we been told? Not much; we do not know whether the signal is voltage, current, or gravitational field strength and we cannot, for example, calculate power delivered to the speaker even if we know the impedance of the speaker.

System parameters also are usually dimensioned quantities specified by a number *and a unit*. For example, the parameter K of a system whose output $y(t)$ for input $x(t)$ is given by $y(t) = Kx(t)$ must be expressed in the unit of $y(t)/x(t)$.

Dimensions and units are more than just extra baggage that must be carried along in order to get meaningful results. They provide powerful checks on correctness of relations and reasonableness of numerical values. A relation that is dimensionally incorrect is incorrect, period. A calculated current of 10^5 amperes through a loudspeaker is unreasonable, indicating that an error has been made in the calculation.

The SI system† of units is used in this book. Dimensions, SI units, and prefixes used in this book are given in Tables 1-1 and 1-2. Consistent use of certain symbols is helpful to students and practicing engineers alike. In this book we abide by certain conventions insofar as possible and reasonable. These conventions are pointed out where the need for them arises.

†The abbreviation SI is for the French Système Internationale d'Unités. An excellent discussion of dimensions and units is given in Kraus and Carver, chap. 1 and app. A-1. (References are listed in Appendix G.)

Table 1-1 Dimensions and SI units

Dimension or quantity	Name of SI unit	Symbol for SI unit	Dimension or quantity	Name of SI unit	Symbol for SI unit
Acceleration	meter/second²	m/s²	Inductance	henry	H
Angle	radian	rad	Length†	meter	m
Angular frequency	radian/second	rad/s	Mass†	kilogram	kg
Angular acceleration	radian/second²	rad/s²	Moment (torque)	newton-meter	Nm
Capacitance	farad	F	Power	watt	W
Charge	coulomb	C	Pressure	newton/meter²	N/m²
Current†	ampere	A	Resistance	ohm	Ω
Energy (work)	joule	J	Temperature†	Kelvin	K
Frequency	hertz	Hz	Time†	second	s
Force	newton	N	Velocity	meter/second	m/s
Impedance	ohm	Ω	Voltage	volt	V

†Fundamental unit in SI system.

4 INTRODUCTION TO SYSTEM ANALYSIS

Table 1-2 SI prefixes

Prefix	Abbreviation	Magnitude
giga	G	10^9
mega	M	10^6
kilo	k	10^3
milli	m	10^{-3}
micro	μ	10^{-6}
nano	n	10^{-9}
pico	p	10^{-12}

1-2 ELEMENTARY SIGNALS

In this section we define signal models for a step, a rectangular pulse, an impulse, a sinusoid, and an exponential pulse. These five elementary signals, particularly the step and the sinusoid, are used widely as test inputs for specifying system performance. They are also used in mathematical descriptions of more complicated signals.

1-2A Step

The *unit step function* is denoted by $u(\alpha)$ and defined by†

$$u(\alpha) = \begin{cases} 0 & \alpha \leq 0 \\ 1 & \alpha > 0 \end{cases} \qquad (1\text{-}1)$$

Figure 1-4 shows a graph of $u(\alpha)$ versus α. The *step signal* of Fig. 1-5 is described by

†The unit step function $u(\alpha)$ often is defined to be unity for $\alpha = 0$ and sometimes is defined to be $\frac{1}{2}$ for $\alpha = 0$. We prefer $u(0) = 0$ because this simplifies using step functions to describe signals that change abruptly (but continuously) in response to an event that occurs at a specified time, e.g., closing a switch at $t = 0$.

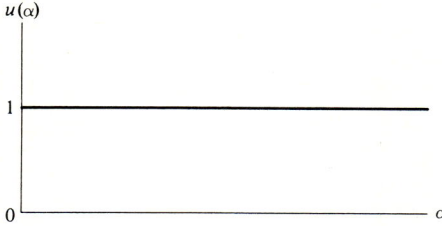

Figure 1-4 Unit step function.

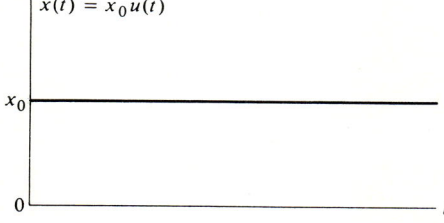

Figure 1-5 Step signal.

ELEMENTS OF SYSTEM ANALYSIS 5

$$x(t) = x_0 u(t) \qquad (1\text{-}2)$$

where t is time and x_0 is the *amplitude* of the step. Time t may be expressed in any convenient unit because the value of $u(t)$ depends only on whether t is positive or not. Amplitude x_0 is expressed in the unit of $x(t)$.

Example 1-1 In the circuit of Fig. 1-6 the switch is moved from contact A to contact B at $t = 0$. The voltage $v(t)$ is given by

$$v(t) = v_0 u(t)$$

where $v_0 = 5$ V.

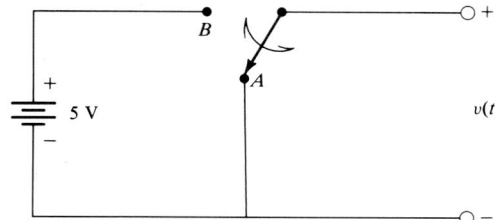

Figure 1-6 Circuit of Example 1-1.

1-2B Rectangular Pulse

The *rectangular function* is denoted by $r(\alpha)$ and defined by

$$r(\alpha) = \begin{cases} 1 & 0 < \alpha \le 1 \\ 0 & \text{otherwise} \end{cases} \qquad (1\text{-}3)$$

Figure 1-7 shows a graph of $r(\alpha)$ versus α. The *rectangular pulse* of Fig. 1-8 is described by

$$x(t) = x_0 r\left(\frac{t}{\tau}\right) \qquad (1\text{-}4)$$

where x_0 is the *amplitude* of the pulse and τ is the *duration* of the pulse. The unit of x_0 is the unit of $x(t)$, and the unit of τ is the unit of t.

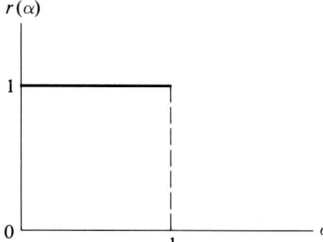

Figure 1-7 Rectangular function.

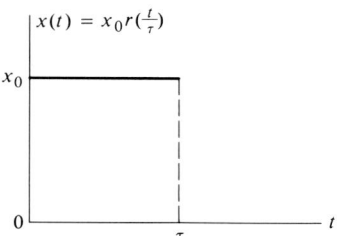

Figure 1-8 Rectangular pulse.

Example 1-2 In the circuit of Fig. 1-6 the switch is moved from contact A to contact B at $t = 0$ and from contact B back to contact A at $t = 10$ μs. The voltage $v(t)$ is given by

$$v(t) = v_0 r\left(\frac{t}{\tau}\right)$$

where $v_0 = 5$ V and $\tau = 10$ μs.

1-2C Impulse

The *delta function*† is denoted by $\delta(\alpha)$ and defined‡ by

$$\delta(\alpha) = \frac{du(\alpha)}{d\alpha} \tag{1-5}$$

where $u(\alpha)$ is the unit step function. The step function $u(\alpha)$ is dimensionless, and the operator $d/d\alpha$ has the dimension of α^{-1}; consequently the delta function $\delta(\alpha)$ has the dimension of α^{-1}.

The two most important properties of the delta function are expressed by the relations

$$\delta(\alpha) = 0, \quad \alpha \neq 0 \tag{1-6}$$

and

$$\int_{-\infty}^{\infty} \delta(\alpha)\, d\alpha = 1 \tag{1-7}$$

Equation (1-6) follows from (1-5) because the derivative (the slope) of the unit step function $u(\alpha)$ is zero except for $\alpha = 0$. Equation (1-7) is derived from (1-5) as follows:

$$\int_{-\infty}^{\infty} \delta(\alpha)\, d\alpha = \int_{-\infty}^{\infty} d[u(\alpha)] = u(\infty) - u(-\infty) = 1 - 0 = 1$$

According to (1-6) and (1-7), the delta function is nonzero at only a single point ($\alpha = 0$), yet it has unit area. This peculiar property will become more understandable in applications. For now, it is sufficient to think of the delta function as a spike having nearly zero width, nearly infinite amplitude, and unit area. The delta function is represented graphically by an arrow, as shown in Fig. 1-9.

An *impulse* $x(t)$ is described by

$$x(t) = \mathbf{a}\delta(t) \tag{1-8}$$

and is represented graphically as shown in Fig. 1-10. The quantity **a** in (1-8) is called the *strength* of the impulse $x(t)$. The unit of $\delta(t)$ is that of t^{-1}. The unit of strength **a** is that of $tx(t)$. For example, if $x(t)$ is voltage in volts and t is time in milliseconds, the appropriate unit for **a** is volt-milliseconds (Vms). The strength **a** is *not* the amplitude (height) of the impulse $x(t)$; it is the *area* bounded by the impulse and the time axis because, from (1-7),

---

†Also called the dirac delta, after the English physicist P. A. M. Dirac (1902–   ), and the unit impulse. We reserve the term "impulse" for a *signal* whose amplitude is given by a delta function of time t [see (1-8)].
‡This definition has been known to cause apoplexy in mathematicians. Nonetheless, its consequences are consistent with those of a rigorous treatment, and it allows us to avoid a great deal of tedious mathematics.

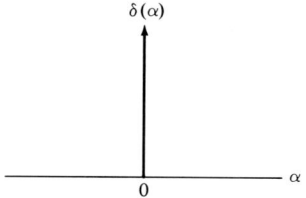

Figure 1-9 Graphical representation of a delta function.

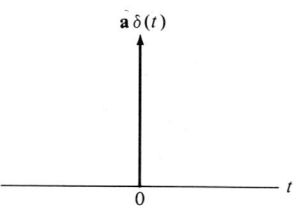

Figure 1-10 Graphical representation of an impulse.

$$\int_{-\infty}^{\infty} a\delta(t)\, dt = a \int_{-\infty}^{\infty} \delta(t)\, dt = a \tag{1-9}$$

Example 1-3 A billiard ball at rest is struck by a cue ball whose momentum before the collision is p_0. After the collision the cue ball is at rest. Describe the force on the cue ball as a function of time.

SOLUTION We assume that the collision is instantaneous and that it occurs at $t = 0$. The momentum of the cue ball is given by

$$p_c(t) = p_0[1 - u(t)]$$

By Newton's law† the force on the cue ball is given by

$$f_c(t) = \frac{dp_c(t)}{dt} = -p_0 \delta(t)$$

Note that this result is dimensionally correct because the dimension of momentum p_0 is force × time and the dimension of $\delta(t)$ is time^{-1}

1-2D Sinusoid

The *sinusoidal signal* of Fig. 1-11 is described by

$$x(t) = x_0 \cos \frac{2\pi t}{T} \tag{1-10}$$

†Force is rate of change of momentum; thus $f = dp/dt$, where f is force, p is momentum, and t is time. For $p = mv$, where m is a fixed mass and v is velocity, the relation $f = dp/dt$ becomes the familiar $f = ma$, where a is acceleration.

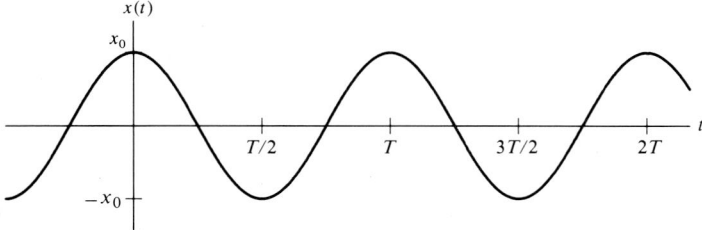

Figure 1-11 Sinusoidal signal described by (1-10).

where x_0 [unit of $x(t)$] is the *peak amplitude* of the sinusoid and T (unit of t) is the *period* of the sinusoid.

In engineering a sinusoid usually is described in terms of *frequency f*, defined by

$$f = \frac{1}{T} \tag{1-11}$$

or in terms of *angular frequency* ω, defined by

$$\omega = 2\pi f = \frac{2\pi}{T} \tag{1-12}$$

Thus, the sinusoid of (1-10) is described by

$$x(t) = x_0 \cos 2\pi f t \tag{1-13}$$

or by

$$x(t) = x_0 \cos \omega t \tag{1-14}$$

The SI unit of frequency f is the hertz (Hz), with 1 Hz = 1 s^{-1}. The SI unit of angular frequency ω is the radian per second with 2π rad/s = 1 Hz.

Example 1-4 The sinusoid of Fig. 1-12 is described by

$$i(t) = i_0 \cos 2\pi f t$$

where $f = 1/0.002 = 500$ Hz and $i_0 = 50$ mA.

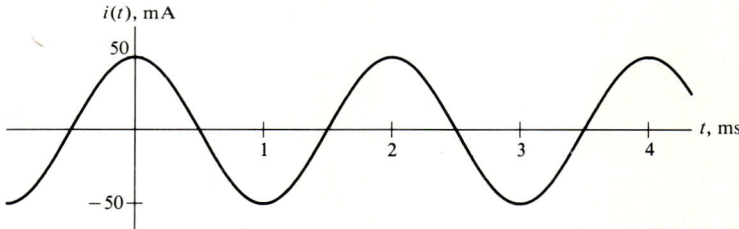

Figure 1-12 Sinusoidal signal of Example 1-4.

1-2E Exponential Pulse

The *exponential pulse* of Fig. 1-13 is described by

$$x(t) = x_0 e^{-t/\tau} u(t) \tag{1-15}$$

The quantity x_0 [unit of $x(t)$] is called the *initial amplitude* of the pulse. It is the amplitude of the exponential for $t = 0^+$.† The quantity τ (unit of t) is called the *time constant* of the exponential. In any interval of duration $n\tau$ the amplitude of the exponential pulse of (1-15) decreases by the factor e^{-n}; that is,

$$x(t + n\tau) = e^{-n} x(t) \tag{1-16}$$

†The symbol 0^+ (0^-) denotes a very small positive (negative) time; thus $x(0^+)$ is the value of $x(t)$ immediately after $t = 0$.

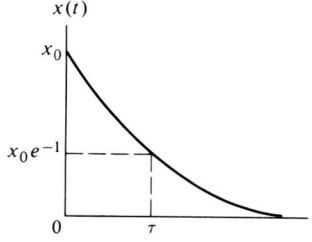

Figure 1-13 Exponential pulse described by (1-15).

In particular, for $t = 0^+$ and $n = 3$

$$x(3\tau) = e^{-3}x(0^+) \approx 0.05x_0$$

The amplitude of the pulse drops to about 5 percent of the initial amplitude in a time equal to three time constants. Often we regard three time constants as the duration of an exponential pulse even though the amplitude of the pulse is nonzero for all $t > 0$.

Example 1-5 The exponential pulse of Fig. 1-14 is described by

$$w(t) = w_0 e^{-t/\tau} u(t)$$

where $w_0 = 5$ mV and $\tau = 1$ ms.

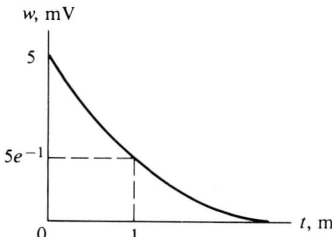

Figure 1-14 Exponential pulse of Example 1-5.

1-3 ELEMENTARY SYSTEMS

In this section we define several elementary systems. These systems, also called *elements*, are described by relatively simple input-output relations. Such elements in system analysis are analogous to resistors, capacitors, and other circuit elements in circuit analysis.

The relation between an input $x(t)$ and corresponding output $y(t)$ of an elementary system has the form

$$y(t) = \text{operation on } x(t) \qquad (1\text{-}17)$$

This relation is called the *transfer characteristic* of the element. For example, the transfer characteristic of the audio amplifier of Fig. 1-2 is $y(t) = Kx(t)$.

For organizational purposes we divide elementary systems into three classes: static elements (also called memoryless or instantaneous elements), dynamic elements, and arithmetic elements. A *static element* is one whose output at any instant depends only on the value of the input at the same instant. A *dynamic element* is one whose output at an instant t depends on values of the input at one or more instants different from t. An *arithmetic element* is an element that adds, subtracts, multiplies, or divides signals.

1-3A Static Elements

An element whose output $y(t)$ for input $x(t)$ is given by

$$y(t) = Kx(t) \tag{1-18}$$

is called a *proportion element*. A proportion element is a static element because the value of an output at time t depends only on the value of the corresponding input at the same time t. The parameter K of the transfer characteristic for a proportion element is called *gain*, regardless of its magnitude. The unit of gain K in (1-18) is the unit of $y(t)/x(t)$; for example, if $x(t)$ is voltage in volts and $y(t)$ is current in milliamperes, the appropriate unit for K is milliamperes per volt. An electronic amplifier with gain K, an ideal transformer with turns ratio K, and a gear train with gear ratio K are examples of components often modeled as proportion elements. A proportion element is represented by the symbol (block diagram) of Fig. 1-15.

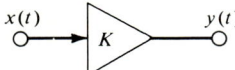

Figure 1-15 Symbol for a proportion element.

Example 1-6 For the gear train shown in Fig. 1-16a the gear ratio is N_1/N_2, where N_1 is the number of teeth on the primary gear and N_2 the number of teeth on the secondary gear. Instantaneous angular displacement of the secondary gear is given by

$$\theta_2 = \frac{N_1}{N_2} \theta_1$$

where θ_1 is instantaneous angular displacement of the primary gear. This relation has the form of (1-18). The gear train can be modeled as a proportion element having input θ_1, output θ_2, and gain N_1/N_2 rad/rad, as shown in Fig. 1-16b.

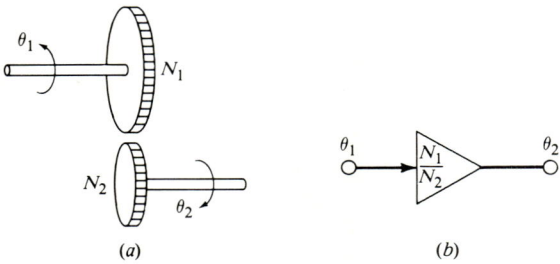

Figure 1-16 Gear train of Example 1-6: (*a*) schematic diagram, (*b*) block diagram.

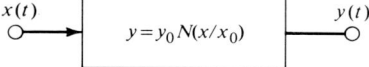

Figure 1-17 Symbol for a static element.

Transfer characteristics for many important static elements have the form

$$y(t) = y_0 N\left[\frac{x(t)}{x_0}\right] \quad (1\text{-}19)$$

where $x(t)$ is an input, $y(t)$ is the corresponding output, x_0 is a parameter having the unit of $x(t)$, y_0 is a parameter having the unit of $y(t)$, and N is a single-valued real function. Such an element is represented by the symbol of Fig. 1-17.

Example 1-7 The circuit diagram symbol for a semiconductor diode is shown in Fig. 1-18a. Forward current $i(t)$ is related to forward voltage $v(t)$ by

$$i(t) = i_0(e^{v(t)/v_0} - 1)$$

where v_0 has the unit of $v(t)$ and i_0 has the unit of $i(t)$. This relation has the form of (1-19) with x replaced by v, y replaced by i, and $N(\alpha) = e^\alpha - 1$. The diode is a static element having input $v(t)$ and output $i(t)$ given by the relation (transfer characteristic) above, as illustrated in Fig. 1-18b.

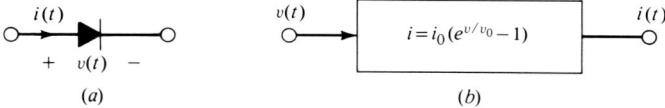

Figure 1-18 Diode of Example 1-7: (a) schematic diagram, (b) block diagram.

To carry this example a little further, we let $v_0 = 25$ mV, $i_0 = 1.0$ μA, and $v(t) = v_1 r(t/\tau)$, where $v_1 = 250$ mV. The corresponding output $i(t)$ is obtained as follows. For $t \leq 0$, $v(t) = 0$ and

$$i(t) = i_0(e^0 - 1) = 0 \quad t \leq 0$$

For $0 < t \leq \tau$, $v(t) = v_1$ and

$$i(t) = i_0(e^{v_1/v_0} - 1) = 22 \text{ mA} \quad 0 < t \leq \tau$$

For $t > \tau$, $v(t) = 0$ and $i(t) = 0$. The output is described for all time by

$$i(t) = 22r\left(\frac{t}{\tau}\right) \quad \text{mA}$$

Several static elements are common enough to deserve special mention. One of these is the proportion element. Others are the *soft limiter*, the *hard limiter*, the *half-wave rectifier*, the *full-wave rectifier*, the *comparator*, and the *square-law rectifier*, all of whose transfer characteristics are given in Table 1-3 (another important static element, called a quantizer, is described in Chap. 6).

12 INTRODUCTION TO SYSTEM ANALYSIS

Table 1-3 Transfer characteristics for six static elements

Element	Description	Graph
Soft limiter	$y = \begin{cases} -y_0 & x < -x_0 \\ y_0 x/x_0 & |x| \leq x_0 \\ y_0 & x > x_0 \end{cases}$	
Hard limiter	$y = \begin{cases} -y_0 & x < 0 \\ 0 & x = 0 \\ y_0 & x > 0 \end{cases}$	
Half-wave rectifier	$y = \begin{cases} 0 & x \leq 0 \\ y_0 x/x_0 & x > 0 \end{cases}$	
Full-wave rectifier	$y = y_0 \left| \dfrac{x}{x_0} \right|$	
Comparator	$y = \begin{cases} 0 & x \leq x_0 \\ y_0 & x > x_0 \end{cases}$	
Square-law rectifier	$y = y_0 \left(\dfrac{x}{x_0} \right)^2$	

Example 1-8 The parameters of a certain soft limiter are $x_0 = 2.5$ mV and $y_0 = 1.0$ V (see Table 1-3). Plot the output of the limiter for input

$$x(t) = x_1 \cos \omega_0 t$$

where $x_1 = 5.0$ mV and $f_0 = 1.0$ kHz.

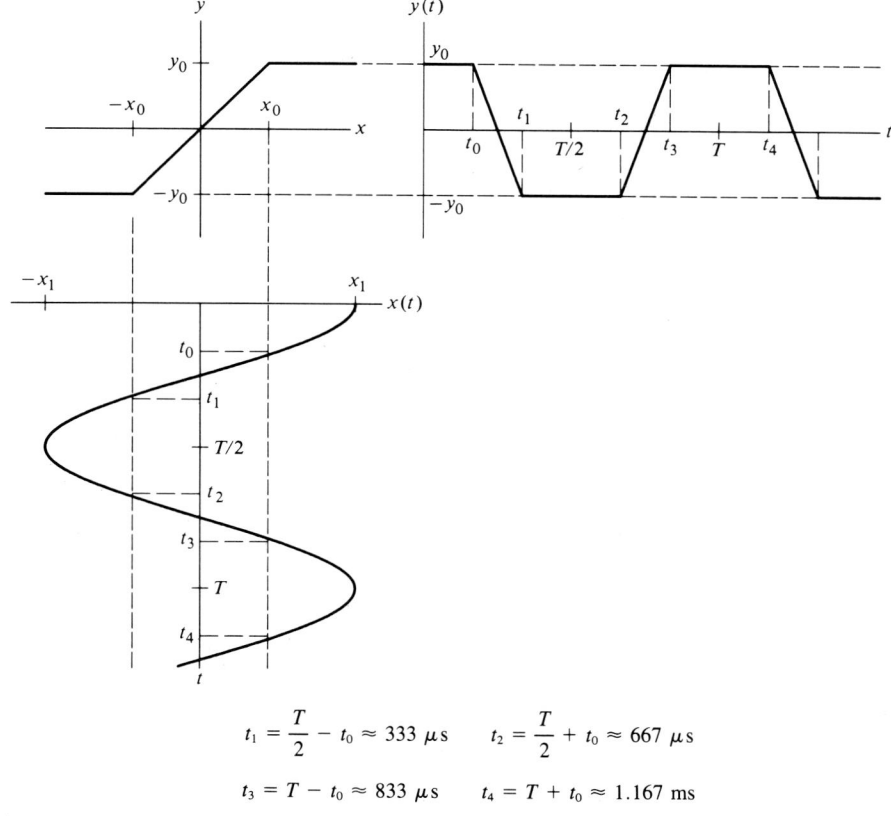

$$t_1 = \frac{T}{2} - t_0 \approx 333 \ \mu s \qquad t_2 = \frac{T}{2} + t_0 \approx 667 \ \mu s$$

$$t_3 = T - t_0 \approx 833 \ \mu s \qquad t_4 = T + t_0 \approx 1.167 \ ms$$

Figure 1-19 Input $x(t)$, transfer characteristic, and output $y(t)$ of the soft limiter of Example 1-8.

SOLUTION Refer to Fig. 1-19. The output $y(t)$ is limited to y_0 at time t_0 given (implicitly) by

$$x_1 \cos \omega_0 t_0 = x_0$$

which yields

$$t_0 = \omega_0^{-1} \cos^{-1} \frac{x_0}{x_1} = \frac{\pi}{3\omega_0} = \frac{1}{6f_0} = 167 \ \mu s$$

Other times at which limiting occurs are found using symmetry and periodicity of $x(t)$; this is illustrated in Fig. 1-19, which also shows a graph of the output $y(t)$.

1-3B Dynamic Elements

A *delay element* is a dynamic element whose transfer characteristic has the form

$$y(t) = x(t - t_0) \tag{1-20}$$

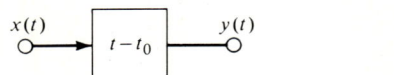
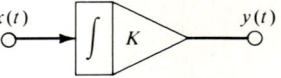

Figure 1-20 Symbol for a delay element. **Figure 1-21** Symbol for an integrator.

where $x(t)$ is an input, $y(t)$ is the corresponding output, and the quantity t_0 is called *delay time*. Input and output have the same unit, and delay time has the unit of time t. A delay element is a dynamic element because an output for time t depends on the input for time $t - t_0$ which (for $t_0 \neq 0$) is different from t. A delay element is represented by the symbol of Fig. 1-20.

Example 1-9 The output $y(t)$ of a delay element described by (1-20) for input

$$x(t) = x_0 e^{-t/\tau} u(t)$$

is given by

$$y(t) = x_0 e^{-(t-t_0)/\tau} u(t - t_0)$$

Some delay is inherent in all systems because signals propagate with finite speed. Propagation delay is significant in many systems; e.g., in radar and sonar systems propagation delay is used to measure distance to a target. Also, delay is often introduced intentionally; e.g., in many oscilloscopes a signal to be displayed is delayed after triggering the sweep so that the leading edges of fast pulses will be visible on the screen.

An *integration element* (an *integrator*) is a dynamic element whose transfer characteristic has the form

$$y(t) = K \int_{-\infty}^{t} x(\alpha) \, d\alpha \tag{1-21}$$

where $x(t)$ is an input, $y(t)$ is an output, and K is a parameter having the unit of y/xt. For example, if $x(t)$ is voltage in volts, $y(t)$ is current in milliamperes, and t is in seconds, the appropriate unit for K is milliamperes per volt-second [mA/(Vs)]. An integrator is represented by the symbol of Fig. 1-21.

Example 1-10 The output of an integrator described by (1-21) for step input

$$x(t) = x_0 u(t)$$

is given by

$$y(t) = K \int_{-\infty}^{t} x_0 u(\alpha) \, d\alpha$$

For $t \leq 0$, the integration is over only nonpositive values of α, where $u(\alpha) = 0$. Therefore,

$$y(t) = 0 \quad t \leq 0$$

For $t > 0$, the integration from $\alpha = 0$ to $\alpha = t$ is over positive values of α, where $u(\alpha) = 1$. Therefore,

$$y(t) = K \int_{0}^{t} x_0 \, d\alpha = K x_0 t \quad t > 0$$

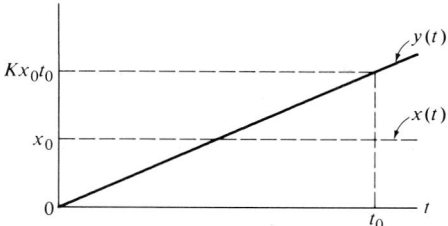

Figure 1-22 Output $y(t)$ of an integrator for step input $x(t) = x_0 u(t)$.

The output is described for all time by

$$y(t) = Kx_0 t u(t)$$

Figure 1-22 shows graphs of the input $x(t)$ and output $y(t)$.

Example 1-11 The output of an integrator described by (1-21) for input

$$x(t) = x_0 e^{-t/\tau} u(t)$$

is given by

$$y(t) = K \int_{-\infty}^{t} x_0 e^{-\alpha/\tau} u(\alpha)\, d\alpha$$

For $t \leq 0$, the integration is over only nonpositive values of α, where $u(\alpha) = 0$. Therefore,

$$y(t) = 0 \qquad t \leq 0$$

For $t > 0$, the integration from $\alpha = 0$ to $\alpha = t$ is over positive values of α, where $u(\alpha) = 1$. Therefore,

$$y(t) = K \int_{0}^{t} x_0 e^{-\alpha/\tau}\, d\alpha = K\tau x_0 (1 - e^{-t/\tau}) \qquad t > 0$$

The output is described for all time by

$$y(t) = K\tau x_0 (1 - e^{-t/\tau}) u(t)$$

Figure 1-23 shows graphs of the input $x(t)$ and the output $y(t)$.

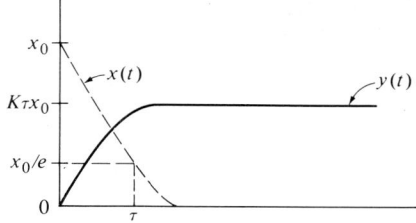

Figure 1-23 Output $y(t)$ of an integrator for input $x(t) = x_0 e^{-t/\tau} u(t)$.

Example 1-12 We wish to show that under certain simplifying assumptions the circuit of Fig. 1-24 is an electronic integrator having transfer characteristic

$$y(t) = K \int_{-\infty}^{t} x(\alpha)\, d\alpha \qquad \text{where } K = -\frac{1}{RC}$$

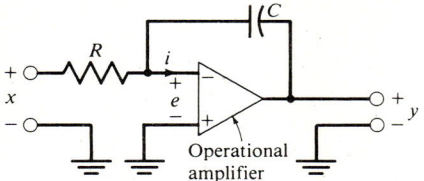

Figure 1-24 Electronic integrator of Example 1-12.

The simplifying assumptions referred to above are the following:

1. The gain of the operational amplifier is so large (typically 10^5 V/V) that $e(t)$ is negligibly small as long as $y(t)$ is within limits imposed by the power supply (typically $|y(t)| < 15$ V).
2. The input impedance of the operational amplifier is so large (typically 1 MΩ) that the current $i(t)$ drawn by the operational amplifier is negligibly small.

Writing Kirchhoff's current law for the node between the resistor and the capacitor gives

$$\frac{e(t) - x(t)}{R} + C\frac{d[e(t) - y(t)]}{dt} + i(t) = 0$$

Under the assumptions above, $e(t) \approx 0$, $i(t) \approx 0$, and the expression above simplifies to

$$\frac{dy(t)}{dt} = -\frac{1}{RC}x(t)$$

Integrating this relation gives

$$y(t) = -(RC)^{-1}\int_{-\infty}^{t} x(\alpha)\,d\alpha$$

as was to be shown.

A *differentiation* element (a *differentiator*) is a dynamic element whose transfer characteristic has the form

$$y(t) = K\frac{dx(t)}{dt} \tag{1-22}$$

where $x(t)$ is an input and $y(t)$ is the corresponding output. The parameter K has the unit of $t\,y(t)/x(t)$; for example, if $x(t)$ is voltage in volts, $y(t)$ is current in milliamperes, and t is in seconds, the appropriate unit for K is milliampere-seconds per volt (mAs/V). A differentiator is represented by the symbol of Fig. 1-25.

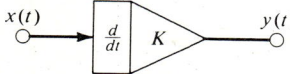

Figure 1-25 Symbol for a differentiator.

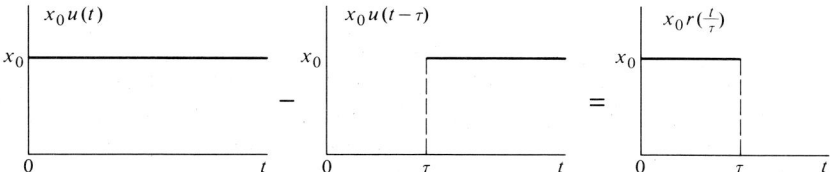

Figure 1-26 Representation of a rectangular pulse as a difference of two step signals.

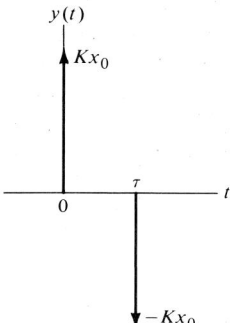

Figure 1-27 Output $y(t)$ of a differentiator for input $x(t) = x_0 r(t/\tau)$.

Example 1-13 Find the output of a differentiator described by (1-22) for input

$$x(t) = x_0 r\left(\frac{t}{\tau}\right)$$

Solution The rectangular pulse $x(t)$ can be expressed as the difference of two step signals (Fig. 1-26); thus

$$x(t) = x_0 r\left(\frac{t}{\tau}\right) = x_0[u(t) - u(t - \tau)]$$

Using (1-5) gives

$$y(t) = K x_0 [\delta(t) - \delta(t - \tau)]$$

Figure 1-27 shows a graph of the output $y(t)$.

Example 1-14 Find the output of a differentiator described by (1-22) for input

$$x(t) = x_0 e^{-t/\tau} u(t)$$

Solution Using the rule for differentiating a product gives

$$y(t) = K x_0 \frac{d}{dt}[e^{-t/\tau} u(t)] = K x_0 \left[e^{-t/\tau} \frac{du(t)}{dt} + \frac{de^{-t/\tau}}{dt} u(t) \right]$$

$$= K x_0 [e^{-t/\tau} \delta(t) - \tau^{-1} e^{-t/\tau} u(t)]$$

Since the delta function is zero for $t \ne 0$, we can replace the exponential by $e^0 = 1$ in the first term of the expression above. This gives

18 INTRODUCTION TO SYSTEM ANALYSIS

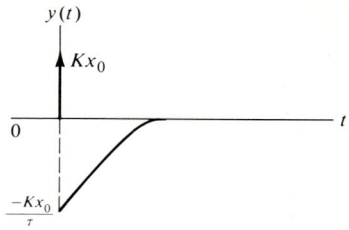

Figure 1-28 Output $y(t)$ of a differentiator for input $x(t) = x_0 e^{-t/\tau} u(t)$.

$$y(t) = Kx_0[\delta(t) - \tau^{-1}e^{-t/\tau}u(t)]$$

Figure 1-28 shows a graph of the output $y(t)$.

A *compression element* is a dynamic element whose transfer characteristic has the form

$$y(t) = x(\eta t) \qquad (1\text{-}23)$$

where $x(t)$ is an input, $y(t)$ is the corresponding output, and η is a dimensionless positive parameter called *compression ratio*. An input $x(t)$ and output $y(t)$ have the same unit. A compression element is represented by the symbol of Fig. 1-29.

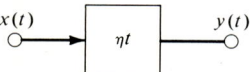

Figure 1-29 Symbol for a compression element.

Example 1-15 The output of a compression element described by (1-23) for input

$$x(t) = x_0 r\left(\frac{t}{\tau}\right)$$

(a rectangular pulse) is given by

$$y(t) = x(\eta t) = x_0 r\left(\frac{\eta t}{\tau}\right) = x_0 r\left(\frac{t}{\tau'}\right) \qquad \text{where } \tau' = \tau/\eta$$

The output is also a rectangular pulse, but with duration $\tau' = \tau/\eta$. Figure 1-30 shows graphs of the output $y(t)$ for $\eta > 1$ and $\eta < 1$. The duration of the output is smaller than that of the input for $\eta > 1$ (compression) and larger than that of the input for $\eta < 1$ (expansion).

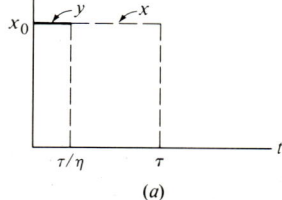
(a)

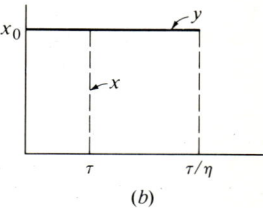
(b)

Figure 1-30 Output $y(t)$ of a compression element for input $x(t) = x_0 r(t/\tau)$ and for (a) $\eta > 1$ (compression) and (b) $0 < \eta < 1$ (expansion).

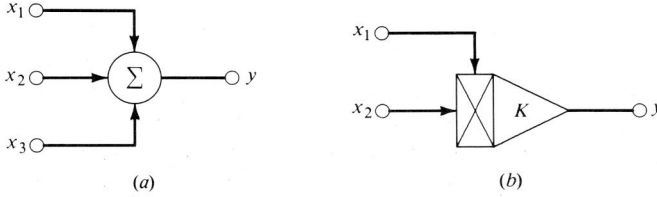

Figure 1-31 Symbols for (*a*) addition and (*b*) multiplication of signals.

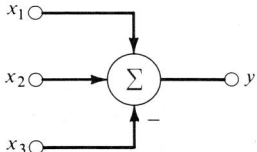

Figure 1-32 Symbol for $y(t) = x_1(t) + x_2(t) - x_3(t)$.

1-3C Addition and Multiplication of Signals

Figure 1-31 shows symbols for addition and multiplication of signals. The symbol for addition also is used to denote subtraction, in which case minus signs are shown beside the appropriate arrowheads, as illustrated in Fig. 1-32. Division of one signal by another is rarely encountered and is not treated in this book.

For an addition element (an adder), all inputs and the output have the same unit. For a multiplication element (a multiplier), inputs and output may have different units. In Fig. 1-31*b* the output is given by

$$y(t) = Kx_1(t)x_2(t)$$

The parameter K has the unit of $y(t)/x_1(t)x_2(t)$; for example, if $x_1(t)$ is current in milliamperes, $x_2(t)$ is voltage in volts, and $y(t)$ is voltage in millivolts, the appropriate unit for K is mV/(mAV) = A^{-1}.

Example 1-16 In Fig. 1-33 let

$$x_1(t) = 2r\left(\frac{t}{\tau}\right) \quad \text{V} \qquad x_2(t) = \cos 2\pi f_0 t \quad \text{V}$$

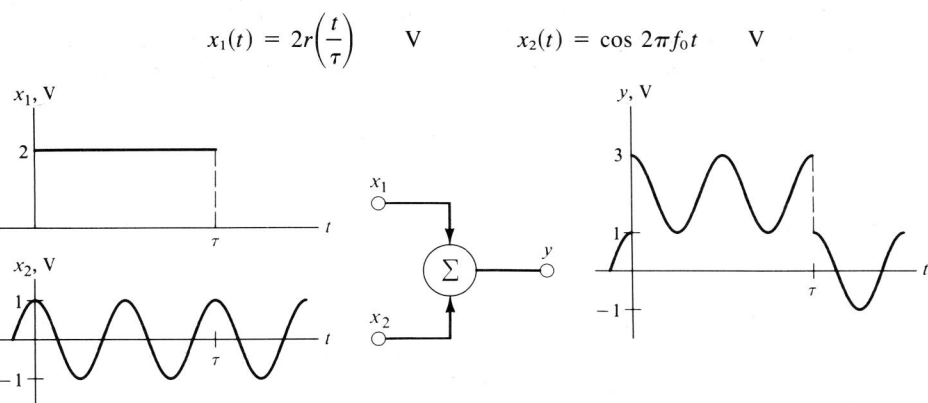

Figure 1-33 System of Example 1-16.

where $f_0 = 1/\tau$. The output $y(t)$ is given by

$$y(t) = x_1(t) + x_2(t) = 2r\left(\frac{t}{\tau}\right) + \cos 2\pi f_0 t \qquad \text{V}$$

Example 1-17 In Fig. 1-34 $K = 1 \text{ V}^{-1}$, and the inputs $x_1(t)$ and $x_2(t)$ are as defined in Example 1-16. The output $y(t)$ is given by

$$y(t) = 2(\cos 2\pi f_0 t)r\left(\frac{t}{\tau}\right) \qquad \text{V}$$

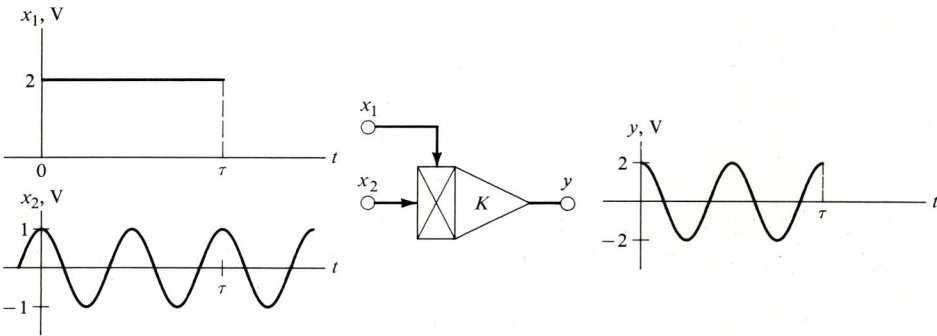

Figure 1-34 System of Example 1-17.

1-4 BLOCK DIAGRAMS

Elementary systems defined above can be connected together to form series, parallel, and feedback systems. In this section we describe these three connections and the forms of corresponding transfer characteristics.

1-4A Series Connections

Figure 1-35 shows examples of *series connections* (also called *cascade* or *tandem connections*). The output of one element is the input to the next. We can obtain the transfer characteristic (or an output) of a series connection by treating each element in turn, as illustrated in the following examples.

Example 1-18 We can obtain the output $z(t)$ of the system of Fig. 1-35a for input $x(t) = x_0 r(t/\tau)$ as follows. First we find the output of the first (proportion) element

$$y(t) = Kx(t) = Kx_0 r\left(\frac{t}{\tau}\right)$$

This is the input to the second (delay) element. Then the output of the delay element (the output of the system) is

$$z(t) = y(t - t_0) = Kx_0 r\left(\frac{t - t_0}{\tau}\right)$$

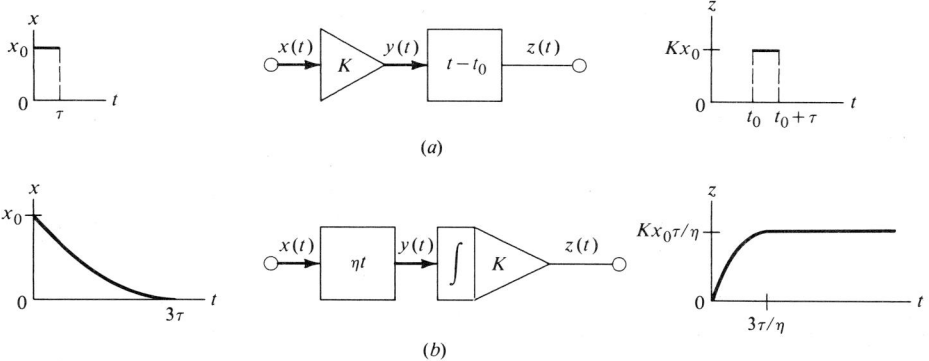

Figure 1-35 Systems in series (see Examples 1-18 and 1-19).

Example 1-19 We can obtain the output $z(t)$ of the system of Fig. 1-35b for input $x(t) = x_0 e^{-t/\tau} u(t)$ as follows. First we find the output of the compression element

$$y(t) = x(\eta t) = x_0 e^{-\eta t/\tau} u(\eta t) = x_0 e^{-t/\tau'} u(t)$$

where $\tau' = \tau/\eta$ and we have used the fact that (for $\eta > 0$) $u(\eta t) = u(t)$. This is the input to the integrator. Then the output of the integrator (the output of the system) is

$$z(t) = K \int_{-\infty}^{t} y(\alpha)\, d\alpha = K x_0 \tau'(1 - e^{-t/\tau'}) u(t)$$

where we have used the result of Example 1-11.

1-4B Parallel Connections

A parallel connection results when two or more operations are performed on an input and the resulting signals are applied to an element that accepts multiple inputs. Addition and multiplication elements allow parallel connections, as illustrated in the next two examples.

Example 1-20 In Fig. 1-36 let $x(t) = x_0 u(t)$. The inputs to the multiplier are

$$v_1(t) = x(t) \qquad v_2(t) = x(t - t_0)$$

The output is

$$w(t) = K v_1(t) v_2(t) = K x(t) x(t - t_0) = K x_0^2 u(t) u(t - t_0) = K x_0^2 u(t - t_0)$$

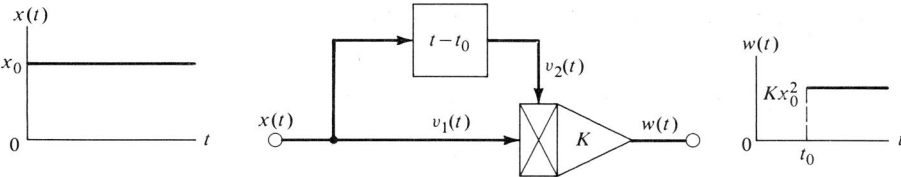

Figure 1-36 Systems in parallel (see Example 1-20).

22 INTRODUCTION TO SYSTEM ANALYSIS

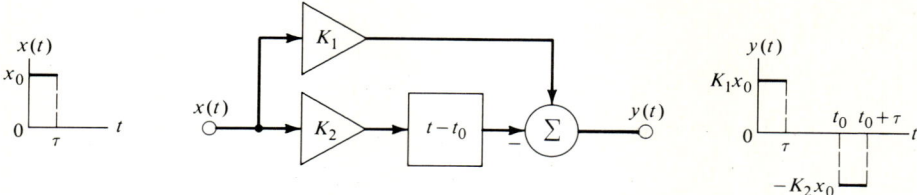

Figure 1-37 Systems in parallel (see Example 1-21).

Example 1-21 In Fig. 1-37 let $x(t) = x_0 r(t/\tau)$. The output is

$$y(t) = K_1 x(t) - K_2 x(t - t_0) = K_1 x_0 r\left(\frac{t}{\tau}\right) - K_2 x_0 r\left(\frac{t - t_0}{\tau}\right)$$

1-4C Feedback Connections

A feedback connection is one in which the input of at least one element is obtained from a point downstream of the element, i.e., from the output of the element or from the output of any other element closer to the overall system output. Figure 1-38 shows typical feedback connections. The next example illustrates a recursive method for finding an output of a feedback system containing only a proportion element, a delay element, and an adder.

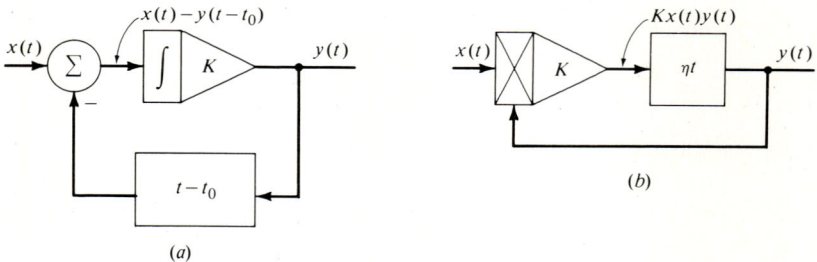

Figure 1-38 Examples of feedback systems.

Example 1-22 We wish to obtain the output $y(t)$ of the system of Fig. 1-39 for input $x(t) = x_0 r(t/\tau)$, where $\tau < t_0$. we assume that $y(t) = 0$ for $t < 0$.

From Fig. 1-39, an input $x(t)$ and corresponding output $y(t)$ are related by

$$y(t) = x(t) + Ky(t - t_0) \tag{1-24}$$

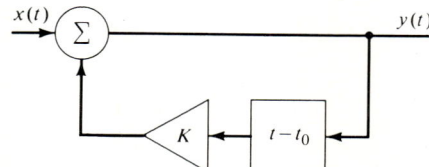

Figure 1-39 Feedback system of Example 1-22.

ELEMENTS OF SYSTEM ANALYSIS 23

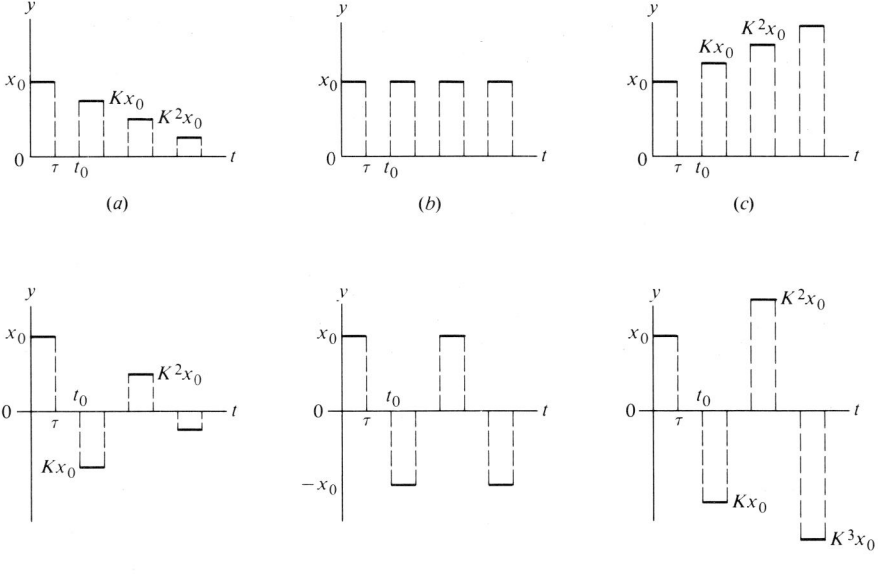

Figure 1-40 Output of the system of Fig. 1-39 for input $x(t) = x_0 r(t/\tau)$, where $\tau < t_0$ and (a) $0 < K < 1$, (b) $K = 1$, (c) $K > 1$, (d) $-1 < K < 0$, (e) $K = -1$, (f) $K < -1$.

This relation contains $y(t)$ and $y(t - t_0)$. We seek a relation (a transfer characteristic) of the form

$$y(t) = \text{operations on } x(t)$$

Such a relation can be obtained by repeated use of (1-24), as follows. Replacing t by $t - t_0$ in (1-24) gives

$$y(t - t_0) = x(t - t_0) + Ky(t - 2t_0) \qquad (1\text{-}25a)$$

Similarly,

$$y(t - 2t_0) = x(t - 2t_0) + Ky(t - 3t_0) \qquad (1\text{-}25b)$$

and so on. Using (1-25a) to eliminate $y(t - t_0)$ from (1-24) gives

$$y(t) = x(t) + Kx(t - t_0) + K^2 y(t - 2t_0) \qquad (1\text{-}26)$$

Using (1-25b) to eliminate $y(t - 2t_0)$ from (1-26) gives

$$y(t) = x(t) + Kx(t - t_0) + K^2 x(t - 2t_0) + K^3 y(t - 3t_0) \qquad (1\text{-}27)$$

Continuing in this manner, we find by induction that

$$y(t) = x(t) + Kx(t - t_0) + K^2 x(t - 2t_0) + K^3 x(t - 3t_0) + \cdots$$

This relation can be written compactly as

$$y(t) = \sum_{n=0}^{\infty} K^n x(t - nt_0) \qquad (1\text{-}28)$$

An output is a sum of scaled and delayed replicas of the input.
For input $x(t) = x_0 r(t/\tau)$, (1-28) gives

$$y(t) = \sum_{n=0}^{\infty} K^n x_0 r\left(\frac{t - nt_0}{\tau}\right)$$

Figure 1-40 shows graphs of this output for six different values of K.

1-4D Feedforward and Feedback Systems

A system consisting entirely of series and parallel connections is called a *feedforward system*. We can always obtain an explicit relation (a transfer characteristic) of the form

Output = operations on input

for a feedforward system. Thus, we can find an output of a feedforward system simply by performing a series of operations on the input. Obtaining an output of a feedback system is generally much more difficult, essentially because an input to at least one element in a feedback system depends in part on the output of that element. Consequently, the relation between input and output of a feedback system usually is implicit in the output; i.e., the relation between input and output of a feedback system usually has the form

Output = operations on input and output

To illustrate these remarks we consider the systems of Fig. 1-41. The system of Fig. 1-41a is a feedforward system whose transfer characteristic is

$$z(t) = y_0\left[\frac{x(t)}{x_0}\right]^2 + K \int_{-\infty}^{t} x(\alpha)\, d\alpha$$

This gives an output $z(t)$ explicitly (as the result of operations on the input alone). For the feedback system of Fig. 1-41b we find

$$z(t) = K \int_{-\infty}^{t} \left\{ x(\alpha) + y_0\left[\frac{z(\alpha)}{z_0}\right]^2 \right\} d\alpha$$

This gives an output implicitly (as the result of operations on the input and output). We cannot reduce this relation to an explicit expression for $z(t)$; that is, we cannot obtain a transfer characteristic for the system of Fig. 1-41b.

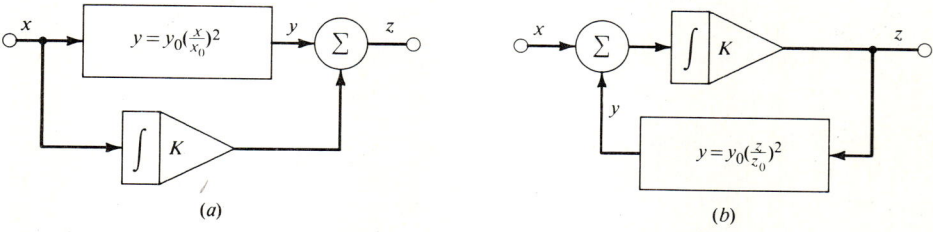

Figure 1-41 Examples of (a) a feedforward system and (b) a feedback system.

1-5 FUNDAMENTAL CONCEPTS

In this section we describe five fundamental properties of systems: realizability, stability, transient and steady-state response, fidelity, and sensitivity to parameter variations. Virtually all methods of describing or specifying system performance refer ultimately to one or more of these five properties. We refer to these properties often in the remainder of this book.

1-5A Realizability

A *realizable* transfer characteristic is one that describes a cause-and-effect relation, i.e., one for which the present value of an output (an effect) depends only on past and present values of the input (the cause). This idea can be expressed as follows. A realizable transfer characteristic gives a causal output for any input, where a causal output is one that does not anticipate the input that produces it.

Example 1-23 Consider the transfer characteristic

$$y(t) = x(t + t_0)$$

where $x(t)$ is an input, $y(t)$ is the corresponding output, and $t_0 > 0$. This transfer characteristic is nonrealizable because it gives a noncausal output for any input. For example, let

$$x(t) = x_0 r\left(\frac{t}{\tau}\right)$$

Figure 1-42 shows graphs of this input and the corresponding output. The output pulse is noncausal because it appears earlier than the input.

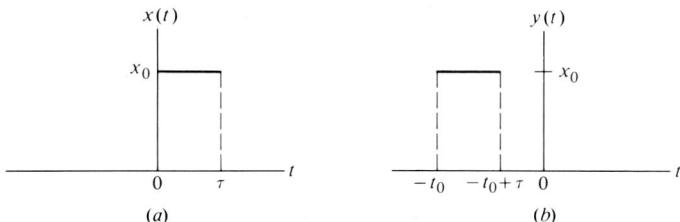

Figure 1-42 (*a*) Input $x(t)$ and (*b*) output $y(t)$ of the nonrealizable system of Example 1-23.

Example 1-24 The transfer characteristic for a compression element is

$$y(t) = x(\eta t)$$

where $x(t)$ is an input and $y(t)$ is the corresponding output. For $\eta > 1$, the transfer characteristic is nonrealizable. To show this we let $x(t) = x_0 r(t/\tau)$, for which

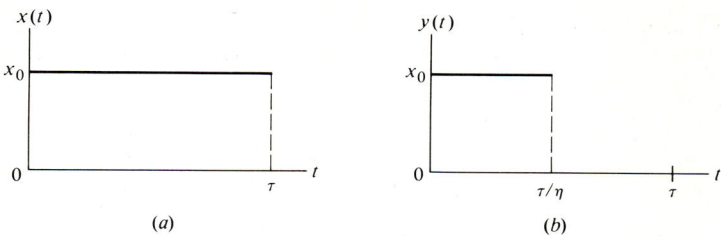

Figure 1-43 (*a*) Input $x(t)$ and (*b*) output $y(t)$ of the compression element of Example 1-24.

$$y(t) = x_0 r\left(\frac{t}{\tau'}\right)$$

with $\tau' = \tau/\eta$. Figure 1-43 shows graphs of this input and the corresponding output. The output pulse is noncausal because its trailing edge appears earlier than the trailing edge of the input. True compression implies prior knowledge of the duration of an input.

It is not always easy to find whether a transfer characteristic is realizable or not. A transfer characteristic may give causal outputs for some inputs and noncausal outputs for others. The best input to use to test for causality is a rectangular pulse. Others can give misleading results, as illustrated in the next example.

Example 1-25 The output of the compression element of Example 1-24 for input $x(t) = x_0 u(t)$ is

$$y(t) = x_0 u(\eta t)$$

where $\eta > 1$. This simplifies to

$$y(t) = x_0 u(t)$$

because the value of $u(\alpha)$ depends only on whether α is positive or not. This output is causal because it appears no earlier than the input that produces it. Without further study we might conclude wrongly that the compression element is realizable. Using a rectangular pulse for the test input leads to the correct conclusion, as shown in Example 1-24.

Some nonrealizable transfer characteristics are nonetheless desirable. For example, in radar (or sonar) it is desirable to transmit long pulses in order to get echoes that are sufficiently energetic for detection of small or distant targets; on the other hand, it is desirable to receive short pulses in order to obtain a good estimate of distance to a target. A radar system can achieve both objectives by transmitting long pulses and compressing echoes. This requires a realizable compression element.

Often, a realizable transfer characteristic can be obtained from a nonrealizable one by incorporating delay. An example is using a tape recorder to achieve compression. We can record a signal at 9.5 cm/s, play back the recorded signal at 19 cm/s, and thereby achieve a 2-to-1 ($\eta = 2$) compression of the signal. The delay necessary for realizing this particular compression is the time between the beginning of the recording and the beginning of the playback. The necessary delay increases with the duration of a signal to be compressed. In general, introducing delay can circumvent non-realizability only for a limited class of inputs; e.g., a tape recorder can compress only

ELEMENTS OF SYSTEM ANALYSIS **27**

signals having durations less than or equal to the maximum possible recording time (determined by recording speed and tape length).

Example 1-26 To illustrate how introducing delay can make a nonrealizable transfer characteristic realizable for a limited class of inputs we consider the system of Fig. 1-44, where $\eta > 1$ (compression). We wish to determine a condition on t_0 such that the system can achieve compression (with delay) of a rectangular pulse

$$x(t) = x_0 r\left(\frac{t}{\tau}\right)$$

The output for input $x(t)$ above is given by

$$y(t) = x_0 r\left[\frac{\eta(t - t_0)}{\tau}\right] = x_0 r\left[\frac{t - t_0}{\tau'}\right] \quad \text{where } \tau' = \tau/\eta$$

The trailing (rightmost) edge of the output occurs for $(t - t_0)/\tau' = 1$ or for $t = t_0 + \tau/\eta$, which must equal or exceed τ (the time for which the trailing edge of the input occurs) if the system is to give a causal output for input $x(t)$ above. This means that delay t_0, compression ratio η, and pulse duration τ must satisfy

$$t_0 + \frac{\tau}{\eta} \geq \tau$$

which implies

$$t_0 \geq \tau - \frac{\tau}{\eta}$$

Figure 1-45a shows the input $x(t)$ and the output $y(t)$ for a case where $t_0 < \tau - \tau/\eta$. The trailing edge of the output occurs earlier than the trailing edge of the input (the output is

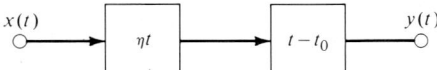

Figure 1-44 System of Example 1-26.

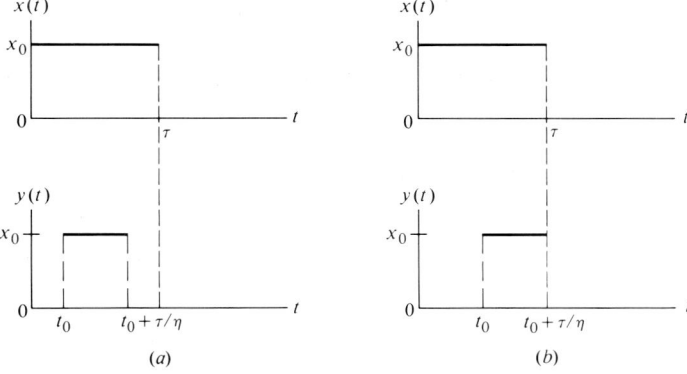

Figure 1-45 Output of the system of Example 1-26 for input $x(t) = x_0 r(t/\tau)$: (a) noncausal output, where $t_0 < \tau - \tau/\eta$, and (b) causal output, where $t_0 = \tau - \tau/\eta$.

noncausal). Figure 1-45b shows the input $x(t)$ and the output $y(t)$ for a case where $t_0 = \tau - \tau/\eta$. The trailing edge of the output and the trailing edge of the input occur at the same time (the output is causal). For $t_0 > \tau - \tau/\eta$, the trailing edge of the output occurs later than the trailing edge of the input.

A realizable transfer characteristic is merely one that does not violate our suppositions regarding cause and effect. A realizable transfer characteristic is not necessarily a feasible one because feasibility also depends on cost, size, legality, current technology, and other considerations. Nonetheless, realizability is certainly necessary. A systems engineer must be aware of this requirement, especially in using design procedures where an optimum transfer characteristic is derived mathematically from performance criteria without regard to causality. Often the result is a nonrealizable transfer characteristic which subsequently is made realizable (for inputs of interest) by introducing delay.

1-5B Transient and Steady-State Response

Any signal can be separated into a transient component, which vanishes as time t approaches infinity, and a steady-state component, which persists (does not vanish) as time t approaches infinity.

Example 1-27 Consider the signal

$$x(t) = x_0 u(t) - x_0 e^{-t/\tau} u(t) \qquad \text{where } \tau > 0$$

The second term, which vanishes as t approaches infinity, constitutes a transient component of $x(t)$. The first term, which persists as t approaches infinity, constitutes a steady-state component of $x(t)$.

In general, a signal does not have a unique transient component; e.g., the signal of Example 1-27 can be expressed as

$$x(t) = x_0 r\left(\frac{t}{\tau}\right) + x_0 u(t - \tau) - x_0 e^{-t/\tau} u(t)$$

in which case we might identify the first and the third terms as "the" transient components of the signal (because both vanish as t approaches infinity). In almost all applications where separation of a response into transient and steady-state components is useful, the steady-state component has the form of the associated input for large time t; for example, if the input is a step, the steady-state output is a constant and if the input is a sinusoid, the steady-state output is a sinusoid of the same frequency. Thus, we separate an output into transient and steady-state components by first obtaining the steady-state output; subsequently, we obtain the transient components by subtracting the steady-state output from the (total) output.

For many systems some specifications refer mainly to transient components and some refer mainly to steady-state components of an output (for a particular test input). For example, top speed of a car is a steady-state specification, whereas acceleration time from 0 to 60 mi/h is a transient specification (here the test input is a step increase in fuel flow). Separating an output into transient and steady-state components often helps distinguish those parameters of the system in question which influence mainly steady-state performance from those which influence mainly transient performance. For example, wind resistance may have considerable influence on the top speed of a car but little influence on the acceleration time from 0 to 60 mi/h. Conversely, the mass of a car has considerable influence on the acceleration time but little influence on top speed.

There also may be system parameters that have considerable influence on both steady-state and transient components of a response. It is helpful to identify those parameters which affect both components in the same way; e.g., increasing engine power improves both top speed and acceleration time of a car. It also is helpful to identify parameters that affect steady-state performance and transient performance in opposite ways; e.g., increasing transmission gear ratio may decrease the acceleration time (an improvement) and decrease top speed (a reduction in performance).

1-5C Stability

Essentially, a stable system is one whose output remains under control of the input at all times. For example, we may say that a public-address system is stable as long as the output of the system is an amplified version of a singer's voice. It can become unstable if the singer wanders (microphone in hand) too close to a loudspeaker; the ensuing loud squeal has an intensity and pitch which are unrelated to the intensity and pitch of the singer's voice.

There are several formal (mathematical) definitions of stable operation. The one used in this book, called the *bounded-input–bounded-output* (BIBO) definition, is based on the definition of a bounded signal.

A *bounded signal* is one whose magnitude never exceeds some preassigned (finite) value; i.e., a signal $x(t)$ is bounded if and only if there is a constant M_x such that

$$|x(t)| \leq M_x \qquad -\infty < t < \infty$$

Example 1-28 The signal

$$x(t) = x_0(\cos \omega t) u(t)$$

is bounded because $|x(t)| \leq x_0$ for all time t. In contrast, the signal

$$x(t) = x_0 \frac{t}{\tau} u(t)$$

is unbounded because for any preassigned constant M_x there is a time $t > M_x |\tau/y_0|$ for which $|x(t)|$ exceeds M_x. Figure 1-46 illustrates these results.

The BIBO stability condition defines a *stable system* as one that gives a bounded output for any bounded input, a *conditionally stable* system as one that gives bounded outputs for some (but not all) bounded inputs, and an *unstable system* as one that gives an unbounded output for every nonzero bounded input.

Example 1-29 A proportion element with input $x(t)$ and output $y(t) = Kx(t)$ is stable because the output is bounded for every bounded input.

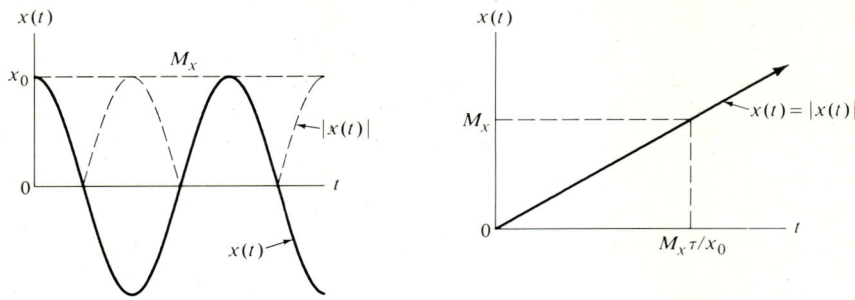

Figure 1-46 Bounded and unbounded signals of Example 1-28.

Example 1-30 An integrator is conditionally stable. The input

$$x(t) = x_0 e^{-t/\tau} u(t)$$

is bounded. The corresponding output

$$y(t) = K \int_{-\infty}^{t} x(\alpha)\, d\alpha = K\tau x_0 (1 - e^{-t/\tau}) u(t)$$

is also bounded. However, for the bounded input

$$x(t) = x_0 u(t)$$

the output

$$y(t) = K x_0 t u(t)$$

is unbounded, as shown in Example 1-28. These results show that an integrator gives bounded outputs for some bounded inputs and unbounded outputs for other bounded inputs. Therefore an integrator is conditionally stable.

Example 1-31 The system of Fig. 1-47 is unstable. Any nonzero input $w(t)$ causes the signal $x(t)$ to be positive for some time interval which in turn causes the output $y(t)$ of the first integrator to be positive for all time. Consequently, the output $z(t)$ of the second integrator increases monotonically with time. The output $z(t)$ for any nonzero input $w(t)$ is unbounded, and the system is unstable.

For a specific case, let $w(t) = w_0 r(t/\tau)$. We find

$$x(t) = x_0 r\left(\frac{t}{\tau}\right) \qquad y(t) = K_1 x_0 \left[tr\left(\frac{t}{\tau}\right) + \tau u(t - \tau) \right]$$

and

$$z(t) = K_1 K_2 x_0 \left[0.5 t^2 r\left(\frac{t}{\tau}\right) + (\tau t - 0.5\tau^2) u(t - \tau) \right]$$

as shown in Fig. 1-48.

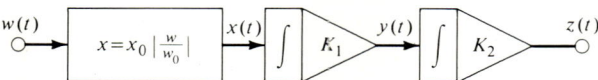

Figure 1-47 An unstable system.

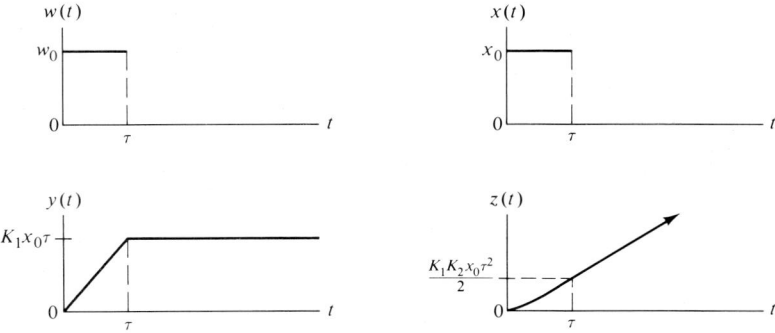

Figure 1-48 Output of the system of Fig. 1-47 for input $x(t) = x_0 r(t/\tau)$.

Stability is almost always important in analysis and design of feedback systems. In a feedback system a signal that is fed back may reinforce an input and cause the output to increase without bound (mathematically) or oscillate, as when a singer holding a microphone gets too close to a loudspeaker.

Example 1-32 For what values of K is the system of Example 1-22 at least conditionally stable (not unstable)?

SOLUTION Refer to Fig. 1-40. The output $y(t)$ is bounded for $|K| < 1$. Therefore the system is at least conditionally stable for $|K| < 1$.

To show that a system is stable (or unstable) we must show that *every* bounded input produces a bounded (or unbounded) output. This may be difficult or even impossible because it may be difficult or impossible to find a transfer characteristic (an explicit expression for an output) for the system. In this book we limit study of stability to feedforward systems, simple feedback systems that can be treated using recursive methods (as in Example 1-22), and the class of linear stationary systems treated in Chap. 2. In practice, extensive computer simulation is often used for studying stability.

1-5D Fidelity

Some systems are designed to scale or to transmit certain signals without introducing appreciable distortion, i.e., without causing any significant change in the waveforms of the signals. For example, an output of a high-fidelity recording system should sound like the original performance.

A system having input $x(t)$ and output $y(t)$ is said to provide *distortionless transmission* if the transfer characteristic of the system has the form

$$y(t) = Kx(t - t_0) \quad \text{where } K \neq 0$$
$$t_0 \geq 0 \qquad (1\text{-}29)$$

Such a system is equivalent (at its terminals) to a series connection of a delay element and a proportion element, as shown in Fig. 1-49.

32 INTRODUCTION TO SYSTEM ANALYSIS

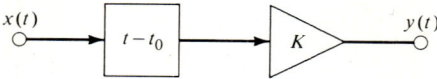

Figure 1-49 Block diagram for a distortionless system.

According to this definition, a system introduces no distortion if an output is proportional to a delayed replica of the input. This is intuitively reasonable; e.g., for a high-fidelity recording system neither delay between recording and playback nor loudness different from that of the original performance is perceived as distortion by a listener.

Instantaneous distortion (or instantaneous error) introduced by a system is defined as shown in Fig. 1-50. The signal $x(t)$ is an appropriate test input, $y_a(t)$ is the actual output (the output of a system under test), $y_i(t)$ is the ideal (distortionless) output, and $v(t)$ is instantaneous distortion.

Any measure of the size of instantaneous distortion is called a *distortion measure*. A distortion measure that is sometimes used is *integrated squared error* (ISE), defined by

$$\text{ISE} = \int_{-\infty}^{\infty} v^2(t)\, dt \qquad (1\text{-}30)$$

where $v(t)$ is instantaneous distortion.

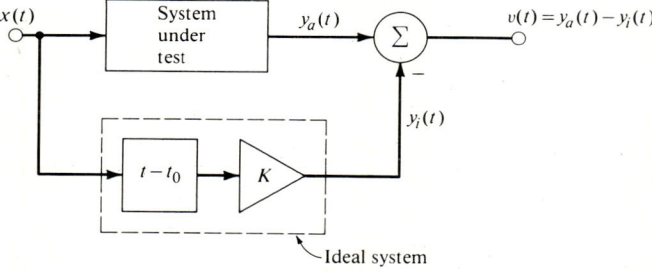

Figure 1-50 Definition of instantaneous distortion.

Example 1-33 The ideal transfer characteristic for a certain system is given by (1-29) with $K = 10$ and $t_0 = 0$. The actual output of the system for step input $x(t) = x_0 u(t)$ is

$$y_a(t) = 10x_0(1 - e^{-t/\tau})u(t)$$

Calculate the corresponding ISE.

SOLUTION From (1-29), the ideal output is

$$y_i(t) = 10x_0 u(t)$$

The instantaneous distortion is

$$v(t) = y_a(t) - y_i(t) = -10x_0 e^{-t/\tau} u(t)$$

From (1-30), the ISE is

$$\text{ISE} = \int_0^\infty (-10 x_0 e^{-t/\tau})^2 \, dt$$

$$= 100 x_0^2 \int_0^\infty e^{-2t/\tau} \, dt = 50 x_0^2 \tau$$

The ISE is proportional to the time constant τ. Decreasing the time constant τ decreases the ISE.

For a quantitative description of the distortion introduced by a system we must choose a distortion measure, e.g., ISE, and a test input and specify appropriate values for K and t_0 in (1-29). There are few general rules for these choices. We treat distortion further in subsequent chapters where more meaningful discussion is possible.

1-5E Sensitivity to Parameter Variations

Imperfect quality control, environmental effects, component aging, and imperfect modeling conspire to make parameters of an actual system different from their design values. An intelligent designer foresees this and takes steps to desensitize performance of a system to anticipated variations of parameters. The designer must be able to calculate sensitivity of performance to variations of each parameter.

Any sensible performance measure for a system, e.g., ISE, is a function of parameters of the system, e.g., gains. Let $p_1, p_2, \ldots, p_k$ denote design values for the parameters of a system and let ϕ_0 denote the corresponding (design) value of a performance measure for the system. Then

$$\phi_0 = \phi(p_1, p_2, \ldots, p_k)$$

Let $\Delta p_1, \Delta p_2, \ldots, \Delta p_k$ denote small deviations of the parameters from their design values. The corresponding deviation of the performance measure from its design value is given (approximately) by†

$$\Delta \phi \approx \frac{\partial \phi}{\partial p_1} \Delta p_1 + \frac{\partial \phi}{\partial p_2} \Delta p_2 + \cdots + \frac{\partial \phi}{\partial p_k} \Delta p_k$$

We consider the parameters one at a time. Let all parameters except p_n have their design values. Then $\Delta p_i = 0$ for $i \neq n$, and

$$\Delta \phi \approx \frac{\partial \phi}{\partial p_n} \Delta p_n$$

gives the change $\Delta \phi$ of the performance measure corresponding to a small change Δp_n in the parameter p_n alone. The fractional change $\Delta \phi / \phi_0$ is usually more meaningful than the actual change $\Delta \phi$. From the last expression above we obtain

$$\frac{\Delta \phi}{\phi_0} \approx \frac{\partial \phi}{\partial p_n} \frac{\Delta p_n}{\phi_0}$$

†This is the rule for calculating the total differential of a function ϕ of several variables $p_1, p_2, \ldots, p_k$ (see Thomas and Finney, p. 605).

Likewise, the fractional change in p_n is usually more meaningful than the actual change Δp_n. Multiplying the right side of the last relation by p_n/p_n gives

$$\frac{\Delta \phi}{\phi_0} \approx \left(\frac{\partial \phi}{\partial p_n} \frac{p_n}{\phi_0}\right) \frac{\Delta p_n}{p_n} \tag{1-31}$$

Equation (1-31) gives the fractional change $\Delta \phi / \phi$ of a performance measure ϕ for a system in terms of the fractional change $\Delta p_n / p_n$ of a parameter p_n of the system. The quantity in parentheses is defined as the *sensitivity* of the performance measure ϕ to variations of the parameter p_n. We denote this quantity by the symbol $S_\phi(p_n)$; thus

$$S_\phi(p_n) = \frac{\partial \phi}{\partial p_n} \frac{p_n}{\phi_0} \tag{1-32}$$

From (1-31) we have also that

$$S_\phi(p_n) \approx \frac{\Delta \phi / \phi}{\Delta p_n / p_n} \tag{1-33}$$

The sensitivity $S_\phi(p_n)$ is approximately the ratio of the fractional (or percent) change in the performance measure ϕ to the fractional (or percent) change in the parameter p_n. Equation (1-32) provides a means of calculating a sensitivity. Equation (1-33) provides a means of interpreting or measuring a sensitivity.

Example 1-34 For the system of Fig. 1-51

$$y(t) = \frac{K_1}{1 + K_1 K_2} x(t)$$

Let overall gain be the performance measure of interest; thus

$$\phi(K_1, K_2) = \frac{K_1}{1 + K_1 K_2}$$

Calculate the sensitivities of overall gain ϕ to variations in K_1 and K_2 for $K_1 = 10^4$ and $K_2 = 0.1$.

SOLUTION From (1-32)

$$S_\phi(K_1) = \frac{\partial \phi}{\partial K_1} \frac{K_1}{\phi} = \frac{1 + K_1 K_2 - K_1 K_2}{(1 + K_1 K_2)^2} \frac{K_1}{K_1/(1 + K_1 K_2)} = \frac{1}{1 + K_1 K_2} \approx 10^{-3}$$

and

$$S_\phi(K_2) = \frac{\partial \phi}{\partial K_2} \frac{K_2}{\phi} = -\frac{K_1 K_2}{1 + K_1 K_2} \approx -1$$

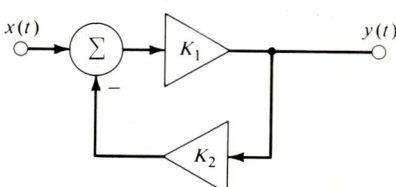

Figure 1-51 System of Example 1-34.

The magnitude of $S_\phi(K_2)$ is much larger than the magnitude of $S_\phi(K_1)$ (provided $K_1 K_2 \gg 1$, as in this example). Therefore overall gain ϕ is much more sensitive to variations of K_2 than to variations of K_1. A 1 percent change in K_2 causes a 1 percent change in ϕ, whereas a 1 percent change in K_1 causes only a 0.001 percent change in ϕ. This means that the proportion element in the feedback path is the critical component of this system. If it is necessary for overall gain ϕ to be within 1 percent of its design value, K_2 must be within 1 percent of its design value. In contrast, overall gain is relatively insensitive to variations of K_1 (so long as $K_1 K_2 \gg 1$).

Equation (1-31) is an approximation. Sensitivities can be used to determine which parameters of a system are critical, but in general they cannot be used to calculate actual variations of a performance measure ϕ unless variations of all parameters are very small or ϕ is a linear function of the parameters. To calculate actual variations of a performance measure ϕ we must calculate ϕ for many possible sets of values of the associated parameters. Such a calculation often requires use of a computer.

SUMMARY

A *signal* is a measurable quantity that varies with time. A *system* is a cause-and-effect relation between two or more signals. Signals identified as causes are called *inputs*. Signals identified as effects are called *outputs*. A system can be represented by a *block diagram*, in which the system is represented by a box, inputs are represented by arrows entering the box, outputs are represented by arrows leaving the box, and the relation between inputs and outputs is given by an expression written in the box. Signals and parameters of systems are dimensioned quantities. They are described at any instant by a number and a unit. The SI system of units is used in this book.

System analysis is the separation of systems into components for further study. The main objective of system analysis is to determine whether a proposed system will perform as intended and if not, why not, so the design can be corrected.

The further study referred to above usually entails finding the output of a system for one or more test inputs. Five *elementary signals* often used as test inputs are defined in Table 1-4.

In analysis, a system is represented by an interconnection of simpler subsystems. Several *elementary systems* (elements) often used in such representations are defined in Table 1-5. These elements can be connected in *series, parallel,* and *feedback* configurations to form more complicated systems or subsystems.

Elementary systems (and many nonelementary systems) are described by transfer characteristics. A *transfer characteristic* is a relation of the form

$$\text{Output} = \text{operation(s) on input}$$

which gives an output explicitly. A system is *static* if its output at time t depends only on the input at that same time. A system is *dynamic* if its output at time t depends on values of the input for one or more times different from t.

A *feedforward* system contains no feedback connection (contains only series and parallel connections). A transfer characteristic can be found for any feedforward system. A *feedback system* contains one or more feedback connections. Only special kinds

Table 1-4 Elementary signals

Name	Description	Graph
Step	$x(t) = x_0 u(t)$	
Rectangular pulse	$x(t) = x_0 r\left(\dfrac{t}{\tau}\right)$	
Impulse	$x(t) = \mathbf{a}\delta(t)$	
Sinusoid	$x(t) = x_0 \cos \dfrac{2\pi t}{T}$	
Exponential pulse	$x(t) = x_0 e^{-t/\tau} u(t)$	

of feedback systems can be described by transfer characteristics. In general, the relation between input and output for a feedback system is an implicit relation of the form

$$\text{Output} = \text{operation(s) on input and output}$$

Analysis of such a system is much more difficult than analysis of a system described by a transfer characteristic. Calculating an output of such a system often requires numerical methods. In this book (except for Chap. 9) we consider only systems that can be described by transfer characteristics.

The following properties of systems are referred to in descriptions (or specifications) of system performance:

Table 1-5 Elementary systems

Name	Transfer characteristic	Symbol
Proportion	$y(t) = Kx(t)$	$x(t) \to [K] \to y(t)$
Static	$y(t) = y_0 N\left[\dfrac{x(t)}{x_0}\right]$	$x(t) \to [y = y_0 N(\tfrac{x}{x_0})] \to y(t)$
Delay	$y(t) = x(t - t_0)$	$x(t) \to [t - t_0] \to y(t)$
Integration	$y(t) = K \displaystyle\int_{-\infty}^{t} x(\alpha)\, d\alpha$	$x(t) \to [\int\ K] \to y(t)$
Differentiation	$y(t) = K \dfrac{dx(t)}{dt}$	$x(t) \to [\tfrac{d}{dt}\ K] \to y(t)$
Compression	$y(t) = x(\eta t)$	$x(t) \to [\eta t] \to y(t)$
Addition	$y(t) = x_1(t) + x_2(t)$	$x_1(t), x_2(t) \to [\Sigma] \to y(t)$
Multiplication	$y(t) = Kx_1(t)x_2(t)$	$x_1(t), x_2(t) \to [\bowtie\ K] \to y(t)$

Realizability, which implies that the system gives a causal output for any input of interest

Stability, which implies that the system gives a bounded output for any bounded input

Fidelity, which implies that the system provides (nearly) distortionless transmission for inputs of interest

Sensitivity to parameter variations, which tells whether performance remains satisfactory for anticipated deviations of parameters from their design values

Transient and steady-state response, which often help identify parameters that are important to particular components of a performance specification

Concepts and notation introduced in this chapter are used throughout the remainder

PROBLEMS

1-1 Plot $x(t)$ versus time t. Assume $t_0 > 0$ and $\tau > 0$.

(a) $x(t) = x_0 u(t_0 - t)$

(b) $x(t) = \left(\dfrac{x_0 t}{\tau}\right) r\left(\dfrac{t}{\tau}\right)$

(c) $x(t) = x_0 \left(1 + \dfrac{t}{\tau}\right) r\left(1 + \dfrac{t}{\tau}\right)$

(d) $x(t) = x_0 r\left(\dfrac{t}{\tau}\right) \cos 2\pi f t \quad f = \dfrac{1}{\tau}$

(e) $x(t) = x_0 \sum_{n=0}^{3} K^n u(t - nt_0) \quad K = 0.5$

(f) $x(t) = a \sum_{n=0}^{2} K^n \delta(t - nt_0) \quad K = 0.8$

1-2 Use the elementary signals of Table 1-4 to describe the waveforms of Fig. P1-2. Assume that the signals are zero for times not shown.

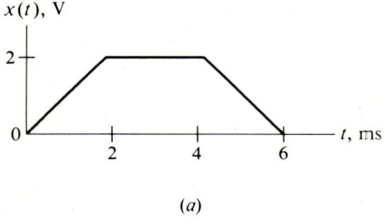
(a)

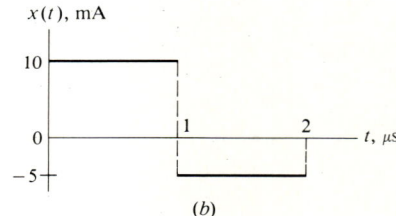
(b)

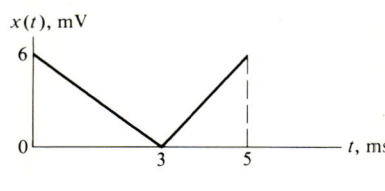
(c)

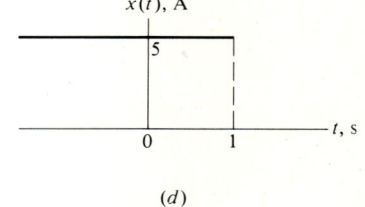
(d)

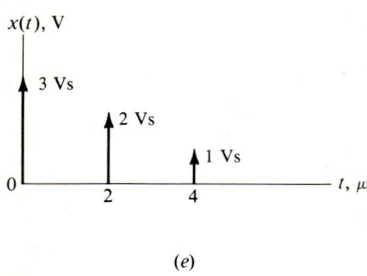
(e)

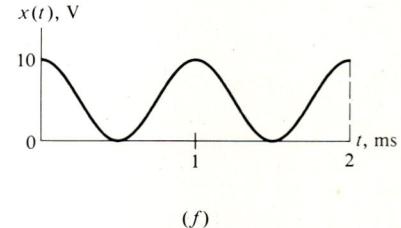
(f)

Figure P1-2

1-3 Does $x_0 u(t) u(\tau - t)$ equal $x_0 r(t/\tau)$?

1-4 Show that the definition

$$\delta(\alpha) = \lim_{\tau \to 0}\left[\frac{1}{\tau} r\left(\frac{\alpha}{\tau}\right)\right]$$

is consistent with (1-6) and (1-7).

1-5 Show that the definition

$$\delta(\alpha) = \lim_{\tau \to 0}\left[\frac{1}{\tau} e^{-\alpha/\tau} u(\alpha)\right]$$

is consistent with (1-6) and (1-7).

1-6 Show that

$$\int_a^b \delta(\alpha)\, d\alpha = u(b) - u(a)$$

Use this result to find

$$I = \int_a^b \delta(\alpha - \alpha_0)\, d\alpha$$

for $a < b$ and

(a) $\alpha_0 < a$ (b) $\alpha_0 = a$ (c) $a < \alpha_0 < b$ (d) $\alpha_0 = b$ (e) $\alpha_0 > b$

1-7 Show that

(a) $\displaystyle\int_{-\infty}^{\infty} \delta(a\alpha - b)\, d\alpha = \frac{1}{|a|}$ (b) $\displaystyle\int_{-\infty}^{\infty} x(\alpha)\delta(\alpha - \alpha_0)\, d\alpha = x(\alpha_0)$

(c) $\displaystyle\int_{-\infty}^{\infty} x(\alpha)\delta(a\alpha - b)\, d\alpha = \frac{1}{|a|} x\left(\frac{b}{a}\right)$

1-8 An instrument called a frequency counter measures the frequency of a sinusoid by counting the number N of zero crossings in a fixed interval of duration T.

   (a) Find the frequency of a sinusoid that gives $N = 100$ counts in an interval having duration $T = 2$ ms.

   (b) Is this measurement exact? What is the maximum error in part (a)?

   (c) For $T = 2$ ms what is the lowest frequency that can be measured to within 1 percent?

1-9 Plot the output $y(t)$ versus t of each system below for the specified input $x(t)$. Refer to Table 1-3 for definitions of the transfer characteristics.

   (a) Hard limiter with $y_0 = 5$ V, $x(t) = x_0 \cos \omega t$ with $f = 1$ kHz.

   (b) Soft limiter with $x_0 = 1$ mA, $y_0 = 5$ mV, $x(t) = x_1 \cos \omega t$ with $x_1 = 2$ mA and $f = 1$ kHz.

   (c) Half-wave rectifier with $x_0 = y_0 = 1$ V, $x(t) = x_1 \cos \omega t$ with $x_1 = 120$ V and $f = 60$ Hz.

   (d) Full-wave rectifier with $x_0 = y_0 = 1$ V, $x(t) = x_1 \cos \omega t$ with $x_1 = 120$ V and $f = 60$ Hz.

   (e) Square-law rectifier with $x_0 = 1$ mV and $y_0 = 5$ mA, $x(t) = x_1 \cos \omega t$ with $x_1 = 5$ mV and $f = 5$ kHz.

   (f) Comparator with $x_0 = 1.5$ V and $y_0 = 5$ V, $x(t) = x_1 r(t/\tau) + x_2 \cos \omega t$ with $x_1 = 3$ V, $\tau = 1$ ms, $x_2 = 1$ V, and $f = 4$ kHz.

1-10 Figure P1-10 shows the transfer characteristic for a so-called dead-band gain element. Plot the output of this element versus time for input

$$x(t) = 2 \cos 2\pi f t \quad \text{V}$$

where $f = 1$ kHz.

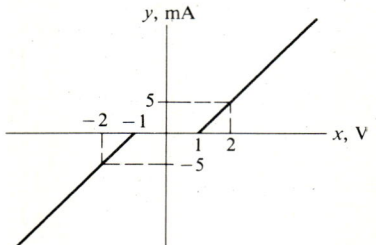

Figure P1-10

1-11 The transfer characteristic for a certain static element is

$$y(t) = y_0 \tanh \frac{x(t)}{x_0}$$

where $x(t)$ is an input and $y(t)$ is the corresponding output. Describe how this element could be modeled (approximated) for input $x(t) = A \cos \omega t$ for (a) $A \ll x_0$ and (b) $A \gg x_0$.

1-12 The static element of Prob. 1-11 is approximated as a soft limiter having parameters x_0 and y_0 (see Table 1-3). Find (numerically) the maximum absolute error of this approximation for $x_0 = y_0 = 1$ V.

1-13 Plot the output of each element:

(a) Delay element with delay time $t_0 = 250$ μs and input

$$x(t) = 5 \cos \omega t \quad \text{V}$$

with $f = 1$ kHz.

(b) Integrator with gain $K = 2$ V/(mAs) and input

$$x(t) = 5r\left(\frac{t}{\tau}\right) \quad \text{mA}$$

with $\tau = 500$ ms.

(c) Integrator with gain $K = 5$ s^{-1} and input

$$x(t) = \mathbf{a}\delta(t) - \mathbf{a}\delta(t - t_0)$$

with $\mathbf{a} = 2$ Vs and $t_0 = 5$ ms.

(d) Integrator with gain $K = 2$ s^{-1} and input

$$x(t) = x_0 r\left(\frac{t}{\tau}\right) - \mathbf{a}\delta(t - \tau)$$

with $x_0 = 5$ V, $\tau = 2$ s, and $\mathbf{a} = 10$ Vs.

(e) Differentiator with gain $K = 2$ Vs/mA and input

$$x(t) = 5(\cos \omega t) r\left(\frac{t}{\tau}\right) \quad \text{mA}$$

with $f = 1$ kHz and $\tau = 1.25$ ms.

(f) Differentiator with gain $K = 2$ s and input
$$x(t) = 50[u(t - t_0) - u(t - 2t_0)] \quad \text{mA}$$
with $t_0 = 10$ μs.

(g) Compression element with compression ratio $\eta = 2$ and input
$$x(t) = 5e^{-t/\tau}u(t) \quad \text{V}$$
with $\tau = 2$ ms.

(h) Compression element with compression ratio $\eta = 4$ and input
$$x(t) = 5r\left(\frac{t}{\tau}\right) \cos \omega t \quad \text{V}$$
with $\tau = 10$ ms and $f = 500$ Hz.

1-14 Explain why a differentiator is a dynamic element.

1-15 Give units for all signals and parameters appearing in the systems of Fig. P1-15a to g.

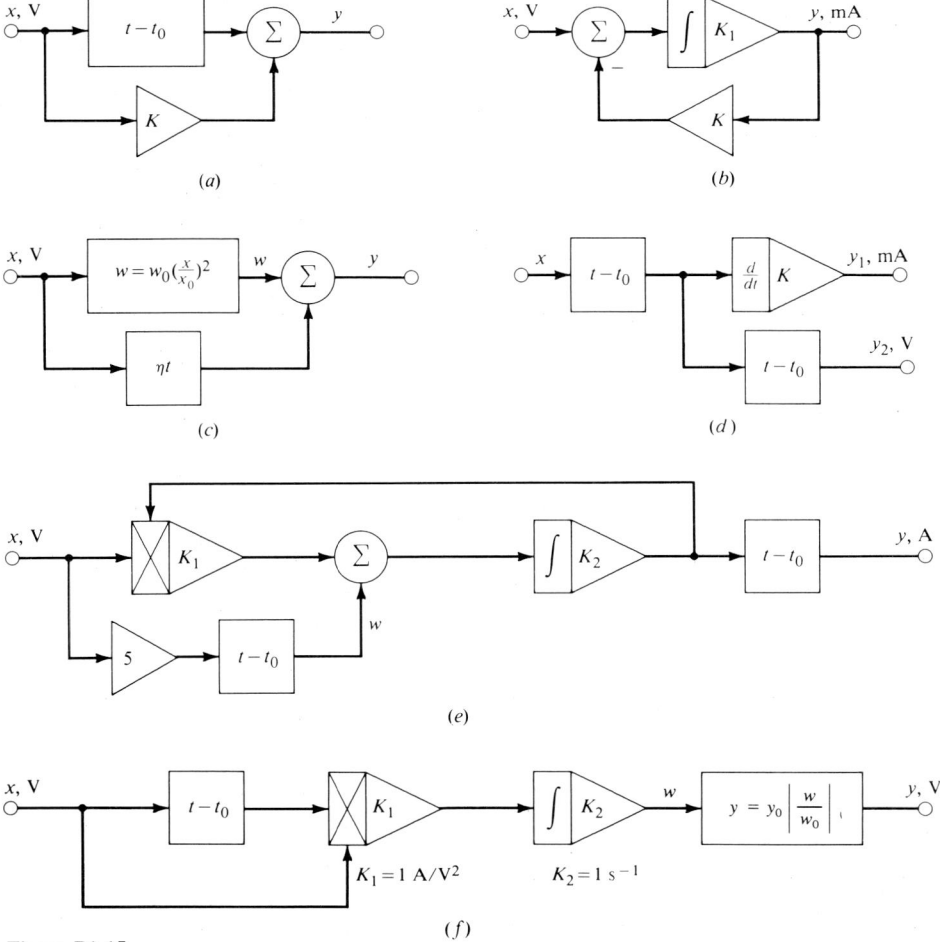

Figure P1-15

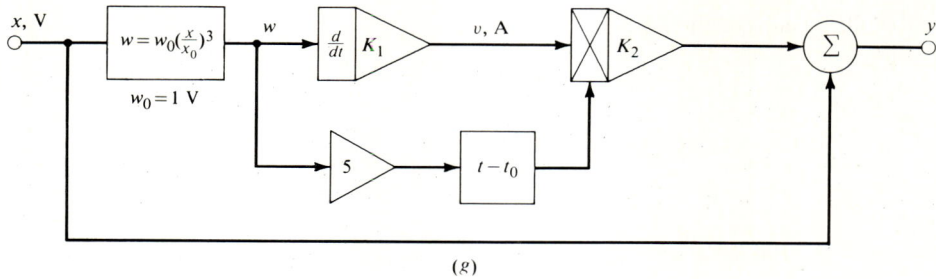

(g)

Figure P1-15 *Continued*

1-16 Find the output of the system of Fig. P1-16, where $K = 1 \text{ V}^{-1}$, $t_0 = 250 \ \mu\text{s}$, $x_1(t) = 5 \cos \omega_1 t$ V, and $x_2(t) = 4 \cos \omega_2 t$ V, with $f_1 = 1$ kHz and $f_2 = 3$ kHz.

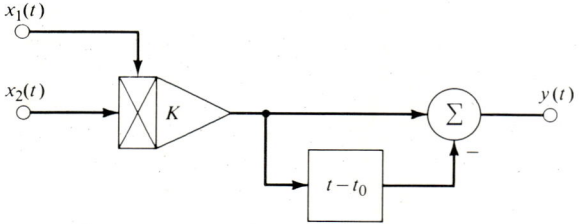

Figure P1-16

1-17 Find the output of the system of Fig. P1-17, where $K = 1 \text{ s}^{-1}$, $K_1 = 1$, $t_0 = 1$ s, and $x(t) = 2u(t)$ V.

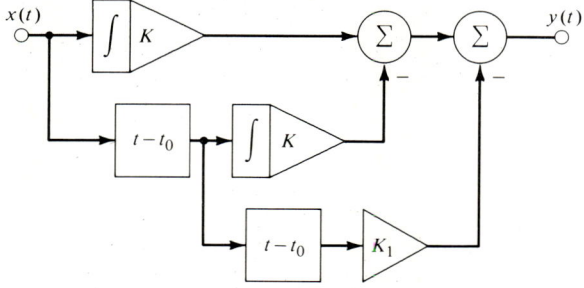

Figure P1-17

1-18 Find the output of the system of Fig. P1-18, where $K = 1 \text{ s}^{-1}$ and $x(t) = 2 \cos \omega t$ V with $f = 5$ kHz.

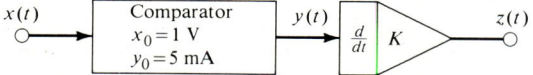

Figure P1-18

1-19 Find the output of the system of Fig. P1-19, where $K = 1$, $t_0 = 1$ ms, and $x(t) = 1 + \cos \omega t + \cos 2\omega t$ V with $f = 500$ Hz.

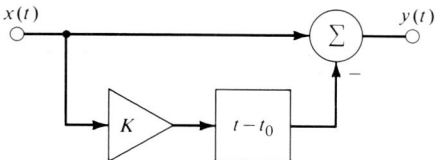

Figure P1-19

1-20 Find the output of the system of Fig. P1-20, where $K_1 = 1$ V^{-1}, $K_2 = 2$, $x_1(t) = 2 \cos \omega_1 t$ V, and $x_2(t) = \cos \omega_2 t$ V with $f_1 = 1$ kHz and $f_2 = 500$ Hz.

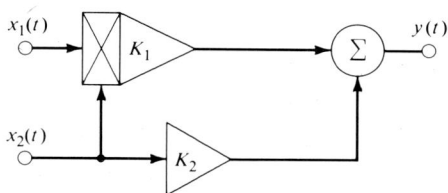

Figure P1-20

1-21 Find the output of the system of Fig. P1-21, where $K_1 = 1$ V^{-1}, $K_2 = 8\pi$ s^{-1}, $x_1(t) = 2 \cos \omega_1 t$ V, and $x_2(t) = \cos \omega_2 t$ V with $f_1 = 5$ kHz and $f_2 = 500$ kHz.

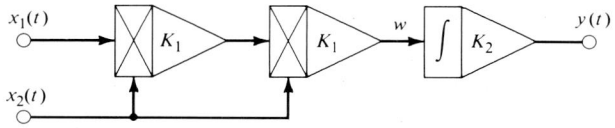

Figure P1-21

1-22 Find the output of the system of Fig. P1-22, where $t_0 = 2$ μs, $K_n = (0.5)^n$, and $x(t) = 10\, r(t/\tau)$ V with $\tau = 1$ μs.

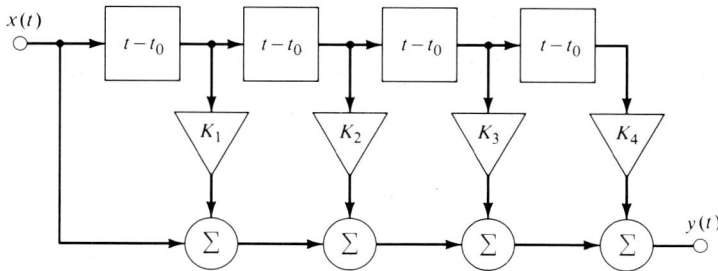

Figure P1-22

1-23 Find the output of the system of Fig. P1-23, where $t_0 = 2$ ms, $K = 0.5$, and $x(t) = 4r(t/\tau)$ V with $\tau = 1$ ms.

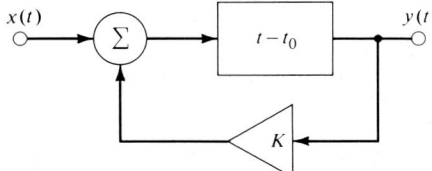

Figure P1-23

44 INTRODUCTION TO SYSTEM ANALYSIS

1-24 Find the output of the system of Fig. P1-24, where $t_0 = 2$ ms, $K = 0.5$, and $x(t) = 4r(t/\tau)$ V with $\tau = 1$ ms.

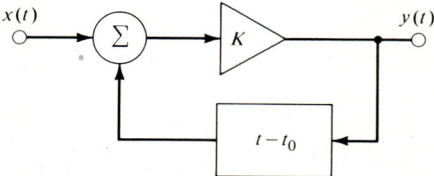

Figure P1-24

1-25 Find the output of the system of Fig. P1-25, where $t_0 = 2$ ms and $x(t) = 5r(t/\tau)$ V with $\tau = 1$ ms.

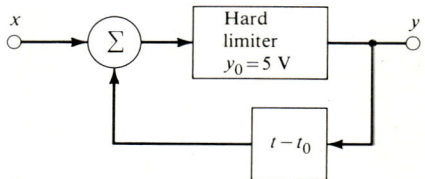

Figure P1-25

1-26 Find the output of the system of Fig. P1-26, where $K_1 = 0.5$, $K_2 = 0.8$, $t_0 = 2$ ms, and $x(t) = 2r(t/\tau)$ mA with $\tau = 1$ ms.

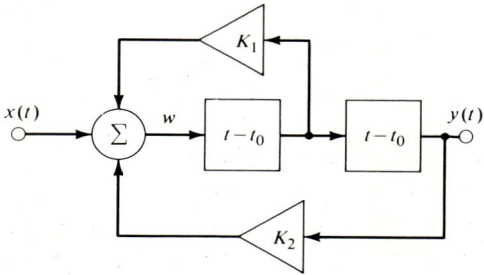

Figure P1-26

1-27 Give block diagrams for systems whose outputs for input $x(t) = u(t)$ V are shown in Fig. P1-27a to d.

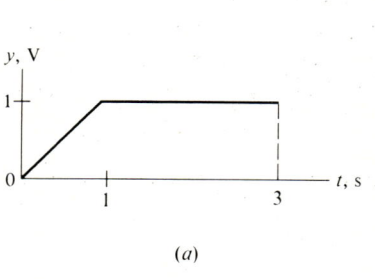

(a)

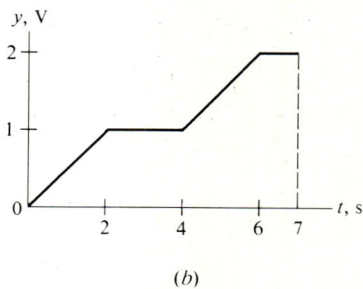

(b)

Figure P1-27

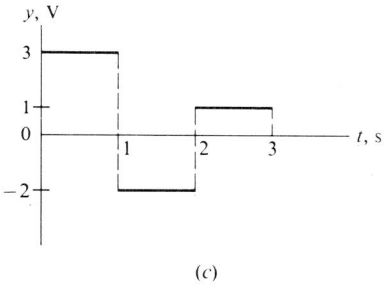
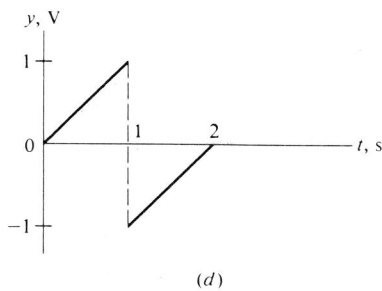

(c) (d)

Figure P1-27 *Continued*

1-28 Show that proportion, delay, integration, and differentiation elements commute; i.e., show that the transfer characteristic of a series connection of these four elements is independent of the order in which they are connected. *Hint:* It is sufficient to show that these elements commute in pairs.

1-29 For the system of Fig. P1-29 $K_1 = 1$ s, $K_2 = 1$ s^{-1}, and

$$x(t) = x_0(\cos \omega t)u(t)$$

Show that the output $v(t)$ is proportional to the frequency of the input $x(t)$ (in steady state). Why is this system impractical? How could it be made more practical?

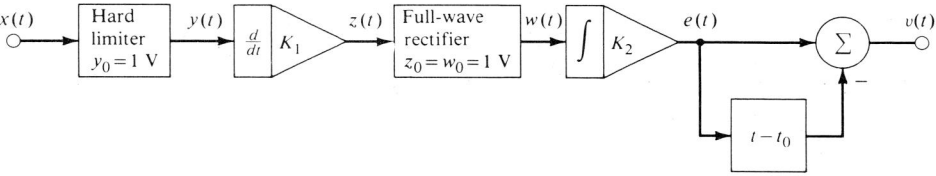

Figure P1-29

1-30 According to the result of Prob. 1-28, the systems of Fig. P1-30 are equivalent. Does this equivalence hold for $x(t) = x_0$ (independent of time t)? Why not? What proviso should be attached to the result of Prob. 1-28?

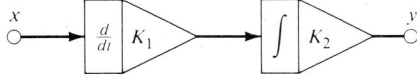

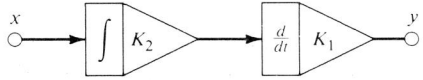

Figure P1-30

1-31 Does the result of Prob. 1-28 imply that *physical* components represented by proportion, delay, integration, and differentiation elements commute? Under what conditions?

1-32 A pulse $x(t)$ travels along a coaxial cable with speed $c = 3 \times 10^8$ m/s. It is attenuated by the factor 0.1 in each 1-km length of cable. Draw a block diagram representing a 3-km length of the cable and plot the output for input

$$x(t) = 5r\left(\frac{t}{\tau}\right) \quad V$$

with $\tau = 2$ μs.

1-33 At $t = 0$ a radar transmits a 500-ns pulse which travels at 3×10^8 m/s. The pulse is reflected by an airplane, and the leading edge of the reflected pulse (the echo) is received by the radar at time $t = 6$ μs. Draw a block diagram representing these events. How far is the airplane from the antenna?

1-34 A sonar system transmits a 500-ms pulse. Since the same transducer is used for transmitting and receiving, the sonar is "deaf" during transmission. Find the minimum range for which a target can be detected. The speed of sound in seawater is $c = 1.5$ km/s.

1-35 The playback head on a certain tape recorder is 1.5 cm downstream from the recording head. When monitoring a recording, the playback is delayed relative to the source (the signal being recorded). Find the delay time for recording speeds of 19, 9.5, and 4.75 cm/s.

1-36 Find all positive values of t_0 for which the systems of Fig. P1-36 give a causal output for input $x(t) = x_0 r(t/\tau)$ and compression ratio $\eta = 2$.

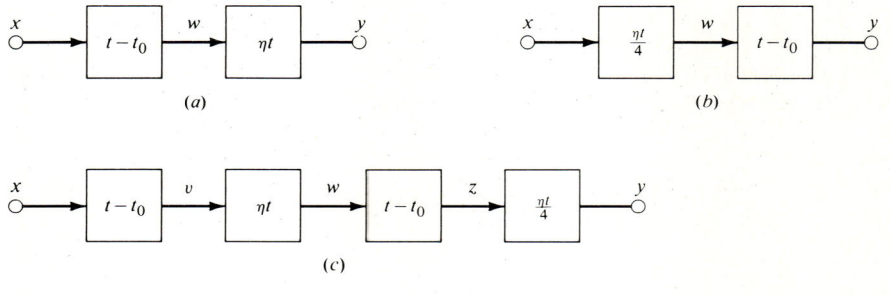

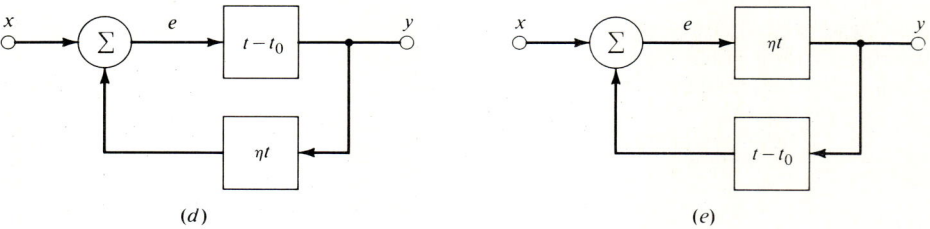

Figure P1-36

1-37 Determine whether a compression element with compression ratio $0 < \eta < 1$ (corresponding to expansion) is realizable for input $x(t) = x_0 r(t/\tau)$.

1-38 Determine whether an element whose output $y(t)$ for input $x(t)$ is given by $y(t) = x(\eta t)$ with $\eta < 0$ is realizable for input $x(t) = x_0 r(t/\tau)$.

1-39 It is desired to realize compression for an exponential pulse $x(t) = x_0 e^{-t/\tau} u(t)$. Can this be done by introducing delay (a) strictly or (b) practically? Justify your answers.

1-40 Two tape recorders are to be used to achieve 2-to-1 compression, as shown in Fig. P1-40. The tape is fed from the supply reel of the first recorder through the head assembly of the first

ELEMENTS OF SYSTEM ANALYSIS 47

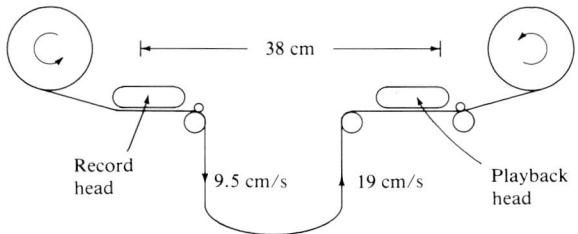

Figure P1-40

recorder, then through the head assembly of the second recorder onto the takeup reel on the second recorder. The first recorder begins recording at $t = 0$ and runs at 9.5 cm/s. The second recorder begins playback at $t = t_1$ and runs at 19 cm/s. Draw a block diagram representing this operation and find the minimum value of t_1 that allows 2-to-1 compression of a 60-s signal. Assume that the tape is taut at $t = 0$.

1-41 See Fig. P1-41. At $t = 0$ a radar transmits a pulse $x(t) = x_0 r(t/\tau) \cos \omega t$. The pulse propagates at 3×10^8 m/s. It is reflected by an object whose distance from the radar at $t = 0$ is d_0. The object is moving toward the radar with constant velocity v.

(a) Show that the echo received by the radar is a compressed and delayed version of the transmitted pulse (ignore attenuation). Draw a block diagram for the events described and express all delays and compression ratios as functions of v and d_0.

(b) For $d_0 = 1$ km and $v = 500$ m/s find the duration of the longest pulse that can be compressed as described above. What prevents (interrupts!) compression of a longer pulse?

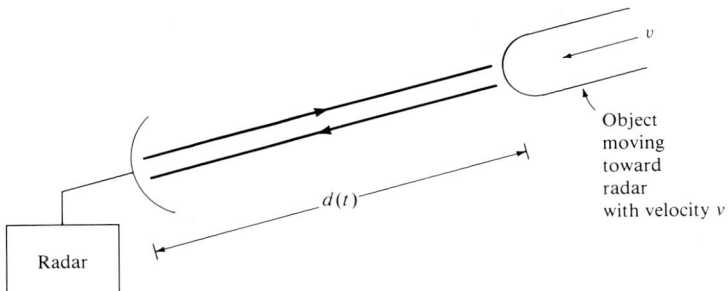

Figure P1-41

1-42 Find the steady-state and transient components of $x(t)$.

(a) $x(t) = x_0 e^{-t/\tau} u(t) + x_1 (\cos \omega t) u(t)$ (b) $x(t) = a\delta(t) - x_0 e^{-t/\tau} u(t)$

(c) $x(t) = x_0 \left[\dfrac{t}{\tau} r\left(\dfrac{t}{\tau}\right) + u(t - \tau) \right]$ (d) $x(t) = x_0 \left[\dfrac{t}{\tau} u(t) - \dfrac{t - \tau}{\tau} u(t - \tau) \right]$

(e) $x(t) = x_0 \sum_{n=0}^{\infty} 2^{-n} r\left(\dfrac{t - n\tau}{\tau}\right)$

1-43 For the systems of Fig. P1-43, $x(t) = x_0 r(t/\tau)$, with $\tau = 500$ μs. Find all values of the unspecified parameter for which the output $y(t)$ is bounded.

1-44 Figure P1-44a shows a schematic diagram for a public-address system, where feedback from the speaker to the microphone is possible. Figure P1-44b shows a simplified model for the system. The model and the actual system are related as follows:

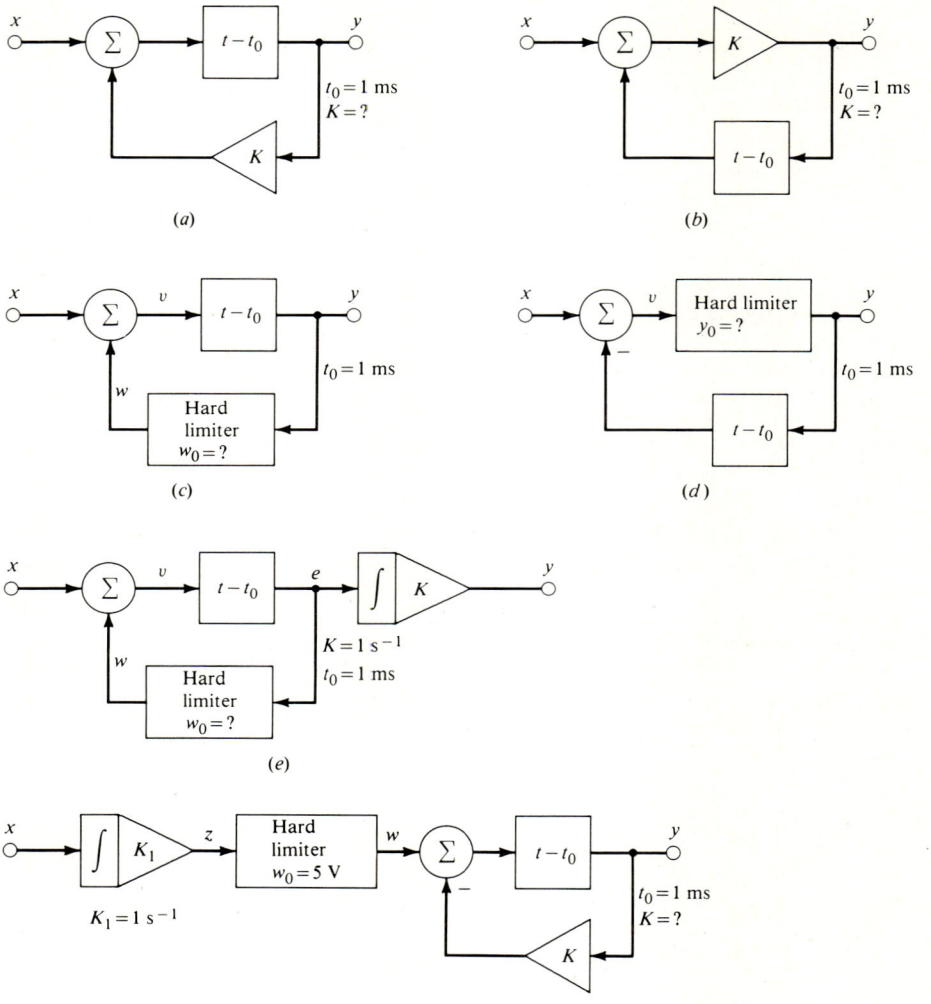

Figure P1-43

1. The soft limiter represents the amplifier. Its input $x(t)$ and output $v(t)$ are voltages. The amplifier is linear with gain v_0/x_0 so long as $|x(t)| < x_0$. When $|x(t)| > x_0$, the amplifier is overdriven and its output is limited (by the power supply) to $\pm v_0$. For this problem, $x_0 = 2.5$ mV and $v_0 = 25$ V.
2. The proportion element following the amplifier (limiter) represents the speaker. Its output $y(t)$ is incremental (from atmospheric) pressure, where $K_1 = 0.1$ N/(m²V).
3. The delay represents time required for propagation of an acoustic wave from the speaker to the microphone, and the proportion element following the delay element accounts for attenuation of the wave in air and for the efficiency and radiation pattern of the speaker. The

ELEMENTS OF SYSTEM ANALYSIS 49

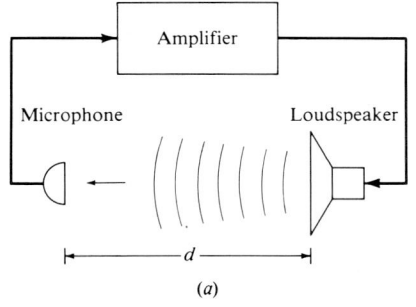

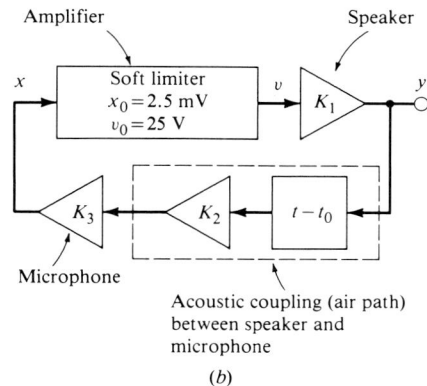

Figure P1-44 (*a*) Physical system; (*b*) block diagram.

quantity d denotes distance from the speaker to the microphone. Let $K_2 = A/d$, $t_0 = d/c$, where $c = 335$ m/s (speed of sound in air) and $A = 20$ mm.
4. The gain K_3 of the proportion element representing the microphone accounts for the efficiency and pattern of the microphone, where $K_3 = 0.1$ V/(N/m²).

At $t = 0$, a noise pulse $v(t) = v_0 r(t/\tau)$, with $\tau = 1$ ms, appears at the output of the amplifier. Find the largest distance d for which the system oscillates (is unstable), the period of the oscillation, and the amplitude of the oscillation at the amplifier output. What important effects does this analysis neglect?

1-45 For each system of Fig. P1-45:

(*a*) Obtain an expression for the ISE for input $x(t) = x_0 r(t/\tau)$. Assume the ideal transfer characteristic is $y(t) = x(t)$.

(*b*) Obtain expressions for the sensitivities of the ISE to variations of K_1, K_2, and K_3 using the values given in Fig. P1-45 as the design values for these parameters.

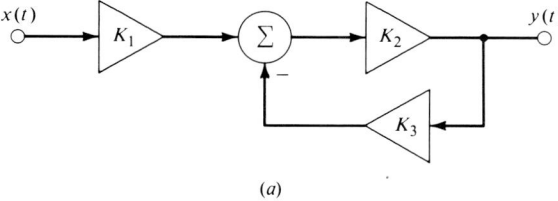

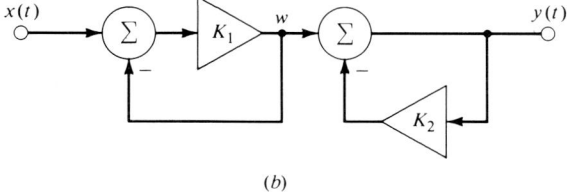

Figure P1-45 (*a*) $K_1 = K_3 = 1$, $K_2 = 10$; (*b*) $K_1 = 10$, $K_2 = 0.01$.

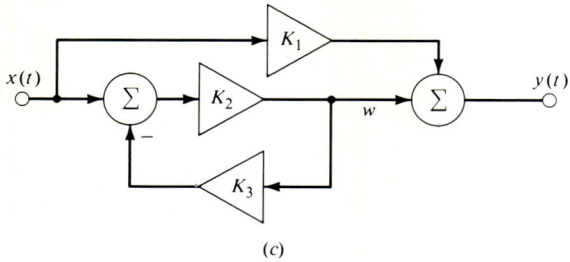

Figure P1-45 *Continued* (c) $K_1 = 0.5$, $K_2 = 100$, $K_3 = 2$.

(c) Use (1-33) to calculate the percent change of ISE for a 2 percent change of each parameter.

CHAPTER
TWO

LINEAR STATIONARY SYSTEMS

A linear stationary system is one that obeys superposition and has constant parameters. An example is an electric circuit composed of fixed resistors, capacitors, and inductors. Many other systems (or components of systems) such as electric motors and electronic amplifiers are linear and stationary under normal operating conditions.

In this chapter we describe the fundamental properties (superposition and time invariance) of linear stationary systems. We then show that any linear stationary system, whether feedforward or feedback, can be represented by a transfer characteristic (a convolution integral) which gives an output explicitly as the result of an operation on the input. We show how the transfer characteristic of a system can be obtained from a differential equation describing the system and how the transfer characteristic is interpreted in terms of realizability, stability, fidelity, and transient and steady-state response.

2-1 DEFINITION OF A LINEAR STATIONARY SYSTEM

In this section we describe superposition and time invariance, which are the defining properties of linear stationary systems. All methods for analyzing linear stationary systems are based on these important properties.

2-1A Linear Systems and Superposition

A *linear system* is one that obeys the following *principle of superposition:*

> If input $x_1(t)$ (acting alone) produces output $y_1(t)$ and input $x_2(t)$ (acting alone) produces output $y_2(t)$, then input
> $$x(t) = a_1 x_1(t) + a_2 x_2(t)$$

produces output

$$y(t) = a_1 y_1(t) + a_2 y_2(t)$$

Example 2-1 Show that the system (integrator) of Fig. 2-1 is linear.

SOLUTION The transfer characteristic of the system is

$$y(t) = K \int_{-\infty}^{t} x(\alpha) \, d\alpha$$

The output for input

$$x(t) = a_1 x_1(t) + a_2 x_2(t)$$

is given by

$$y(t) = K \int_{-\infty}^{t} [a_1 x_1(\alpha) + a_2 x_2(\alpha)] \, d\alpha$$

$$= a_1 K \int_{-\infty}^{t} x_1(\alpha) \, d\alpha + a_2 K \int_{-\infty}^{t} x_2(\alpha) \, d\alpha$$

Comparing the last two terms (individually) with the transfer characteristic of the system shows that the output above has the form

$$y(t) = a_1 y_1(t) + a_2 y_2(t)$$

where $y_1(t)$ is the output for input $x_1(t)$ and $y_2(t)$ is the output for input $x_2(t)$. Therefore the system is linear.

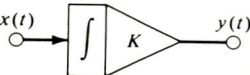

Figure 2-1 System of Example 2-1.

Example 2-2 Show that the system of Fig. 2-2 is nonlinear.

SOLUTION The transfer characteristic of the system is

$$y(t) = y_0 \left[\frac{x(t)}{x_0} \right]^2$$

The output for input

$$x(t) = a_1 x_1(t) + a_2 x_2(t)$$

is given by

$$y(t) = y_0 \left[\frac{a_1 x_1(t) + a_2 x_2(t)}{x_0} \right]^2$$

$$= a_1^2 y_0 \left[\frac{x_1(t)}{x_0} \right]^2 + a_2^2 y_0 \left[\frac{x_2(t)}{x_0} \right]^2 + 2 a_1 a_2 y_0 \frac{x_1(t) x_2(t)}{x_0^2}$$

This does not have the form

$$y(t) = a_1 y_1(t) + a_2 y_2(t)$$

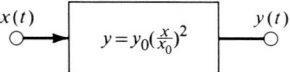

Figure 2-2 System of Example 2-2.

where $y_1(t)$ is the output for input $x_1(t)$ and $y_2(t)$ is the output for input $x_2(t)$. Therefore the system is nonlinear.

Superposition can be extended by induction to inputs having more than two terms. The output of a linear stationary system for input

$$x(t) = a_1 x_1(t) + a_2 x_2(t) + \cdots + a_n x_n(t)$$

is given by

$$y(t) = a_1 y_1(t) + a_2 y_2(t) + \cdots + a_n y_n(t)$$

where $y_i(t)$ is the output for input $x_i(t)$ ($i = 1, 2, \ldots, n$). Superposition is especially helpful when all terms of an input have the same mathematical form, as illustrated in the next example.

Example 2-3 Find the output $y(t)$ of the system of Fig. 2-3 for input

$$x(t) = 5x_0 \cos \omega_1 t + 4x_0 \cos \omega_2 t - 3x_0 \cos \omega_3 t$$

SOLUTION Since the system is linear, we can use superposition. The input is a linear combination of terms of the form

$$x_i(t) = x_0 \cos \omega_i t$$

for $i = 1, 2, 3$. The corresponding term of the output is

$$y_i(t) = K_1 x_0 \cos \omega_i t - K_2 \omega_i x_0 \sin \omega_i t$$

for $i = 1, 2, 3$. By superposition the output is

$$\begin{aligned} y(t) &= 5y_1(t) + 4y_2(t) - 3y_3(t) \\ &= 5(K_1 x_0 \cos \omega_1 t - K_2 \omega_1 x_0 \sin \omega_1 t) \\ &\quad + 4(K_1 x_0 \cos \omega_2 t - K_2 \omega_2 x_0 \sin \omega_2 t) \\ &\quad - 3(K_1 x_0 \cos \omega_3 t - K_2 \omega_3 x_0 \sin \omega_3 t) \end{aligned}$$

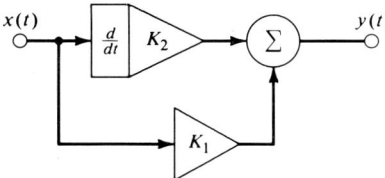

Figure 2-3 System of Example 2-3.

Any system consisting entirely of linear subsystems is a linear system (see Probs. 2-5 to 2-7). Proportion, delay, compression, integration, differentiation, and addition elements are linear (see Prob. 2-3), and any system (block diagram) consisting entirely of these elements is a linear system. Consequently, we can identify a linear

2-1B Stationary Systems and Time Invariance

A *stationary system* is one that obeys the following *principle of time invariance:*

If input $x(t)$ produces output $y(t)$, then input $x(t - t_0)$ produces output $y(t - t_0)$.

This implies that a stationary system commutes with a delay element, as illustrated in Fig. 2-4.

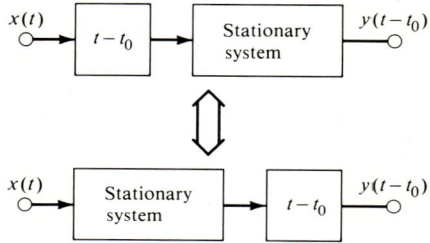

Figure 2-4 The principle of time invariance.

Example 2-4 Show that the system of Fig. 2-5a is stationary.

SOLUTION To show that the system is stationary we show that the system commutes with a delay element. The system is stationary if the signal $y_1(t)$ of Fig. 2-5b is identical to the signal $y_2(t)$ of Fig. 2-5c. For the system of Fig. 2-5b,

$$y_1(t) = K_1 \int_{-\infty}^{t-t_0} x(\alpha)\, d\alpha - K_2 x(t - t_0)$$

For the system of Fig. 2-5c,

$$y_2(t) = K_1 \int_{-\infty}^{t} x(\alpha - t_0)\, d\alpha - K_2 x(t - t_0)$$

Changing the variable of integration in the last integral from α to $\beta = \alpha - t_0$ gives

$$y_2(t) = K_1 \int_{-\infty}^{t-t_0} x(\beta)\, d\beta - K_2 x(t - t_0)$$

Since this expression for $y_2(t)$ is identical to the expression for $y_1(t)$ above, the system is stationary.

Example 2-5 Determine whether a compression element is stationary.

SOLUTION We determine whether a compression element commutes with a delay element. For the system of Fig. 2-6a,

$$y_1(t) = x[\eta(t - t_0)] = x(\eta t - \eta t_0)$$

LINEAR STATIONARY SYSTEMS 55

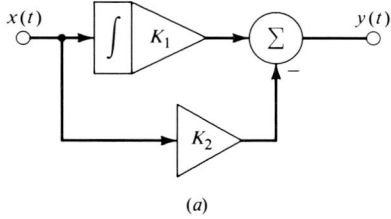

(a)

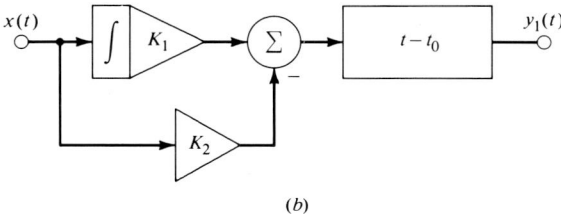

(b)

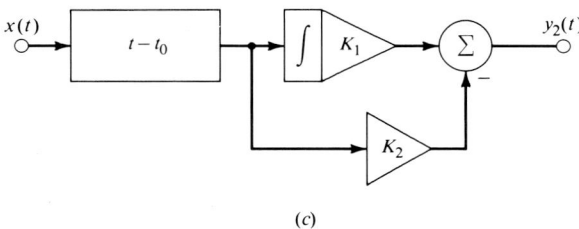

(c)

Figure 2-5 System of Example 2-4.

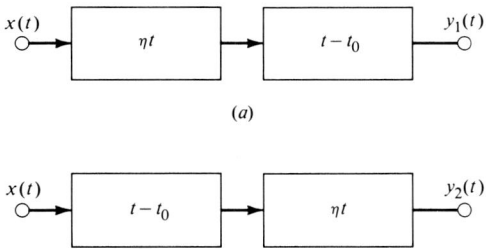

(a)

(b)

Figure 2-6 System of Example 2-5.

For the system of Fig. 2-6b,

$$y_2(t) = x(\eta t - t_0)$$

In general (for $\eta \neq 1$), $y_1(t) \neq y_2(t)$. Therefore a compression element is nonstationary.

Example 2-6 Determine whether a proportion element having time-varying gain $K(t)$ is stationary.

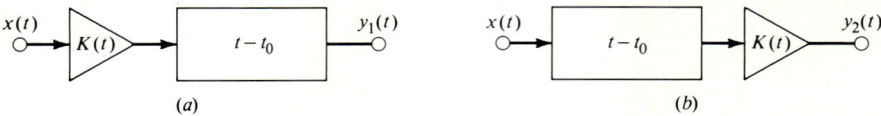

Figure 2-7 System of Example 2-6.

SOLUTION The transfer characteristic for a proportion element having time-varying gain $K(t)$ is

$$y(t) = K(t)x(t)$$

where $x(t)$ is an input and $y(t)$ is an output. We wish to determine whether the element commutes with a delay element, as illustrated in Fig. 2-7. For the system of Fig. 2-7a,

$$y_1(t) = K(t - t_0)x(t - t_0)$$

For the system of Fig. 2-7b,

$$y_2(t) = K(t)x(t - t_0) \neq y_1(t)$$

Therefore a proportion element having time-varying gain is nonstationary. This example illustrates the fact that a system having a time-varying parameter is likely to be nonstationary.

Any system that consists entirely of stationary subsystems is stationary (see Prob. 2-7). Of the elementary systems defined in Sec. 1-3 only a compression element is inherently nonstationary. Addition elements and multiplication elements cause a system to be nonstationary only if they represent components having time-varying parameters. Any system considered in this book is stationary if it contains no compression element or time-varying parameter.

2-1C Linear Stationary Systems

A *linear stationary system* is one that obeys both superposition and time invariance. The output of a linear stationary system for input

$$x(t) = a_1 x_1(t - t_1) + a_2 x_2(t - t_2) + \cdots$$

is given by

$$y(t) = a_1 y_1(t - t_1) + a_2 y_2(t - t_2) + \cdots$$

where $y_i(t)$ is the output for input $x_i(t)$ ($i = 1, 2, \ldots$).

Example 2-7 The output of a certain linear stationary system for input $x_1(t) = x_0 u(t)$ is

$$y_1(t) = y_0 e^{-t/\tau_1} u(t)$$

Find the output for input

$$x(t) = 5x_0 r\left(\frac{t}{\tau_2}\right)$$

SOLUTION The input $x(t)$ can be expressed as

$$x(t) = 5x_0 u(t) - 5x_0 u(t - \tau_2) = 5x_1(t) - 5x_1(t - \tau_2)$$

where $x_1(t) = x_0 u(t)$. By superposition and time invariance, the output is given by

$$y(t) = 5y_1(t) - 5y_1(t - \tau_2)$$

where $y_1(t)$ is the output for input $x_1(t) = x_0 u(t)$; thus

$$y(t) = 5y_0 e^{-t/\tau_1} u(t) - 5y_0 e^{-(t-\tau_2)/\tau_1} u(t - \tau_2)$$

A system is linear and stationary if it contains only linear stationary subsystems and has no time-varying parameters. Therefore, a system is linear and stationary if it contains only proportion, delay, integration, differentiation, and addition elements and has no time-varying parameters. We can identify a linear stationary system by inspecting a block diagram describing the system.

Example 2-8 Find the output $y(t)$ of the system of Fig. 2-8 for input

$$x(t) = 5x_0 r\left(\frac{t}{\tau}\right) - 2x_0 r\left(\frac{t - t_0}{\tau}\right)$$

SOLUTION By inspection of the block diagram, the system is linear and stationary. Therefore we can use superposition and time invariance. The input can be expressed as

$$x(t) = 5x_0 u(t) - 5x_0 u(t - \tau) - 2x_0 u(t - t_0) + 2x_0 u(t - t_0 - \tau)$$

This has the form

$$x(t) = 5x_1(t) - 5x_1(t - \tau) - 2x_1(t - t_0) + 2x_1(t - t_0 - \tau)$$

where $x_1(t) = x_0 u(t)$. Using superposition and time invariance, we can express the output as

$$y(t) = 5y_1(t) - 5y_1(t - \tau) - 2y_1(t - t_0) + 2y_1(t - t_0 - \tau)$$

where $y_1(t)$ is the output for input $x_1(t)$. From Fig. 2-8,

$$y_1(t) = K_1 x_0 \int_{-\infty}^{t} u(\alpha) \, d\alpha - K_2 x_0 u(t)$$

$$= K_1 x_0 t u(t) - K_2 x_0 u(t) = x_0 (K_1 t - K_2) u(t)$$

Therefore,

$$y(t) = 5x_0 (K_1 t - K_2) u(t) - 5x_0 [K_1 (t - \tau) - K_2] u(t - \tau)$$
$$- 2x_0 [K_1 (t - t_0) - K_2] u(t - t_0) + 2x_0 [K_1 (t - t_0 - \tau) - K_2] u(t - t_0 - \tau)$$

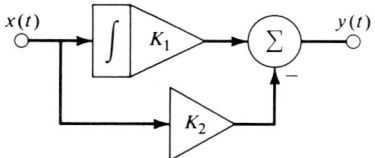

Figure 2-8 System of Example 2-8.

2-2 CONVOLUTION

In this section we develop a convolution integral which gives an output of a linear stationary system explicitly as the result of an operation on an input. This shows that

2-2A Superposition Integral

A preliminary relation needed in the development of the convolution integral below is

$$x(t) = \int_{-\infty}^{\infty} x(\lambda)\delta(t - \lambda)\, d\lambda \qquad (2\text{-}1)$$

This relation can be derived from properties (1-6) and (1-7) of the delta function as follows. We note that

$$x(\lambda)\delta(t - \lambda) = x(t)\delta(t - \lambda)$$

because $\delta(t - \lambda) = 0$ except for $\lambda = t$. This implies that

$$\int_{-\infty}^{\infty} x(\lambda)\delta(t - \lambda)\, d\lambda = x(t) \int_{-\infty}^{\infty} \delta(t - \lambda)\, d\lambda$$

because $x(t)$ is not a function of λ. From (1-7), the last integral equals unity. Equation (2-1) follows.

We can obtain (2-1) another way. A signal can be approximated by a sum of rectangular pulses, as illustrated in Fig. 2-9, where

$$x(t) \approx \sum_{n=-\infty}^{\infty} x(n\Delta\lambda) r\left(\frac{t - n\Delta\lambda}{\Delta\lambda}\right)$$

Multiplying and dividing the summand by $\Delta\lambda$ gives

$$x(t) \approx \sum_{n=-\infty}^{\infty} x(n\Delta\lambda) \frac{1}{\Delta\lambda} r\left(\frac{t - n\Delta\lambda}{\Delta\lambda}\right) \Delta\lambda$$

To make the approximation exact we let $\Delta\lambda \rightarrow 0$. This gives

$$x(t) = \int_{-\infty}^{\infty} x(\lambda) \lim_{\Delta\lambda \to 0} \left[\frac{1}{\Delta\lambda} r\left(\frac{t - \lambda}{\Delta\lambda}\right)\right] d\lambda$$

But (see Prob. 1-4)

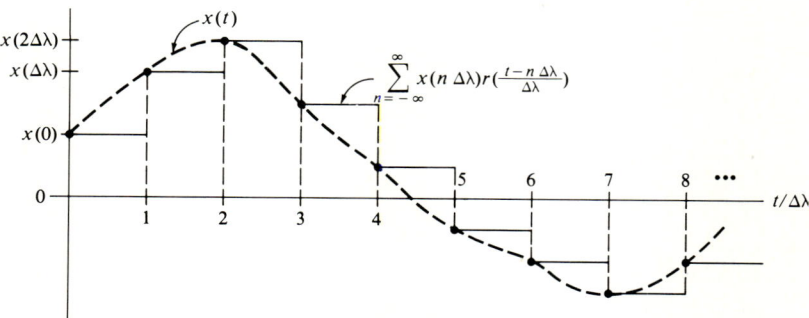

Figure 2-9 Approximating a signal by a superposition of narrow rectangular pulses.

$$\lim_{\Delta\lambda \to 0}\left[\frac{1}{\Delta\lambda} r\left(\frac{t-\lambda}{\Delta\lambda}\right)\right] = \delta(t-\lambda)$$

Equation (2-1) follows.

To interpret (2-1) we use the definition of an integral to express the right side of (2-1) as

$$x(t) = \lim_{\Delta\lambda \to 0}\left[\sum_{n=-\infty}^{\infty} x(n\Delta\lambda)\delta(t - n\Delta\lambda)\Delta\lambda\right]$$

This shows that (2-1) expresses a signal $x(t)$ as a superposition (a sum) of impulses, where the nth impulse is delayed by $n\Delta\lambda$ (occurs for $t = n\Delta\lambda$) and has strength $x(n\Delta\lambda)\Delta\lambda$. Consequently, we refer to the right side of (2-1) as a *superposition integral*.

2-2B Impulse Response and Convolution

The impulse response of a linear stationary system is defined as the function $h(t)$, where $y(t) = \mathbf{a}h(t)$ is the output of the system for input $x(t) = \mathbf{a}\delta(t)$. This definition is illustrated in Fig. 2-10. The unit of strength $\mathbf{a}$ is the unit of xt, and the unit of an impulse response $h(t) = y(t)/\mathbf{a}$ is the unit of y/xt.† For example, if $x(t)$ is voltage in volts, $y(t)$ is current in milliamperes, and time t is expressed in seconds, then the unit of $h(t)$ is milliamperes per volt-second. In this book the symbol $h(t)$ (with subscripts if necessary) always denotes an impulse response.

From the superposition integral (2-1) we can show that the output $y(t)$ of a linear stationary system for any input $x(t)$ is given by a *convolution integral*

$$y(t) = \int_{-\infty}^{\infty} x(\lambda)h(t - \lambda)\,d\lambda \qquad (2\text{-}2)$$

where $h(t)$ is the impulse response for the system. To show this, we write (2-1) as

$$x(t) = \lim_{\Delta\lambda \to 0}\left[\sum_{n=-\infty}^{\infty} x(n\Delta\lambda)\delta(t - n\Delta\lambda)\Delta\lambda\right]$$

Each term in the sum has the form $\mathbf{a}_n\delta(t - t_n)$, where $\mathbf{a}_n = x(n\Delta\lambda)\Delta\lambda$ and $t_n = n\Delta\lambda$. By time invariance and the definition of $h(t)$, the output for input $\mathbf{a}_n\delta(t - t_n)$ is

$$\mathbf{a}_n h(t - t_n) = x(n\Delta\lambda)h(t - n\Delta\lambda)\Delta\lambda$$

By superposition, the complete response is

$$y(t) = \lim_{\Delta\lambda \to 0}\left[\sum_{n=-\infty}^{\infty} x(n\Delta\lambda)h(t - n\Delta\lambda)\Delta\lambda\right]$$

Equation (2-2) follows.

†Impulse response often is defined as the output of a system for input $\delta(t)$. This definition is dimensionally incorrect because (in general) $\delta(t)$ does not have the dimension of an input and $h(t)$ does not have the dimension of an output.

Figure 2-10 Definition of impulse response $h(t)$.

Example 2-9 The impulse response of a certain linear stationary system is

$$h(t) = \frac{1}{\tau} e^{-t/\tau} u(t)$$

Find the output $y(t)$ for input

$$x(t) = x_0 u(t)$$

SOLUTION From (2-2),

$$y(t) = \int_{-\infty}^{\infty} \frac{x_0}{\tau} u(\lambda) e^{-(t-\lambda)/\tau} u(t - \lambda) \, d\lambda$$

The function $u(t - \lambda)$ is equal to 1 for $t - \lambda > 0$ (for $\lambda < t$) and is equal to 0 otherwise. Thus,

$$y(t) = \int_{-\infty}^{t} \frac{x_0}{\tau} e^{-(t-\lambda)/\tau} u(\lambda) \, d\lambda$$

The function $u(\lambda)$ is equal to 1 if $\lambda > 0$ and is equal to 0 otherwise; thus,

$$y(t) = \begin{cases} 0 & t \leq 0 \\ \int_0^t \frac{x_0}{\tau} e^{-(t-\lambda)/\tau} \, d\lambda & t > 0 \end{cases}$$

This can be written

$$y(t) = u(t) \int_0^t \frac{x_0}{\tau} e^{-(t-\lambda)/\tau} \, d\lambda$$

Using the law of exponents to factor the exponential gives

$$y(t) = \frac{x_0}{\tau} e^{-t/\tau} u(t) \int_0^t e^{\lambda/\tau} \, d\lambda$$

Performing the integration yields

$$y(t) = x_0 e^{-t/\tau} u(t)(e^{t/\tau} - 1)$$

which simplifies to

$$y(t) = x_0(1 - e^{-t/\tau}) u(t)$$

An alternative form of the convolution integral is sometimes useful. Changing the variable of integration in (2-2) from λ to $\eta = t - \lambda$ gives (see Prob. 2-14)

$$y(t) = \int_{-\infty}^{\infty} x(t - \eta) h(\eta) \, d\eta \tag{2-3}$$

Equation (2-3) is easier to use than (2-2) when the expression for $x(t)$ is simpler than that for $h(t)$.

Example 2-10 The impulse response of a certain linear stationary system is

$$h(t) = \frac{1}{\tau} e^{-t/\tau} (\sin \omega t) u(t)$$

Find the output $y(t)$ for input

$$x(t) = x_0 u(t)$$

SOLUTION Since the input $x(t)$ is simpler than the impulse response $h(t)$, we use (2-3), which gives

$$y(t) = \int_{-\infty}^{\infty} x_0 u(t-\eta) \frac{1}{\tau} e^{-\eta/\tau}(\sin \omega \eta) u(\eta)\, d\eta$$

Replacing $u(t-\eta)$ by unity for $\eta < t$ and by zero for $\eta > t$ gives

$$y(t) = \frac{x_0}{\tau} \int_{-\infty}^{t} e^{-\eta/\tau}(\sin \omega \eta) u(\eta)\, d\eta$$

Replacing $u(\eta)$ by unity for $\eta > 0$ and by zero for $\eta < 0$ gives

$$y(t) = \frac{x_0}{\tau} u(t) \int_{0}^{t} e^{-\eta/\tau}(\sin \omega \eta)\, d\eta$$

Performing the integration yields

$$\int_{0}^{t} e^{-\eta/\tau}(\sin \omega \eta)\, d\eta = \frac{\tau[\omega\tau - e^{-t/\tau}(\sin \omega t + \omega\tau \cos \omega t)]}{1 + (\omega\tau)^2}$$

Thus the output is

$$y(t) = \frac{x_0[\omega\tau - e^{-t/\tau}(\sin \omega t + \omega\tau \cos \omega t)]u(t)}{1 + (\omega\tau)^2}$$

Sometimes an impulse response contains a delta function. In such a case, we use the integration formula (see Prob. 1-6)

$$\int_{a}^{b} \delta(\alpha)\, d\alpha = u(b) - u(a) \qquad (2\text{-}4)$$

Example 2-11 The impulse response of a certain linear stationary system is

$$h(t) = \delta(t) - \frac{1}{\tau} e^{-t/\tau} u(t)$$

Find the output $y(t)$ for input $x(t) = x_0 u(t)$.

SOLUTION Equation (2-3) gives

$$y(t) = \int_{-\infty}^{\infty} x_0 u(t-\eta)\left[\delta(\eta) - \frac{1}{\tau} e^{-\eta/\tau} u(\eta)\right] d\eta$$

We write this as

$$y(t) = y_1(t) - y_2(t)$$

where

$$y_1(t) = x_0 \int_{-\infty}^{\infty} u(t-\eta)\delta(\eta)\, d\eta$$

and

$$y_2(t) = \int_{-\infty}^{\infty} \frac{x_0}{\tau} e^{-\eta/\tau} u(\eta) u(t-\eta)\, d\eta$$

Replacing $u(t - \eta)$ by unity for $\eta < t$ and by zero for $\eta > t$ gives

$$y_1(t) = x_0 \int_{-\infty}^{t} \delta(\eta)\, d\eta \quad \text{and} \quad y_2(t) = \int_{-\infty}^{t} \frac{x_0}{\tau} e^{-\eta/\tau} u(\eta)\, d\eta$$

From (2-4),

$$y_1(t) = x_0[u(t) - u(-\infty)] = x_0 u(t)$$

From Example 2-9,

$$y_2(t) = x_0(1 - e^{-t/\tau}) u(t)$$

The output is

$$y(t) = y_1(t) - y_2(t) = x_0 e^{-t/\tau} u(t)$$

The significance of the convolution integral (2-2) lies mainly in two facts: (1) deriving (2-2) requires only superposition and time invariance. This means that any linear stationary system (including feedback systems) can be described by a transfer characteristic (a convolution integral). (2) The transfer characteristic of a linear stationary system is completely determined by the impulse response of the system. A linear stationary system is described completely by its output for a single, simple input (an impulse). Consequently, analysis of a linear stationary system is much easier than analysis of a nonlinear or nonstationary feedback system which (in general) has no transfer characteristic and is not described completely in terms of its output for a single input.

2-2C Graphical Interpretation of Convolution

The convolution integral

$$y(t) = \int_{-\infty}^{\infty} x(\lambda) h(t - \lambda)\, d\lambda$$

has a simple graphical interpretation: at any instant t_0 an output $y(t_0)$ is the net area† bounded by the product $x(\lambda) h(t_0 - \lambda)$ and the λ axis. Below, we illustrate graphical interpretation of convolution for a system with input

$$x(t) = x_0 u(t)$$

and impulse response

$$h(t) = \frac{1}{\tau} e^{-t/\tau} u(t)$$

The output at a particular time t_0 is given by

$$y(t_0) = \int_{-\infty}^{\infty} x(\lambda) h(t_0 - \lambda)\, d\lambda$$

†The area is negative where the product is negative.

We interpret this relation graphically as follows:

1. We plot $x(\lambda)$ versus λ, as shown in Fig. 2-11a.
2. We plot $h(t_0 - \lambda)$ versus λ. The simplest way is to note that
$$h(t_0 - \lambda) = h[-(\lambda - t_0)]$$
Thus, to obtain the graph of $h(t_0 - \lambda)$ versus λ we reflect $h(\lambda)$ in the vertical axis and shift the result t_0 units to the right (to the left if $t_0 < 0$). Figure 2-11b shows a graph of $h(t_0 - \lambda)$.
3. We plot the product $x(\lambda)h(t_0 - \lambda)$ versus λ. The result is shown in Fig. 2-11c. Note that it is drawn for $t_0 > 0$. For $t_0 < 0$, there is no value of λ for which both $x(\lambda)$ and $h(t_0 - \lambda)$ are nonzero. Therefore, if $t_0 < 0$, the product $x(\lambda)h(t_0 - \lambda)$ is zero for all λ.
4. We calculate the area bounded by $x(\lambda)h(t_0 - \lambda)$ and the λ axis. This is the shaded area in Fig. 2-11c; thus (see Example 2-9)
$$y(t_0) = \frac{x_0}{\tau} u(t_0) \int_0^{t_0} e^{-(t_0-\lambda)/\tau} d\lambda$$
$$= x_0(1 - e^{-t_0/\tau})u(t_0)$$
as shown in Fig. 2-11d.

We can regard Fig. 2-11c as one frame (taken at $t = t_0$) from a motion picture of a convolution operation. In this motion picture the function $x(\lambda)$ does not change from one frame to the next because $x(\lambda)$ is not a function of time t. As time goes on (from one frame to the next), the function $h(t - \lambda)$ moves to the right along the λ axis. At

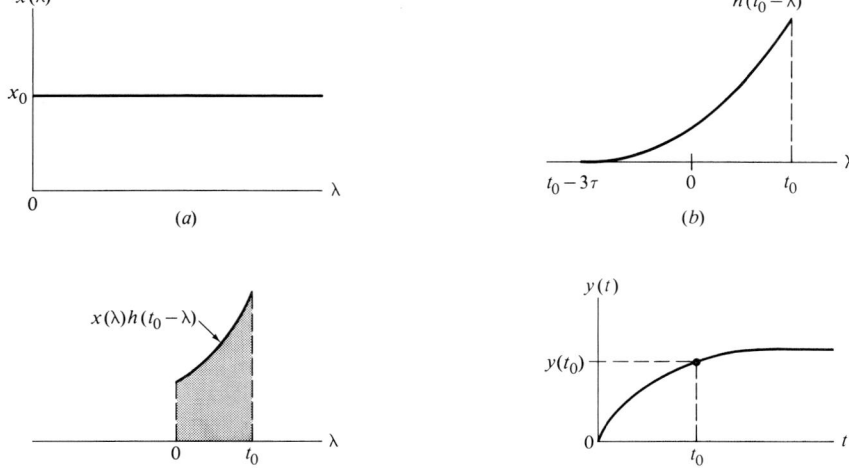

Figure 2-11 Graphical interpretation of convolution.

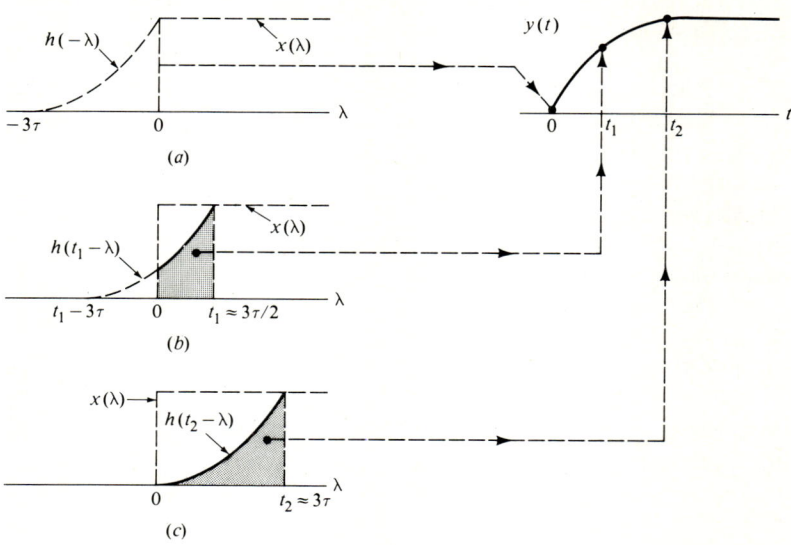

Figure 2-12 Graphical interpretation of convolution.

each instant the output $y(t)$ is the net area bounded by $x(\lambda)h(t - \lambda)$ and the λ axis. Three frames of this motion picture are shown in Fig. 2-12. In Fig. 2-12a, $t = 0$, and the right edge of $h(t - \lambda) = h(-\lambda)$ is at $\lambda = 0$. The functions $x(\lambda)$ and $h(-\lambda)$ do not overlap, the product $x(\lambda)h(t - \lambda)$ is zero, and the output $y(t)$ is zero. In Fig. 2-12b, $t = t_1 = 3\tau/2$ (about half of the duration of the impulse response), and the output is

$$y(t_1) = y\left(\frac{3\tau}{2}\right) = x_0(1 - e^{-3/2}) = 0.78x_0$$

In Fig. 2-12c, $t = t_2 = 3\tau$ and the output is

$$y(t_2) = y(3\tau) = x_0(1 - e^{-3}) = 0.95x_0$$

Graphical interpretation of (2-2) enables us to visualize a convolution operation. This in turn helps make an impulse response a meaningful description of a linear stationary system. For example, we can infer several things from Fig. 2-12:

1. At any instant t the output $y(t)$ depends only on earlier values of the input $x(t)$; that is, $y(t)$ depends only on values of $x(\lambda)$ over $-\infty < \lambda < t$. This shows that $y(t)$ is causal and that a system having impulse response $h(t) = (1/\tau)e^{-t/\tau}u(t)$ is realizable.
2. The contribution to $y(t)$ of an input value $x(t_a)$ decreases exponentially with $t - t_a$ (decreases as $e^{-(t-t_a)/\tau}$). This means the system has a finite memory of past inputs.
3. For $t > 3\tau$, the output has almost reached its steady-state value $y(\infty) = x_0$. The

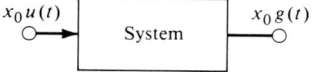

Figure 2-13 Definition of step response $g(t)$.

duration of the transient component of the output is equal to the duration of the impulse response.

4. For a step input, the steady-state component of the output is proportional to the area of the impulse response. This means that the response to a step input is bounded (the system is at least conditionally stable) if the area of the impulse response is finite.

2-2D Step Response

The *step response* of a linear stationary system is the function $g(t)$, where $y(t) = x_0 g(t)$ is the output for input $x_0 u(t)$. This definition is illustrated in Fig. 2-13. In this book the symbol $g(t)$ (with subscripts if necessary) always denotes a step response.

Example 2-12 Find the step response of the system of Fig. 2-14.

SOLUTION From Fig. 2-14,

$$w(t) = K_1 \int_{-\infty}^{t} x(\alpha)\, d\alpha - K_1 \int_{-\infty}^{t-t_0} x(\alpha)\, d\alpha$$

and

$$y(t) = K_2 \frac{dw(t)}{dt} = K_1 K_2 [x(t) - x(t - t_0)]$$

For $x(t) = x_0 u(t)$ this gives

$$y(t) = K_1 K_2 x_0 [u(t) - u(t - t_0)] = K_1 K_2 x_0 r\left(\frac{t}{t_0}\right)$$

Therefore, the step response is

$$g(t) = \frac{y(t)}{x_0} = K_1 K_2 r\left(\frac{t}{t_0}\right)$$

An impulse response is the derivative of the associated step response; i.e.,

$$h(t) = \frac{dg(t)}{dt} \qquad (2\text{-}5)$$

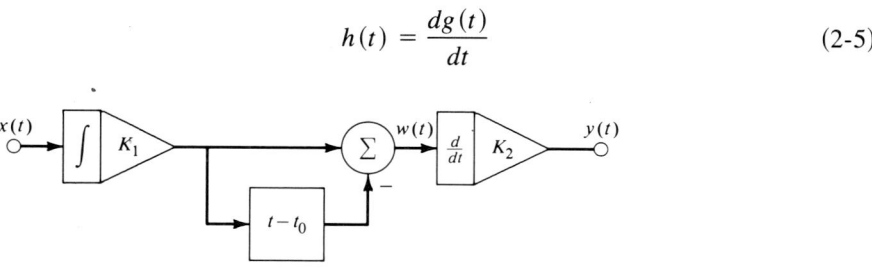

Figure 2-14 System of Example 2-12.

To show this, we use (2-3) with $x(t) = x_0 u(t)$; thus

$$y(t) = x_0 g(t) = \int_{-\infty}^{\infty} h(\eta) x_0 u(t - \eta) \, d\eta$$

Canceling x_0 and replacing $u(t - \eta)$ by unity for $\eta < t$ and by zero for $\eta > t$ gives

$$g(t) = \int_{-\infty}^{t} h(\eta) \, d\eta \qquad (2\text{-}6)$$

Differentiating this relation with respect to t yields (2-5).

Example 2-13 Find the impulse response of the system of Example 2-12.

SOLUTION From Example 2-12, the step response of the system is

$$g(t) = K_1 K_2 [u(t) - u(t - t_0)]$$

Using (2-5) gives

$$h(t) = K_1 K_2 [\delta(t) - \delta(t - t_0)]$$

Step response is important because step signals are widely used as test inputs and because (2-5) often provides the best means of obtaining the impulse response of a system.

2-2E Response to a Rectangular Pulse

A rectangular pulse can be expressed as

$$x_0 r\left(\frac{t}{\tau}\right) = x_0 u(t) - x_0 u(t - \tau) \qquad (2\text{-}7a)$$

Using this expression, superposition, and time invariance, we can express the output $y(t)$ of a linear stationary system for input $x(t) = x_0 r(t/\tau)$ as

$$y(t) = x_0 g(t) - x_0 g(t - \tau) \qquad (2\text{-}7b)$$

where $g(t)$ is the step response of the system. For example, the output of the system of Example 2-12 for input $x(t) = x_0 r(t/\tau)$ is

$$y(t) = K_1 K_2 x_0 \left[r\left(\frac{t}{t_0}\right) - r\left(\frac{t - \tau}{t_0}\right) \right]$$

2-2F Block Diagrams

By using convolution, any linear stationary system, no matter how complicated, can be described by a transfer characteristic. Therefore, any linear stationary system can be represented by a single block, as shown in Fig. 2-15. Conventionally, only the impulse response is written in the box. It is understood that the transfer characteristic is given by (2-2).

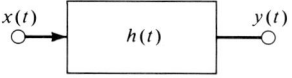

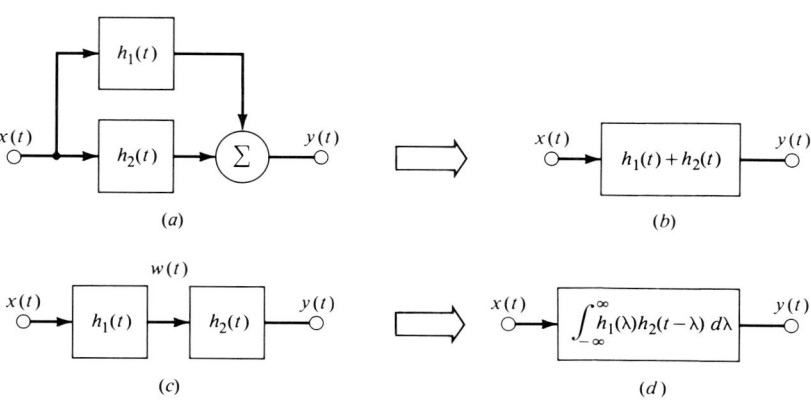

Figure 2-15 Symbol for a linear stationary system having impulse response $h(t)$.

Figure 2-16 Reduction of connections of linear stationary systems to single blocks: (a) and (b) parallel; (c) and (d) series.

Any system composed entirely of linear stationary elements is a linear stationary system. Therefore, the parallel and series connections of Fig. 2-16 are linear and stationary. Each can be reduced to a single block. (Procedures for reducing a feedback system to a single block are given in Chaps. 4 and 5.) For the system of Fig. 2-16a, input $x(t) = \mathbf{a}\delta(t)$ produces output $y(t) = \mathbf{a}h_1(t) + \mathbf{a}h_2(t)$. Therefore, the impulse response of the system is

$$h(t) = h_1(t) + h_2(t)$$

The parallel connection of Fig. 2-16a reduces to a single block having impulse response $h(t)$ above, as shown in Fig. 2-16b. For the system of Fig. 2-16c, the output of the first block for input $\mathbf{a}\delta(t)$ is (by the definition of impulse response)

$$w(t) = \mathbf{a}h_1(t)$$

By convolution, the output of the second block (the output of the system) is

$$y(t) = \mathbf{a}\int_{-\infty}^{\infty} h_1(\lambda)h_2(t - \lambda)\, d\lambda$$

It follows that the impulse response of the system is

$$h(t) = \int_{-\infty}^{\infty} h_1(\lambda)h_2(t - \lambda)\, d\lambda \tag{2-8}$$

The series connection of Fig. 2-16c reduces to a single block having impulse response $h(t)$ above, as shown in Fig. 2-16d.

Example 2-14 Find the impulse response of the system of Fig. 2-17, where

68 INTRODUCTION TO SYSTEM ANALYSIS

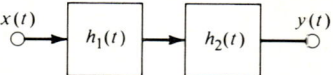

Figure 2-17 System of Example 2-14.

$$h_1(t) = \frac{1}{\tau_1} e^{-t/\tau_1} u(t), \qquad h_2(t) = \frac{1}{\tau_2} u(t)$$

SOLUTION From (2-8) the impulse response is

$$h(t) = \int_{-\infty}^{\infty} h_1(\lambda) h_2(t - \lambda) \, d\lambda = (\tau_1 \tau_2)^{-1} \int_{-\infty}^{\infty} e^{-\lambda/\tau_1} u(\lambda) u(t - \lambda) \, d\lambda$$

$$= \tau_2^{-1} (1 - e^{-t/\tau_1}) u(t)$$

2-2G Commutativity

Linear stationary systems in series commute. The transfer characteristic of a series connection of two linear stationary systems in series is independent of the order in which the systems are connected, as illustrated in Fig. 2-18.

To prove commutativity we first find the impulse response $h_a(t)$ for the series connection of Fig. 2-18a. From (2-8),

$$h_a(t) = \int_{-\infty}^{\infty} h_1(\lambda) h_2(t - \lambda) \, d\lambda$$

Next we find the impulse response $h_b(t)$ for the same systems connected in the reverse order (Fig. 2-18b). From (2-8),

$$h_b(t) = \int_{-\infty}^{\infty} h_2(\eta) h_1(t - \eta) \, d\eta \qquad (2\text{-}9)$$

Changing the variable of integration from η to $\lambda = t - \eta$ in (2-9) yields (2-8). This shows that the systems of Fig. 2-18a and b have the same overall impulse response and the same overall transfer characteristic; i.e., two linear stationary systems in series commute. It follows by induction that the transfer characteristic of a series connection of any number of linear stationary systems is independent of the order in which the systems are connected. Commutativity of linear stationary systems often simplifies analysis and design of systems (because it allows rearrangement of subsystems). Note that commutativity applies to mathematical models of linear stationary systems, not necessarily to physical systems; e.g., an amplifier and a turntable in an audio system do not commute, even though linear stationary models for these components do commute. Note also that nonlinear and nonstationary systems (models) do not commute, either with other nonlinear and nonstationary systems or with linear stationary systems.

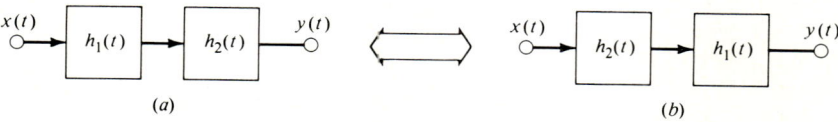

Figure 2-18 The commutativity property of linear stationary systems.

LINEAR STATIONARY SYSTEMS 69

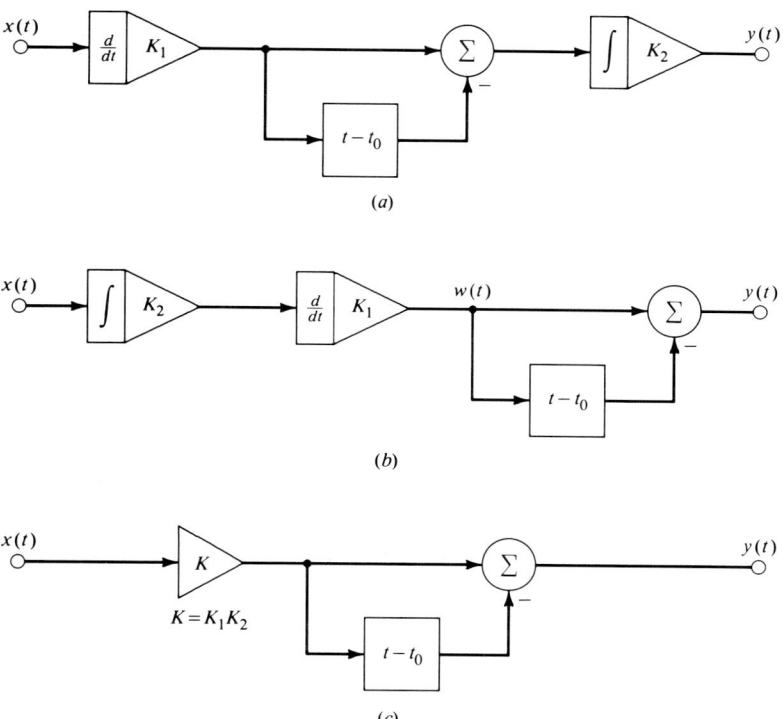

Figure 2-19 System of Example 2-15.

A series connection of systems (models) can be reordered freely only if all systems are linear and stationary.

Example 2-15 Find the step response of the system of Fig. 2-19a.

SOLUTION Using commutativity to reorder the subsystems gives the block diagram of Fig. 2-19b, where

$$w(t) = K_1 \frac{d}{dt} K_2 \int_{-\infty}^{t} x(\alpha)\, d\alpha = K_1 K_2 x(t)$$

This shows that the series connection of the differentiator and the integrator is equivalent to a proportion element having gain $K_1 K_2$, as shown in Fig. 2-19c. The step response of the system is

$$g(t) = K_1 K_2 [u(t) - u(t - t_0)] = K_1 K_2 r\left(\frac{t}{t_0}\right)$$

2-3 INTERPRETATION OF IMPULSE RESPONSE

The transfer characteristic of a linear stationary system is determined by the impulse response of the system. Consequently, many important properties of a linear stationary

70 INTRODUCTION TO SYSTEM ANALYSIS

system can be deduced from the impulse response of the system. We describe below how realizability, stability, and fidelity are reflected in the impulse response of a linear stationary system.

2-3A Realizability

A linear stationary system is realizable if and only if its impulse response is zero for $t < 0$. This follows from (2-2), which gives the output $y(t)$ for input $x(t)$ as

$$y(t) = \int_{-\infty}^{\infty} x(\lambda) h(t - \lambda) \, d\lambda$$

If $h(t) = 0$ for $t < 0$, then $h(t - \lambda) = 0$ for $\lambda > t$ and

$$y(t) = \int_{-\infty}^{t} x(\lambda) h(t - \lambda) \, d\lambda$$

The only values of the input $x(\lambda)$ that are used in finding $y(t)$ are those for which $-\infty < \lambda < t$. This shows that $y(t)$ is causal (that the system is realizable) if $h(t) = 0$ for $t < 0$.

Example 2-16 The impulse response

$$h(t) = \frac{1}{\tau} r\left(\frac{t - t_0}{\tau}\right)$$

describes a realizable system if $t_0 \geq 0$ (Fig. 2-20a) and a nonrealizable system if $t_0 < 0$ (Fig. 2-20b).

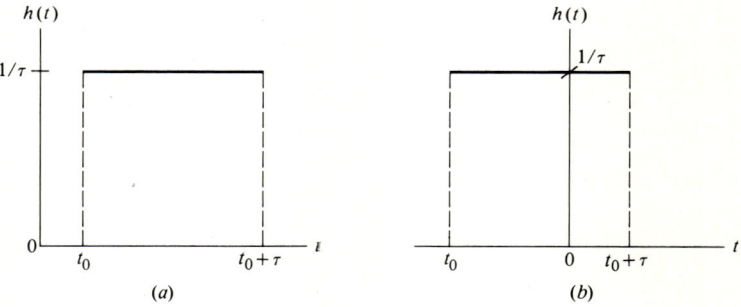

Figure 2-20 Impulse response of Example 2-16: (a) $t_0 > 0$; (b) $t_0 < 0$.

The realizability condition $h(t) = 0$ for $t < 0$ applies only to a linear stationary system because a nonlinear or nonstationary system is not described completely by its response to an impulse.

Example 2-17 A compression element having input $x(t)$ and output $y(t)$ is described by the transfer characteristic

$$y(t) = x(\eta t) \quad \text{where } \eta > 1$$

The "impulse response"

$$h(t) = \delta(\eta t)$$

equals zero for $t < 0$. Nonetheless, a compression element is nonrealizable, as shown in Example 1-30. The condition $h(t) = 0$ for $t < 0$ is insufficient in this case because a compression element is nonstationary (see Example 2-5).

2-3B Stability

A linear stationary system is stable if and only if its impulse response $h(t)$ is absolutely integrable, i.e., if and only if

$$\int_{-\infty}^{\infty} |h(\eta)| \, d\eta < \infty \qquad (2\text{-}10)$$

Otherwise the system is either conditionally stable or unstable.

To show that (2-10) is a sufficient condition for stability we must show that it implies that any bounded input produces a bounded output. From (2-3), the output $y(t)$ for input $x(t)$ is given by

$$y(t) = \int_{-\infty}^{\infty} x(t - \eta) h(\eta) \, d\eta$$

The triangle inequality† gives

$$|y(t)| \le \int_{-\infty}^{\infty} |x(t - \eta)| \, |h(\eta)| \, d\eta$$

Consequently, the condition $|x(t)| \le M_x$, where M_x is a positive constant, implies that

$$|y(t)| \le M_x \int_{-\infty}^{\infty} |h(\eta)| \, d\eta$$

This shows that $y(t)$ is bounded if $x(t)$ is bounded and (2-10) holds. Therefore, (2-10) is a sufficient condition for stability.

To show that (2-10) is a necessary condition for stability we need only find one system for which it does not hold and for which one bounded input produces an unbounded output. Consider a system for which $h(t) = (1/\tau) u(t)$. In this case the integral

$$\int_{-\infty}^{\infty} |h(\eta)| \, d\eta = \left| \frac{1}{\tau} \right| \int_{0}^{\infty} d\eta$$

†The (generalized) triangle inequality

$$\left| \int_{-\infty}^{\infty} f(\alpha) \, d\alpha \right| \le \int_{-\infty}^{\infty} |f(\alpha)| \, d\alpha$$

is a generalization of the inequality $|a + b| \le |a| + |b|$.

72 INTRODUCTION TO SYSTEM ANALYSIS

does not exist; thus, (2-10) does not hold. Furthermore, the output for input $x(t) = x_0 u(t)$ is

$$y(t) = \int_{-\infty}^{\infty} \frac{x_0}{\tau} u(t - \eta) u(\eta) \, d\eta = \frac{x_0}{\tau} u(t) \int_0^t d\eta = \frac{x_0 t}{\tau} u(t)$$

which is unbounded. Therefore, (2-10) is a necessary condition for stability.

Example 2-18 The impulse response for a certain linear stationary system is

$$h(t) = \frac{1}{\tau} e^{-t/\tau} u(t)$$

Find the values of τ for which this system is stable.

SOLUTION We have

$$\int_{-\infty}^{\infty} |h(\eta)| \, d\eta = \int_{-\infty}^{\infty} \left| \frac{1}{\tau} e^{-\eta/\tau} u(\eta) \right| d\eta = \left| \frac{1}{\tau} \right| \int_0^{\infty} e^{-\eta/\tau} \, d\eta = -\frac{\tau}{|\tau|} e^{-\eta/\tau} \Big|_0^{\infty}$$

This gives

$$\int_{-\infty}^{\infty} |h(\eta)| \, d\eta = \begin{cases} 1 & \tau > 0 \\ \infty & \tau \leq 0 \end{cases}$$

Therefore, the system is stable for $\tau > 0$.

Example 2-19 Find the values of K for which the system of Fig. 2-21 is stable.

SOLUTION From Example 1-22, the impulse response of the system is

$$h(t) = \sum_{n=0}^{\infty} K^n \delta(t - nt_0)$$

Figure 2-22 shows graphs of $|h(t)|$ for $|K| < 1$ and $|K| > 1$. The graphs suggest that the system is stable for $|K| < 1$ and unstable for $|K| > 1$. Below, we show that this is the case.

Using $h(t)$ above in (2-10) gives

$$\int_{-\infty}^{\infty} |h(\eta)| \, d\eta = \int_{-\infty}^{\infty} \left| \sum_{n=0}^{\infty} K^n \delta(\eta - nt_0) \right| d\eta$$

The functions $\delta(\eta), \delta(\eta - t_0), \ldots$ do not overlap. Therefore,

$$\left| \sum_{n=0}^{\infty} K^n \delta(\eta - nt_0) \right| = \sum_{n=0}^{\infty} |K^n \delta(\eta - nt_0)|$$

The functions $\delta(\eta), \delta(\eta - t_0), \ldots$ are positive. Therefore,

$$\sum_{n=0}^{\infty} |K^n \delta(\eta - nt_0)| = \sum_{n=0}^{\infty} |K^n| \delta(\eta - nt_0)$$

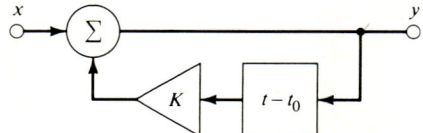

Figure 2-21 System of Example 2-19.

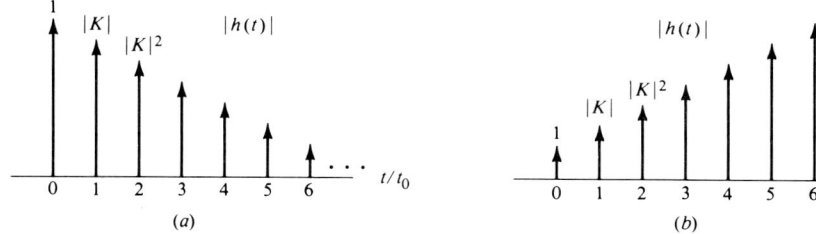

Figure 2-22 Magnitude of the impulse response of the system of Fig. 2-21: (a) $|K| < 1$; (b) $|K| > 1$.

Hence we have

$$\int_{-\infty}^{\infty} |h(\eta)| \, d\eta = \int_{-\infty}^{\infty} \sum_{n=0}^{\infty} |K^n| \delta(\eta - nt_0) \, d\eta$$

$$= \sum_{n=0}^{\infty} |K|^n \int_{-\infty}^{\infty} \delta(\eta - nt_0) \, d\eta = \sum_{n=0}^{\infty} |K|^n$$

This is a geometric series (see Thomas and Finney, pp. 754–755) with ratio $|K|$. For $|K| < 1$, the series converges:

$$\sum_{n=0}^{\infty} |K|^n = \frac{1}{1-|K|} \qquad |K| < 1$$

For $|K| > 1$ the series diverges. This shows that the system of Fig. 2-21 is stable for $|K| < 1$. The reader can show that the system is conditionally stable for $|K| = 1$ by finding the output for a step input and the output for input $x_0 r(t/\tau)$ with $\tau < t_0$ (see Example 1-22).

2-3C Fidelity

From (1-29) a distortionless system is a linear stationary system having impulse response

$$h(t) = K\delta(t - t_0) \qquad (2\text{-}11)$$

The parameters K and t_0 are arbitrary. Gain (or attenuation) and delay do not introduce distortion, no matter how much they may degrade other measures of performance.

An impulse rarely is used as a test input for calculating or measuring distortion because of certain mathematical difficulties. For example, from (2-11) the ideal (distortionless) output for input $x(t) = \mathbf{a}\delta(t)$ is

$$y_i(t) = K\mathbf{a}\delta(t - t_0)$$

In this case, the integral on the right side of (1-30) (the defining relation for integrated squared error) does not exist.† Nonetheless, the impulse response of a system provides some qualitative information about distortion introduced by the system. In the context of convolution, the significant feature of a distortionless system is that the impulse response is a delta function [see (2-11)]. The more nearly the impulse response of a system approximates a delta function the smaller the distortion introduced by the

†An integral of $\delta^2(\alpha)$ is undefined if $\alpha = 0$ lies between the limits of integration.

system. Loosely, a linear stationary system provides high-fidelity (low-distortion) transmission for an input if (1) the impulse response of the system is much narrower than the narrowest significant feature (peak or wiggle) of the input and (2) the area bounded by the impulse response and the time axis is nonzero (a function approximating a delta function must have nonzero area).

To illustrate this idea we consider a system having impulse response

$$h(t) = \frac{1}{\tau} e^{-t/\tau} u(t)$$

with $\tau > 0$ (for stability). This function approaches a delta function as τ approaches zero (see Prob. 1-5). Suppose this system is required to provide high-fidelity transmission for input

$$x(t) = x_0 r\left(\frac{t}{t_0}\right) = x_0 u(t) - x_0 u(t - t_0)$$

From (2-7) and Example 2-9, the corresponding output is

$$y(t) = x_0(1 - e^{-t/\tau})u(t) - x_0[1 - e^{-(t-t_0)/\tau}]u(t - t_0)$$

Figure 2-23 shows graphs of the impulse response $h(t)$, the input $x(t)$, and the output $y(t)$ for two values of τ. For $\tau = t_0$ (Fig. 2-23a), the width of the impulse response

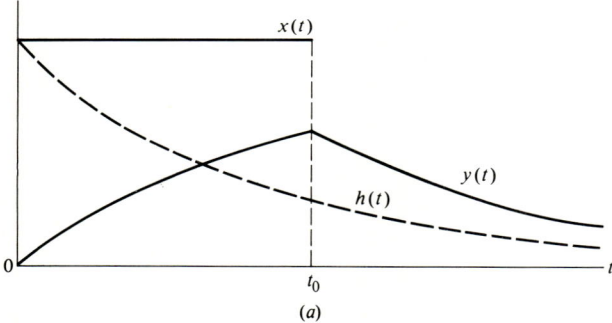

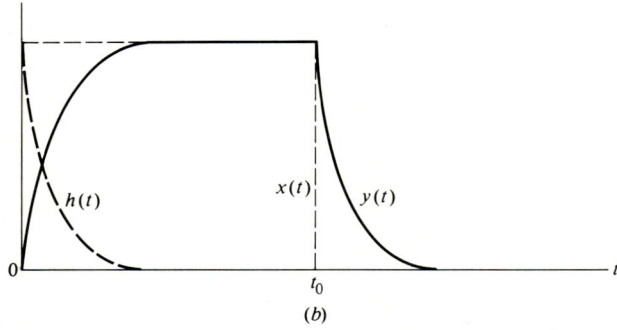

Figure 2-23 An illustration of conditions for high-fidelity transmission: (a) $\tau = t_0$; (b) $\tau = t_0/10$.

LINEAR STATIONARY SYSTEMS 75

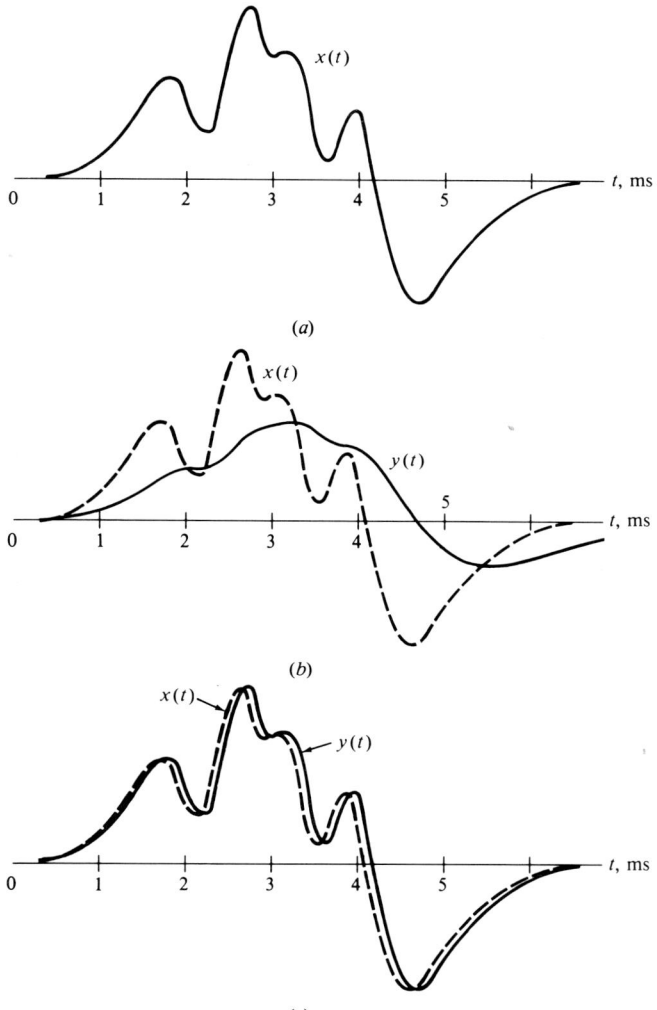

Figure 2-24 An illustration of conditions for high-fidelity transmission: (a) input, (b) output for $\tau = 1.0$ ms, (c) output for $\tau = 0.1$ ms.

$h(t)$ is comparable to that of the narrowest significant feature of the input $x(t)$. The output $y(t)$ does not look much like the input $x(t)$ (the distortion is high). For $\tau = t_0/10$ (Fig. 2-23b), the width of the impulse response $h(t)$ is smaller than the width of the narrowest significant feature of the input, and the output looks more like the input $x(t)$ (the distortion is lower than for $\tau = t_0$). For $\tau = t_0/100$, the output $y(t)$ would be practically indistinguishable from the input $x(t)$ on the scale used in Fig. 2-23.

Suppose this same system is to provide high-fidelity transmission for the input $x(t)$ of Fig. 2-24a. The width of the narrowest significant feature (wiggle) of $x(t)$ is approximately 1 ms. For high-fidelity transmission we require $\tau \ll 1$ ms. Figures 2-24b and c show graphs of the input $x(t)$ (dashed curves) and the output $y(t)$ (solid curves) for two values of τ. For $\tau = 1$ ms (Fig. 2-24b), there is considerable distortion; the fast wiggles in $x(t)$ are almost completely suppressed. For $\tau = 100$ μs (Fig. 2-24c), there is relatively little distortion. For $\tau = 10$ μs, the output would be indistinguishable from the input on the scale used in Fig. 2-24.

2-4 SYSTEMS DESCRIBED BY DIFFERENTIAL EQUATIONS

Many important systems are described by differential equations of the form

$$\frac{d^n y}{dt^n} + a_{n-1}\frac{d^{n-1} y}{dt^{n-1}} + \cdots + a_0 y = b_m \frac{d^m x}{dt^m} + \cdots + b_0 x \qquad (2\text{-}12)$$

where $x = x(t)$ is an input and $y = y(t)$ is the corresponding output. Lumped-constant electric circuits and lumped-constant mechanical systems are examples of such systems.

The order n of the highest derivative of an output $y(t)$ in (2-12) is called the *order* of the equation (the order of the system). The order of a system is one of the most significant attributes of the system. The order m of the highest derivative of an input $x(t)$, though also significant, has no name. The quantities $a_{n-1}, a_{n-2}, \ldots, a_0, b_m, \ldots, b_0$ are called the *coefficients* of (2-12). In a particular case, values of the coefficients are determined by parameters of the system (e.g., gains and time constants) described by (2-12). By convention (in this book), the coefficient of $d^n y/dt^n$ is unity. This causes no loss of generality. We can always enforce this convention by dividing a differential equation by the coefficient of the highest derivative of the output.

For $n > 0$, an output $y(t)$ is given implicitly by (2-12). Consequently, (2-12) is not easy to interpret. For example, we cannot tell at a glance whether a system described by such a differential equation provides high-fidelity transmission for a particular input. However, since a system described by (2-12) is linear and stationary (see Probs. 2-35 and 2-36), such a system also is described by the transfer characteristic

$$y(t) = \int_{-\infty}^{\infty} x(\lambda) h(t - \lambda)\, d\lambda$$

where $h(t)$ is the impulse response of the system. The transfer characteristic of a system described by (2-12) can be obtained by solving (2-12) for $y(t)$ with $x(t) = \mathbf{a}\delta(t)$. Subsequently, important properties of the system, e.g., fidelity, can be ascertained by examining the impulse response, as described in the previous section.

In this section we give examples of systems (block diagrams) described by (2-12). We then describe a procedure for obtaining the impulse response of such a system and illustrate the procedure using practical systems. In the examples given we consider only first- and second-order systems ($n = 1$ and $n = 2$). These are by far the most important cases (1) because a great many important physical systems (or components of systems) are adequately modeled as first- and second-order systems and (2) because any system described by (2-12) can be decomposed into first- and second-order subsystems. Methods for doing this are described in Chaps. 4 and 5.

2-4A Obtaining a Differential Equation from a Block Diagram

We begin by illustrating how an equation having the form of (2-12) is obtained from a block diagram describing a linear stationary system.

Example 2-20 Obtain a differential equation describing the system of Fig. 2-25.

SOLUTION From Fig. 2-25

$$v(t) = K_1 x(t) + K_2 \frac{dx(t)}{dt} - K_4 \frac{dy(t)}{dt}$$

and
$$y(t) = K_3 v(t)$$

Combining the last two relations and rearranging terms gives

$$K_3 K_4 \frac{dy}{dt} + y = K_2 K_3 \frac{dx}{dt} + K_1 K_3 x$$

Dividing by $K_3 K_4$ gives

$$\frac{dy}{dt} + (K_3 K_4)^{-1} y = K_2 K_4^{-1} \frac{dx}{dt} + K_1 K_4^{-1} x$$

This equation has the form of (2-12), with $n = m = 1$, $a_0 = (K_3 K_4)^{-1}$, $b_1 = K_2 K_4^{-1}$, and $b_0 = K_1 K_4^{-1}$.

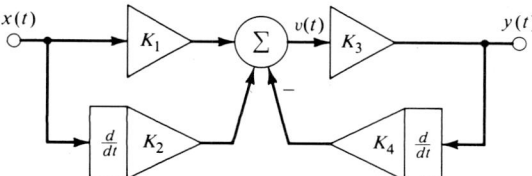

Figure 2-25 System of Example 2-20.

Example 2-21 Obtain a differential equation describing the system of Fig. 2-26.

SOLUTION From the figure,

$$y(t) = K_1 \int_{-\infty}^{t} v(\alpha) \, d\alpha = K_1 \int_{-\infty}^{t} [x(\alpha) - K_2 y(\alpha)] \, d\alpha$$

Differentiating this relation once with respect to t and rearranging terms gives

$$\frac{dy}{dt} + K_1 K_2 y = K_1 x$$

which has the form of (2-12) with $n = 1$, $m = 0$, $a_0 = K_1 K_2$, and $b_0 = K_1$.

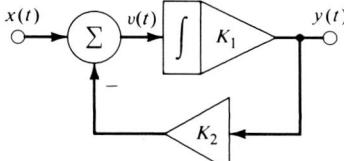

Figure 2-26 System of Example 2-21.

78 INTRODUCTION TO SYSTEM ANALYSIS

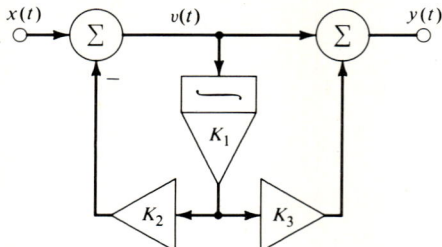

Figure 2-27 System of Example 2-22.

Example 2-22 Obtain a differential equation describing the system of Fig. 2-27.

SOLUTION From the figure,

$$v = x - K_1 K_2 \int_{-\infty}^{t} v(\alpha) \, d\alpha \qquad y = v + K_1 K_3 \int_{-\infty}^{t} v(\alpha) \, d\alpha$$

To simplify notation we define

$$\int_{-\infty}^{t} v(\alpha) \, d\alpha = w(t)$$

Thus

$$v + K_1 K_2 w = x \qquad v + K_1 K_3 w = y$$

Solving these last relations simultaneously for v and w gives

$$w = \frac{x - y}{K_1(K_2 - K_3)} \qquad v = \frac{K_2 y - K_3 x}{K_2 - K_3}$$

From the definition of $w(t)$ above,

$$v = \frac{dw}{dt}$$

Using the expressions above for w and v in this relation gives

$$\frac{K_2 y - K_3 x}{K_2 - K_3} = \frac{dx/dt - dy/dt}{K_1(K_2 - K_3)}$$

Rearranging terms yields

$$\frac{dy}{dt} + K_1 K_2 y = \frac{dx}{dt} + K_1 K_3 x$$

This has the form of (2-12) with $n = m = 1$, $a_0 = K_1 K_2$, $b_1 = 1$, and $b_0 = K_1 K_3$.

Example 2-23 Obtain a differential equation describing the system of Fig. 2-28.

SOLUTION From the figure,

$$y(t) = K_2 \int_{-\infty}^{t} v(\alpha) \, d\alpha \qquad v(t) = K_1 \int_{-\infty}^{t} w(\alpha) \, d\alpha$$

Differentiating these relations once with respect to t gives

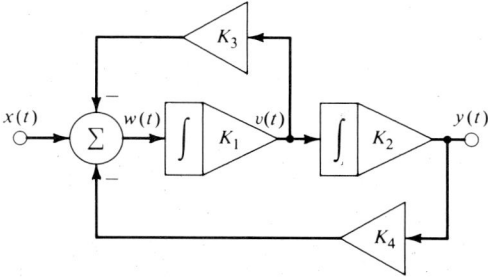

Figure 2-28 System of Example 2-23.

$$v = \frac{1}{K_2}\frac{dy}{dt} \qquad w = \frac{1}{K_1}\frac{dv}{dt} = \frac{1}{K_1 K_2}\frac{d^2 y}{dt^2}$$

From Fig. 2-28 we also have

$$w(t) = x(t) - K_3 v(t) - K_4 y(t)$$

Using the two previous relations to replace $w(t)$ and $v(t)$ and rearranging terms gives

$$\frac{d^2 y}{dt^2} + K_1 K_3 \frac{dy}{dt} + K_1 K_2 K_4 y = K_1 K_2 x$$

This has the form of (2-12) with $n = 2$, $m = 0$, $a_1 = K_1 K_3$, $a_0 = K_1 K_2 K_4$, and $b_0 = K_1 K_2$.

Next, we describe a procedure for obtaining the impulse response of a system described by (2-12) with $m = 0$ (no differentiation of an input). Subsequently, we extend the procedure to systems for which $m > 0$.

2-4B Impulse Response for $m = 0$ (No Differentiation of an Input)

For $m = 0$, (2-12) simplifies to

$$\frac{d^n y}{dt^n} + a_{n-1}\frac{d^{n-1} y}{dt^{n-1}} + \cdots + a_0 y = b_0 x \qquad (2\text{-}13)$$

We obtain the impulse response of a system described by (2-13) by the following indirect method. We obtain the step response $g(t)$ of the system and then we obtain the impulse response $h(t)$ from the step response using (2-5); thus

$$h(t) = \frac{dg(t)}{dt}$$

This indirect method has two advantages over obtaining $h(t)$ directly: (1) it is easier to solve (2-13) with $x(t) = x_0 u(t)$ than with $x(t) = a\delta(t)$, and (2) the method yields both step response and impulse response with little more work than it takes to find the step response alone.

To obtain an equation for the step response $g(t)$ of a system described by (2-13) we take $x(t) = x_0 u(t)$ and $y(t) = x_0 g(t)$. This gives

$$\frac{d^n g}{dt^n} + a_{n-1}\frac{d^{n-1} g}{dt^{n-1}} + \cdots + a_0 g = b_0 u \qquad (2\text{-}14)$$

For $a_0 \neq 0$, the solution of this equation is given by†

$$g(t) = b_0 a_0^{-1}[1 + g_c(t)]u(t) \tag{2-15}$$

where $g_c(t)$ is the complete solution of the equation‡

$$\frac{d^n g_c}{dt^n} + a_{n-1}\frac{d^{n-1} g_c}{dt^{n-1}} + \cdots + a_0 g_c = 0 \tag{2-16}$$

The function $g_c(t)$ obtained by solving (2-16) is called the *complementary function* for a system described by (2-13). By a theorem on differential equations, the complementary function consists of n linearly independent terms each satisfying (2-16). At least one of these terms has the form ce^{st}, where c and s are constants. Substituting ce^{st} for $g_c(t)$ in (2-16) gives

$$(s^n + a_{n-1}s^{n-1} + \cdots + a_0)ce^{st} = 0$$

The function e^{st} cannot be zero. To obtain a nontrivial solution we require $c \neq 0$. Dividing the equation above by ce^{st} gives

$$s^n + a_{n-1}s^{n-1} + \cdots + a_0 = 0 \tag{2-17}$$

The function ce^{st} satisfies (2-16) if s is any solution of (2-17). By the fundamental theorem of algebra, (2-17) has n solutions, which we denote by $p_1, p_2, \ldots, p_n$. Equation (2-17) is called the *characteristic equation* for a system described by (2-13). The left side of (2-17) is called the *characteristic polynomial* for the system. The roots of the characteristic polynomial (the solutions of the characteristic equation) are called the *characteristic roots* of the system. A characteristic equation can be written in factored form as

$$(s - p_1)(s - p_2) \cdots (s - p_n) = 0 \tag{2-18}$$

The complete solution $g_c(t)$ of (2-16) is constructed from the characteristic roots $p_1, p_2, \ldots, p_n$ as follows:

1. For each nonrepeated (distinct) root p include a term of the form

$$ce^{pt}$$

2. For each k-fold-repeated root p include a term of the form

$$(c_1 + c_2 t + \cdots + c_k t^{k-1})e^{pt}$$

For example, suppose the characteristic equation for a particular system is

$$(s - p_1)(s - p_2)(s - p_3)^2 = 0$$

where p_1, p_2, and p_3 are all different. In this case there are two nonrepeated roots (p_1 and p_2) and one 2-fold-repeated root (p_3). The complete solution of (2-16) is

†Most calculus textbooks treat methods for solving (2-14) in detail; we are content to outline the procedure.
‡In mathematics books (2-16) is called the *homogeneous equation* for a system described by (2-14).

$$g_c(t) = c_1 e^{p_1 t} + c_2 e^{p_2 t} + (c_3 + c_4 t)e^{p_3 t}$$

The solution contains $n = 4$ linearly independent terms, as required.

A complementary function contains n constants $c_1, c_2, \ldots, c_n$ (see, for example, the last equation above). So far as (2-16) is concerned, these constants are arbitrary. To determine them in particular applications we need n conditions on the complementary function $g_c(t)$. We can derive these conditions by requiring the step response $g(t)$ to be causal, i.e., the system to be realizable. First we require

$$g(t) = 0 \quad \text{for} \quad t < 0 \tag{2-19a}$$

and

$$\frac{d^i g}{dt^i} = 0 \quad \text{for} \quad t < 0 \tag{2-19b}$$
$$i = 1, 2, \cdots$$

Otherwise Taylor's expansion for $g(t)$ about some point $t_a < 0$ would be nonzero for some negative time t (the step response would be noncausal). Next, we note from (2-14) that the step response and its first $n - 1$ derivatives are continuous everywhere. Otherwise a delta function would result from at least one of the differentiations on the left side of (2-14). This cannot be allowed because there is no delta function on the right side of (2-14) (only a delta function can cancel a delta function). In particular, the step response and its first $n - 1$ derivatives are continuous at $t = 0$, so that

$$g(0^+) = g(0^-) \tag{2-20a}$$

where $t = 0^+$ ($t = 0^-$) denotes an infinitesimal positive (negative) time, and

$$\left.\frac{d^i g}{dt^i}\right|_{t=0^+} = \left.\frac{d^i g}{dt^i}\right|_{t=0^-} \quad \text{for } i = 1, 2, \ldots, n - 1 \tag{2-20b}$$

Combining (2-19) and (2-20) yields

$$g(0^+) = 0 \tag{2-21a}$$

and

$$\left.\frac{d^i g}{dt^i}\right|_{t=0^+} = 0 \quad \text{for } i = 1, 2, \ldots, n - 1 \tag{2-21b}$$

Equations (2-21) constitute n conditions on the step response $g(t)$ of a realizable system described by (2-13). These conditions are easier to apply if they are recast as conditions on the associated complementary function $g_c(t)$. From (2-15),

$$g(t) = b_0 a_0^{-1}[1 + g_c(t)]u(t)$$

The function $u(t)$ equals unity for $t = 0^+$. Therefore (2-21a) implies

$$g_c(0^+) = -1 \tag{2-22a}$$

All derivatives of $u(t)$ equal zero for $t = 0^+$. Therefore (2-21b) implies

$$\left.\frac{d^i g_c}{dt^i}\right|_{t=0^+} = 0 \quad \text{for } i = 1, 2, \ldots, n - 1 \tag{2-22b}$$

The conditions expressed by (2-22) determine the n constants $c_1, c_2, \ldots, c_n$ that appear in the complementary function for (2-13). For example, consider a second-order system having the complementary function

$$g_c(t) = c_1 e^{p_1 t} + c_2 e^{p_2 t}$$

To find the values of c_1 and c_2 we use (2-22) with $n = 2$; thus

$$g_c(0^+) = c_1 + c_2 = -1 \quad \text{and} \quad \left.\frac{dg_c}{dt}\right|_{t=0^+} = p_1 c_1 + p_2 c_2 = 0$$

Solving these equations for c_1 and c_2 gives

$$c_1 = \frac{p_2}{p_1 - p_2} \qquad c_2 = \frac{-p_1}{p_1 - p_2}$$

We can summarize the development above by outlining a procedure for obtaining the step response $g(t)$ and the impulse response $h(t)$ of a system described by (2-13):

1. Obtain the characteristic roots $p_1, p_2, \ldots, p_n$ by solving the characteristic equation

$$s^n + a_{n-1} s^{n-1} + \cdots + a_0 = 0$$

2. Form the complementary function $g_c(t)$ as follows:
   a. For each nonrepeated characteristic root p include a term of the form

$$c e^{pt}$$

   b. For each k-fold-repeated characteristic root p include a term of the form

$$(c_1 + c_2 t + \cdots + c_k t^{k-1}) e^{pt}$$

3. Determine the n constants $c_1, c_2, \ldots, c_n$ that appear in the complementary function $g_c(t)$ using the conditions

$$g_c(0^+) = -1 \qquad \left.\frac{d^i g_c}{dt^i}\right|_{t=0^+} = 0 \qquad \text{for } i = 1, 2, \ldots, n-1$$

4. The step response is given by

$$g(t) = b_0 a_0^{-1} [1 + g_c(t)] u(t)$$

5. The impulse response is given by

$$h(t) = \frac{dg(t)}{dt}$$

For $g(t)$ given above, this yields

$$h(t) = b_0 a_0^{-1} [1 + g_c(t)] \delta(t) + b_0 a_0^{-1} \left[\frac{dg_c(t)}{dt}\right] u(t)$$

The delta function is zero for $t \neq 0$. From (2-22a) the quantity $1 + g_c(t)$ is zero for $t = 0$. Consequently the first term equals zero for all time.† The impulse response is given by

$$h(t) = b_0 a_0^{-1} \frac{dg_c(t)}{dt} u(t) \qquad (2\text{-}23)$$

In Secs. 2-4C and 2-4D we illustrate this procedure using first- and second-order systems described by (2-13). In Sec. 2-4E we extend the procedure to systems described by (2-12), where $m > 0$.

2-4C First-Order Systems

We seek expressions for the step response and the impulse response of a first-order system described by

$$\frac{dy}{dt} + a_0 y = b_0 x \qquad (2\text{-}24)$$

where $x(t)$ is an input and $y(t)$ is the corresponding output. The steps are numbered in agreement with those of the procedure given above.

Step 1 The characteristic equation is

$$s + a_0 = 0$$

Thus, there is one characteristic root, given by $p_1 = -a_0$.

Step 2 The complementary function is

$$g_c(t) = ce^{-a_0 t}$$

Step 3 From (2-22), $g_c(0^+) = c = -1$.

Step 4 From (2-15), the step response is

$$g(t) = b_0 a_0^{-1}(1 - e^{-a_0 t})u(t) \qquad (2\text{-}25)$$

Step 5 From (2-23) the impulse response is

$$h(t) = b_0 e^{-a_0 t} u(t) \qquad (2\text{-}26)$$

Figure 2-29 shows graphs of the step response $g(t)$ and the impulse response $h(t)$ for $a_0 > 0$.

†Actually, the quantity $[1 + g_c(t)]\delta(t)$ is indeterminate (is of the form $0 \cdot \infty$) for $t = 0$. Dropping this term from the expression for $h(t)$ above is justified by the fact that the term contributes nothing to an output; i.e.,

$$\int_{-\infty}^{\infty} x(t - \lambda)[1 + g_c(\lambda)]\delta(\lambda) \, d\lambda = 0$$

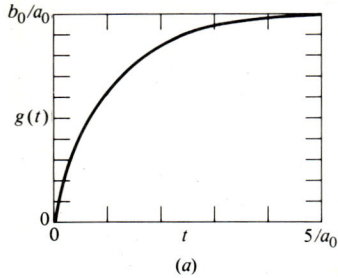 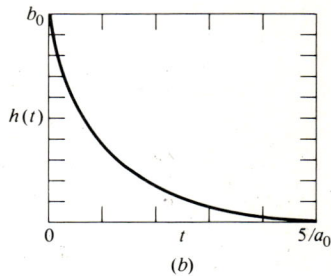

Figure 2-29 (a) Step response $g(t)$ and (b) impulse response $h(t)$ of a first-order system.

The steady-state component of the step response is

$$g_{ss}(t) = b_0 a_0^{-1}$$

Since the steady-state component of the output $y(t)$ for input $x_0 u(t)$ is $b_0 a_0^{-1} x_0$, the quantity $b_0 a_0^{-1}$ is called the *dc gain* of the system. The transient component of the step response is

$$g_t(t) = -b_0 a_0^{-1} e^{-a_0 t} u(t)$$

The time constant associated with the transient component is $1/a_0$. For $t = 3/a_0$, the step response (or the output for a step input) has reached about 95 percent of its steady-state value.

The impulse response $h(t)$ decays exponentially if $a_0 > 0$ and grows exponentially if $a_0 < 0$. Therefore, the system is stable if $a_0 > 0$ and unstable if $a_0 < 0$ (see Example 2-18). For $a_0 = 0$, the impulse response is $h(t) = b_0 u(t)$ and the system described by (2-24) is an integrator. Thus, for $a_0 = 0$, the system is conditionally stable (see Example 1-30).

The duration of the impulse response is proportional to $\tau = 1/a_0$. As τ is decreased (as a_0 is increased), the impulse response becomes narrower and distortion introduced by the system decreases. However, unless b_0 is proportional to a_0, the gain of the system also decreases as a_0 is increased. In practice, to increase the fidelity of transmission provided by this system we would increase both a_0 (to make the impulse response narrower) and b_0 (to maintain reasonable signal strength at the output).

The impulse response $h(t)$ is zero for $t < 0$, and so the system is realizable. This is no surprise since realizability is built into our procedure for finding $h(t)$ by the conditions (2-22). In practice, a differential equation describing a system usually is obtained from physical laws (e.g., Kirchhoff's laws) and a diagram of the system (e.g., a circuit diagram) where it is clear that the differential equation describes a physical (realizable) system. Equations (2-22) are logical consequences of that fact.

Example 2-24 Obtain expressions for the step response and the impulse response of the electric circuit of Fig. 2-30.

SOLUTION Applying Kirchhoff's current law to the node joining the resistor to the capacitor yields

$$C \frac{dy}{dt} + \frac{y - x}{R} = 0$$

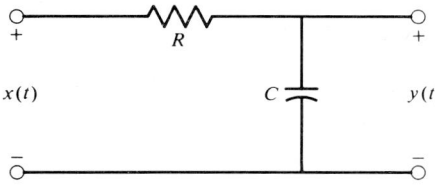

Figure 2-30 Electric circuit of Example 2-24.

Rearranging terms gives

$$\frac{dy}{dt} + \frac{1}{\tau} y = \frac{1}{\tau} x$$

where $\tau = RC$. This has the form of (2-24), with $a_0 = b_0 = 1/\tau$. From (2-25), the step response is

$$g(t) = (1 - e^{-t/\tau}) u(t)$$

From (2-26), the impulse response is

$$h(t) = \frac{1}{\tau} e^{-t/\tau} u(t)$$

The system (circuit) is stable because $\tau = RC > 0$ (resistance R and capacitance C are positive). On physical grounds the circuit must be stable because it is passive (contains no source of energy).

Decreasing either R or C decreases the time constant $\tau = RC$. This speeds up the step response (decreases the duration of the transient). The steady-state component of the step response (the dc gain of the system) is independent of the time constant.

Decreasing τ makes the impulse response proportionately narrower and taller, but the area bounded by $h(t)$ and the time axis is independent of τ. As $\tau \to 0$, $h(t) \to \delta(t)$ and the system becomes distortionless with gain $= 1$ and delay $= 0$. The circuit provides high-fidelity transmission for any input whose narrowest significant feature (peak or wiggle) is much wider than $h(t)$. For example, an input of the form

$$x(t) = A_0 + A_1 \cos \omega_1 t + \cdots + A_n \cos \omega_n t \qquad 0 < \omega_1 < \cdots < \omega_n$$

is transmitted without much distortion if $\tau \ll 2\pi/\omega_n$. As another example, a rectangular pulse

$$x(t) = x_0 r\left(\frac{t}{t_0}\right)$$

is transmitted without much distortion if the duration of the pulse is much larger than the time constant (see Fig. 2-23).

Example 2-25 The motor and load in Fig. 2-31 constitute a system with input $i(t)$ (current in amperes) and output $S(t)$ (angular speed in radians per second). Obtain the output $S(t)$ for step input $i(t) = i_0 u(t)$. Describe how torque constant K (Nm/A), moment of inertia $J[\text{Nm}/(\text{rad}/\text{s}^2)]$, and coefficient of viscous friction $F[\text{Nm}/(\text{rad}/\text{s})]$ influence transient and steady-state response.

SOLUTION As indicated in Fig. 2-31, the torque produced by the motor is

$$T(t) = Ki(t)$$

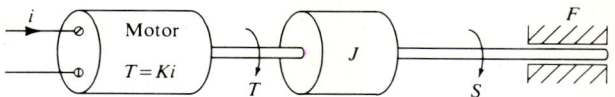

Figure 2-31 System of Example 2-25.

where $i(t)$ is armature current and K is the torque constant for the motor. From Newton's law for angular motion, torque $T(t)$ and angular speed $S(t)$ are related by

$$T = J\frac{dS}{dt} + FS$$

Combining these equations and rearranging terms gives

$$\frac{dS}{dt} + \frac{F}{J}S = \frac{K}{J}i$$

which has the form of (2-24), with $a_0 = F/J$ and $b_0 = K/J$. From (2-25) the output $S(t)$ for input $i(t) = i_0 u(t)$ is given by

$$S(t) = i_0 g(t) = \frac{Ki_0}{F}(1 - e^{-Ft/J})u(t)$$

Figure 2-32 shows a graph of angular speed $S(t)$ versus time t. The steady-state speed Ki_0/F is directly proportional to K, inversely proportional to F, and independent of J. Increasing K (using a bigger motor) or decreasing F (using roller bearings instead of sleeve bearings) increases steady-state speed, as one would expect. The moment of inertia J, which is directly proportional to mass of the load, has no influence on steady-state speed.

The time constant of the response is $\tau = J/F$. Since τ is directly proportional to J, a heavy load accelerates more slowly than a lighter one, as one would expect. The time constant is inversely proportional to the coefficient of friction F. Therefore, transient response can be speeded up by increasing F (by adding viscous damping) and increasing K or i_0 proportionately (to maintain the same steady-state speed). Although this wastes power (the power dissipated by the added viscous friction), it is sometimes the most economical way to improve transient response.

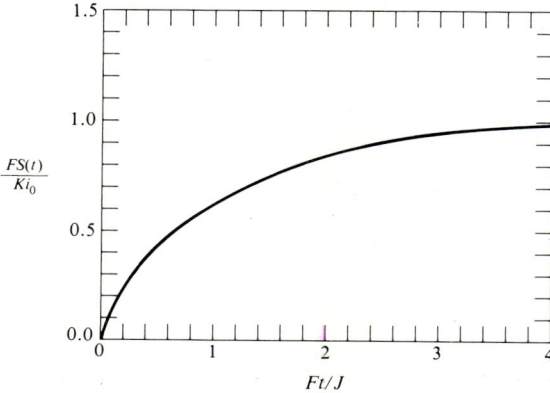

Figure 2-32 A graph of the output of the system of Example 2-25.

2-4D Second-Order Systems

Below we obtain expressions for step response and impulse response of a second-order system described by

$$\frac{d^2y}{dt^2} + a_1\frac{dy}{dt} + a_0 y = b_0 x \tag{2-27}$$

where $x(t)$ is an input and $y(t)$ is the corresponding output. The steps that follow are numbered in agreement with those of the procedure given in Sec. 2-4B.

Step 1 The characteristic equation is

$$s^2 + a_1 s + a_0 = 0$$

There are two characteristic roots, given by

$$p_1 = \frac{-a_1 + \sqrt{a_1^2 - 4a_0}}{2} \qquad p_2 = \frac{-a_1 - \sqrt{a_1^2 - 4a_0}}{2} \tag{2-28}$$

Step 2 For $a_1^2 \neq 4a_0$, the roots p_1 and p_2 are distinct. In that case, the complementary function is

$$g_c(t) = c_1 e^{p_1 t} + c_2 e^{p_2 t} \tag{2-29}$$

For $a_1^2 = 4a_0$, the roots p_1 and p_2 are identical (there is one repeated root). In that case, the complementary function is

$$g_c(t) = (c_1 + c_2 t) e^{pt} \tag{2-30}$$

where $p = -a_1/2$. In what follows, we treat the case $a_1^2 \neq 4a_0$ (distinct roots) in detail. Then we simply give the results (step response and impulse response) for the case $a_1^2 = 4a_0$ (repeated roots), leaving the derivation of those results as an exercise.

Step 3 From (2-22) and (2-29),

$$g_c(0^+) = c_1 + c_2 = -1 \qquad \left.\frac{dg_c(t)}{dt}\right|_{t=0^+} = p_1 c_1 + p_2 c_2 = 0$$

Solving these equations for c_1 and c_2 gives (for $p_1 \neq p_2$)

$$c_1 = \frac{p_2}{p_1 - p_2} \qquad c_2 = \frac{-p_1}{p_1 - p_2}$$

Using these relations in (2-29) gives

$$g_c(t) = \frac{p_2 e^{p_1 t} - p_1 e^{p_2 t}}{p_1 - p_2} \tag{2-31}$$

Step 4 From (2-15), the step response is

$$g(t) = b_0 a_0^{-1}\left(1 + \frac{p_2 e^{p_1 t} - p_1 e^{p_2 t}}{p_1 - p_2}\right)u(t) \tag{2-32}$$

Step 5 From (2-23), the impulse response is

$$h(t) = \frac{b_0 a_0^{-1}(p_1 p_2 e^{p_1 t} - p_1 p_2 e^{p_2 t}) u(t)}{p_1 - p_2}$$

From (2-28), $p_1 p_2 = a_0$; thus

$$h(t) = \frac{b_0}{p_1 - p_2}(e^{p_1 t} - e^{p_2 t}) u(t) \tag{2-33}$$

Equations (2-32) and (2-33) give the step response and the impulse response of a second-order system described by (2-27) having either real or complex distinct characteristic roots, i.e., for a system described by (2-27) with $a_1^2 - 4a_0 \neq 0$, in which case p_1 and p_2 given by (2-28) are different. These expressions, though valid, are cumbersome if the characteristic roots are complex. We wish to put these expressions in forms more appropriate for cases where $a_1^2 - 4a_0 < 0$ (where the characteristic roots are a complex-conjugate pair). Let $p_1 = p$ and $p_2 = p^*$, where†

$$p = \frac{-a_1 + j\sqrt{4a_0 - a_1^2}}{2} \qquad p^* = \frac{-a_1 - j\sqrt{4a_0 - a_1^2}}{2} \tag{2-34}$$

The complementary function given by (2-31) becomes

$$g_c(t) = \frac{p^* e^{pt} - p e^{p^* t}}{p - p^*} = \frac{\operatorname{Im} p^* e^{pt}}{\operatorname{Im} p} \tag{2-35}$$

where we have used the relation $\alpha - \alpha^* = 2j \operatorname{Im} \alpha$. To simplify notation we introduce the quantities

$$\sigma = \operatorname{Re} p = -\frac{a_1}{2} \qquad \omega = \operatorname{Im} p = \frac{\sqrt{4a_0 - a_1^2}}{2} \tag{2-36}$$

Then the characteristic root p is expressed in rectangular form as

$$p = \sigma + j\omega \tag{2-37}$$

and in polar form as

$$p = re^{j\theta} \tag{2-38}$$

where‡

$$r = \sqrt{\sigma^2 + \omega^2} \qquad \theta = \operatorname{Tan}^{-1}(\omega, \sigma) \tag{2-39}$$

†Throughout this book $j = \sqrt{-1}$.
‡The function $\operatorname{Tan}^{-1}(\omega, \sigma)$ is the *four-quadrant inverse tangent function* defined by

$$\operatorname{Tan}^{-1}(\omega, \sigma) = \begin{cases} \tan^{-1}\dfrac{\omega}{\sigma} & \sigma > 0 \\ 0 & \sigma = \omega = 0 \\ \pi + \tan^{-1}\dfrac{\omega}{\sigma} & \sigma < 0 \end{cases}$$

where $\tan^{-1} \alpha$ is the two-quadrant inverse tangent function whose range is $-\pi/2 \leq \tan^{-1} \alpha \leq \pi/2$. A four-quadrant inverse tangent function $\operatorname{Tan}^{-1}(\omega, \sigma)$ can be obtained using a pocket calculator to convert the quantity $\sigma + j\omega$ from rectangular to polar form. This gives $\sigma + j\omega = re^{j\theta}$, where $\theta = \operatorname{Tan}^{-1}(\omega, \sigma)$.

Using (2-37) and (2-38) for p in (2-35) gives

$$g_c(t) = \omega^{-1} \operatorname{Im}(re^{-j\theta}e^{(\sigma+j\omega)t}) = r\omega^{-1}e^{\sigma t} \operatorname{Im} e^{j(\omega t - \theta)}$$

Using Euler's identity† $e^{j\alpha} \equiv \cos \alpha + j \sin \alpha$ gives

$$g_c(t) = r\omega^{-1}e^{\sigma t} \sin(\omega t - \theta) \tag{2-40}$$

Using (2-40) for $g_c(t)$ in (2-32) gives

$$g(t) = b_0 a_0^{-1}[1 + r\omega^{-1}e^{\sigma t} \sin(\omega t - \theta)]u(t) \tag{2-41a}$$

where

$$\sigma = \frac{-a_1}{2} \qquad \omega = \frac{\sqrt{4a_0 - a_1^2}}{2} \qquad r = \sqrt{\sigma^2 + \omega^2} \qquad \theta = \operatorname{Tan}^{-1}(\omega, \sigma)$$

$$\tag{2-41b}$$

To obtain an expression for impulse response $h(t)$ we use (2-33) with $p_1 = p$, $p_2 = p^*$; thus

$$h(t) = b_0 \frac{(\operatorname{Im} e^{pt})u(t)}{\operatorname{Im} p}$$

Using (2-37) and (2-38) gives

$$h(t) = b_0 \omega^{-1} \operatorname{Im} e^{(\sigma+j\omega)t}u(t) = b_0 \omega^{-1} e^{\sigma t}(\sin \omega t)u(t) \tag{2-42a}$$

where

$$\sigma = \frac{-a_1}{2} \qquad \omega = \frac{\sqrt{4a_0 - a_1^2}}{2} \tag{2-42b}$$

Equations (2-32), (2-33), (2-41), and (2-42) give the step response and the impulse response for a second-order system described by (2-27). Equations (2-32) and (2-41) are equivalent, as are (2-33) and (2-42). Equations (2-32) and (2-33) are most appropriate for a system having real, distinct characteristic roots (where $a_1^2 > 4a_0$). Equations (2-41) and (2-42) are most appropriate for a system having complex-conjugate characteristic roots p, $p^* = \sigma \pm j\omega$ (where $a_1^2 < 4a_0$). The step response $g(t)$ and the impulse response $h(t)$ of a system having real, equal roots (where $a_1^2 = 4a_0$) are given by (see Prob. 2-41)

$$g(t) = b_0 a_0^{-1}[1 - (1 - pt)e^{pt}]u(t) \tag{2-43}$$

and

$$h(t) = b_0 a_0^{-1} p^2 t e^{pt} u(t) \tag{2-44}$$

where $p = -a_1/2$. The expressions obtained above are collected in Table 2-1 for easy reference.

Figure 2-33 shows graphs of step response $g(t)$ and impulse response $h(t)$ versus time t for a (stable) second-order system described by (2-27). For a system having real, negative characteristic roots (Fig. 2-33a) the step response rises monotonically from

†Named after Swiss mathematician Leonhard Euler (1707–1783), one of the founders of modern analysis.

Table 2-1 Step response $g(t)$ and impulse response $h(t)$ for a second-order system whose input $x(t)$ and output $y(t)$ are related by

$$\frac{d^2y}{dt^2} + a_1\frac{dy}{dt} + a_0 y = b_0 x$$

Overdamped system ($a_1^2 > 4a_0$)

$$g(t) = b_0 a_0^{-1}\left(1 + \frac{p_2 e^{p_1 t} - p_1 e^{p_2 t}}{p_1 - p_2}\right) u(t)$$

$$h(t) = \frac{b_0}{p_1 - p_2}(e^{p_1 t} - e^{p_2 t}) u(t)$$

$$p_1 = \frac{-a_1 + \sqrt{a_1^2 - 4a_0}}{2} \qquad p_2 = \frac{-a_1 - \sqrt{a_1^2 - 4a_0}}{2}$$

Critically damped system ($a_1^2 = 4a_0$)

$$g(t) = b_0 a_0^{-1}[1 - (1 - pt)e^{pt}] u(t)$$

$$h(t) = b_0 a_0^{-1} p^2 t e^{pt} u(t) \qquad p = -\frac{a_1}{2}$$

Underdamped system ($a_1^2 < 4a_0$)

$$g(t) = b_0 a_0^{-1}[1 + r\omega^{-1} e^{\sigma t} \sin(\omega t - \theta)] u(t)$$

$$h(t) = b_0 \omega^{-1} e^{\sigma t} (\sin \omega t) u(t)$$

$$\sigma = \frac{-a_1}{2} \qquad \omega = \frac{\sqrt{4a_0 - a_1^2}}{2} \qquad r = \sqrt{\sigma^2 + \omega^2} \qquad \theta = \mathrm{Tan}^{-1}(\omega, \sigma)$$

$g(0) = 0$ toward its steady-state value $g(\infty) = b_0/a_0$. The impulse response increases initially and then approaches zero (decays) as t increases. If the characteristic roots are distinct (and real), the system is said to be *overdamped*. If the characteristic roots are identical, the system is said to be *critically damped*.

For a system having complex-conjugate characteristic roots p, p^* (Fig. 2-33b), the step response and the impulse response contain damped sinusoids. Such a system is said to be *underdamped*. The step response exhibits overshoot and ringing as it approaches its steady-state value $g(\infty) = b_0/a_0$. The magnitude of the overshoot and the duration of the ringing are determined by the relative magnitudes of $\sigma = \mathrm{Re}\,p$ and $\omega = \mathrm{Im}\,p$. For $\sigma > \omega$, overshoot is very small, ringing is unnoticeable, and the system is said to be *heavily damped*. For $\sigma \approx \omega$, overshoot is slight (about 5 percent of steady-state amplitude), ringing dies quickly, and the system is said to be *moderately damped*. For $\sigma \ll \omega$, overshoot is large, ringing persists for one or more cycles, and the system is said to be *lightly damped*. For $\sigma = 0$, overshoot equals 100 percent of steady-state amplitude, ringing persists indefinitely, and the system is said to be *undamped*.

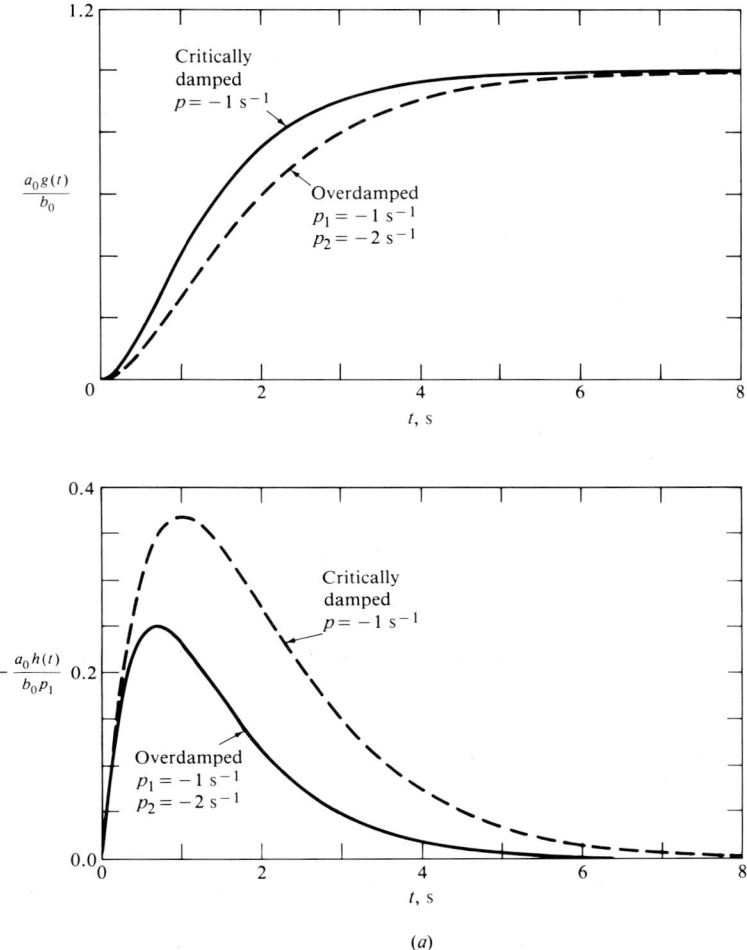

Figure 2-33 Step response $g(t)$ and impulse response $h(t)$ of a second-order system.

From a practical viewpoint, underdamped systems are more important than overdamped and critically damped systems. Overdamped systems are less important partly because they are too sluggish for most applications and partly because an overdamped system can be treated as either a series or parallel connection of first-order systems. (Methods for doing this are described in subsequent chapters; see also Prob. 2-32.) Critically damped systems cannot be built because critical damping requires identical characteristic roots. Parameters of physical systems are too imprecise to allow such a specification. Critically damped systems serve mainly as benchmark systems because critical damping gives the smallest rise time without overshoot. Many systems (e.g., many automatic control systems) are designed to be moderately damped in order to

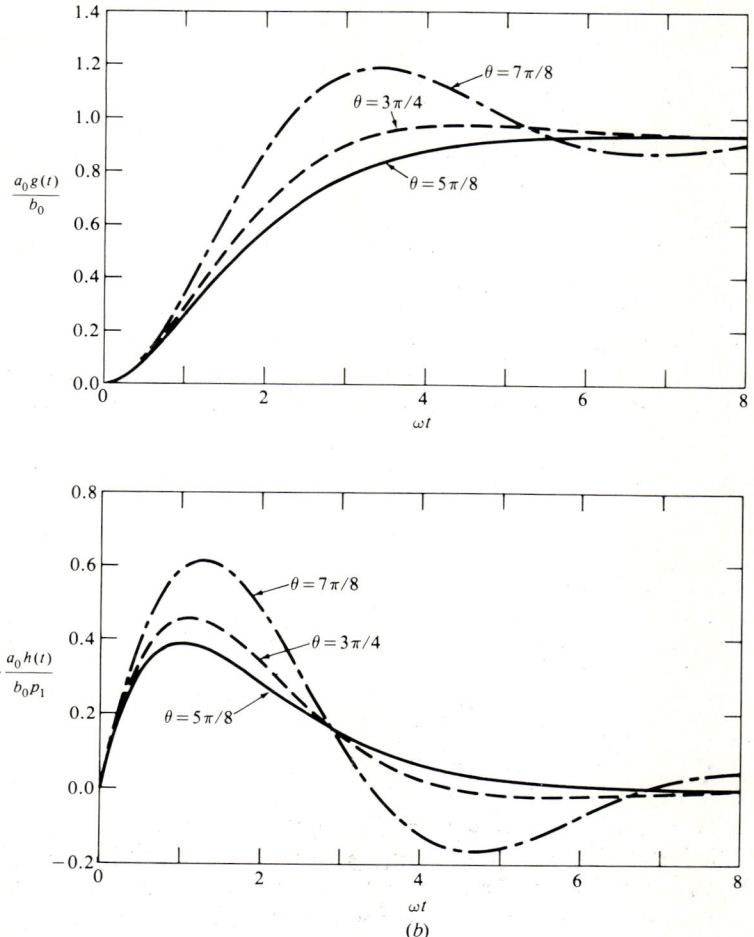

Figure 2-33 *Continued*

achieve a fast response (small rise time) with tolerable overshoot. Heavily damped systems exhibit little overshoot but are relatively sluggish. Lightly damped systems give smaller rise times than moderately damped systems but exhibit more overshoot and more persistent ringing. Generally, large overshoot and persistent ringing are undesirable; however, some systems (e.g., circuits used in electronic oscillators) are designed to produce such a response.

Example 2-26 Obtain the step response and the impulse response for the circuit of Fig. 2-34a, where $L = 400$ mH, $C = 100$ nF, and (a) $R = 500$ Ω, (b) $R = 1$ kΩ, (c) $R = 2$ kΩ.

SOLUTION Kirchhoff's current law gives

$$\frac{1}{L}\int_{-\infty}^{t}[y(\alpha) - x(\alpha)]d\alpha + \frac{y(t)}{R} + C\frac{dy(t)}{dt} = 0$$

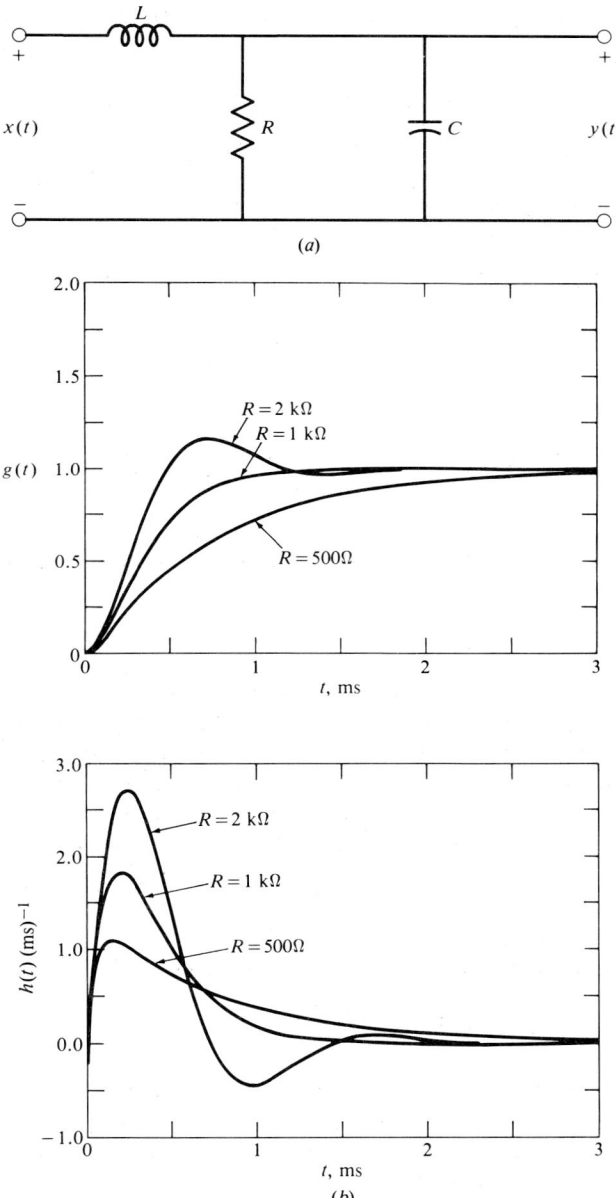

Figure 2-34 (a) Electric circuit of Example 2-26; (b) step response and impulse response.

Differentiating this relation once with respect to t and rearranging terms gives

$$\frac{d^2y}{dt^2} + \frac{1}{RC}\frac{dy}{dt} + \frac{1}{LC}y = \frac{1}{LC}x$$

This has the form of (2-27), with $a_1 = 1/RC$ and $a_0 = b_0 = 1/LC$.

(a) For $R = 500\ \Omega$, the characteristic roots are

$$p_1 = -\frac{a_1}{2} + \frac{\sqrt{a_1^2 - 4a_0}}{2} = -1.34 \times 10^3\ \text{s}^{-1}$$

$$p_2 = -\frac{a_1}{2} - \frac{\sqrt{a_1^2 - 4a_0}}{2} = -18.7 \times 10^3\ \text{s}^{-1}$$

The roots are real and distinct, and the system is overdamped. From Table 2-1, the step response is

$$g(t) = (1 - 1.08e^{p_1 t} + 0.08e^{p_2 t})u(t)$$

and the impulse response is

$$h(t) = (1.44 \times 10^3)(e^{p_1 t} - e^{p_2 t})u(t) \quad \text{s}^{-1}$$

(b) For $R = 1\ \text{k}\Omega$, the characteristic roots are

$$p_1 = p_2 = p = -5 \times 10^3\ \text{s}^{-1}$$

and the system is critically damped. From Table 2-1, the step response is

$$g(t) = [1 - (1 - pt)e^{pt}]u(t)$$

and the impulse response is

$$h(t) = p^2 t e^{pt} u(t)$$

(c) For $R = 2\ \text{k}\Omega$, the characteristic roots are

$$p = (-2.50 + 4.33j) \times 10^3\ \text{s}^{-1} \qquad p^* = (-2.50 - 4.33j) \times 10^3\ \text{s}^{-1}$$

The roots are a complex-conjugate pair, and the system is underdamped. From Table 2-1, the step response is

$$g(t) = [1 + 1.15e^{\sigma t} \sin(\omega t - \theta)]u(t)$$

and the impulse response is

$$h(t) = (5.77 \times 10^3)e^{\sigma t}(\sin \omega t)u(t) \quad \text{s}^{-1}$$

where $\sigma = -2.50 \times 10^3\ \text{s}^{-1}$, $\omega = 4.33 \times 10^3\ \text{s}^{-1}$, and $\theta = \text{Tan}^{-1}(\omega, \sigma) = 2.09$ rad.

Figure 2-34b shows graphs of the step response $g(t)$ and impulse response $h(t)$ for the three cases considered above. In all three cases, the magnitude of the impulse response decreases exponentially with time; therefore, all three impulse responses are absolutely integrable and the system is stable. On physical grounds, we know that the system is stable for any value of resistance R because the system is a passive electric circuit (one containing no energy source).

All three step responses have the same steady-state amplitude (dc gain) $g(\infty) = 1$. This is evident from the circuit diagram because in steady state (for step input) the voltage across the inductor equals zero (because the current in the inductor is constant). That the dc gain is unity is also evident from the differential equation describing the circuit: in steady state, derivatives of the output (for step input) equal zero, in which case the differential equation simplifies to $y = x$.

For $R < 1\ \text{k}\Omega$, the system is overdamped. For $R = 1\ \text{k}\Omega$, the system is critically damped. For $R > 1\ \text{k}\Omega$, the system is underdamped. Increasing R makes the impulse response and the step response become more oscillatory because increasing R decreases the power dissipated by the resistor (the energy removed from the circuit per unit time), allowing the inductor and capacitor to exchange portions of their stored energies for a longer time. For $R \to \infty$, the resistor dissipates no energy (because current in the resistor approaches zero), the inductor and capacitor exchange their stored energies indefinitely, and the impulse response and step response exhibit undamped (sustained) oscillation.

If the system is overdamped, increasing R decreases rise time (speeds up the response to a step or pulse input). If the system is underdamped, increasing R decreases the time for which the step response first reaches its steady-state amplitude but also increases overshoot and duration of ringing. The fastest rise without overshoot is obtained when the system is critically damped, but a faster rise with (usually) tolerable overshoot is obtained when the system is moderately underdamped.

2-4E Equations Containing Derivatives of an Input

The procedure given above applies to systems described by (2-12) with $m = 0$ (no derivatives of an input). Here we extend the procedure to systems described by (2-12) for $m > 0$.

For $m > 0$, (2-12) can be written as two equations;

$$v = b_m \frac{d^m x}{dt^m} + \cdots + b_0 x \tag{2-45}$$

and

$$\frac{d^n y}{dt^n} + \cdots + a_0 y = v \tag{2-46}$$

This implies that a system described by (2-12) can be treated as a series connection of two systems, as shown in Fig. 2-35. One is a feedforward system having input $x(t)$ and output $v(t)$; the other is a feedback system having input $v(t)$ and output $y(t)$. These systems commute because both are linear and stationary. This is illustrated in Fig. 2-36, where

$$\frac{d^n w}{dt^n} + \cdots + a_0 w = x \tag{2-47}$$

and

$$y = b_m \frac{d^m w}{dt^m} + \cdots + b_0 w \tag{2-48}$$

Consequently, the step response and the impulse response of a system described by (2-12) (for $m > 0$) can be obtained as follows:

1. Obtain the step response $g'(t)$ of the feedback system described by (2-47). This can be done using the procedure described above because no derivatives of an input $x(t)$ appear in (2-47).

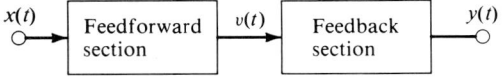

Figure 2-35 How a system described by (2-12) can be analyzed as a series connection of feedforward and feedback subsystems.

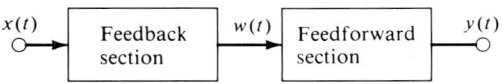

Figure 2-36 Using commutativity to reorder the systems of Fig. 2-35.

2. Obtain the step response $g(t)$ of the overall system by taking $g'(t)$ as the input to the feedforward system described by (2-48). This gives

$$g = b_m \frac{d^m g'}{dt^m} + \cdots + b_0 g' \qquad (2\text{-}49)$$

3. Finally, obtain the impulse response of the overall system using

$$h(t) = \frac{dg(t)}{dt} \qquad (2\text{-}50)$$

Example 2-27 Obtain the step response and the impulse response of a first-order system described by

$$\frac{dy}{dt} + a_0 y = b_1 \frac{dx}{dt} + b_0 x$$

SOLUTION The steps below are numbered in agreement with those of the procedure given above.

Step 1 We obtain the step response $g'(t)$ of the feedback section from

$$\frac{d^n g'}{dt^n} + \cdots + a_0 g' = u(t)$$

From (2-15),

$$g'(t) = a_0^{-1}(1 - e^{-a_0 t})u(t)$$

Step 2 We obtain the step response $g(t)$ of the overall system using (2-49) with $m = 1$. We find

$$\frac{dg'}{dt} = a_0^{-1}(1 - e^{-a_0 t})\delta(t) + e^{-a_0 t}u(t)$$

$$= e^{-a_0 t}u(t)$$

and thus

$$g(t) = b_1 e^{-a_0 t}u(t) + b_0 a_0^{-1}(1 - e^{-a_0 t})u(t)$$

$$= [b_0 a_0^{-1} - (b_0 a_0^{-1} - b_1)e^{-a_0 t}]u(t) \qquad (2\text{-}51)$$

Step 3 From (2-50), the impulse response is

$$h(t) = [b_0 a_0^{-1} - (b_0 a_0^{-1} - b_1)e^{-a_0 t}]\delta(t) + a_0(b_0 a_0^{-1} - b_1)e^{-a_0 t}u(t)$$

$$= b_1 \delta(t) + (b_0 - a_0 b_1)e^{-a_0 t}u(t) \qquad (2\text{-}52)$$

Figure 2-37 shows graphs of the step response and the impulse response versus time.

The steady-state component of the step response is $g(\infty) = b_0/a_0$. This is also evident from the differential equation describing the system because in steady state (for step input) both dx/dt and dy/dt are zero and the equation simplifies to $y = (b_0/a_0)x$. The time constant of the step response (and the impulse response) is $1/a_0$. The duration of the transient component of the step response is proportional to $1/a_0$. It is independent of b_0 and of b_1.

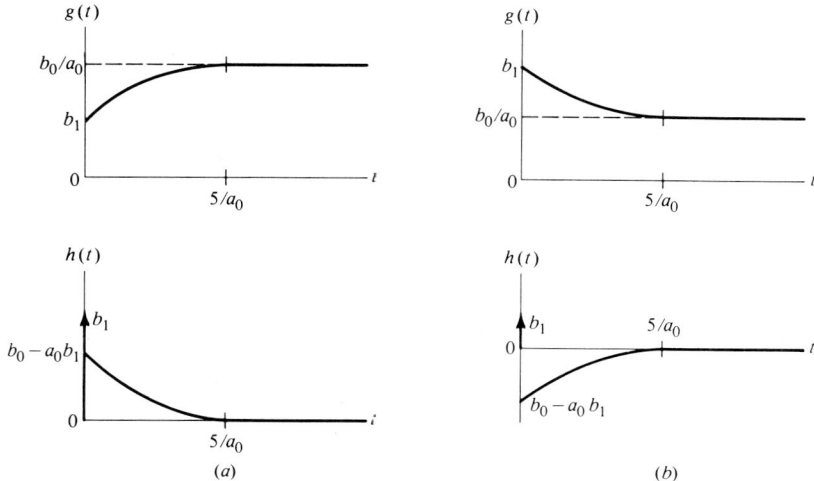

Figure 2-37 Graphs of the step response $g(t)$ and the impulse response $h(t)$ of the system of Example 2-27: (a) $b_0/a_0 > b_1$; (b) $b_0/a_0 < b_1$.

The system is stable if $a_0 < 1$ because in that case the impulse response is absolutely integrable. The coefficients b_0, b_1 have no bearing on stability.

Note that the impulse response of the system of Example 2-27 contains a delta function. In general, the impulse response of a system described by (2-12) contains a delta function if $m = n$ and contains derivatives of a delta function if $m > n$. In this book we apply the procedures given above only to systems described by (2-12) with $m \leq n$.

2-4F Third- and Higher-Order Systems

The procedures given above for obtaining step response and impulse response of systems described by (2-12) are applicable to systems of arbitrary order; however, characteristic roots of systems of order greater than 4 must be obtained numerically because there is no algebraic method for finding roots of polynomials of degree greater than 4. In practice, characteristic roots of systems of order greater than 2 often are obtained numerically because expressions for the roots of third- and fourth-degree polynomials are cumbersome. Computer programs for finding roots (both real and complex) of a polynomial are available at virtually all installations for scientific computing (see, for example, *System/360 Scientific Subroutine Package*, pp. 181–183).

Once the characteristic roots of a system are known, the coefficients $c_1, c_2, \ldots, c_n$ appearing in the complementary function for the system can be found by solving the n simultaneous equations given by (2-22). This too can be done numerically using any one of several widely available computer programs. In practice, obtaining the step response and the impulse response of a system of relatively high order is not much more difficult than for a system of order 1 or 2.

From an engineering viewpoint, a bigger problem than obtaining the step response (or the impulse response) of a high-order system is that of *specifying* the step response of such a system because of the number of parameters involved. For example, specifying the step response of even a third-order system described by (2-12) is difficult if all seven coefficients $(a_0, a_1, a_2, b_0, b_1, b_2, b_3)$ are free because each coefficient may affect more than one property of the response and because seven independent performance specifications are necessary to determine the seven coefficients. It is difficult to come up with seven independent, meaningful performance specifications for even a relatively complex system. In designing a complex system one is almost forced to separate the system into simpler subsystems, each having a manageable number (say, two or three) of free parameters and its own set of performance specifications. That is why system analysis is such an important part of system design.

First- and second-order systems described by (2-12) are important mainly because (1) many important components of physical systems can be modeled (under normal operating conditions) as first- or second-order linear stationary systems and (2) a system of order greater than 2 can be analyzed as a series or parallel connection of first- and second-order systems. These ideas are pursued further in subsequent chapters.

2-4G Stability

To conclude this chapter we describe a method for finding whether a system described by (2-12) is stable. Recall that a linear stationary system having impulse response $h(t)$ is stable if and only if $h(t)$ is absolutely integrable, i.e., if and only if

$$\int_{-\infty}^{\infty} |h(\eta)| \, d\eta < \infty$$

The impulse response of a system described by (2-12) is a sum of *linearly independent* terms. Since there is no cancellation (for all time) of any one term by any or all of the other terms, the system is stable if and only if each term is absolutely integrable. For $m = n$ in (2-12) one term of the impulse response is a delta function, which is absolutely integrable. Every other term has the form $ct^k e^{pt} u(t)$, where p is a characteristic root and k is a nonnegative integer. Such a function is absolutely integrable if and only if Re $p < 0$. It follows that a system described by (2-12) is stable if and only if all its characteristic roots have negative real parts.

Example 2-28 A system having input $x(t)$ and output $y(t)$ is described by the differential equation

$$\frac{d^2y}{dt^2} + a_1 \frac{dy}{dt} + a_0 y = b_1 \frac{dx}{dt} + b_0 x$$

where $a_1 = 4 \text{ s}^{-1}$ and a_0 depends on an adjustable parameter. Find all values of a_0 for which the system is stable.

SOLUTION The characteristic equation for the system is

$$s^2 + a_1 s + a_0 = 0$$

The two characteristic roots are given by

$$p_1 = \frac{-a_1 + \sqrt{a_1^2 - 4a_0}}{2} \qquad p_2 = \frac{-a_1 - \sqrt{a_1^2 - 4a_0}}{2}$$

We consider the cases $a_1^2 < 4a_0$ (complex roots) and $a_1^2 \geq 4a_0$ (real roots) separately.
For $a_1^2 < 4a_0$, the characteristic roots are complex. The real part of both is $-a_1/2$. In this case the system is stable because $-a_1/2 = -2 \text{ s}^{-1}$ is negative. Therefore the system is stable if the roots are complex, which requires $a_1^2 < 4a_0$ or $a_0 > 4 \text{ s}^{-2}$.
For $a_1^2 \geq 4a_0$, the roots are real. Clearly, p_2 is negative, and p_1 is negative if $a_1 > \sqrt{a_1^2 - 4a_0}$. This requires $a_0 > 0$, which implies the condition $a_0 > 4 \text{ s}^{-2}$ obtained above (for complex roots). Therefore the system is stable if and only if $a_0 > 0$.

SUMMARY

A *linear stationary system* obeys both *superposition* and *time invariance*. The output of a linear stationary system for input

$$x(t) = a_1 x_1(t - t_1) + a_2 x_2(t - t_2) + \cdots + a_n x_n(t - t_n)$$

is given by

$$y(t) = a_1 y_1(t - t_1) + a_2 y_2(t - t_2) + \cdots + a_n y_n(t - t_n)$$

where $y_i(t)$ is the output for input $x_i(t)$ ($i = 1, 2, \ldots, n$).

The transfer characteristic for a series connection of linear stationary systems is independent of the order in which the systems are connected. The systems of a series connection of linear stationary systems can be reordered freely (to simplify analysis); however, nonlinear and nonstationary systems do not commute, either with each other or with linear stationary systems.

The *impulse response* of a linear stationary system is the function $h(t)$, where $y(t) = ah(t)$ is the output for input $x(t) = a\delta(t)$. The *step response* is a function $g(t)$, where $y(t) = x_0 g(t)$ is the output for input $x(t) = x_0 u(t)$. The impulse response is the derivative of the step response [see (2-5)] and the step response is the integral of the impulse response [see (2-6)]. The unit of $h(t)$ is that of $yx^{-1}t^{-1}$, and the unit of $g(t)$ is that of yx^{-1}, where $y(t)$ is the output for input $x(t)$.

The output $y(t)$ of a linear stationary system for input $x(t)$ is given explicitly by a *convolution integral*

$$y(t) = \int_{-\infty}^{\infty} x(\lambda) h(t - \lambda) \, d\lambda$$

where $h(t)$ is the impulse response of the system. This means that any linear stationary system can be described by a transfer characteristic. The transfer characteristic is determined completely by the impulse response $h(t)$; consequently all important properties of a linear stationary system can be deduced from the impulse response of the system, as follows:

1. A linear stationary system is *realizable* if its impulse response is zero for $t < 0$.
2. A linear stationary system is *stable* if its impulse response is absolutely integrable.
3. A linear stationary system provides *high-fidelity transmission* for input $x(t)$ if its impulse response approximates a delta function compared to the narrowest significant feature of $x(t)$.
4. The *duration of a transient response* is the duration of the impulse response.

5. The *steady-state output* for a step input is proportional to the area bounded by the impulse response and the time axis.

Using superposition and time invariance, we can express the output of a linear stationary system for input

$$x(t) = x_0 r\left(\frac{t}{\tau}\right) = x_0 u(t) - x_0 u(t - \tau)$$

(the pulse response of the system) as

$$y(t) = x_0 g(t) - x_0 g(t - \tau)$$

where $g(t)$ is the step response of the system.

Often, a linear stationary system is represented by the block diagram of Fig. 2-15, which denotes the convolution relation (2-2). Note that the block diagram of Fig. 2-15 does *not* imply that $y(t) = h(t)x(t)$.

Section 2-4 gives a procedure for finding the step response and the impulse response of a linear stationary system described by the differential equation (2-12). The order of the highest derivative of an output $y(t)$ is called the *order* of the system. The most important systems of this kind are *first-* and *second-order systems*. The impulse response for a first-order system (with $m = 0$) is a real exponential. The impulse response for a second-order system (with $m = 0$) is a damped sinusoid if the system is *underdamped* (if the characteristic roots are complex), is of the form $te^{pt}u(t)$ if the system is *critically damped* (if the characteristic roots are real and equal), and consists of two real exponentials if the system is *overdamped* (if the characteristic roots are real and distinct). The most important second-order systems are underdamped systems because an overdamped system can be analyzed as a series or parallel connection of first-order systems. An underdamped system cannot be analyzed as a series or parallel connection of (realizable) first-order systems because the corresponding first-order differential equations have complex coefficients.

The impulse response for a linear stationary system described by (2-12) consists of terms of the form $t^k e^{pt} u(t)$, where k is a nonnegative integer and p is a characteristic root, and (if $m = n$) a delta function. A delta function is absolutely integrable. A term of the form $t^k e^{pt} u(t)$ is absolutely integrable if Re $p < 0$. Therefore, *a linear stationary system is stable if and only if all its characteristic roots have negative real parts*.

PROBLEMS

2-1 Use the definitions in Sec. 2-1 to find whether the systems of Fig. P2-1a to f are linear and stationary.

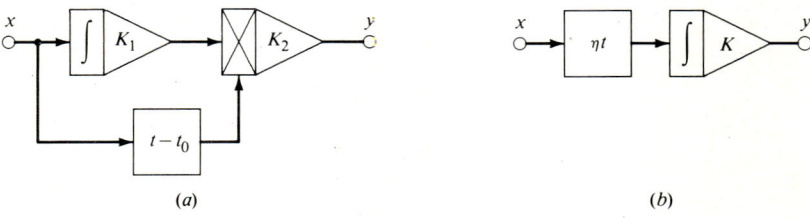

(a)   (b)

Figure P2-1

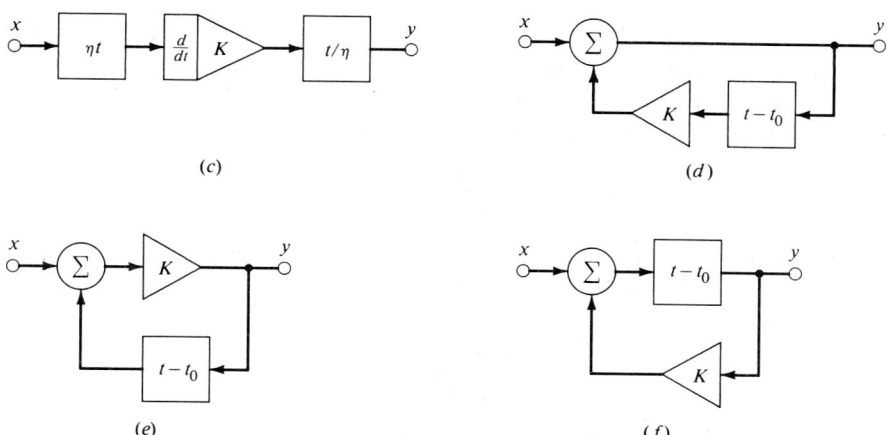

Figure P2-1 *Continued*

2-2 Obtain the impulse response $h(t)$, the step response $g(t)$, and the output $y(t)$ for input $x(t) = x_0 r(t/t_0)$ for the systems of Fig. P2-2. Sketch graphs of $h(t)$, $g(t)$, and $y(t)$ versus t. Assume that $0 < K < 1$.

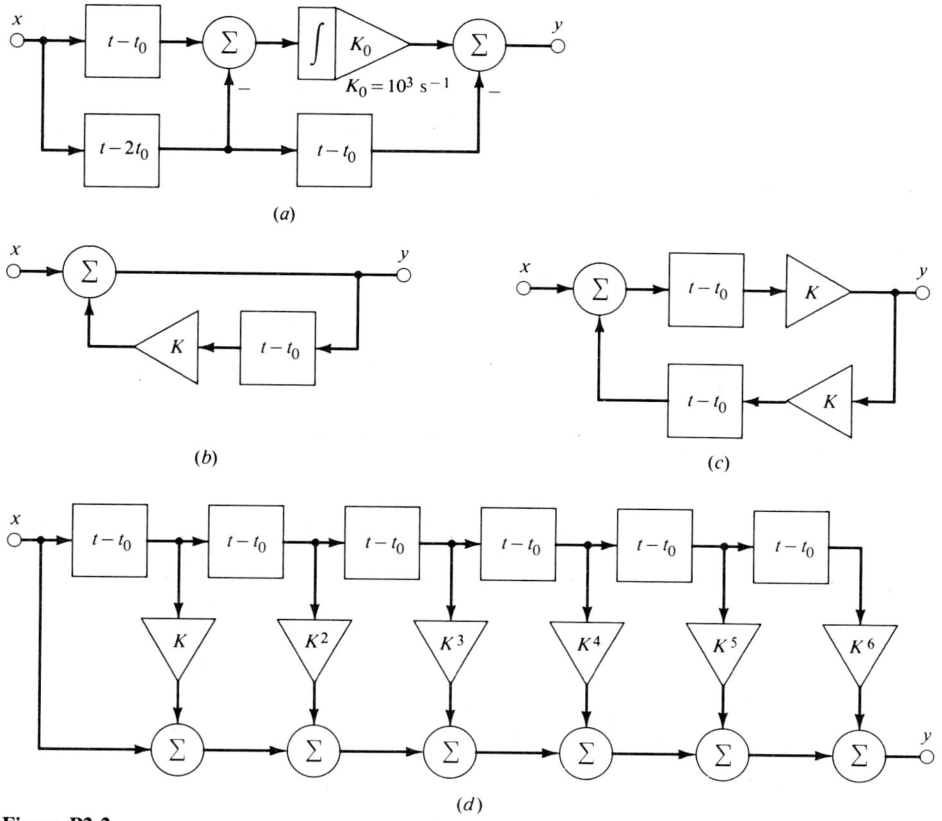

Figure P2-2

2-3 Show that proportion, delay, integration, and differentiation elements are linear and stationary.

2-4 Show that a static element described by (1-19) is stationary.

2-5 Use the definitions to show that a series connection of two linear stationary systems is a linear stationary system.

2-6 Use the definitions to show that a parallel connection (through addition) of two linear stationary systems is a linear stationary system.

2-7 Use the results of Probs. 2-5 and 2-6 to argue that any feedforward system consisting entirely of linear stationary elements is linear and stationary.

2-8 The output of a certain linear stationary system for input $x_1(t) = x_0 u(t)$ is

$$y_1(t) = 5x_0(1 - e^{-t/\tau})u(t)$$

Find and plot the output for input

$$x(t) = 2x_c r\left(\frac{t}{5\tau}\right)$$

2-9 Use superposition to determine the outputs $y_1(t)$ and $y_2(t)$ of the system of Fig. P2-9, where $t_0 = 1$ ms, $t_1 = 2/3$ ms, and

$$x(t) = x_0(\cos \omega_1 t + \cos \omega_2 t)$$

with $f_1 = 1$ kHz, $f_2 = 1.5$ kHz, and $x_0 = 100$ mV.

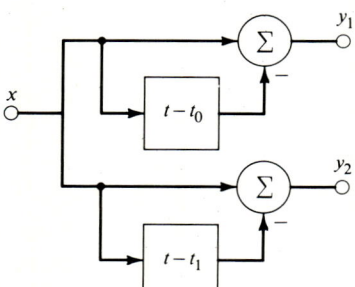

Figure P2-9

2-10 The output of a certain linear stationary system for input $x_1(t) = x_0 r(t/\tau)$ is

$$y_1(t) = x_0\left[\frac{t}{\tau}r\left(\frac{t}{\tau}\right) + u(t - \tau)\right]$$

Find and plot the output $y(t)$ for input

$$x(t) = x_0 r\left(\frac{t}{2\tau}\right)$$

2-11 The output of a certain linear stationary system for input $x_1(t) = a\delta(t)$ is

$$y_1(t) = y_0 \frac{\tau - t}{\tau} r\left(\frac{t}{\tau}\right)$$

where $y_0 = 1$ V and $\tau = 5$ ms.

(a) Draw a block diagram for a linear stationary system that gives output $y_1(t)$ for input $x_1 = \mathbf{a}\delta(t)$.

(b) Plot the output $y(t)$ for input

$$x(t) = 2\mathbf{a}\delta(t) + 3\mathbf{a}\delta(t - t_0) - 5\mathbf{a}\delta(t - 2t_0) \qquad \text{where } t_0 = \frac{\tau}{2}$$

2-12 Use convolution to obtain the outputs of systems having the following impulse responses $h(t)$ and inputs $x(t)$:

(a) $h(t) = \dfrac{1}{\tau} e^{-t/\tau} u(t) \qquad x(t) = x_0 r\left(\dfrac{t}{5\tau}\right)$

(b) $h(t) = \dfrac{1}{\tau} r\left(\dfrac{t}{\tau}\right) \qquad x(t) = x_0 u(t)$

(c) $h(t) = \delta(t) + \delta(t - \tau) \qquad x(t) = x_0 r\left(\dfrac{t}{\tau}\right)$

(d) $h(t) = \dfrac{1}{\tau} e^{-t/\tau} u(t) \qquad x(t) = x_0 \cos \omega t$

(e) $h(t) = \dfrac{1}{\tau} e^{-t/\tau} u(t) \qquad x(t) = x_0 e^{-t/\tau} u(t)$

(f) $h(t) = \dfrac{1}{\tau_1} r\left(\dfrac{t}{\tau_1}\right) \qquad x(t) = x_0 r\left(\dfrac{t}{\tau_2}\right)$

where (i) $\tau_1 < \tau_2$, (ii) $\tau_1 = \tau_2$, (iii) $\tau_1 > \tau_2$.

2-13 Obtain and plot the impulse response and the step response for (a) a proportion element, (b) a delay element, and (c) an integrator.

2-14 Derive (2-3) from (2-2).

2-15 Obtain the impulse response of each system of Fig. P2-15. Find all values of the parameter K for which the system is stable. Assume $t_0 > 0$.

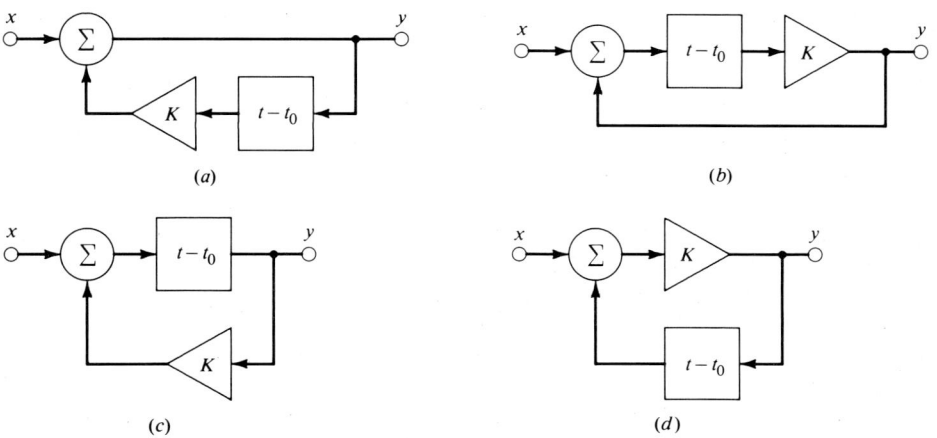

Figure P2-15

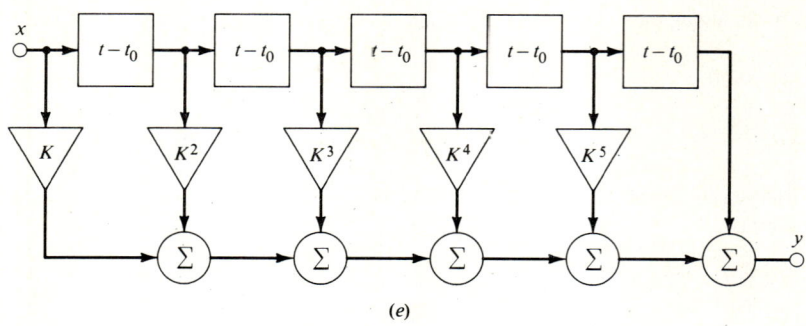

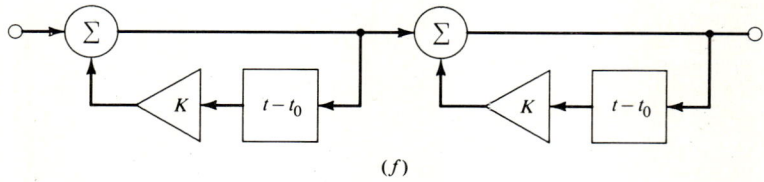

(f)

Figure P2-15 *Continued*

2-16 Use (2-6) to obtain step responses from the following impulse responses:

(a) $h(t) = \dfrac{1}{\tau} e^{-t/\tau} u(t)$
(b) $h(t) = \delta(t) - \dfrac{1}{\tau} e^{-t/\tau} u(t)$

(c) $h(t) = \left(\dfrac{1}{\tau_1} e^{-t/\tau_1} + \dfrac{1}{\tau_2} e^{-t/\tau_2} \right) u(t)$
(d) $h(t) = \dfrac{1}{\tau} (\sin \omega t) u(t)$

2-17 Use (2-5) to obtain impulse responses from the following step responses:

(a) $g(t) = (1 - e^{-t/\tau}) u(t)$
(b) $g(t) = e^{-t/\tau} u(t)$
(c) $g(t) = (1 - e^{-t/\tau} \cos \omega t) u(t)$
(d) $g(t) = (2 - e^{-t/\tau_1} - e^{-t/\tau_2}) u(t)$

2-18 Use the superposition integral (2-1) and integration by parts to show that a signal $x(t)$ can be expressed as a superposition of step signals; thus

$$x(t) = \int_{-\infty}^{\infty} \dfrac{dx(\lambda)}{d\lambda} u(t - \lambda) \, d\lambda$$

provided $x(-\infty) = 0$.

2-19 Use the result of Prob. 2-18 to show that the output $y(t)$ of a linear stationary system for input $x(t)$ is given by

$$y(t) = \int_{-\infty}^{\infty} \dfrac{dx(\lambda)}{d\lambda} g(t - \lambda) \, d\lambda$$

where $g(t)$ is the step response of the system.

2-20 Use the result of Prob. 2-19 to obtain the output of a system having step response $g(t) = (t/\tau) u(t)$ for input $x(t) = x_0 r(t/\tau)$.

2-21 Find the impulse responses of the systems in the dashed boxes in Fig. P2-21, where

$$h_1(t) = \frac{1}{\tau_1} e^{-t/\tau_1} \qquad h_2(t) = \frac{1}{\tau_2} e^{-t/\tau_2} \qquad h_3(t) = \frac{1}{\tau_3} e^{-t/\tau_3} \qquad \tau_1 < \tau_2 < \tau_3$$

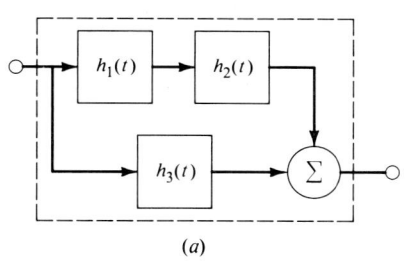

(a)

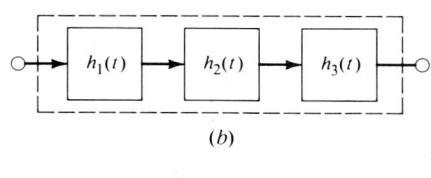

(b)

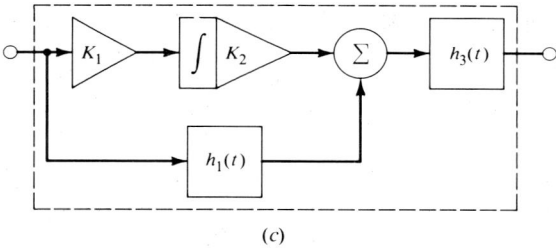

(c)

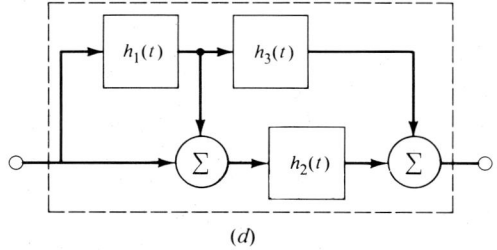

(d)

Figure P2-21

2-22 Use convolution to show that the output of a linear stationary system for sinusoidal input is a sinusoid whose frequency equals the frequency of the input. *Hint:* Use (2-3) and the identity

$$\cos(\alpha - \beta) \equiv \cos\alpha \cos\beta + \sin\alpha \sin\beta$$

2-23 Use a graphical representation of convolution to obtain a graph of the output of a linear stationary system having impulse response $h(t) = (1/\tau_1)r(t/\tau_1)$ for input $x(t) = x_0 r(t/\tau_2)$, where (a) $\tau_2 < \tau_1$, (b) $\tau_2 = \tau_1$, and (c) $\tau_2 > \tau_1$. Under what conditions does the system provide high-fidelity transmission for input $x(t)$ above?

2-24 Under what conditions on K and t_0 does the system of Fig. P2-24 provide high-fidelity transmission for input $x(t) = x_0 r(t/\tau)$?

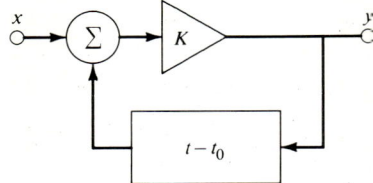

Figure P2-24

2-25 The impulse response of a certain linear stationary system is $h(t) = (1/\tau)e^{-t/\tau}u(t)$. What can be said of the class of inputs for which this system provides high-fidelity transmission?

2-26 The impulse response of a certain linear stationary system is $h(t) = \delta(t) - \tau^{-1}e^{-t/\tau}u(t)$. Sketch a graph of the output of this system for input $x(t) = x_0 r(t/t_0)$ for $t_0 = \tau$. Describe how fidelity of the output depends on the relative magnitudes of t_0 and τ.

2-27 The impulse response of a certain linear stationary system is $h(t) = (1/\tau)e^{-t/\tau}(\sin \omega t)u(t)$. Find all values of τ for which the system is stable. Under what conditions does the system provide high-fidelity transmission for input $x(t) = x_0 r(t/\tau)$?

2-28 For the system of Fig. P2-28, $K_1 = 5$ mA/V and $K_2 = 1$ Vs/A.
(a) Obtain a differential equation describing the system.
(b) Obtain the step response and the impulse response of the system.
(c) Sketch a graph of the output $y(t)$ for input $x(t) = 5r(t/\tau)$ V, where $\tau = 5$ ms.
(d) Calculate the ISE for input $x(t)$.
(e) Calculate the sensitivities of ISE to variations of K_1 and K_2. Interpret the results of the calculation.

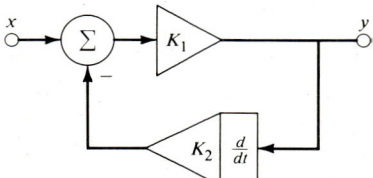

Figure P2-28

2-29 Repeat Prob. 2-28 for the system of Fig. P2-29, where $K_1 = 1$ V/A, $K_2 = 0.1$ A/(Vs), and $x(t) = 5r(t/\tau)$ mA, with $\tau = 5$ s.

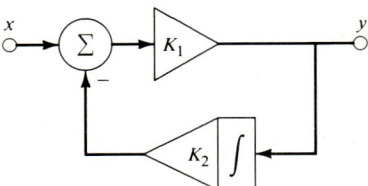

Figure P2-29

2-30 Obtain differential equations describing the systems of Fig. P2-30 and obtain the step response and the impulse response of each system. Describe how the parameter K_2 affects fidelity of transmission, assuming all other parameters are fixed and positive.

2-31 The step response of a certain linear stationary system is

$$g(t) = (1 - 2e^{p_1 t} + 3e^{p_2 t} - 2e^{p_3 t})u(t)$$

where $p_1 = -1$ s^{-1}, $p_2 = -2$ s^{-1}, and $p_3 = -3$ s^{-1}. Draw a block diagram for the system using only proportion, addition, and integration elements.

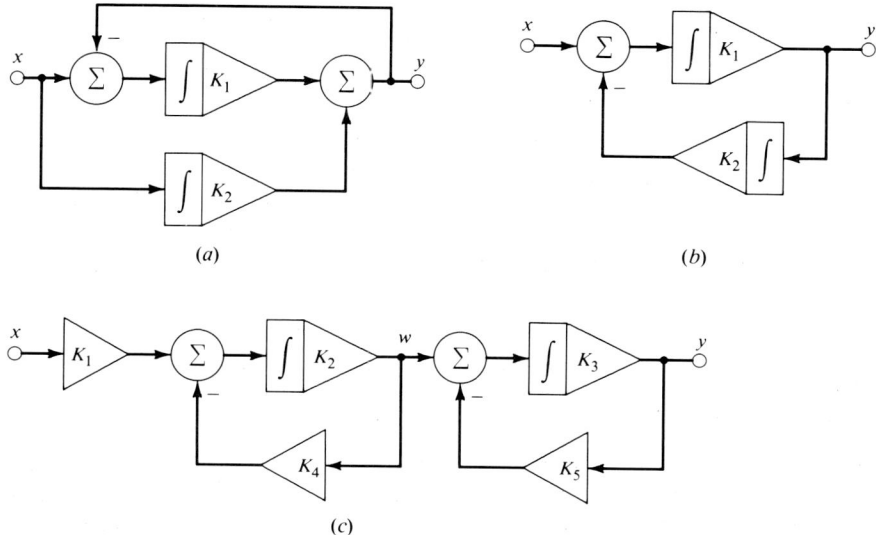

Figure P2-30

2-32 A system is described by the differential equation

$$\frac{d^2y}{dt^2} + a_1\frac{dy}{dt} + a_0 y = b_0 x$$

where $x = x(t)$ denotes an input and $y = y(t)$ denotes the corresponding output. Show that the system can be analyzed as a parallel connection of two first-order systems, provided $a_1^2 > 4a_0$.

2-33 A first-order system is described by the differential equation

$$\frac{dy}{dt} + a_0 y = b_1\frac{dx}{dt}$$

where $a_0 = 2 \text{ s}^{-1}$ and $b_1 = 10$. Find the output $y(t)$ for input $x(t) = x_0 e^{-t/\tau} u(t)$, where $x_0 = 5$ V and (a) $\tau = 1$ s and (b) $\tau = 500$ ms.

2-34 A second-order system is described by the differential equation

$$\frac{d^2y}{dt^2} + a_1\frac{dy}{dt} + a_0 y = b_0 x$$

where $a_1 = 5 \text{ s}^{-1}$, $a_0 = 6 \text{ s}^{-2}$, and $b_0 = 1 \text{ V/(mAs}^2)$. Find the output $y(t)$ for input $x(t) = x_0 e^{-t/\tau} u(t)$, where $x_0 = 5$ mA and $\tau = 1$ s.

2-35 Show that a system described by (2-12) is linear. *Hint:* Let y_i satisfy (2-12) for input x_i ($i = 1, 2, \ldots$). Show that $a_1 y_1 + a_2 y_2$ satisfies (2-12) for input $a_1 x_1 + a_2 x_2$.

2-36 Show that a system described by (2-12) is stationary. *Hint:* Use the fact that delay and differentiation commute.

2-37 Verify by direct substitution that

$$y(t) = \mathbf{a}h(t) = \frac{\mathbf{a}}{\tau} e^{-t/\tau} u(t)$$

satisfies the differential equation

$$\tau \frac{dy}{dt} + y = x$$

for $x(t) = \mathbf{a}\delta(t)$.

2-38 Explain why the system of Fig. P2-38 is nonlinear.

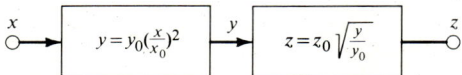

Figure P2-38

2-39 Consider a second-order linear stationary system described by (2-13) with $n = 2$.
  (a) Use l'Hospital's rule to show that in the limit as $p_1 \to p_2$, (2-32) yields (2-43).
  (b) Use l'Hospital's rule to show that in the limit as $\omega \to 0$, (2-41) yields (2-43).

2-40 A differential equation describing an underdamped second-order system often is written

$$\frac{d^2y}{dt^2} + 2\zeta\omega_0 \frac{dy}{dt} + \omega_0^2 y = b_0 x$$

where ζ (zeta) is *damping ratio* and ω_0 is *undamped natural frequency*.
  (a) Express the characteristic roots (assumed complex) of the system in terms of ζ and ω_0.
  (b) Find all values of ζ for which the system is actually underdamped. Does ω_0 have any bearing on damping?
  (c) Find the value of b_0 for which the dc gain of the system equals unity.
  (d) Obtain the impulse response of the system for $\zeta = 0$. Why is ω_0 called the undamped natural frequency?

2-41 Use the procedure given in Sec. 2-4C to obtain the step response and the impulse response of a (critically damped) system described by (2-27) for $a_1^2 = 4a_0$.

2-42 The output of a certain linear stationary system for input $w(t)$ is denoted by $z(t)$. Show that the output of the system for input

$$x(t) = K \int_{-\infty}^{t} w(\alpha)\,d\alpha$$

is given by

$$y(t) = K \int_{-\infty}^{t} z(\alpha)\,d\alpha$$

CHAPTER
THREE
RESPONSE TO SINUSOIDAL EXCITATION

Sinusoidal signals are widely used as test inputs for specifying system performance. For example, frequency response, harmonic distortion, and intermodulation distortion of an audio amplifier all describe the output of the amplifier for one or more sinusoidal inputs. There are two more or less obvious reasons why sinusoidal signals are especially useful test inputs: (1) Signals in certain systems (such as power systems and some communication systems) are nearly sinusoidal — at least over certain intervals. (2) Sinusoids are easy to generate. The two most compelling reasons for using sinusoids as test inputs are more subtle: (1) The output of a stable linear stationary system for a sinusoidal input is a sinusoid whose frequency is the same as that of the input. (2) Any signal can be expressed as a sum of sinusoids (see Chap. 4). These two facts and superposition make it possible to infer the output of a stable linear stationary system for any input from the response of the system to sinusoidal inputs (from the frequency response of the system). For example, fidelity of an output of an audio amplifier can be judged from the frequency response of the amplifier even though music signals are not sinusoidal.

In this chapter we describe methods for finding outputs of systems for one or more sinusoidal inputs. In Chap. 4 we extend these methods to nonsinusoidal inputs.

3-1 LINEAR STATIONARY SYSTEMS

In this section we describe an algebraic method for obtaining the output of a sinusoidally excited linear stationary system. The method is essentially identical to the phasor method for obtaining outputs of sinusoidally excited electric circuits. We start with a review of amplitude-phase and exponential representations of sinusoidal signals.

3-1A Amplitude-Phase Form for a Sinusoidal Signal

A sinusoidal signal $x(t)$ is represented in *amplitude-phase form* as

$$x(t) = A \cos(\omega t + \theta) \qquad (3\text{-}1)$$

where A = *peak amplitude*, unit of x
ω = *angular frequency*, rad/s
θ = *initial phase*, rad

We adopt the convention that peak amplitude A is nonnegative; thus, $A \geq 0$. This convention can be enforced without loss of generality using the identity

$$-\cos \alpha \equiv \cos(\alpha + \pi)$$

For example, we can write

$$x(t) = -5 \cos\left(\omega t - \frac{\pi}{4}\right) \quad \text{mA}$$

as

$$x(t) = 5 \cos\left(\omega t + \frac{3\pi}{4}\right) \quad \text{mA}$$

This is a sinusoid having peak amplitude 5 mA and initial phase $3\pi/4$ rad.

3-1B Exponential Form for a Sinusoidal Signal

The sinusoidal signal of (3-1) can be represented in *exponential form* as

$$x(t) = Xe^{j\omega t} + X^*e^{-j\omega t} \qquad (3\text{-}2)$$

where X is a complex quantity called *complex amplitude* (unit of x). Usually the exponential form of (3-2) is more convenient than the amplitude-phase form of (3-1) for analysis of linear stationary systems.

To obtain the exponential form for a sinusoidal signal from the amplitude-phase form for the signal (to obtain X from A and θ) we use Euler's identity

$$\cos \alpha \equiv \tfrac{1}{2}e^{j\alpha} + \tfrac{1}{2}e^{-j\alpha}$$

Applying this identity to the right side of (3-1) with $\alpha = \omega t + \theta$ gives the right side of (3-2) with

$$X = \tfrac{1}{2}Ae^{j\theta} \qquad (3\text{-}3)$$

Equation (3-3) gives complex amplitude X in polar form. From (3-3), peak amplitude A equals twice the magnitude of complex amplitude X and initial phase θ equals the angle of complex amplitude X. To obtain the peak amplitude A and initial phase θ from the complex amplitude X of a sinusoid expressed in exponential form we use

$$A = 2|X| \qquad \theta = \angle X \qquad (3\text{-}4)$$

where X is the complex amplitude of the sinusoid.†

---

†In this book the symbol $\angle$ denotes "angle of"; for example, $\angle(a + jb) = \text{Tan}^{-1}(b, a)$ and $\angle e^{j\theta} = \theta$.

Example 3-1 The sinusoidal signal

$$x(t) = 10 \cos\left(\omega t - \frac{\pi}{4}\right) \quad \text{V}$$

is expressed in exponential form as

$$x(t) = Xe^{j\omega t} + X^*e^{-j\omega t}$$

where $X = 5e^{-j\pi/4}$ V.

Example 3-2 A sinusoidal signal $x(t)$ expressed in exponential form as

$$x(t) = Xe^{j\omega t} + X^*e^{-j\omega t}$$

with $X = 4e^{j3\pi/4}$ mA, is expressed in amplitude-phase form as

$$x(t) = 8 \cos\left(\omega t + \frac{3\pi}{4}\right) \quad \text{mA}$$

3-1C Transfer Function for a Linear Stationary System

From (2-3), the output $y(t)$ of a linear stationary system for input $x(t)$ is given by the convolution integral

$$y(t) = \int_{-\infty}^{\infty} h(\eta) x(t - \eta) \, d\eta$$

where $h(t)$ is the impulse response for the system. The output of a linear stationary system for input

$$x(t) = Xe^{j\omega t}$$

is given by

$$y(t) = X \int_{-\infty}^{\infty} h(\eta) e^{j\omega(t-\eta)} \, d\eta = Xe^{j\omega t} \int_{-\infty}^{\infty} h(\eta) e^{-j\omega \eta} \, d\eta$$

The last integral is independent of time t. It is a function of only the frequency ω of the input $x(t) = Xe^{j\omega t}$. We denote the integral by $H(j\omega)$;† thus

$$H(j\omega) = \int_{-\infty}^{\infty} h(\eta) e^{-j\omega \eta} \, d\eta \tag{3-5}$$

With this definition the previous equation can be written

$$y(t) = Ye^{j\omega t} \quad \text{where } Y = H(j\omega)X \tag{3-6}$$

A function $H(j\omega)$ given by (3-5) is called the *transfer function* for a system having impulse response $h(t)$.‡ The unit of the transfer function for a system having input $x(t)$ and output $y(t)$ is the unit of $y(t)/x(t)$, for example, volts per milliampere. If $y(t)$ and

---

†We use $H(j\omega)$ rather than $H(\omega)$ to be consistent with other notation introduced in subsequent chapters.
‡In this book the symbol $H(j\omega)$ (subscripted if necessary) always denotes a transfer function. Do not confuse transfer *function* with transfer *characteristic*.

$x(t)$ have the same dimension, the transfer function is dimensionless. In such a case, it is good practice to associate a dimensionless unit, e.g., volts per volt, with the transfer function in order to keep track of dimensions and units of input and output.

An impulse response $h(t)$ contains terms of the form $t^k e^{st}$ and (possibly) an impulse. For the integral in (3-5) to exist, each such term in $h(t)$ must go to zero as t goes to infinity. This requires that all characteristic roots of the system have negative real parts. Therefore, for a system to have a transfer function, the system must be stable.

Equation (3-6) shows that the output of a stable linear stationary system for input $x(t) = X e^{j\omega t}$ is also an exponential having angular frequency ω. Equation (3-6) also shows that the complex amplitude of the output is the product of the complex amplitude of the input and the transfer function for angular frequency ω (the frequency of the input).

Equation (3-6) gives the output of a stable linear stationary system for input $X e^{j\omega t}$. To obtain the output $y(t)$ of the system for sinusoidal input $x(t) = A \cos(\omega t + \theta)$ we first express the input $x(t)$ in exponential form as

$$x(t) = X e^{j\omega t} + X^* e^{-j\omega t}$$

where $X = (A/2) e^{j\theta}$. Both terms on the right side of this expression have the same form. The first has complex amplitude X and angular frequency ω. The second has complex amplitude X^* and angular frequency $-\omega$. Using (3-6) and superposition yields

$$y(t) = H(j\omega) X e^{j\omega t} + H(-j\omega) X^* e^{-j\omega t}$$

This can be written

$$y(t) = Y e^{j\omega t} + Y^* e^{-j\omega t} \qquad \text{where } Y = H(j\omega) X \qquad (3\text{-}7)$$

Equation (3-7) gives the exponential form of the output. From (3-4), the amplitude-phase form of the output is

$$y(t) = B \cos(\omega t + \psi) \qquad \text{where } B = 2|Y| \qquad (3\text{-}8)$$
$$\psi = \angle Y$$

We can summarize the developments above by giving a procedure for finding an output of a sinusoidally excited, stable linear stationary system:

1. Express the input in exponential form as

$$A \cos(\omega t + \theta) = X e^{j\omega t} + X^* e^{-j\omega t} \qquad \text{where } X = \tfrac{1}{2} A e^{j\theta}$$

2. Obtain the transfer function for the system. From (3-5),

$$H(j\omega) = \int_{-\infty}^{\infty} h(\eta) e^{-j\omega \eta} \, d\eta$$

3. Express the output in exponential form as

$$y(t) = Y e^{j\omega t} + Y^* e^{-j\omega t} \qquad \text{where } Y = H(j\omega) X$$

4. Express the output in amplitude-phase form as

$$y(t) = B\cos(\omega t + \psi) \qquad \text{where } B = 2|Y|$$
$$\psi = \angle Y$$

Example 3-3 The impulse response of a certain linear stationary system is

$$h(t) = \frac{K}{\tau} e^{-t/\tau} u(t)$$

where $\tau = 1$ ms and $K = 10^3$ V/A. Find the output $y(t)$ for input

$$x(t) = 10\cos\left(\omega_0 t + \frac{\pi}{4}\right) \qquad \text{mA}$$

where $\omega_0 = 2$ krad/s.

SOLUTION *Step 1* The peak amplitude and the initial phase of the input are $A = 10$ mA and $\theta = \pi/4$ rad, respectively. The complex amplitude of the input is

$$X = \tfrac{1}{2} A e^{j\theta} = 5 e^{j\pi/4} \qquad \text{mA}$$

Step 2 The transfer function of the system is

$$H(j\omega) = \int_{-\infty}^{\infty} h(\eta) e^{-j\omega\eta} \, d\eta = \int_{-\infty}^{\infty} \frac{K}{\tau} e^{-\eta/\tau} u(\eta) e^{-j\omega\eta} \, d\eta$$

$$= \int_{0}^{\infty} \frac{K}{\tau} e^{-(1+j\omega\tau)\eta/\tau} \, d\eta = \frac{K}{\tau}\left[-\frac{e^{-(1+j\omega\tau)\eta/\tau}}{(1+j\omega\tau)/\tau}\right]_{\eta=0}^{\eta=\infty} = \frac{K}{1+j\omega\tau}$$

Step 3 The complex amplitude of the output is

$$Y = H(j\omega_0)X = \frac{0.005 K e^{j\pi/4}}{1+j\omega_0\tau} = \frac{(5\times 10^{-3})(10^3) e^{j\pi/4}}{1+j(2\times 10^3)(10^{-3})}$$

$$= \frac{5 e^{j\pi/4}}{\sqrt{5}\, e^{1.11j}} = \sqrt{5}\, e^{-0.32j} \qquad \text{V}$$

Step 4 The peak amplitude and the initial phase of the output are

$$B = 2|Y| = 2\sqrt{5} \text{ V} \qquad \text{and} \qquad \psi = \angle Y = -0.32 \text{ rad}$$

respectively. The output is expressed in amplitude-phase form as

$$y(t) = 2\sqrt{5}\cos(\omega_0 t - 0.32) \qquad \text{V}$$

where $\omega_0 = 2$ krad/s

3-1D Gain and Phase Shift

The development above shows that the output $y(t)$ of a stable linear stationary system for input

$$x(t) = A\cos(\omega_0 t + \theta)$$

is given by
$$y(t) = |H(j\omega_0)|A \cos[\omega_0 t + \theta + \angle H(j\omega_0)] \quad (3\text{-}9)$$
where $H(j\omega)$ is the transfer function of the system. To obtain the output we scale the amplitude of the input by $|H(j\omega)|$ and increase the initial phase of the input by $\angle H(j\omega_0)$. In other words, a sinusoidally excited, stable linear stationary system can be regarded as a system that introduces a frequency-dependent gain and a frequency-dependent phase shift. We denote *gain* by $\Gamma(\omega)$ and *phase shift* by $\phi(\omega)$, where
$$\Gamma(\omega) = |H(j\omega)| \qquad \phi(\omega) = \angle H(j\omega) \quad (3\text{-}10)$$
With these definitions (3-9) can be written
$$y(t) = \Gamma(\omega_0)A \cos[\omega_0 t + \theta + \phi(\omega_0)] \quad (3\text{-}11)$$

Example 3-4 The transfer function for a certain linear stationary system is
$$H(j\omega) = \frac{K(1 + j\omega/\omega_2)}{(1 + j\omega/\omega_1)(1 + j\omega/\omega_3)}$$
where $K = 100$ V/V, $\omega_1 = 1$ krad/s, $\omega_2 = 2$ krad/s, and $\omega_3 = 3$ krad/s. Find the output $y(t)$ for input
$$x(t) = 5\cos\left(\omega_0 t - \frac{\pi}{3}\right) \quad V$$
where $\omega_0 = 1$ krad/s.

SOLUTION From (3-10), the gain is
$$\Gamma(\omega) = |H(j\omega)| = \left|\frac{K(1 + j\omega/\omega_2)}{(1 + j\omega/\omega_1)(1 + j\omega/\omega_3)}\right|$$
$$= \frac{|K||1 + j\omega/\omega_2|}{|1 + j\omega/\omega_1||1 + j\omega/\omega_3|} = \frac{|K|\sqrt{1 + (\omega/\omega_2)^2}}{\sqrt{1 + (\omega/\omega_1)^2}\sqrt{1 + (\omega/\omega_3)^2}}$$

For $\omega = \omega_0 = 1$ krad/s, this gives
$$\Gamma(\omega_0) = \frac{100\sqrt{1 + (\tfrac{1}{2})^2}}{\sqrt{1 + 1^2}\sqrt{1 + (\tfrac{1}{3})^2}} = 75 \text{ V/V}$$

From (3-10) the phase shift is
$$\phi(\omega) = \angle H(j\omega) = \angle K + \angle\left(1 + \frac{j\omega}{\omega_2}\right) - \angle\left(1 + \frac{j\omega}{\omega_1}\right) - \angle\left(1 + \frac{j\omega}{\omega_3}\right)$$
$$= 0 + \text{Tan}^{-1}\left(\frac{\omega}{\omega_2}, 1\right) - \text{Tan}^{-1}\left(\frac{\omega}{\omega_1}, 1\right) - \text{Tan}^{-1}\left(\frac{\omega}{\omega_3}, 1\right)$$

For $\omega = \omega_0$, this gives
$$\phi(\omega_0) = \text{Tan}^{-1}(0.50, 1.00) - \text{Tan}^{-1}(1.00, 1.00) - \text{Tan}^{-1}(0.33, 1.00)$$
$$= 0.46 - 0.79 - 0.32 = -0.65 \text{ rad}$$

From (3-11), the output is

$$y(t) = \Gamma(\omega_0)A \cos[\omega_0 t + \theta + \phi(\omega_0)]$$

where $A = 5$ V, $\theta = -\pi/3$ rad, and $\Gamma(\omega_0)$ and $\phi(\omega_0)$ are as given above; thus

$$y(t) = 375 \cos(\omega_0 t - 1.70) \quad \text{V}$$

3-1E Constant (dc) Signals

It is often convenient to treat a constant (dc) signal $x(t) = x_0$ as a sinusoid having frequency $\omega = 0$. For $\omega = 0$, (3-1) gives the amplitude-phase representation of a dc signal as

$$x_0 = A \cos \theta \tag{3-12a}$$

In order to abide by the convention that peak amplitude is nonnegative we define

$$A = |x_0| \qquad \theta = \angle x_0 = \begin{cases} 0 & x_0 \geq 0 \\ \pi & x_0 < 0 \end{cases} \tag{3-12b}$$

For example, the signal $x(t) = 5$ V has peak amplitude 5 V and initial phase 0, whereas the signal $z(t) = -5$ V has peak amplitude 5 V and initial phase π rad.

The exponential form for a dc signal is defined differently from that for a sinusoidal $(\omega \neq 0)$ signal. For $\omega = 0$, $e^{j\omega t} = 1$, and (3-2) gives $x(t) = X + X^*$; however, it is unnecessarily cumbersome to represent a single constant (x_0) as the sum of two other constants $(X + X^*)$. We define the complex amplitude of a dc signal as the signal itself; thus the exponential representation of a dc signal $x(t) = x_0$ is just

$$X = x_0 \tag{3-13}$$

For a dc signal, (3-12) and (3-13) give

$$X = Ae^{j\theta} \qquad \begin{aligned} A &= |X| \\ \theta &= \angle X \end{aligned} \tag{3-14}$$

Note that the relation between peak amplitude and complex amplitude for a dc signal is different (by a factor of 2) from that for a sinusoidal $(\omega \neq 0)$ signal.

The main advantage of treating a dc signal as a zero-frequency sinusoid lies in the fact that we can use the transfer function of a system (for $\omega = 0$) to obtain the output of the system for a dc input. The output of a system having transfer function $H(j\omega)$ for input $x(t) = x_0$ is given by

$$y(t) = H(0)x_0 \tag{3-15}$$

The quantity $H(0)$ is called the *dc gain* of the system.

Example 3-5 The output $y(t)$ of the system of Example 3-4 for input $x(t) = x_0 = -5$ mV is given by $y(t) = H(0)x_0$, where $H(j\omega)$ is the transfer function of the system. From Example 3-4

$$H(0) = \frac{K(1 + j0)}{(1 + j0)(1 + j0)} = K = 100 \text{ V/V}$$

Thus $y(t) = (100)(-0.005) = -500$ mV.

116 INTRODUCTION TO SYSTEM ANALYSIS

Equation (3-5) provides the most direct route to the transfer function of a system described by an impulse response. Using (3-5) to obtain the transfer function of a system described by a block diagram or a differential equation is tedious because it entails first finding the impulse response of the system. In Secs. 3-1F and 3-1G we show how the transfer function of a system can be obtained directly from a differential equation or a block diagram describing the system.

3-1F Systems Described by Differential Equations

We seek the transfer function of a stable linear stationary system whose output $y(t)$ and input $x(t)$ are related by a differential equation of the form

$$\frac{d^n y}{dt^n} + a_{n-1}\frac{d^{n-1}y}{dt^{n-1}} + \cdots + a_0 y = b_m \frac{d^m x}{dt^m} + \cdots + b_0 x \qquad (3\text{-}16)$$

Because the system is linear and stationary, the output for input $x(t) = Xe^{j\omega t}$ is given by $y(t) = H(j\omega)Xe^{j\omega t}$ [see (3-6)]. Substituting $Xe^{j\omega t}$ for $x(t)$ and $H(j\omega)Xe^{j\omega t}$ for $y(t)$ in (3-16) gives

$$[(j\omega)^n + a_{n-1}(j\omega)^{n-1} + \cdots + a_0]H(j\omega)Xe^{j\omega t} = [b_m(j\omega)^m + \cdots + b_0]Xe^{j\omega t}$$

This yields

$$H(j\omega) = \frac{b_m(j\omega)^m + \cdots + b_0}{(j\omega)^n + a_{n-1}(j\omega)^{n-1} + \cdots + a_0} \qquad (3\text{-}17)$$

Example 3-6 The input $x(t)$ and output $y(t)$ of a certain system are currents in amperes related by

$$\frac{d^3 y}{dt^3} + a_2 \frac{d^2 y}{dt^2} + a_1 \frac{dy}{dt} + a_0 y = b_1 \frac{dx}{dt} + b_0 x$$

where $a_0 = b_0 = 24 \times 10^9 \text{ s}^{-3}$, $a_1 = 26 \times 10^6 \text{ s}^{-2}$, $a_2 = 9 \times 10^3 \text{ s}^{-1}$, and $b_1 = 24 \times 10^6 \text{ s}^{-2}$. Find the output for input

$$x(t) = 20 \cos\left(\omega_0 t - \frac{\pi}{2}\right) \quad \text{mA}$$

where $\omega_0 = 5$ krad/s.

SOLUTION From (3-17), the transfer function of the system is

$$H(j\omega) = \frac{b_1(j\omega) + b_0}{(j\omega)^3 + a_2(j\omega)^2 + a_1(j\omega) + a_0} = \frac{j\omega b_1 + b_0}{j(a_1\omega - \omega^3) + a_0 - a_2\omega^2} \quad \text{A/A}$$

For $\omega = \omega_0 = 5$ krad/s, this gives

$$H(j\omega_0) = \frac{j(5000)(24 \times 10^6) + 24 \times 10^9}{j(5000)(26 \times 10^6 - 25 \times 10^6) + 24 \times 10^9 - (9 \times 10^3)(25 \times 10^6)}$$

$$= \frac{24 + 120j}{-201 + 5j} = 0.61e^{-1.75j} \text{ A/A}$$

From this (polar-form) expression for $H(j\omega_0)$, we have immediately that the gain and phase shift for $\omega = \omega_0$ are

$$\Gamma(\omega_0) = 0.61 \text{ A/A} \qquad \phi(\omega_0) = -1.75 \text{ rad}$$

It follows from (3-11) that the output is

$$y(t) = (0.61)(0.02) \cos\left(\omega_0 t - \frac{\pi}{2} - 1.75\right) \qquad \text{A}$$

$$= 12.2 \cos(\omega_0 t - 3.31) = 12.2 \cos(\omega_0 t + 2.97) \qquad \text{mA}$$

3-1G Block-Diagram Reduction

A system having transfer function $H(j\omega)$ can be represented by a *block diagram*, as shown in Fig. 3-1, where $Y = H(j\omega)X$. A block diagram consisting of series, parallel, and feedback connections of such blocks can be reduced to a single block by an algebraic process called *block-diagram reduction*. The transfer function of the single block thus obtained is the transfer function of the system represented by the original block diagram. Block-diagram reduction is based on a few simple rules for simplifying series, parallel, and feedback connections and for reordering blocks and addition elements or blocks and pickoff points. We develop these rules below and give examples of their use.

Figure 3-2a shows a parallel connection of two systems. The complex amplitude of an output is given by

$$Y = [H_1(j\omega) + H_2(j\omega)]X$$

where X is the complex amplitude of the corresponding input. This implies that the transfer function of the parallel connection of Fig. 3-2a is

$$H(j\omega) = H_1(j\omega) + H_2(j\omega) \qquad (3\text{-}18)$$

as shown in Fig. 3-2b.

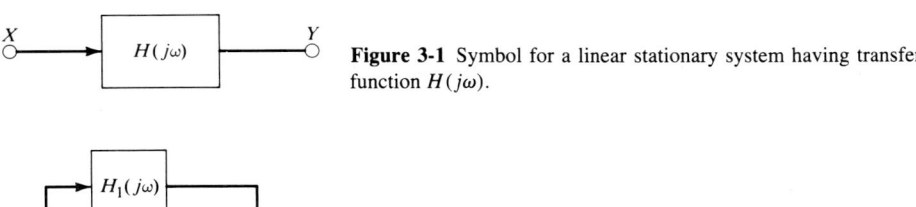

Figure 3-1 Symbol for a linear stationary system having transfer function $H(j\omega)$.

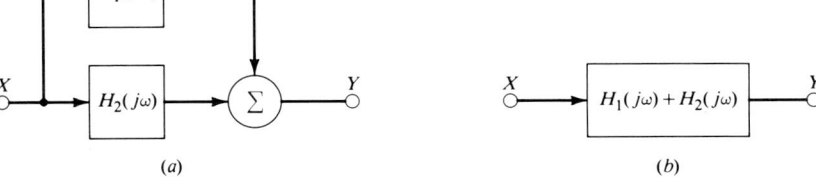

Figure 3-2 Reduction of a parallel connection to a single block: (*a*) Original and (*b*) reduced system.

Figure 3-3 Reduction of a series connection to a single block: (a) original and (b) reduced system.

Figure 3-3a shows a series connection of two systems. The complex amplitude of an output is given by

$$Y = H_2(j\omega)W = H_2(j\omega)H_1(j\omega)X$$

where X is the complex amplitude of the corresponding input. This implies that the transfer function of the series connection of Fig. 3-3a is

$$H(j\omega) = H_2(j\omega)H_1(j\omega) \qquad (3\text{-}19)$$

as shown in Fig. 3-3b.

Figure 3-4a shows a feedback connection of two blocks. The complex amplitude of an output is given by

$$Y = H_1(j\omega)E = H_1(j\omega)[X \pm H_2(j\omega)Y]$$

which yields

$$Y = \frac{H_1(j\omega)}{1 \mp H_1(j\omega)H_2(j\omega)}X$$

This implies that the transfer function of the feedback connection of Fig. 3-4a is

$$H(j\omega) = \frac{H_1(j\omega)}{1 \mp H_1(j\omega)H_2(j\omega)} \qquad (3\text{-}20)$$

as shown in Fig. 3-4b. Note that the negative sign in (3-20) corresponds to the positive sign in Fig. 3-4a and vice versa.

The feedback system of Fig. 3-4a may not be stable, even if the constituent systems represented by $H_1(j\omega)$ and $H_2(j\omega)$ are stable. If the feedback system is not stable, it cannot be described by the transfer function of (3-20) because the output for input $Xe^{j\omega t}$ is not given by $H(j\omega)Xe^{j\omega t}$. We can determine whether a system is stable or not as follows:

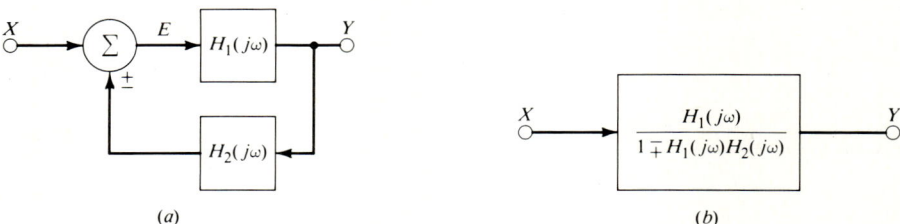

Figure 3-4 Reduction of a feedback system to a single block: (a) Original and (b) reduced system.

1. Express the transfer function (obtained by block-diagram reduction or from a differential equation) as a ratio of two polynomials in $j\omega$, as in (3-17).
2. By inspecting the denominator of the transfer function identify the coefficients $a_0, a_1, \ldots, a_n$ in the differential equation describing the system [see (3-16)].
3. Find the characteristic roots for the system by solving the characteristic equation

$$s^n + a_{n-1}s^{n-1} + \cdots + a_0 = 0$$

The system is stable if and only if all its characteristic roots have negative real parts.

Block-diagram reduction can be used to obtain the transfer function of a system composed of any number of subsystems in series, parallel, and feedback. Block-diagram reduction is often the easiest way to find the transfer function of a system or to obtain a differential equation describing a system [by working backward from (3-17) to (3-16)]. Subsequently, the impulse response can be obtained using the procedure given in Sec. 2-4. We describe easier methods for obtaining the impulse response of a system in Chaps. 4 and 5.

Example 3-7 Obtain the transfer function of the system of Fig. 3-5a.

SOLUTION We use block-diagram reduction. Figure 3-5b to d shows steps in the reduction. First, we reduce the series connection of H_1 and H_2 to a single block according to (3-19). Next, we reduce the feedback connection containing $H_1 H_2$ in the forward path and H_3 in the feedback path to a single block according to (3-20). Finally, we reduce the parallel

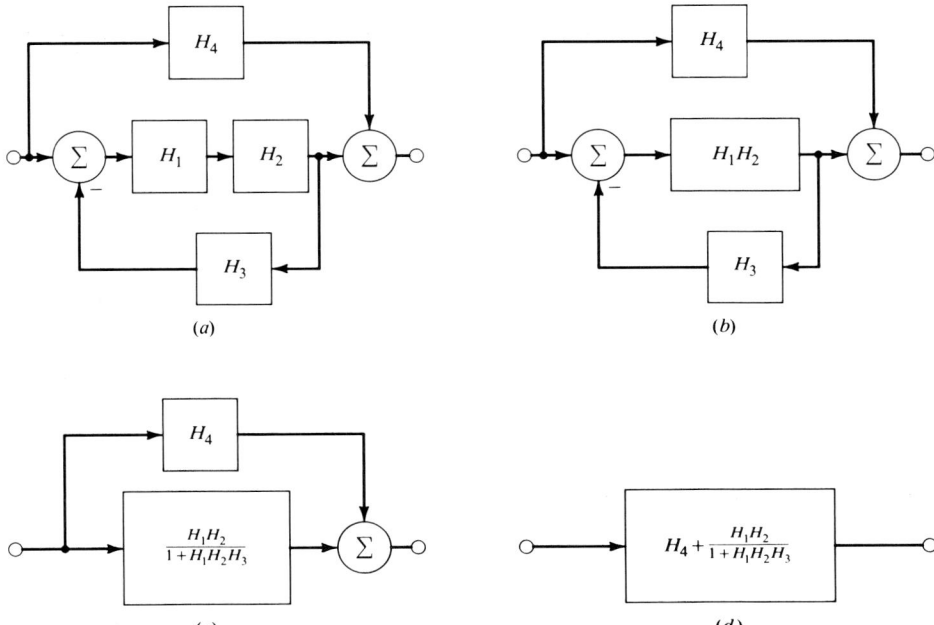

Figure 3-5 System of Example 3-7.

120 INTRODUCTION TO SYSTEM ANALYSIS

connection to a single block according to (3-18). The transfer function of the system is given by

$$H = H_4 + \frac{H_1 H_2}{1 + H_1 H_2 H_3}$$

The rules given in (3-18) to (3-20) can be used to reduce a system consisting entirely of simple series, parallel and feedback subsystems. It is often necessary to reorder blocks and addition elements or blocks and pickoff points before applying those rules. Figures 3-6 and 3-7 show how addition elements and pickoff points are moved from the input of a block to the output of the block and vice versa.

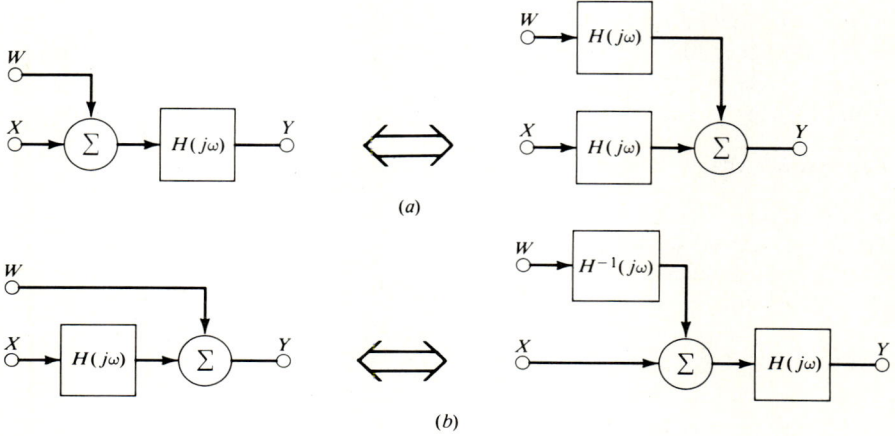

Figure 3-6 Rules for exchanging the order of systems and addition elements.

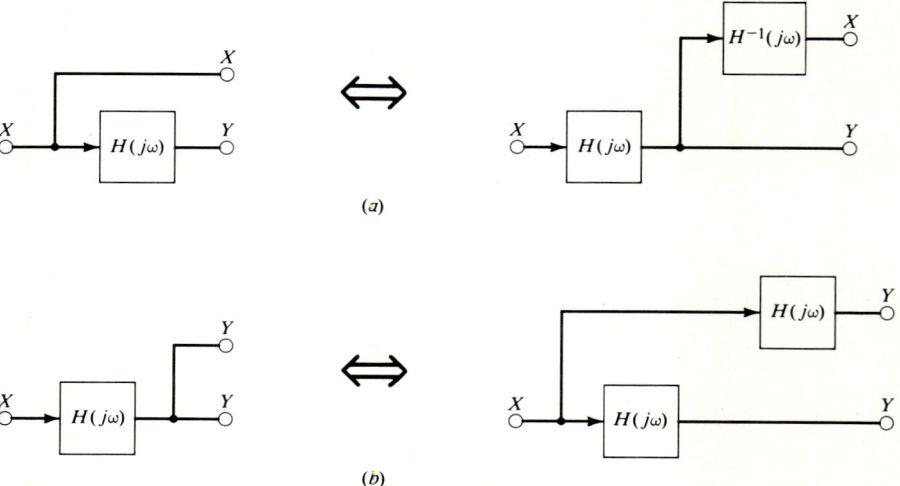

Figure 3-7 Rules for exchanging the order of systems and pickoff points.

RESPONSE TO SINUSOIDAL EXCITATION **121**

Example 3-8 Obtain the transfer function of the system of Fig. 3-8a.

SOLUTION The system does not consist entirely of simple series, parallel, and feedback subsystems because of the addition element between H_2 and H_3. The addition element can

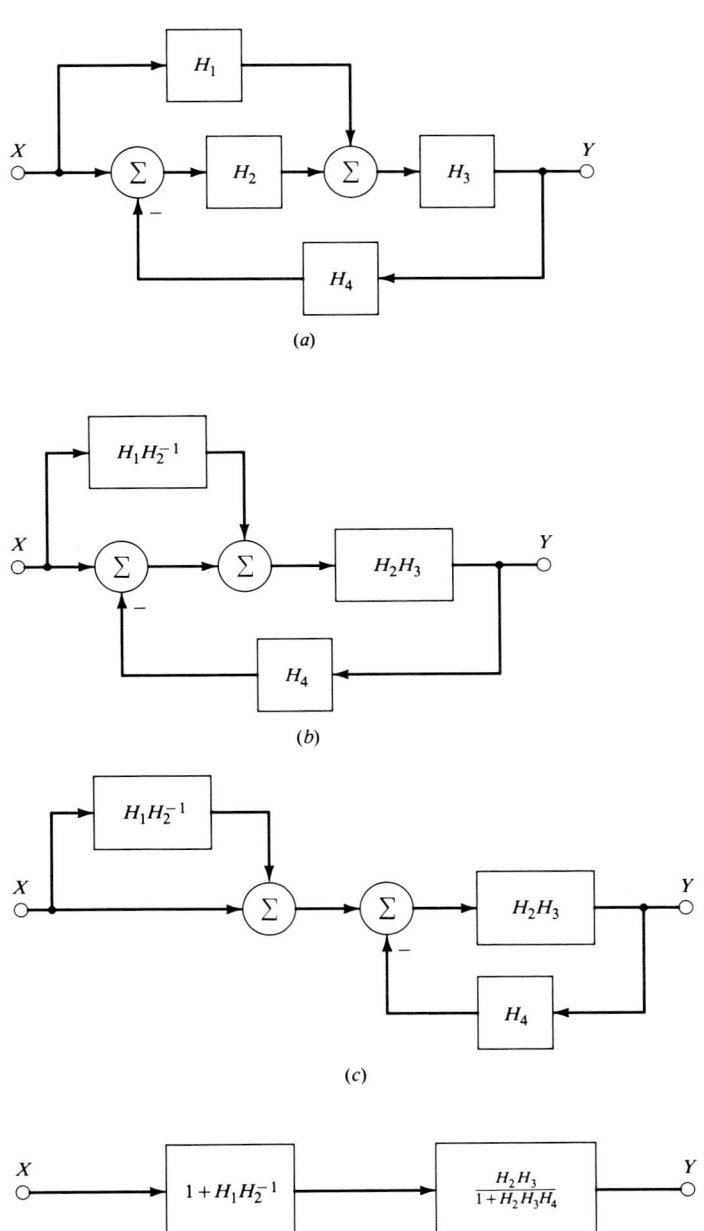

Figure 3-8 System of Example 3-8.

be moved outside the feedback loop as follows: Move the addition element from the output to the input of H_2 using the rule given in Fig. 3-6b; this gives the block diagram of Fig. 3-8b. Reverse the order of the two addition elements in Fig. 3-8b (addition is commutative); this gives the block diagram of Fig. 3-8c, which consists of a parallel subsystem and a feedback subsystem connected in series. Figure 3-8d shows the last step of the reduction. The transfer function of the system is given by

$$H = \frac{(1 + H_1 H_2^{-1}) H_2 H_3}{1 - H_2 H_3 H_4}$$

Example 3-9 Obtain the transfer function of the system of Fig. 3-9a.

SOLUTION The system does not consist entirely of simple series, parallel, and feedback subsystems because of the pickoff point between H_2 and H_3. Move the pickoff point outside the feedback loop, using the rule in Fig. 3-7a. This gives the block diagram of Fig. 3-9b,

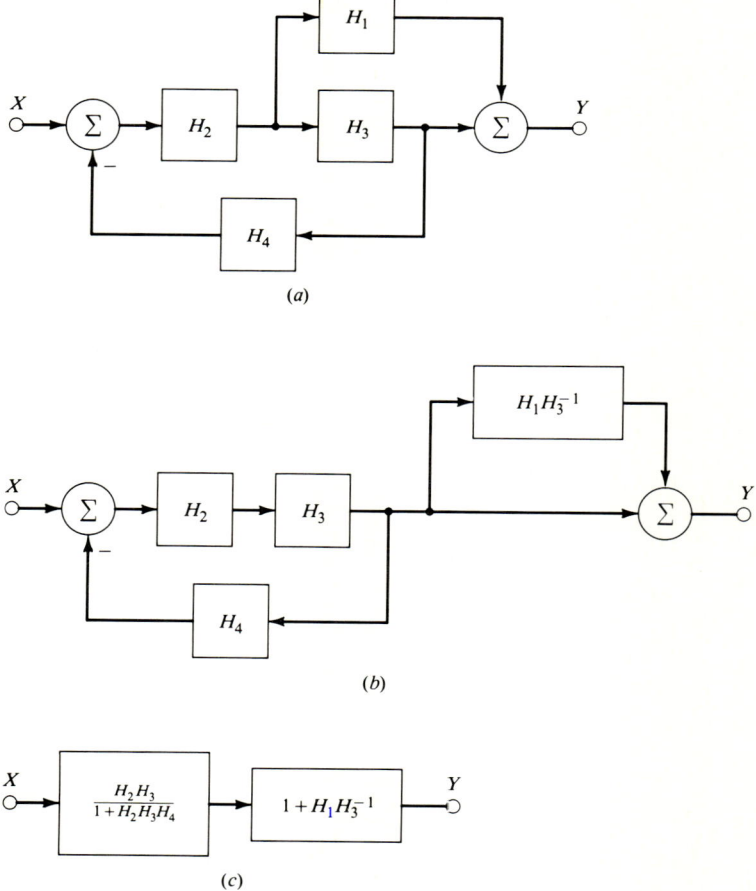

Figure 3-9 System of Example 3-9.

which consists entirely of series, parallel, and feedback subsystems. Figure 3-9c shows the last step in the reduction. The transfer function is given by

$$H = \frac{H_2 H_3 (1 + H_1 H_3^{-1})}{1 + H_2 H_3 H_4}$$

Example 3-10 Obtain a differential equation describing the system of Fig. 3-10, where

$$H_1(j\omega) = \frac{K_1}{1 + j\omega/\omega_1} \qquad H_2(j\omega) = \frac{K_2}{1 + j\omega/\omega_2}$$

with $K_1 = 10^3$, $\omega_1 = 1$ krad/s, $K_2 = 10$, and $\omega_2 = 2$ krad/s. Determine whether the system is stable.

SOLUTION Block-diagram reduction yields the transfer function

$$H = \frac{H_1}{1 + H_1 H_2}$$

Substituting the functions above for H_1, H_2 gives

$$H(j\omega) = \frac{K_1(1 + j\omega/\omega_2)}{(1 + j\omega/\omega_1)(1 + j\omega/\omega_2) + K_1 K_2} = \frac{K_1 \omega_1 (\omega_2 + j\omega)}{(\omega_1 + j\omega)(\omega_2 + j\omega) + K_1 K_2 \omega_1 \omega_2}$$

which can be written

$$H(j\omega) = \frac{b_1(j\omega) + b_0}{(j\omega)^2 + a_1(j\omega) + a_0}$$

where

$$b_1 = K_1 \omega_1 = 1 \times 10^6 \text{ s}^{-1}$$
$$b_0 = K_1 \omega_1 \omega_2 = 2 \times 10^9 \text{ s}^{-2}$$
$$a_1 = \omega_1 + \omega_2 = 3 \times 10^3 \text{ s}^{-1}$$
$$a_0 = (1 + K_1 K_2) \omega_1 \omega_2 = 2 \times 10^{10} \text{ s}^{-2}$$

A differential equation describing the system is

$$\frac{d^2 y}{dt^2} + a_1 \frac{dy}{dt} + a_0 y = b_1 \frac{dx}{dt} + b_0 x$$

To check for stability we find the characteristic roots for the system. The characteristic equation for the system is

$$s^2 + a_1 s + a_0 = 0$$

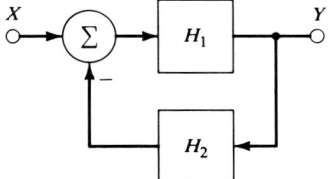

Figure 3-10 System of Example 3-10.

and the characteristic roots are

$$p_1, p_2 = \frac{-a_1 \pm \sqrt{a_1^2 - 4a_0}}{2} = (-1.5 \pm 141j) \times 10^3 \text{ s}^{-1}$$

The system is stable because the real parts of the characteristic roots are negative.

Block-diagram reduction can be applied to a block diagram whose constituents (individual blocks) are described by transfer functions. To convert a block diagram whose constituents are described by transfer characteristics (or impulse responses) into one whose constituents are described by transfer functions, we find the transfer functions of the individual blocks. A block described by an impulse response $h(t)$ is replaced by one described by the transfer function $H(j\omega)$ given by (3-5). A block described by a differential equation given by (3-16) is replaced by one described by a transfer function given by (3-17). A block representing an elementary system is replaced according to Table 3-1. Below, we derive the entries in this table.

Proportion element with gain K The transfer characteristic is $y(t) = Kx(t)$, where $x(t)$ is an input and $y(t)$ is the corresponding output. The output for input $x(t) = Xe^{j\omega t}$ is given by $y(t) = KXe^{j\omega t}$. This implies that the transfer function for a proportion element having gain K is $H(j\omega) = K$, as given in Table 3-1.

Differentiator with gain K The transfer characteristic is $y(t) = K\,dx(t)/dt$, where $x(t)$ is an input and $y(t)$ is the corresponding output. This has the form of (3-16) with

Table 3-1 Frequency-domain symbols for elementary linear stationary systems

Element	Transfer characteristic	Transfer function	Symbol
Proportion	$y(t) = Kx(t)$	K	$X \to [K] \to Y$
Differentiation	$y(t) = K\dfrac{dx(t)}{dt}$	$j\omega K$	$X \to [j\omega K] \to Y$
Integration	$y(t) = K\displaystyle\int_{-\infty}^{t} x(\alpha)\,d\alpha$	$\dfrac{K}{j\omega}$	$X \to [\frac{K}{j\omega}] \to Y$
Delay	$y(t) = x(t - t_0)$	$e^{-j\omega t_0}$	$X \to [e^{-j\omega t_0}] \to Y$
Addition	$y(t) = w(t) + x(t)$	$Y = W + X$	$W, X \to [\Sigma] \to Y$

$n = 0$, $a_0 = 1$, $m = 1$, $b_1 = K$, and $b_0 = 0$. From (3-17), the transfer function for a differentiator is $H(j\omega) = j\omega K$, as given in Table 3-1.

Integrator with gain K The transfer characteristic is

$$y(t) = K \int_{-\infty}^{t} x(\alpha)\, d\alpha$$

where $x(t)$ is an input and $y(t)$ the corresponding output. Differentiating this relation once with respect to time gives

$$\frac{dy(t)}{dt} = Kx(t)$$

This has the form of (3-17), with $n = 1$, $a_0 = 0$, $m = 0$, and $b_0 = K$. From (3-7), the transfer function of an integrator (see Prob. 3-17) is $H(j\omega) = K/j\omega$, as given in Table 3-1.

Delay element with delay time t_0 The transfer characteristic is $y(t) = x(t - t_0)$, where $x(t)$ is an input and $y(t)$ the corresponding output. For input $x(t) = Xe^{j\omega t}$, this gives

$$y(t) = Xe^{j\omega(t-t_0)} = Xe^{-j\omega t_0} e^{j\omega t}$$

Comparing this relation with (3-6) shows that the transfer function for a delay element having delay time t_0 is $e^{-j\omega t_0}$, as given in Table 3-1.

Addition element The output of an addition element (an adder) for inputs $w(t) = We^{j\omega t}$ and $x(t) = Xe^{j\omega t}$ is $y(t) = (X + W)e^{j\omega t} = Ye^{j\omega t}$, where $Y = X + W$. This shows that the complex amplitude of the output is the sum of the complex amplitudes of the inputs provided both inputs have the same frequency. If the inputs have different frequencies, the output cannot be represented as a single exponential.

Example 3-11 Obtain the output of the system of Fig. 3-11a for input $x(t) = 2\cos(\omega_0 t + \pi/2)$ V, where $K_1 = 2 \times 10^{-4}$, $t_0 = 25\ \mu\text{s}$, $K_2 = 2\pi\ \text{s}^{-1}$, $K_3 = 5 \times 10^3$, and $\omega_0 = 10\pi$ krad/s.

SOLUTION Referring to Table 3-1, we first transform the block diagram of Fig. 3-11a to that of Fig. 3-11b. We then use block-diagram reduction (Figs. 3-11c and d) to obtain the transfer function $H(j\omega)$ of the system. This gives

$$H(j\omega) = \frac{K_2 + j\omega K_1}{K_2 K_3 + j\omega e^{j\omega t_0}}$$

We then calculate gain and phase shift of the system for the frequency of the input. For $\omega = \omega_0 = 10\pi$ krad/s we obtain

$$H(j\omega_0) = \frac{2\pi + j(\pi \times 10^4)(2 \times 10^{-4})}{(2\pi)(5000) + j(\pi \times 10^4)e^{j\pi(10^4)(25\times 10^{-6})}} = 370e^{-0.393j} \quad \mu\text{V/V}$$

The gain and phase shift are

$$\Gamma(\omega_0) = 370\ \mu\text{V/V} \qquad \phi(\omega_0) = -0.393\text{ rad}$$

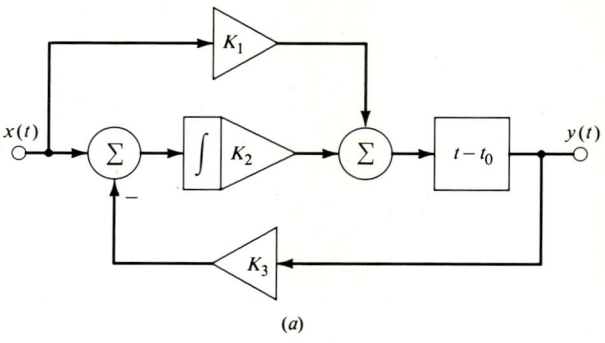

(a)

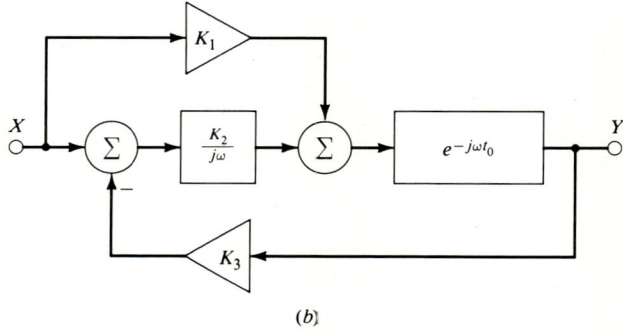

(b)

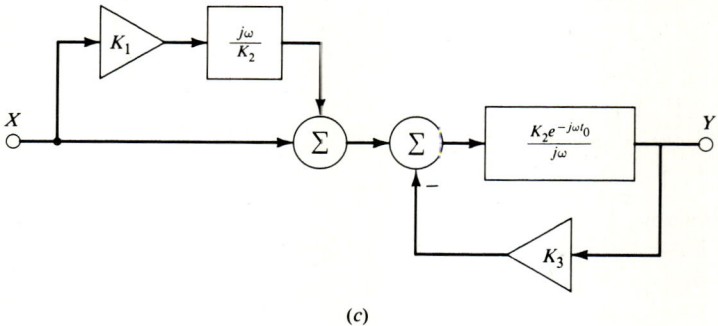

(c)

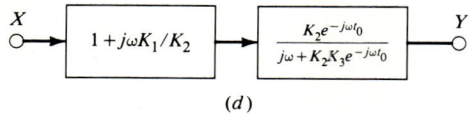

(d)

Figure 3-11 System of Example 3-11.

The output is

$$y(t) = (370)(2)\cos\left(\omega_0 t + \frac{\pi}{2} - 0.393\right) = 740\cos(\omega_0 t + 1.18) \quad \mu V$$

3-1H Superposition

We use superposition to obtain the output of a linear stationary system for an input consisting of several sinusoidal terms. We use the methods described above to obtain the output for each term considered individually. The complete output is the sum of the individual outputs thus obtained.

Example 3-12 Obtain the output $y(t)$ of a system having transfer function

$$H(j\omega) = \frac{10}{1 + j\omega\tau} \quad V/V$$

for input

$$x(t) = 5 + 4\cos\omega_0 t + 3\cos\left(2\omega_0 t - \frac{\pi}{2}\right) \quad V$$

where $\tau = 1$ ms and $\omega_0 = 1$ krad/s.

SOLUTION From (3-10), gain and phase shift of the system are given by

$$\Gamma(\omega) = \frac{10}{\sqrt{1 + (\omega\tau)^2}} \quad V/V \qquad \phi(\omega) = -\text{Tan}^{-1}(\omega\tau, 1)$$

The output for the dc term $x_0 = 5$ V is

$$y_0 = H(0)x_0 = (10)(5) = 50 \text{ V}$$

The output for the term $x_1(t) = 4\cos\omega_0 t$ V is

$$y_1(t) = \Gamma(\omega_0)4\cos[\omega_0 t + \phi(\omega_0)] = 28.3\cos\left(\omega_0 t - \frac{\pi}{4}\right) \quad V$$

The output for the term $x_2(t) = 3\cos(2\omega_0 t - \pi/2)$ V is

$$y_2(t) = \Gamma(2\omega_0)3\cos\left[2\omega_0 t - \frac{\pi}{2} + \phi(2\omega_0)\right] = 13.4\cos(2\omega_0 t - 2.68) \quad V$$

Superposition gives

$$y(t) = y_0 + y_1(t) + y_2(t)$$

$$= 50 + 28.3\cos\left(\omega_0 t - \frac{\pi}{4}\right) + 13.4\cos(2\omega_0 t - 2.68) \quad V$$

3-2 NONLINEAR STATIC SYSTEMS

The transfer characteristic for a nonlinear static system has the form

$$y = y_0 N\left(\frac{x}{x_0}\right) \tag{3-21}$$

where $x = x(t)$ is an input, $y = y(t)$ is the corresponding output, and N is a nonlinear function. A nonlinear system cannot be described by a transfer function because (1) the output for sinusoidal input is not a sinusoid having the frequency of the input and (2) a nonlinear system does not obey superposition. Nonetheless, sinusoids often are used as test inputs for nonlinear systems; e.g., specifications on harmonic distortion and intermodulation distortion for an audio amplifier refer to the output of the amplifier (regarded as a nonlinear system) for one or two sinusoidal inputs.

Sometimes, it is adequate to calculate an output of a nonlinear static system for a sinusoidal input using (3-21) directly; thus, the output for input $x(t) = A \cos(\omega t + \theta)$ is given by

$$y(t) = y_0 N\left[\frac{A}{x_0} \cos(\omega t + \theta)\right]$$

Often, however, the resulting expression is too cumbersome to be of much use. We then seek other, more easily interpreted expressions for an output of a sinusoidally excited nonlinear system.

In this section we show that an output of a sinusoidally excited nonlinear static system described by (3-21) can be expressed as a *harmonic series* of the form

$$y(t) = B_0 \cos \theta_0 + B_1 \cos(\omega_0 t + \theta_1) + B_2 \cos(2\omega_0 t + \theta_2) + \cdots \quad (3\text{-}22)$$

where the fundamental frequency ω_0 equals the frequency of the input. This is especially helpful when (as is often the case) an output of a nonlinear static system is the input of a linear stationary system (because then we can use the methods described in the previous section to find the output of the linear stationary system). It is also helpful for interpreting an output of a nonlinear static system alone, e.g., for describing distortion introduced by the system.

3-2A Polynomial Transfer Characteristics

First we consider a nonlinear static system whose output $y(t)$ for input $x(t)$ is given by

$$y = y_0\left[d_0 + d_1 \frac{x}{x_0} + d_2\left(\frac{x}{x_0}\right)^2 + \cdots + d_k\left(\frac{x}{x_0}\right)^k\right] \quad (3\text{-}23)$$

This transfer characteristic has the form of (3-21), where

$$N(\alpha) = d_0 + d_1\alpha + d_2\alpha^2 + \cdots + d_k\alpha^k$$

i.e., where $N(\alpha)$ is a polynomial of degree k in $\alpha = x/x_0$. We show below that a sinusoidally excited system whose transfer characteristic has the form of (3-23) gives an output having the form of (3-22). Then we extend that result to systems having other than polynomial transfer characteristics.

The output $y(t)$ of a system described by (3-23) for input

$$x(t) = A \cos \omega_0 t$$

is given by

$$y(t) = y_0\left[d_0 + d_1 \frac{A}{x_0} \cos \omega_0 t + d_2\left(\frac{A}{x_0}\right)^2 \cos^2 \omega_0 t + \cdots + d_k\left(\frac{A}{x_0}\right)^k \cos^k \omega_0 t\right]$$

To express the output as a harmonic series we apply the identity

$$\cos^m \alpha \equiv 2^{-m} \sum_{i=0}^{m} \frac{m! \cos(|2i - m|\alpha)}{i!(m-i)!} \quad (3\text{-}24)$$

to each term of the series above; e.g.,

$$\cos^2 \alpha \equiv \tfrac{1}{2} + \tfrac{1}{2} \cos 2\alpha \qquad \cos^3 \alpha \equiv \tfrac{3}{4} \cos \alpha + \tfrac{1}{4} \cos 3\alpha$$

Example 3-13 The transfer characteristic of a certain system is

$$y = y_0 \left[\frac{x}{x_0} - \frac{1}{3}\left(\frac{x}{x_0}\right)^3 \right]$$

where $x_0 = 1$ mA and $y_0 = 5$ mV. Obtain the output $y(t)$ for input

$$x(t) = 5 \cos \omega_0 t \quad \text{mA}$$

where $\omega_0 = 2$ krad/s. Express the output as a harmonic series having fundamental frequency ω_0.

SOLUTION The output is given by

$$y(t) = y_0(5 \cos \omega_0 t - \tfrac{125}{3} \cos^3 \omega_0 t)$$

Using the identity $\cos^3 \alpha \equiv \tfrac{3}{4} \cos \alpha + \tfrac{1}{4} \cos 3\alpha$ gives

$$y(t) = y_0(5 \cos \omega_0 t - \tfrac{125}{4} \cos \omega_0 t - \tfrac{125}{12} \cos 3\omega_0 t)$$
$$= -131 \cos \omega_0 t - 52.1 \cos 3\omega_0 t \quad \text{mV}$$

Example 3-14 Obtain the output $y(t)$ of the system of Fig. 3-12 for input $x(t) = 10 \cos \omega_0 t$ mA, where $x_0 = 2$ mA, $v_0 = 1$ V, $K = 1$ ms, and $\omega_0 = 1$ krad/s. Express the output as a harmonic series having fundamental frequency ω_0.

SOLUTION The output of the nonlinear (square-law) element is given by

$$v(t) = v_0 \left(\frac{10 \cos \omega_0 t}{2}\right)^2 = 25 v_0 \cos^2 \omega_0 t$$

Using the identity $\cos^2 \alpha \equiv \tfrac{1}{2} + \tfrac{1}{2} \cos 2\alpha$ gives

$$v(t) = 12.5 v_0 + 12.5 v_0 \cos 2\omega_0 t$$

The transfer function for the differentiator is (see Table 3-1)

$$H(j\omega) = j\omega K$$

where $K = 1$ ms. The associated gain and phase shift are given by

$$\Gamma(\omega) = |j\omega K| = \omega K \qquad \phi(\omega) = \angle(j\omega K) = \frac{\pi}{2} \quad \text{rad}$$

The output of the differentiator is

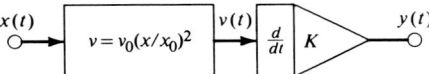

Figure 3-12 System of Example 3-14.

130 INTRODUCTION TO SYSTEM ANALYSIS

$$y(t) = 12.5H(0)v_0 + 12.5\Gamma(2\omega_0)v_0 \cos[2\omega_0 t + \phi(2\omega_0)]$$

$$= (12.5)(0)v_0 + (12.5)(2\omega_0 K)v_0 \cos\left(2\omega_0 t + \frac{\pi}{2}\right)$$

$$= 25.0 \cos\left(2\omega_0 t + \frac{\pi}{2}\right) \quad \text{V}$$

This system is a frequency doubler. Its output for sinusoidal input is a sinusoid whose frequency equals twice the frequency of the input.

In principle, we can obtain a good polynomial approximation for any practical nonlinear transfer characteristic of the form $y = y_0 N(x/x_0)$; for example, by truncating Maclaurin's or Taylor's expansion (Thomas and Finney, p. 789) for the function $N(\alpha)$ or by the method of least squares. We can use Maclaurin's or Taylor's expansion only if the transfer characteristic is sufficiently smooth. Otherwise we must resort to other methods. We show how a polynomial approximation for a nonlinear static transfer characteristic is obtained using Maclaurin's or Taylor's expansion in Secs. 3-2B and 3-2C, respectively. Other methods for obtaining polynomial approximations generally require use of a digital computer.

3-2B Maclaurin's Expansion

Maclaurin's expansion for a function $N(\alpha)$ is given by

$$N(\alpha) = d_0 + d_1\alpha + d_2\alpha^2 + \cdots \quad \text{where } d_n = \frac{1}{n!} \left.\frac{d^n N(\alpha)}{d\alpha^n}\right|_{\alpha=0} \quad (3\text{-}25)$$

Section A-7 (Appendix A) gives Maclaurin's expansions for the most common elementary functions. We can use Maclaurin's expansion to obtain a polynomial approximating a static transfer characteristic under two conditions: (1) the transfer characteristic must be smooth enough for the necessary derivatives to exist and (2) the series given by (3-25) must converge for all values of $\alpha = x/x_0$ for which the approximation is used.

Example 3-15 Use Maclaurin's expansion to obtain a third-degree polynomial approximating the transfer characteristic

$$y = y_0 \tanh \frac{x}{x_0}$$

where $x_0 = 25.0$ mA and $y_0 = 5.00$ V. Use the approximation to obtain the output for input

$$x(t) = A \cos \omega_0 t$$

where $A = 5.00$ mA and $\omega_0 = 1.00$ krad/s.

SOLUTION Maclaurin's expansion for $\tanh \alpha$ is

$$\tanh \alpha = \alpha - \frac{\alpha^3}{3} + \frac{2\alpha^5}{15} - \cdots$$

The series converges for $|\alpha| < \pi/2$. The largest value of $|\alpha|$ for which the series is used in this example is $5.00/25.00 = 0.20$. A third-degree polynomial approximating the transfer characteristic is obtained by keeping only the first two terms of the series above; thus

$$y \approx y_0 \left[\frac{x}{x_0} - \frac{1}{3}\left(\frac{x}{x_0}\right)^3 \right]$$

$$= 0.20 y_0 \cos \omega_0 t - \tfrac{1}{3}(0.20)^3 y_0 (\tfrac{3}{4} \cos \omega_0 t + \tfrac{1}{4} \cos 3\omega_0 t)$$

$$= 990 \cos \omega_0 t - 3.33 \cos 3\omega_0 t \quad \text{mV}$$

The third harmonic is insignificant for the precision with which parameters are specified in this example, and for practical purposes the output is

$$y(t) = 990 \cos \omega_0 t \quad \text{mV}$$

Maclaurin's expansion for an even (odd) function contains only even (odd) powers of the argument; i.e., if $N(-\alpha) = N(\alpha)$, Maclaurin's series for $N(\alpha)$ has the form

$$N(\alpha) = d_0 + d_2 \alpha^2 + d_4 \alpha^4 + \cdots$$

and if $N(-\alpha) = -N(\alpha)$, Maclaurin's series for $N(\alpha)$ has the form

$$N(\alpha) = d_1 \alpha + d_3 \alpha^3 + d_5 \alpha^5 + \cdots$$

This means that a polynomial approximation obtained by truncating Maclaurin's series for an even (odd) transfer characteristic contains only even (odd) powers of an input. The same is true of polynomial approximations obtained in other (reasonable) ways, e.g., using least squares. Also, any realistic static transfer characteristic can be approximated as accurately as desired by a polynomial. Further, (3-24) shows that even (odd) powers of a sinusoidal signal contain only even (odd) harmonics of the signal. It follows from these observations that the output of a sinusoidally excited static system contains only even harmonics of the input if the transfer characteristic is even and only odd harmonics of the input if the transfer characteristic is odd. If the transfer characteristic is neither even nor odd, the output contains (in general) all harmonics of a sinusoidal input. For example, the output of the system of Example 3-13 contains only odd harmonics of the input, and the output of the system of Example 3-14 contains only even harmonics of the input.

3-2C Taylor's Expansion

Maclaurin's expansion is a special case of Taylor's expansion

$$N(\alpha + \alpha_0) = d_0 + d_1 \alpha + d_2 \alpha^2 + \cdots \quad \text{where } d_n = \frac{1}{n!} \left. \frac{d^n N(\alpha)}{d\alpha^n} \right|_{\alpha=\alpha_0} \quad (3\text{-}26)$$

We use Maclaurin's series where an input contains no dc component and Taylor's expansion when an input has the form

$$x(t) = x_{\text{dc}} + A \cos \omega_0 t$$

i.e., when an input contains a dc bias x_{dc}. For present purposes the function $N(\alpha)$ represents a static transfer characteristic, α_0 represents the dc component of an input, and α represents the ac component of an input. Figure 3-13a shows a case where Maclaurin's series can be used to approximate a transfer characteristic and Fig. 3-13b shows a case where Taylor's series should be used.

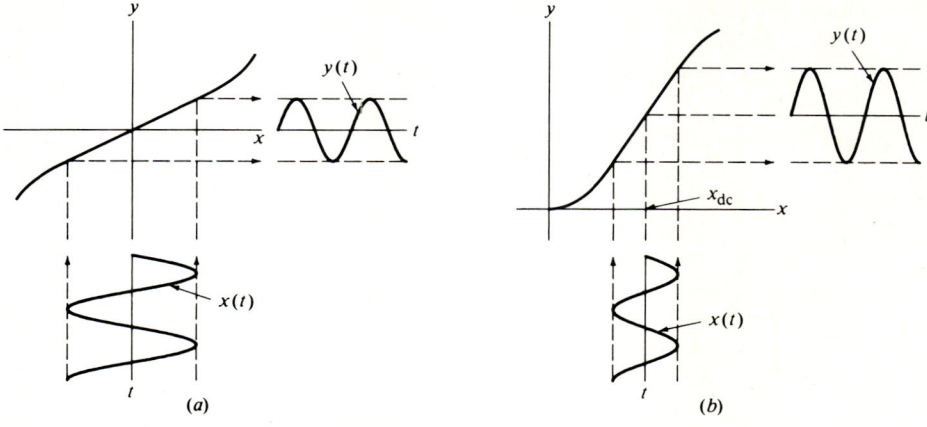

Figure 3-13 An illustration of (a) Unbiased and (b) biased operation of a nonlinear static system.

Example 3-16 Use a third-degree polynomial to approximate the transfer characteristic

$$y = y_0(e^{x/x_0} - 1)$$

where $x_0 = 25.0$ mV and $y_0 = 100$ mA. Use the approximation to obtain the output $y(t)$ for input

$$x(t) = A_0 + A_1 \cos \omega_0 t$$

where $A_0 = 10.0$ mV, $A_1 = 5.00$ mV, and $f_0 = \omega_0/2\pi = 10.00$ kHz.

SOLUTION Since the input has a dc component, we use Taylor's series to obtain a polynomial approximating the transfer characteristic of the system. From (3-26), a polynomial approximating $e^{\alpha + \alpha_0}$ in a neighborhood of α_0 is

$$e^{\alpha_0 + \alpha} = e^{\alpha_0} + e^{\alpha_0}\alpha + \tfrac{1}{2}e^{\alpha_0}\alpha^2 + \tfrac{1}{6}e^{\alpha_0}\alpha^3 = e^{\alpha_0}(1 + \alpha + \tfrac{1}{2}\alpha^2 + \tfrac{1}{6}\alpha^3)$$

Using this approximation for e^α in the transfer characteristic above gives

$$y \approx y_0[e^{\alpha_0}(1 + \alpha + \tfrac{1}{2}\alpha^2 + \tfrac{1}{6}\alpha^3) - 1]$$

where $\alpha_0 = A_0/x_0 = 0.4$ and $\alpha = (A_1/x_0) \cos \omega_0 t = 0.20 \cos \omega_0 t$; thus

$$y(t) \approx y_0(e^{0.40} - 1) + y_0 e^{0.40}[0.20 \cos \omega_0 t \\ + (0.50)(0.20)^2 \cos^2 \omega_0 t + (0.17)(0.20)^3 \cos^3 \omega_0 t]$$

Using (3-24) to express $\cos^2 \omega_0 t$ and $\cos^3 \omega_0 t$ as linear combinations of harmonics of $\cos \omega_0 t$ gives

$$y(t) \approx B_0 + B_1 \cos \omega_0 t + B_2 \cos 2\omega_0 t + B_3 \cos 3\omega_0 t$$

where $B_0 = y_0\{e^{0.40}[1 + (0.25)(0.20)^2] - 1\} = 50.7$ mA
$B_1 = y_0 e^{0.40}[0.20 + (0.125)(0.20)^3] = 30.0$ mA
$B_2 = 0.25 y_0 e^{0.40}(0.20)^2 = 1.49$ mA
$B_3 = 0.04 y_0 e^{0.40}(0.20)^3 = 47.7$ μA

Considering the precision with which parameters are given in the statement of the problem, the third harmonic is insignificant. To a good approximation,

$$y(t) = 50.7 + 30.0 \cos \omega_0 t + 1.50 \cos 2\omega_0 t \quad \text{mA}$$

For any particular input peak amplitude, we can approximate any practical transfer characteristic by a polynomial.† Error incurred by approximating a transfer characteristic by a polynomial generally decreases as the degree of the polynomial is increased. As a rule, such an approximation is useful for pencil-and-paper calculations only if the degree k of the resulting polynomial is small (say, $k \leq 5$). Also, error of a polynomial approximation generally increases as the peak amplitude of an input is increased. Consequently, polynomial approximations of low degree are generally useful only for smooth transfer characteristics and small inputs. Using a polynomial of low degree to calculate an output of a nonlinear static element is called a *small-signal calculation*. Cases where an input is so large or a transfer characteristic is so irregular that polynomial approximation is impractical require *large-signal methods*. Such methods include graphical methods, which do not give an output as a harmonic series, and other methods which do give an output as a harmonic series. In Chap. 4 we describe a method for large-signal analysis which gives an output of a nonlinear static system as a harmonic series without requiring polynomial (or any other) approximation.

3-2D Harmonic Distortion

Since a distortionless system is a linear stationary system, the output of a distortionless system for sinusoidal input is a sinusoid having the frequency of the input. On the other hand, the output of a nonlinear static system for sinusoidal input is a harmonic series. The output may contain a sinusoid having the frequency of the input (the fundamental or first harmonic), but it also contains higher harmonics of the input. This suggests that an output of a sinusoidally excited nonlinear static system can be analyzed as a distortionless component (the first harmonic) plus a distortion component (all other harmonics). Strictly speaking, this is incorrect because all odd powers of the input contribute to the first harmonic; nonetheless, it is a useful way to describe distortion introduced by a nonlinear static system if (as is usually the case) contributions of higher powers of the input to the fundamental are much smaller than the fundamental. The amplitude of a harmonic (above the first) expressed as a fraction or percentage of the amplitude of the fundamental (first harmonic) is often used to describe distortion introduced by a nonlinear static system. Distortion calculated or measured in this way is called *harmonic distortion*.

Example 3-17 The transfer characteristic for a certain system is

$$y = y_0 \left[\frac{x}{x_0} - \frac{1}{3}\left(\frac{x}{x_0}\right)^3 + \frac{1}{5}\left(\frac{x}{x_0}\right)^5 \right]$$

where $x = x(t)$ is an input, $y = y(t)$ is the corresponding output, $x_0 = 2.00$ mV, and

†This is obvious if $N(\alpha)$ in (3-21) is defined by an infinite series. Otherwise approximating $N(\alpha)$ by a polynomial usually requires least squares or other methods.

$y_0 = 20.0$ V. Calculate the third-harmonic distortion D_3 and the fifth-harmonic distortion D_5, defined by

$$D_3, \% = \frac{100 B_3}{B_1} \qquad D_5, \% = \frac{100 B_5}{B_1}$$

where B_1 is the peak amplitude of the fundamental (first harmonic), B_3 is the peak amplitude of the third harmonic, and B_5 is the peak amplitude of the fifth harmonic in the output for input

$$x(t) = A \cos \omega_0 t$$

where $A = 1.00$ mV.

SOLUTION The output is given (approximately) by

$$y(t) = y_0 \left[\frac{A}{x_0} \cos \omega_0 t - \frac{1}{3} \left(\frac{A}{x_0} \right)^3 \cos^3 \omega_0 t + \frac{1}{5} \left(\frac{A}{x_0} \right)^5 \cos^5 \omega_0 t \right]$$

where $A/x_0 = 1.00/2.00 = 0.50$. Using the identity (3-24) for $\cos^3 \alpha$ and $\cos^5 \alpha$ gives

$$y(t) = B_1 \cos \omega_0 t + B_3 \cos(3\omega_0 t + \pi) + B_5 \cos 5\omega_0 t$$

where $B_1 = y_0 |0.50 - (0.33)(0.75)(0.50)^3 + (0.20)(0.63)(0.50)^5| = 9.45$ V
$B_3 = y_0 |-(0.33)(0.25)(0.50)^3 + (0.20)(0.31)(0.50)^5| = 168$ mV
$B_5 = y_0 |(0.20)(0.06)(0.50)^5| = 8$ mV

The third- and fifth-harmonic distortions are

$$D_3 = \frac{(100)(0.168)}{9.45} = 1.79\% \qquad D_5 = \frac{(100)(0.008)}{9.45} = 0.08\%$$

3-2E Linearization

Virtually all physical systems are nonlinear. A linear model for a physical system represents an approximation that is valid only so long as signals in the system are sufficiently small; e.g., the transfer characteristic of an operational amplifier is that of a soft limiter (Fig. 3-14a). If an input $x(t)$ satisfies $|x(t)| \leq x_0$ or, equivalently, if an output $y(t)$ satisfies $|y(t)| \leq y_0$, the amplifier can be modeled as a proportion element having gain $K = y_0/x_0$ (Fig. 3-14b). For a typical operational amplifier, gain K is about 10^5 and maximum output y_0 (determined by power-supply voltage) is about 15 V. Consequently, we can model a typical operational amplifier as a proportion element only if an input satisfies $|x(t)| \leq 150$ μV. Often, a nonlinear element in a system, e.g., an operational amplifier, is modeled by a linear element, e.g., a proportion element. This is referred to as *linearizing* the element (or system). Linearization is possible if the element is intended to operate linearly, in which case one objective of analysis is to determine limits on signals in the system such that linear operation is assured. For example, the circuit of Fig. 1-24 is an integrator only if the operational amplifier operates linearly. The operational amplifier operates linearly only if every input $x(t)$ satisfies $|x(t)| \leq V_0/K$, where V_0 is the supply voltage and K is the gain of the amplifier. Linearization is unreasonable for analysis of a system that is intended to

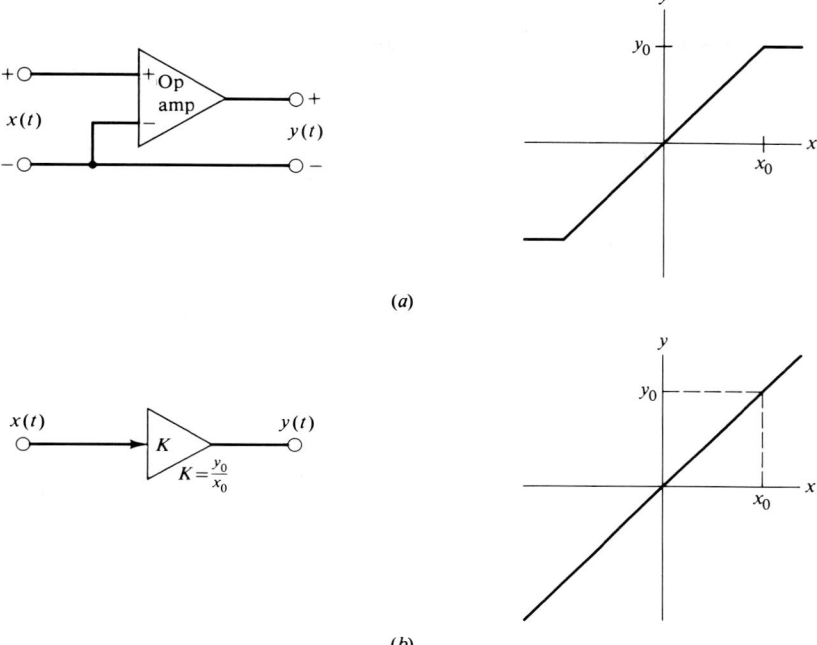

Figure 3-14 Linearization of a nonlinear static system: (*a*) operational amplifier and associated nonlinear transfer characteristic, (*b*) linear model.

operate nonlinearly; e.g., one ordinarily would not linearize the rectifying circuit in a power supply.

For the operational amplifier of Fig. 3-14 there is a sharply defined boundary between linear and nonlinear operation. The amplifier is linear for $|x(t)| \leq x_0$, and the amplifier is nonlinear if $|x(t)|$ exceeds x_0 at any instant. In many practical cases no such sharp boundary exists. Frequently an element (or system) is almost linear for very small inputs, a little less linear for larger inputs, and so on. In such cases an objective of analysis is to determine departure from linearity as a function of input amplitude (or as a function of the amplitude of some other signal in the system). This requires some measure of departure from linearity, e.g., third-harmonic distortion.

Example 3-18 For inputs of interest, the output $y(t)$ of a certain audio amplifier for input $x(t)$ is given by

$$y = y_0 \tanh \frac{x}{x_0}$$

where $x_0 = 25$ mV and $y_0 = 25$ V. Find the maximum peak amplitude A for input $x(t) = A \cos \omega_0 t$ if third-harmonic distortion introduced by the amplifier must not exceed 0.1 percent.

SOLUTION We approximate the transfer characteristic of the amplifier by a polynomial. Because the specified third-harmonic distortion is so small, we assume that a fifth-degree polynomial is adequate. Truncating Maclaurin's series for tanh α yields

$$y \approx y_0 \left[\frac{x}{x_0} - \frac{1}{3}\left(\frac{x}{x_0}\right)^3 + \frac{2}{15}\left(\frac{x}{x_0}\right)^5 \right]$$

where $x = A \cos \omega_0 t$. Using the identity (3-24) gives

$$y(t) = B_1 \cos \omega_0 t + B_3 \cos 3\omega_0 t + B_5 \cos 5\omega_0 t$$

where
$$|B_1| = y_0|R - 0.250R^3 + 0.083R^5|$$
$$|B_3| = y_0|-0.083R^3 + 0.042R^5|$$
$$R = \frac{A}{x_0}$$

Third-harmonic distortion is given by $D_3 = 100|B_3/B_1|$ (percent). We require $D_3 \leq 0.1$ percent. This implies that

$$\log \frac{|-0.083R^3 + 0.042R^5|}{|R - 0.250R^3 + 0.083R^5|} \leq -3$$

Figure 3-15 shows a graph of the quantity on the left versus $R = A/x_0$. We see from the graph that $D_3 \leq 0.1$ percent implies that $A/x_0 \leq 0.110$, which for $x_0 = 25$ mV gives $A \leq 2.75$ mV. This is the largest peak amplitude of a sinusoidal input for which third-harmonic distortion does not exceed 0.1 percent. The peak amplitude of the corresponding output is approximately $B = y_0 A/x_0 = 2.75$ V. The amplifier can deliver only about $\frac{1}{2}$ W to an 8-Ω speaker with no more than 0.1 percent third-harmonic distortion. The transfer characteristic implies that the power supply could support output swings as large as ± 25 V, which means that about 40 W could be delivered to an 8-Ω speaker. The amplifier does not make very good use of available power.

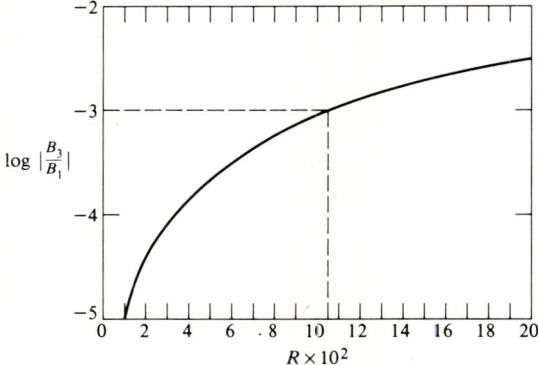

Figure 3-15 Third-harmonic distortion versus input amplitude for the system of Example 3-18.

3-2F Intermodulation Distortion

The output of a nonlinear static system for input

$$x(t) = A_1 \cos \omega_1 t + A_2 \cos \omega_2 t$$

consisting of two sinusoidal components contains harmonics of each component and also *intermodulation components* of the form

$$\cos[(m_1\omega_1 \pm m_2\omega_2)t]$$

RESPONSE TO SINUSOIDAL EXCITATION **137**

where m_1, m_2 are integers. The root-mean-square (rms) amplitude of an intermodulation component expressed as a fraction or percentage of rms amplitude of the fundamental components is sometimes used to describe distortion introduced by a nonlinear system. This measure of distortion is called *intermodulation distortion*.

Example 3-19 The transfer characteristic for a certain system is

$$y = y_0 \left[\frac{x}{x_0} - \frac{1}{3}\left(\frac{x}{x_0}\right)^3 \right]$$

where $x = x(t)$ is an input, $y = y(t)$ is the corresponding output, $x_0 = 5.00$ mV, and $y_0 = 1.00$ V. We wish to calculate intermodulation distortion introduced by the system for input

$$x(t) = A_1 \cos \omega_1 t + A_2 \cos \omega_2 t$$

where $A_1 = 1.00$ mV and $A_2 = 2.00$ mV. To simplify notation we define $B = A_1/x_0 = 0.20$ and $C = A_2/x_0 = 0.40$. The output is given by

$$y(t) = y_0[B \cos \omega_1 t + C \cos \omega_2 t - \tfrac{1}{3}(B^3 \cos^3 \omega_1 t + 3B^2 C \cos^2 \omega_1 t \cos \omega_2 t$$
$$+ 3BC^2 \cos \omega_1 t \cos^2 \omega_2 t + C^3 \cos^3 \omega_2 t)]$$

Using the identity (3-24) yields

$$y(t) = y_0[B \cos \omega_1 t + C \cos \omega_2 t - \tfrac{1}{4}B^3 \cos \omega_1 t - \tfrac{1}{12}B^3 \cos 3\omega_1 t$$
$$- \tfrac{1}{2}B^2 C(1 + \cos 2\omega_1 t) \cos \omega_2 t - \tfrac{1}{2}C^2 B(1 + \cos 2\omega_2 t) \cos \omega_1 t$$
$$- \tfrac{1}{4}C^3 \cos \omega_2 t - \tfrac{1}{12}C^3 \cos 3\omega_2 t]$$

This simplifies to

$$y(t) = D_{10} \cos \omega_1 t + D_{30} \cos 3\omega_1 t + D_{01} \cos \omega_2 t + D_{03} \cos 3\omega_2 t$$
$$+ D_{12} \cos \omega_1 t \cos 2\omega_2 t + D_{21} \cos 2\omega_1 t \cos \omega_2 t$$

where

$$D_{10} = y_0(B - \tfrac{1}{4}B^3 - \tfrac{1}{2}BC^2) = 180 \text{ mV} \qquad D_{30} = y_0(-\tfrac{1}{12}B^3) = -670 \ \mu\text{V}$$
$$D_{01} = y_0(C - \tfrac{1}{4}C^3 - \tfrac{1}{2}B^2 C) = 380 \text{ mV} \qquad D_{03} = y_0(-\tfrac{1}{12}C^3) = -5.30 \text{ mV}$$
$$D_{12} = y_0(-\tfrac{1}{2}BC^2) = -16 \text{ mV} \qquad D_{21} = y_0(-\tfrac{1}{2}B^2 C) = -8.0 \text{ mV}$$

Using the identity

$$\cos \alpha \cos \beta \equiv \tfrac{1}{2}\cos(\alpha + \beta) + \tfrac{1}{2}\cos(\alpha - \beta)$$

gives

$$y(t) = D_{10} \cos \omega_1 t + D_{30} \cos 3\omega_1 t + D_{01} \cos \omega_2 t + D_{03} \cos 3\omega_2 t$$
$$+ 0.5D_{12} \cos[(\omega_1 + 2\omega_2)t] + 0.5D_{12} \cos[(\omega_1 - 2\omega_2)t]$$
$$+ 0.5D_{21} \cos[(2\omega_1 + \omega_2)t] + 0.5D_{21} \cos[(2\omega_1 - \omega_2)t]$$

The first and third terms represent the desired (distortionless) output. The second and the fourth terms represent harmonic distortion. The last four terms represent intermodulation distortion. The rms amplitude of the desired component is given by†

†The mean-squared amplitude of a sum of sinusoids having different frequencies is the sum of the mean-squared amplitudes of the sinusoids; the rms amplitude is the square root of the mean-squared amplitude.

138 INTRODUCTION TO SYSTEM ANALYSIS

$$\sqrt{2}\,A_{rms} = \sqrt{D_{10}^2 + D_{01}^2} = 420 \text{ mV}$$

The rms amplitude of the intermodulation component is given by

$$\sqrt{2}\,B_{rms} = \sqrt{(0.5D_{12})^2 + (0.5D_{12})^2 + (0.5D_{21})^2 + (0.5D_{21})^2} = 12.6 \text{ mV}$$

The total intermodulation distortion is

$$\text{IMD} = \frac{100 B_{rms}}{A_{rms}} = \frac{(100)(12.6)}{420} = 3.00\%$$

Since intermodulation distortion is more difficult to calculate, more difficult to measure, and more difficult to interpret than harmonic distortion, intermodulation distortion is used less often for describing performance; however, both harmonic and intermodulation distortion are usually given as performance measures for high-fidelity audio equipment. Generally, harmonic distortion is somewhat more tolerable than intermodulation distortion in audio equipment, because (loosely) harmonic distortion simply enriches harmonics already present whereas intermodulation distortion introduces frequencies not present in a source. For example, plucking two strings of a guitar produces two harmonic series, one for each note. The structure of these series distinguishes one guitar from another. Harmonic distortion changes the relative amplitudes of the components of each series but introduces no new frequencies. If it is not too severe, harmonic distortion alone simply makes one guitar sound like a different guitar. On the other hand, intermodulation distortion mixes the two harmonic series and introduces frequencies that should not be produced by *any* guitar. The result (if audible) is very disturbing to many listeners.

3-3 INTRODUCTION TO FREQUENCY-DOMAIN ANALYSIS

In Secs. 3-1 and 3-2 we describe methods for calculating outputs of sinusoidally excited systems. These methods are very important, partly because a few systems are sinusoidally excited (or nearly so) in actual use, but mainly because it turns out that any signal can be represented as a sum (or superposition) of sinusoids. Methods described above (especially those of Sec. 3-1) are therefore much more general than they first appear.

In Chap. 4 we show how a signal is expressed as a sum of sinusoids using Fourier analysis (Fourier series and Fourier integral). In this section we introduce some concepts that make Fourier analysis easier to appreciate, use, and understand. It is difficult to exaggerate the importance of concepts introduced in this section and treated further in Chap. 4. Without them it is virtually impossible to understand operation of many commonplace and important systems (e.g., radio, television, and telephone).

3-3A Amplitude and Phase Spectrum

In Chap. 4 we show how a signal $x(t)$ can be described over any finite interval $t_0 \leq t < t_0 + T$ by a harmonic series of the form

$$x(t) = A_0 \cos \theta_0 + A_1 \cos(\omega_0 t + \theta_1) + A_2 \cos(2\omega_0 t + \theta_2) + \cdots$$

$$= \sum_{n=0}^{\infty} A_n \cos(n\omega_0 t + \theta_n) \tag{3-27}$$

where the fundamental frequency ω_0 equals $2\pi/T$. The starting point t_0 and the duration T of the interval are arbitrary. For all practical purposes, such a series can describe a signal for all time.

A signal $x(t)$ described by a harmonic series as in (3-27) is determined completely (is defined for all time) by specifying frequency, peak amplitude, and initial phase of each term. It is a common and useful practice to present these quantities as shown in Fig. 3-16. Figure 3-16a defines the *amplitude spectrum* of the signal of (3-27). Figure 3-16b defines the *phase spectrum* of the signal of (3-27). The amplitude spectrum and the phase spectrum of a signal are referred to collectively as the *spectrum* of the signal. The unit of an amplitude spectrum is the unit of the associated signal, and the unit of a phase spectrum in this book is the radian.

The amplitude spectrum and the phase spectrum of a signal are functions of frequency. We denote the amplitude spectrum for a signal $x(t)$ described by (3-27) by $A_x(f)$, and we denote the phase spectrum by $\theta_x(f)$, where

$$A_x(f) = \begin{cases} A_n & f = nf_0 \\ 0 & f \neq nf_0 \end{cases} \quad \theta_x(f) = \begin{cases} \theta_n & f = nf_0 \\ 0 & f \neq nf_0 \end{cases} \quad n = 0, 1, 2, \ldots \tag{3-28}$$

Together, an amplitude spectrum and a phase spectrum completely determine a signal; i.e., specifying $A_x(f)$ and $\theta_x(f)$ for all values of frequency f is equivalent to specifying $x(t)$ for all values of time t.

Example 3-20 Obtain the spectrum of the signal

$$x(t) = -15 + 10 \cos\left(\omega_0 t - \frac{\pi}{4}\right) + 5 \cos\left(2\omega_0 t + \frac{3\pi}{4}\right) \quad \text{V}$$

where $f_0 = 5$ kHz.

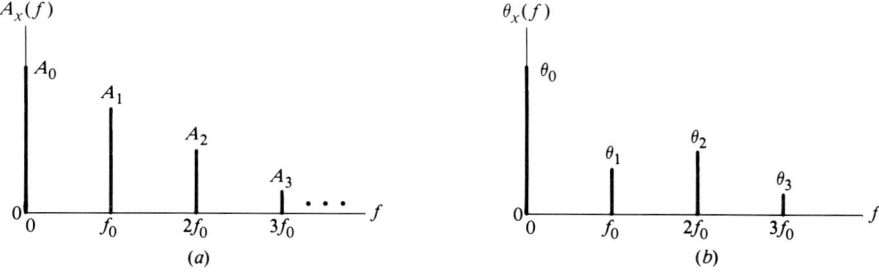

Figure 3-16 Graphs of (*a*) the amplitude spectrum and (*b*) the phase spectrum of the signal of (3-27). The vertical lines connecting coordinates (nf_0, A_n) to the horizontal axis have no mathematical meaning. They often are included to make the shape of a spectrum stand out.

140 INTRODUCTION TO SYSTEM ANALYSIS

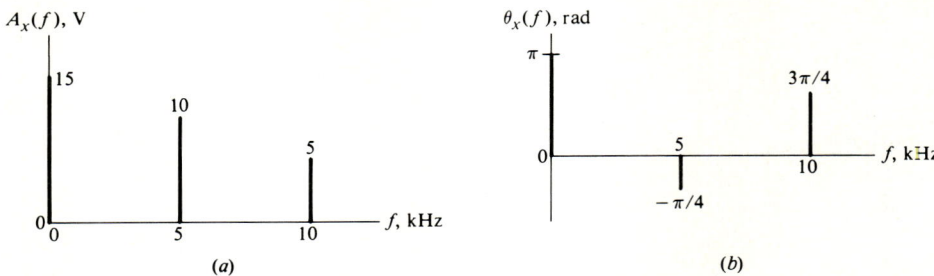

Figure 3-17 Spectrum of the signal of Example 3-20.

SOLUTION We first put the expression above in standard (amplitude-phase) form by expressing the dc component as

$$-15 \text{ V} = 15 \cos \pi \quad \text{V}$$

The peak amplitudes of the components of $x(t)$ are $A_0 = 15$ V, $A_1 = 10$ V, and $A_2 = 5$ V. The initial phases are $\theta_0 = \pi$ rad, $\theta_1 = -\pi/4$ rad, and $\theta_2 = 3\pi/4$ rad. We have

$$A_x(f) = \begin{cases} 15 \text{ V} & f = 0 \\ 10 \text{ V} & f = 5 \text{ kHz} \\ 5 \text{ V} & f = 10 \text{ kHz} \\ 0 & \text{all other } f \end{cases} \qquad \theta_x(f) = \begin{cases} \pi \text{ rad} & f = 0 \\ -\dfrac{\pi}{4} \text{ rad} & f = 5 \text{ kHz} \\ \dfrac{3\pi}{4} \text{ rad} & f = 10 \text{ kHz} \\ 0 & \text{all other } f \end{cases}$$

Figure 3-17 shows graphs of the amplitude spectrum and the phase spectrum.

Example 3-21 Plot (as a function of time t) the signal $x(t)$ whose spectrum is shown in Fig. 3-18.

SOLUTION From (3-27) and Fig. 3-18, we have

$$x(t) = 5 + 2.5 \cos\left(\omega_0 t - \frac{\pi}{2}\right) \quad \text{mA}$$

where $f_0 = 1$ kHz. Figure 3-19 shows a graph of $x(t)$ versus time.

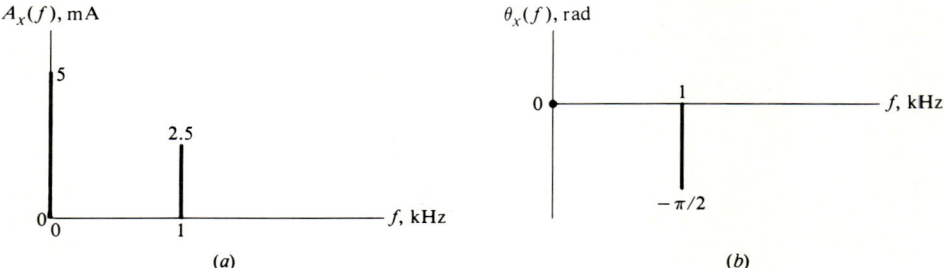

Figure 3-18 Spectrum of the signal of Example 3-21.

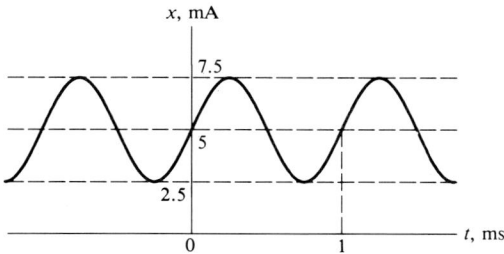

Figure 3-19 Waveform of the signal of Example 3-21.

3-3B Complex (Two-Sided) Spectrum

The signal of (3-27) can be expressed in exponential form as

$$x(t) = \cdots + X_2^* e^{-j2\omega_0 t} + X_1^* e^{-j\omega_0 t} + X_0 + X_1 e^{j\omega_0 t} + X_2 e^{j2\omega_0 t} + \cdots \quad (3\text{-}29)$$

where, from (3-14) and (3-4),

$$X_0 = A_0 \cos\theta_0 \quad \text{and} \quad X_n = \tfrac{1}{2} A_n e^{j\theta_n} \quad \text{for } n = 1, 2, 3, \ldots$$

With the definition

$$X_{-n} = X_n^* \quad (3\text{-}30)$$

we can write (3-29) more compactly as

$$x(t) = \sum_{n=-\infty}^{\infty} X_n e^{jn\omega_0 t} \quad (3\text{-}31)$$

A signal $x(t)$ is determined completely by specifying the frequency and complex amplitude of each term on the right side of (3-31). It is useful to do so by giving complex amplitude as a function of frequency called the complex spectrum of the signal. The *complex spectrum* of a signal $x(t)$ is denoted by $C_x(f)$ and defined by

$$C_x(f) = \begin{cases} X_n & f = nf_0 \\ 0 & f \neq nf_0 \end{cases} \quad n = 0, \pm 1, \pm 2, \ldots \quad (3\text{-}32)$$

The unit of a complex spectrum is the unit of the associated signal.

The complex spectrum $C_x(f)$ for a signal described by (3-31) is presented graphically by plotting the magnitude and the angle of $C_x(f)$, as illustrated in Fig. 3-20. The function $|C_x(f)|$ of Fig. 3-20a is called the *two-sided amplitude spectrum* of a signal having complex spectrum $C_x(f)$ [a signal described by (3-31)]. The function $\angle C_x(f)$ of Fig. 3-20b is called the *two-sided phase spectrum* of a signal having complex spectrum $C_x(f)$. The two-sided amplitude and phase spectra of a signal are referred to collectively as the *complex spectrum* or the *two-sided spectrum* of the signal. Wherever confusion is possible, we refer to the amplitude and phase spectra of Fig. 3-16 as *one-sided spectra*.

The relations between the complex spectrum $C_x(f)$, the (one-sided) amplitude spectrum $A_x(f)$, and the (one-sided) phase spectrum $\theta_x(f)$ of a signal $x(t)$ are

142 INTRODUCTION TO SYSTEM ANALYSIS

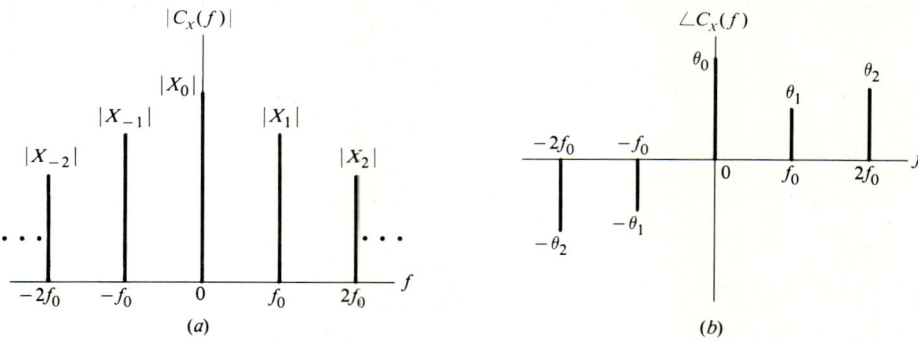

Figure 3-20 Graphs of (a) the magnitude and (b) the angle of the complex spectrum of the signal of (3-31).

$$A_x(f) = \begin{cases} |C_x(f)| & f = 0 \\ 2|C_x(f)| & f > 0 \end{cases} \qquad \theta_x(f) = \angle C_x(f) \qquad f \geq 0 \qquad (3\text{-}33a)$$

$$C_x(f) = \begin{cases} \frac{1}{2}A_x(f)e^{-j\theta_x(f)} & f < 0 \\ A_x(f)e^{j\theta_x(f)} & f = 0 \\ \frac{1}{2}A_x(f)e^{j\theta_x(f)} & f > 0 \end{cases} \qquad (3\text{-}33b)$$

These relations show that a two-sided amplitude spectrum is symmetric about $f = 0$ (is an even function of frequency) and (if we ignore the phase of a dc component) that a two-sided phase spectrum is antisymmetric about $f = 0$ (is an odd function of frequency). For $f = 0$, a one-sided spectrum and the associated two-sided spectrum are identical. For $f > 0$, a one-sided amplitude spectrum equals twice the associated two-sided spectrum, and a one-sided phase spectrum is identical to the associated two-sided phase spectrum.

Example 3-22 Figure 3-21 shows graphs of the two-sided amplitude spectrum and the two-sided phase spectrum of the signal

$$x(t) \sum_{n=-2}^{2} X_n e^{jn\omega_0 t}$$

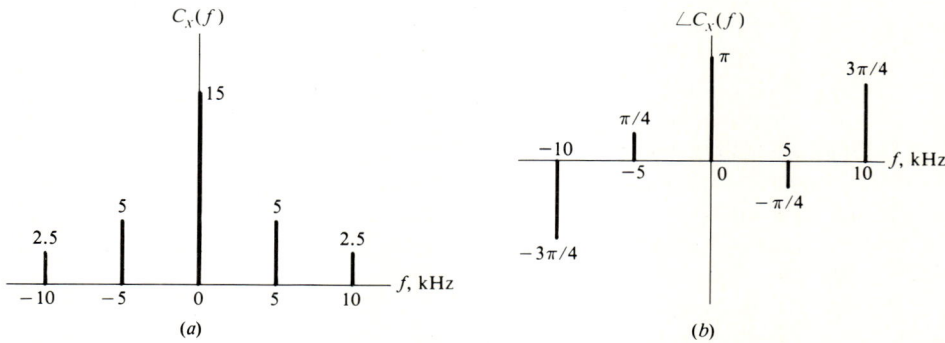

Figure 3-21 Two-sided spectrum of the signal of Example 3-22.

where $f_0 = 5$ kHz, $X_0 = -15$ V, $X_1 = 5e^{-j\pi/4}$ V, and $X_2 = 2.5e^{j3\pi/4}$ V. Note that $X_{-1} = X_1^*$ and $X_{-2} = X_2^*$.

In (3-27) a signal $x(t)$ is expressed as a sum of sinusoids. In the associated spectrum of Fig. 3-16 the independent variable f is frequency of sinusoidal oscillation (number of cycles per second), which is a nonnegative quantity. Consequently, a one-sided spectrum is defined only for nonnegative frequencies. In (3-31) a signal $x(t)$ is expressed as a sum of complex exponentials, where each exponential $Xe^{j2\pi ft}$ can be represented by a point orbiting the origin of a complex plane. In the associated (two-sided) spectrum the independent variable f is angular velocity (revolutions per second) of the point. Since this frequency can be either positive (counterclockwise motion) or negative (clockwise motion), a complex spectrum is defined for both positive and negative frequencies.

The idea that a signal has both positive- and negative-frequency components often causes confusion. There is a tendency to attach physical significance to negative-frequency components of a signal when in fact there is none. Indeed, neither the positive- nor the negative-frequency components of a signal are physical quantities. The two-sidedness of a complex spectrum is simply a result of a mathematical relation (Euler's identity) between a real sinusoid and a conjugate pair of complex exponentials. No physical system can separate a real sinusoid into its positive- and negative-frequency components because both components are complex whereas signals produced by a physical system are real.

3-3C Time Domain and Frequency Domain

A spectrum completely determines the waveform of a signal through either (3-27) or (3-31). In Chap. 4 we show that the converse is also true. The spectrum of a signal can be obtained from the waveform of the signal even though the signal is not expressed (originally) as a superposition of sinusoids. Therefore, a signal can be described either in the *time domain* by its waveform (a function of time) or in the *frequency domain* by its spectrum (a function of frequency). Similarly, a system can be described in the time domain by describing how the waveform of an output is obtained from the waveform of an input, or it can be described in the frequency domain by describing how the spectrum of an output is obtained from the spectrum of an input. For example, a transfer characteristic is a time-domain description of a system, whereas a transfer function is a frequency-domain description of a system.

Time- and frequency-domain descriptions of a signal or system contain the same information about the signal or system but present it in different ways. A frequency-domain description often provides much more insight than the associated time-domain description. Indeed, some systems that are easily understood in the frequency domain are almost incomprehensible in the time domain. Such a system, called a single-sideband transmission system, is described in Chap. 4.

In the remainder of this section we introduce frequency-domain analysis of linear stationary systems and signal-transmission systems. Conscientious study of these topics here will be rewarded in Chap. 4, where a mathematically more advanced treatment of frequency-domain analysis is given.

3-3D Frequency-Domain Analysis of Linear Stationary Systems

By using superposition the output of a linear stationary system for input

$$x(t) = \sum_{n=-\infty}^{\infty} X_n e^{jn\omega_0 t}$$

can be expressed as

$$y(t) = \sum_{n=-\infty}^{\infty} H(jn\omega_0) X_n e^{jn\omega_0 t}$$

where $H(j\omega)$ is the transfer function of the system. This can be written

$$y(t) = \sum_{n=-\infty}^{\infty} Y_n e^{jn\omega_0 t} \quad \text{where } Y_n = H(jn\omega_0)X_n \quad (3\text{-}34)$$

The complex spectrum of the output is given by

$$C_y(f) = \begin{cases} Y_n = H(j2\pi f)X_n & f = nf_0 \\ 0 & f \neq nf_0 \end{cases}$$

This shows that the complex (two-sided) spectrum of the output of a linear stationary system for input $x(t)$ is given by

$$C_y(f) = H(j2\pi f)C_x(f) \quad (3\text{-}35)$$

where $C_x(f)$ is the complex spectrum of the input. Taking the magnitude of (3-35) and the angle of (3-35) gives

$$|C_y(f)| = |H(j2\pi f)||C_x(f)| \quad \angle C_y(f) = \angle H(j2\pi f) + \angle C_x(f) \quad (3\text{-}36)$$

Using (3-33) and (3-10) in (3-36) yields

$$A_y(f) = \Gamma(2\pi f)A_x(f) \quad \theta_y(f) = \phi(2\pi f) + \theta_x(f) \quad (3\text{-}37)$$

where $\Gamma(2\pi f)$ is the gain of the system of $\phi(2\pi f)$ is the phase shift of the system. This shows that the amplitude spectrum of the output is the product of the amplitude spectrum of the input and the gain of the system and that the phase spectrum of the output is the sum of the phase spectrum of the input and the phase shift of the system.

Example 3-23 The transfer function of a certain system is

$$H(j\omega) = \frac{1}{1 + j(\omega/\omega_1)} \quad \text{mA/V}$$

where $f_1 = \omega_1/2\pi = 1$ kHz. The two-sided spectrum of an input $x(t)$ is given in Fig. 3-22. Obtain the complex spectrum of the corresponding output $y(t)$. Give the time-domain description of the output $y(t)$ in amplitude-phase form.

SOLUTION Table 3-2 gives values of $C_x(f)$ obtained from Fig. 3-22 and values of $C_y(f)$ calculated using (3-35). Table 3-3 gives the one-sided spectrum of the output, obtained from the two-sided spectrum $C_y(f)$ using (3-33). The output is expressed in amplitude-phase form by

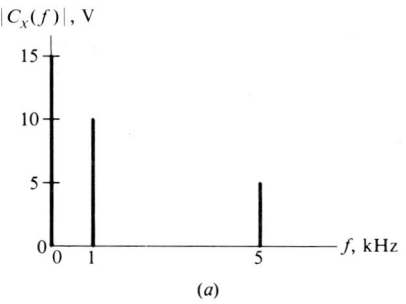 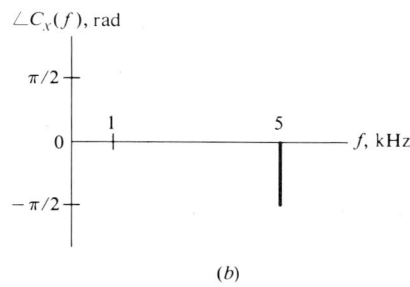

Figure 3-22 Spectrum of the signal of Example 3-23.

Table 3-2 Quantities calculated in Example 3-23

f, kHz	$C_x(f)$, V	$H(j2\pi f)$, mA/V	$C_y(f)$, mA
−5.00	$5.00e^{j\pi/2}$	$0.196e^{1.37j}$	$0.980e^{2.94j}$
−1.00	10.0	$0.707e^{0.785j}$	$7.07e^{0.785j}$
0.00	15.0	1.00	15.0
1.00	10.0	$0.707e^{-0.785j}$	$7.07e^{-0.785j}$
5.00	$5.00e^{-j\pi/2}$	$0.196e^{-1.37j}$	$0.980e^{-2.94j}$

Table 3-3 One-sided spectrum of the output for Example 3-23

f, kHz	$A_y(f)$, mA	$\theta_y(f)$, rad
0.00	15.0	0.00
1.00	14.1	−0.785
5.00	1.96	−2.94

$$y(t) = 15.0 + 14.1 \cos(\omega_0 t - 0.785) + 1.96 \cos(5\omega_0 t - 2.94) \quad \text{mA}$$

where $f_0 = 1$ kHz.

Equation (3-35) is the fundamental relation for frequency-domain analysis of linear stationary systems. It is to be compared with the convolution relation (2-2), which is the fundamental relation for time-domain analysis of linear stationary systems. Multiplication is simpler than convolution, both computationally and conceptually. It is easier to perform the multiplication indicated by (3-35) than it is to perform the convolution indicated by (2-2). It is easier to visualize a spectrum given by (3-35) without actually performing the indicated multiplication than it is to visualize a waveform given by (2-2) without actually performing the indicated convolution. As a consequence, the frequency-domain description of a linear stationary system (the transfer function of the system) is often much easier to interpret than the corresponding time-domain description (the impulse response of the system).

3-3E Linear Distortion

The transfer characteristic of a *distortionless system* has the form

$$y(t) = Kx(t - t_0)$$

where $y(t)$ is the output for input $x(t)$. A distortionless system is a linear stationary system. Distortion introduced because a system fails to be linear, e.g., harmonic distortion, is called *nonlinear distortion*. Distortion introduced because a system, though linear, fails to have a transfer characteristic of the form given above is called *linear distortion*. In this section we describe frequency-domain analysis of linear distortion.

The transfer function for a distortionless system described by the transfer characteristic above is

$$H_0(j\omega) = Ke^{-j\omega t_0}$$

The associated gain $\Gamma_0(2\pi f)$ and phase shift $\phi_0(2\pi f)$ are given by

$$\Gamma_0(2\pi f) = |H_0(j2\pi f)| = |K| \qquad \phi_0(2\pi f) = \angle K - 2\pi f t_0$$

Figure 3-23 shows graphs of gain $\Gamma_0(2\pi f)$ and phase shift $\phi_0(2\pi f)$ versus frequency f. In frequency-domain analysis of linear distortion the most important properties of a distortionless system are that its gain is *independent of frequency* and that its phase shift is a *linear function of frequency* having phase intercept $\phi(0) = 0$ (for $K \geq 0$) or $\phi(0) = \pi$ (for $K < 0$). Phase shift can always be interpreted modulo 2π rad (by casting out integral multiples of 2π rad) because $\cos(\alpha \pm 2\pi) = \cos \alpha$. Therefore, the phase shift of a distortionless system is a linear function of frequency whose phase intercept equals zero modulo π.

No physical system provides distortionless transmission for all inputs. At best, a physical system can provide *high-fidelity (low-distortion) transmission* for a limited class of inputs. In time-domain language, a high-fidelity system is one whose impulse response approximates a delta function for inputs of interest. In frequency-domain language, a high-fidelity system is one whose gain is nearly independent of frequency and whose phase shift is a linear function of frequency having phase intercept zero (modulo π) where the spectrum of an input is significantly different from zero.

Figure 3-24 illustrates application of the frequency-domain conditions for high-fidelity transmission. The solid curves represent actual gain and phase shift. The dashed lines represent linear approximations to phase shift in frequency bands of interest. The system of Fig. 3-24a provides high-fidelity transmission for any input whose spectrum is confined to the band $f < f_c$. The gain of the system is nearly independent of frequency for $f < f_c$, the phase shift is nearly linear for $f < f_c$, and the phase intercept for the linear approximation equals π rad. The system of Fig. 3-24b provides high-fidelity transmission for any input whose spectrum is confined to the band $f > f_c$. The gain of the system is nearly independent of frequency for $f > f_c$, the phase shift is nearly linear for $f > f_c$, and the phase intercept for the linear approximation equals zero. The system of Fig. 3-24c provides high-fidelity transmission for any input whose spectrum is confined either to the band $f < f_a$ or to the band $f > f_b$. In both these bands

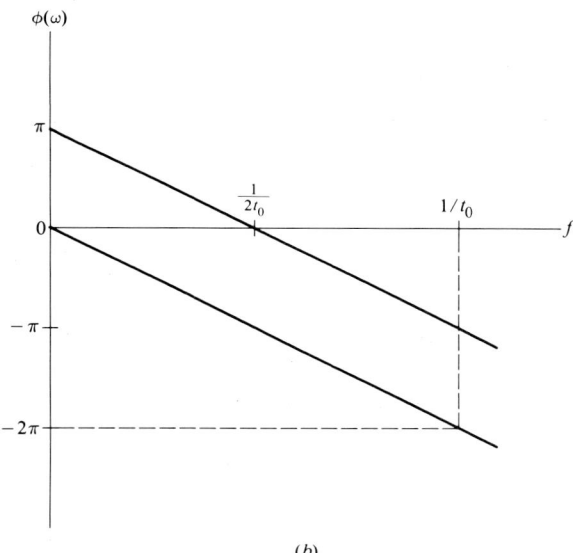

Figure 3-23 Graphs of (a) gain versus frequency and (b) phase shift versus frequency for a distortionless system.

the gain is nearly independent of frequency, the phase shift is nearly linear, and the phase intercept is either π rad (for $f < f_a$) or zero (for $f > f_b$). The system does not provide high-fidelity transmission for an input whose spectrum has components in both these bands, even if the input has no components in the band $f_a < f < f_b$, because the linear approximations to the phase shifts in the bands $f < f_a$ and $f > f_b$ have different phase intercepts. Finally, the system of Fig. 3-24d does not provide high-fidelity transmission for any input, even though its gain is nearly independent of frequency and its phase shift is nearly linear for $f_a < f < f_b$, because the phase intercept for a linear approximation to the phase shift over $f_a < f < f_b$ is not zero (modulo π rad).

148 INTRODUCTION TO SYSTEM ANALYSIS

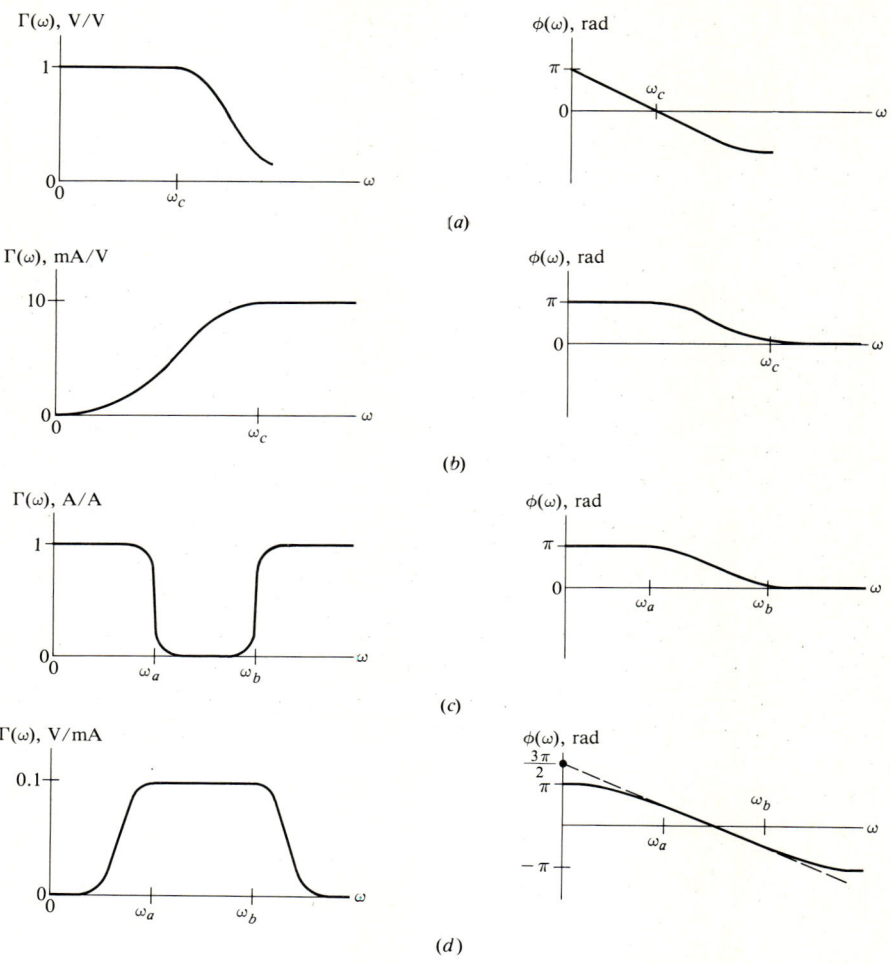

Figure 3-24 An illustration of conditions on gain and phase shift for high-fidelity transmission.

Example 3-24 In this example, we obtain conditions under which a system having transfer function

$$H(j\omega) = \frac{1}{1 + j\omega/\omega_c} \quad \text{V/V}$$

provides high-fidelity transmission, and then we illustrate effects of distortion on the waveform of a particular input.

The gain and the phase shift of the system are given by

$$\Gamma(\omega) = \frac{1}{\sqrt{1 + (\omega/\omega_c)^2}} \quad \text{V/V} \qquad \phi(\omega) = -\tan^{-1}\frac{\omega}{\omega_c} \quad \text{rad}$$

where we can use the two-quadrant inverse tangent function because the real part of $1 + j\omega/\omega_c$ is positive. Figures 3-25a and b show graphs of gain and phase shift versus ω/ω_c.

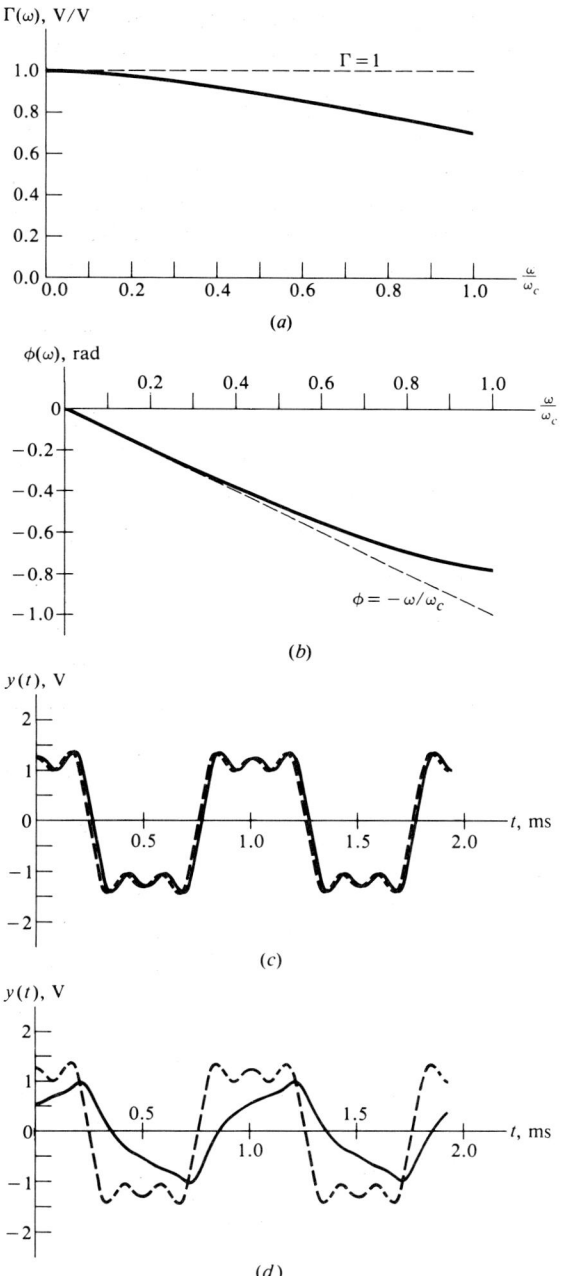

Figure 3-25 Graphs of (a) gain and (b) phase shift versus frequency for the system of Example 3-24 (solid curves) and for a comparable distortionless system (dashed curves). (c) and (d) Waveforms of an input (dashed curve) and corresponding output (solid curve) for the system of Example 3-24. In (c), $\omega_c = 20\omega_0$; in (d), $\omega_c = \omega_0$.

For $\omega/\omega_c \ll 1$ (for $\omega \ll \omega_c$), the gain and phase shift can be approximated by

$$\Gamma(\omega) \approx 1 \qquad \phi(\omega) \approx -\frac{\omega}{\omega_c}$$

These approximations are obtained by dropping all but the lowest terms from Maclaurin's expansions for $\Gamma(\omega)$ and $\phi(\omega)$. For $\omega \ll \omega_c$, gain is nearly independent of frequency and phase shift is nearly linear with intercept $\phi(0) = 0$. For inputs confined to a band $f \ll f_c$, the transfer function for the system is approximately

$$H(j\omega) = \Gamma(\omega)e^{j\phi(\omega)} \approx e^{-j\omega/\omega_c}$$

This shows that the output $y(t)$ for input $x(t)$ is given (approximately) by

$$y(t) \approx x(t - t_0)$$

where $t_0 = 1/\omega_c$. The system provides high-fidelity transmission for any signal whose spectrum is confined to a band $0 < f \ll f_c$. Conversely, a signal having significant components outside that band suffers significant distortion on transmission through the system. Figures 3-25c and d illustrate these remarks for input

$$x(t) = 15 \cos \omega_0 t - 5 \cos 3\omega_0 t + 3 \cos 5\omega_0 t \qquad V$$

The spectrum of $x(t)$ is confined entirely to the band $f \leq 5f_0$. Figure 3-25c shows the corresponding output for $f_c = 20f_0$. The spectrum of $x(t)$ is confined to the band $f \leq f_c/4$; in this band gain is approximately $\Gamma(\omega) \approx 1$, and phase shift is approximately $\phi(\omega) \approx -\omega/\omega_c = -\omega t_0$, with

$$t_0 = \frac{1}{\omega_c} = \frac{1}{20\omega_0} = \frac{T}{40\pi}$$

The delay t_0 is less than 1 percent of the period T of $x(t)$ and is barely discernible in Fig. 3-25. For $f_c = 20\omega_0$, the waveform of the output is nearly identical to the waveform of the input, as we would expect. Figure 3-25d shows the output for $f_c = f_0$. In this case, significant components of $x(t)$ lie outside the band $f < f_c$, and considerable distortion is evident in the waveform of the output.

The condition $\phi(0) = 0$ (modulo π rad) can be ignored in many practical applications for reasons which are given at appropriate places in the sequel. Here we wish only to point out that the reason this condition can be ignored is *not* that distortion introduced when this condition fails to hold is negligible. Indeed, such distortion can be severe, as illustrated in Fig. 3-26. Figure 3-26a shows graphs of gain and phase shift for a particular system. For $f > f_c$, gain is independent of frequency and phase shift is a linear function of frequency; however, since the phase intercept equals $\pi/2$ rad, the system fails to meet the condition $\phi(0) = 0$ (modulo π). The solid curve in Fig. 3-26b shows the output of this system for input

$$x(t) = A(\cos \omega_0 t - \tfrac{1}{3} \cos 3\omega_0 t + \tfrac{1}{5} \cos 5\omega_0 t - \cdots - \tfrac{1}{15} \cos 15\omega_0 t)$$

(shown by the dashed curve), where $f_0 \gg f_c$. The spectrum of this input lies entirely in the band $f > f_c$, where gain is nearly independent of frequency and phase shift is nearly linear, yet the signal suffers significant distortion because the phase-intercept condition is not met.

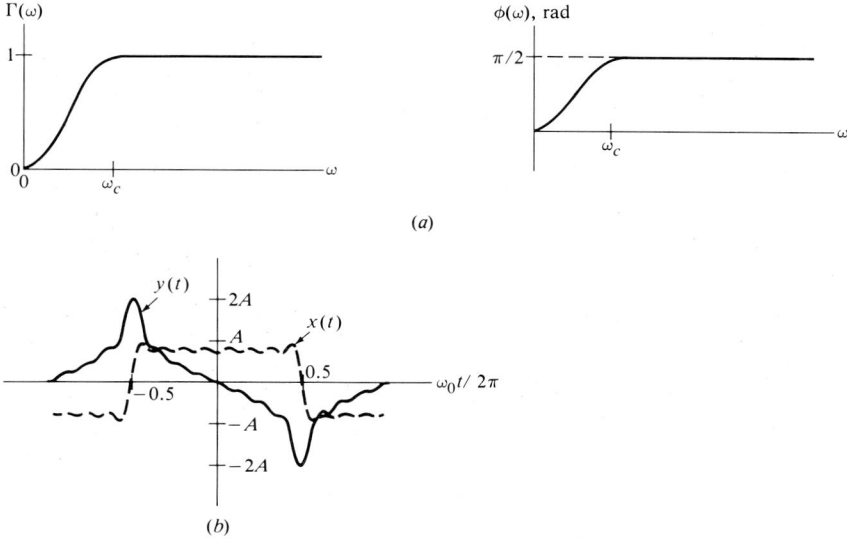

Figure 3-26 An illustration of phase-intercept distortion.

3-3F Filters

A *filter* is a linear stationary system designed to pass (transmit) sinusoidal signals whose frequencies lie in some intervals called *passbands* and stop (attenuate completely) sinusoidal signals whose frequencies lie in other intervals called *stopbands*. Filters are used in all kinds of systems, particularly in signal-transmission systems, where they are used to separate one signal from two or more that are present simultaneously, e.g., in the channel selector of a television receiver.

Table 3-4 gives symbols and graphs of gain versus frequency for four *ideal filters*. In their passbands these filters have unit (frequency-independent) gain. For the present we assume that these ideal filters introduce no phase shift. Therefore, each of these ideal filters provides distortionless transmission for an input whose spectrum lies entirely in a passband. In their stopbands these filters have zero gain. Therefore, each of these filters stops a signal whose spectrum lies entirely in a stopband.

Example 3-25 Find the output of the ideal bandpass filter of Fig. 3-27a for input

$$x(t) = \sum_{n=0}^{\infty} A_n \cos(n\omega_0 t + \theta_n)$$

where $f_0 = 10$ kHz, $\theta_n = -0.2n$ rad, and $A_n = 4/\sqrt{1 + n^2}$ V.

SOLUTION Figure 3-27b shows the amplitude spectrum of the input and the gain of the filter on one frequency axis. The filter passes the 30-kHz component of the input $x(t)$ and stops all other components of the input. The output is

$$y(t) = A_3 \cos(3\omega_0 t + \theta_3) = 1.26 \cos(3\omega_0 t - 0.60) \quad \text{V}$$

where $3f_0 = 30$ kHz.

Table 3-4 Ideal filters

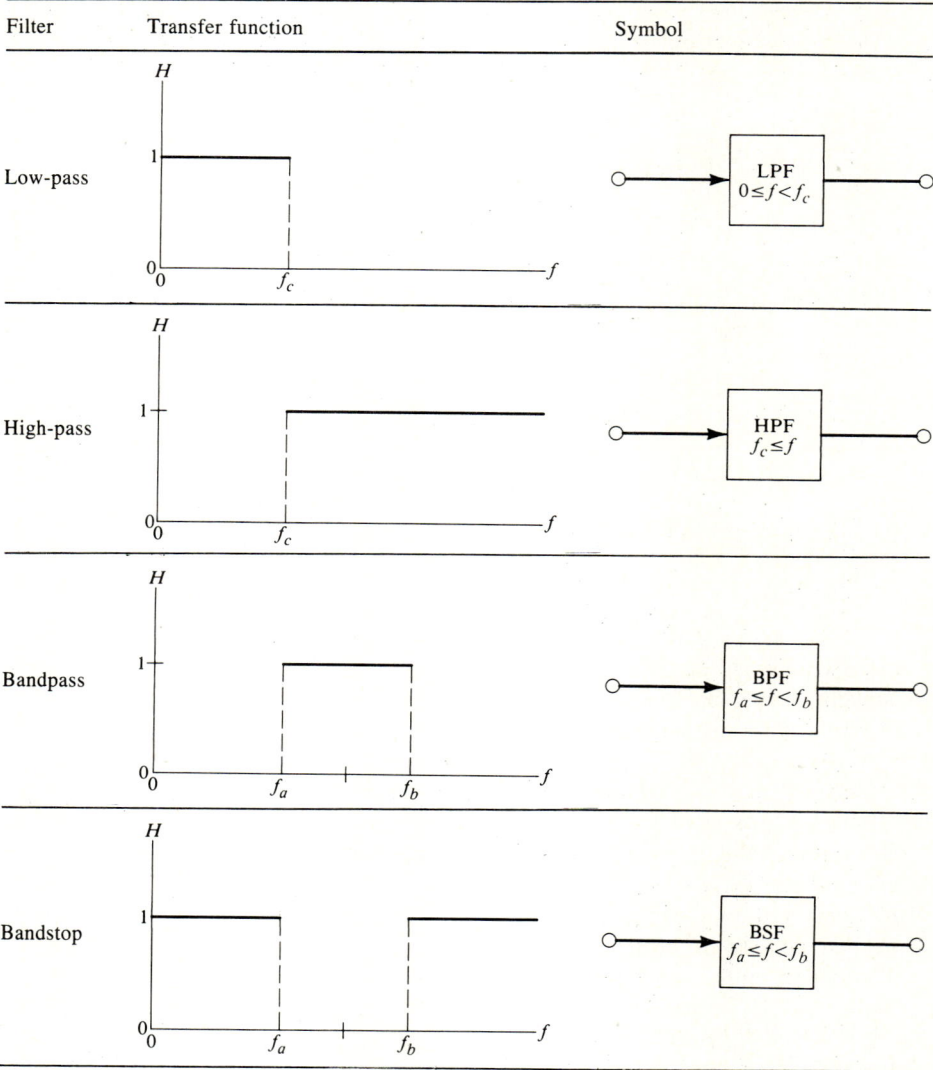

In theory, ideal filters are nonrealizable, for reasons given in Chap. 4. In practice, they usually can be approximated as accurately as necessary using realizable transfer functions. In any event, ideal filters are often used in first-cut analysis and design of systems and for rough calculations of various performance measures.

3-3G Spectrum Translation and Modulation

One of the main purposes of a signal-transmission system is to match a signal to be transmitted to an available transmission medium through which the signal must pass.

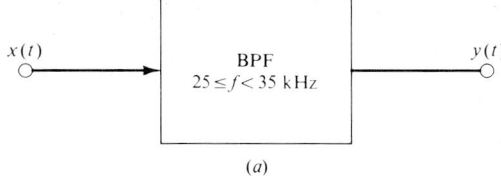

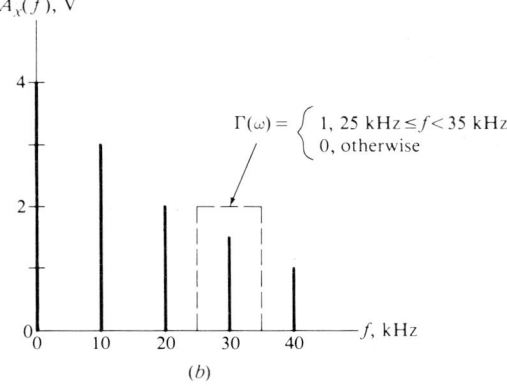

Figure 3-27 System and input of Example 3-25.

Another is to make it possible for a number of different signals to share a single transmission medium; e.g., in commercial (standard-broadcast) radio, signals representing speech and music are transmitted through space. In commercial radio systems, the matching problem is that the spectra of (audible) speech and music signals lie mainly below about 20 kHz, whereas economical radio transmission occurs at much higher frequencies. The sharing problem arises because a number of different radio stations (not to mention television stations, police radio, amateur radio, airport control radio, radar systems, and a host of others) all use the same transmission medium. Both the matching problem and the sharing problem described above are solved by translating the spectrum of a signal before transmission. For example, a commercial AM radio station translates signals representing speech and music to a center frequency somewhere between about 500 kHz and 1.6 MHz. In any particular geographic area different radio stations use different center frequencies to avoid interfering with each other. Radio receivers use filters to separate a particular radio signal from the many that appear simultaneously at the antenna and reverse the translation used at the radio station to obtain an audible signal. In this section we introduce the mathematical basis for spectrum translation.

An effect of multiplying a signal by a complex exponential $e^{j\omega_c t}$ is to translate the complex (two-sided) spectrum of the signal — to the right if ω_c is positive and to the left if ω_c is negative. Multiplying a signal $x(t)$ by $e^{j\omega_c t}$ multiplies each component of the signal by $e^{j\omega_c t}$. For any particular component $X_n e^{jn\omega_0 t}$, this gives

$$e^{j\omega_c t} X_n e^{jn\omega_0 t} = X_n e^{j(\omega_c + n\omega_0)t}$$

which is simply the original component translated from $f = nf_0$ to $f = f_c + nf_0$, as

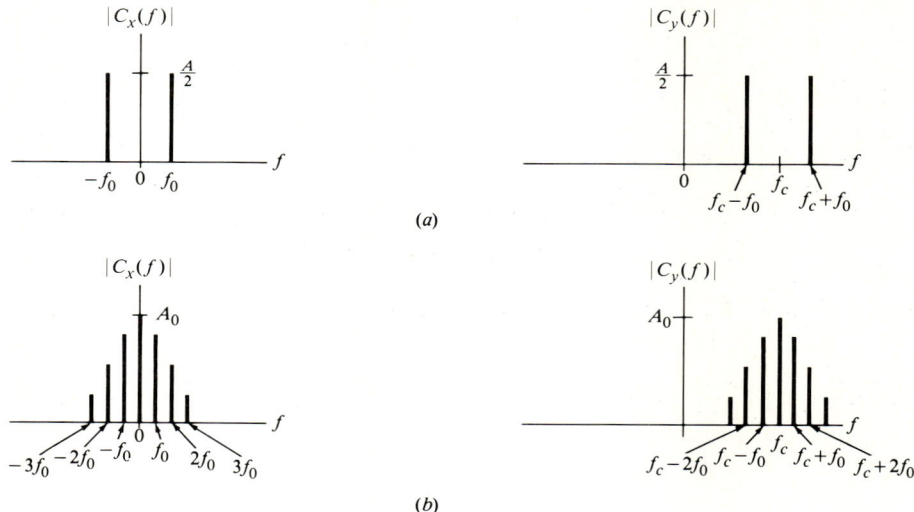

Figure 3-28 Spectrum translation for (a) a sinusoid and (b) a signal having several sinusoidal components.

illustrated in Fig. 3-28a. Because the same thing happens to every component of $x(t)$, the entire complex spectrum is translated, as shown in Fig. 3-28b.

The rule for *spectrum translation* can be expressed as follows: The complex spectrum of

$$y(t) = x(t)e^{j\omega_c t} \tag{3-38a}$$

is given by

$$C_y(f) = C_x(f - f_c) \tag{3-38b}$$

where $C_x(f)$ is the complex spectrum of $x(t)$. Note that (ideal) spectrum translation, as presented above, is associated with complex (two-sided) spectra. A signal $y(t)$ given by (3-38a), being complex, does not have a one-sided spectrum.

True spectrum translation cannot be achieved in a physical system because it requires multiplication of a signal by a complex quantity; however, spectrum translation can be achieved for all practical purposes by a realizable operation called *linear modulation*, where a signal is multiplied by a sinusoidal signal. Consider the system of Fig. 3-29, where

$$y(t) = KBx(t) \cos \omega_c t \tag{3-39a}$$

Using Euler's identity gives

$$y(t) = \tfrac{1}{2} KB [x(t) e^{j\omega_c t} + x(t) e^{-j\omega_c t}]$$

From (3-38), the complex spectrum of $y(t)$ above is given by

$$C_y(f) = \tfrac{1}{2} KB [C_x(f - f_c) + C_x(f + f_c)] \tag{3-39b}$$

where $C_x(f)$ is the complex spectrum of $x(t)$. To obtain the complex spectrum of $y(t)$ we translate the spectrum of $x(t)$ to the right by f_c and to the left by f_c and form the sum

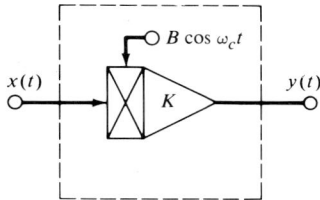

Figure 3-29 System for linear modulation.

of the translated spectra (scaled by $KB/2$). If the frequency f_c is large enough to ensure that the terms $C_x(f - f_c)$ and $C_x(f + f_c)$ do not overlap, the one-sided spectrum of $y(t)$ is a translated version of the two-sided spectrum of $x(t)$.

If f_c is so large that $C_x(f - f_c)$ and $C_x(f + f_c)$ in (3-39) do not overlap, the system of Fig. 3-29 is called a *linear modulator*. An input $x(t)$ is called the *modulating signal*, the sinusoid $B \cos \omega_c t$ is called the *carrier*, and the output $y(t)$ is called the *modulated signal*.

Example 3-26 In the system of Fig. 3-29 let $f_c = 10$ kHz, $B = 4$ mA, and $K = 1$ mA^{-1}. Obtain the spectrum of the output $y(t)$ for input

$$x(t) = 15 + 20 \cos\left(\omega_0 t - \frac{\pi}{4}\right) \quad \text{V}$$

where $f_0 = 1$ kHz.

SOLUTION Figure 3-30a shows the two-sided spectrum of the input (the modulating signal) $x(t)$. Figures 3-30b and c show the terms $C_x(f - f_c)$ and $C_x(f + f_c)$ appearing in (3-39). Figure 3-30d shows the two-sided spectrum of the output (the modulated signal) $y(t)$. The quantities $C_x(f - f_c)$ and $C_x(f + f_c)$ do not overlap because f_c is larger than the highest frequency (f_0) represented in $x(t)$. Consequently, the one-sided spectrum of the modulated signal (which is essentially the right half of the two-sided spectrum of the modulated signal) is a translated version of the two-sided spectrum of $x(t)$. For all practical purposes, the system of this example achieves spectrum translation.

Example 3-27 In the system of Fig. 3-29 let $f_c = 1$ kHz, $B = 4$ mA, and $K = 1$ mA^{-1}. Obtain the spectrum of the modulated signal $y(t)$ for input

$$x(t) = 15 + 20 \cos\left(\omega_0 t - \frac{\pi}{4}\right) \quad \text{V}$$

where $f_0 = 1$ kHz.

SOLUTION This example is just like the previous one except that the frequency of the carrier does not exceed the highest frequency present in the input $x(t)$. Figures 3-31a and b show the terms $C_x(f - f_c)$ and $C_x(f + f_c)$ that appear in (3-39). Figure 3-31c shows the two-sided spectrum of the output $y(t)$. Because the terms $C_x(f - f_c)$ and $C_x(f + f_c)$ overlap, the right half of $C_y(f)$ is not just $C_x(f - f_c)$. This system does not achieve frequency translation. Note that where the terms $C_x(f - f_c)$ and $C_x(f + f_c)$ overlap, addition of complex quantities is required; we cannot simply add the magnitudes of these terms to obtain $C_y(f)$; thus, for $f = 0$,

$$C_y(f) = 20(e^{-j\pi/4} + e^{j\pi/4}) = 28.3 \text{ V}$$

not $C_y(f) = 20 + 20 = 40$ V.

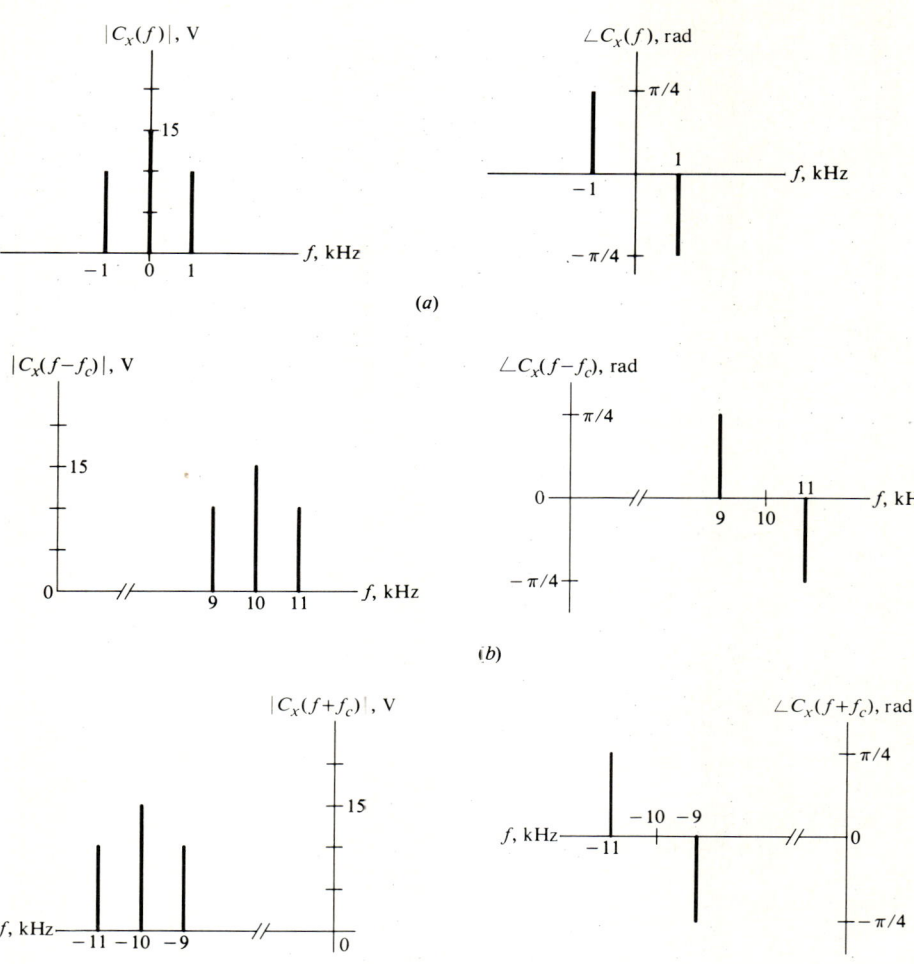

Figure 3-30 Solution of Example 3-26.

Recovering a modulating signal from a modulated signal is called *demodulation*. Figure 3-32 shows a system for demodulating a linearly modulated signal $y(t)$, where

$$y(t) = K_1 B_1 x(t) \cos \omega_c t$$

The output of the multiplier is given by

$$z(t) = K_2 B_2 y(t) \cos \omega_c t = Dx(t) \cos^2 \omega_c t$$

where $D = K_1 B_1 K_2 B_2$. Using the identity $2 \cos^2 \alpha \equiv 1 + \cos 2\alpha$ gives

$$z(t) = \tfrac{1}{2} Dx(t) + \tfrac{1}{2} Dx(t) \cos 2\omega_c t$$

The signal $z(t)$ above consists of two terms: the first is proportional to the modulating

RESPONSE TO SINUSOIDAL EXCITATION **157**

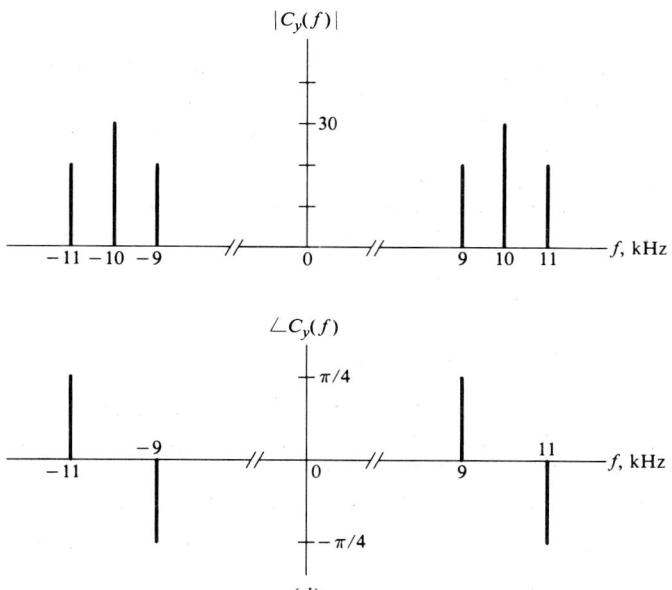

(d)

Figure 3-30 *Continued*

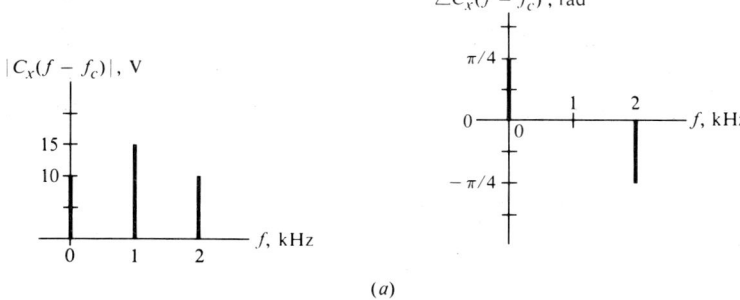

(a)

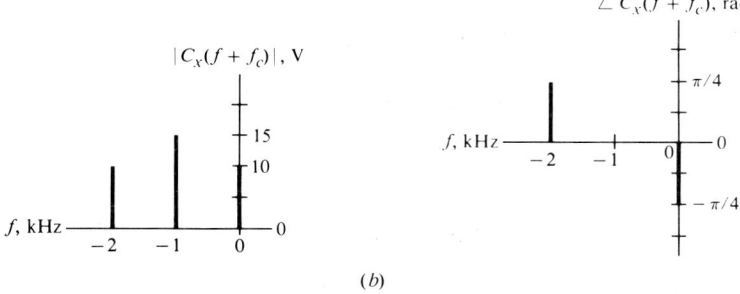

(b)

Figure 3-31 Spectrum of the signal of Example 3-27.

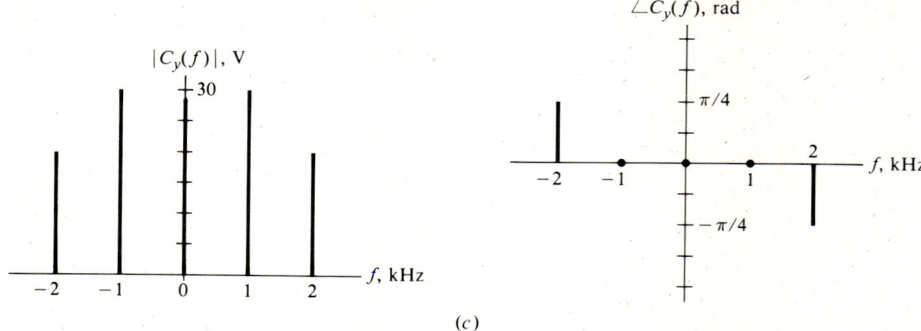

(c)

Figure 3-31 Continued

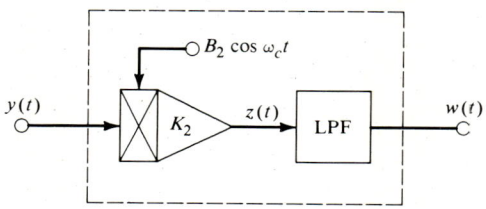

Figure 3-32 System for linear demodulation.

signal $x(t)$, and the other is a linearly modulated signal having carrier frequency $2f_c$. From (3-39), the complex spectrum of $z(t)$ is given by

$$C_z(f) = \tfrac{1}{2}DC_x(f) + \tfrac{1}{4}D[C_x(f - 2f_c) + C_x(f + 2f_c)]$$

where $C_x(f)$ is the complex spectrum of the modulating signal. Figure 3-33 shows the spectrum of $z(t)$ above for

$$x(t) = A_0 \cos \theta_0 + A_1 \cos(\omega_0 t + \theta_1) + A_2 \cos(2\omega_0 t + \theta_2) + \cdots$$
$$+ A_n \cos(n\omega_0 t + \theta_n)$$

for a case where $f_c > nf_0$. The low-pass filter (assumed ideal) passes components of $z(t)$ in the band $f < f_c$ and stops components lying outside that band, as indicated by

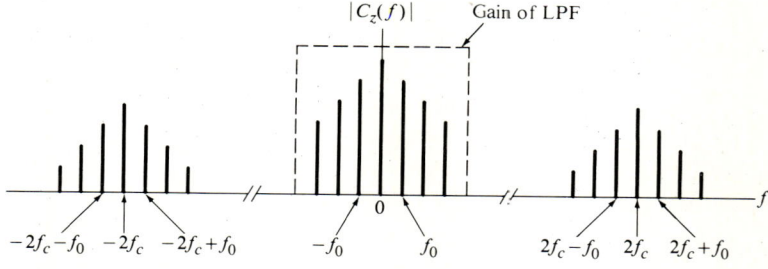

Figure 3-33 Spectrum of the signal $z(t)$ in the demodulator of Fig. 3-32.

the dashed lines in Fig. 3-33; thus components represented by $C_x(f)$ are passed through the filter, and components represented by $C_x(f \pm 2f_c)$ are stopped by the filter. The output of the filter (the output of the demodulator) is

$$w(t) = \tfrac{1}{2}Dx(t)$$

The output is proportional to the modulating signal $x(t)$. A system consisting of the linear modulator of Fig. 3-29 in series with the demodulator of Fig. 3-32 provides distortionless transmission for any signal whose spectrum is confined to the band $f \leq f_c$, where f_c is the carrier frequency used in the system.

3-4 SUMMARY

This chapter treats three main topics: Sec. 3-1 treats analysis of sinusoidally excited linear stationary systems, Sec. 3-2 describes polynomial approximation for calculating outputs of sinusoidally excited nonlinear static systems, and Sec. 3-3 introduces frequency-domain analysis of systems.

Amplitude-Phase and Exponential Representation of a Sinusoid

A sinusoidal signal can be expressed in *amplitude-phase* or *exponential form;* thus

$$A\cos(\omega t + \theta) = Xe^{j\omega t} + X^*e^{-j\omega t}$$

where (for $\omega \neq 0$)

$$X = \tfrac{1}{2}Ae^{j\theta}$$

For a constant (dc) signal,

$$x_0 = A\cos\theta = X \quad \text{where } A = |x_0| \quad \text{and} \quad \theta = \begin{cases} 0 & x_0 \geq 0 \\ \pi \text{ rad} & x_0 < 0 \end{cases}$$

Peak amplitude A is a nonnegative quantity. Peak amplitude A and complex amplitude X are given in the unit of the signal, and initial phase θ is given in radians.

Transfer Function

The output of a stable linear stationary system for input $x(t) = Xe^{j\omega t}$ is given by

$$y(t) = Ye^{j\omega t} \qquad Y = H(j\omega)X$$

The *transfer function* $H(j\omega)$ can be obtained in several ways:

1. For a system described by an impulse response, use (3-5).
2. For a system described by a differential equation, use (3-17).
3. For a system described by a block diagram, use block-diagram reduction (see Table 3-1 and Figs. 3-2 to 3-4, 3-6, and 3-7).
4. For a system described by a transfer characteristic, it often is easiest to use $H(j\omega) = Y/X$, where $Ye^{j\omega t}$ is the output for input $Xe^{j\omega t}$.

Gain and Phase Shift

The output of a stable linear stationary system for input

$$x(t) = A \cos(\omega_0 t + \theta)$$

is given by

$$y(t) = \Gamma(\omega_0) A \cos[\omega_0 t + \theta + \phi(\omega_0)] \tag{3-40a}$$

where *gain* $\Gamma(\omega)$ and *phase shift* $\phi(\omega)$ are given by

$$\Gamma(\omega) = |H(j\omega)| \qquad \phi(\omega) = \angle H(j\omega) \tag{3-40b}$$

To calculate gain and phase shift express a transfer function in polar form; thus

$$H(j\omega) = \Gamma(\omega) e^{j\phi(\omega)}$$

Superposition

The output of a stable linear stationary system for an input having the form

$$x(t) = A_0 \cos \theta_0 + A_1 \cos(\omega_0 t + \theta_1) + A_2 \cos(2\omega_0 t + \theta_2) + \cdots$$

is given by

$$y(t) = B_0 \cos \psi_0 + B_1 \cos(\omega_0 t + \psi_1) + B_2 \cos(2\omega_0 t + \psi_2) + \cdots$$

$$B_n = \Gamma(n\omega_0) A_n \qquad \psi_n = \theta_n + \phi(n\omega_0)$$

where $\Gamma(\omega)$ is the gain of the system and $\phi(\omega)$ is the phase shift of the system.

Polynomial Approximation for Nonlinear Static Systems

A sufficiently smooth nonlinear static transfer characteristic can be approximated for small inputs by a polynomial. The polynomial can be obtained by truncating Maclaurin's or Taylor's expansion for the transfer characteristic. Use Maclaurin's expansion for ac inputs (see Sec. A-7). Use Taylor's expansion for inputs having a dc bias [see (3-26)]. Write the approximate transfer characteristic as

$$y = y_0 N\left(\frac{x}{x_0}\right) = y_0 N(\alpha)$$

where $x = x(t)$ is an input, $y = y(t)$ is the corresponding output, x_0 (unit of x) and y_0 (unit of y) are constants, and

$$N(\alpha) = d_0 + d_1 \alpha + d_2 \alpha^2 + \cdots + d_k \alpha^k$$

is a polynomial in $\alpha = x(t)/x_0$. Then use the identity (3-24) to express powers of $\cos \omega_0 t$ as harmonics of $\cos \omega_0 t$. The resulting harmonic series contains only odd harmonics if $N(\alpha)$ is odd, only even harmonics if $N(\alpha)$ is even, and both odd and even harmonics if $N(\alpha)$ is neither odd nor even. The highest harmonic represented in the output is the kth, where k is the degree of the polynomial.

Harmonic Distortion

The output of a nonlinear static system for input $x(t) = A \cos \omega_0 t$ has the form

$$y(t) = B_0 \cos \psi_0 + B_1 \cos(\omega_0 t + \psi_1) + B_2 \cos(2\omega_0 t + \psi_2) + \cdots$$
$$+ B_k \cos(k\omega_0 t + \psi_k)$$

The *nth-harmonic distortion* introduced by the system is defined by

$$D_n = \frac{100 B_n}{B_1} \%$$

As a rule, use a polynomial whose degree is at least $n + 1$ for calculating nth-harmonic distortion. If nth-harmonic distortion is greater than about 1 percent, repeat the calculation using a polynomial of higher degree.

Amplitude and Phase Spectrum

A signal is described as a harmonic series in amplitude-phase form by

$$x(t) = \sum_{n=0}^{\infty} A_n \cos(n\omega_0 t + \theta_n)$$

The *amplitude spectrum* $A_x(f)$ and the *phase spectrum* $\theta_x(f)$ of the signal are defined by

$$A_x(f) = \begin{cases} A_n & f = nf_0 \\ 0 & \text{all other } f \end{cases} \qquad \theta_x(f) = \begin{cases} \theta_n & f = nf_0 \\ 0 & \text{all other } f \end{cases}$$

Complex Spectrum

A signal is expressed as a harmonic series in exponential form as

$$x(t) = \sum_{n=-\infty}^{\infty} X_n e^{jn\omega_0 t}$$

The *complex spectrum* of the signal is defined by

$$C_x(f) = \begin{cases} X_n & f = nf_0 \\ 0 & \text{all other } f \end{cases}$$

Time Domain and Frequency Domain

A signal is described in the time domain by its waveform and in the frequency domain by its spectrum. A system is described in the time domain by a relation between the waveforms of an input and output and in the frequency domain by a relation between the spectra of an input and output. The three most important frequency-domain relations and their time-domain counterparts are

$$C_y(f) = H(j\omega)C_x(f) \Leftrightarrow y(t) = \int_{-\infty}^{\infty} x(\lambda)h(t-\lambda)\,d\lambda$$

$$C_y(f) = Ke^{-j\omega t_0}C_x(f) \Leftrightarrow y(t) = Kx(t-t_0)$$

$$C_y(f) = C_x(f - f_c) \Leftrightarrow y(t) = e^{j\omega_c t}x(t)$$

The first pair of relations shows that convolution in the time domain corresponds to multiplication in the frequency domain. This largely accounts for the relative ease of frequency-domain analysis of linear stationary systems. The second pair of relations shows that high-fidelity transmission of a signal requires frequency-independent gain and linear phase shift. This often makes frequency-domain analysis of distortion preferable to time-domain analysis of distortion. The third pair of relations is the basis for operation of a great many signal-transmission systems, where a signal is translated to center frequency f_c before transmission.

Concepts, relations, and procedures presented in this chapter are generalized in subsequent chapters. The mathematics necessary for the generalizations is somewhat more complicated than that used in this chapter, but the fundamental ideas are the same.

PROBLEMS

3-1 Express the following signals in amplitude-phase form:

(a) $x(t) = 10 - 5\cos\left(\omega t - \dfrac{\pi}{2}\right)$ mA

(b) $x(t) = -20 + 10\sin \omega t$ V

(c) $x(t) = 3\cos \omega t + 4\sin \omega t$ mV

(d) $x(t) = 5e^{-j\omega t} + 5e^{j\omega t}$ V

(e) $x(t) = 2 + (1+j)e^{-j\omega t} + (1-j)e^{j\omega t}$ mA

(f) $x(t) = je^{j(\omega t - \pi/3)} - je^{-j(\omega t - \pi/3)}$ mV

(g) $x(t) = \dfrac{10e^{j\omega t}}{1-j} + \dfrac{10e^{-j\omega t}}{1+j} - 5$ mA

3-2 Express the following signals in exponential form:

(a) $x(t) = 10 + 5\cos\left(\omega t - \dfrac{\pi}{4}\right)$ V

(b) $x(t) = 20 + 10\cos \omega_1 t + 5\cos\left(\omega_2 t - \dfrac{\pi}{2}\right)$ V

(c) $x(t) = -20 + 10\sin \omega t$ mA

(d) $x(t) = 10 - 3\cos \omega t + 3\sin \omega t$ V

(e) $x(t) = 5 + 10\cos \omega_1 t + 10\cos\left(\omega_1 t - \dfrac{\pi}{4}\right) - 10\sin\left(\omega_2 t + \dfrac{\pi}{4}\right)$ V

3-3 Obtain the transfer functions of the systems having the following impulse responses:

(a) $h(t) = \delta(t) - \dfrac{1}{\tau}e^{-t/\tau}u(t)$ (b) $h(t) = \dfrac{t}{\tau^2}e^{-t/\tau}u(t)$

(c) $h(t) = \dfrac{1}{\tau} e^{-t/\tau} (\sin \omega t) u(t)$ (d) $h(t) = \left(\dfrac{1}{\tau_1} e^{-t/\tau_1} + \dfrac{1}{\tau_2} e^{-t/\tau_2}\right) u(t)$

(e) $h(t) = \dfrac{1}{\tau} r\left(\dfrac{t}{\tau}\right)$ (f) $h(t) = \omega_0 r\left(\dfrac{t}{\tau}\right) \cos \omega_0 t$

(g) $h(t) = \omega_0 r\left(\dfrac{t}{\tau}\right) \sin \omega_0 t$ (h) $h(t) = 4\delta(t) + 3\delta(t - t_0) + 2\delta(t - 2t_0)$

3-4 Obtain the outputs of the systems having the following transfer functions for input

$$x(t) = 5 + 4 \cos \omega_0 t + 3 \cos\left(2\omega_0 t - \dfrac{\pi}{2}\right) \quad \text{V}$$

(a) $H(j\omega) = \dfrac{100}{1 + 4j(\omega/\omega_0)}$ mA/V

(b) $H(j\omega) = \dfrac{1 - 10j(\omega/\omega_0)}{1 + 5j(\omega/\omega_0)}$ V/V

(c) $H(j\omega) = \dfrac{j\omega\omega_0}{-100\omega^2 + j\omega\omega_0 + 100\omega_0^2}$ V/V

(d) $H(j\omega) = \dfrac{j(\omega/\omega_0)}{1 + j(\omega/\omega_0)}$ V/V

(e) $H(j\omega) = \dfrac{2}{[1 + j(\omega/\omega_0)][2 + j(\omega/\omega_0)]}$ mA/V

(f) $H(j\omega) = \dfrac{5}{0.5 + j(\omega/\omega_0)}$ V/V

(g) $H(j\omega) = 5e^{-j\pi\omega/\omega_0}$ mA/V

3-5 Plot gain and phase shift versus normalized frequency f/f_0 for each system of Prob. 3-4.

3-6 Obtain the outputs of the systems described by the differential equations below for input

$$x(t) = 5 + 5 \cos \omega_0 t \quad \text{V}$$

where $\omega_0 = 1$ rad/s.

(a) $\dfrac{dy}{dt} + 5\omega_0 y = \dfrac{dx}{dt}$ (b) $\dfrac{dy}{dt} + 5\omega_0 y = 10 \dfrac{dx}{dt} + 2\omega_0 x$

(c) $\dfrac{d^2y}{dt^2} + 3\omega_0 \dfrac{dy}{dt} + 2\omega_0^2 y = 5 \dfrac{d^2x}{dt^2} + 20\omega_0 \dfrac{dx}{dt} + 15\omega_0^2 x$

(d) $\dfrac{d^2y}{dt^2} + 3\omega_0 \dfrac{dy}{dt} + 2\omega_0^2 y = \omega_0 \dfrac{dx}{dt}$ (e) $\dfrac{d^3y}{dt^3} + 3\omega_0 \dfrac{d^2y}{dt^2} + 3\omega_0^2 \dfrac{dy}{dt} + 2\omega_0^3 y = 10 \dfrac{d^3x}{dt^3}$

3-7 Obtain the outputs of the systems of Fig. P3-7 for input $x(t) = 5 + 4 \cos \omega_0 t$ V, where $\omega_0 = 10$ krad/s.

3-8 Show that a second-order system having transfer function

$$H(j\omega) = \dfrac{1}{(1 + j\omega/\omega_1)(1 + j\omega/\omega_2)}$$

where $\omega_1 \neq \omega_2$, can be analyzed as either a series connection or a parallel connection of two first-order systems. *Hint:* To obtain the parallel connection find A and B such that

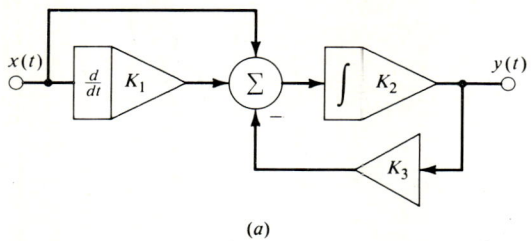

(a)

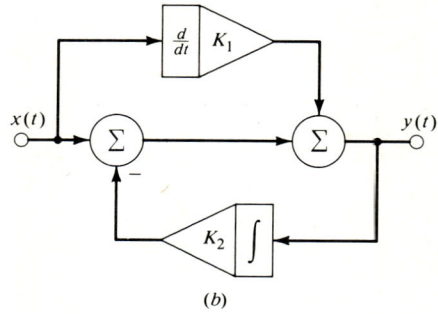

(b)

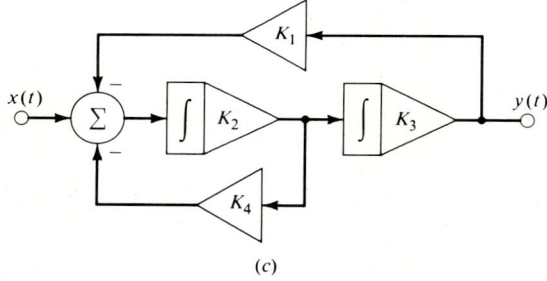

(c)

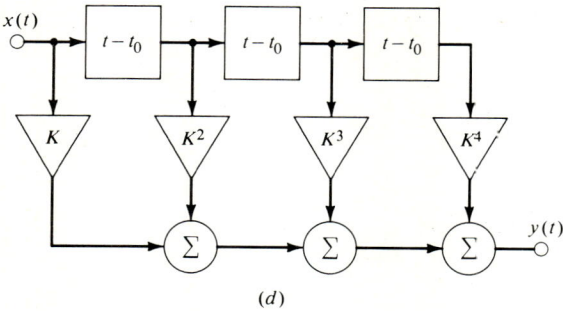

(d)

Figure P3-7 (a) $K_1 = 100\ \mu s$, $K_2 = 10^3$ mA/Vs, $K_3 = 1$ V/mA; (b) $K_1 = 100\ \mu s$, $K_2 = 10^3\ s^{-1}$; (c) $K_1 = 10$, $K_2 = 20\ s^{-1}$, $K_3 = 5 \times 10^5\ s^{-1}$, $K_4 = 100$; (d) $K = 0.05$, $t_0 = 10\pi/3$ ms;

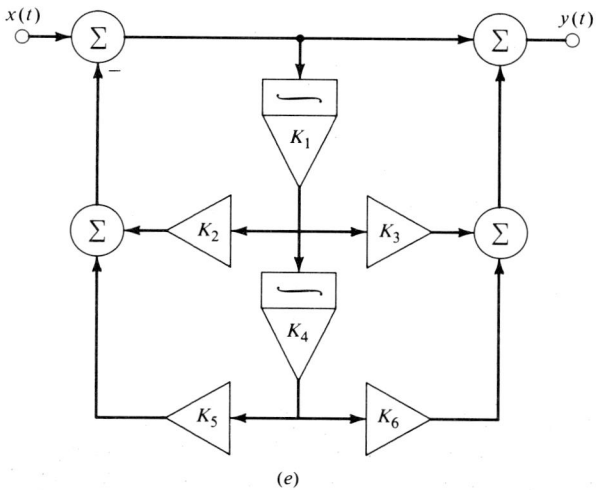

(e)

Figure P3-7 *Continued* (e) $K_1 = K_4 = 10^4 \text{ s}^{-1}$, $K_2 = 1$, $K_5 = K_6 = 2$.

$$H(j\omega) = \frac{A}{1 + j\omega/\omega_1} + \frac{B}{1 + j\omega/\omega_2}$$

3-9 Show that a system having transfer function

$$H(j\omega) = \frac{1}{1 + j\omega/\omega_0}$$

can be represented by a feedback connection of elementary systems.

3-10 Obtain differential equations describing the systems of Prob. 3-4.

3-11 Obtain differential equations describing the systems of Prob. 3-3.

3-12 Obtain block diagrams containing only elementary systems for the systems of Prob. 3-4.

3-13 Show that for $\tau_1 \neq \tau_2$, the system of part (d) of Prob. 3-3 can be analyzed as either a series connection or a parallel connection of first-order systems (see Prob. 3-8).

3-14 Obtain a block diagram containing only elementary systems for the system of Part (c) of Prob. 3-3.

3-15 Use block-diagram reduction to obtain the transfer functions of the systems of Fig. P3-15. Express each transfer function in terms of $H_1, H_2, \ldots$.

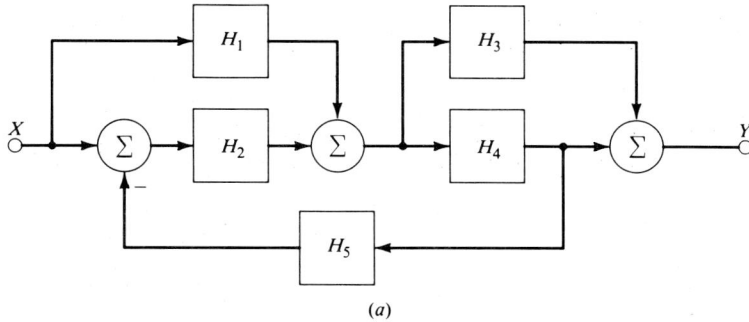

(a)

Figure P3-15

166 INTRODUCTION TO SYSTEM ANALYSIS

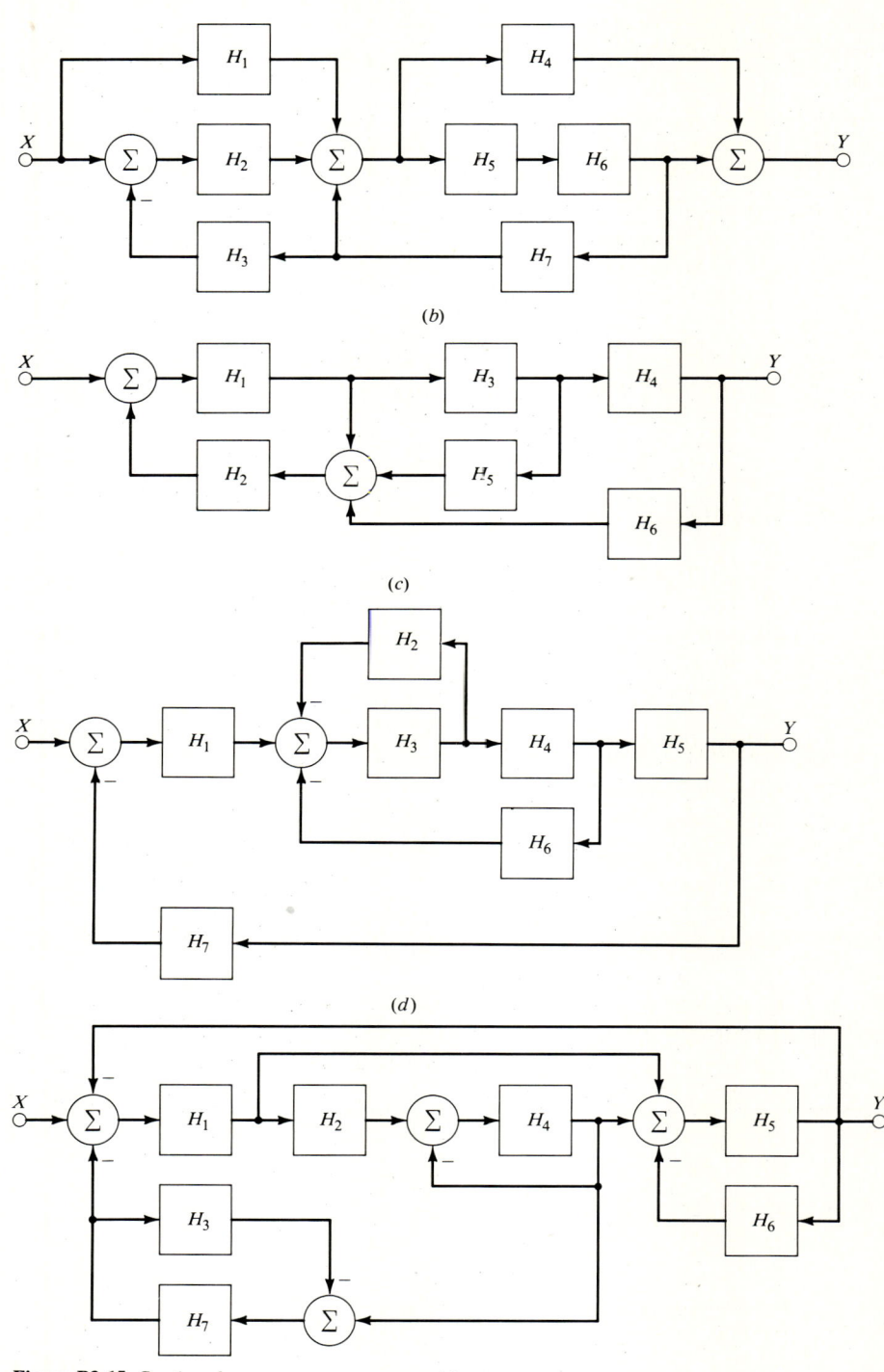

Figure P3-15 *Continued*

3-16 Figure P3-16 shows graphs of a sinusoidal input $x(t)$ and the corresponding output $y(t)$ of a linear stationary system. Find the gain and the phase shift of the system for the frequency of the input.

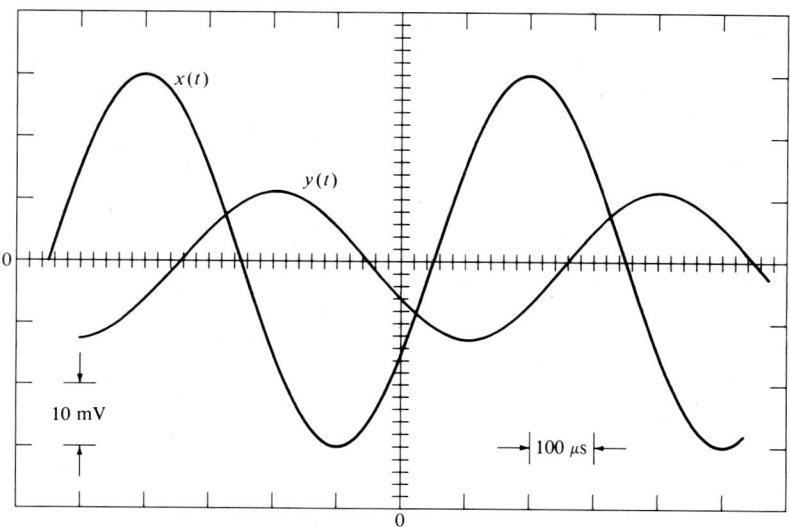

Figure P3-16

3-17 The impulse response of an ideal integrator is

$$h(t) = Ku(t)$$

The inpulse response of a practical integrator is more accurately modeled by

$$h(t) = Ke^{-t/\tau}u(t)$$

where τ is large (on the order of a few hours) but finite.

(a) Obtain expressions for the outputs of the ideal integrator and the practical integrator described above for input $x(t) = x_0 u(t)$.

(b) For $\tau = 3600$ s find the largest time for which the output of the practical integrator for input $x(t) = x_0 u(t)$ differs from that of the ideal integrator by no more than 1 percent (of the output of the ideal integrator).

(c) Show that the transfer function of the practical integrator is

$$H(j\omega) = \frac{K\tau}{1 + j\omega\tau}$$

(d) Plot gain versus frequency for both the ideal and the practical integrator on the same axes. For $\tau = 3600$ s find the smallest frequency for which the gain of the practical integrator differs from that of the ideal integrator by no more than 1 percent (of the gain of the ideal integrator).

3-18 Use Maclaurin's expansion to obtain third-degree polynomials approximating the transfer characteristics given below. Use the approximations to obtain the outputs of the systems for input $x(t) = A \cos \omega_0 t$, where $A = 500$ mV, $f_0 = 1$ kHz, $x_0 = 1$ V, and $y_0 = 5$ V. Express the outputs as harmonic series in amplitude-phase form.

(a) $y = y_0 \tan^{-1} \dfrac{x}{x_0}$ (b) $y = y_0 \pi^{-1} \sin^{-1} \dfrac{x}{x_0}$ (c) $y = y_0 \dfrac{x/x_0}{\sqrt{1+(x/x_0)^2}}$

(d) $y = y_0 \ln\left(1 + \dfrac{x}{x_0}\right)$ (e) $y = y_0 \operatorname{erf} \dfrac{x}{x_0}$ where $\operatorname{erf} \alpha = \dfrac{2}{\sqrt{\pi}} \displaystyle\int_0^\alpha e^{-\lambda^2} d\lambda$

3-19 Use Taylor's expansion to obtain third-degree polynomials approximating the transfer characteristics below. Use the approximations to obtain the outputs for input

$$x(t) = A_0 + A_1 \cos \omega_0 t$$

where $A_0 = 1$ V, $A_1 = 200$ mV, $f_0 = 1$ kHz, $x_0 = 500$ mV, and $y_0 = 100$ mA. Express the outputs as harmonic series in amplitude-phase form.

(a) $y = y_0 e^{x/x_0}$ (b) $y = y_0 \ln\left(1 + \dfrac{x}{x_0}\right)$ (c) $y = y_0 \cosh \dfrac{x}{x_0}$

3-20 Many introductory calculus textbooks give Taylor's expansion as

$$N(\alpha) = d_0 + d_1(\alpha - \alpha_0) + d_2(\alpha - \alpha_0)^2 + \cdots \qquad \text{where } d_n = \dfrac{1}{n!} \dfrac{d^n N(\alpha)}{dt^n}\bigg|_{\alpha = \alpha_0}$$

Show that this form of Taylor's expansion is equivalent to (3-26).

3-21 Obtain a small-signal linear model for the system of Fig. P3-21. Give conditions on the input $x(t)$ under which the model is valid.

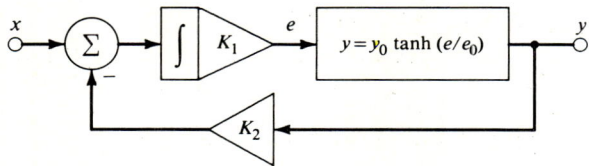

Figure P3-21

3-22 The circuit of Fig. 1-24 is to be used to integrate a periodic signal $x(t)$. The (small-signal) gain of the operational amplifier is $K = 10^5$ V/V. For linear operation the output $y(t)$ must satisfy the relation $|y(t)| \le 10$ V. Find the maximum allowable peak amplitude A as a function of period T for which the circuit functions as an integrator for the inputs of Fig. P3-22.

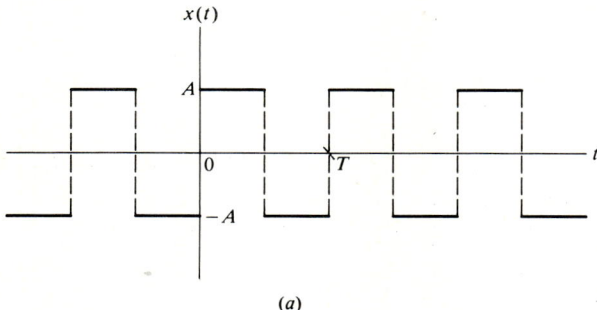

(a)

Figure P3-22

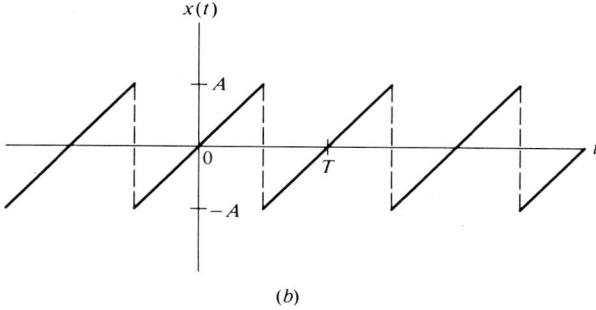

(b)

Figure P3-22 *Continued*

3-23 The output $y(t)$ of a certain system for input $x(t)$ is given by

$$y = y_0 \tan^{-1} \frac{x}{x_0}$$

where $x_0 = 20$ mV and $y_0 = 50$ V. Use a fifth-degree polynomial approximation for the transfer characteristic to calculate the percent third-harmonic distortion introduced by the system for input

$$x(t) = 5 \cos \omega_0 t \quad \text{mV}$$

How would you check the result of this calculation?

3-24 For the system of Fig. P3-24, $x_0 = 1$ V, $y_0 = w_0 = 5$ mA, and $x(t) = x_0 \cos \omega_0 t$, with $f_0 = 1$ kHz. Obtain the output $z(t)$ for

(a) $H(j\omega) = \dfrac{10}{1 + j(\omega/\omega_0)}$ V/mA (b) $H(j\omega) = \dfrac{10}{1 - j(\omega_0/\omega)}$ V/mA

(c) $H(j\omega) = \dfrac{-\omega_0^2}{\omega^2 - 10 j\omega\omega_0 - \omega_0}$ V/mA

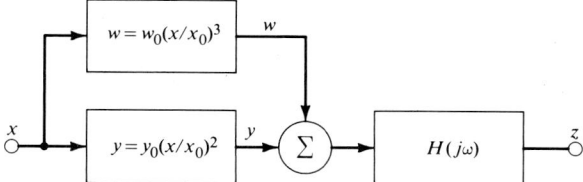

Figure P3-24

3-25 For each of the static elements of Table 1-3 give the frequencies represented in the output for sinusoidal input having frequency f_0.

3-26 Plot the one- and the two-sided spectrum (both amplitude and phase) for each of the following signals.

(a) $x(t) = -10 + 4 \cos\left(\omega_0 t - \dfrac{\pi}{4}\right) + 2 \cos\left(2\omega_0 t - \dfrac{\pi}{4}\right)$ V

where $f_0 = 2$ kHz.

170 INTRODUCTION TO SYSTEM ANALYSIS

(b) $\quad x(t) = \sum_{n=0}^{5} \dfrac{10}{\sqrt{1+n^2}} \cos\left(n\omega_0 t - \dfrac{n\pi}{5}\right) \quad$ mA

where $f_0 = 500$ Hz.

(c) $\quad x(t) = 5 + 5 \sum_{n=1}^{8} \dfrac{\sin(n\pi/2)}{n\pi/2} \cos n\omega_0 t \quad$ V

where $f_0 = 1$ kHz.

3-27 Express the signals whose spectra are shown in Fig. P3-27 as harmonic series in amplitude-phase form.

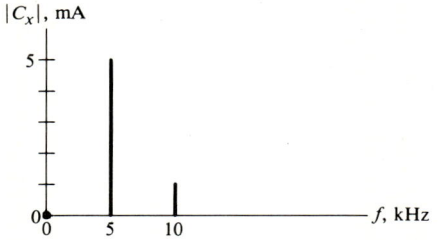
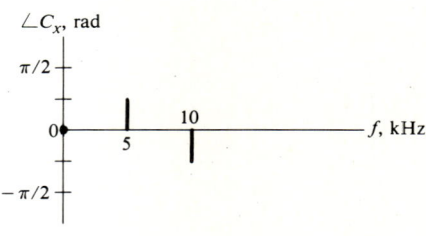

(a)

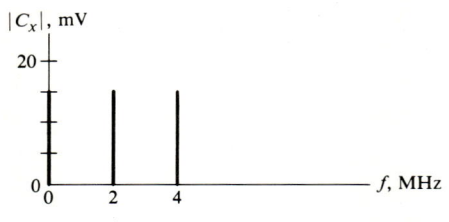
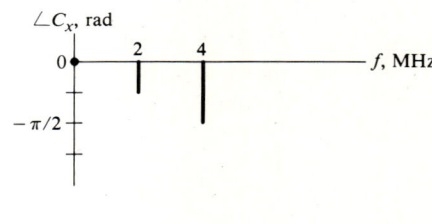

(b)

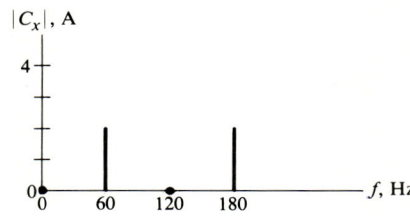
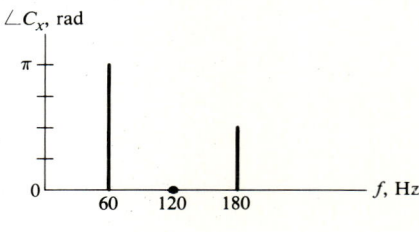

(c)

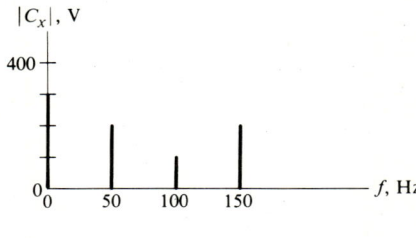
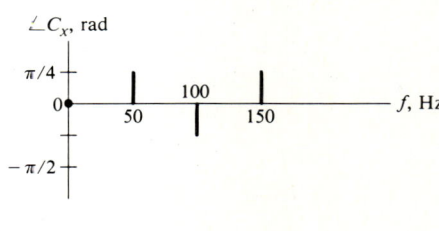

(d)

Figure P3-27

3-28 Sketch the waveforms of the signals whose spectra are shown in Fig. P3-28.

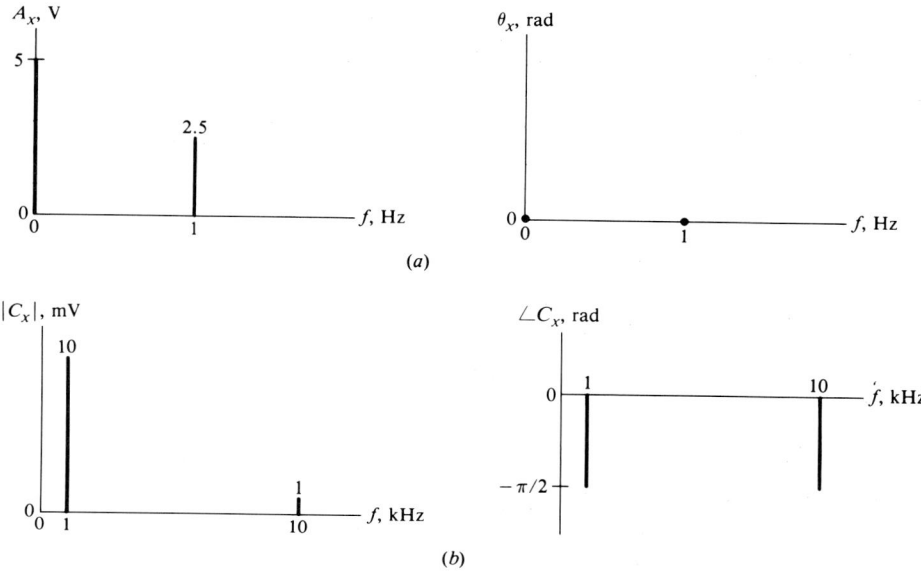

Figure P3-28

3-29 Plot the two-sided spectra of the outputs of the systems of Fig. P3-7 for the input having the two-sided spectrum of Fig. P3-29.

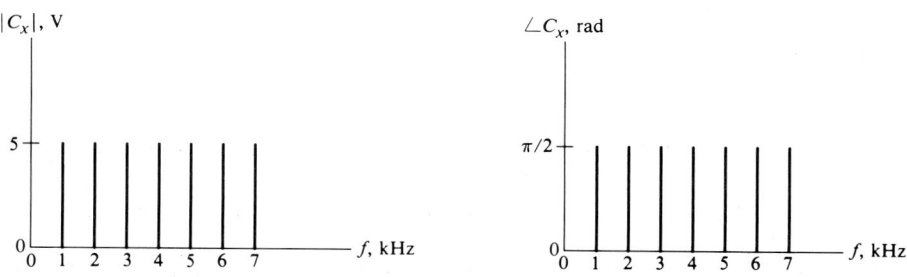

Figure P3-29

3-30 Describe conditions on the waveform and the spectrum of an input $x(t)$ under which the systems of Fig. P3-30 provide high-fidelity transmission for $x(t)$.

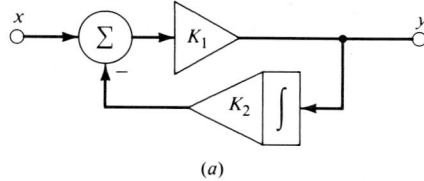

Figure P3-30 (a) $K_1 = 10^3$, $K_2 = 1 \text{ s}^{-1}$.

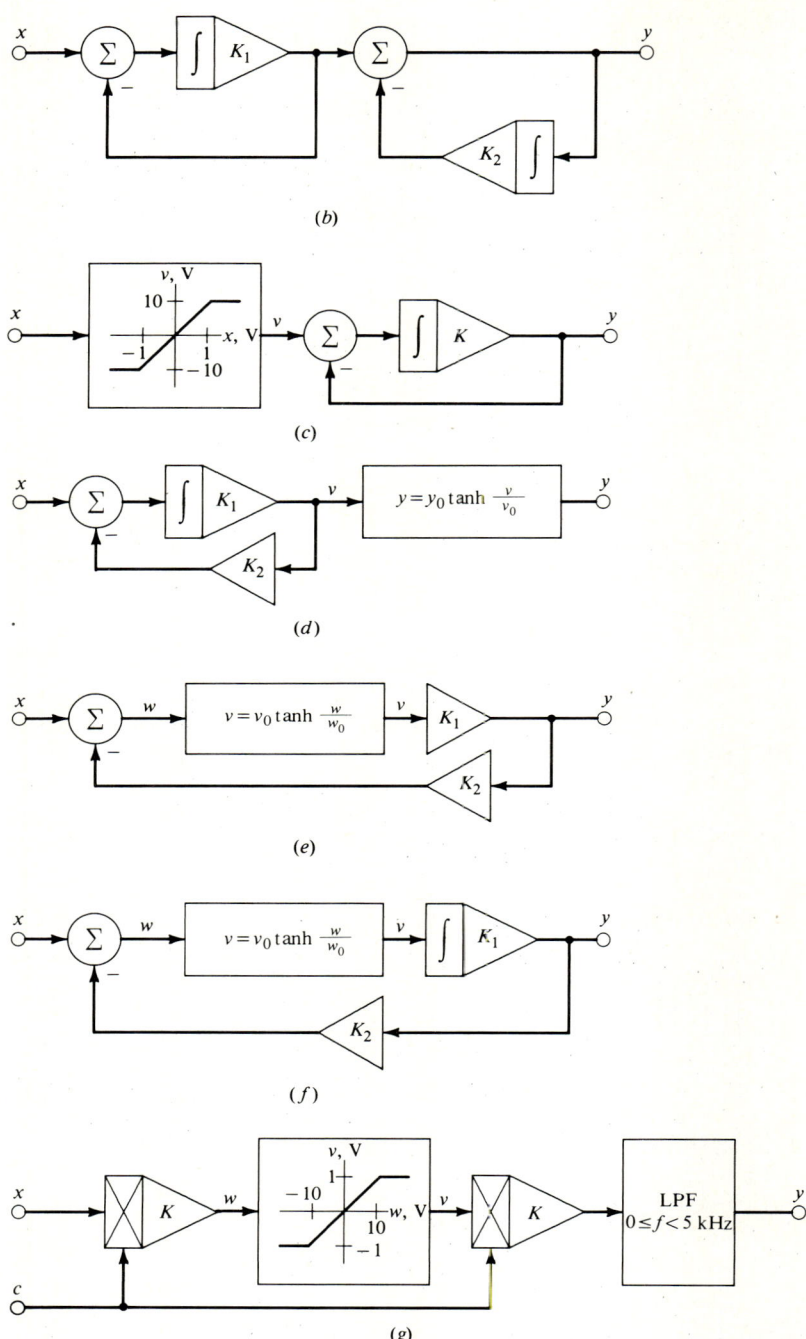

Figure P3-30 *Continued* (b) $K_1 = 10^3 \text{ s}^{-1}$, $K_2 = 10^5 \text{ s}^{-1}$; (c) $K = 10^4 \text{ s}^{-1}$; (d) $K_1 = 10^5 \text{ s}^{-1}$, $K_2 = 0.1$, $y_0 = 10$ mA, $v_0 = 10$ V; (e) $v_0 = w_0 = 1$ V, $K_1 = 100$, $K_2 = 0.1$; (f) $v_0 = w_0 = 1$ V, $K_4 = 10^4 \text{ s}^{-1}$, $K_2 = 0.1$; (g) $K = 2 \text{ V}^{-1}$, $c(t) = 5 \cos \omega_0 t$ V with $f_0 = 1$ MHz.

3-31 Use third-degree polynomial approximations to calculate third-harmonic distortion introduced by the systems of Prob. 3-18.

3-32 The output of a certain nonlinear static system for input $x(t) = A \cos \omega_0 t$ is found (by measurement) to be

$$y(t) \approx \begin{cases} 10 \cos \omega_0 t & \text{mV} & \text{for } A = 1 \text{ mV} \\ 125 \cos \omega_0 t + 8.33 \cos 3\omega_0 t & \text{mV} & \text{for } A = 10 \text{ mV} \end{cases}$$

Obtain a third-degree polynomial that approximates the transfer characteristic of the system.

3-33 In the modulator-demodulator system of Fig. P3-33 the carriers used for modulation and demodulation differ in initial phase by θ. Let $K = 2/B$,

$$x(t) = A \cos \omega_0 t \qquad w(t) = B \cos \omega_c t \qquad v(t) = B \cos(\omega_c t + \theta)$$

where $f_0 = 1$ kHz and $f_c = 1$ MHz. Obtain an expression for the demodulator output. Describe how the carrier phase difference θ affects operation of the system.

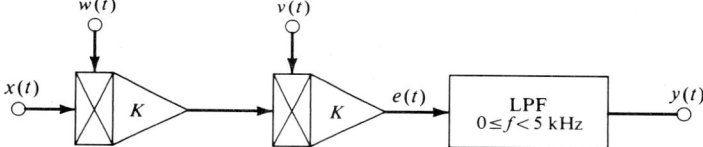

Figure P3-33

3-34 In the modulator-demodulator system of Fig. P3-33, the carriers used for modulation and demodulation differ in frequency by $\Delta\omega$. Let $K = 2/B$,

$$x(t) = A \cos \omega_0 t \qquad w(t) = B \cos \omega_c t \qquad v(t) = B \cos[(\omega_c + \Delta\omega)t]$$

where $f_0 = 1$ kHz and $f_c = 1$ MHz. Obtain an expression for the demodulator output. Describe how the carrier frequency difference $\Delta\omega$ affects operation of the system.

3-35 Figure P3-35 shows a simplified block diagram for a single-sideband signal-transmission system. Let

$$x(t) = A \cos \omega_0 t$$

with $f_0 < W$ and

$$c_T(t) = c_R(t) = B \cos \omega_c t$$

with $f_c > 2W$. Show that the output $y(t)$ is proportional to the input $x(t)$.

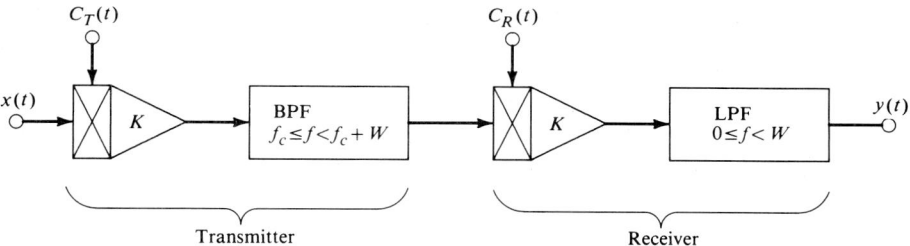

Figure P3-35

3-36 Figure P3-36 shows a simplified block diagram for a servomechanism (a position-control system). The input $v(t)$ represents 60-Hz interference (hum) introduced by the power amplifier driving the motor. The parameters K_1 and K_2 are individually adjustable over $10 \le K_1 \le 200$ V/rad, $10 \le K_2 \le 50$ rad/V, subject to the constraint $K_1 K_2 = 200$. Find the values of K_1 and K_2 that minimize contribution of hum to output position θ_C.

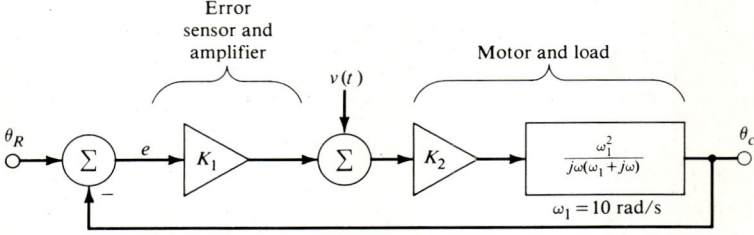

Figure P3-36

3-37 For the system (servomechanism) of Fig. P3-36, where $v(t) = 0$, find the value of $K_1 K_2 = K$ for which the signal (the position error) $e(t)$ does not exceed 1 percent of the peak amplitude of a sinusoidal input θ_R having frequency $\omega_0 = 10$ rad/s.

CHAPTER
FOUR

FOURIER SERIES, FOURIER INTEGRAL, AND FOURIER TRANSFORMATION

In this chapter we show how a signal can be represented as a superposition of sinusoids; thus we show how the spectrum of a signal is obtained from the waveform of the signal and vice versa. Subsequently, we show how important properties of a system, e.g., stability and fidelity, are determined from a frequency-domain description of the system.

The chapter is organized as follows. In Sec. 4-1 we describe a Fourier series,[†] a harmonic series that can represent any signal over any finite interval. In Sec. 4-2 we introduce the Fourier integral, which can describe a signal over an infinite interval. In Sec. 4-3 we describe the Fourier transformation, which is essentially a formalism for using Fourier series and Fourier integrals. The Fourier transformation allows compact presentation of important principles, e.g., relations between time- and frequency-domain descriptions of signals and systems. In Sec. 4-4 we show how important properties of systems and signals are examined in the frequency domain. Finally, in Sec. 4-5, we illustrate frequency-domain analysis by its application to systems for filtering, modulation, and automatic control.

4-1 FOURIER SERIES

Any realistic periodic waveform can be produced (mathematically, at least) by distorting a sinusoidal waveform. For example, the system of Fig. 4-1 produces a triangular wave, a square wave, and a sawtooth wave in response to a sinusoidal input

[†]After Jean Baptiste Joseph Fourier (1768–1830), French physicist and mathematician, who used harmonic series to study heat conduction in solids.

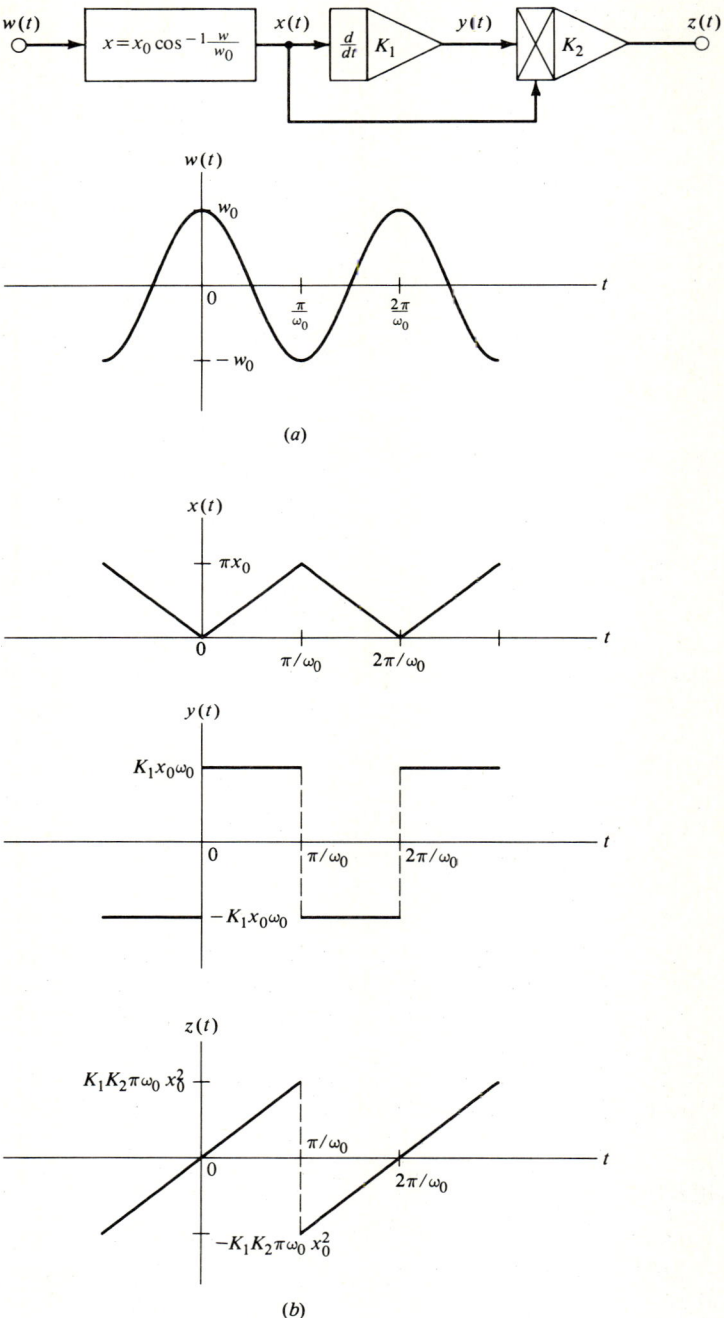

Figure 4-1 An illustration of how a periodic signal can be derived from a sinusoidal signal.

$w(t) = w_0 \cos \omega_0 t$. This suggests that any periodic signal can be described by a harmonic series (because the output of a sinusoidally excited static system can be described by a harmonic series). For example, the triangular wave $x(t)$ produced by the system of Fig. 4-1 can be described by a harmonic series because the triangular wave is the output of a sinusoidally excited static element; thus

$$x(t) = A_0 \cos \theta_0 + A_1 \cos(\omega_0 t + \theta_1) + A_2 \cos(2\omega_0 t + \theta_2) + \cdots$$

where the peak amplitudes A_n and the initial phases θ_n ($n = 0, 1, 2, \ldots$) can be determined using methods described in Sec. 3-2. Similarly, the square wave $y(t)$ produced by the system of Fig. 4-1 also can be described by a harmonic series; thus

$$y(t) = B_0 \cos \psi_0 + B_1 \cos(\omega_0 t + \psi_1) + B_2 \cos(2\omega_0 t + \psi_2) + \cdots$$

where $B_n = K_1 n \omega_0 A_n$ and $\psi_n = \theta_n + \pi/2$. Finally, the sawtooth wave $z(t)$ produced by the system of Fig. 4-1 can be described by a harmonic series because the product $z(t) = K_2 x(t) y(t)$ consists of terms of the form

$$\cos(n\omega_0 t + \theta_n) \cos(m\omega_0 t + \psi_m) = 0.5 \cos[(n + m)\omega_0 t + \theta_n + \psi_m]$$
$$+ 0.5 \cos[(n - m)\omega_0 t + \theta_n - \psi_m]$$

These terms are harmonics of $\cos \omega_0 t$ because n and m are integers. Other systems can be devised for obtaining other periodic waveforms from a sinusoidal waveform (see Prob. 4-1). This does not prove that *any* periodic waveform can be so obtained, but at least it makes that claim plausible.

4-1A Fourier Coefficients

We assume† that any realistic periodic signal $x(t)$ can be described by a harmonic series expressed in exponential form as

$$x(t) = \sum_{n=-\infty}^{\infty} X_n e^{jn\omega_0 t} \qquad (4\text{-}1a)$$

We wish to show how the fundamental frequency ω_0 and the complex amplitudes $X_0, X_{\pm 1}, X_{\pm 2}, \ldots$ are determined from the waveform of a signal $x(t)$ described by (4-1a). Subsequently, the signal can be expressed (if necessary) in amplitude-phase form as

$$x(t) = \sum_{n=0}^{\infty} A_n \cos(n\omega_0 t + \theta_n) \qquad (4\text{-}1b)$$

where
$$A_n = \begin{cases} |X_0| & n = 0 \\ 2|X_n| & n \neq 0 \end{cases} \qquad \theta_n = \angle X_n \qquad (4\text{-}2)$$

Since the period of the series must equal the period of the signal represented by the series, the *fundamental frequency* of the harmonic series describing a signal $x(t)$ is

†Proof is omitted because it is largely irrelevant to practical application of Fourier series; see Churchill, chaps. 4 and 5.

given by

$$f_0 = \frac{1}{T} \quad (4\text{-}3)$$

where T is the period of the signal.

We obtain a formula for the complex amplitudes of the series by the following trick. We multiply (4-1a) by $e^{-jk\omega_0 t}$ and integrate the product over one period (from $t = t_0$ to $t = t_0 + T$). This gives

$$\int_{t_0}^{t_0+T} x(t) e^{-jk\omega_0 t} \, dt = \int_{t_0}^{t_0+T} \sum_{n=-\infty}^{\infty} X_n e^{j(n-k)\omega_0 t} \, dt$$

where $\omega_0 = 2\pi/T$. Reversing the order of the integration and the summation on the right side gives

$$\int_{t_0}^{t_0+T} x(t) e^{-jk\omega_0 t} \, dt = \sum_{n=-\infty}^{\infty} X_n \int_{t_0}^{t_0+T} e^{j(n-k)\omega_0 t} \, dt$$

But (see Prob. 4-2)

$$\int_{t_0}^{t_0+T} e^{j(n-k)\omega_0 t} \, dt = \begin{cases} T & n = k \\ 0 & n \neq k \end{cases} \quad (4\text{-}4)$$

Consequently, the only nonzero term in the sum above is the one for $n = k$, which equals TX_k. It follows that

$$X_k = \frac{1}{T} \int_{t_0}^{t_0+T} x(t) e^{-jk\omega_0 t} \, dt \quad k = 0, \pm 1, \pm 2, \ldots \quad (4\text{-}5)$$

Equations (4-3) and (4-5) show how the fundamental frequency and the complex amplitudes of the harmonic series describing a periodic signal $x(t)$ are obtained from the waveform of the signal. Complex amplitudes obtained from (4-5) for a signal $x(t)$ are called *Fourier coefficients* for the signal $x(t)$. A harmonic series obtained from (4-3) and (4-5) is called a *Fourier series*. Equation (4-1a), where ω_0 is given by (4-3) and X_k is given by (4-5), gives a Fourier series for $x(t)$ in exponential form. Equation (4-1b), where A_k and θ_k are obtained from X_k using (4-2), is a Fourier series for $x(t)$ in amplitude-phase form.

From (3-32), the complex spectrum of a signal $x(t)$ described by the exponential Fourier series of (4-1a) is given by

$$C_x(f) = \begin{cases} X_k & f = kf_0 \\ 0 & f \neq kf_0 \end{cases} \quad k = 0, \pm 1, \pm 2, \ldots$$

Fourier's coefficients for a signal constitute the complex spectrum of the signal. Consequently, (4-3) and (4-5) provide a path from the time-domain description (the waveform) of a periodic signal to the frequency-domain description (the spectrum) of the signal and (4-1) provides a path from the frequency domain to the time domain.

Example 4-1 In this example we obtain the Fourier series for the periodic signal (the rectangular pulse train) $x(t)$ of Fig. 4-2.

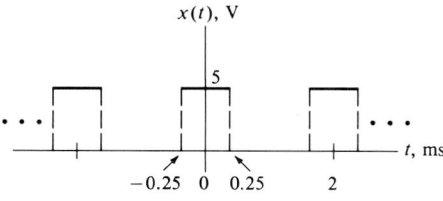

Figure 4-2 Rectangular pulse train of Example 4-1.

The period of the signal $x(t)$ is $T = 2$ ms. From (4-3), the fundamental frequency of the Fourier series for $x(t)$ is

$$f_0 = \frac{1}{0.002} = 500 \text{ Hz}$$

The Fourier coefficients for $x(t)$ are given by (4-5), where $T = 2$ ms, $\omega_0 = \pi$ krad/s, and t_0 is arbitrary (see Prob. 4-3). A convenient value for t_0 is† $t_0 = -T/2 = -1$ ms, in which case $t_0 + T = 1$ ms. From (4-5) and Fig. 4-2,

$$X_k = \frac{1}{T} \int_{-T/2}^{T/2} x(t) e^{-jk\omega_0 t} \, dt \quad \text{where} \quad x(t) = \begin{cases} 0 & -T/2 < t \leq -\tau/2 \\ x_0 & -\tau/2 < t \leq \tau/2 \\ 0 & \tau/2 < t \leq T/2 \end{cases}$$

with $x_0 = 5$ V and $\tau = 500$ μs; thus

$$X_k = \frac{1}{T} \int_{-\tau/2}^{\tau/2} x_0 e^{-jk\omega_0 t} \, dt$$

Performing the integration yields

$$X_k = \begin{cases} \dfrac{x_0 \tau}{T} & k = 0 \\ \dfrac{x_0(e^{-jk\omega_0 \tau/2} - e^{jk\omega_0 \tau/2})}{-jk\omega_0 T} = \dfrac{2x_0}{T} \dfrac{\sin(k\omega_0 \tau/2)}{k\omega_0} & k \neq 0 \end{cases}$$

This can be written compactly as

$$X_k = \frac{x_0 \tau}{T} \text{sa} \frac{k\omega_0 \tau}{2} = \frac{x_0 \tau}{T} \text{sa} \frac{k\pi\tau}{T}$$

where sa α denotes the *sampling function*, defined by

$$\text{sa } \alpha = \begin{cases} 1 & \alpha = 0 \\ \dfrac{\sin \alpha}{\alpha} & \alpha \neq 0 \end{cases} \quad (4\text{-}6)$$

The exponential form of the Fourier series for the rectangular pulse train of Fig. 4-2 is

$$x(t) = \sum_{n=-\infty}^{\infty} \frac{x_0 \tau}{T} \text{sa} \frac{n\pi\tau}{T} e^{jn\omega_0 t}$$

where $x_0 = 5$ V, $T = 2$ ms, $\tau = 500$ μs, and $\omega_0 = \pi$ krad/s. The peak amplitudes and the initial phases for the associated amplitude-phase series can be obtained using (4-2).

†Although t_0 is arbitrary, the integration in (4-5) is usually easier for some values of t_0 than others (see Prob. 4-4).

Table 4-1 Fourier coefficients for the rectangular pulse train of Fig. 4-2

n	f, kHz	X_n, V	A_n, V	θ_n, rad
0	0.00	1.25	1.25	0.00
1	0.50	1.13	2.25	0.00
2	1.00	0.80	1.59	0.00
3	1.50	0.38	0.75	0.00
4	2.00	0.00	0.00	0.00
5	2.50	−0.23	0.45	3.14
6	3.00	−0.27	0.53	3.14
7	3.50	−0.16	0.32	3.14
8	4.00	0.00	0.00	0.00

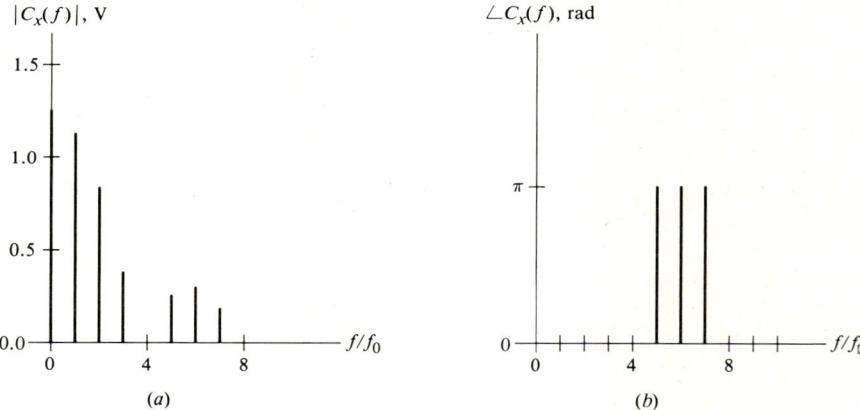

Figure 4-3 Spectrum of the rectangular pulse train of Fig. 4-2.

Table 4-1 gives the complex amplitudes for the exponential series and the peak amplitudes and initial phases for the amplitude-phase series for $n = 0, 1, \ldots, 8$. Figure 4-3 shows a graph of the two-sided spectrum of the rectangular pulse train $x(t)$.

Using (4-1b) and Table 4-1, we can write the first few terms of the series as

$$x(t) = 1.25 + 2.25 \cos \omega_0 t + 1.59 \cos 2\omega_0 t + 0.75 \cos 3\omega_0 t$$
$$+ 0.45 \cos(5\omega_0 t + \pi) + 0.53 \cos(6\omega_0 t + \pi) + \cdots \quad \text{V}$$

where $\omega_0 = 3.14$ krad/s.

Appendix B gives the Fourier coefficients for several periodic waveforms. Using Appendix B (or a similar table of Fourier coefficients) avoids the tedious integration required by (4-5).

We assume above that a Fourier series can describe any signal (any measurable function of time). This is equivalent to assuming that a Fourier series converges to the associated signal waveform at every instant. In Sec. 4-1B we give conditions under which this is actually true.

4-1B Convergence of a Fourier Series

It can be shown that a Fourier series for a periodic signal $x(t)$ converges to (is equal to) the signal at every instant if the signal satisfies the following conditions:†

1. The signal is absolutely integrable over one period; i.e., for a signal $x(t)$ having period T,

$$\int_0^T |x(t)|\, dt < \infty$$

2. The signal has a finite number of extrema (maxima and minima) in one period.
3. The signal is continuous.

We can forget about conditions 1 and 2 because they are met by any realistic signal model; however, several useful signal models, e.g., a rectangular pulse train, fail to meet condition 3. We wish to describe convergence of a Fourier series for a signal having one or more jump discontinuities.

A Fourier series converges to the associated signal wherever the signal is continuous. In the immediate neighborhood of a jump discontinuity the series converges to values different from those of the signal, as follows. Let $\hat{x}(t)$ denote a Fourier series for a signal $x(t)$; then

$$\hat{x}(t) = \sum_{n=-\infty}^{\infty} X_n e^{jn\omega_0 t}$$

where the coefficients $X_0, X_{\pm 1}, X_{\pm 2}, \ldots$ are given by (4-5). Let $x(t)$ have a jump discontinuity for $t = t_1$, so that

$$x(t_1^-) \neq x(t_1^+)$$

where $x(t_1^-)$ and $x(t_1^+)$ denote the left- and right-hand limits

$$x(t_1^-) = \lim_{t \to t_1} x(t) \qquad x(t_1^+) = \lim_{t_1 \leftarrow t} x(t)$$

For t near t_1, the Fourier series $\hat{x}(t)$ for the signal $x(t)$ converges to the following values (Stackgold, pp. 134–136):

$$\hat{x}(t_1^-) \approx x(t_1^-) + 0.09[x(t_1^-) - x(t_1^+)] \qquad (4\text{-}7a)$$

$$\hat{x}(t_1) = 0.50[x(t_1^-) + x(t_1^+)] \qquad (4\text{-}7b)$$

$$\hat{x}(t_1^+) \approx x(t_1^+) + 0.09[x(t_1^+) - x(t_1^-)] \qquad (4\text{-}7c)$$

†These are sufficient conditions for convergence of a Fourier series. The necessary conditions are unknown. (See Stackgold, pp. 128–137.)

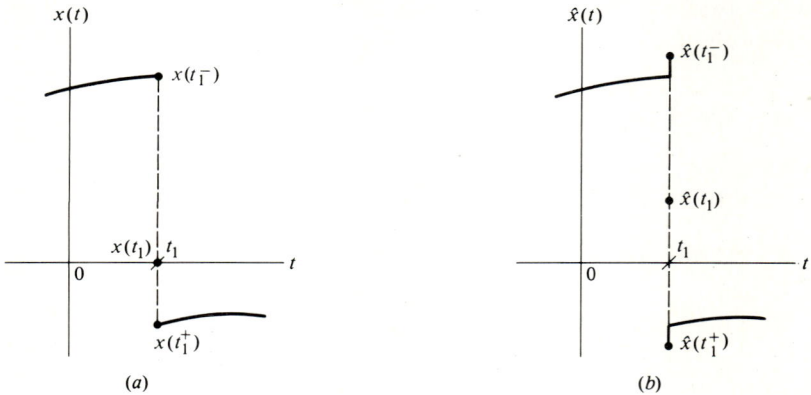

Figure 4-4 Convergence of Fourier's series for a signal near a jump discontinuity of the signal.

Figure 4-4 illustrates these relations. Figure 4-4a shows a signal $x(t)$ having a jump discontinuity at $t = t_1$. Figure 4-4b shows a Fourier series for this signal. For $t = t_1^-$, the series overshoots the signal by about 9 percent of the jump, in agreement with (4-7a). For $t = t_1$, the series equals the average value of $x(t_1^+)$ and $x(t_1^-)$, in agreement with (4-7b). For $t = t_1^+$, the series undershoots the signal by about 9 percent of the jump, in agreement with (4-7c).

The differences expressed by (4-7) between a signal and its Fourier series are insignificant. No physical system can distinguish between two signals that differ only over infinitesimal intervals. For all practical purposes, a signal and its Fourier series are identical at every instant, even if the signal is discontinuous.

Failure of a Fourier series for a discontinuous signal to converge to the signal near a jump discontinuity is reflected by a finite (truncated) Fourier series of the form

$$\hat{x}(t) = \sum_{k=-K}^{K} X_k e^{jk\omega_0 t}$$

Figure 4-5 shows graphs of two finite Fourier series ($K = 7$ and $K = 50$) for the rectangular pulse train of Fig. 4-2. Because the series approaches the values given by (4-7), it overshoots the signal by about 9 percent of x_0 on the high sides of the discontinuities and undershoots the signal by about 9 percent of x_0 on the low sides of the discontinuities. Increasing K from 7 to 50 crowds the overshoots and undershoots closer to the discontinuities and improves convergence at points distant from the discontinuities but does not decrease (by much) the amplitudes of the overshoots and undershoots. This oscillatory behavior of a finite Fourier series for a discontinuous signal in a neighborhood of a discontinuity is called *Gibbs' phenomenon*.† The oscillations themselves are called *Gibbs' oscillations*. For theoretical work, where we can use infinite Fourier series to describe signals (or functions), Gibbs' phenomenon is insignificant; but Gibbs' oscillations can be troublesome when a discontinuous signal is approximated by a finite Fourier series. In such cases it may be necessary to eliminate

---

†After Josiah Willard Gibbs (1839–1903), an American mathematician who contributed extensively to thermodynamics, chemistry, and statistical mechanics.

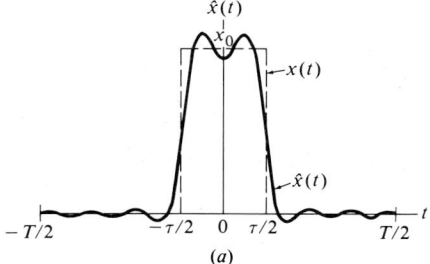
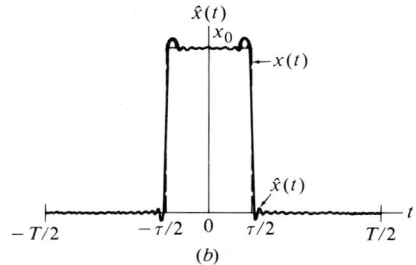

Figure 4-5 Partial sums of Fourier's series for the rectangular pulse train of Fig. 4-2 including harmonics through (a) the seventh and (b) the fiftieth; $\tau = 500\ \mu s$, $T = 2$ ms, and $x_0 = 5$ V.

(or decrease) the oscillations by using steep but continuous curves to replace jump discontinuities of a signal (or function) to be approximated by a finite Fourier series.

4-1C Applications in System Analysis

The examples below illustrate application of Fourier series to finding outputs of linear stationary systems and nonlinear static systems. The methods are essentially identical to those described in Chap. 3. The only new procedure is using Fourier's series (instead of polynomial approximation, for example) to obtain harmonic series describing inputs and outputs.

Example 4-2 Figure 4-6 shows a system and its input $x(t)$. Obtain a harmonic series describing the output $y(t)$. Express the output as a harmonic series in amplitude-phase form.

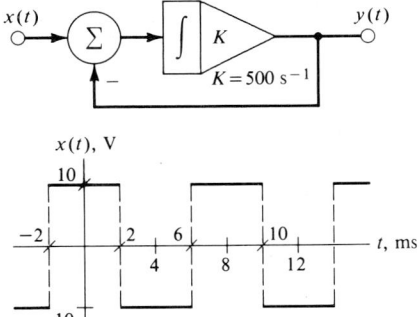

Figure 4-6 System and input of Example 4-2.

SOLUTION The system is linear and stationary. We can obtain the output $y(t)$ by expressing the input as a harmonic series, obtaining the transfer function of the system, and using superposition.

The period of the input is $T = 8$ ms. A harmonic (Fourier) series describing the input is

$$x(t) = \sum_{n=-\infty}^{\infty} X_n e^{jn\omega_0 t}$$

From (4-3), the fundamental frequency of the series is

$$f_0 = \frac{1}{T} = \frac{1}{0.008} = 125 \text{ Hz}$$

From Appendix B, the Fourier coefficients for the input (square wave) are given by

$$X_n = \begin{cases} 0 & n = 0 \\ x_0 \text{ sa } \dfrac{n\pi}{2} & n = \pm 1, \pm 2, \pm 3, \ldots \end{cases}$$

where $x_0 = 10$ V. Using block-diagram reduction to obtain the transfer function of the system gives

$$H(j\omega) = \frac{K}{K + j\omega}$$

where $K = 500$ s^{-1}. By superposition, the output is given by

$$y(t) = \sum_{n=-\infty}^{\infty} Y_n e^{jn\omega_0 t} \qquad \text{where } Y_n = H(jn\omega_0)X_n$$

The output is expressed in amplitude-phase form as

$$y(t) = \sum_{n=0}^{\infty} B_n \cos(n\omega_0 t + \psi_n) \qquad \text{where } B_n = \begin{cases} |Y_0| & n = 0 \\ 2|Y_n| & n \neq 0 \end{cases} \qquad \psi_n = \angle Y_n$$

Table 4-2 gives Fourier coefficients for the input, values of the transfer function $H(j\omega)$, and Fourier coefficients for the output for $n = 1, 2, \ldots, 13$. Referring to Table 4-2 and using the relations above for B_n and ψ_n, we can write the Fourier series for the output as

Table 4-2 Quantities calculated in Example 4-2
All even-numbered amplitudes for the input equal zero

n	X_n, V	$H(jn\omega_0)$, V/V	Y_n, V
1	6.37	$0.54e^{-1.00j}$	$3.42e^{-1.00j}$
3	−2.12	$0.21e^{-1.36j}$	$0.44e^{1.78j}$
5	1.27	$0.13e^{-1.44j}$	$0.16e^{-1.44j}$
7	−0.91	$0.09e^{-1.48j}$	$0.08e^{1.66j}$
9	0.71	$0.07e^{-1.50j}$	$0.05e^{-1.50j}$
11	−0.58	$0.06e^{-1.51j}$	$0.03e^{1.63j}$
13	0.49	$0.05e^{-1.52j}$	$0.02e^{-1.52j}$

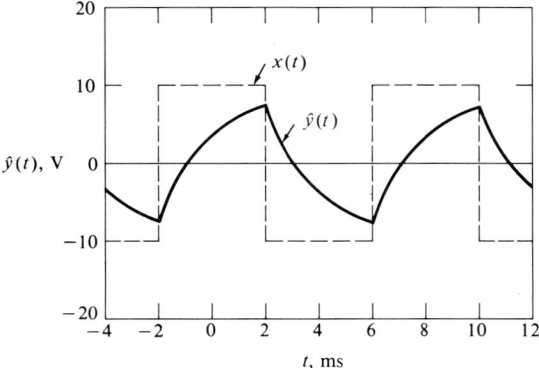

Figure 4-7 Partial sum of Fourier's series for the output of the system of Example 4-2.

$$y(t) = 6.84 \cos(\omega_0 t - 1.00) + 0.88 \cos(3\omega_0 t + 1.78) + 0.32 \cos(5\omega_0 t - 1.44)$$
$$+ 0.16 \cos(7\omega_0 t + 1.66) + 0.10 \cos(9\omega_0 t - 1.50)$$
$$+ 0.06 \cos(11\omega_0 t + 1.63) + 0.04 \cos(13\omega_0 t - 1.52) + \cdots \quad \text{V}$$

where $\omega_0 = 2\pi f_0 = 785$ rad/s. Figure 4-7 shows a graph of the partial sum (finite Fourier series)

$$\hat{y}(t) = \sum_{n=0}^{13} B_n \cos(n\omega_0 t + \psi_n)$$

which includes harmonics through the thirteenth.

Example 4-3 Obtain the output of the system of Fig. 4-8 for input

$$x(t) = A \cos \omega_0 t$$

where $A = 25$ V and $f_0 = 60$ Hz. Express the output in amplitude-phase form.

SOLUTION The output of the full-wave rectifier is given by

$$y(t) = B|\cos \omega_0 t|$$

where $B = y_0|A/x_0| = 100$ mA. Figure 4-9 shows a graph of the signal $y(t)$. The fundamental frequency of $y(t)$ is $f_0 = 60$ Hz.† From Appendix B, Fourier's coefficients for the signal (the full-wave-rectified sinusoid) $y(t)$ are given by

†We could also use a series having fundamental frequency $2\omega_0$ because $y(t)$ contains only even harmonics of $\cos \omega_0 t$; see Prob. 4-22.

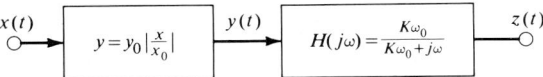

Figure 4-8 System of Example 4-3; $x_0 = 500$ mV, $y_0 = 2$ mA, $\omega_0 = 120\pi$ rad/s, and $K = 0.10$.

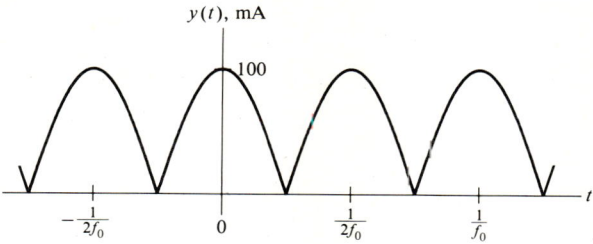

Figure 4-9 Rectified sinusoid $y(t)$ in the system of Fig. 4-8.

$$Y_n = \tfrac{1}{4}B[1 + (-1)^n]\left[\operatorname{sa}\frac{(n+1)\pi}{2} + \operatorname{sa}\frac{(n-1)\pi}{2}\right]$$

Table 4-3 gives the first few Fourier coefficients (complex amplitudes) and associated peak amplitudes and initial phases for $y(t)$. The amplitude-phase form of Fourier's series for $y(t)$ is

$$y(t) = 63.66 + 42.44 \cos 2\omega_0 t + 8.49 \cos(4\omega_0 t + \pi) + \cdots \quad \text{mA}$$

Having expressed $y(t)$ as a sum of sinusoids, we can use superposition to determine $z(t)$. This gives

$$z(t) = \sum_{n=-\infty}^{\infty} Z_n e^{jn\omega_0 t} = \sum_{n=0}^{\infty} D_n \cos(n\omega_0 t + \lambda_n)$$

where
$$Z_n = H(jn\omega_0)Y_n = \frac{0.10\omega_0}{0.10\omega_0 + jn\omega_0}Y_n = \frac{0.10}{0.10 + jn}Y_n$$

Table 4-4 gives complex amplitudes, peak amplitudes, and initial phases for the first few terms of Fourier's series for $z(t)$. The series can be expressed in amplitude-phase form as

$$z(t) = 63.66 + 2.12 \cos(2\omega_0 t - 1.52) + 0.21 \cos(4\omega_0 t + 1.60) + \cdots \quad \text{mA}$$

Figure 4-10 shows a graph of $z(t)$ obtained from the partial sum (finite Fourier series)

$$z(t) = \sum_{n=0}^{8} D_n \cos(n\omega_0 t + \lambda_n)$$

which includes harmonics through the eighth.

Table 4-3 Fourier coefficients for the full-wave rectified sinusoid $y(t)$ of Fig. 4-9

All odd-numbered amplitudes equal zero

n	Y_n, mA	B_n, mA	ψ_n, rad
0	63.66	63.66	0.00
2	21.22	42.44	0.00
4	−4.24	8.49	3.14

Table 4-4 Fourier coefficients for the output $z(t)$ of the system of Fig. 4-8

All odd-numbered amplitudes equal zero

n	Z_n, mA	D_n, mA	λ_n, rad
0	63.66	63.66	0.00
2	1.06	2.12	−1.52
4	−0.11	0.21	1.60

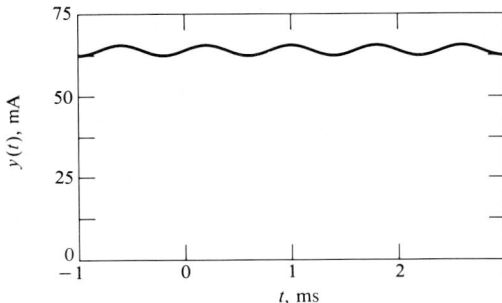

Figure 4-10 Partial sum of Fourier's series for the output $z(t)$ of the system of Fig. 4-8.

Example 4-4 Figure 4-11 shows the transfer characteristic for a certain nonlinear amplifier, where x denotes instantaneous amplitude of an input and y denotes instantaneous amplitude of an output. Calculate the third-harmonic distortion (percent) introduced by the amplifier for input

$$x(t) = A \cos \omega_0 t$$

where $A = 2.20$ mV and $f_0 = 1$ kHz.

SOLUTION The output of the amplifier is the clipped sinusoid $y(t)$ of Fig. 4-12. From Appendix B, the complex amplitudes of the first and third harmonics of $y(t)$ are given by

$$Y_1 = \frac{b}{2} - \frac{b\tau}{T}\left(\operatorname{sa}\frac{2\pi\tau}{T} + 1\right) + \frac{2a\tau}{T}\operatorname{sa}\frac{\pi\tau}{T}$$

$$Y_3 = \frac{2a\tau}{T}\operatorname{sa}\frac{3\pi\tau}{T} - \frac{b\tau}{T}\left(\operatorname{sa}\frac{4\pi\tau}{T} + \operatorname{sa}\frac{2\pi\tau}{T}\right)$$

where $\dfrac{\tau}{T} = \dfrac{1}{\pi}\cos^{-1}\dfrac{a}{b}$

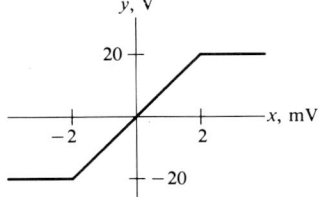

Figure 4-11 Transfer characteristic of the nonlinear amplifier of Example 4-4.

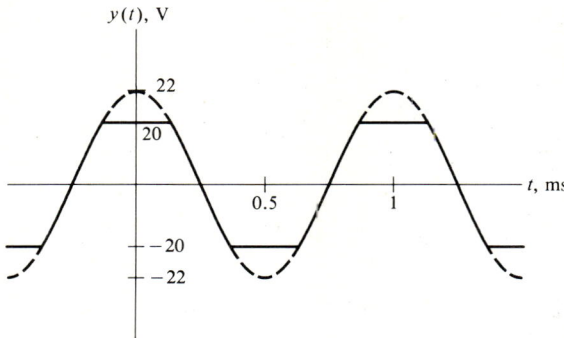

Figure 4-12 Output of the nonlinear amplifier of Fig. 4-11 for a sinusoidal input.

From Figs. 4-11 and 4-12, $a = 20$ V and $b = 22$ V; thus

$$\frac{\tau}{T} = \frac{1}{\pi} \cos^{-1} \frac{20}{22} = 0.14$$

whence $Y_1 = 11 - (3.01)(1.88) + (5.47)(0.97) = 10.65$
$Y_3 = (5.47)(0.75) - (3.01)(0.58 + 0.88) = -0.29$
The third-harmonic distortion is

$$D_3 = \frac{100|Y_3|}{|Y_1|} = 2.92\%$$

In Examples 4-3 and 4-4 harmonic series representing outputs of nonlinear elements are obtained directly from the waveforms of the outputs using the table of Fourier coefficients in Appendix B. In these examples, at least, this table-lookup operation is a welcome alternative to the polynomial-approximation procedure described in Sec. 3-2 because high-degree polynomials would be needed to describe the transfer characteristics with sufficient accuracy. On the other hand, using polynomial approximation is often easier (for pencil-and-paper calculations) than using (4-5). For example, consider a system having transfer characteristic

$$y = y_0 \tanh \frac{x}{x_0}$$

where $x = x(t)$ is an input and $y = y(t)$ the corresponding output. For this system it is much easier to calculate third-harmonic distortion (for a small input) using polynomial approximation than using (4-5) because the integral

$$Y_n = \frac{1}{T} \int_{t_0}^{t_0+T} y_0 \tanh\left(\frac{A}{x_0} \cos \omega_0 t\right) e^{-jn\omega_0 t} \, dt$$

cannot be expressed in simple form. As a rule (for pencil-and-paper calculations), using polynomial approximation is appropriate for small-signal calculations involving reasonably smooth transfer characteristics and using Fourier coefficients is appropriate for large-signal calculations involving piecewise-linear transfer characteristics. Other cases, e.g., large-signal calculations involving smooth transfer characteristics, usually

4-1D Interpretation of a Fourier Series

A Fourier series expresses a signal (a waveform) as a sum of sinusoids. The Fourier coefficients for a signal constitute the complex spectrum of the signal, which shows the relative importance of each sinusoidal component of the signal. Where the complex spectrum of a signal is relatively large, the associated sinusoidal components contribute greatly to the waveform of the signal. Where the complex spectrum of a signal is relatively small, the associated sinusoidal components contribute only slightly to the waveform of the signal. In brief, the waveform of a signal is determined predominantly by the larger sinusoidal components of the signal. Figure 4-13 illustrates this for a triangular pulse train $x(t)$. Figure 4-13a shows a graph of the waveform $x(t)$, and Fig. 4-13b shows the complex spectrum (Fourier coefficients) $C_x(f)$ of $x(t)$. Figures 4-13c and d show waveforms given by the finite (truncated) Fourier series

$$\hat{x}(t) = \sum_{n=-K}^{K} X_n e^{jn\omega_0 t} = \sum_{n=-K}^{K} C_x(nf_0) e^{jn\omega_0 t}$$

obtained by discarding components lying above Kf_0. For $K = 32$, the waveform of the truncated Fourier series is indistinguishable from that of the actual signal (on the scale used in Fig. 4-13).

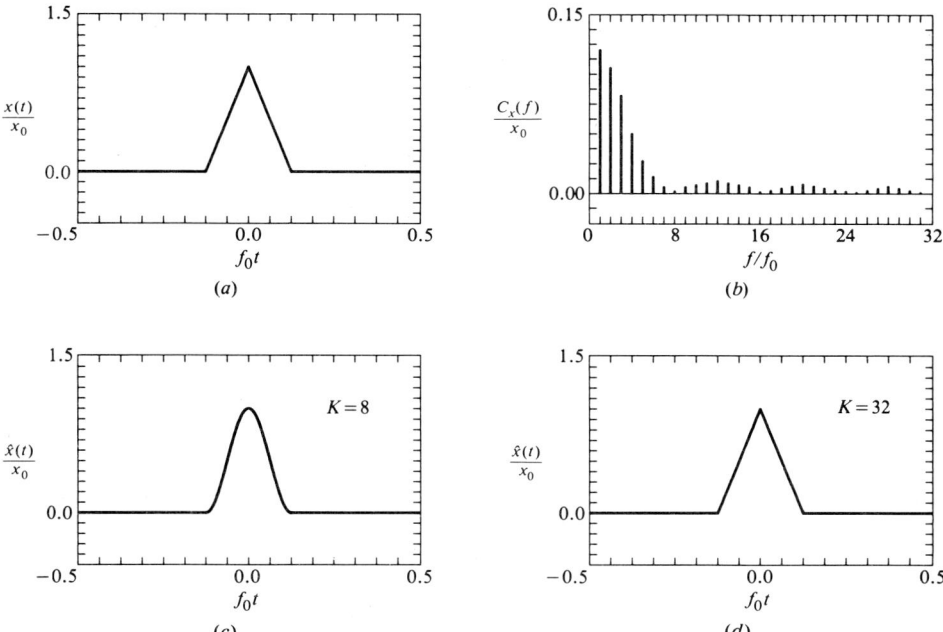

Figure 4-13 Finite Fourier series for a triangular wave. See text.

In almost every practical application the significant components of a signal occupy a finite band on the frequency axis. The physical reason is that every signal we observe is transmitted to us through some system or medium having finite passband. The mathematical reason is that a Fourier series converges for any realistic signal model, which implies that the Fourier coefficients decrease with increasing harmonic number. The width of the band on the positive-frequency axis for which the spectrum of a signal is significantly different from zero is called the *bandwidth* of the signal. For example, we might say that the bandwidth of the triangular pulse train of Fig. 4-13 is $32f_0$, where f_0 is the fundamental frequency of the signal. Bandwidth is the most important frequency-domain property of a signal. The definition and significance of bandwidth are discussed more fully in subsequent sections.

4-2 FOURIER INTEGRAL

Above, we show how a periodic signal can be described by a Fourier series (a harmonic series). This extends the frequency-domain methods of Chap. 3 to periodically excited systems. In this section we describe the Fourier integral, which represents a nonperiodic signal as a superposition of sinusoids. This extends frequency-domain methods to a wide variety of practical problems. To introduce our development of the Fourier integral we show how a Fourier series can be used to describe (approximately) a nonperiodic signal.

4-2A Fourier Series for a Nonperiodic Signal

A Fourier series is periodic. The period of a Fourier series is $1/f_0$, where f_0 is the fundamental frequency of the series. Conversely, the fundamental frequency of a Fourier series is $f_0 = 1/T$, where T is the duration of the interval $t_0 \leq t < t_0 + T$ in the formula

$$X_n = \frac{1}{T} \int_{t_0}^{t_0+T} x(t) e^{-jn\omega_0 t} \, dt$$

whereby the coefficients of the series are determined from the waveform of a signal described by the series. The series describes the signal $x(t)$ over that interval and describes the periodic extension of the signal outside that interval. If the signal is periodic, and if T is an integral multiple of the period of $x(t)$, the series describes the signal for all time. If $x(t)$ is nonperiodic, the series describes the signal only over the interval $t_0 \leq t < t_0 + T$. See Fig. 4-14, where the dashed curve represents a nonperiodic signal $x(t)$ and the solid curve represents a Fourier series whose coefficients are obtained from $x(t)$ using the formula above. For $t_0 \leq t < t_0 + T$, the series and the signal are identical. Outside that interval, the series and the signal are different because the series is periodic and the signal is not.

The starting point t_0 and the duration T of the interval of integration in the formula for the Fourier coefficients must be finite, but otherwise they are arbitrary. In any practical application we can always choose t_0 early enough and T large enough to make the periodicity of the associated series inconsequential, i.e., such that the series describes a signal for all times of interest.

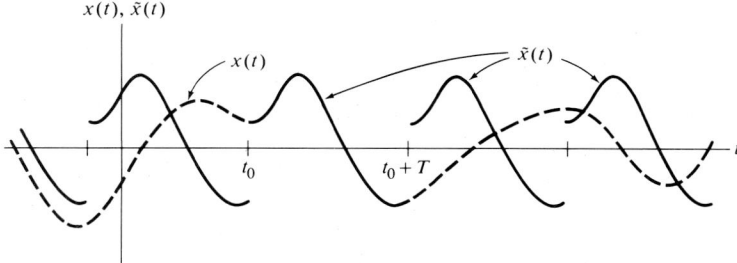

Figure 4-14 Representing a nonperiodic signal $x(t)$ over an interval $t_0 \leq t < t_0 + T$ by a Fourier series $\tilde{x}(t)$.

Example 4-5 We use a Fourier series to obtain the response of a system to a single rectangular pulse (a nonperiodic signal). The transfer function of the system is

$$H(j\omega) = \frac{1}{1 + j\omega\tau}$$

where $\tau = 1$ ms. This is the transfer function of a low-pass first-order RC circuit having time constant $\tau = RC$. The input is a (single) rectangular pulse described by

$$x(t) = x_0 r\left(\frac{t + \tau/2}{\tau}\right)$$

where we have centered the pulse on $t = 0$ to simplify the expression for the Fourier coefficients for $x(t)$.

To determine the response $y(t)$ to the single pulse $x(t)$ we determine the response $\tilde{y}(t)$ to a periodic pulse train $\tilde{x}(t)$, where each pulse is identical to $x(t)$ and where consecutive pulses are far enough apart for the system to respond to any one pulse as if that pulse were the only one present. This means that the period T of the pulse train $\tilde{x}(t)$ must be large enough to give the system time to relax completely between consecutive pulses; i.e., the period T must be much larger than the time constant τ of the system. Under this condition the nonperiodic output $y(t)$ for the nonperiodic input $x(t)$ is given (to a good approximation) by

$$y(t) = \begin{cases} \tilde{y}(t) & -\frac{\tau}{2} \leq t < -\frac{\tau}{2} + T \\ 0 & \text{all other } t \end{cases}$$

The necessary calculations are carried out as follows:

1. We obtain a Fourier series $\tilde{x}(t)$ describing the input $x(t)$ over an interval $-\tau/2 \leq t < -\tau/2 + T$; thus

$$\tilde{x}(t) = \sum_{n=-\infty}^{\infty} X_n e^{jn\omega_0 t}$$

where (from Appendix B) $X_n = (x_0 \tau/T) \, \mathrm{sa}(n\pi\tau/T)$.

2. We obtain the output $\tilde{y}(t)$ for input $\tilde{x}(t)$ above. This gives

$$\tilde{y}(t) = \sum_{n=-\infty}^{\infty} Y_n e^{jn\omega_0 t}$$

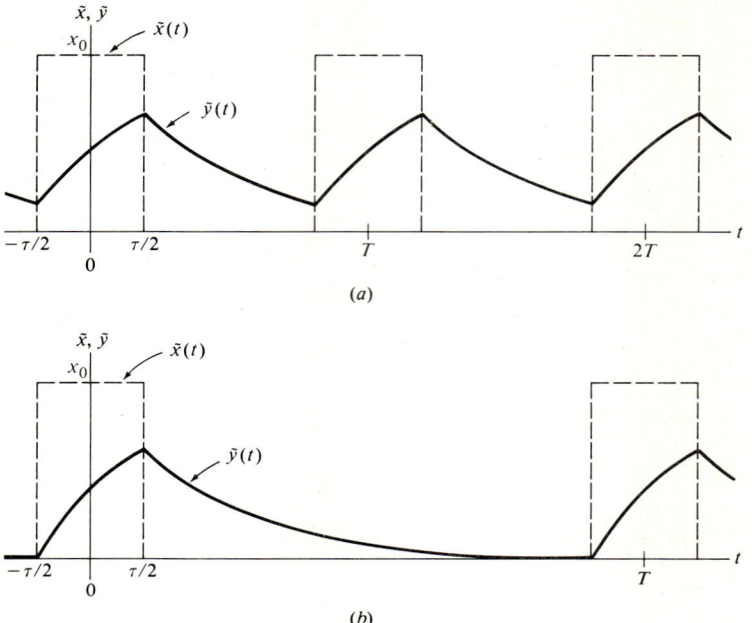

Figure 4-15 Using a Fourier series to obtain the output of a system for a nonperiodic input (see Example 4-5); (a) $T = 2.5\tau$; (b) $T = 5\tau$.

where

$$Y_n = H(jn\omega_0)X_n = \left(\frac{x_0\tau}{T}\right)\frac{\text{sa}(n\pi\tau/T)}{1 + jn\omega_0\tau}$$

3. We plot a graph of the periodic signal $\tilde{y}(t)$ using a finite number of terms from the Fourier series above.

Figure 4-15 shows graphs of $\tilde{y}(t)$ for two values of period T. For $T = 2.5\tau$ (Fig. 4-15a), the system does not have time to relax between consecutive pulses of the periodic input $\tilde{x}(t)$; that is, the system is still responding to the nth pulse when the $(n + 1)$th pulse comes along. In this case a single period of $\tilde{y}(t)$ is not a good approximation to the output for a single pulse. For $T = 5\tau$ (Fig. 4-15b), the system relaxes almost completely between consecutive pulses of $\tilde{x}(t)$; that is, it responds to each pulse as if that pulse were the only one present. Consequently, for $T = 5\tau$ a single cycle of $\tilde{y}(t)$ is a good approximation to the waveform of the response $y(t)$ to a single pulse.

This mathematical procedure of using the response to a periodic pulse train to determine the response to a single pulse is analogous to the experimental procedure of using a pulse generator and an oscilloscope to observe the response of an electric circuit to a single pulse. The pulse generator actually produces a periodic pulse train, and the oscilloscope actually displays the periodic response of the circuit to that pulse train. We obtain the waveform of the response to a single pulse by making the period of the pulse generator output much larger than the largest time constant of the system

FOURIER SERIES, FOURIER INTEGRAL, AND FOURIER TRANSFORMATION 193

(so that the system relaxes between pulses) and adjusting the trigger and sweep circuits of the oscilloscope such that consecutive periods of the output are superimposed on the screen of the oscilloscope (to give a steady, observable trace).

A logical extension of the mathematical procedure described above is to allow the period of a Fourier series to approach infinity, thereby obtaining a description of a signal for all time. This leads to the Fourier integral, as we show below.

4-2B Fourier Integral

The Fourier series for a signal $x(t)$ over $-T/2 \leq t < T/2$ is

$$x(t) = \sum_{n=-\infty}^{\infty} X_n e^{jn\omega_0 t} \quad (4\text{-}8)$$

where $f_0 = 1/T$ and

$$X_n = \frac{1}{T} \int_{-T/2}^{T/2} x(t) e^{-jn\omega_0 t} dt \quad (4\text{-}9)$$

Substituting the right side of (4-9) for X_n in (4-8) gives

$$x(t) = \sum_{n=-\infty}^{\infty} \left[\frac{1}{T} \int_{-T/2}^{T/2} x(t) e^{-jn\omega_0 t} dt \right] e^{jn\omega_0 t}$$

Substituting $\Delta\omega$ for $\omega_0 = 2\pi/T$ gives

$$x(t) = \frac{1}{2\pi} \sum_{n=-\infty}^{\infty} \left[\int_{-T/2}^{T/2} x(t) e^{-jn\Delta\omega t} dt \right] e^{jn\Delta\omega t} \Delta\omega$$

Finally, taking the limit as $T \to \infty$ (as $\Delta\omega \to 0$) yields

$$x(t) = \frac{1}{2\pi} \int_{-\infty}^{\infty} X(j\omega) e^{j\omega t} d\omega \quad (4\text{-}10)$$

where

$$X(j\omega) = \int_{-\infty}^{\infty} x(t) e^{-j\omega t} dt \quad (4\text{-}11)$$

The right side of (4-10), where $X(j\omega)$ is given by (4-11), is the *Fourier integral* for a signal $x(t)$. It is analogous to the Fourier series for a signal $x(t)$. The quantity $X(j\omega)$ defined by (4-11), called the *spectral density* of the signal $x(t)$, is analogous to the complex spectrum (consisting of Fourier coefficients) for a signal $x(t)$. It is important to note that the unit of the spectral density of a signal $x(t)$ is the unit of $tx(t)$ or $x(t)/f$; for example, if $x(t)$ is voltage in volts and t is in seconds, the unit of $X(j\omega)$ is volt-seconds or volts per hertz. This is why $X(j\omega)$ is called a spectral *density* (and not a spectrum).

Example 4-6 This example illustrates the limiting process used above to obtain the Fourier integral from the Fourier series for a particular signal (a rectangular pulse). Figure 4-16 shows a Fourier series and the associated complex spectrum (which happens to be real) for a rectangular pulse having duration τ. The series actually describes a periodic pulse train $\tilde{x}(t)$, which in turn describes $x(t)$ over an interval $-T/2 \leq t < T/2$. Note that doubling T

194 INTRODUCTION TO SYSTEM ANALYSIS

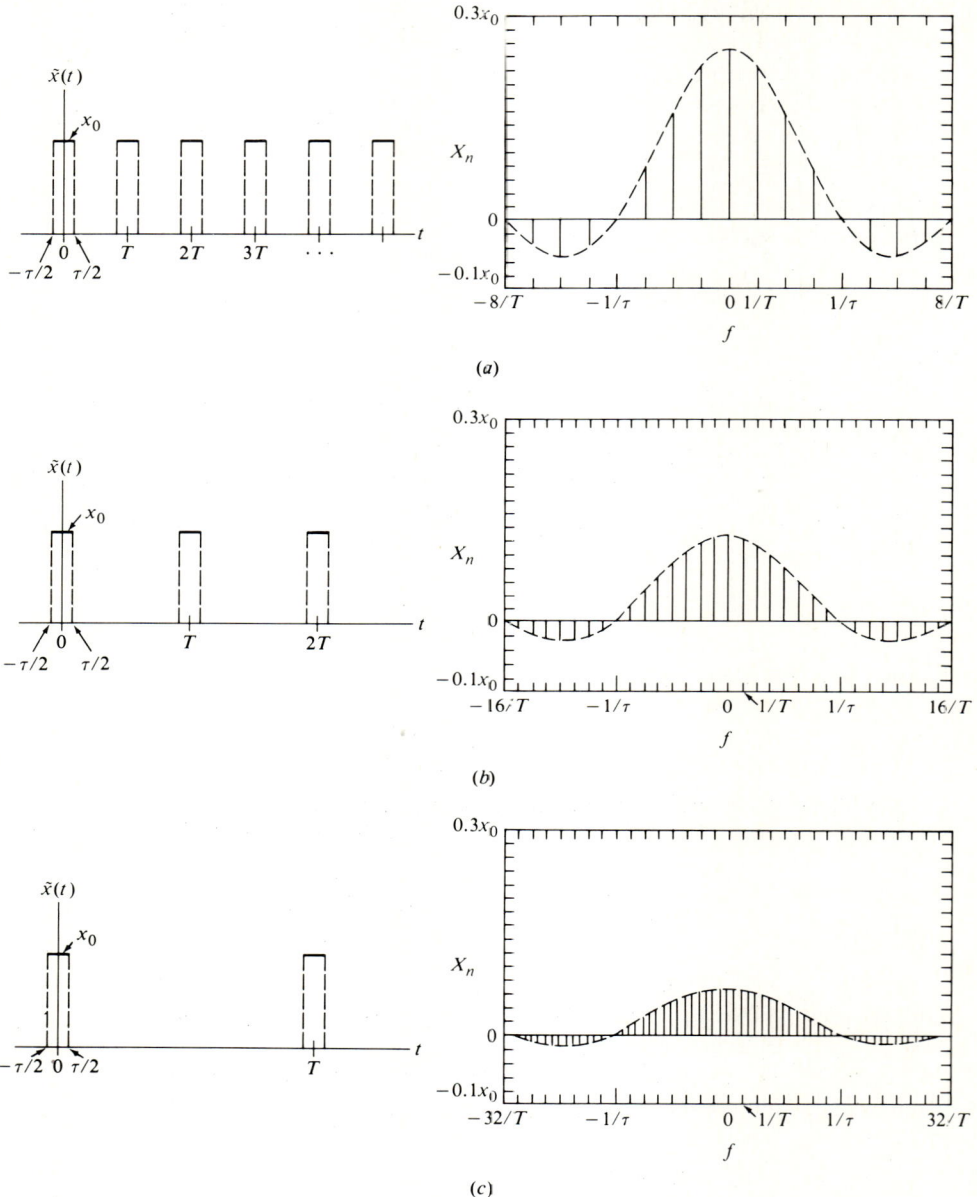

Figure 4-16 Spectrum of a rectangular pulse train having period T and pulse duration τ for (a) $T = 4\tau$, (b) $T = 8\tau$, and (c) $T = 16\tau$.

halves the separation (in frequency) of adjacent components and halves the peak amplitudes of all components. This suggests that as $T \to \infty$, both the separation and the amplitudes of the sinusoidal components of $\tilde{x}(t)$ become infinitesimal; i.e., as $T \to \infty$, $\Delta f = 1/T \to df$, $n\pi/T \to \omega/2$, and

$$X_n = \frac{x_0 \tau}{T} \operatorname{sa} \frac{n\pi\tau}{T} \longrightarrow x_0 \tau \operatorname{sa} \frac{\omega\tau}{2} df$$

This implies that the spectral density for

$$x(t) = x_0 r\left(\frac{t + \tau/2}{\tau}\right)$$

is

$$X(j\omega) = x_0 \tau \operatorname{sa} \frac{\omega\tau}{2}$$

This can be verified using (4-11). The spectral density of the pulse $x(t)$ above is

$$X(j\omega) = \int_{-\infty}^{\infty} x_0 r\left(\frac{t + \tau/2}{\tau}\right) e^{-j\omega t} dt = x_0 \int_{-\tau/2}^{\tau/2} e^{-j\omega t} dt = x_0 \tau \operatorname{sa} \frac{\omega\tau}{2}$$

4-2C Convergence of the Fourier Integral

The Fourier integral for a signal $x(t)$ converges to the signal at every instant if the following conditions are met:

1. The square of the signal is integrable; i.e.,

$$\int_{-\infty}^{\infty} x^2(t) dt < \infty$$

2. The signal has a finite number of extrema in any finite interval.
3. The signal is continuous.

Although condition 2 is met by any realistic signal model, many useful signal models fail to meet condition 1, e.g., a sinusoid, condition 3, e.g., a rectangular pulse, or both, e.g., a step. Some signals that fail to meet condition 1 can still be described by the Fourier integral if delta functions are allowed to appear in the associated spectral densities. We treat such signals in the sequel. Here we discuss convergence of the Fourier integral for a signal that meets conditions 1 and 2 but not 3.

It can be shown that the Fourier integral for a discontinuous signal converges to the signal where the signal is continuous and satisfies (4-7) in the neighborhood of a jump discontinuity. This means that the Fourier integral for a discontinuous signal is physically indistinguishable from the signal itself because no physical system can distinguish between two signals that differ only over a finite number of infinitesimal intervals. However, if a discontinuous signal $x(t)$ is approximated by a *finite Fourier integral* $\hat{x}(t)$ such that

$$\hat{x}(t) = \frac{1}{2\pi} \int_{-\omega_0}^{\omega_0} X(j\omega) e^{j\omega t} d\omega$$

then the approximation exhibits Gibbs' oscillations near each discontinuity. Sometimes Gibbs' oscillations are troublesome. Usually they can be eliminated by using steep but continuous curves to replace jump discontinuities of a signal before calculating its spectral density.

4-2D Amplitude-Phase Form of the Fourier Integral

We note in passing that the Fourier integral for a signal $x(t)$ can be written (see Prob. 4-33)

$$x(t) = \frac{1}{2\pi} \int_0^\infty a_x(\omega) \cos[\omega t + \theta_x(\omega)] \, d\omega \qquad (4\text{-}12a)$$

where $a_x(\omega) = \begin{cases} 0 & \omega < 0 \\ |X(j\omega)| & \omega = 0 \\ 2|X(j\omega)| & \omega > 0 \end{cases} \qquad \theta_x(\omega) = \begin{cases} 0 & \omega < 0 \\ \angle X(j\omega) & \omega \geq 0 \end{cases} \qquad (4\text{-}12b)$

Equation (4-12a) gives the *amplitude-phase* form of the Fourier integral for a signal $x(t)$. The quantity $a_x(\omega)$ is the *one-sided amplitude spectral density* for the signal $x(t)$ and the quantity $\theta_x(\omega)$ is the *one-sided phase spectrum* for the signal $x(t)$. The amplitude-phase form of the Fourier integral is used only rarely in practice and not at all in this book.

4-2E Interpretation of Spectral Density

A spectral density is analogous to a complex spectrum (to a set of Fourier coefficients), as can be seen by writing the Fourier integral for a signal $x(t)$ as

$$x(t) = \lim_{\Delta\omega \to 0} \sum_{n=-\infty}^{\infty} X(jn\,\Delta\omega) e^{jn\Delta\omega t} \frac{\Delta\omega}{2\pi}$$

If we ignore the limiting operation, this sum looks like a Fourier series having fundamental frequency $\Delta f = \Delta\omega/2\pi$ and coefficients (complex amplitudes)

$$X_n = X(jn\,2\pi\,\Delta f)\Delta f \qquad n = 0, \pm 1, \pm 2, \cdots$$

The quantity X_n is a complex amplitude. The quantity $X(jn\,2\pi\,\Delta f)\Delta f$ is an area bounded by the spectral density $X(jn\,2\pi\,\Delta f)$ and the frequency axis. An *area* bounded by a spectral density and the frequency axis can be interpreted as the complex amplitude of a component of the associated signal. The *value* of a spectral density *at a point* has no physical significance.

Loosely speaking, the Fourier integral expresses a signal as a sum of sinusoids. The spectral density of a signal shows the relative contributions of the sinusoidal components to the sum. Where the spectral density of a signal is relatively large, the associated sinusoidal components contribute greatly to the waveform of the signal. Where the spectral density of a signal is relatively small, the associated sinusoidal components contribute relatively little to the waveform of the signal.

Example 4-7 This example illustrates the above interpretation of spectral density as it applies to the spectral density of a constant (dc) signal.

From (4-11), the spectral density of a dc signal $x(t) = x_0$ is

$$X(j\omega) = x_0 \int_{-\infty}^{\infty} e^{-j\omega t} \, dt$$

The integral diverges. To obtain a useful expression for the spectral density $X(j\omega)$ we rewrite the relation in the form

$$X(j\omega) = x_0 \lim_{\tau \to \infty} \left(\int_{-\tau/2}^{\tau/2} e^{-j\omega t} dt \right)$$

Performing the indicated integration gives

$$X(j\omega) = x_0 \lim_{\tau \to \infty} \left(\tau \operatorname{sa} \frac{\omega \tau}{2} \right)$$

To interpret this expression for the spectral density $X(j\omega)$ we refer to Fig. 4-17, which shows a graph of the quantity $\tau \operatorname{sa}(\omega\tau/2)$ versus ω. Note that increasing τ makes the function $\tau \operatorname{sa}(\omega\tau/2)$ narrower and taller. Furthermore,

$$\int_{-\infty}^{\infty} \tau \left(\operatorname{sa} \frac{\omega\tau}{2} \right) d\omega = 2\pi$$

so that the area bounded by $\tau \operatorname{sa}(\omega\tau/2)$ and the ω axis is independent of τ. Therefore, the limit above yields a delta function. The spectral density of a dc signal $x(t) = x_0$ is given by

$$X(j\omega) = x_0 \int_{-\infty}^{\infty} e^{-j\omega t} dt = 2\pi x_0 \delta(\omega) \tag{4-13}$$

This result is consistent with our interpretation of spectral density. Since a dc signal can be regarded as a zero-frequency sinusoid, we expect the spectral density of a dc signal to be concentrated at $f = 0$. The spectral density above equals zero for all f except $f = 0$. Also, the unit of the spectral density above is the unit of the dc signal per unit frequency because the unit of the delta function is that of f^{-1}. Furthermore, the amplitude of the dc signal is equal to the area bounded by the spectral density above and the frequency axis because

$$\int_{-\infty}^{\infty} X(j2\pi f) df = \int_{-\infty}^{\infty} 2\pi x_0 \delta(2\pi f) df = x_0 \int_{-\infty}^{\infty} \delta(\alpha) d\alpha = x_0$$

Finally, using the spectral density above in the Fourier integral gives

$$\frac{1}{2\pi} \int_{-\infty}^{\infty} 2\pi x_0 \delta(\omega) d\omega = x_0 \int_{-\infty}^{\infty} \delta(\omega) d\omega = x_0$$

which is the correct result.

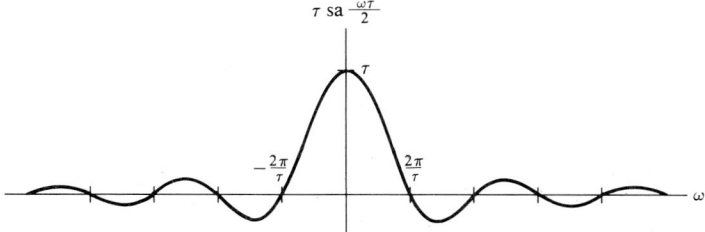

Figure 4-17 Graph of $\tau \operatorname{sa}(\omega\tau/2)$ versus ω.

A spectral density can be treated, e.g., plotted, either as a function of frequency f or as a function of angular frequency ω. This is often confusing, especially if a spectral density at hand contains a delta function. In this context it is important to recognize that $\delta(\omega) = \delta(2\pi f)$ and $\delta(f)$ have different strengths when both are regarded as functions of frequency f or when both are regarded as functions of angular frequency ω. From the properties of a delta function, we have

$$\int_{-\infty}^{\infty} \delta(\omega)\, d\omega = 1 \quad \text{and} \quad \int_{-\infty}^{\infty} \delta(f)\, df = 1$$

But

$$\int_{-\infty}^{\infty} \delta(\omega)\, df = \frac{1}{2\pi} \int_{-\infty}^{\infty} \delta(\omega)\, d\omega = \frac{1}{2\pi}$$

and

$$\int_{-\infty}^{\infty} \delta(f)\, d\omega = 2\pi \int_{-\infty}^{\infty} \delta(f)\, df = 2\pi \qquad (4\text{-}14)$$

These relations show that a delta function $\delta(f)$ has strength 2π when it is regarded as a function of angular frequency ω and that a delta function $\delta(\omega)$ has strength $(2\pi)^{-1}$ when it is regarded as a function of frequency (f). The relations (4-14) also show that

$$\delta(\omega) = \delta(2\pi f) = \frac{1}{2\pi} \delta(f) \qquad (4\text{-}15)$$

This is simply a consequence of the difference of scale between frequency f and angular frequency ω; for example, the fact that $\delta(\omega)$ has strength 1 on an angular-frequency (ω) axis and has strength $(2\pi)^{-1}$ on a frequency f axis simply reflects the relation

$$(1 \text{ rad/s})^{-1} = (2\pi \text{ Hz})^{-1} = (2\pi)^{-1} \text{ Hz}$$

Note also that the natural unit of a spectral density for, say, a voltage is volts per *hertz*, not volts per radian per second. This is because the Fourier integral can be written

$$x(t) = \frac{1}{2\pi} \int_{-\infty}^{\infty} X(j\omega) e^{j\omega t}\, d\omega = \int_{-\infty}^{\infty} X(j2\pi f) e^{j2\pi f t}\, df$$

where we have used the relations $\omega = 2\pi f$, $d\omega = 2\pi\, df$.

It is instructive to compare the spectral density of a dc signal with the complex spectrum of the dc signal. The complex spectrum of a dc signal $x(t) = x_0$ is

$$C_x(f) = \begin{cases} x_0 & f = 0 \\ 0 & f \neq 0 \end{cases}$$

whereas the spectral density of the signal is [see (4-15)]

$$X(j2\pi f) = 2\pi x_0 \delta(2\pi f) = x_0 \delta(f) \qquad (4\text{-}16)$$

Figure 4-18 shows graphical representations of these quantities. The amplitude of a dc signal is given by the *value* of the complex spectrum of the signal for $f = 0$ and by the *integral* of the spectral density of the signal over a small interval containing $f = 0$ (or $\omega = 0$).

Figure 4-18 Graphical representation of (a) the complex spectrum and (b) the spectral density of a dc signal $x(t) = x_0$.

4-2F Applications in System Analysis

Writing the Fourier integral as

$$x(t) = \lim_{\Delta\omega \to 0} \left[\frac{1}{2\pi} \sum_{n=-\infty}^{\infty} X(jn\,\Delta\omega) e^{jn\,\Delta\omega t}\,\Delta\omega \right]$$

shows that the Fourier integral for a signal $x(t)$ expresses the signal as a sum of exponentials. Each exponential in the sum has the form

$$x_n(t) = \frac{1}{2\pi} X(jn\,\Delta\omega) e^{jn\,\Delta\omega t}\,\Delta\omega$$

The output of a linear stationary system for this input $x_n(t)$ is

$$y_n(t) = \frac{1}{2\pi} H(jn\,\Delta\omega) X(jn\,\Delta\omega) e^{jn\,\Delta\omega t}\,\Delta\omega$$

where $H(jn\,\Delta\omega)$ is the transfer function of the system. By superposition, the output of the system for input $x(t)$ is

$$y_n(t) = \lim_{\Delta\omega \to 0} \left[\frac{1}{2\pi} \sum_{n=-\infty}^{\infty} H(jn\,\Delta\omega) X(jn\,\Delta\omega) e^{jn\,\Delta\omega t}\,\Delta\omega \right]$$

$$= \frac{1}{2\pi} \int_{-\infty}^{\infty} H(j\omega) X(j\omega) e^{j\omega t}\,d\omega \qquad (4\text{-}17)$$

This is the Fourier integral for the output $y(t)$. The spectral density of the output is

$$Y(j\omega) = H(j\omega) X(j\omega) \qquad (4\text{-}18)$$

This relation is essentially identical to the relation

$$C_y(f) = H(j2\pi f) C_x(f)$$

between the complex spectra (Fourier coefficients) of an input and output.

Example 4-8 In this example we use the Fourier integral to obtain the output of a system having transfer function

$$H(j\omega) = \frac{1}{1 + j\omega\tau}$$

and (dc) input $x(t) = x_0$. This problem is very easy to solve using methods described in Chap. 3. The dc gain of the system is $H(0) = 1$, and so the output for the dc input $x(t) = x_0$ is the dc signal $y(t) = H(0)x_0 = x_0$. Using the Fourier integral for this problem is like using a sledgehammer to drive a tack; we do so only to illustrate the method in a simple setting.

From (4-16), the spectral density of the input is

$$X(j\omega) = 2\pi x_0 \delta(\omega)$$

From (4-18), the spectral density of the output is

$$Y(j\omega) = H(j\omega)X(j\omega) = \frac{2\pi x_0 \delta(\omega)}{1 + j\omega\tau}$$

The waveform of the output is given by the Fourier integral

$$y(t) = \frac{1}{2\pi} \int_{-\infty}^{\infty} \frac{2\pi x_0 \delta(\omega)}{1 + j\omega\tau} e^{j\omega t} \, d\omega = x_0 \int_{-\infty}^{\infty} \frac{\delta(\omega)}{1 + j\omega\tau} e^{j\omega t} \, d\omega$$

This yields

$$y(t) = x_0 \int_{-\infty}^{\infty} \frac{\delta(\omega)}{1 + j0} e^0 \, d\omega = x_0 \int_{-\infty}^{\infty} \delta(\omega) \, d\omega = x_0$$

as expected. This example illustrates application of the Fourier integral to a very simple problem; problems more deserving of such elegant treatment are treated below. First, however, we describe the Fourier transformation, which is a powerful formalism for applying the Fourier integral to system analysis.

4-3 FOURIER TRANSFORMATION

The introduction to the Fourier integral in Sec. 4-2 is intended to be heuristic. In this section we describe a mathematical formalism called the *Fourier transformation,* which allows more compact treatment of frequency-domain system analysis. We start by describing what is meant in mathematics by a transformation.

4-3A Transformations

Certain mathematical operations are simplified by transformations. A familiar example is using logarithms for exponentiation, as illustrated by the flowchart of Fig. 4-19. Instead of calculating $c = a^b$ directly in the "number domain," e.g., by repeated multiplication if b is an integer, we can transform the problem to the "logarithm domain," whereby raising a to the bth power is transformed into multiplying the logarithm of a by b. Subsequently, the quantity $c = a^b$ is obtained by the inverse transformation (the antilogarithm) $c = e^{b \ln a}$.

Figure 4-20 shows a flowchart for obtaining an output of a linear stationary system. Comparing Fig. 4-20 with Fig. 4-19 shows that using Fourier's integral to obtain an output of a linear stationary system is analogous to using logarithms to perform exponentiation. The output $y(t)$ of a system for input $x(t)$ can be obtained directly in the time domain by convolution of $x(t)$ and the impulse response $h(t)$ of the system or by transforming the problem to the frequency domain, whereby the waveform

FOURIER SERIES, FOURIER INTEGRAL, AND FOURIER TRANSFORMATION **201**

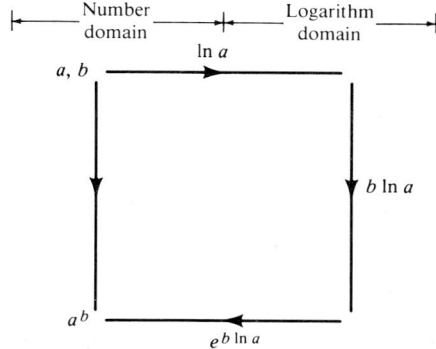

Figure 4-19 Flowchart for using logarithms to calculate $c = a^b$.

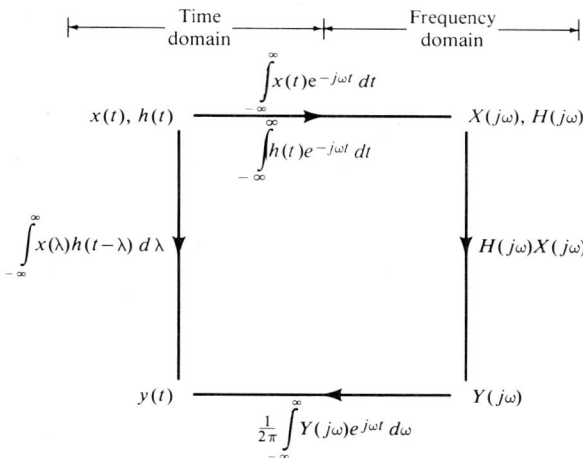

Figure 4-20 Flowchart for using the Fourier transformation to obtain an output of linear stationary system having impulse response $h(t)$.

$x(t)$ is transformed into the spectral density $X(j\omega)$ and convolution of $x(t)$ and $h(t)$ is transformed into multiplication of $X(j\omega)$ by the transfer function $H(j\omega)$ of the system. This gives the spectral density $Y(j\omega)$ of the output. Then the waveform $y(t)$ of the output can be obtained using the inverse transformation (the Fourier integral) of (4-17). We describe frequency-domain analysis by considering (4-10) to be the defining relation for a transformation called the *Fourier transformation*.

4-3B Definition of the Fourier Transformation

The *Fourier transform* of a function $e(t)$ is denoted by $E(j\omega)$ and defined by

$$E(j\omega) = \int_{-\infty}^{\infty} e(t)e^{-j\omega t}\, dt \tag{4-19}$$

The *inverse Fourier transform* of a function $E(j\omega)$ is given by

$$e(t) = \frac{1}{2\pi} \int_{-\infty}^{\infty} E(j\omega)e^{j\omega t}\, d\omega \tag{4-20}$$

The Fourier transformation defined by (4-19) is analogous to the transformation $A = \ln a$, where a is a number and A is its natural logarithm. The inverse Fourier transformation given by (4-20) recovers a function $e(t)$ from its Fourier transform $E(j\omega)$. It is analogous to the inverse transformation $a = e^A$, which recovers a number a from its natural logarithm A.

To show (formally) that the operation indicated by (4-20) is the inverse of that indicated by (4-19) we must show that

$$\frac{1}{2\pi} \int_{-\infty}^{\infty} \int_{-\infty}^{\infty} e(t') e^{-j\omega t'} \, dt' \, e^{j\omega t} \, d\omega = e(t)$$

Reversing the order of the integrations gives

$$\frac{1}{2\pi} \int_{-\infty}^{\infty} e(t') \int_{-\infty}^{\infty} e^{-j\omega t'} e^{j\omega t} \, d\omega \, dt' = \frac{1}{2\pi} \int_{-\infty}^{\infty} e(t') \int_{-\infty}^{\infty} e^{-j(t'-t)\omega} \, d\omega \, dt'$$

From (4-13), the inner integral gives

$$\int_{-\infty}^{\infty} e^{-j(t'-t)\omega} \, d\omega = 2\pi \delta(t' - t)$$

It follows that

$$\frac{1}{2\pi} \int_{-\infty}^{\infty} e(t') \int_{-\infty}^{\infty} e^{-j(t'-t)\omega} \, d\omega \, dt' = \int_{-\infty}^{\infty} e(t') \delta(t' - t) \, dt'$$

$$= e(t) \int_{-\infty}^{\infty} \delta(t' - t) \, dt' = e(t)$$

as was to be shown.

Example 4-9 The Fourier transform of a delta function

$$e(t) = \mathbf{a}\delta(t)$$

is

$$\int_{-\infty}^{\infty} \mathbf{a}\delta(t) e^{-j\omega t} \, dt = \mathbf{a} \int_{-\infty}^{\infty} \delta(t) \, dt = \mathbf{a}$$

It follows that the inverse Fourier transform of a frequency-independent quantity $E(j\omega) = \mathbf{a}$ is $e(t) = \mathbf{a}\delta(t)$.

4-3C Operator Notation

To simplify notation we use the operator $\mathscr{F}$ to denote Fourier transformation and the operator $\mathscr{F}^{-1}$ to denote inverse Fourier transformation. With this notation (4-19) and (4-20) are written as

$$E(j\omega) = \mathscr{F}\{e(t)\} \qquad (4\text{-}21)$$

and

$$e(t) = \mathscr{F}^{-1}\{E(j\omega)\} \qquad (4\text{-}22)$$

respectively. An operator is simply a shorthand expression for a mathematical operation; thus $\mathscr{F}\{e(t)\}$ denotes multiplying $e(t)$ by $e^{-j\omega t}$ and integrating over all t, and $\mathscr{F}^{-1}\{E(j\omega)\}$ denotes multiplying $E(j\omega)$ by $(2\pi)^{-1}e^{j\omega t}$ and integrating over all ω. Using $\mathscr{F}$ and $\mathscr{F}^{-1}$ to denote Fourier transformation and inverse Fourier transformation, respectively, is like using d/dt to denote differentiation, where

$$\frac{d}{dt}[e(t)] = \lim_{\Delta t \to 0} \frac{e(t + \Delta t) - e(t)}{\Delta t}$$

Example 4-10 From Example 4-9,

$$\mathscr{F}\{a\delta(t)\} = a \qquad \mathscr{F}^{-1}\{a\} = a\delta(t)$$

Example 4-11 The Fourier transform of a rectangular function is given by

$$\mathscr{F}\left\{r\left(\frac{t}{\tau}\right)\right\} = \int_{-\infty}^{\infty} r\left(\frac{t}{\tau}\right)e^{-j\omega t}\,dt = \int_{0}^{\tau} e^{-j\omega t}\,dt = -\frac{1}{j\omega}e^{-j\omega t}\bigg|_{0}^{\tau} = \frac{1 - e^{-j\omega\tau}}{j\omega}$$

Thus we have

$$\mathscr{F}\left\{r\left(\frac{t}{\tau}\right)\right\} = \frac{1 - e^{-j\omega\tau}}{j\omega} \qquad \mathscr{F}^{-1}\left\{\frac{1 - e^{-j\omega\tau}}{j\omega}\right\} = r\left(\frac{t}{\tau}\right)$$

It is often convenient to express the Fourier transform of a rectangular function in another form, as follows. We can write the transform above as

$$\frac{1 - e^{-j\omega\tau}}{j\omega} = \frac{e^{-j\omega\tau/2}[2j\,\sin(\omega\tau/2)]}{j\omega}$$

Multiplying numerator and denominator by $\tau/2$ gives

$$\mathscr{F}\left\{r\left(\frac{t}{\tau}\right)\right\} = \tau\,\text{sa}\,\frac{\omega\tau}{2}\,e^{-j\omega\tau/2}$$

4-3D Linearity of the Fourier Transformation

Because integration is a linear operation, it follows from (4-19) and (4-20) that Fourier transformation and inverse Fourier transformation are *linear operations;* i.e.,

$$\mathscr{F}\{a_1e_1(t) + a_2e_2(t) + \cdots\} = a_1\mathscr{F}\{e_1(t)\} + a_2\mathscr{F}\{e_2(t)\} + \cdots \qquad (4\text{-}23)$$

and

$$\mathscr{F}^{-1}\{a_1E_1(j\omega) + a_2E_2(j\omega) + \cdots\} = a_1\mathscr{F}^{-1}\{E_1(j\omega)\} + a_2\mathscr{F}^{-1}\{E_2(j\omega)\} + \cdots \qquad (4\text{-}24)$$

Example 4-12 From (4-23), the Fourier transform of

$$x(t) = \mathbf{a}\delta(t) + x_0 r\left(\frac{t}{\tau}\right)$$

is given by

$$X(j\omega) = \mathscr{F}\left\{\mathbf{a}\delta(t) + x_0 r\left(\frac{t}{\tau}\right)\right\} = \mathbf{a}\mathscr{F}\{\delta(t)\} + x_0\mathscr{F}\left\{r\left(\frac{t}{\tau}\right)\right\}$$

From Examples 4-10 and 4-11,

$$\mathcal{F}\{\delta(t)\} = 1 \qquad \mathcal{F}\left\{r\left(\frac{t}{\tau}\right)\right\} = \tau \, \text{sa} \, \frac{\omega\tau}{2} e^{-j\omega\tau/2}$$

Thus

$$X(j\omega) = \mathbf{a} + x_0\tau \, \text{sa} \, \frac{\omega\tau}{2} e^{-j\omega\tau/2}$$

4-3E Interpretation of the Fourier Transformation

Comparing (4-19) with (4-11) shows that *the Fourier transform of a signal is the spectral density of the signal.* Comparing (4-20) with (4-10) shows that *the waveform of a signal is the inverse Fourier transform of the spectral density of the signal;* thus

$$X(j\omega) = \mathcal{F}\{x(t)\} \qquad x(t) = \mathcal{F}^{-1}\{X(j\omega)\} \qquad (4\text{-}25)$$

Example 4-13 From Example 4-11, the Fourier transform of a rectangular function is

$$\mathcal{F}\left\{r\left(\frac{t}{\tau}\right)\right\} = \tau \, \text{sa} \, \frac{\omega\tau}{2} e^{-j\omega\tau/2}$$

Therefore, the spectral density of a rectangular pulse $x(t) = x_0 r(t/\tau)$ is

$$X(j\omega) = x_0\tau \, \text{sa} \, \frac{\omega\tau}{2} e^{-j\omega\tau/2}$$

Comparing (4-19) with (3-5) shows that *the transfer function of a linear stationary system is the Fourier transform of the impulse response of the system.* Conversely, *the impulse response of a linear stationary system is the inverse Fourier transform of the transfer function of the system;* thus

$$H(j\omega) = \mathcal{F}\{h(t)\} \qquad h(t) = \mathcal{F}^{-1}\{H(j\omega)\} \qquad (4\text{-}26)$$

Comparing (4-20) with (4-17) shows that the output $y(t)$ of a linear stationary system for input $x(t)$ is given by

$$y(t) = \mathcal{F}^{-1}\{H(j\omega)X(j\omega)\} \qquad (4\text{-}27a)$$

where $X(j\omega)$ is the spectral density (the Fourier transform) of the input and $H(j\omega)$ is the transfer function of the system. The frequency-domain procedure for finding an output of a linear stationary system can be expressed compactly using operator notation as

$$y(t) = \mathcal{F}^{-1}\{H(j\omega)\mathcal{F}\{x(t)\}\} \qquad (4\text{-}27b)$$

This expression symbolizes the following procedure for obtaining the output $y(t)$ of a linear stationary system for input $x(t)$:

1. Obtain the Fourier transform (the spectral density) $X(j\omega)$ of the input $x(t)$.
2. Obtain the transfer function $H(j\omega)$ of the system, e.g., by block-diagram reduction.

3. Multiply the Fourier transform of the input by the transfer function.
4. Obtain the waveform of the output $y(t)$ by taking the inverse Fourier transform of $H(j\omega)X(j\omega)$.

Example 4-14 Obtain the output of the system of Fig. 4-21 for (dc) input $x(t) = x_0$.

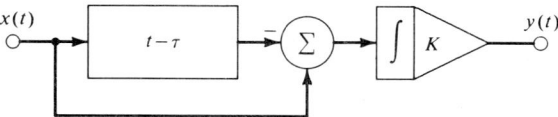

Figure 4-21 System of Example 4-14.

SOLUTION **Step 1** From (4-13), the Fourier transform of the input is

$$X(j\omega) = 2\pi x_0 \delta(\omega)$$

Step 2 By block-diagram reduction, the transfer function of the system is

$$H(j\omega) = K\tau \text{ sa } \frac{\omega\tau}{2} e^{-j\omega\tau/2}$$

Step 3 The Fourier transform of the output is

$$Y(j\omega) = 2\pi K\tau x_0 \text{ sa } \frac{\omega\tau}{2} e^{-j\omega\tau/2} \delta(\omega)$$

Because $\delta(\omega) = 0$ for $\omega \neq 0$, the Fourier transform above simplifies to

$$Y(j\omega) = 2\pi K\tau x_0 \text{ sa } 0 \, e^0 \delta(\omega) = 2\pi K\tau x_0 \delta(\omega)$$

Step 4 Taking the inverse Fourier transform gives

$$y(t) = K\tau x_0 \mathcal{F}^{-1}\{2\pi\delta(\omega)\} = K\tau x_0$$

4-3F Use of Tables

The most difficult steps in using the Fourier transformation to obtain an output of a system are the integrations necessary for finding Fourier transforms and inverse Fourier transforms. These steps are made much easier by using a table of Fourier transforms, just as using Fourier series is made easier by using a table of Fourier coefficients. It is permissible to use such a table for finding both Fourier transforms *and inverse Fourier transforms* because the Fourier transform of a function is unique if it exists. The uniqueness of a Fourier transform follows from the fact that an integral, e.g., the integral defining the Fourier transform of a function, is unique. A short table of Fourier transforms is given in Table C-1. More extensive tables are available (e.g., in Campbell and Foster), but Table C-1 is adequate for problems considered in this book.

Two of the transforms given in Table C-1 are derived in examples given above. Others are derived subsequently, but first we show how to use Table C-1 to find the output of a linear stationary system.

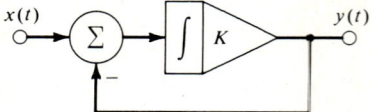

Figure 4-22 System of Example 4-15.

Example 4-15 Obtain the output of the system of Fig. 4-22 for input $x(t) = x_0 e^{-t/\tau} u(t)$.

SOLUTION We follow the procedure given above, using Table C-1 to obtain the necessary transforms. From Table C-1, the Fourier transform of the input is

$$X(j\omega) = \mathcal{F}\{x_0 e^{-t/\tau} u(t)\} = \frac{x_0 \tau}{1 + j\omega\tau}$$

Using block-diagram reduction to obtain the transfer function of the system gives

$$H(j\omega) = \frac{1}{1 + j\omega\tau_0}$$

where $\tau_0 = 1/K$. From (4-27),

$$y(t) = \mathcal{F}^{-1}\left\{\frac{x_0 \tau}{(1 + j\omega\tau)(1 + j\omega\tau_0)}\right\}$$

From Table C-1,

$$y(t) = \begin{cases} \dfrac{x_0 \tau (e^{-t/\tau} - e^{-t/\tau_0}) u(t)}{\tau - \tau_0} & \tau \neq \tau_0 \\ \dfrac{x_0 t e^{-t/\tau} u(t)}{\tau} & \tau = \tau_0 \end{cases}$$

The frequency-domain procedure just illustrated is much easier than the corresponding time-domain procedure (convolution), but the computational advantages of the Fourier transformation should not be overemphasized. After all, we could also simplify time-domain calculations by tabulating impulse responses and convolution integrals. The Fourier transformation is important in system engineering partly because it transforms certain difficult operations into simpler ones e.g., convolution to multiplication, but mainly because it transforms these operations in such a way as to provide considerable insight into certain signal-processing operations.

4-3G Operational Properties of the Fourier Transformation

The Fourier transformation establishes correspondences between operations on a function and other operations on the Fourier transform of the function. These correspondences between time- and frequency-domain operations are called *operational properties* of the Fourier transformation. The most useful operational properties of the Fourier transformation in system analysis are given in Table C-2. Below, we derive and interpret some of these properties.

One of the most important operational properties of the Fourier transformation is the *convolution property*, expressed by

$$\mathcal{F}\left\{\int_{-\infty}^{\infty} e_1(\lambda) e_2(t - \lambda)\, d\lambda\right\} = \mathcal{F}\{e_1(t)\}\mathcal{F}\{e_2(t)\} \qquad (4\text{-}28)$$

According to this relation, convolution of functions corresponds to multiplication of the transforms of the functions. A special case of this relation, where $e_1(t)$ and $e_2(t)$ represent a signal and an impulse response, is derived by heuristic arguments in Sec. 4-2F. Here we give a more formal derivation. From (4-19),

$$\mathcal{F}\left\{\int_{-\infty}^{\infty} e_1(\lambda)e_2(t-\lambda)\,d\lambda\right\} = \int_{-\infty}^{\infty}\int_{-\infty}^{\infty} e_1(\lambda)e_2(t-\lambda)\,d\lambda\, e^{-j\omega t}\,dt$$

Reversing the order of the integrations and using the change of variable $t' = t - \lambda$ gives

$$\int_{-\infty}^{\infty} e_1(\lambda) \int_{-\infty}^{\infty} e_2(t-\lambda)e^{-j\omega t}\,dt\,d\lambda = \int_{-\infty}^{\infty} e_1(\lambda) \int_{-\infty}^{\infty} e_2(t')e^{-j\omega(t'+\lambda)}\,dt'\,d\lambda$$

$$= \int_{-\infty}^{\infty} e_1(\lambda)e^{-j\omega\lambda} \int_{-\infty}^{\infty} e_2(t')e^{-j\omega t'}\,dt'\,d\lambda$$

By definition, the inner integral (over t') is the Fourier transform of $e_2(t)$; thus the expression above simplifies to

$$\int_{-\infty}^{\infty} e_1(\lambda)e^{-j\omega\lambda}\mathcal{F}\{e_2(t)\}\,d\lambda = \mathcal{F}\{e_2(t)\}\int_{-\infty}^{\infty} e_1(\lambda)e^{-j\omega\lambda}\,d\lambda$$

The last integral is the Fourier transform of $e_1(t)$. Equation (4-28) follows.

The convolution property of the Fourier transformation provides the mathematical basis for frequency-domain analysis of linear stationary systems. For $e_1(t) = x(t)$ (an input) and $e_2(t) = h(t)$ (an impulse response), (4-28) gives the output $y(t)$ of a linear stationary system as

$$\mathcal{F}\{y(t)\} = \mathcal{F}\{x(t)\}\mathcal{F}\{h(t)\} \quad \text{or} \quad y(t) = \mathcal{F}^{-1}\{\mathcal{F}\{x(t)\}\mathcal{F}\{h(t)\}\}$$

This result is obtained by more heuristic arguments in Sec. 4-2F [see (4-18)].

Another important operational property of the Fourier transformation is the *frequency-translation property,* expressed by

$$\mathcal{F}\{e(t)e^{j\omega_c t}\} = E[j(\omega - \omega_c)] \tag{4-29}$$

where $E(j\omega) = \mathcal{F}\{e(t)\}$. According to this property, multiplying a signal by an exponential $e^{j\omega_c t}$ corresponds to translating the Fourier transform (the spectral density) of the signal. This property is a generalization of the relation (3-38) for complex spectra. The frequency-translation property provides the mathematical basis for analysis and design of many important signal-transmission systems. A formal proof of the frequency-translation property follows. From (4-19),

$$\mathcal{F}\{e(t)e^{j\omega_c t}\} = \int_{-\infty}^{\infty} e(t)e^{j\omega_c t}e^{-j\omega t}\,dt = \int_{-\infty}^{\infty} e(t)e^{-j(\omega-\omega_c)t}\,dt$$

The last integral is simply the right side of (4-19) with ω replaced by $\omega - \omega_c$. Equation (4-29) follows.

Example 4-16 Obtain the spectral density of the signal

$$x(t) = x_0 r\left(\frac{t}{\tau}\right)e^{j\omega_c t}$$

SOLUTION From the frequency-translation property, the Fourier transform (the spectral density) of $x(t)$ is given by

$$X(j\omega) = x_c E[j(\omega - \omega_c)]$$

where $E(j\omega)$ is the Fourier transform of $r(t/\tau)$. From Table C-1, $E(j\omega) = \tau \operatorname{sa}(\omega\tau/2)e^{-j\omega\tau/2}$; thus

$$X(j\omega) = x_0 \tau \operatorname{sa} \frac{(\omega - \omega_c)\tau}{2} e^{-j(\omega-\omega_c)\tau/2}$$

Example 4-17 Obtain the waveform of a signal having spectral density

$$X(j\omega) = 2\pi x_0 [\delta(\omega - \omega_c) + \delta(\omega + \omega_c)]$$

SOLUTION Because inverse Fourier transformation is a linear operation, we have

$$x(t) = 2\pi x_0 \mathscr{F}^{-1}\{\delta(\omega - \omega_c)\} + 2\pi x_0 \mathscr{F}^{-1}\{\delta(\omega + \omega_c)\}$$

From the frequency-translation property,

$$x(t) = 2\pi x_0 e^{j\omega_c t} \mathscr{F}^{-1}\{\delta(\omega)\} + 2\pi x_0 e^{-j\omega_c t} \mathscr{F}^{-1}\{\delta(\omega)\}$$
$$= 4\pi x_0 \cos \omega_c t \, \mathscr{F}^{-1}\{\delta(\omega)\} = 2x_0 \cos \omega_c t$$

Other useful operational properties of the Fourier transformation are the *delay property*

$$\mathscr{F}\{e(t - t_0)\} = E(j\omega)e^{-j\omega t_0} \qquad (4\text{-}30)$$

the *compression-expansion* property

$$\mathscr{F}\{e(\eta t)\} = \frac{1}{|\eta|} E\left(\frac{j\omega}{\eta}\right) \qquad (4\text{-}31)$$

the *differentiation property*

$$\mathscr{F}\left\{\frac{de(t)}{dt}\right\} = j\omega E(j\omega) \qquad (4\text{-}32)$$

and the *integration property*

$$\mathscr{F}\left\{\int_{-\infty}^{t} e(\alpha)\,d\alpha\right\} = \frac{1}{j\omega} E(j\omega) \qquad (4\text{-}33)$$

Deriving these four properties from (4-19) is left as an exercise.

Example 4-18 Obtain the spectral density (the Fourier transform) of the signal

$$x(t) = x_0 r\left(\frac{t + \tau/2}{\tau}\right)$$

SOLUTION The signal $x(t)$ can be written

$$x(t) = x_0 e(t - t_0)$$

where $e(t) = r(t/\tau)$ and $t_0 = -\tau/2$. From the delay property (4-30),

$$X(j\omega) = \mathscr{F}\{x(t)\} = x_0 E(j\omega)e^{-j\omega t_0} = x_0 E(j\omega)e^{j\omega\tau/2}$$

From Table C-1,

$$E(j\omega) = \mathcal{F}\{e(t)\} = \tau \text{ sa} \frac{\omega\tau}{2} e^{-j\omega\tau/2}$$

Thus

$$X(j\omega) = x_0 \tau \text{ sa} \frac{\omega\tau}{2}$$

Example 4-19 Obtain the spectral density of the signal

$$y(t) = K \frac{dx(t)}{dt} \quad \text{where } x(t) = x_0 e^{-t/\tau} u(t)$$

SOLUTION From the differentiation property (4-32),

$$Y(j\omega) = \mathcal{F}\{y(t)\} = j\omega K X(j\omega)$$

where, from Table C-1,

$$\mathcal{F}\{x(t)\} = x_0 \mathcal{F}\{e^{-t/\tau} u(t)\} = \frac{x_0 \tau}{1 + j\omega\tau}$$

Thus

$$Y(j\omega) = \frac{j\omega K \tau x_0}{1 + j\omega\tau}$$

Alternatively, we can do the indicated differentiation first, as follows:

$$y(t) = \frac{K \, dx(t)}{dt} = -\frac{K x_0}{\tau} e^{-t/\tau} u(t) + K x_0 \delta(t)$$

whence

$$Y(j\omega) = -\frac{K x_0}{\tau} \mathcal{F}\{e^{-t/\tau} u(t)\} + K x_0 \mathcal{F}\{\delta\}(t)\} = -\frac{K x_0}{1 + j\omega\tau} + K x_0 = \frac{j\omega K \tau x_0}{1 + j\omega\tau}$$

as before.

The compression-expansion property (4-31) is also called the *duration-bandwidth property* because it shows that duration and bandwidth of a signal are inversely related; i.e., increasing the duration of a signal $(0 < \eta < 1)$ decreases the bandwidth of the signal, and decreasing the duration of a signal $(\eta > 1)$ increases the bandwidth of the signal. Figure 4-23 illustrates this relation for the signal $x(t) = x_0 e^{-t/\tau} u(t)$. The spectral density of $x(t)$ is

$$X(j\omega) = x_0 \mathcal{F}\{e^{-t/\tau} u(t)\} = \frac{x_0 \tau}{1 + j\omega\tau}$$

The effective duration of $x(t)$ is proportional to τ and the effective width of $X(j\omega)$ is inversely proportional to τ.

Special care is required in using the differentiation property (4-32) to obtain the Fourier transform of a function $e(t)$ having nonzero average value, e.g., a signal

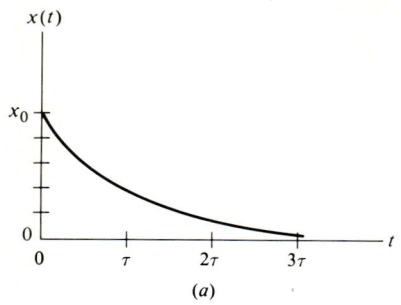
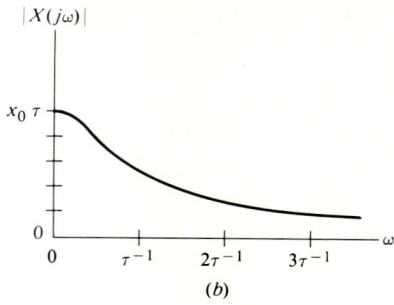

Figure 4-23 Waveform and amplitude spectral density of an exponential pulse, illustrating the duration-bandwidth relation (4-31): (a) $x(t) = e^{-t/\tau}u(t)$; (b) $X(j\omega) = |\mathcal{F}\{x(t)\}|$.

having a dc component. For example, consider the constant function $e(t) = 1$. Equation (4-32) gives

$$\mathcal{F}\left\{\frac{d}{dt}(1)\right\} = \mathcal{F}\{0\} = 0 = j\omega \mathcal{F}\{1\}$$

which implies that $\mathcal{F}\{1\} = 0$, whereas the (correct) Fourier transform of $e(t) = 1$ is

$$\mathcal{F}\{1\} = 2\pi\delta(\omega)$$

The fallacy in using (4-32) in reverse, as above, to obtain the Fourier transform of a function having nonzero average value is the implicit assumption that a function $e(t)$ is determined uniquely by its derivative $de(t)/dt$. This is not the case. A function is determined by its derivative only to within an additive constant; i.e., if e_0 is constant (is independent of t), then

$$\frac{d}{dt}[e(t) + e_0] = \frac{de(t)}{dt}$$

is independent of e_0. Consequently, we cannot use (4-32) to obtain the Fourier transform of a function $e(t)$ from the Fourier transform of $de(t)/dt$ if $e(t)$ has nonzero average value. In such a case, we must use the more general relation

$$\mathcal{F}\{e(t)\} = \frac{1}{j\omega}\mathcal{F}\left\{\frac{de(t)}{dt}\right\} + 2\pi e_0 \delta(\omega) \tag{4-34a}$$

where e_0 is the average value of $e(t)$, defined by

$$e_0 = \lim_{T \to \infty} \frac{1}{T} \int_{-T/2}^{T/2} e(t)\, dt \tag{4-34b}$$

Example 4-20 Obtain the spectral density of the step signal $x(t) = x_0 u(t)$.

SOLUTION The spectral density of $x(t)$ is given by

$$X(j\omega) = \mathcal{F}\{x(t)\} = x_0 \mathcal{F}\{u(t)\}$$

The unit step function $u(t)$ has average value u_0 given by

FOURIER SERIES, FOURIER INTEGRAL, AND FOURIER TRANSFORMATION **211**

$$u_0 = \lim_{T \to \infty} \frac{1}{T} \int_{-T/2}^{T/2} u(t)\, dt = \lim_{T \to \infty} \frac{1}{T} \int_0^{T/2} dt = \lim_{T \to \infty} \frac{1}{T} \frac{T}{2} = \frac{1}{2}$$

From (4-34),

$$\mathcal{F}\{u(t)\} = \frac{1}{j\omega} \mathcal{F}\left\{\frac{du(t)}{dt}\right\} + \pi\delta(\omega) = \frac{1}{j\omega} + \pi\delta(\omega)$$

Thus the spectral density of $x(t) = x_0 u(t)$ is

$$X(j\omega) = \frac{x_0}{j\omega} + \pi x_0 \delta(\omega)$$

The delta function arises from the dc component of the step signal $x(t)$. The other term arises from the transition of the step from $x(t) = 0$ to $x(t) = x_0$ at $t = 0$.

4-3H Fourier Transform of a Step Function

In Example 4-20 we show that the Fourier transform of the unit step function is

$$\mathcal{F}\{u(t)\} = \frac{1}{j\omega} + \pi\delta(\omega) \tag{4-35}$$

Although this relation is correct, the delta function appearing in (4-35) is a nuisance in many manipulations. The relation

$$\mathcal{F}\{u(t)\} = \frac{1}{j\omega} \tag{4-36}$$

which in effect ignores the average value of $u(t)$, is used more often than (4-35). Indeed, (4-36) is the relation given in Table C-1. We wish to resolve this apparent conflict and to show that (4-36) can be used in virtually all practical applications.

In system engineering the Fourier transform of a function usually represents either the spectral density of a signal or the transfer function of a system. Therefore the Fourier transform of a unit step arises either in the spectral density of an (ideal) step signal or in the transfer function of an (ideal) integrator. Neither ideal step signals nor ideal integrators exist. A step signal is a model for a signal that is approximately constant for a long (but finite) time. Similarly, an integrator is a model for a system whose impulse response is approximately constant for a long (but finite) time. Nothing is lost in any practical application by treating a step function as an exponential function having a very large (but finite) time constant; thus

$$u(t) = \begin{cases} 0 & t \le 0 \\ \lim_{\tau \to \infty} e^{-t/\tau} & t > 0 \end{cases}$$

For τ finite (no matter how large) this function has zero average value. Formally, the Fourier transform of the "practical step function" $u(t)$ defined by the relation above is given by

$$\mathcal{F}\{u(t)\} = \lim_{\tau \to \infty} \int_{-\infty}^{\infty} e^{-t/\tau} e^{-j\omega t}\, dt = \lim_{\tau \to \infty} \frac{\tau}{1 + j\omega\tau} = \frac{1}{j\omega}$$

212 INTRODUCTION TO SYSTEM ANALYSIS

In view of this result, we consider (4-36) an acceptable alternative to (4-35). This amounts to adopting the relations

$$\mathcal{F}\{u(t)\} = \frac{1}{j\omega} \qquad \mathcal{F}^{-1}\left\{\frac{1}{j\omega}\right\} = u(t)$$

as *definitions*. This special treatment of the unit step function causes no difficulty if it is used consistently.

Example 4-21 To illustrate the discussion above we determine the output of a system for input $x(t) = x_0 u(t)$ twice, first using (4-35) and then using (4-36). The transfer function of the system is

$$H(j\omega) = \frac{1}{1 + j\omega\tau}$$

From (4-27), the output $y(t)$ for input $x(t) = x_0 u(t)$ is given by

$$y(t) = \mathcal{F}^{-1}\{H(j\omega)\mathcal{F}\{x(t)\}\}$$

Case 1 Using (4-35) gives

$$y(t) = x_0 \mathcal{F}^{-1}\left\{\frac{1}{1 + j\omega\tau}\left[\frac{1}{j\omega} + \pi\delta(\omega)\right]\right\} = x_0 \mathcal{F}^{-1}\left\{\frac{1}{j\omega(1 + j\omega\tau)} + \frac{\pi\delta(\omega)}{1 + j\omega\tau}\right\}$$

$$= x_0 \mathcal{F}^{-1}\left\{\frac{1}{j\omega(1 + j\omega\tau)} + \pi\delta(\omega)\right\}$$

where the last relation follows from the previous one because the delta function equals zero for $\omega \neq 0$. Expanding the first term in the braces in partial fractions gives†

$$y(t) = x_0 \mathcal{F}^{-1}\left\{\frac{1}{j\omega} - \frac{\tau}{1 + j\omega\tau} + \pi\delta(\omega)\right\} = x_0 \mathcal{F}^{-1}\left\{\frac{1}{j\omega} + \pi\delta(\omega)\right\} - x_0 \mathcal{F}^{-1}\left\{\frac{\tau}{1 + j\omega\tau}\right\}$$

$$= x_0 u(t) - x_0 e^{-t/\tau} u(t)$$

Case 2 We solve the problem again, this time using (4-36). From (4-27),

$$y(t) = \mathcal{F}^{-1}\{H(j\omega)\mathcal{F}\{x(t)\}\}$$

Using (4-36) gives

$$y(t) = x_0 \mathcal{F}^{-1}\left\{\frac{1}{j\omega(1 + j\omega\tau)}\right\} = x_0 \mathcal{F}^{-1}\left\{\frac{1}{j\omega}\right\} - x_0 \mathcal{F}^{-1}\left\{\frac{\tau}{1 + j\omega\tau}\right\}$$

Using (4-36) and Table C-1 gives

$$y(t) = x_0 u(t) - x_0 e^{-t/\tau} u(t)$$

as before. This illustrates the fact that (4-36) gives correct results if it is used consistently (i.e., for both transforms and inverse transforms in any one problem).

†The partial fraction expansion used here (see Thomas and Finney, pp. 356–362) is

$$\frac{1}{\alpha(1 + a\alpha)} = \frac{1}{\alpha} - \frac{a}{1 + a\alpha}$$

In this book we use (4-36) exclusively. All entries in Table C-1 are consistent with (4-36). Any transform in Table C-1 having $j\omega$ as a factor in its denominator may lead to incorrect results if used with (4-35) but will lead to correct results if used with (4-36).

4-3I Fourier Transform of a Periodic Signal

Above, we treat periodic and nonperiodic signals differently. A periodic signal is described in the time domain by a Fourier (harmonic) series and in the frequency domain by a complex spectrum, whereas a nonperiodic signal is described in the time domain by a Fourier integral and in the frequency domain by a spectral density. Actually, a Fourier integral can describe both periodic and nonperiodic signals. From Table C-1,

$$\mathscr{F}\{e^{jn\omega_0 t}\} = 2\pi\delta(\omega - n\omega_0)$$

Using this relation and the linearity of the Fourier transformation, we find that the Fourier transform (the spectral density) of a periodic signal $x(t)$ described by an exponential Fourier series can be expressed as

$$X(j\omega) = \mathscr{F}\left\{\sum_{n=-\infty}^{\infty} X_n e^{jn\omega_0 t}\right\} = 2\pi \sum_{n=-\infty}^{\infty} X_n \delta(\omega - n\omega_0) \qquad (4\text{-}37)$$

where
$$X_n = \frac{1}{T}\int_{t_0}^{t_0+T} x(t)e^{-jn\omega_0 t}\,dt \qquad T = \frac{2\pi}{\omega_0}$$

is the nth Fourier coefficient for $x(t)$. This shows that Fourier's series can be treated as a special case of Fourier's integral. This fact is important in some theoretical developments because it implies that relations derived from (4-19) and (4-20) apply to both nonperiodic and periodic signals. But in practical applications it is usually much easier to use a Fourier series for describing periodic signals, reserving the Fourier integral for describing nonperiodic signals.

4-3J Normalized Energy, Normalized Power, and Bandwidth

The *normalized energy* of a signal $x(t)$ is denoted by J_x and defined by

$$J_x = \int_{-\infty}^{\infty} x^2(t)\,dt \qquad (4\text{-}38)$$

For a signal $x(t)$ applied to a dissipative load, normalized energy is proportional to actual energy dissipated in the load; e.g., if $x(t)$ is current in a resistor having resistance R, energy dissipated (converted to heat) is given by RJ_x. The unit of normalized energy is the unit of $x^2(t)t$, for example, ampere-squared-seconds, which is also the unit of actual energy dissipated per unit load, e.g., joules per ohm. Normalized energy is so called because (for a signal applied to a dissipative load) it is actual energy dissipated *per unit load*.

We can calculate the normalized energy of a signal from the waveform of the signal using the definition (4-38). Alternatively, we can calculate the normalized energy of a signal from the spectral density of the signal using Rayleigh's theorem,† which is expressed by the relation

$$\int_{-\infty}^{\infty} x^2(t)\, dt = \frac{1}{2\pi} \int_{-\infty}^{\infty} |X(j\omega)|^2\, d\omega \qquad (4\text{-}39)$$

where $X(j\omega) = \mathcal{F}\{x(t)\}$ is the spectral density of $x(t)$. Rayleigh's theorem can be proved as follows. Using Fourier's integral to replace one $x(t)$ in the integral on the left side of (4-39) gives

$$\int_{-\infty}^{\infty} x(t) \frac{1}{2\pi} \int_{-\infty}^{\infty} X(j\omega)e^{j\omega t}\, d\omega\, dt = \frac{1}{2\pi} \int_{-\infty}^{\infty} X(j\omega) \int_{-\infty}^{\infty} x(t)e^{j\omega t}\, dt\, d\omega$$

For a real signal $x(t)$ the inner integral on the right is

$$\int_{-\infty}^{\infty} x(t)e^{j\omega t}\, dt = X(-j\omega) = X^*(j\omega)$$

Equation (4-39) follows because $X(j\omega)X^*(j\omega) = |X(j\omega)|^2$.

Example 4-22 In this example we calculate the normalized energy of an exponential pulse $x(t) = x_0 e^{-t/\tau} u(t)$ using the definition (4-38) and then again using Rayleigh's theorem (4-39).
From (4-38),

$$J_x = \int_{-\infty}^{\infty} x_0^2 (e^{-t/\tau})^2 u^2(t)\, dt = x_0^2 \int_0^{\infty} e^{-2t/\tau}\, dt$$

This yields

$$J_x = \frac{x_0^2 \tau}{2}$$

The spectral density (the Fourier transform) of $x(t)$ is

$$X(j\omega) = \frac{x_0 \tau}{1 + j\omega\tau}$$

Using Rayleigh's theorem (4-39) gives

$$J_x = \frac{1}{2\pi} \int_{-\infty}^{\infty} \frac{(x_0\tau)^2\, d\omega}{1 + (\omega\tau)^2} = \frac{x_0^2 \tau}{2\pi} \tan^{-1} \omega\tau \bigg|_{-\infty}^{\infty} = \frac{x_0^2 \tau}{2}$$

as before.

Because $X(-j\omega) = X^*(j\omega)$, (4-39) can be written

$$J_x = \int_{-\infty}^{\infty} |X(j2\pi f)|^2\, df = \int_0^{\infty} 2|X(j2\pi f)|^2\, df$$

The quantity $2|X(j2\pi f)|^2$ is called the *energy density spectrum* of the associated signal $x(t)$. In this book the energy density spectrum of a signal $x(t)$ is denoted by $E_x(f)$; thus

†After Lord Rayleigh (John William Strutt) (1842–1919), British physicist and mathematician.

$$E_x(f) = 2|X(j2\pi f)|^2 \qquad f \geq 0 \qquad (4\text{-}40)$$

Example 4-23 Obtain the energy density spectrum of a rectangular pulse $x(t) = x_0 r(t/\tau)$.

SOLUTION From Table C-1, the spectral density of the pulse is

$$X(j\omega) = x_0 \tau \text{ sa} \frac{\omega \tau}{2} e^{-j\omega\tau/2}$$

From (4-40), the energy density spectrum of the pulse is

$$E_x(f) = 2|X(j2\pi f)|^2 = 2(x_0\tau)^2 \text{ sa}^2 \pi f\tau$$

Figure 4-24 shows a graph of the pulse $x(t)$ versus time and a graph of the energy density spectrum $E_x(f)$ versus frequency.

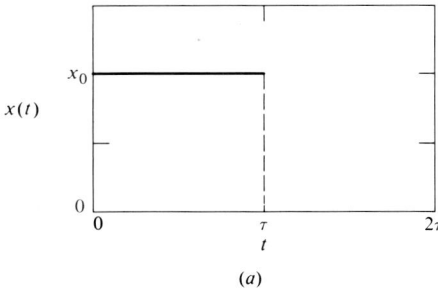

(a)

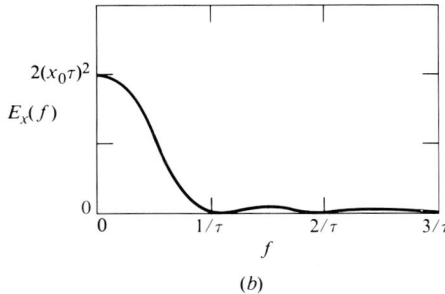

(b)

Figure 4-24 Signal of Example 4-23; (a) waveform; (b) energy density spectrum.

The energy density spectrum of a signal describes how normalized energy is distributed among the spectral components of the signal. By Rayleigh's theorem, the (total) normalized energy of a signal $x(t)$ is given by

$$J_x = \int_0^\infty E_x(f) \, df \qquad (4\text{-}41a)$$

This implies that the normalized energy of components in a band $f_1 \leq f < f_2$ is given by

$$J_x(f_1 \leq f < f_2) = \int_{f_1}^{f_2} E_x(f) \, df \qquad (4\text{-}41b)$$

The contribution of components to (total) normalized energy is large where the energy density spectrum is large and small where the energy density spectrum is small. For example, the energy density spectrum of Fig. 4-24 shows that most of the normalized energy of a rectangular pulse is contributed by components in the band $0 \leq f < 1/\tau$.

Example 4-24 Find the fraction of the normalized energy of an exponential pulse $x(t) = x_0 e^{-t/\tau} u(t)$ that is contributed by components in the band $0 \leq f < 1/2\pi\tau$.

SOLUTION The energy density spectrum of the exponential pulse (see Example 4-22) is

$$E_x(f) = \frac{2(x_0\tau)^2}{1 + (2\pi\tau f)^2}$$

Figure 4-25 shows a graph of $E_x(f)$. From (4-41), the normalized energy contributed by components in the band $0 \leq f < 1/2\pi\tau$ is given by

$$J_x\left(0 \leq f < \frac{1}{2\pi\tau}\right) = \int_0^{1/2\pi\tau} E_x(f)\, df$$

as illustrated by the shaded area in Fig. 4-25. This yields

$$J_x\left(0 \leq f < \frac{1}{2\pi\tau}\right) = 2x_0^2 \tau (2\pi)^{-1} \tan^{-1} 2\pi f \tau \Big|_0^{1/2\pi\tau}$$

$$= 2x_0^2 \tau (2\pi)^{-1} (\tan^{-1} 1 - \tan^{-1} 0) = 0.25 x_0^2 \tau$$

From Example 4-22, the (total) normalized energy of the exponential pulse $x(t)$ is $0.5 x_0^2 \tau$. Therefore half of the normalized energy of $x(t)$ is contributed by components in the band $0 \leq f < 1/2\pi\tau$.

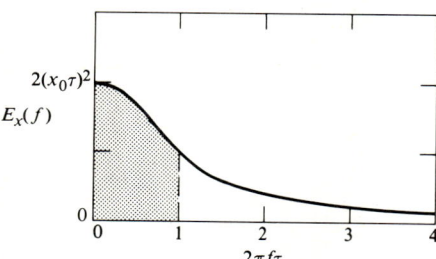

Figure 4-25 Energy density spectrum of the exponential pulse of Example 4-24.

Rayleigh's theorem is applicable only to a signal having finite normalized energy; i.e., to a signal $x(t)$ that satisfies

$$\int_{-\infty}^{\infty} x^2(t)\, dt < \infty$$

Many useful signal models, e.g., sinusoidal signals, have infinite normalized energy. Rayleigh's theorem and the concepts of normalized energy and energy density spectrum are inapplicable to such signals; however, many signals having infinite normalized energy have finite normalized power, where normalized power is normalized energy averaged over all time. The *normalized power* of a signal $x(t)$ is denoted (in this book) by L_x and defined by

$$L_x = \lim_{T \to \infty} \frac{1}{T} \int_{-T/2}^{T/2} x^2(t)\, dt \tag{4-42}$$

The normalized power of a signal is just the mean-squared amplitude of the signal, which in turn is the square of the rms amplitude of the signal (Hayt and Kemmerly, pp. 342–343). For a signal $x(t)$ applied to a dissipative load, normalized power L_x is proportional to actual power dissipated; e.g., if $x(t)$ is voltage across a resistor having resistance R, the actual power dissipated is L_x/R, where L_x is the normalized power of $x(t)$. The unit of normalized power of a signal $x(t)$ is the unit of x^2 (e.g., volts squared) which is also the unit of power dissipated per unit load (e.g., watts per ohm). Normalized power is so called because (for a signal applied to a dissipative load) it is actual power dissipated *per unit load*.

Example 4-25 Obtain an expression for the normalized average power of a sinusoidal signal $x(t) = A \cos(\omega t + \theta)$ $(\omega \neq 0)$.

SOLUTION From (4-42),

$$L_x = \lim_{T \to \infty} \frac{1}{T} \int_{-T/2}^{T/2} A^2 \cos^2(\omega t + \theta)\, dt$$

Using the identity $2 \cos^2 \alpha \equiv 1 + \cos 2\alpha$ gives

$$L_x = \lim_{T \to \infty} \frac{A^2}{2T} \int_{-T/2}^{T/2} [1 + \cos(2\omega t + 2\theta)]\, dt = \lim_{T \to \infty} \frac{A^2}{2T}(T) = \tfrac{1}{2}A^2 \qquad \omega \neq 0$$

The normalized power of the sinusoid is just the square of the rms amplitude $x_{\text{rms}} = A/\sqrt{2}$. It is independent of the frequency and the initial phase of the sinusoid.

A signal having finite normalized energy is called an *energy signal*. The normalized power of an energy signal equals zero. A signal having finite (and nonzero) normalized power is called a *power signal*. The normalized energy of a power signal is infinite. So far as we know, all real-world signals are energy signals; however, many useful signal models describe power signals. Fortunately, we can define a power density spectrum for a power signal. The power density spectrum of a power signal is analogous to the energy density spectrum of an energy signal.

The *power density spectrum* of a power signal $x(t)$ is denoted by $p_x(f)$ and defined by

$$p_x(f) = \lim_{\tau \to \infty} \left[\frac{1}{\tau} 2|X_\tau(j2\pi f)|^2 \right] \qquad f \geq 0 \tag{4-43}$$

where

$$X_\tau(j\omega) = \int_{-\tau/2}^{\tau/2} x(t) e^{-j\omega t}\, dt \tag{4-44}$$

The reasoning behind this definition is as follows. The quantity $2|X_\tau(j2\pi f)|^2$ is the energy density spectrum of $x(t)$ (over $-\tau/2 \leq t < \tau/2$). The total normalized energy of the segment is

$$J'_x = \int_0^\infty 2|X_\tau(j2\pi f)|^2 \, df$$

and the normalized power (the average normalized energy) of the segment is

$$\frac{1}{\tau} J'_x = \frac{1}{\tau} \int_0^\infty 2|X_\tau(j2\pi f)|^2 \, df$$

Therefore, the normalized power of the signal $x(t)$ (for $-\infty \leq t < \infty$) is given by

$$L_x = \int_0^\infty \lim_{\tau \to \infty} \left[\frac{1}{\tau} 2|X_\tau(j2\pi f)|^2\right] df$$

The integrand is called the power density spectrum of $x(t)$ because it describes how much of the total normalized power of $x(t)$ is contributed by components in any specified band. A power density spectrum describes how the total normalized power of a power signal is distributed among the spectral components of the signal, just as an energy density spectrum describes how the total normalized energy of an energy signal is distributed among the spectral components of the signal. The total normalized power of a signal $x(t)$ having power density spectrum $p_x(f)$ is given by

$$L_x = \int_{0^-}^\infty p_x(f) \, df \quad (4\text{-}45)$$

where the lower limit of integration is taken as $f = 0^-$ to make it clear that the strength of a delta function $\delta(f)$ appearing in $p_x(f)$ is to be included in the integral. The normalized power of components in a band $f_1 \leq f < f_2$ is given by

$$L_x(f_1 \leq f < f_2) = \int_{f_1^-}^{f_2} p_x(f) \, df \quad (4\text{-}46)$$

where the lower limit of integration is taken as $f = f_1^-$ to make it clear that the strength of a delta function $\delta(f - f_1)$ appearing in $p_x(f)$ is to be included in the integral.

Example 4-26 Obtain the power density spectrum of a dc signal $x(t) = x_0$.

SOLUTION We use the definition (4-43). First, we obtain the spectral density of a segment of the signal. This gives

$$X_\tau(j\omega) = \int_{-\tau/2}^{\tau/2} x_0 e^{-j\omega t} \, dt = x_0 \tau \, \text{sa} \, \frac{\omega\tau}{2}$$

The power density spectrum of the signal is

$$p_x(f) = \lim_{\tau \to \infty} \left[\frac{2}{\tau}|X_\tau(j2\pi f)|^2\right] = 2x_0^2 \lim_{\tau \to \infty} \left(\tau \, \text{sa}^2 \frac{\omega\tau}{2}\right)$$

As τ approaches infinity, the width of the function $\tau \, \text{sa}^2(\omega\tau/2)$ approaches zero and the amplitude of the function $\tau \, \text{sa}^2(\omega\tau/2)$ (for $\omega = 0$) approaches infinity. Furthermore, the area bounded by the function $\tau \, \text{sa}^2(\omega\tau/2)$ and the ω axis, given by

$$\int_0^\infty \tau \, \text{sa}^2 \frac{\omega\tau}{2} \, d\omega = 2 \int_0^\infty \text{sa}^2 \frac{\omega\tau}{2} \, d\frac{\omega\tau}{2} = \pi$$

(see Sec. A-5) is independent of τ. Therefore, the limit above yields a delta function having strength π. The power density spectrum of a dc signal $x(t) = x_0$ [see (4-15)] is

$$p_x(f) = x_0^2 2\pi\delta(\omega) = x_0^2 \delta(f)$$

This is the expected result. The normalized power is concentrated at $f = 0$, as it should be for a dc signal. Also, the total normalized power is

$$L_x = \int_{0^-}^\infty p_x(f) \, df = x_0^2 \int_{0^-}^\infty \delta(f) \, df = x_0^2$$

which is correct for a dc signal having amplitude x_0. Finally, because $\delta(f)$ has the unit of $1/f$, the unit of $p_x(f)$ above is the unit of x_0^2 per unit frequency, as it should be for a power density spectrum.

Proceeding as in Example 4-26, we can show that the power density spectrum of a sinusoidal signal $x(t) = A \cos(\omega_0 t + \theta)$ is

$$p_x(f) = \frac{A^2}{2} \delta(f - f_0)$$

The normalized power is concentrated at $f = f_0$, and the total normalized power is $A^2/2$, as expected. Furthermore, we can show that the normalized power of a sum of sinusoids (having different frequencies) is equal to the sum of the normalized powers of the sinusoids. It follows that the power density spectrum of a periodic signal $x(t)$ described by a harmonic (Fourier) series

$$x(t) = \sum_{n=0}^\infty A_n \cos(n\omega_0 t + \theta_n) \tag{4-47}$$

is

$$p_x(f) = A_0^2 \delta(f) + \frac{1}{2} \sum_{n=1}^\infty A_n^2 \delta(f - nf_0) \tag{4-48}$$

The total normalized power of a periodic signal $x(t)$ is given by

$$L_x = \lim_{\tau \to \infty} \frac{1}{\tau} \int_{-\tau/2}^{\tau/2} x^2(t) \, dt = A_0^2 + \frac{1}{2} \sum_{n=1}^\infty A_n^2 = \sum_{n=-\infty}^\infty |X_n|^2 \tag{4-49}$$

This relation is known as *Parseval's theorem*.†

Example 4-27 Obtain the power density spectrum of the exponential pulse train $x(t)$ of Fig. 4-26a, where $x_0 = 20$ V and $\tau = 1$ ms.

SOLUTION From Appendix B, Fourier's coefficients for $x(t)$ are given by

$$X_n = \frac{x_0 \tau}{T} \frac{1 - e^{-T/\tau}}{1 + j2\pi n\tau/T} \approx \frac{x_0 \tau}{T} \frac{1}{1 + jn\pi/2}$$

†After the French mathematician Marc-Antoine Parseval (1755–1836).

220 INTRODUCTION TO SYSTEM ANALYSIS

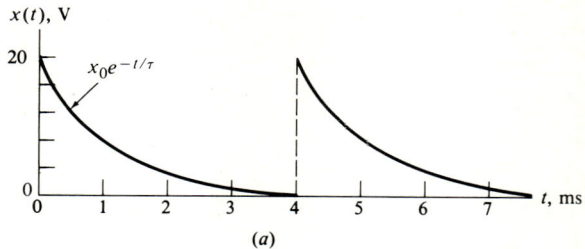

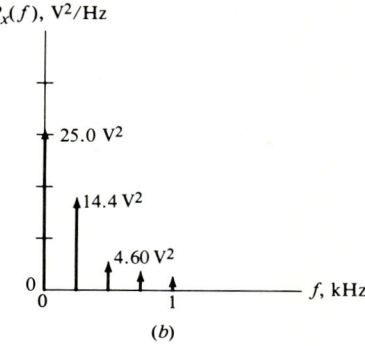

Figure 4-26 Signal of Example 4-27: (a) waveform; (b) power density spectrum.

where $T = 4$ ms. From (4-48), the power density spectrum of $x(t)$ is

$$p_x(f) = \left(\frac{x_0\tau}{T}\right)^2 \delta(f) + 2\left(\frac{x_0\tau}{T}\right)^2 \sum_{n=1}^{\infty} \frac{1}{1 + (n\pi/2)^2} \delta(f - nf_0)$$

Figure 4-26b shows a graphical representation of $p_x(f)$.

In any practical application the significant spectral components of a signal occupy a band having finite width on a frequency axis; i.e., virtually all the normalized energy or power of a signal is concentrated in components having frequencies in a band $f_1 \leq f < f_2$. We refer to this band as the *band occupied by the signal,* and we refer to the width $f_2 - f_1$ of the band as the *bandwidth* of the signal. For example, Fig. 4-24 shows that most of the normalized energy of a rectangular pulse is concentrated in the band $0 \leq f < 1/\tau$, where τ is the duration of the pulse. We say that the pulse occupies the band $0 \leq < f < 1/\tau$ and that the pulse has bandwidth $1/\tau$. There are various quantitative definitions of bandwidth. For example, the bandwidth of a rectangular pulse can be defined as the width W of a band $0 \leq f < W$ that accounts for a specified fraction of the total normalized energy of the pulse. According to this definition, the bandwidth W of a rectangular pulse $x(t) = x_0 r(t/\tau)$ is given by

$$\int_0^W E_x(f)\,df = 2(x_0\tau)^2 \int_0^W (\text{sa}^2\,\pi f\tau)\,df = \alpha J_x \qquad (4\text{-}50)$$

where αJ_x is some specified fraction, say 95 percent, of the total normalized energy

$$J_x = \int_{-\infty}^{\infty} x^2(t)\, dt = x_0^2 \tau$$

Each of the various quantitative definitions of bandwidth finds some use, but, by and large, engineers use the term bandwidth rather loosely, in a manner suggested by the shape of an energy density spectrum or power density spectrum. For example, most engineers would say that the bandwidth of a rectangular pulse is $W = 1/\tau$, where τ is the duration of the pulse, because this definition is suggested by the shape of the energy density spectrum and because the duration of the pulse is a meaningful parameter. It is significant that a large fraction of the normalized energy of the pulse is accounted for by components lying in the band $0 \leq f < W$, but the fact that that large fraction equals 0.90 is more or less incidental. Similarly, most system engineers would say that the bandwidth of an exponential pulse is $W = 1/2\pi\tau$, where τ is the time constant of the pulse. This definition is suggested by the shape of the energy density spectrum and by the fact that the time constant of an exponential pulse is itself a meaningful parameter, not by any prior specification on normalized energy accounted for by components in the band $0 \leq f < W$. In fact, only half of the normalized energy of an exponential pulse is accounted for by components in that band. As a rule, when we refer to the bandwidth of a signal, we mean simply some *conventional or convenient measure of the width of the energy (or power) density spectrum of the signal*.

The term bandwidth also is used to describe systems. In that case, bandwidth refers to the *width of the band over which the transfer function of the system approximates a desired transfer function*. For example, bandwidth of a bandpass filter refers to the width of the passband of the filter; bandwidth of a bandstop filter usually refers to the width of the stopband of the filter; and bandwidth of a practical differentiator refers to the width of the band over which the transfer function of the practical differentiator approximates that of an ideal differentiator. For systems, as for signals, the term bandwidth usually is used rather loosely. For example, we ordinarily regard the parameter f_c as the bandwidth of a low-pass filter having transfer function

$$H(j\omega) = \frac{1}{1 + j\omega/\omega_c} = \frac{1}{1 + jf/f_c}$$

even though the filter passes only half the normalized power of a sinusoidal input having frequency f_c (this implies that the system introduces significant distortion in transmitting a signal having significant components near but below f_c). As another example, a system having transfer function

$$H(j\omega) = \frac{jf/f_c}{1 + jf/f_c}$$

can be used to approximate a differentiator because for $f \ll f_c$, $H(j\omega) \approx jf/f_c$, which is the transfer function of an ideal differentiator. Such a system is often described as a differentiator having bandwidth f_c, even though the transfer function of the system is significantly different from jf/f_c for some frequencies less than f_c.

4-4 FREQUENCY-DOMAIN SYSTEM ANALYSIS

We show above how any signal of practical importance can be expressed as a superposition of sinusoids using Fourier's integral. This implies that many properties of a linear stationary system can be determined from the transfer function of the system. In this section we show how realizability, fidelity, and other properties of a linear stationary system are determined from the transfer function of the system. In Sec. 4-5 we show how these important concepts arise in designing filters, signal-transmission systems, and automatic control systems.

4-4A Stability

We wish to show how to determine whether a system having a rational transfer function of the form

$$H(j\omega) = \frac{b_m(j\omega)^m + b_{m-1}(j\omega)^{m-1} + \cdots + b_0}{(j\omega)^n + a_{n-1}(j\omega)^{n-1} + \cdots + a_0} \quad (4\text{-}51)$$

is stable. A system described in the frequency domain by the transfer function of (4-51) is described in the time domain by the differential equation

$$\frac{d^n y}{dt^n} + a_{n-1}\frac{d^{n-1} y}{dt^{n-1}} + \cdots + a_0 y = b_m \frac{d^m x}{dt^m} + b_{m-1}\frac{d^{m-1} x}{dt^{m-1}} + \cdots + b_0 x$$

where $x = x(t)$ is an input and $y = y(t)$ is the corresponding output. The system is stable if and only if all characteristic roots have negative real parts. The characteristic roots of the system are the roots of the characteristic polynomial

$$c(s) = s^n + a_{n-1}s^{n-1} + \cdots + a_0$$

If any characteristic root has nonnegative real part, the system is unstable.† The characteristic polynomial is just the denominator of the transfer function above with $j\omega$ replaced by s.

Example 4-28 Determine whether the system of Fig. 4-27 is stable.

SOLUTION First, we obtain the transfer function of the system. Transforming the block diagram to the frequency domain and using block-diagram reduction yields

$$H(j\omega) = \frac{K_1 K_2}{(j\omega)^2 + K_1 K_3(j\omega) + K_1 K_2 K_4}$$

By inspection of the denominator of $H(j\omega)$, the characteristic equation of the system is

$$s^2 + K_1 K_3 s + K_1 K_2 K_4 = 0$$

where $K_1 K_3 = 5 \times 10^3$ s^{-1} and $K_1 K_2 K_4 = 4 \times 10^6$ s^{-2}. The characteristic roots are $p_1 = -1 \times 10^3$ s^{-1} and $p_2 = -4 \times 10^3$ s^{-1}. The system is stable because both characteristic roots have negative real parts (both are real and negative).

---
†In theory, a system having a simple (nonrepeated) imaginary characteristic root is conditionally stable. In practice, such a system is regarded as unstable or at least as "insufficiently stable" because parameters of a physical system cannot be specified with sufficient accuracy to guarantee conditional stability.

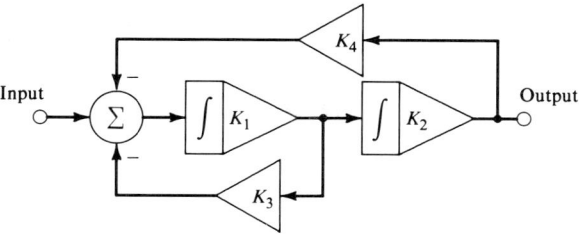

Figure 4-27 System of Example 4-28; $K_1 = K_2 = 10^3 \text{ s}^{-1}$, $K_3 = 5$, $K_4 = 4$.

4-4B Realizability

Frequency-domain conditions for realizability sometimes are confused with the condition for stability. It sometimes is asserted that a linear stationary system is nonrealizable if any characteristic root of the system has positive real part. The correct assertion is that such a system is either unstable or nonrealizable. To explain why, we consider the impulse response

$$h(t) = -pe^{pt}u(-t)$$

where $p > 0$. Figure 4-28 shows a graph of $h(t)$ versus t. A system having this impulse response is stable because $h(t)$ is absolutely integrable; however, the system is nonrealizable because $h(t) \neq 0$ for $t < 0$. With these facts in mind, we next examine this system in the frequency domain.

The transfer function of a system having impulse response $h(t)$ above is

$$H(j\omega) = \mathcal{F}\{h(t)\} = \frac{p}{j\omega - p}$$

This shows that p is the characteristic root of the system. Surprisingly, the transfer function can be realized, as shown by the block diagram of Fig. 4-29, even though the transfer function is derived from a nonrealizable impulse response; however, the resulting system is unstable because the characteristic root p is positive.

We can conclude from this discussion that a transfer function $H(j\omega) = p/(j\omega - p)$, where $p > 0$, represents two different systems: one stable but nonrealizable, the other realizable but unstable. In general, if a characteristic root obtained from a transfer function $H(j\omega)$ has positive real part, the transfer function $H(j\omega)$

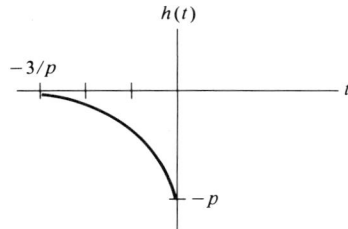

Figure 4-28 Impulse response of a nonrealizable, stable system.

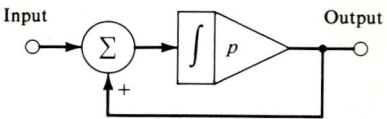

Figure 4-29 System having transfer function $H(j\omega) = p/(j\omega - p)$.

describes either a stable, nonrealizable system or a realizable, unstable system.† In practice, this ambiguity usually causes no difficulty because both possibilities are equally unacceptable. In this book, we assume that any rational transfer function of the form

$$H(j\omega) = \frac{b_m(j\omega)^m + b_{m-1}(j\omega)^{m-1} + \cdots + b_0}{(j\omega)^n + a_{n-1}(j\omega)^{n-1} + \cdots + a_0} \qquad (4\text{-}52)$$

is realizable. With this assumption, a characteristic root having positive real part implies instability, not nonrealizability. To justify this assumption we need only exhibit a realization of the transfer function $H(j\omega)$ above. The reader can show that the transfer function of the system of Fig. 4-30 is given by (4-52) (for $m \leq n$).

A general frequency-domain condition for realizability of a transfer function $H(j\omega)$ (not necessarily rational) is as follows. A transfer function $H(j\omega)$ that satisfies

$$\int_{-\infty}^{\infty} |H(j\omega)|^2 \, d\omega < \infty \qquad (4\text{-}53a)$$

is realizable if and only if

$$\int_{-\infty}^{\infty} \frac{|\log |H(j\omega)||}{1 + \omega^2} \, d\omega < \infty \qquad (4\text{-}53b)$$

†An exception to this rule arises if the system is to be passive (is to contain no energy sources); e.g., if a transfer function is to be realized as a passive electric circuit (one consisting only of resistors, capacitors, and inductors), a characteristic root having positive real part implies nonrealizability, not instability, because a passive electric circuit cannot be unstable.

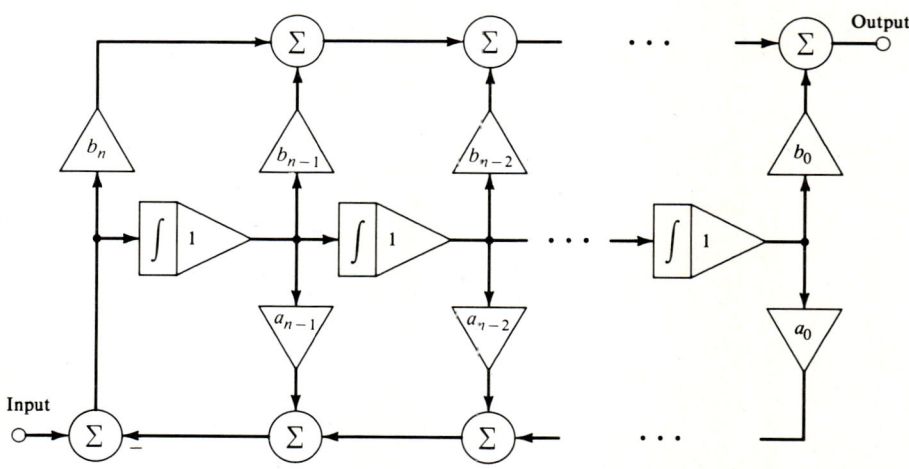

Figure 4-30 System having the transfer function of (4-52), showing that a rational transfer function is realizable.

This is called the *Paley-Wiener condition* for realizability. Its proof involves concepts beyond the scope of this book (see Papoulis, pp. 215–217).

Equation (4-53b) represents a necessary and sufficient condition for realizability of $H(j\omega)$ only if $H(j\omega)$ satisfies (4-53a). If $H(j\omega)$ fails to satisfy (4-53a), then (4-53b) is neither necessary nor sufficient. For example, the Paley-Wiener condition shows that an ideal low-pass filter is nonrealizable because the transfer function of such a filter satisfies (4-53a) and fails to satisfy (4-53b). On the other hand, the Paley-Wiener condition cannot be invoked to show that an ideal high-pass filter is nonrealizable (though this is indeed true) because the transfer function of an ideal high-pass filter fails to satisfy (4-53a).

The Paley-Wiener condition is an important contribution to the theory of linear stationary systems, but it is of little practical value. For one thing, it is inapplicable to transfer functions that fail to satisfy (4-53a). This renders the Paley-Wiener condition useless in many cases. For another, the Paley-Wiener condition is a condition on only the magnitude of a transfer function (the gain of a system). It cannot guarantee that a particular *complex* transfer function is realizable. At most, the Paley-Wiener condition can guarantee that a gain $\Gamma(\omega)$ can be associated with some unspecified phase shift $\phi(\omega)$ such that the transfer function $H(j\omega) = \Gamma(\omega)e^{j\phi(\omega)}$ is realizable. The Paley-Wiener condition suggests no method for finding the phase shift $\phi(\omega)$. Finally, it is often much more difficult to apply the Paley-Wiener condition to a transfer function $H(j\omega)$ than it is to obtain the associated impulse response $h(t) = \mathcal{F}^{-1}\{H(j\omega)\}$ and use the condition $h(t) = 0$ for $t < 0$.

Example 4-29 Show that the ideal low-pass transfer function

$$H(j\omega) = \begin{cases} 1 & |\omega| < \omega_0 \\ 0 & |\omega| \geq \omega_0 \end{cases}$$

is nonrealizable.

SOLUTION The associated impulse response is

$$h(t) = \mathcal{F}^{-1}\{H(j\omega)\} = \frac{1}{2\pi}\int_{-\omega_0}^{\omega_0} e^{j\omega t}\, d\omega = 2f_0 \text{ sa } \omega_0 t$$

where $f_0 = \omega_0/2\pi$. The transfer function is nonrealizable because the associated impulse response $h(t)$ is nonzero for $t < 0$. This shows that the particular complex transfer function $H(j\omega)$ above is nonrealizable. The Paley-Wiener condition can be used to show that *any* ideal low-pass transfer function is nonrealizable.

4-4C Linear Distortion

Virtually all signals requiring high-fidelity transmission, e.g., signals representing speech, music, television video, and data, are so-called *baseband signals* whose spectral densities are concentrated near the origin on a frequency axis. Consequently, almost all systems for high-fidelity transmission are on the whole (from source to destination) *low-pass systems*. The following discussion of linear distortion is therefore limited to low-pass systems. In Secs. 4-5B to 4-5D we show how the results are applied to systems containing bandpass subsystems. To avoid cluttering our treatment of linear distortion with things that are irrelevant (i.e., sign and magnitude of dc gain) we assume that the low-pass systems considered here have unit dc gain. This causes no loss of generality because we can always write the transfer function of a low-pass system as

$$H(j\omega) = H(0)\frac{H(j\omega)}{H(0)} = KH'(j\omega)$$

where $H'(j\omega) = H(j\omega)/H(0)$ is a system having unit dc gain. The scale factor $K = H(0)$ has no bearing on distortion introduced by the system.

The transfer function of a distortionless system having unit dc gain is

$$H_0(j\omega) = e^{-j\omega t_0}$$

The gain and phase shift of the system are

$$\Gamma_0(\omega) = 1 \qquad \phi_0(\omega) = -\omega t_0$$

The gain is independent of frequency and the phase shift is a linear function of frequency having phase intercept $\phi(0) = 0$. In the frequency domain, linear distortion can be separated into two components. Distortion introduced because a system has frequency-dependent gain is called *amplitude distortion*.† Distortion introduced because a system has nonlinear phase shift is called *phase distortion*. This analysis of linear distortion is often useful. For example, it has been determined experimentally that intelligibility of speech is relatively insensitive to phase distortion. This makes some systems for speech transmission less expensive than they would be otherwise because it costs less to build a system providing frequency-independent gain than one providing both frequency-independent gain and linear phase shift.

The delay introduced by a distortionless system having phase shift $\phi_0(\omega) = -\omega t_0$ is given by $t_0 = -d\phi_0(\omega)/d\omega$. In general, the quantity

$$D(\omega) = -\frac{d\phi(\omega)}{d\omega} \tag{4-54}$$

is called the *delay* of a system having phase shift $\phi(\omega)$. For a distortionless system, delay $D(\omega)$ is independent of frequency and equals actual delay introduced by the system. In general (for a system having nonlinear phase shift), delay $D(\omega)$ is frequency-dependent. A system is said to introduce *delay distortion* if the delay $D(\omega)$ of the system is frequency-dependent. In principle, phase distortion and delay distortion are simply two different descriptions of one component of linear distortion; however, in practical applications, delay distortion is used almost exclusively. There are mainly two reasons for preferring delay distortion to phase distortion as a measure of linear distortion: (1) it is easier to describe departure of delay from a frequency-independent value than it is to describe departure of phase shift from a linear function of frequency, and (2) if delay distortion introduced by a system is small, i.e., if delay $D(\omega)$ is almost frequency-independent, the quantity $D(0)$ can often be interpreted as actual delay introduced by a system. Indeed, in virtually all applications where delay distortion is

---

†This terminology is poor because it suggests that amplitude distortion is somehow dependent on the amplitude of an input. A better term would be gain distortion, but since "amplitude distortion" is entrenched, we must learn to live with it.

an important measure of performance, $D(\omega)$ is approximately frequency-independent and the quantity $D(0)$ is approximately equal to actual delay.

Example 4-30 To illustrate these remarks we describe linear distortion introduced by a system having transfer function

$$H(j\omega) = \frac{1}{1 + j\omega/\omega_0}$$

Figure 4-31 shows graphs of the associated gain $\Gamma(\omega)$, phase shift $\phi(\omega)$, and delay $D(\omega)$, where

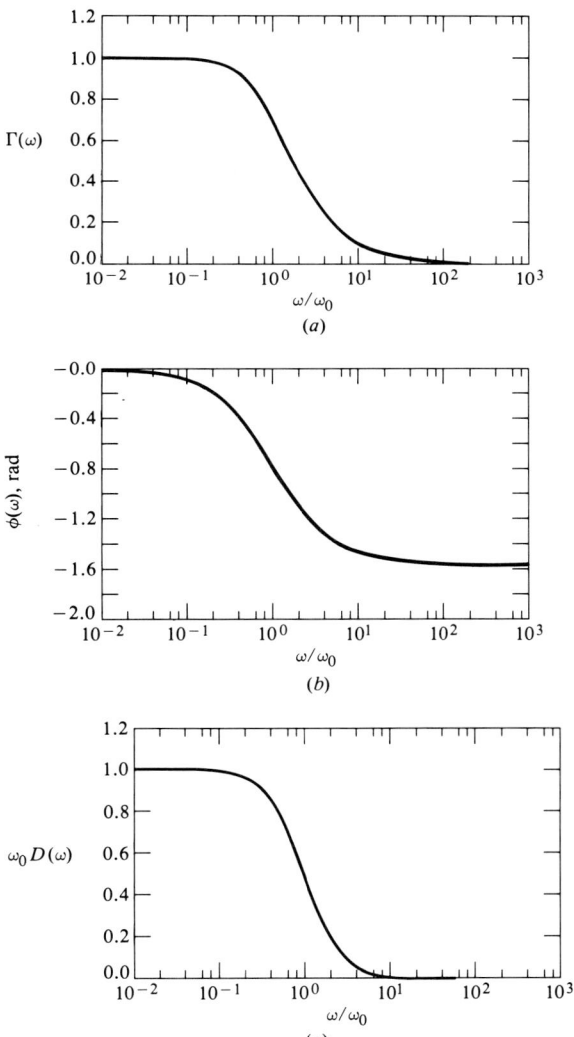

Figure 4-31 Gain, phase shift, and delay of the system of Example 4-30.

$$\Gamma(\omega) = |H(j\omega)| = \frac{1}{\sqrt{1 + (\omega/\omega_0)^2}} \qquad \phi(\omega) = \angle H(j\omega) = -\tan^{-1}\frac{\omega}{\omega_0}$$

$$D(\omega) = -\frac{d\phi(\omega)}{d\omega} = \frac{\omega_0}{\omega^2 + \omega_0^2}$$

For $\omega \ll \omega_0$, gain $\Gamma(\omega)$ and delay $D(\omega)$ are nearly independent of frequency; thus

$$\Gamma(\omega) \approx 1 \qquad D(\omega) \approx \omega_0^{-1} \qquad \omega \ll \omega_0$$

This means that the system introduces little distortion for any input occupying a band $0 \leq f \ll f_0$. The output $y(t)$ for an input $x(t)$ occupying a band $0 \leq f \ll f_0$ is given (approximately) by

$$y(t) \approx x(t - \omega_0^{-1})$$

To illustrate the discussion above we find the output of the system for input $x(t) = x_0 r(t/\tau)$. Note that the input $x(t)$ occupies (mainly) the band $0 \leq f < 1/\tau$ (see **Fig. 4-24**). The output is given by

$$y(t) = \mathcal{F}^{-1}\{H(j\omega)\mathcal{F}\{x(t)\}\}$$

where (from Table C-1)

$$\mathcal{F}\{x(t)\} = \frac{x_0(1 - e^{-j\omega\tau})}{j\omega}$$

This yields

$$y(t) = \mathcal{F}^{-1}\left\{\frac{x_0(1 - e^{-j\omega\tau})}{j\omega(1 + j\omega/\omega_c)}\right\}$$

$$= x_0 \mathcal{F}^{-1}\left\{\frac{1}{j\omega(1 + j\omega/\omega_0)}\right\} - x_0 \mathcal{F}^{-1}\left\{\frac{e^{-j\omega\tau}}{j\omega(1 + j\omega/\omega_0)}\right\}$$

Using the delay property of the Fourier transform gives

$$y(t) = x_0 \mathcal{F}^{-1}\left\{\frac{1}{j\omega(1 + j\omega/\omega_0)}\right\} - x_0 \mathcal{F}^{-1}\left\{\frac{1}{j\omega(1 + j\omega/\omega_0)}\right\}_{t \to t-\tau}$$

$$= x_0(1 - e^{-\omega_0 t})u(t) - x_0(1 - e^{-\omega_0(t-\tau)})u(t - \tau)$$

Figure 4-32 shows graphs of the amplitude spectral density of the input $x(t)$, the gain and the delay of the system, and the waveform of the output $y(t)$ for $f_0 = 10/\tau$ and for $f_0 = 1/\tau$. For $f_0 = 10/\tau$, the spectral density of $x(t)$ lies well within the nominal passband $0 \leq f < f_0$ of the system. In this case, the system introduces little distortion (the waveform of the output is a good approximation to the waveform of the input). The output is given to a good approximation by $y(t) \approx x_0 r[(t - t_0)/\tau]$, where $t_0 = 1/\omega_0$ and $\tau = 10/\omega_0$. The delay t_0 is imperceptible because it is a small fraction of the pulse duration. For $f_0 = 1/\tau$, the spectral density of the input $x(t)$ extends outside the passband of the system. In this case, the system introduces considerable distortion. The waveform of the output $y(t)$ is appreciably different from the waveform of the input.

Next, we describe (separately) typical effects of amplitude distortion and delay distortion (phase distortion). To do this we describe the outputs of several systems for the input (the triangular pulse) $x(t)$ of Fig. 4-33. The nominal bandwidth of the pulse $x(t)$ is $W_x = 2/\tau$.

FOURIER SERIES, FOURIER INTEGRAL, AND FOURIER TRANSFORMATION

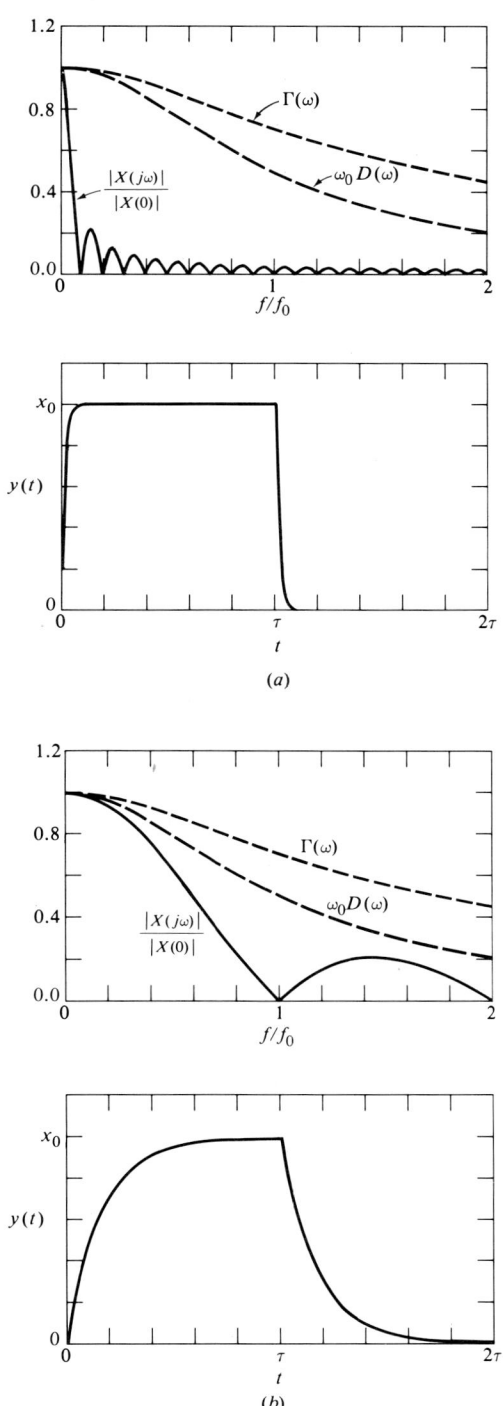

Figure 4-32 Linear distortion of a rectangular pulse having duration τ on transmission by a system having transfer function $H(j\omega) = 1/(1 + j\omega/\omega_0)$: (a) $f_0 = 10/\tau$, (b) $f_0 = 1/\tau$.

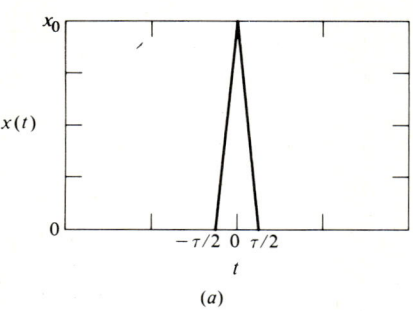 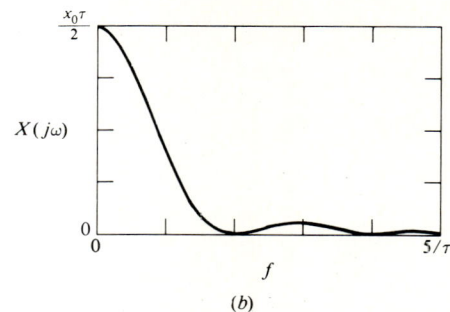

Figure 4-33 Triangular pulse used to illustrate effects of amplitude and phase distortion: (a) waveform, (b) spectral density.

Figure 4-34 shows graphs of the output $y(t)$ of a system having transfer function

$$H(j\omega) = \frac{1}{\sqrt{1 + (\omega/\omega_0)^2}}$$

for the input $x(t)$ of Fig. 4-33. The graphs of the output are normalized to unit peak amplitude to facilitate comparison with the input waveform of Fig. 4-33. This system introduces amplitude distortion but no delay distortion (the delay of the system is zero for all ω). The nominal bandwidth of the system is $W_s = f_0$. In Fig. 4-34a the system bandwidth is 2.5 times the input bandwidth, and little distortion is evident in the waveform of the output. In Fig. 4-34b the system bandwidth is half the input bandwidth, and some distortion is evident (the waveform of the output is a slightly rounded version of the waveform of the input). In Fig. 4-34c the system bandwidth is one-fifth of the input bandwidth, and considerable distortion is evident in the waveform of the output. This figure illustrates typical effects of amplitude distortion introduced by drooping gain (by inadequate bandwidth) on the waveform of a symmetric pulse: sharp features of the pulse are rounded (smoothed), and the pulse is smeared (spread out), but the pulse retains its symmetry.

Another kind of amplitude distortion arises from *passband ripple,* where the gain of a system is alternately above and below its average value (in the passband). Passband ripple is often exhibited by systems composed of several subsystems in series or parallel, e.g., when a number of tuned RLC circuits are connected in parallel to obtain a bandpass filter. To illustrate effects of passband ripple we use the input of Fig. 4-33 and a system having transfer function

$$H(j\omega) = 1 + \alpha \cos \omega t_0 \tag{4-55}$$

where $0 \leq \alpha \leq 1$. Figure 4-35 shows a graph of this transfer function. The peak amplitude of the ripple equals α, and the period (in hertz) of the ripple equals $1/t_0$. We assume that the bandwidth of the system is infinite in order to show effects of passband ripple apart from those of a drooping gain (or inadequate bandwidth).

The output of the system of Fig. 4-35 for input $x(t)$ is given by

$$y(t) = \mathcal{F}^{-1}\{(1 + \alpha \cos \omega t_0)X(j\omega)\}$$

FOURIER SERIES, FOURIER INTEGRAL, AND FOURIER TRANSFORMATION 231

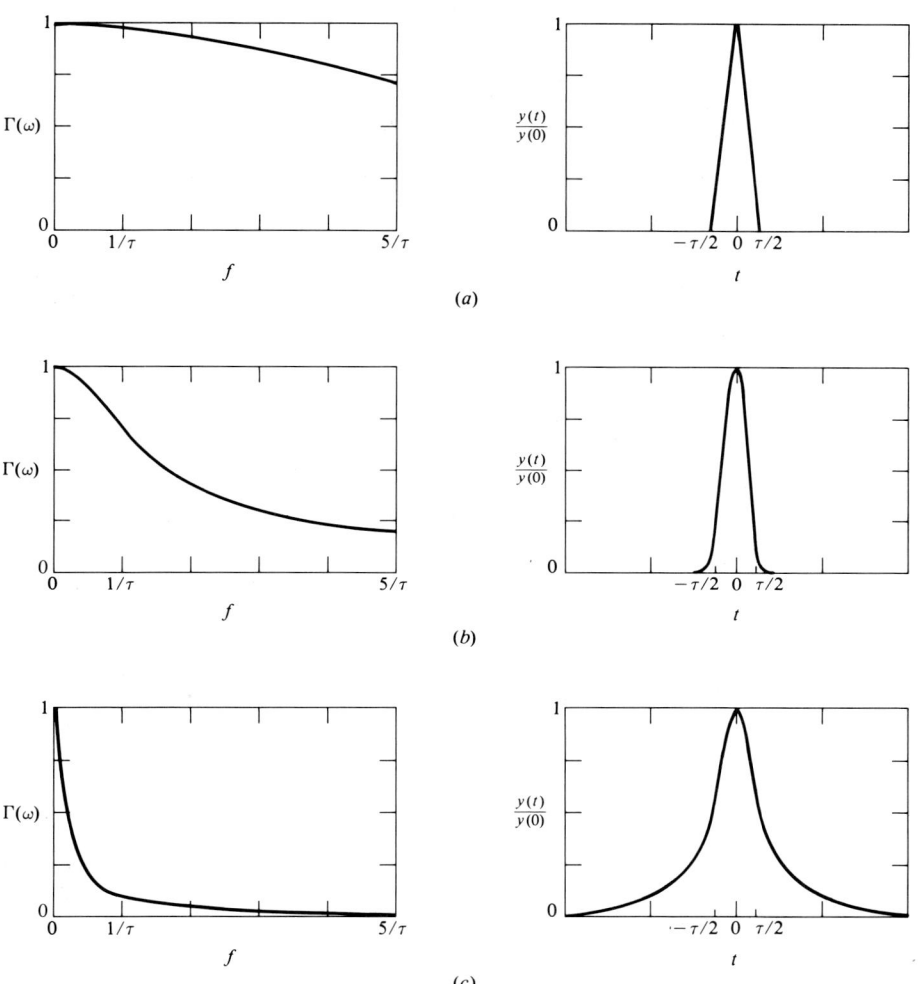

Figure 4-34 Amplitude distortion of a triangular pulse caused by inadequate bandwidth (drooping gain): (a) $f_0 = 5/\tau$, (b) $f_0 = 1/\tau$, (c) $f_0 = 1/10\tau$.

where $X(j\omega) = \mathcal{F}\{x(t)\}$ is the spectral density of $x(t)$. Using Euler's identity and the linearity of the Fourier transform gives

$$y(t) = \mathcal{F}^{-1}\{X(j\omega)\} + \frac{\alpha}{2}\mathcal{F}^{-1}\{X(j\omega)e^{j\omega t_0}\} + \frac{\alpha}{2}\mathcal{F}^{-1}\{X(j\omega)e^{-j\omega t_0}\}$$

Using the delay property of the Fourier transformation gives

$$y(t) = x(t) + \frac{\alpha}{2}x(t - t_0) + \frac{\alpha}{2}x(t + t_0)$$

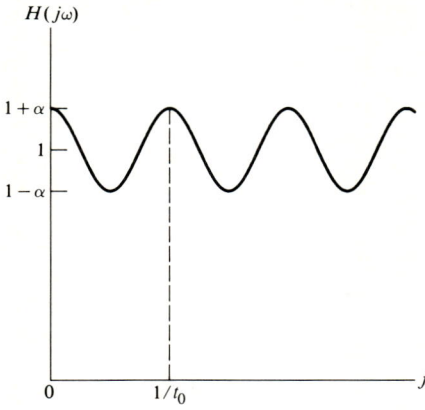

Figure 4-35 Transfer function having passband ripple.

The output $y(t)$ equals the input $x(t)$ plus so-called *paired echoes* of the form $x(t \pm t_0)$. The paired echoes constitute instantaneous distortion introduced by passband ripple. The amplitude of the paired echoes is $\alpha/2$, where α is the peak amplitude of the passband ripple. The separation of the paired echoes from the central pulse $x(t)$ equals t_0, where $1/t_0$ is the period of the passband ripple. Figure 4-36 illustrates these results for the input (triangular pulse) of Fig. 4-33 and for several values of α and t_0.

Next we describe effects of delay (phase) distortion arising from frequency-dependent delay (nonlinear phase shift) using the input (triangular pulse) of Fig. 4-33 and a system having gain $\Gamma(\omega) = 1$ (no amplitude distortion) and phase shift

$$\phi(\omega) = -\tan^{-1} \frac{\omega}{\omega_0}$$

The delay of the system is given by

$$D(\omega) = -\frac{d\phi}{d\omega} = \frac{\omega_0}{\omega^2 + \omega_0^2}$$

The delay is frequency-dependent because the phase shift is a nonlinear function of frequency. Figure 4-37 shows graphs of delay versus frequency and corresponding graphs of the output $y(t)$ versus time. The system introduces delay distortion because the delay departs from $D(0) = 1/\omega_0$ within the band $0 \leq f \leq 2/\tau$ occupied by the input $x(t)$.

For $f_0 = 5/\tau$ (Fig. 4-37a), delay is nearly independent of frequency over the band occupied by the signal, and very little distortion is evident in the waveform of the output. Also, since $D(0)$ is a small fraction of the duration of the pulse $x(t)$, the pulse suffers little actual delay. For $f_0 = 1/\tau$ (Fig. 4-37b), the delay $D(\omega)$ departs significantly from $D(0)$ over the band occupied by the input, and some distortion is apparent in the waveform of the output $y(t)$, i.e., the sharp corners of the input are rounded, and there is a small undershoot on the leading edge of the pulse; however, the pulse retains most of its original form. For $f_0 = 0.1/\tau$ (Fig. 4-37c), the delay is a strong function of frequency within the band occupied by the input, and considerable dis-

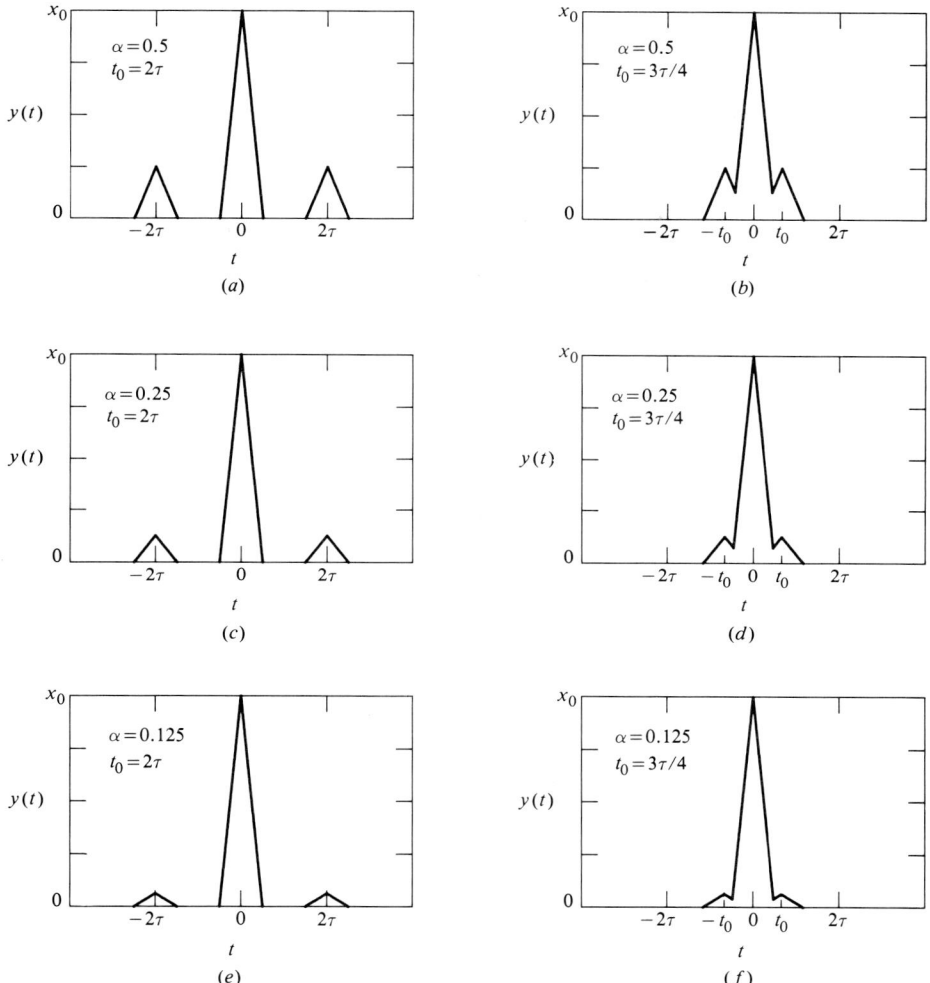

Figure 4-36 Paired-echo distortion of a triangular pulse caused by passband ripple.

tortion is apparent in the waveform of the output. The corners of the triangular pulse are completely rounded, the pulse is smeared (widened), and the symmetry of the pulse is destroyed.

Figures 4-34, 4-36, and 4-37 illustrate typical effects of linear distortion. Amplitude distortion arising from drooping gain or inadequate system bandwidth has a smoothing effect because it attenuates high-frequency components necessary for describing sharp peaks and fast transitions. This smoothing effect causes smearing of a pulse on transmission through a system whose bandwidth is smaller than the bandwidth of the pulse. Delay distortion (phase distortion) also causes smearing, but by a different mechanism. For a sum of sinusoids (spectral components) to describe a pulse, the sinusoids must interfere constructively over the interval (on a time axis) occupied by

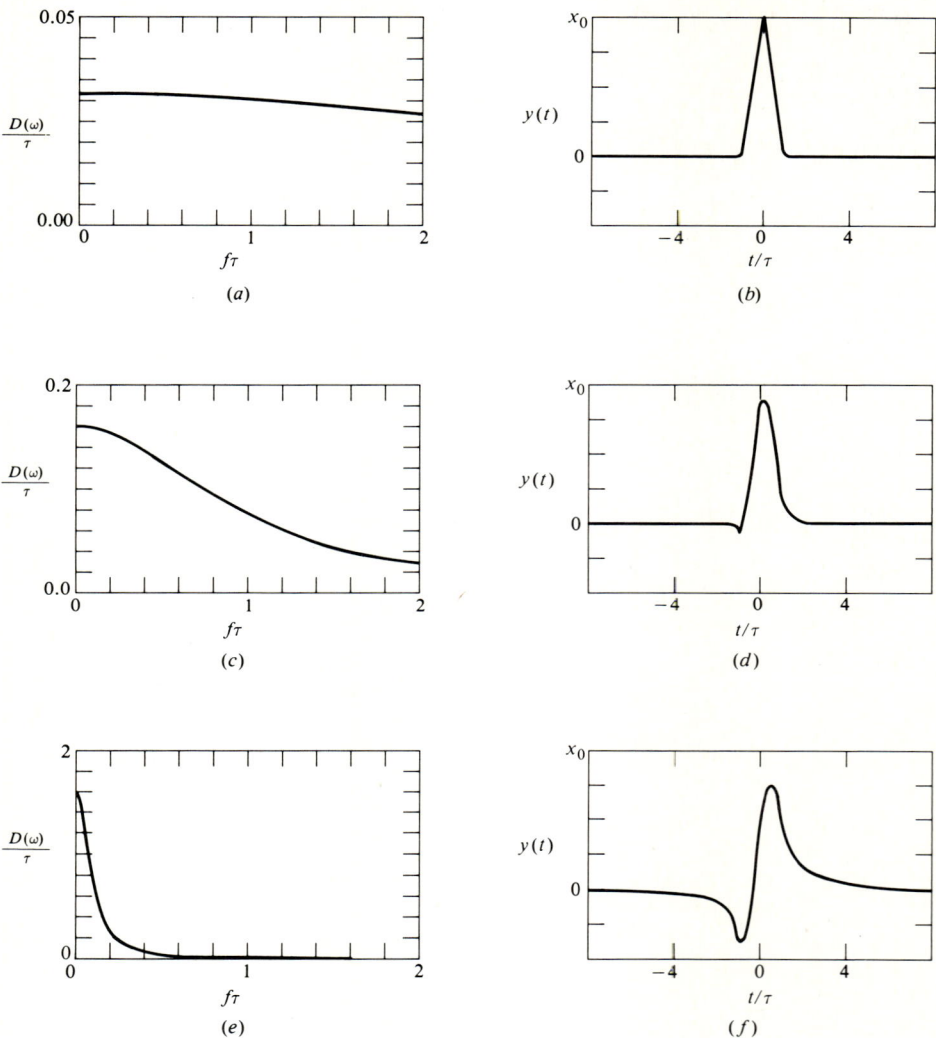

Figure 4-37 Phase (delay) distortion of a triangular pulse caused by nonlinear phase shift (frequency-dependent delay): (a) $f_0 = 5/\tau$, (b) $f_0 = 1/\tau$, (c) $f_0 = 1/10\tau$.

the pulse and destructively elsewhere; i.e., the sinusoids must pile up in the interval occupied by the pulse and must cancel elsewhere. Therefore the relative phases of the spectral components of a pulse are critical. Delay distortion (phase distortion) changes the relative phases of spectral components. This causes the spectral components of a pulse to interfere constructively (to pile up) outside the interval occupied by the pulse and to interfere destructively (to cancel) inside the interval occupied by the pulse. This gives rise to smearing, overshoots and undershoots, and asymmetry.

4-4D Transient Response

In practice, transient response of a system usually refers to the transient component of the output of the system for a step input. In this section we describe how various properties of a step response are reflected in the transfer function of the system.

Consider a low-pass system having the transfer function

$$H_2(j\omega) = \frac{1}{1 + j\omega/\omega_2}$$

The step response of the system is

$$g_2(t) = \mathscr{F}^{-1}\left\{\frac{H_2(j\omega)}{j\omega}\right\} = (1 - e^{-\omega_2 t})u(t)$$

Figure 4-38 shows graphs of the gain $|H_2(j\omega)|$, the step response $g_2(t)$, and the pulse response $y_2(t)$, defined by

$$y_2(t) = g_2(t) - g_2(t - \tau)$$

For present purposes, two things are significant: (1) the rise time of the step response is inversely proportional to the bandwidth of the system (a fast rise requires a large

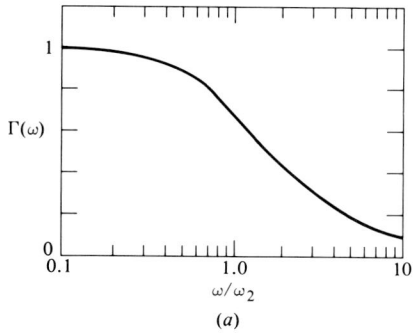
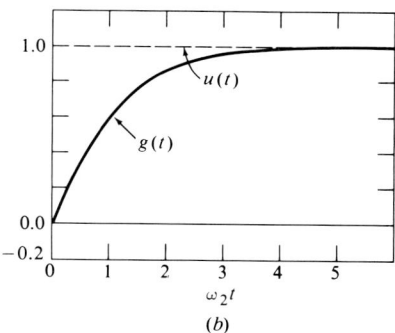
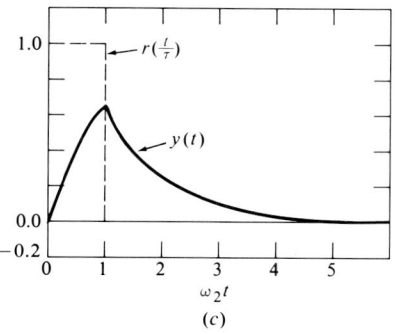

Figure 4-38 (a) Gain, (b) step response, and (c) pulse response of a first-order low-pass system.

236 INTRODUCTION TO SYSTEM ANALYSIS

bandwidth), and (2) the steady-state amplitude of the step response equals the dc gain $H_2(0) = 1$ of the system.

Now consider a high-pass system having the transfer function

$$H_1(j\omega) = \frac{j\omega/\omega_1}{1 + j\omega/\omega_1}$$

The step response of the system is

$$g_1(t) = e^{-\omega_1 t} u(t)$$

Figure 4-39 shows graphs of the gain, the step response, and the pulse response of the system. For present purposes, two things are significant: (1) the rise time equals zero (not surprising because the bandwidth of the passband is infinite), and (2) the dc gain equals zero. The step response sags toward zero with time constant $\tau = 1/\omega_1$. The sag time is inversely proportional to the cutoff frequency ω_1.

Finally, consider a bandpass system having transfer function

$$H(j\omega) = \frac{j\omega/\omega}{(1 + j\omega/\omega_1)(1 + \omega/\omega_2)} = H_1(j\omega)H_2(j\omega)$$

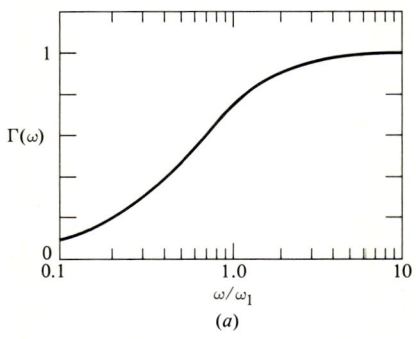
(a)

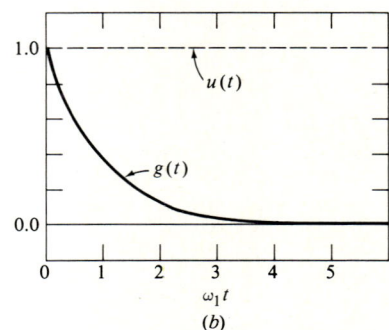
(b)

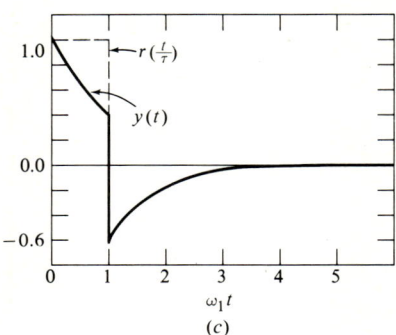
(c)

Figure 4-39 (a) Gain, (b) step response, and (c) pulse response of a first-order high-pass system.

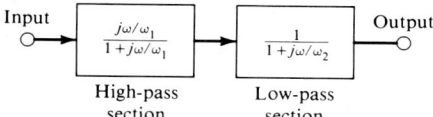

Figure 4-40 Block diagram for an overdamped second-order bandpass system.

where $\omega_2 \gg \omega_1$. This system can be represented by the block diagram of Fig. 4-40, where the subsystems are the low- and high-pass systems described above.

The step response of this bandpass system is

$$g(t) = a(e^{-\omega_1 t} - e^{-\omega_2 t})u(t)$$

where $a = \omega_2/(\omega_2 - \omega_1) \approx 1$ (because $\omega_2 \gg \omega_1$). Figure 4-41 shows graphs of the gain $|H(j\omega)|$ and the step response $g(t)$ for $\omega_2 = 10\omega_1$. We can analyze the step response of the bandpass system in terms of the step responses of the high- and low-pass subsystems as follows (see Fig. 4-40). The rise time of the high-pass subsystem is zero, and the sag time ($\approx 1/\omega_1$) of the high-pass subsystem is much larger than the rise time ($\approx 1/\omega_2$) of the low-pass subsystem. Consequently, the output of the high-pass sub-

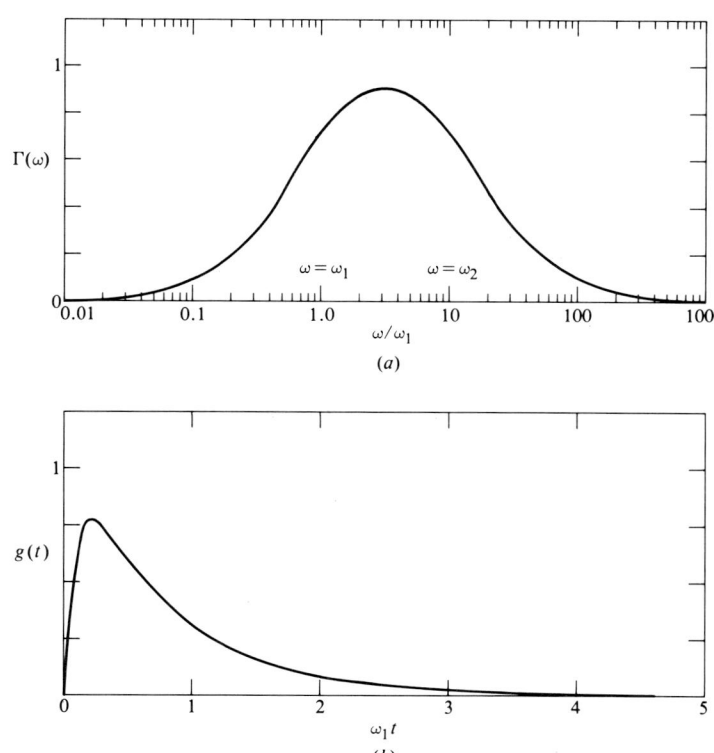

Figure 4-41 (a) Gain and (b) step response of the system of Fig. 4-40.

system (the input of the low-pass subsystem) looks like a step for $0 \leq t < 1/\omega_2$, and the step response of the bandpass system looks (for a short time) like the step response of the low-pass subsystem alone. The rise time of the step response $g(t)$ is proportional to $1/\omega_2$ as $g(t)$ approaches (initially) a value determined by the dc gain of the low-pass subsystem. For times approaching $1/\omega_1$, the output of the high-pass subsystem sags toward zero with time constant $1/\omega_1$. The output of the low-pass subsystem follows this sag because the time constant $1/\omega_2$ of the low-pass subsystem is much smaller than the time constant $1/\omega_1$ of the high-pass subsystem. These observations can be summarized as follows. The step response of a bandpass system having transfer function $H(j\omega)$ above, where $\omega_2 \gg \omega_1$, is characterized by a rise time inversely proportional to the upper cutoff frequency ω_2 and a sag time inversely proportional to the lower cutoff frequency ω_1. These results are approximately true for a high-order bandpass system, provided the upper cutoff frequency is much larger than the lower cutoff frequency.

The gains of the three systems considered above are almost flat (independent of frequency) in the passbands. The gains of many important systems have peaks in the passbands. Such peaks indicate resonance, which gives rise to a ringing step response. Consider a second-order underdamped system having transfer function

$$H(j\omega) = \frac{1}{(j\omega/\omega_0)^2 + 2\zeta j\omega/\omega_0 + 1} \quad (4\text{-}56)$$

where $0 < \zeta < 1$. From Table C-1, the step response of the system is

$$g(t) = [1 + \beta^{-1} e^{-\zeta \omega_0 t} \sin(\beta \omega_0 t - \theta)]u(t) \quad (4\text{-}57a)$$

where

$$\beta = \sqrt{1 - \zeta^2} \qquad \theta = \text{Tan}^{-1}(\beta, -\zeta) \quad (4\text{-}57b)$$

Figure 4-42 shows graphs of the gain versus ω/ω_0 and the step response versus $\omega_0 t$ (these normalizations of the independent variables make it possible to draw graphs corresponding to several values of ζ on single pairs of axes). Significant properties of the gain are the dc gain $\Gamma(0) = 1$, the peak gain Γ_p, the resonant frequency ω_p, and the bandwidth $W \approx \omega_0$. Significant properties of the step response are the steady-state amplitude $g_{ss} = 1$, the rise time t_R, the peak amplitude g_p, the peak time t_p, the ring frequency ω_p, and the settling time t_s. Figure 4-43 illustrates the definitions of these quantities.

The properties of the gain and the step response are determined by the two parameters ζ and ω_0 because these are the only parameters appearing in the expressions for $H(j\omega)$ and $g(t)$ above. The dimensionless parameter ζ is called *damping ratio*, and the parameter ω_0 (unit of ω) is called *undamped natural frequency*. In what follows we obtain expressions for resonant frequency ω_p, rise time t_R, and other quantities of interest in terms of damping ratio ζ and undamped natural frequency ω_0.

Resonant frequency ω_p and *peak gain* Γ_p are determined by the relation

$$\left. \frac{d\Gamma(\omega)}{d\omega} \right|_{\omega = \omega_p} = 0$$

This gives

$$\omega_p = \omega_0 \sqrt{1 - 2\zeta^2} \quad (4\text{-}58)$$

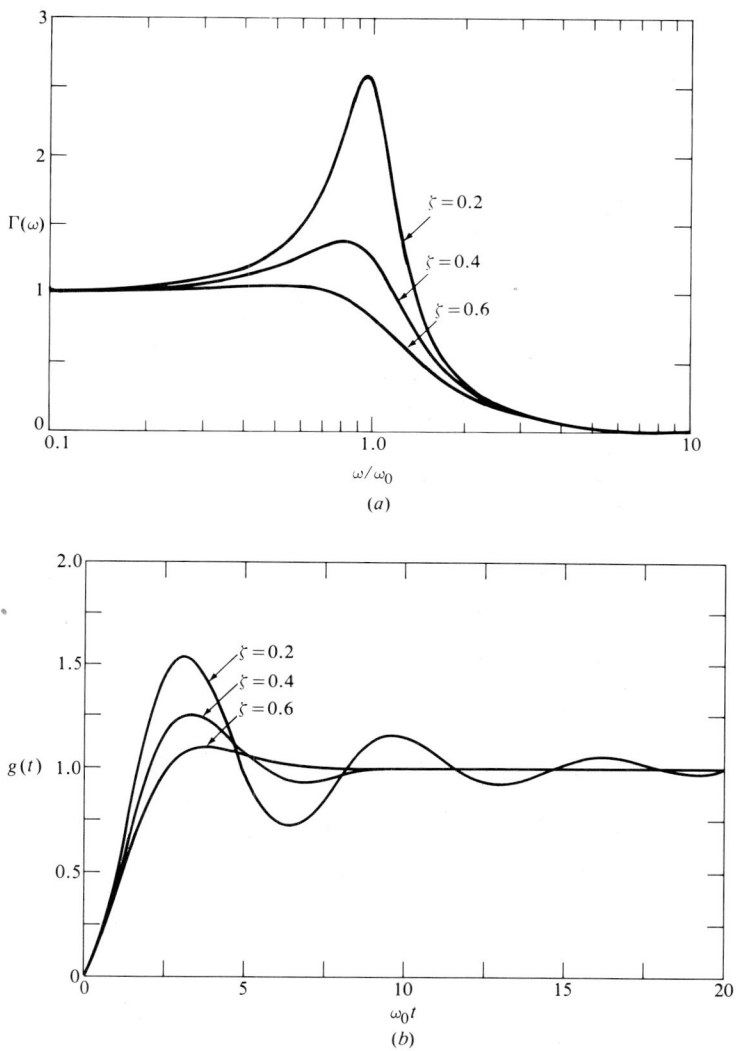

Figure 4-42 (a) Gain and (b) step response of an underdamped second-order system.

and
$$\Gamma_p = \frac{1}{2\zeta\sqrt{1-\zeta^2}} \tag{4-59}$$

From (4-58), resonant frequency ω_p is imaginary for $\zeta > 1/\sqrt{2}$, which means that there is no peak (above $\omega = 0$) in gain $\Gamma(\omega)$ for $\zeta > 1/\sqrt{2}$. Equation (4-59) shows that Γ_p depends only on ζ. Solving (4-59) for ζ^2 gives

$$2\zeta^2 = 1 - \sqrt{1 - \frac{1}{\Gamma_p^2}} \tag{4-60a}$$

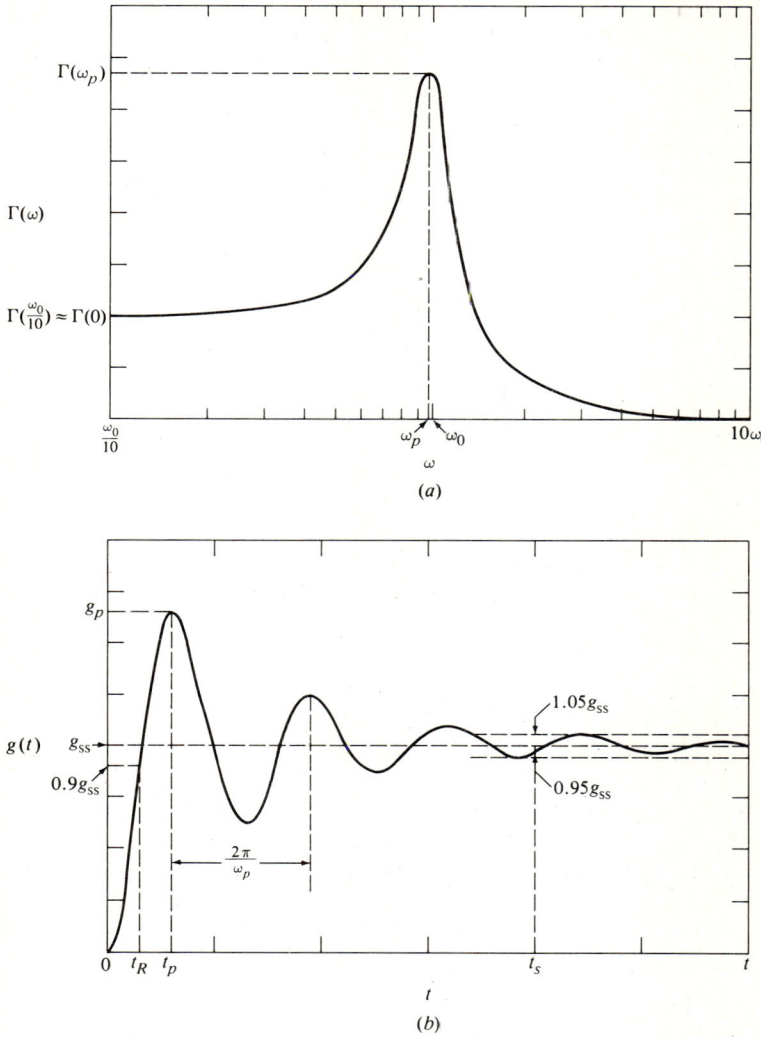

Figure 4-43 Definition of quantities used to describe (*a*) gain and (*b*) step response of an underdamped second-order system.

For Γ_p greater than about 2, this yields

$$\zeta \approx \frac{1}{2\Gamma_p} \qquad (4\text{-}60b)$$

Figure 4-44 shows a graph of damping ratio ζ versus peak gain Γ_p. This graph can be used to obtain Γ_p from ζ (or vice versa) if the approximation (4-60*b*) cannot be used. Once the damping ratio ζ is determined, the undamped natural frequency ω_0 can be calculated using (4-58); thus

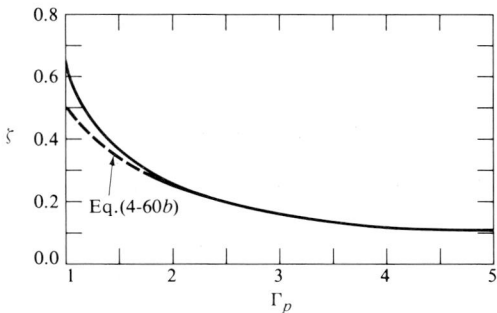

Figure 4-44 Damping ratio ζ versus peak gain Γ_p for an underdamped second-order system.

$$\omega_0 = \frac{\omega_p}{\sqrt{1 - \zeta^2}} \quad (4\text{-}61)$$

The *rise time* for the step response is determined by the definition $g(t_R) = 0.9$ (see Fig. 4-43). This gives

$$e^{-\zeta\omega_0 t_R} \sin(\beta\omega_0 t_R - \theta) = -0.1\beta$$

This equation can be solved numerically for $\omega_0 t_R$ as a function of ζ. Figure 4-45 shows a graph of the solution. For $0 < \zeta < 0.7$, normalized rise time $\omega_0 t_R$ is an almost linear function of damping ratio ζ. A least-squares fit over this interval gives

$$\omega_0 t_R \approx 1.38 + 1.62\zeta \quad 0 < \zeta < 0.7 \quad (4\text{-}62)$$

as shown by the dashed line in Fig. 4-45. Note that for any particular value of ζ, rise time t_R is approximately inversely proportional to bandwidth (undamped natural frequency) ω_0.

Peak time t_p is determined from

$$\left.\frac{dg(t)}{dt}\right|_{t=t_p} = 0$$

This gives

$$\omega_0 t_p = \frac{\pi}{\sqrt{1 - \zeta^2}} \quad (4\text{-}63)$$

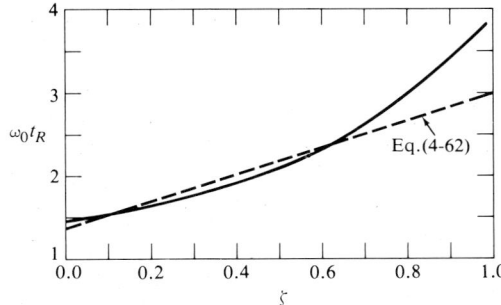

Figure 4-45 Normalized rise time $\omega_0 t_R$ versus damping ratio ζ for an underdamped second-order system.

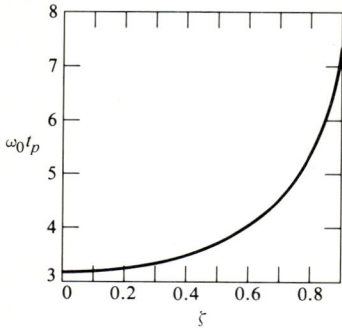

Figure 4-46 Normalized peak time $\omega_0 t_p$ versus damping ratio ζ for an underdamped second-order system.

Figure 4-46 shows a graph of $\omega_0 t_p$ versus ζ. For any particular value of damping ratio ζ, peak time t_p is inversely proportional to bandwidth ω_0. Note that peak time is defined only if $0 \le \zeta < 1$ (only for an underdamped system). For $\zeta = 1$, the system is critically damped. For $\zeta > 1$, the system is overdamped. In either case, the step response approaches its steady-state amplitude from below, and no peak time exists.

The maximum amplitude of the step response is

$$g_p = g(t_p) = 1 + \exp\left(\frac{-\pi\zeta}{\sqrt{1-\zeta^2}}\right)$$

The *percent overshoot* (PO) defined by

$$\text{PO} = 100(g_p - 1) = 100\exp\left(\frac{-\pi\zeta}{\sqrt{1-\zeta^2}}\right) \tag{4-64}$$

is often of interest, e.g., in automatic control systems. Figure 4-47 shows a graph of percent overshoot versus damping ratio. Percent overshoot ranges from 100 percent for $\zeta = 0$ to zero for $\zeta = 1$.

The *settling time* for a second-order system is a time after which the step response is always within a few percent (often 5 percent) of its steady-state amplitude. It is difficult to obtain an exact expression for settling time. It is often approximated as

$$\omega_0 t_s \approx \frac{3}{\zeta} \tag{4-65}$$

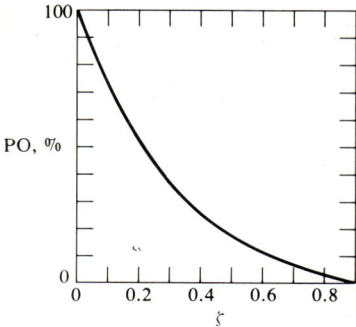

Figure 4-47 Percent overshoot of step response versus damping ratio for an underdamped second-order system.

which represents three time constants for the exponential damping factor $e^{-\zeta\omega_0 t}$ in the expression for step response. Note that for any particular value of ζ, settling time is approximately inversely proportional to bandwidth.

There is no reason to memorize the formulas given above. They can be looked up (or derived) when needed. However, it is worthwhile to learn the following qualitative relationships between gain $\Gamma(\omega)$ and step response $g(t)$ for an underdamped second-order system:

1. The steady-state amplitude of the step response is equal to the dc gain $H(0)$.
2. Rise time, peak time, and settling time vary inversely with bandwidth.
3. Percent overshoot increases if peak gain is increased or if damping ratio is decreased.

Such qualitative relationships are of considerable value in system engineering. They make it possible to infer (at a glance) important time-domain properties of a system from frequency-domain properties of the system and vice versa.

4-4E Approximations for Systems of High Order

Systems of high order can often be approximated (modeled) as first- or second-order systems. For example, consider an overdamped second-order system having the transfer function

$$H(j\omega) = \frac{1}{(1 + j\omega/\omega_1)(1 + j\omega/\omega_2)}$$

where $\omega_2 \gg \omega_1$. This system can be analyzed as a series connection of two first-order low-pass subsystems: one having bandwidth $f_1 = \omega_1/2\pi$ and the other having bandwidth $f_2 = \omega_2/2\pi$. The subsystem having bandwidth f_2 has little effect on an input occupying a band $0 \leq f \ll f_2$ (the subsystem provides almost distortionless transmission for such an input). The transfer function of the system can be approximated as

$$H(j\omega) \approx \frac{1}{1 + j\omega/\omega_1} \qquad 0 \leq f \ll f_2$$

for any input occupying that band.

Similarly, for $\omega_2 \gg \omega_1$ the bandpass transfer function

$$H(j\omega) = \frac{j\omega/\omega_1}{(1 + j\omega/\omega_1)(1 + j\omega/\omega_2)}$$

can be approximated by a first-order high-pass system for inputs occupying a band $0 \leq f \ll f_2$ and by a first-order low-pass system for inputs occupying a band $f_1 \ll f < \infty$.

Finally, a high-order system having complex-conjugate characteristic roots can often be approximated by a second-order underdamped system. For example, consider a third-order system with transfer function

$$H(j\omega) = \frac{\omega_0^2}{(1 + j\omega/\omega_1)(\omega_0^2 + 2j\zeta\omega\omega_0 - \omega^2)}$$

where $\omega_1 \gg \omega_0$ and $\zeta < 1$ (where the second-order section is underdamped). For inputs occupying a band $0 \le f \ll f_1$, the system can be approximated as a second-order system having transfer function

$$H(j\omega) = \frac{\omega_0^2}{\omega_0^2 + 2j\zeta\omega\omega_0 - \omega^2}$$

Example 4-31 Obtain the output of a system having transfer function

$$H(j\omega) = \frac{1}{(1 + j\omega/\omega_1)(1 + j\omega/\omega_2)(1 + j\omega/\omega_3)}$$

for input $x(t) = x_0 r(t/\tau)$, where $f_1 = 5$ kHz, $f_2 = 50$ kHz, $f_3 = 100$ kHz, $x_0 = 5$ V, and $\tau = 1$ ms.

SOLUTION The system can be analyzed as a series connection of three first-order low-pass subsystems:

$$H(j\omega) = \frac{1}{(1 + j\omega/\omega_1)} \frac{1}{(1 + j\omega/\omega_2)} \frac{1}{(1 + j\omega/\omega_3)}$$

The input occupies the band $0 \le f < 1/\tau$, where $1/\tau = 1$ kHz. Therefore the low-pass subsystems represented by the last two factors have little effect on the input. For input $x(t)$ above, the transfer function can be approximated by

$$H(j\omega) \approx \frac{1}{1 + j\omega/\omega_1}$$

The output is given to a good approximation by (see Example 4-30)

$$y(t) \approx x_0(1 - e^{-\omega_1 t})u(t) - x_0(1 - e^{-\omega_1(t-\tau)})u(t - \tau)$$

4-5 APPLICATIONS OF FREQUENCY-DOMAIN ANALYSIS

In this section we describe applications of frequency-domain analysis to systems for filtering, signal transmission, and automatic control. Our purpose is to illustrate how concepts introduced above arise in specific practical applications. We do not attempt a thorough treatment of the applications themselves.

4-5A Low-Pass Butterworth Filter Design

Designing a filter can be separated into two tasks: The first, called *approximation,* is to obtain a realizable transfer function $H(j\omega)$ that approximates the transfer function of an ideal filter. The second, called *synthesis,* is to define a physical system (e.g., an electric circuit) having transfer function $H(j\omega)$. We describe a particular solution to the approximation problem in enough detail to illustrate certain frequency-domain concepts introduced above. We do not discuss synthesis.

Consider the problem of approximating the transfer function of an ideal low-pass filter by a realizable transfer function of the form

$$H(j\omega) = \frac{K}{c(j\omega)} \qquad (4\text{-}66a)$$

where K is a constant and $c(j\omega)$ is a polynomial of the form

$$c(j\omega) = (j\omega)^n + a_{n-1}(j\omega)^{n-1} + \cdots + a_1(j\omega) + a_0 \qquad (4\text{-}66b)$$

The degree n of $c(j\omega)$ is the order of a filter described by (4-66). The order n largely determines how accurately $H(j\omega)$ above approximates an ideal low-pass transfer function. It also largely determines the complexity (the number of components) and thus the cost of a filter described by (4-66); e.g., an electric circuit described by (4-66) contains n energy-storage elements (capacitors and inductors). The approximation problem is to find $c(j\omega)$ such that $H(j\omega)$ describes a realizable, stable low-pass filter that approximates an ideal low-pass transfer function. We begin by describing how realizability and stability are incorporated in the approximation procedure.

We require that $H(j\omega)$ given by (4-66) be realizable and that the realization be stable. Realizability requires only that the coefficients $a_0, a_1, \ldots, a_{n-1}$ be real (see Fig. 4-30). Stability requires that all roots of the characteristic polynomial $c(s)$ have negative real parts. These requirements are incorporated in the approximation procedure described below, as follows. The squared gain of a filter described by (4-66) is given by

$$\Gamma^2(\omega) = |H(j\omega)|^2 = H(j\omega)H^*(j\omega) = \frac{K^2}{c(j\omega)c^*(j\omega)}$$

Because the coefficients of $c(j\omega)$ are real (for realizability), we have $c^*(j\omega) = c(-j\omega)$ and

$$\Gamma^2(\omega) = \frac{K^2}{c(j\omega)c(-j\omega)}$$

Consider the polynomial

$$F(s) = c(s)c(-s)$$

obtained by replacing $j\omega$ by s in the denominator of $\Gamma^2(\omega)$ above. Let $p_0, p_1, p_2, \ldots, p_{n-1}$ denote the roots of $c(s)$. Then the roots of $c(-s)$ are $-p_0, -p_1, \ldots, -p_{n-1}$, and the roots of $F(s)$ are $\pm p_0, \pm p_1, \pm p_2, \ldots, \pm p_{n-1}$. Thus, for every root p_m of $F(s)$ there is also a root $-p_m$. Furthermore, complex roots of $c(s)$ occur in conjugate pairs because the coefficients of $c(s)$ are real. Therefore complex roots of $F(s)$ also occur in conjugate pairs. For every complex root p_m of $F(s)$ there is also a root p_m^*. It follows from these observations that the roots of $F(s)$ exhibit the double symmetry illustrated by Fig. 4-48, where roots of $F(s)$ are represented by points (crosses) in a complex plane. Note in particular that for each root having positive real part there is a mirror-image root having negative real part.

The double symmetry of the roots of $F(s)$ illustrated by Fig. 4-48 is a consequence of the definition $F(s) = c(s)c(-s)$, where $c(s)$ is a polynomial having real coeffi-

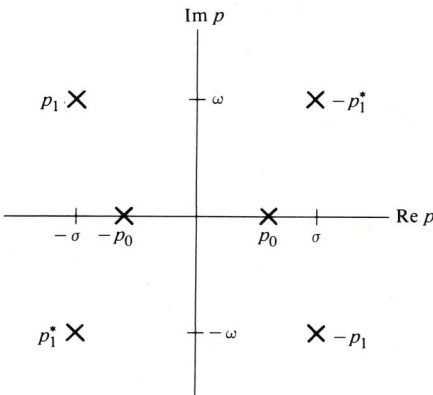

Figure 4-48 An illustration of the double symmetry of the roots of $c(s)c(-s)$, where $c(s)$ is a polynomial having real coefficients.

cients. Conversely, any polynomial $F(s)$ whose roots have the double symmetry illustrated by Fig. 4-48 can be factored as $F(s) = c(s)c(-s)$, where $c(s)$ is a polynomial whose coefficients are real and whose roots all have negative real parts. This factoring is done by associating all roots in the left half of the complex plane with $c(s)$ and all roots in the right half with $c(-s)$. A polynomial $c(s)$ obtained in this manner is the characteristic polynomial of a realizable, stable system. Therefore, we can obtain the transfer function of a realizable, stable low-pass filter as follows:

1. Approximate the squared gain of an ideal low-pass filter using a function of the form

$$\Gamma^2(\omega) = |H(j\omega)|^2 = \left.\frac{K^2}{F(s)}\right|_{s=j\omega} \qquad (4\text{-}67)$$

where $F(s)$ is a polynomial whose roots have the double symmetry illustrated in Fig. 4-48.
2. Factor $F(s)$ as $F(s) = c(s)c(-s)$, where the roots of $c(s)$ are the roots of $F(s)$ having negative real parts.
3. Form the transfer function of the filter as

$$H(j\omega) = \frac{K}{c(j\omega)} \qquad (4\text{-}68)$$

There is an infinite number of polynomials whose roots have the double symmetry illustrated in Fig. 4-48. We can form as many such polynomials as we like simply by laying out doubly symmetric but otherwise arbitrary root patterns in a complex plane. The heart of the approximation problem is to find an appropriate root pattern for $F(s)$ from specifications on gain of a filter. Several solutions to this problem have been found. Below, we describe a solution given by Butterworth.[†]

[†] S. Butterworth, "On the Theory of Filter Amplifiers," *Experimental Wireless and Wireless Engineering*, 1930, pp. 536–541.

Butterworth's solution consists of taking $F(s)$ in (4-67) to be a polynomial of the form

$$F(s) = 1 + (-1)^n \left(\frac{s}{\omega_c}\right)^{2n} \qquad (4\text{-}69)$$

where the order n (a positive integer) and the parameter ω_c are determined from specifications on gain of a low-pass filter. A polynomial given by (4-69) is called a *Butterworth polynomial*. A filter obtained using a Butterworth polynomial for $F(s)$ in (4-67) is called a *Butterworth filter*. In what follows we first show that the roots of a Butterworth polynomial have the necessary double symmetry. Subsequently, we show how n and ω_c in (4-69) are determined from specifications on gain.

The $2n$ roots $p_0, p_1, \ldots, p_{2n-1}$ of a Butterworth polynomial are the solutions of

$$F(p) = 1 + (-1)^n \left(\frac{p}{\omega_c}\right)^{2n} = 0$$

This gives

$$p = \omega_c(-1)^{(n+1)/2n} = j\omega_c(-1)^{1/(2n)}$$

By De Moivre's theorem (see Thomas and Finney, p. 836) the $2n$ roots of -1 are given by

$$(-1)^{1/(2n)} = e^{j(2k+1)\pi/2n} \qquad k = 0, 1, 2, \ldots, 2n-1$$

Therefore the roots of a Butterworth polynomial are given by

$$p_k = j\omega_c e^{j(2k+1)\pi/2n} \qquad k = 0, 1, 2, \ldots, 2n-1 \qquad (4\text{-}70)$$

The roots given by (4-70) lie on a circle having radius ω_c, as illustrated (for $n = 3$) by Fig. 4-49. The roots having negative real parts are given by (4-70) for $k = 0, 1, 2, \ldots, n-1$.

The characteristic polynomial for a low-pass Butterworth filter is

$$c(s) = \frac{(s - p_0)(s - p_1) \cdots (s - p_{n-1})}{(-p_0)(-p_1) \cdots (-p_{n-1})}$$

where $p_0, p_1, \ldots, p_{n-1}$ are given by (4-70). It can be shown using (4-70) that $(-p_0)(-p_1) \cdots (-p_{n-1}) = (-\omega_c)^n$; thus

$$c(s) = (-\omega_c)^{-n}(s - p_0)(s - p_1) \cdots (s - p_{n-1}) \qquad (4\text{-}71)$$

From (4-68), (4-70), and (4-71), the transfer function of a low-pass Butterworth filter is

$$H(j\omega) = \frac{K(-\omega_c)^n}{(j\omega - p_0)(j\omega - p_1) \cdots (j\omega - p_{n-1})} \qquad (4\text{-}72a)$$

where

$$p_k = j\omega_c e^{j(2k+1)\pi/2n} \qquad (4\text{-}72b)$$

From (4-67) and (4-69), the gain of a Butterworth filter is given by

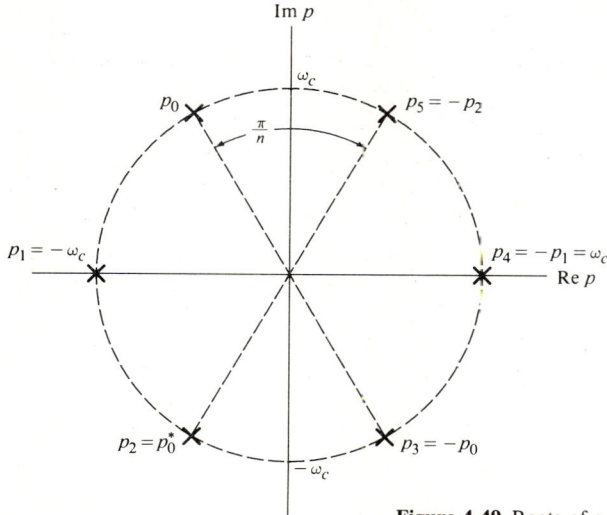

Figure 4-49 Roots of a third-degree Butterworth polynomial.

$$\Gamma(\omega) = \frac{|K|}{\sqrt{1 + (\omega/\omega_c)^{2n}}} = \frac{|K|}{\sqrt{1 + (f/f_c)^{2n}}} \qquad (4\text{-}73)$$

Figure 4-50 shows graphs of gain versus f/f_c. The gain of a Butterworth filter approximates the gain of an ideal low-pass filter having dc gain K and bandwidth f_c. The approximation improves as the order n of the filter is increased. Next we show how the parameters K, n, and ω_c are determined from specifications on gain of a Butterworth filter.

In particular applications the dc gain K, the order n, and the so-called cutoff frequency f_c of a Butterworth filter are determined from specifications on gain versus

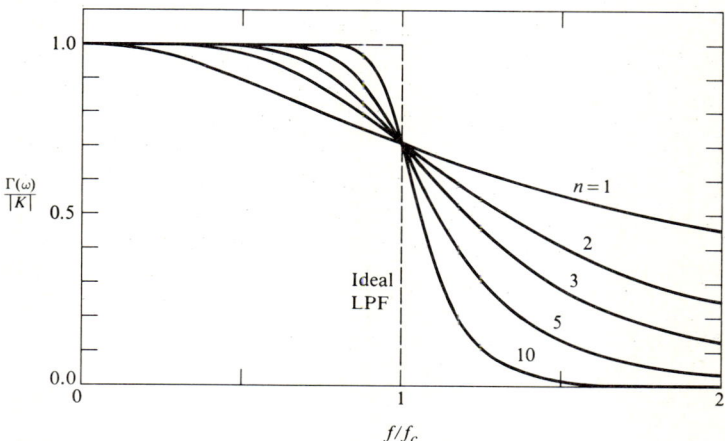

Figure 4-50 Gain of a low-pass Butterworth filter.

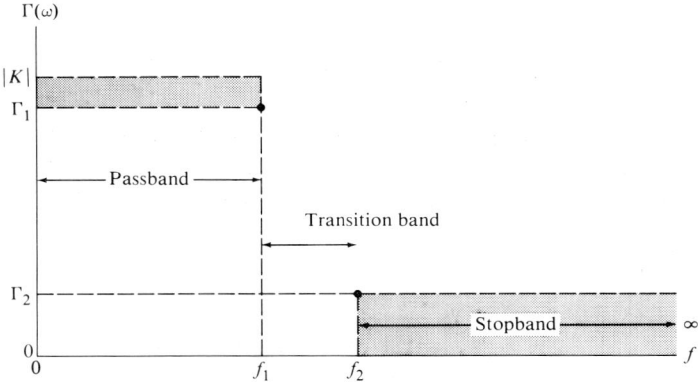

Figure 4-51 Specifications on gain for a low-pass filter.

frequency, as illustrated in Fig. 4-51. The frequency f_1 is called the passband edge. The gain Γ_1 is the minimum gain in the passband. The frequency f_2 is called the stopband edge. The gain Γ_2 is the maximum gain in the stopband. The band $f_1 < f < f_2$ is called the transition band. The filter gain $\Gamma(\omega)$ must lie in the shaded regions.

From Fig. 4-51 and (4-73), we have

$$\left(\frac{f_1}{f_c}\right)^{2n} = \left(\frac{K}{\Gamma_1}\right)^2 - 1 = \gamma_1 \tag{4-74a}$$

and

$$\left(\frac{f_2}{f_c}\right)^{2n} = \left(\frac{K}{\Gamma_2}\right)^2 - 1 = \gamma_2 \tag{4-74b}$$

Dividing (4-74a) by (4-74b) gives $(f_1/f_2)^{2n} = \gamma_1/\gamma_2$, whence

$$n = \operatorname{int} \frac{\ln(\gamma_1/\gamma_2)}{2 \ln(f_1/f_2)} + 1 \tag{4-75a}$$

where int α denotes the largest integer less than α. Once the filter order n is determined, we can use either (4-74a) or (4-74b) to determine the cutoff frequency f_c. Unless the argument of the int function in (4-75a) happens to be an integer, (4-74a) and (4-74b) give different values for f_c. Using (4-74a) to determine f_c yields $\Gamma(2\pi f_1) = \Gamma_1$ and $\Gamma(2\pi f_2) \leq \Gamma_2$, whereas using (4-74b) to determine f_c yields $\Gamma(2\pi f_1) \geq \Gamma_1$ and $\Gamma(2\pi f_2) = \Gamma_2$; that is, determining f_c from (4-74a) yields a filter that just meets the passband specification and surpasses the stopband specification, whereas determining f_c from (4-74b) yields a filter that just meets the stopband specification and surpasses the passband specification. No matter which of the relations (4-74) is used to obtain f_c, a slight variation of a parameter of the filter, e.g., a capacitance, can cause the gain of the filter to fail to meet the specifications because in either case the gain just meets one specification. To make the design more secure we calculate two cutoff frequencies, one from each of the relations (4-74), and use the geometric mean of these two frequencies for f_c. Thus

250 INTRODUCTION TO SYSTEM ANALYSIS

$$f_c = \sqrt{f_{c_1} f_{c_2}} \qquad (4\text{-}75b)$$

where, from (4-74),

$$f_{c_1} = f_1 e^{-(\ln \gamma_1)/2n} \qquad f_{c_2} = f_2 e^{-(\ln \gamma_2)/2n} \qquad (4\text{-}75c)$$

Example 4-32 Obtain the transfer function of a Butterworth filter that meets the following specifications (see Fig. 4-51):

dc gain = 1 V/V = K   Stopband edge = 10 kHz = f_2
Passband edge = 3.0 kHz = f_1 Maximum stopband gain = 0.1 V/V = Γ_2
Minimum passband gain = 0.9 V/V = Γ_1

SOLUTION First we calculate the quantities γ_1, γ_2 defined by (4-74):

$$\gamma_1 = \left(\frac{1}{0.9}\right)^2 - 1 = 0.23 \qquad \gamma_2 = \left(\frac{1}{0.1}\right)^2 - 1 = 99$$

We determine the filter order using (4-75a), which gives

$$n = \text{int} \frac{\ln(\gamma_1/\gamma_2)}{2 \ln(f_1/f_2)} + 1 = \text{int} \frac{\ln(0.23/99)}{2 \ln(3.0/10)} + 1 = \text{int } 2.52 + 1 = 3$$

We use (4-75b) and (4-75c) to determine the cutoff frequency f_c. This gives

$$f_{c_1} = 3.0 \times 10^3 e^{-(\ln 0.24)/6} = 3.8 \text{ kHz} \qquad f_{c_2} = 10 \times 10^3 e^{-(\ln 99)/6} = 4.6 \text{ kHz}$$

and

$$f_c = \sqrt{(3.8)(4.6)} = 4.2 \text{ kHz}$$

Using (4-72b) to determine the characteristic roots of the filter gives

$$p_0 = j\omega_c e^{j\pi/6} = \frac{\omega_c}{2}(-1 + j\sqrt{3}) \qquad p_1 = j\omega_c e^{j3\pi/6} = -\omega_c$$

$$p_2 = j\omega_c e^{j5\pi/6} = \frac{\omega_c}{2}(-1 - j\sqrt{3}) = p_0^*$$

We use (4-72a) to obtain the transfer function of the filter. This gives

$$H(j\omega) = \frac{-K\omega_c^3}{(j\omega + \omega_c)(-\omega^2 + j\omega\omega_c + \omega_c^2)}$$

Finally, we determine the value of K from the specification on dc gain. This gives $H(0) = -K = 1$ V/V, whence $K = -1$ V/V. Thus

$$H(j\omega) = \frac{\omega_c^3}{(j\omega + \omega_c)(-\omega^2 + j\omega\omega_c + \omega_c^2)}$$

$$= \frac{1}{(1 + j\omega/\omega_c)[1 + j\omega/\omega_c - (\omega/\omega_c)^2]} \qquad \text{V/V}$$

The filter can be realized using a series connection of a first-order system and an underdamped second-order system. The gain of the filter is

$$\Gamma(\omega) = \frac{1}{\sqrt{1 - (\omega/\omega_c)^6}} \qquad \text{V/V}$$

the phase shift of the filter is

$$\phi(\omega) = -\tan^{-1}\frac{\omega}{\omega_c} - \tan^{-1}\frac{\omega/\omega_c}{1-(\omega/\omega_c)^2}$$

and the delay of the filter is

$$D(\omega) = -\frac{d\phi(\omega)}{d\omega} = \frac{\omega_c}{\omega^2 + \omega_c^2} + \frac{1}{\omega_c}\frac{1+(\omega/\omega_c)^2}{[1-(\omega/\omega_c)^2]^2 + (\omega/\omega_c)^2}$$

Figure 4-52 shows graphs of the gain $\Gamma(\omega)$ and the delay $D(\omega)$ of the filter, where the dashed lines describe the gain and the delay of a comparable ideal filter. For $f < f_c/2$, the gain and the delay are almost independent of frequency. This indicates that the filter provides high-fidelity transmission for an input occupying a band $0 \leq f < f_c/2$.

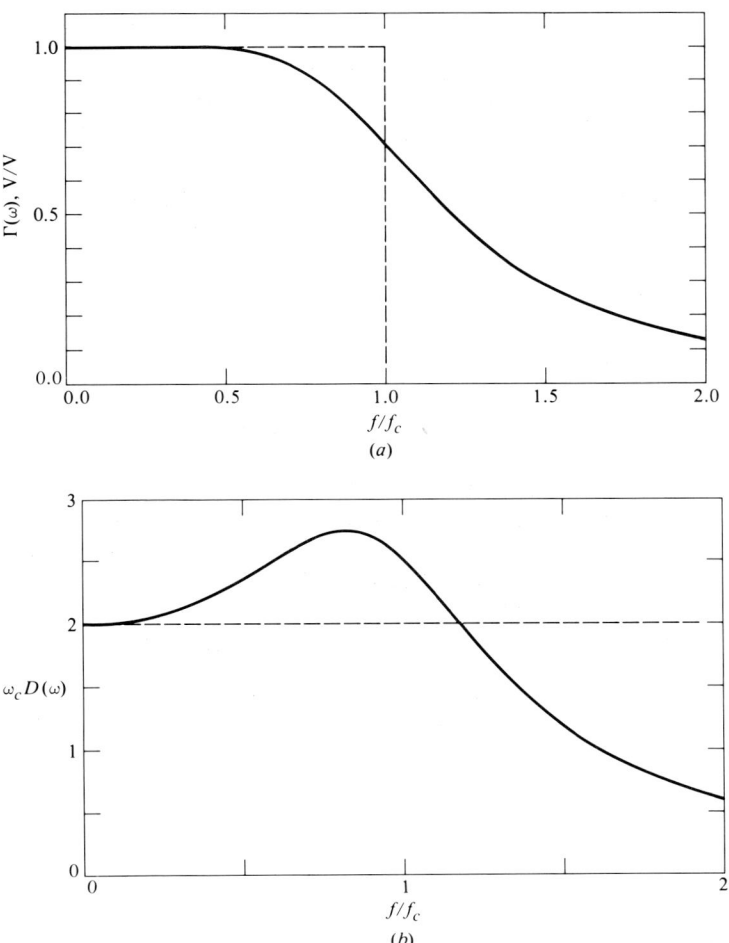

Figure 4-52 Gain and delay of a third-order Butterworth filter.

Figure 4-50 shows that amplitude distortion introduced by a Butterworth filter in the (nominal) passband $0 \leq f < f_c$ decreases as the order of the filter is increased. Unfortunately, delay distortion in the passband increases as the filter order is increased (Fig. 4-53). A high-order Butterworth filter (say, $n \geq 5$) introduces little amplitude distortion, but it introduces considerable delay distortion in its passband. It is often necessary to use (in series with a filter) a system (usually an electric circuit) called a *delay equalizer* whose gain is independent of frequency and whose delay complements the delay of the filter in such a way that the overall delay of the filter and equalizer will be almost independent of frequency. Delay equalization is discussed further in Sec. 5-3I.

In addition to Butterworth filters there are other kinds of filters corresponding to other choices for the function $F(s)$ in (4-67). Some are of lower order than a Butterworth filter for comparable specifications on gain. This means that they can be synthesized using fewer components than a comparable Butterworth filter, which in turn means that they are less expensive. Generally, the reduced complexity and cost of these other filters is accompanied by poorer performance in other respects, e.g., by passband ripple (paired-echo distortion) and greater delay distortion. In some applications keeping passband gain, delay distortion, and stopband gain within acceptable limits using an equalized non-Butterworth filter requires a system (filter and equalizer) whose overall complexity is about the same as that of an equalized Butterworth filter. In other applications where delay (phase) distortion is relatively harmless, e.g., in some systems for speech transmission, certain non-Butterworth filters are significantly less expensive than a comparable Butterworth filter.

4-5B Linear Modulation and Demodulation

Figure 4-54 shows a block diagram for an ideal linear modulator-demodulator. We wish to determine conditions under which the system of Fig. 4-54 provides distortionless transmission for an input (modulating signal) $m(t)$.

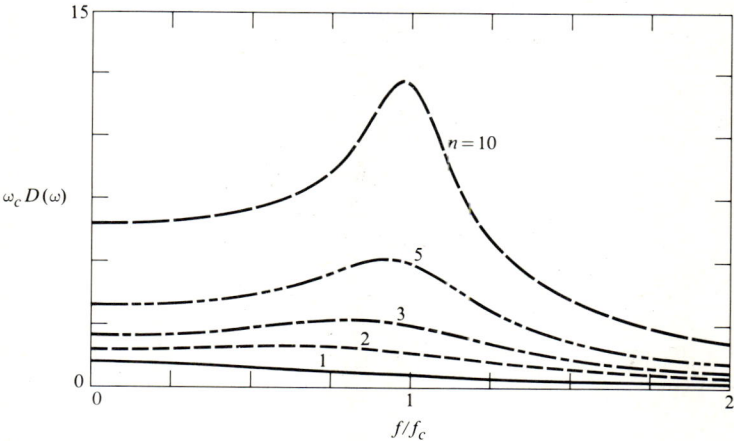

Figure 4-53 Delay of a Butterworth filter.

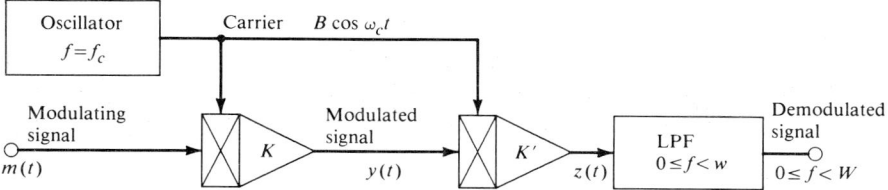

Figure 4-54 Ideal linear modulator-demodulator.

The *modulated signal* $y(t)$ in the system of Fig. 4-54 is

$$y(t) = KBm(t)\cos \omega_c t \tag{4-76a}$$

where $m(t)$ is the *modulating signal* and $B\cos \omega_c t$ is the *carrier*. Using Euler's identity and the frequency-translation property of the Fourier transformation, we find that the spectral density (the Fourier transform) of the modulated signal is

$$Y(j\omega) = \frac{KB}{2} M(j\omega - j\omega_c) + \frac{KB}{2} M(j\omega + j\omega_c) \tag{4-76b}$$

where $M(j\omega)$ is the spectral density of the modulating signal $m(t)$. Figure 4-55 illustrates this relation between a modulating signal and the associated modulated signal. The spectral density of the modulated signal is obtained by translating the spectral density of the modulating signal to the right by f_c, to the left by f_c, and adding the translated spectra (scaled by $KB/2$). Note that the terms $M(j\omega \pm j\omega_c)$ do not overlap if the carrier frequency f_c exceeds the bandwidth W of the modulating signal.

The signal $z(t)$ in the demodulator of Fig. 4-54 is

$$z(t) = K'By(t)\cos \omega_c t \tag{4-77a}$$

Using Euler's identity and the frequency-translation property of the Fourier transformation, we find that the spectral density of $z(t)$ is given by

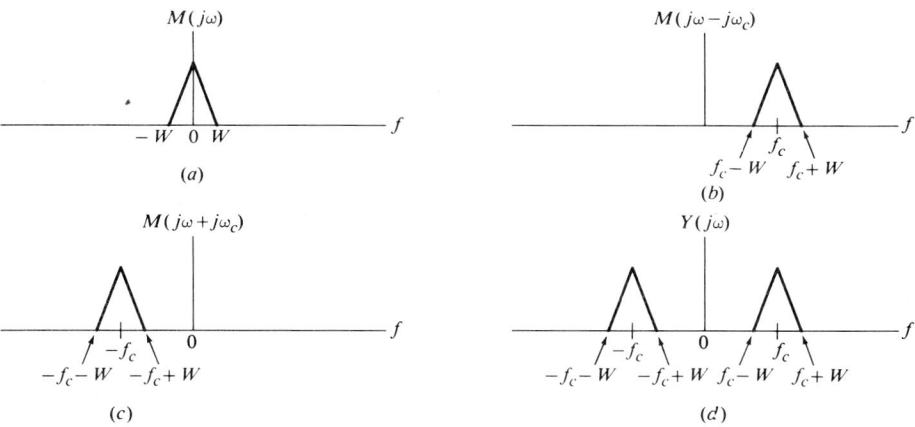

Figure 4-55 An illustration of linear modulation, as in Fig. 4-54: (*a*) spectral density of a modulating signal; (*b*) and (*c*) terms appearing in (4-76); (*d*) spectral density of the modulated signal.

$$Z(j\omega) = \frac{K'B}{2} Y(j\omega - j\omega_c) + \frac{K'B}{2} Y(j\omega + j\omega_c) \qquad (4\text{-}77b)$$

where $Y(j\omega)$ is the spectral density of the modulated signal $y(t)$. Figure 4-56 illustrates this relation. The spectral density of $z(t)$ is obtained by translating the spectral density of the modulated signal to the right by f_c, to the left by f_c, and adding the results (scaled by $K'B/2$). Using (4-76b) to eliminate $Y(j\omega)$ from (4-77b) gives

$$Z(j\omega) = CM(j\omega) + \frac{C}{2}[M(j\omega - 2j\omega_c) + M(j\omega + 2j\omega_c)] \qquad (4\text{-}78)$$

where $C = KK'B^2/2$. If the carrier frequency f_c exceeds the bandwidth W of the modulating signal $m(t)$, the terms $M(j\omega \pm 2j\omega_c)$ do not overlap the term $M(j\omega)$ (see Fig. 4-56d). In this case, the ideal low-pass filter in the demodulator passes

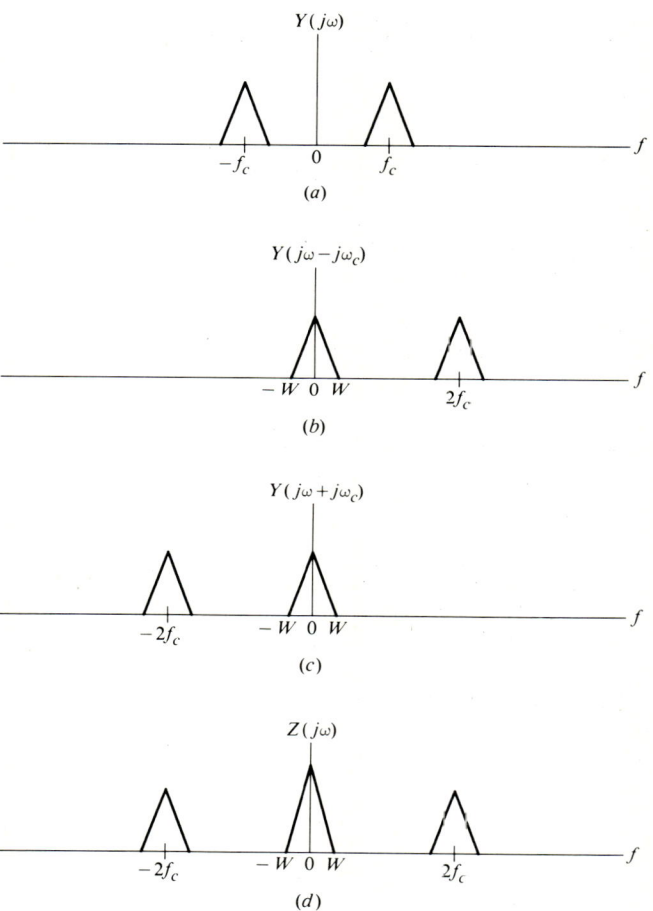

Figure 4-56 An illustration of linear demodulation, as in Fig. 4-54: (*a*) spectral density of received signal; (*b*) and (*c*) terms appearing in (4-77b); (*d*) spectral density of the input to the low-pass filter.

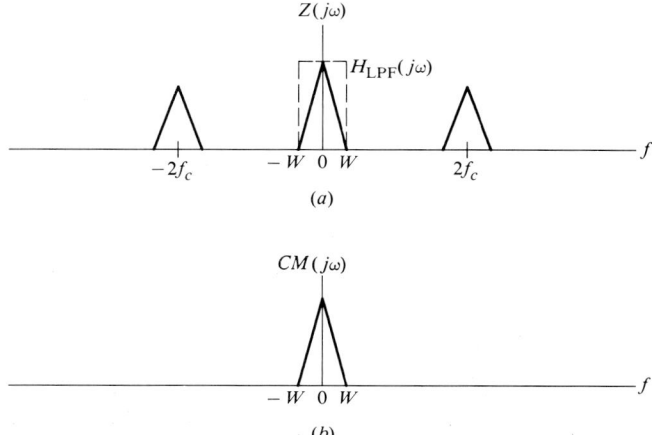

Figure 4-57 Recovering a modulating signal for the signal of Fig. 4-77d by low-pass filtering.

(without distortion) all components associated with $M(j\omega)$ and stops all components associated with $M(j\omega \pm 2j\omega_c)$, as illustrated in Fig. 4-57. For $f_c > W$, the output of the demodulator of Fig. 4-54 is

$$v(t) = \mathcal{F}^{-1}\{CM(j\omega)\} = Cm(t) \qquad f_c > W \qquad (4\text{-}79)$$

Thus the demodulator output $v(t)$ is proportional to the modulator input $m(t)$.

The development above shows that the system of Fig. 4-54 provides distortionless transmission for an input $m(t)$ under the following conditions:

1. The modulating signal is strictly bandlimited to a band $0 \leq f \leq W$; that is, the spectral density of the modulating signal is confined entirely to a band $0 \leq f \leq W$.
2. The carrier frequency exceeds the bandwidth of the modulating signal; that is, $f_c > W$.
3. The carrier (or a signal having the frequency and initial phase of the carrier) is available at the demodulator.
4. The low-pass filter in the demodulator is ideal, having frequency-independent gain and linear phase shift (frequency-independent delay) in the band occupied by the modulating signal.

Next, we describe distortion introduced if these conditions are not met. First, we consider conditions 1 and 2 above. We assume (for this discussion) that conditions 3 and 4 are met. If a modulating signal $m(t)$ is not confined entirely to the band $0 \leq f \leq f_c$, where f_c is the carrier frequency, the terms $M(j\omega \pm 2j\omega_c)$ overlap the (desired) term $M(j\omega)$ in (4-78). The low-pass filter passes not only components associated with $M(j\omega)$ but also some components associated with tails of the terms $M(j\omega \pm 2j\omega_c)$. This causes higher-frequency components of the modulating signal to be muddled at the demodulator output. The severity of this distortion varies, depending on the number and size of the components of $m(t)$ above f_c. In many systems this distortion is negligible; e.g., in ordinary AM radio carrier frequencies are on the order of 1 MHz, bandwidths of modulating signals are on the order of 5 kHz, and there is virtually no overlap of the terms $M(j\omega \pm 2j\omega_c)$ with the term $M(j\omega)$. In systems where carrier frequencies are not so much larger than bandwidths of modulating signals, e.g., in some telephone transmission systems, sharp-cutoff filters are used to limit bandwidths of modulating signals before modulation.

Now consider condition 4, assuming that conditions 1 to 3 are met. For this purpose we determine the overall frequency response of the system of Fig. 4-54, where the low-pass filter is nonideal. We let $\Gamma(\omega)$ and $\phi(\omega)$ denote the gain and the phase shift, respectively, of the low-pass filter, and we determine the output $v(t)$ for input $m(t) = A \cos \omega_m t$. The modulated signal is

$$y(t) = KBA \cos \omega_m t \cos \omega_c t$$

whence
$$z(t) = K'KB'BA \cos \omega_m t \cos^2 \omega_c t$$

Using the identity $2 \cos^2 \alpha \equiv 1 + \cos 2\alpha$ gives

$$z(t) = C \cos \omega_m t + C \cos \omega_m t \cos 2\omega_c t$$

where $C = K'KB'BA/2$. Using the identity

$$2 \cos \alpha \cos \beta \equiv \cos(\alpha + \beta) + \cos(\alpha - \beta)$$

gives

$$z(t) = C \cos \omega_m t + \frac{C}{2} \cos[(2\omega_c + \omega_m)t] + \frac{C}{2} \cos[(2\omega_c - \omega_m)t]$$

We assume that the low-pass filter, though not ideal, effectively eliminates the components at $f = 2f_c \pm f_m$, so that the demodulator output is

$$v(t) = \Gamma(\omega_m)C \cos[\omega_m t + \phi(\omega_m)]$$

Comparing the output $v(t)$ with the input $m(t) = A \cos \omega_m t$ shows that, on the whole, the system of Fig. 4-54 behaves like a linear stationary system whose gain (except for a frequency-independent scale factor) is the gain of the low-pass filter and whose phase shift is the phase shift of the low-pass filter. Therefore the system introduces no distortion if the gain $\Gamma(\omega)$ and the delay $D(\omega) = -d\phi(\omega)/d\omega$ are independent of frequency over the band $0 \le f \le W$ occupied by the signal. Effects of amplitude distortion and delay distortion introduced by a nonideal low-pass filter are exactly those that would arise if the signal $m(t)$ were passed through the low-pass filter alone.

Finally, we examine condition 3. Linear demodulation requires that the carrier (or a replica of the carrier) be available at the demodulator. To show why, we consider the system of Fig. 4-58, where the sinusoid used for demodulation is produced by a so-called *local oscillator*. We assume that the frequency and initial phase of the signal $c'(t)$ produced by the local oscillator are different from the frequency and initial phase of the carrier; thus

$$c(t) = B \cos \omega_c t \qquad \text{and} \qquad c'(t) = B' \cos[(\omega_c + \Delta\omega_c)t + \Delta\theta]$$

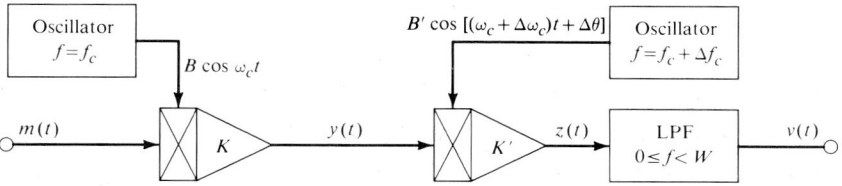

Figure 4-58 Linear modulation and demodulation using a local oscillator in the demodulator.

The fact the carrier and the local oscillator output have different peak amplitudes has no bearing on fidelity of demodulation. We examine effects of frequency error $\Delta\omega_c$ and phase error $\Delta\theta$ separately. For $\Delta\theta = 0$ (no phase error) the output of the local oscillator is

$$c'(t) = B' \cos(\omega_c t + \Delta\omega_c t)$$

and the signal $z(t)$ in the system of Fig. 4-58 is

$$z(t) = c'(t)y(t) = 2Dm(t) \cos \omega_c t \cos(\omega_c t + \Delta\omega_c t)$$

where $2D = K'B'KB$. Using the identity

$$2 \cos \alpha \cos \beta \equiv \cos(\alpha - \beta) + \cos(\alpha + \beta)$$

gives

$$z(t) = Dm(t) \cos(\Delta\omega_c t) + Dm(t) \cos(2\omega_c t + \Delta\omega_c t)$$

If the carrier frequency f_c exceeds the bandwidth of the modulating signal $m(t)$, the output of the ideal low-pass filter (the demodulator output) is

$$v(t) = Dm(t) \cos(\Delta\omega_c t)$$

This shows that frequency error $\Delta\omega_c$ introduces time-varying gain $\cos(\Delta\omega_c t)$. This time-varying gain is intolerable in most applications. For example, suppose the carrier frequency is $f_c = 1$ MHz, a relatively low radio frequency, and that the frequency error is $\Delta f_c = 0.1$ Hz, which is only one-tenth of 1 part per million. In this case the amplitude of the demodulator output varies from zero to $|Dv(t)|$ and back to zero once every 5 s. The effect is that of turning the volume control on the receiver from zero to a normal setting and back to zero 12 times per minute (smoothly).

Similarly, it can be shown that a constant phase error $\Delta\theta$, where $c(t) = B \cos \omega_c t$ and $c'(t) = B' \cos(\omega_c t + \Delta\theta)$, gives

$$v(t) = Dm(t) \cos \Delta\theta$$

The effect of a constant phase error is constant attenuation of the demodulated signal by a factor $\cos \Delta\theta$. It appears that this could be corrected by amplification, but that is impractical for two reasons: (1) A phase error results in more attenuation of the desired output $m(t)$ than of the noise, e.g., static, that invariably accompanies the signal received by the demodulator. If $\Delta\theta$ is near $\pi/2$, the desired output will be buried, e.g., rendered inaudible, by noise. (2) A *constant* phase error is a rarity. Phase error usually varies more or less slowly with time because the phase of the carrier is affected by phenomena associated with the transmission medium, whereas the initial phase of the local oscillator is relatively constant. In practical systems effects of phase error and frequency error are similar. We conclude that the linear modulator-demodulator of Fig. 4-58 performs satisfactorily only if the local oscillator is somehow locked (in frequency and phase) to the carrier $c(t)$.

The simplest way (conceptually) to make a replica of the carrier available to the receiver is to transmit the carrier on a separate channel (Fig. 4-54), but this arrangement is often impossible. For example, if a radio link is used, the modulated signal and the carrier share the same transmission medium (space). Even where separate channels are possible, e.g., where two different cables can be used, transmitting the carrier is undesirable for at least three reasons: (1) it uses power that could otherwise be used to boost the level of the modulating signal; (2) it is expensive to provide a separate channel for the carrier, especially when the cost of the channel is the dominant cost, as for a transoceanic cable; and (3) if separate channels are used, the carrier and the modulating signal may suffer different propagation delays and lose the phase coherence necessary for satisfactory demodulation.

Figure 4-59a shows a widely used method for making a replica of the carrier available for demodulation. A *subcarrier* (a *sub*harmonic of the *carrier*) is added to the

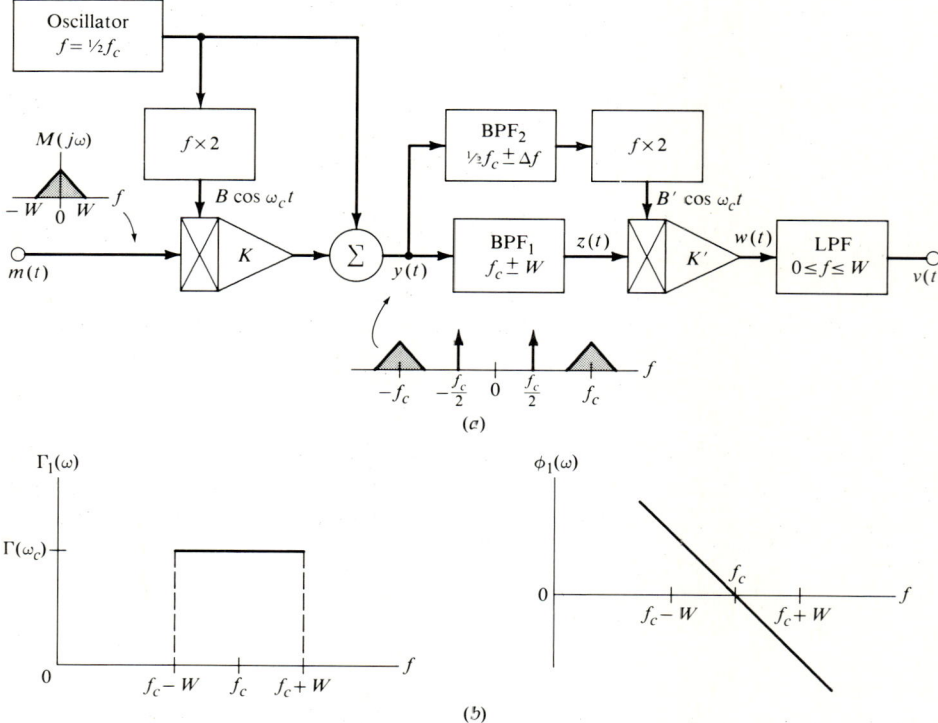

Figure 4-59 Linear modulator-demodulator in which a subharmonic of the carrier is transmitted to the demodulator: (*a*) block diagram; (*b*) gain and phase shift of the ideal bandpass filter BPF_1.

modulated signal, and the sum is transmitted. In the demodulator, bandpass filters are used to separate the modulated signal from the subcarrier and a frequency multiplier is used to reconstruct the carrier. We wish to show that the bandpass filter BPF$_1$ introduces no distortion if its gain and phase shift are given by

$$\Gamma_1(\omega) = \begin{cases} \Gamma_1(\omega_c) & f_c - W \leq f \leq f_c + W \\ 0 & \text{otherwise} \end{cases} \quad \text{and} \quad \phi_1(\omega) = -(\omega - \omega_c)t_1$$

as shown in Fig. 4-59b, where W is the bandwidth of the modulating signal. To show this we obtain the frequency response of the system of Fig. 4-59a under the following assumptions:

1. The modulating signal is strictly bandlimited to $0 \leq f \leq W$.
2. The low-pass filter is ideal, having gain

$$\Gamma_0(\omega) = \begin{cases} \Gamma_0(0) & 0 \leq f \leq W \\ 0 & f > W \end{cases}$$

and phase shift

$$\phi_0(\omega) = -\omega t_0$$

To obtain the frequency response of the system we find an expression for the output $v(t)$ for (sinusoidal) input

$$m(t) = A \cos \omega_m t$$

where $f_m < W$. The modulated signal is

$$y(t) = KBA \cos \omega_m t \cos \omega_c t + B \cos 0.5\omega_c t$$
$$= C \cos[(\omega_c + \omega_m)t] + C \cos[(\omega_c - \omega_m)t] + B \cos 0.5\omega_c t$$

where $C = KBA/2$. The first two terms lie in the passband of the bandpass filter BPF$_1$, where

$$\Gamma_1(\omega_c \pm \omega_m) = \Gamma_1(\omega_c)$$
$$\phi_1(\omega_c \pm \omega_m) = -(\omega_c \pm \omega_m - \omega_c)t_1 = \mp \omega_m t_1$$

The last term (the subcarrier) lies in the stopband of the bandpass filter BPF$_1$, where

$$\Gamma_1(0.5\omega_c) = 0$$

Therefore the output of the bandpass filter BPF$_1$ is given by

$$z(t) = \Gamma_1(\omega_c) C \cos[(\omega_c + \omega_m)t - \omega_m t_1] + \Gamma_1(\omega_c) C \cos[(\omega_c - \omega_m)t + \omega_m t_1]$$

The input of the low-pass filter is given by

$$\begin{aligned} w(t) &= K'B'z(t) \cos \omega_c t \\ &= K'B'\Gamma_1(\omega_c)C \cos[(2\omega_c + \omega_m)t - \omega_m t_1] \\ &\quad + K'B'\Gamma_1(\omega_c)C \cos[(2\omega_c - \omega_m)t + \omega_m t_1] \\ &\quad + 2K'B'\Gamma_1(\omega_c)C \cos(\omega_m t - \omega_m t_1) \end{aligned}$$

where we have used the identity $\cos(-\alpha) \equiv \cos\alpha$ to write
$$\cos(-\omega_m t + \omega_m t_1) = \cos(\omega_m t - \omega_m t_1)$$

The terms having frequencies $2\omega_c \pm \omega_m$ lie outside the passband of the low-pass filter, where
$$\Gamma_0(2\omega_c \pm \omega_m) = 0$$

The term having frequency ω_m lies in the passband of the low-pass filter, where
$$\Gamma_0(\omega_m) = \Gamma_0(0) \qquad \phi_0(\omega_m) = -\omega_m t_0$$

Therefore the output of the low-pass filter (the output of the demodulator) is
$$v(t) = 2K'B'\Gamma_1(\omega_c)\Gamma_0(0)C \cos(\omega_m t - \omega_m t_1 - \omega_m t_0)$$

Recalling that $C = KBA/2$, we have
$$v(t) = K_a A \cos[\omega_m(t - t_a)]$$

where the overall gain $K_a = K'B'KB\Gamma_1(\omega_c)\Gamma_0(0)$ and the overall delay $t_a = t_1 + t_0$ are independent of frequency. This shows that the system of Fig. 4-59 provides distortionless transmission for an input $m(t)$ under the following conditions:

1. The modulating signal $m(t)$ is confined to a band $0 \leq f \leq W$, and the carrier frequency f_c exceeds the bandwidth W of $m(t)$.
2. The gain and the phase shift of the bandpass filter BPF_1 are as given in Fig. 4-59b.
3. The low-pass filter has frequency-independent gain and linear phase shift $\phi_0(\omega) = -\omega t_0$ over the band $0 \leq f \leq W$.

These conditions guarantee distortionless transmission for the system as a whole, such that an output $v(t)$ is a scaled-and-delayed replica of the corresponding input $m(t)$. They do not guarantee distortionless transmission for the individual components of the system. In particular, note that condition 2 does not guarantee that $z(t)$ is an undistorted version of $m(t) \cos \omega_c t$. Indeed, the signal $m(t) \cos \omega_c t$ can suffer considerable distortion on passing through the bandpass filter BPF_1 if the phase intercept $\phi_1(0)$ is not equal to zero (modulo π). What condition 2 actually guarantees is that the *envelope* of $m(t) \cos \omega_c t$ will suffer no distortion at the hands of the filter. This is really all we care about because it is the envelope of $m(t) \cos \omega_c t$ and not the high-frequency wave (the carrier) under the envelope that determines the waveform of the overall output $v(t)$. So-called phase-intercept distortion is unimportant in this application.

Note also that the overall delay introduced by the system is
$$t_a = t_1 + t_0$$

where $\quad t_1 = -\dfrac{d\phi_1(\omega)}{d\omega} = D_1(\omega) \quad$ and $\quad t_0 = -\dfrac{d\phi_0(\omega)}{d\omega} = D_0(\omega)$

are the delays of the bandpass filter BPF_1 and the low-pass filter, respectively. It is interesting to compare values of system delay t_a and propagation delay (in the transmission medium) for a typical system. Typically, t_1 and t_0 are on the order of $1/W$, where W is the bandwidth of the system (the bandwidth of each filter). In an ordinary AM radio receiver $W \approx 5$ kHz, and the system delay t_a is on the order of a few hundred microseconds. The propagation delay (from transmitter to receiver) is $t_p = d/c$, where c is the speed of light ($c \approx 3 \times 10^8$ m/s) and d is distance from the transmitting antenna to the receiving antenna. For $d = 10$ km (about 6 mi), the propagation delay is $t_p = 33$ μs. Thus, in ordinary AM radio, delay introduced by filters in the receiver is typically larger than propagation delay.

4-5C Single-Sideband Modulation

The main drawback of linear modulation is that the bandwidth of a linearly modulated signal is twice the bandwidth of the modulating signal (see Fig. 4-55); i.e., linear modulation uses twice as much channel bandwidth as should be necessary. This bandwidth doubling occurs because (in effect) a linear modulator transmits both the positive- and negative-frequency components of a modulating signal. Figure 4-60 shows the spectral densities of a modulating signal $m(t)$ having bandwidth W and an associated linearly modulated signal having carrier frequency f_c. The spectral density of the modulated signal consists of an *upper sideband* occupying the band $f_c \leq f \leq f_c + W$ and a *lower sideband* occupying the band $f_c - W \leq f \leq f_c$. The upper sideband can be associated with the positive-frequency components of $m(t)$, and the lower sideband can be associated with the negative-frequency components of $m(t)$. Systems for linear modulation also are called *double-sideband* (DSB) *systems* because they transmit both sidebands of a modulated signal.

Transmitting both sidebands of a modulated signal is unnecessary because the associated modulating signal can be reconstructed from either sideband alone. Systems that transmit only one sideband are called *single-sideband* (SSB) *systems*. Figure 4-61 shows a simplified block diagram for an SSB system. It is essentially a DSB system with bandpass *sideband filters* inserted between the linear modulator and the channel and between the channel and the demodulator. The spectral densities shown in Fig. 4-61 illustrate the operation of the system. A modulating signal $m(t)$ is translated (by linear modulation) to center frequency f_c (the carrier frequency). The lower sideband of the resulting DSB signal is removed by the bandpass sideband filter to obtain an SSB signal. A subcarrier whose frequency is half the carrier frequency is added to

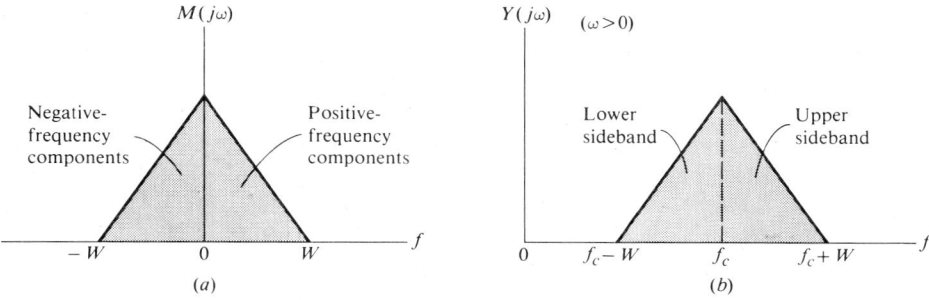

Figure 4-60 Spectral density of a linearly modulated (double-sideband) signal.

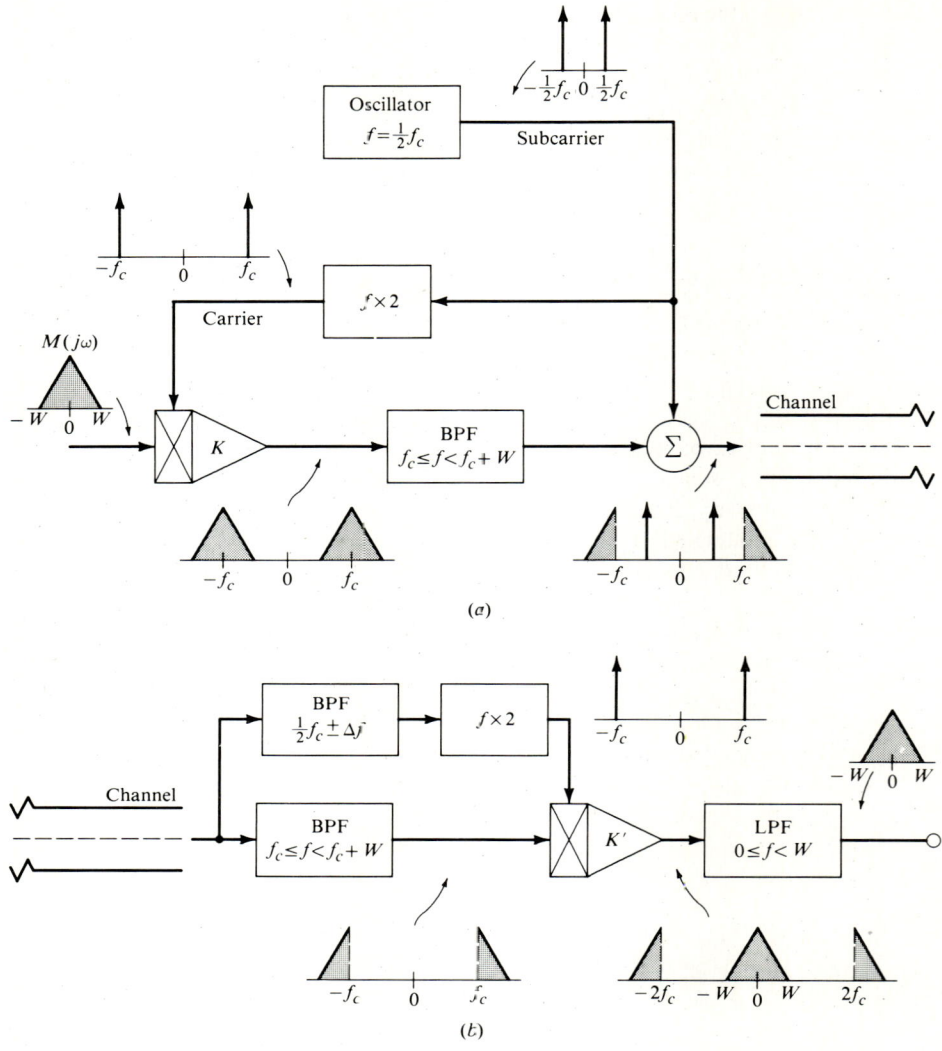

Figure 4-61 Simplified system for single-sideband transmission: (a) modulator; (b) demodulator.

the SSB signal. Adding a subcarrier instead of the carrier itself simplifies extracting the carrier at the receiver because the subcarrier lies outside the band occupied by the SSB signal. The composite signal (SSB signal plus subcarrier) is transmitted through the channel to the demodulator. The demodulator separates the subcarrier and the SSB signal, using bandpass filters, and reconstructs the carrier, using a frequency doubler. Then the demodulator multiplies the SSB signal by the carrier, producing a signal having a baseband component proportional to the modulating signal $m(t)$ plus a high-frequency SSB component occupying the band $2f_c \leq f \leq 2f_c + W$. The low-pass filter stops the high-frequency components and passes the baseband component, giving an output proportional to the modulating signal.

To find the conditions under which the bandpass filters BPF_1 and BPF_2 introduce no distortion we proceed (as before) by finding the overall frequency response of the system. We use a test input $m(t) = A \cos \omega_m t$, where $0 \leq f_m \leq W$. For simplicity we assume that all filters have unit passband gain and zero stopband gain. (Using other than unit passband gain simply introduces a scale factor which has no bearing on fidelity.) We find that the demodulator output is

$$v(t) = K_a A \cos[\omega_m t + \phi_{12}(\omega_c + \omega_m) + \phi_0(\omega_m)]$$

where K_a is independent of frequency, $\phi_{12}(\omega_c + \omega_m)$ is the net phase shift of the sideband filters, and $\phi_0(\omega_m)$ is the phase shift of the low-pass filter. We assume that the low-pass filter is ideal, so that $\phi_0(\omega) = -\omega t_0$. Thus

$$v(t) = K_a A \cos[\omega_m t + \phi_{12}(\omega_c + \omega_m) - \omega_m t_0]$$

For distortionless transmission, overall phase shift $\phi(\omega)$ must be a linear function of frequency having phase intercept $\phi(0) = 0$. Thus

$$\phi_{12}(\omega_c + \omega_m) - \omega_m t_0 = -\omega_m t_a \qquad 0 \leq f_m \leq W$$

This yields

$$\phi_{12}(\omega_c + \omega_m) = -\omega_m (t_a - t_0) \qquad 0 \leq f_m \leq W$$

or (replace $\omega_c + \omega_m$ by ω)

$$\phi_{12}(\omega) = -(\omega - \omega_c)(t_a - t_0) = -(\omega - \omega_c) t_{12} \qquad f_c \leq f \leq f_c + W$$

where $t_{12} = t_a - t_0$ is the delay introduced by the bandpass filters. This shows that for distortionless transmission the combined phase shift of the bandpass filters must be a linear function of frequency intersecting the frequency axis at the *left edge* of the passband. (This is different from the corresponding condition for the DSB system of Fig. 4-59, where the phase shift of the bandpass filter BPF_1 must be a linear function of frequency intersecting the frequency axis at the *center* of the passband.) This condition is hard to meet because phase shifts of most practical bandpass filters have approximately odd symmetry about the center of the passband. Consequently, most SSB systems introduce phase (delay) distortion even if the channel and all filters have linear phase shift. This is why some phase compensation (delay equalization) is usually necessary for SSB transmission of phase-sensitive signals, e.g., data signals.

The main advantage of SSB modulation over DSB modulation is that SSB modulation uses only half as much channel bandwidth. This is significant where many signals must share a single transmission medium. The main disadvantages are the added cost of sideband filters (and delay equalizers in phase-sensitive applications). This cost can be significant because sharp-cutoff sideband filters (and delay equalizers) may be expensive. As a rule, DSB systems are used for transmission over relatively short distances, where the cost of the receiver and transmitter dominate the cost of the system, and SSB systems are used for transmission over long distances, where the cost of the channel dominates the cost of the system.

The development of SSB transmission is a good example of how frequency-domain analysis has led to significant practical innovations that probably would not have been suggested by time-domain analysis. We conclude our discussion of signal-transmission systems by describing another such innovation.

4-5D Frequency-Division Multiplexing

In signal-transmission systems it is usually necessary to provide for simultaneous transmission of several (perhaps many) signals over a single link, e.g., cable, radio link, or optical fiber. For example, all commercial broadcast systems (AM radio, FM radio, and television) share the same transmission medium. Sharing a link can be made possible by *frequency-division multiplexing* (FDM), illustrated in Fig. 4-62. The several modulating signals are translated to nonoverlapping bands (or channels) and added to produce the transmitted signal. At the other end of the link the signals are separated (demultiplexed) by bandpass filters and demodulated as described above.

Frequency-division multiplexing can be accomplished in conjunction with DSB modulation (DSB FDM) or SSB modulation (SSB FDM). More signals can share a link if SSB FDM is used than if DSB FDM is used. On the other hand, an SSB FDM system is more expensive than a comparable DSB FDM system. In practice, a choice between SSB FDM and DSB FDM is decided on the basis of cost, as described above.

4-5E Servomechanisms

Figure 4-63 shows a schematic diagram for a simple dc servomechanism whose input θ_i is angular position of a control knob, e.g., on an antenna rotator, and whose output θ_o is angular position of a load, e.g., an antenna, having moment of inertia J and coefficient of viscous friction F. The purpose of the system is to make the angular position of the load match the angular position of the control knob. Input θ_i and output θ_o are connected mechanically to wipers on potentiometers in an error-detecting circuit, which produces voltage e proportional to position error $\theta_i - \theta_o$. This voltage is applied to a dc motor which drives the load in the direction corresponding to decreasing position error. Ideally, the motor stops when the position error is reduced to zero (when the voltage e is reduced to zero).

Step response often is used to describe or specify performance of a servomechanism, the idea being that a servomechanism should respond quickly, smoothly, and with little or no steady-state error to an abrupt change of desired (input) position. Below, we analyze the servomechanism of Fig. 4-63. We start by obtaining a block diagram of the system. Then we use block-diagram reduction to obtain the overall transfer function of the system, and we use the Fourier transformation to obtain the step response of the system. Finally, we examine the performance (rise time, overshoot, steady-state error, etc.) of the system and we show how performance can be improved by adding tachometer feedback.

We can obtain a block diagram of the system of Fig. 4-63 by obtaining and then connecting together blocks representing individual subsystems. We begin with the error detector. Figure 4-64a shows a graph of the transfer characteristic of the potentiometer bridge circuit. This transfer characteristic is obtained by inspection from the circuit diagram (see Fig. 4-63). We assume that the magnitude of the position error does not exceed 2π rad. With this assumption, error voltage e is proportional to position error $\theta_i - \theta_o$, and the error detector is represented by the block diagram of Fig. 4-64b, where

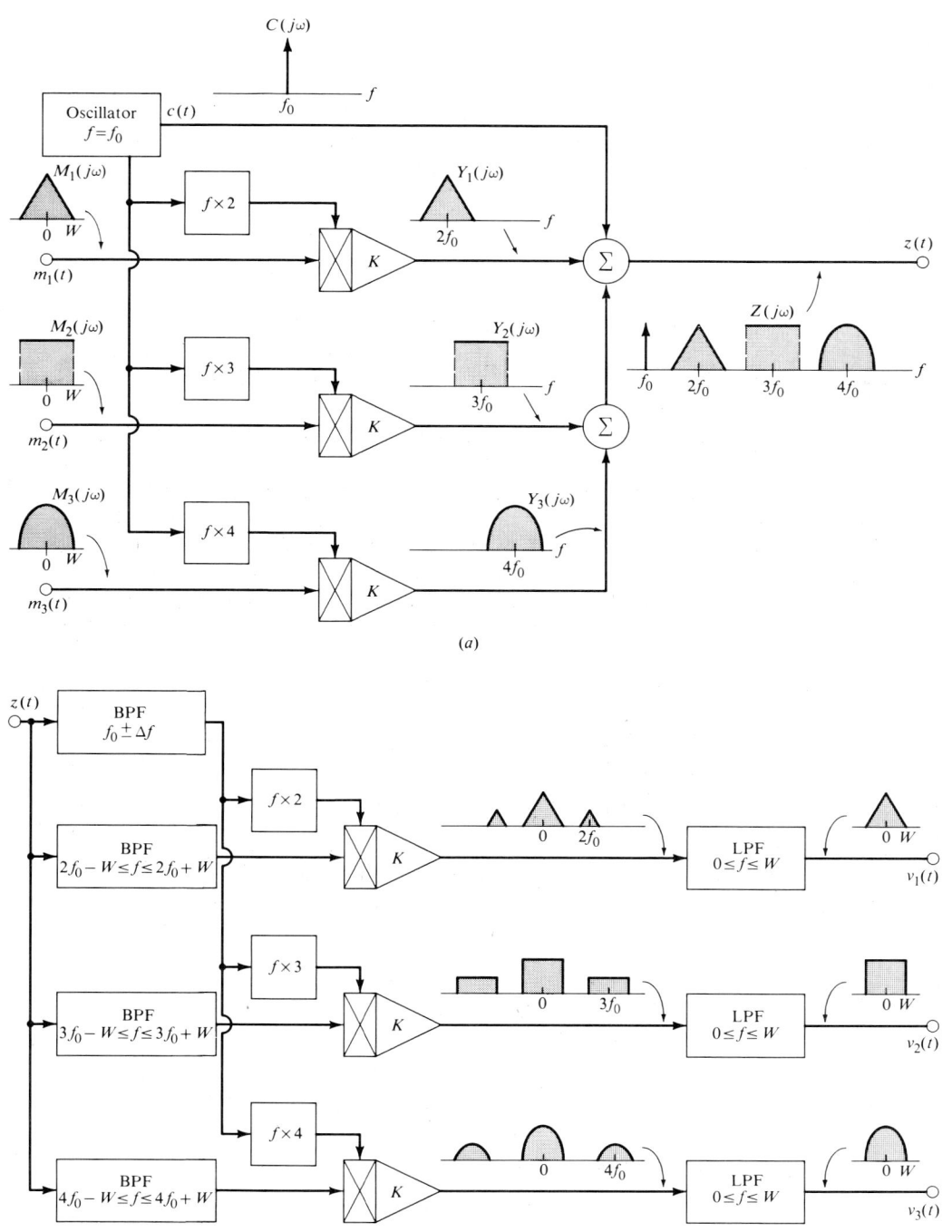

Figure 4-62 System for double-sideband frequency-division multiplex (DSB FDM): (a) multiplexer; (b) demultiplexer.

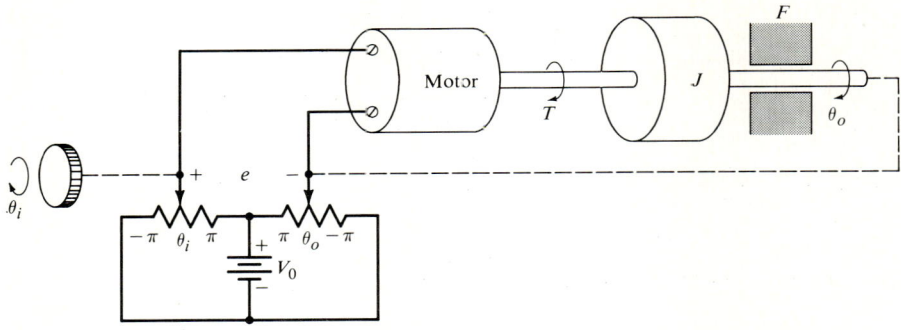

Figure 4-63 A dc servomechanism.

$$e = K_e(\theta_i - \theta_o) \quad \text{with } K_e = \frac{V_0}{2\pi}$$

We assume that the transfer characteristic of the dc motor is

$$T = K_T i_a$$

where i_a is armature current, T is torque, and K_T is a constant called the torque constant of the dc motor. For simplicity we assume that the armature circuit is purely resistive, so that $i_a = e/R$, where R is armature resistance. Thus the transfer characteristic of the motor is

$$T = K_T R^{-1} e = K_m e$$

where $K_m = K_T R^{-1}$. The motor (including the armature circuit) is represented by the block diagram of Fig. 4-65.

The angular position θ_o of the load and the torque T applied to the load are related by the differential equation (Newton's law applied to angular motion)

$$T = J\frac{d^2\theta_o}{dt^2} + F\frac{d\theta_o}{dt}$$

where J is the moment of inertia of the load and F is the coefficient of viscous friction of the load. It follows that the transfer function of the load is

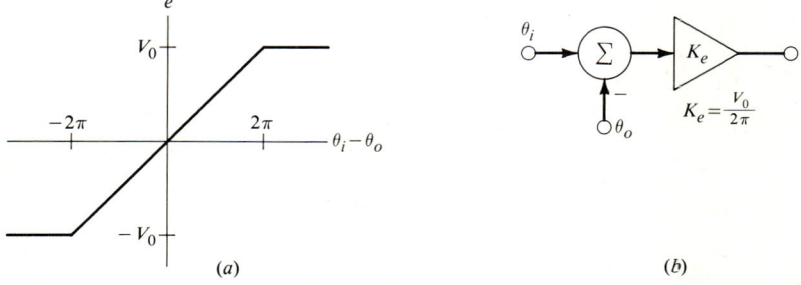

Figure 4-64 (a) Transfer characteristic and (b) block diagram for the error-detecting circuit in the servomechanism of Fig. 4-63.

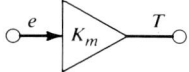

Figure 4-65 Block diagram for the motor in the servomechanism of Fig. 4-63.

$$H_L(j\omega) = \frac{1/J}{(j\omega)^2 + (F/J)j\omega}$$

The load is represented by the block diagram of Fig. 4-66.

Figure 4-67, which shows a block diagram representing the servomechanism of Fig. 4-63, is obtained by connecting the blocks of Figs. 4-64 to 4-66 together. Using block-diagram reduction, we find that the transfer function of the system is

$$H(j\omega) = \frac{K_e K_m / J}{(j\omega)^2 + (F/J)j\omega + K_e K_m / J} \tag{4-80}$$

The servomechanism is a second-order system. In order to use relations given in Sec. 4-4 we write the transfer function above as

$$H(j\omega) = \frac{1}{(j\omega/\omega_0)^2 + 2j\zeta(\omega/\omega_0) + 1} \tag{4-81a}$$

where undamped natural frequency ω_0 and damping ratio ζ are given by

$$\omega_0 = \sqrt{\frac{K_e K_m}{J}} \qquad \zeta = \frac{F}{2J\omega_0} \tag{4-81b}$$

The step response of the system is given by (4-57). Figure 4-42 shows graphs of the step response for various values of damping ratio. Since the step response contains only two parameters (damping ratio ζ and undamped natural frequency ω_0), we can place at most two independent specifications on step response. The two properties of step response of a second-order servomechanism most often specified are percent overshoot PO and rise time t_R. From (4-64),

$$\text{PO} = 100 \exp\left(\frac{-\pi\zeta}{\sqrt{1-\zeta^2}}\right) \tag{4-82a}$$

Solving this relation for damping ratio ζ gives

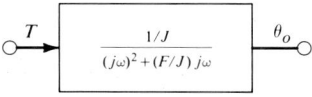

Figure 4-66 Block diagram for the load in the servomechanism of Fig. 4-63.

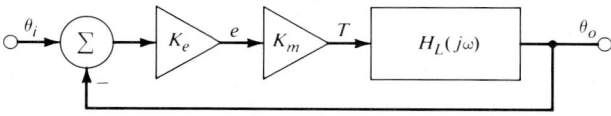

Figure 4-67 Block diagram for the servomechanism of Fig. 4-63.

$$\zeta = \frac{1}{\sqrt{1 + (\pi/\alpha)^2}} \qquad \alpha = \ln\frac{100}{\text{PO}} \qquad (4\text{-}82b)$$

where PO is expressed in percent. Thus damping ratio can be determined from a specification on peak overshoot. From (4-62),

$$\omega_0 \approx \frac{1.38 + 1.62\zeta}{t_R} \qquad (4\text{-}83)$$

Thus, undamped natural frequency can be determined from damping ratio and a specification on rise time.

In practical applications other properties of the step response, e.g., steady-state error and settling time, are also of interest. For the system under consideration, steady-state error equals zero (because dc gain equals unity). We cannot place an independent specification on settling time (in addition to specifications on overshoot and rise time) because a system having only two adjustable parameters (damping ratio and undamped natural frequency) can meet at most two independent specifications.

Once we have determined damping ratio and undamped natural frequency, e.g., from specifications on overshoot and rise time, we must determine corresponding values for the adjustable physical parameters of the system. In all, the system of Fig. 4-67 has four parameters: gain K_e, gain K_m, moment of inertia J, and coefficient of viscous friction F. In (4-81) the parameters K_e and K_m appear only in the product $K_e K_m$. Even though they are independently adjustable, K_e and K_m provide only one free parameter ($K_e K_m$) for purposes of meeting specifications on performance. Further, J and F have minimum values determined by the load. Since the quantities J and F can be increased, e.g., by adding a flywheel and damper, but cannot be decreased, the system has one free parameter ($K_e K_m$) and two partly free parameters (J and F).

Equation (4-81) shows that increasing the moment of inertia J decreases damping ratio ζ. Figure 4-47 shows that decreasing the damping ratio increases percent overshoot. Thus increasing the moment of inertia increases percent overshoot. Further, from (4-81) and (4-83),

$$t_R = \frac{2.76\sqrt{K_e K_m J} + 1.62F}{2K_e K_m} \qquad (4\text{-}84)$$

This shows that increasing the moment of inertia increases rise time t_R. Consequently, increasing the moment of inertia of the load usually degrades the performance of a servomechanism.

We are left with one completely free parameter ($K_e K_m$) and one partly free parameter (F). From (4-81) and Fig. 4-47, increasing $K_e K_m$ increases overshoot and decreases rise time. From (4-81) and Fig. 4-47, increasing F decreases percent overshoot and increases rise time. Because $K_e K_m$ and F affect overshoot and rise time in opposite ways, it sometimes is possible to meet specifications on overshoot and rise time by adding viscous damping and adjusting $K_e K_m$ appropriately.

Example 4-33 The servomechanism of Fig. 4-63 is to position a load having moment of inertia $J = 0.10$ Nm/(rad/s^2) and coefficient of viscous friction $F_L =$

0.40 Nm/(rad/s). Overshoot must not exceed 10 percent, and rise time must not exceed 500 ms. Determine values for $K_e K_m$ and additional viscous friction F_0.

SOLUTION From (4-82) and (4-83), the required values for damping ratio ζ and undamped natural frequency ω_0 are

$$\zeta = 0.59 \qquad \omega_0 = 4.67 \text{ rad/s}$$

Solving (4-81b) for $K_e K_m$ and F gives

$$K_e K_m = J\omega_0^2 = 2.18 \text{ Nm/rad} \qquad F = 2J\omega_0\zeta = 0.55 \text{ Nm/(rad/s)}$$

Because the necessary value for F is greater than the minimum value F_L determined by the load, the specifications can be met using $K_e K_m = 2.18$ Nm/rad and additional viscous damping

$$F_0 = F - F_L = 0.55 - 0.40 = 0.15 \text{ Nm/(rad/s)}$$

Adding viscous damping to meet specifications on step response is impossible if the total damping required is less than the damping of the load alone, e.g., if $F_L = 0.6$ Nm/(rad/s) in Example 4-33. Also, adding viscous damping increases power required of the motor, so that adding damping may also require a larger motor. An alternative to adding viscous (mechanical) damping is to achieve the same effect using electrical damping provided by a tachometer connected as shown in Fig. 4-68a. A tachometer is a dc generator that produces a voltage proportional to angular speed of the shaft. For the tachometer of Fig. 4-68a,

$$v_t = K_t \frac{d\theta_0}{dt}$$

The transfer function of the tachometer is

$$H_t(j\omega) = j\omega K_t$$

The voltage v_t produced by the tachometer is subtracted from the error voltage e, and the difference $v = e - v_t$ is applied to the motor. Figure 4-68b shows a block diagram of the system of Fig. 4-68a. The transfer function of the system is

$$H(j\omega) = \frac{1}{(j\omega/\omega_0)^2 + 2j\zeta\omega/\omega_0 + 1} \tag{4-85a}$$

where

$$\omega_0 = \sqrt{\frac{K_e K_m}{J}} \qquad \zeta = \frac{F}{2J\omega_0} + \frac{K_t K_m}{2J\omega_0} \tag{4-85b}$$

Comparing (4-85b) with (4-81b) shows that tachometer feedback increases damping ratio by $F_t = K_t K_m/2J\omega_0$. Tachometer feedback does not affect undamped natural frequency. There are two advantages of using tachometer feedback instead of mechanical damping: (1) a tachometer generally uses less power than an equivalent damper, and (2) tachometer feedback can either increase ($K_t > 0$) or decrease ($K_t < 0$) damping ratio.

Determining the parameters of the system of Fig. 4-68 from specifications on overshoot and rise time is complicated by the fact that tachometer feedback introduces

270 INTRODUCTION TO SYSTEM ANALYSIS

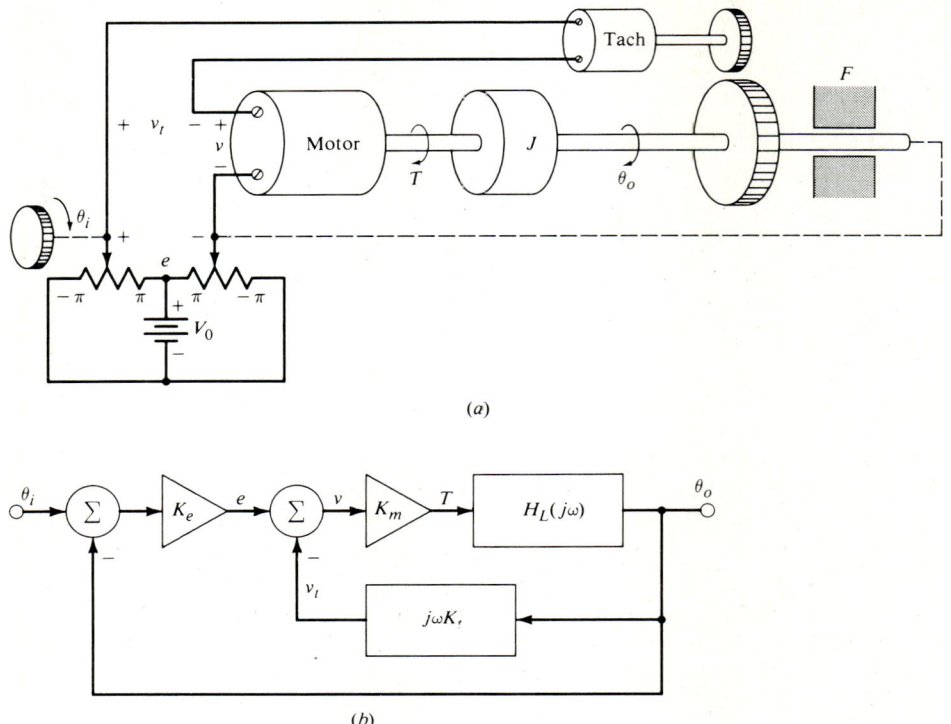

Figure 4-68 Servomechanism with tachometer feedback: (a) schematic diagram, (b) block diagram.

another parameter (K_t) and decouples the parameters K_e and K_m. We must determine three parameters (K_t, K_e, K_m) from only two specifications. We cannot impose a third specification on performance because the step response still contains only two parameters (ζ and ω_0). In practice, the parameter K_m is determined by choosing a motor whose characteristics, e.g., peak torque, are consistent with specifications on rise time and properties of the load; e.g., fast acceleration of a heavy load requires more power than slow acceleration of light load. Subsequently, the parameters K_e and K_t are determined from specifications on damping ratio and overshoot.

> **Example 4-34** The servomechanism of Fig. 4-68 is to position a load having moment of inertia $J = 0.15$ Nm/(rad/s^2) and coefficient of viscous friction $F = 1.00$ Nm/(rad/s). The motor torque constant is $K_T = 0.30$ Nm/A, and the armature resistance is $R = 1.00$ Ω. The overshoot must not exceed 10 percent, and the rise time must not exceed 700 ms. Determine the parameters K_e and K_t.
>
> SOLUTION We have $K_m = K_T R^{-1} = 0.30$ Nm/V. From (4-82) and (4-83), the damping ratio ζ and the undamped natural frequency ω_0 are
>
> $$\zeta = 0.59 \qquad \omega_0 = 3.34 \text{ rad/s}$$
>
> Solving (4-85b) for K_e and K_t gives

$$K_e = \frac{J\omega_0^2}{K_m} = 5.58 \text{ V/rad} \qquad K_t = \frac{2J\omega_0\zeta - F}{K_m} = -1.36 \text{ V/(rad/s)}$$

Note that the specifications cannot be met using added mechanical damping because the effective coefficient of viscous friction is smaller than the coefficient of viscous friction for the load.

Designing servomechanisms is considerably more involved than suggested by the discussion and examples above. A realistic model for a servomechanism usually includes effects omitted from the model of Fig. 4-68, such as nonlinearity of the motor, e.g., velocity and torque saturation; electrical time constants of the motor, e.g., armature inductance; static friction of the motor and load; nonlinearity (saturation) of the error detector; disturbance torques, e.g., wind gusts on an antenna; and a tortional spring constant that accounts for twisting of the shaft during acceleration of the load. Realistic performance specifications often include more complete specifications on step response, e.g., steady-state error and settling time; specifications on responses to other inputs, e.g., disturbance torques; and specifications on sensitivity to parameter variations. Including all significant effects often makes pencil-and-paper design impractical. In practice computer simulation (both analog and digital) is used extensively in designing servomechanisms and other feedback control systems.

SUMMARY

This chapter describes how frequency-domain representations of signals are obtained from waveforms of signals, how frequency-domain representations of signals and systems are used to obtain outputs of systems, and how frequency-domain descriptions of systems are interpreted in terms of realizability, stability, fidelity, and transient and steady-state response.

In Sec. 4-1 we show how a signal $x(t)$ can be described over any finite interval $t_0 \leq t < t_0 + T$ by a *Fourier series*

$$x(t) = \sum_{n=-\infty}^{\infty} X_n e^{jn\omega_0 t}$$

where $\omega_0 = 2\pi/T$ and

$$X_n = \frac{1}{T} \int_{t_0}^{t_0+T} x(t) e^{-jn\omega_0 t} \, dt$$

The quantity ω_0 is the *fundamental frequency* of the series. The quantities $X_0, X_{\pm 1}, X_{\pm 2}, \ldots$ are the *Fourier coefficients* for $x(t)$ over $t_0 \leq t < t_0 + T$.

A Fourier series for a signal is periodic with period T. If the signal $x(t)$ also is periodic with period T, the series describes $x(t)$ for all time. In that case, the Fourier coefficients constitute the *complex spectrum* $C_x(f)$ of the signal; thus

$$C_x(f) = \begin{cases} X_n & f = nf_0 \\ 0 & f \neq nf_0 \end{cases} \qquad n = 0, \pm 1, \pm 2, \ldots$$

The spectrum of the output $y(t)$ of a linear stationary system for periodic input $x(t)$ is given by

$$C_y(f) = H(j2\pi f)C_x(f)$$

where $H(j\omega)$ is the transfer function of the system.

In Sec. 4-2 we generalize a Fourier series by allowing the duration T of the interval of expansion to approach infinity. This yields the Fourier integral

$$x(t) = \frac{1}{2\pi}\int_{-\infty}^{\infty} X(j\omega)e^{j\omega t}\,d\omega$$

where

$$X(j\omega) = \int_{-\infty}^{\infty} x(t)e^{-j\omega t}\,dt$$

is the *spectral density* of the signal $x(t)$. Loosely speaking, the Fourier integral expresses a signal $x(t)$ as a superposition of sinusoids, where the amplitude of a sinusoidal component having angular frequency ω is proportional to $|X(j\omega)|$ and the phase of the component is $\angle X(j\omega)$. By superposition and time invariance, the spectral density of the output $y(t)$ of a linear stationary system for input $x(t)$ is given by

$$Y(j\omega) = H(j\omega)X(j\omega)$$

where $H(j\omega)$ is the transfer function of the system.

In Sec. 4-3 we present the *Fourier transformation*. The *Fourier transform* of a function $e(t)$ is denoted by $\mathcal{F}\{e(t)\} = E(j\omega)$ and defined by

$$\mathcal{F}\{e(t)\} = E(j\omega) = \int_{-\infty}^{\infty} e(t)e^{-j\omega t}\,dt$$

The *inverse Fourier transform* of a function $E(j\omega)$ is denoted by $\mathcal{F}^{-1}\{E(j\omega)\} = e(t)$ and given by

$$\mathcal{F}^{-1}\{E(j\omega)\} = \frac{1}{2\pi}\int_{-\infty}^{\infty} E(j\omega)e^{j\omega t}\,d\omega$$

Fourier series and Fourier integrals can be regarded as special cases of the inverse Fourier transform. The spectral density of a signal is the Fourier transform of the signal. The inverse Fourier transform of a spectral density gives the waveform of the associated signal. Thus the Fourier transformation provides a two-way path between time domain and frequency domain. Using the Fourier transformation, we can show how operations in one domain are transformed to other operations in the other domain. The three most important *operational properties* of the Fourier transformation in system engineering are expressed by the relations

$$\mathcal{F}\left\{\int_{-\infty}^{\infty} x(\lambda)h(t-\lambda)\,d\lambda\right\} = H(j\omega)X(j\omega)$$

$$\mathcal{F}\{x(t)e^{j\omega_c t}\} = X(j\omega - j\omega_c)$$

$$\mathcal{F}\{x(t-t_0)\} = e^{-j\omega t_0}X(j\omega)$$

The first is the fundamental relation for frequency-domain analysis of linear stationary systems, the second is the fundamental relation for frequency-domain analysis of many important modulation systems, and the third makes frequency-domain analysis especially useful where fidelity is important, e.g. in filtering and signal transmission.

In Sec. 4-4 we show how realizability, stability, fidelity, and transient and steady-state response are reflected in the frequency-domain description (the transfer function) of a linear stationary system. We also show that any rational transfer function of the form

$$H(j\omega) = \frac{b_m(j\omega)^m + b_{m-1}(j\omega)^{m-1} + \cdots + b_0}{(j\omega)^n + a_{n-1}(j\omega)^{n-1} + \cdots + a_0}$$

is realizable and that the realization is stable if all roots of the characteristic polynomial $c(s) = s^n + a_{n-1}s^{n-1} + \cdots + a_0$ have negative real parts. We also present the *Paley-Wiener condition for realizability*, which can be applied to nonrational transfer functions although it has little practical value because of drawbacks pointed out in the text.

We show how *linear distortion* is analyzed as *amplitude distortion* and *delay distortion*. A linear stationary system having transfer function $H(j\omega)$ introduces amplitude distortion if gain $\Gamma(\omega) = |H(j\omega)|$ is not independent of frequency over the band occupied by an input. The system introduces delay distortion if *delay* $D(\omega)$ defined by

$$D(\omega) = -\frac{d\phi(\omega)}{d\omega} = -\frac{d\angle H(j\omega)}{d\omega}$$

is not independent of frequency over the band occupied by an input. We describe typical effects of amplitude and delay distortion for a pulse input, including rounding and smearing caused by inadequate bandwidth (drooping gain); paired echoes caused by passband ripple; and overshoots, undershoots, smearing, and asymmetry caused by delay distortion (phase distortion).

Finally, we show how important properties of a step response are reflected in the transfer function of a linear stationary system. The three most important applications of these ideas are to first-order low- and high-pass systems and second-order bandpass systems. For a first-order low-pass system, rise time of step response is inversely proportional to bandwidth, and steady-state amplitude of step response equals dc gain $H(0)$. For a first-order high-pass system, sag time is inversely proportional to the width of the stopband, and steady-state amplitude of step response equals zero. For a second-order bandpass system having almost constant gain (no big peaks) over a wide passband $\omega_1 < \omega < \omega_2$ (where $\omega_1 \ll \omega_2$), rise time varies inversely with bandwidth $\omega_2 - \omega_1 \approx \omega_2$, steady-state amplitude equals zero, and sag time varies inversely with the low cutoff frequency ω_1. For a second-order system having a resonant peak in its passband, rise time, peak time, and settling time all vary inversely with bandwidth and overshoot varies directly with the height of the resonant peak.

In Sec. 4-5 we illustrate important frequency-domain concepts by applications to practical systems for filtering, modulation, and automatic control. We describe one method (Butterworth) for designing a low-pass filter, where realizability and stability arise in the process of obtaining a design procedure. We discuss double- and single-sideband modulation, where fidelity is the dominant performance measure and the frequency-translation property of the Fourier transformation is a powerful tool for analysis. Finally, we describe a simple dc servomechanism, where relations between transient response (step response) and frequency response (transfer function) prove useful for design subject to specifications on rise time and overshoot.

It is difficult to overemphasize the usefulness of concepts presented in this chapter. Frequency-domain analysis is the most widely used method for system analysis and design. It is by far the dominant method for analysis and design of systems for filtering, signal transmission, and signal processing (instrumentation and measurement). In practice, it has at least equal footing with other methods for analysis and design of automatic control systems. Mastery of concepts and procedures presented in this chapter will pay off handsomely in practice.

PROBLEMS

4-1 Draw block diagrams showing how the periodic signals of Fig. P4-1 can be produced by operations on a sinusoidal signal $x(t) = x_0 \cos \omega_0 t$, with $\omega_0 = 2\pi/T$. Show that each of the periodic signals thus obtained can be described by a harmonic series having fundamental frequency ω_0.

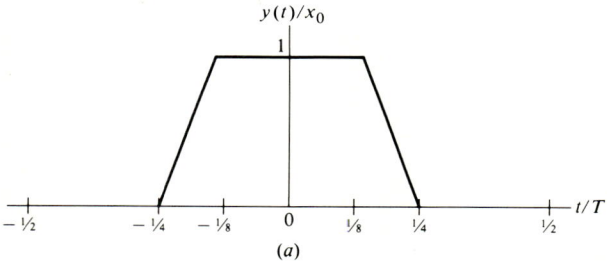

(a)

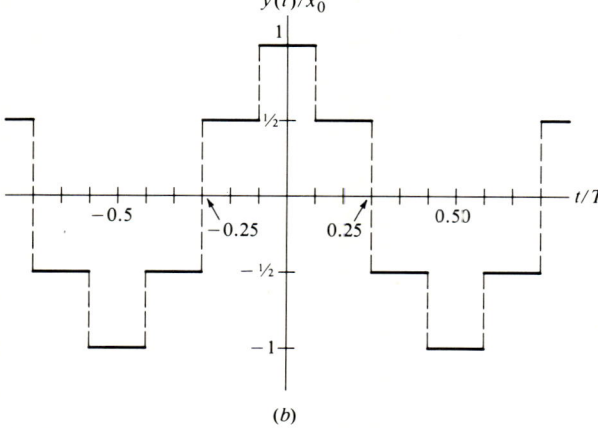

(b)

Figure P4-1

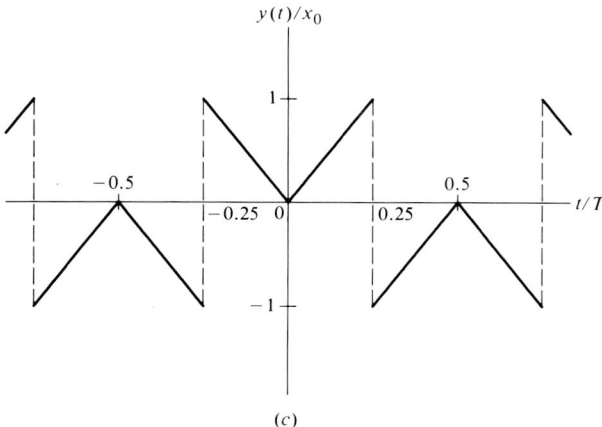

(c)

Figure P4-1 *Continued*

4-2 Show that

$$\int_{t_0}^{t_0+T} e^{j2\pi(n-k)t/T}\,dt = \begin{cases} T & n = k \\ 0 & n \neq k \end{cases}$$

4-3 Show that the Fourier coefficients given by (4-5) are independent of t_0 if $x(t)$ is a periodic signal $x(t)$ having period T. *Hint:* Show that $\partial X_k/\partial t_0 = 0$.

4-4 Obtain the Fourier coefficients for the rectangular pulse train of Example 4-1 using (4-5) with $t_0 = 0$. Why is $t_0 = -T/2$ a better choice?

4-5 Fourier's series for a signal $x(t)$ over $t_0 \leq t < t_0 + T$ can be expressed in *quadrature form* as

$$x(t) = \sum_{n=0}^{\infty} (I_n \cos n\omega_0 t + Q_n \sin n\omega_0 t)$$

where $\omega_0 = 2\pi/T$.

(a) Obtain expressions for I_n and Q_n in terms of the Fourier coefficients given by (4-5).
(b) Obtain expressions for the peak amplitudes and the initial phases of the amplitude-phase form of a Fourier series in terms of I_n and Q_n.

4-6 Use (4-5) to obtain the Fourier coefficients for the periodic signals of Fig. P4-6.

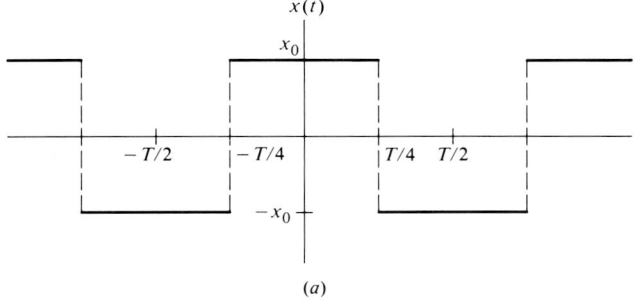

(a)

Figure P4-6

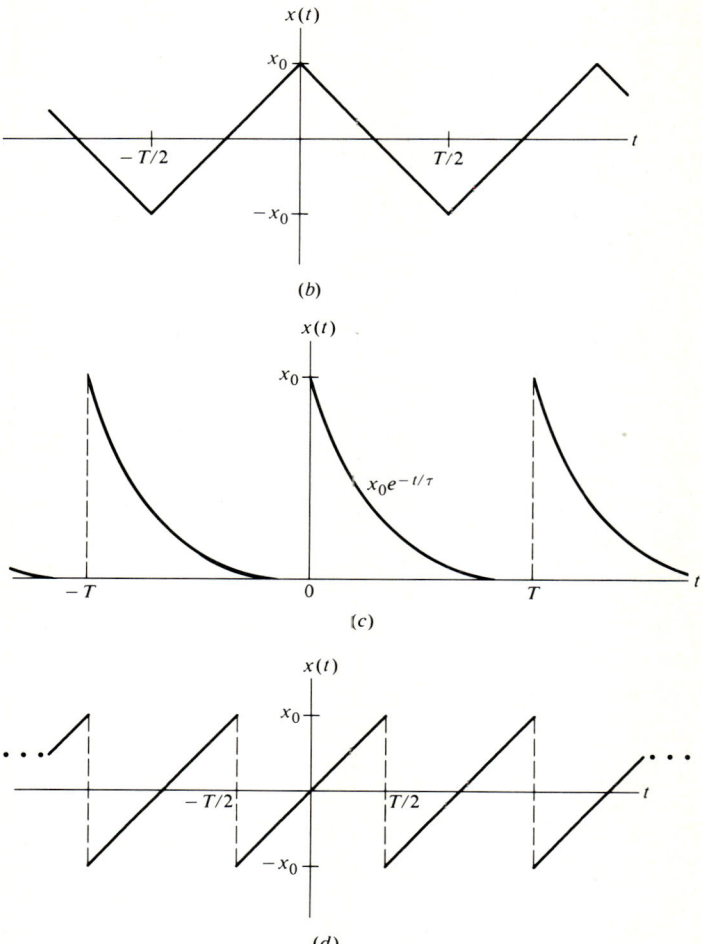

Figure P4-6 *Continued*

4-7 Two periodic signals have periods T_1 and T_2. What is the period of the sum of the signals? What is the period of the product of the signals?

4-8 Two Fourier series have fundamental frequencies f_1 and $f_2 = 2f_1$. What is the fundamental frequency of the sum of these series?

4-9 Two periodic signals have periods T_1 and $T_2 = \sqrt{2}\,T_1$. Is the sum of these signals a periodic signal? If so, find the period.

4-10 In the relations below, $x(t)$ is a periodic signal. Show that $y(t)$ is also a periodic signal and obtain an expression for the Fourier coefficients for $y(t)$ in terms of the Fourier coefficients for $x(t)$.

(a) $y(t) = Kx(t)$ (b) $y(t) = K\dfrac{dx(t)}{dt}$ (c) $y(t) = x(-t)$

(d) $y(t) = x(t - t_0)$ (e) $K\dfrac{dy(t)}{dt} = x(t)$

4-11 A periodic signal $x(t)$ is applied to the input of a compression element having compression ratio η. Express the Fourier coefficients and the fundamental frequency for the output in terms of the corresponding quantities for the input.

4-12 Give a Fourier series in both exponential and amplitude-phase form for the signals whose spectra are shown in Fig. P4-12.

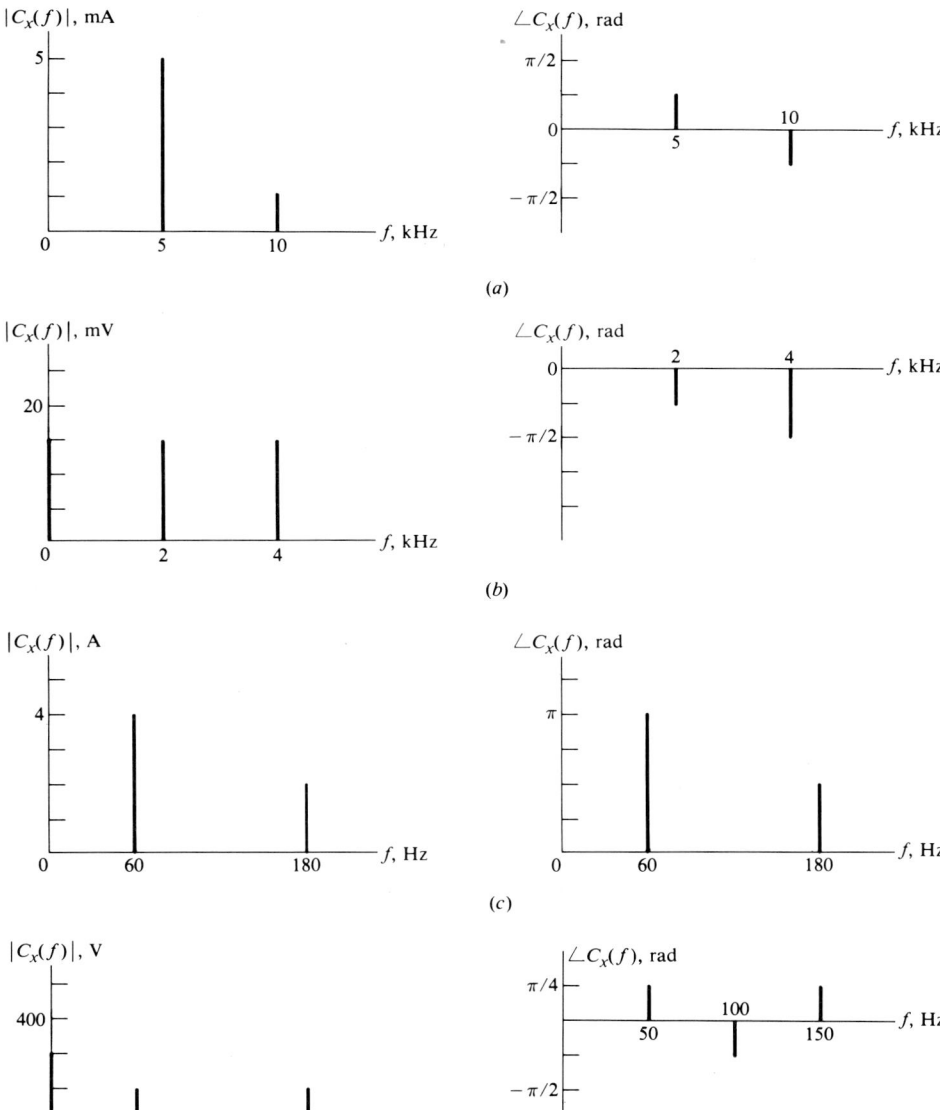

Figure P4-12

278 INTRODUCTION TO SYSTEM ANALYSIS

4-13 Obtain a Fourier series for the output $y(t)$ of each system of Fig. P4-13 for input $x(t) = A \cos \omega_0 t$, where $A = 5$ V and $f_0 = 1$ kHz.

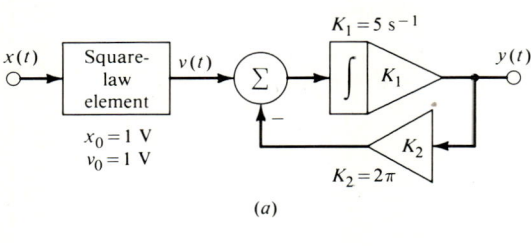

(a)

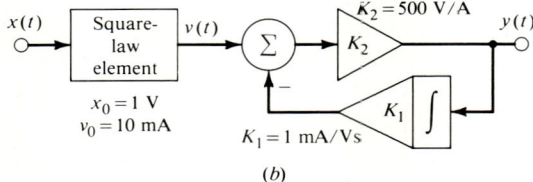

(b)

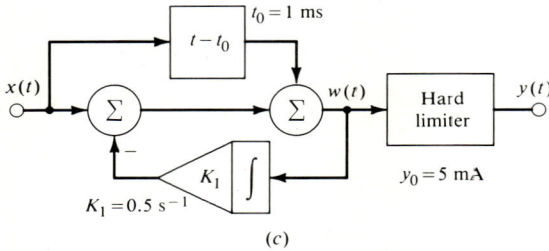

(c)

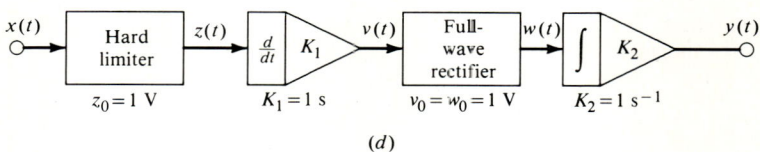

(d)

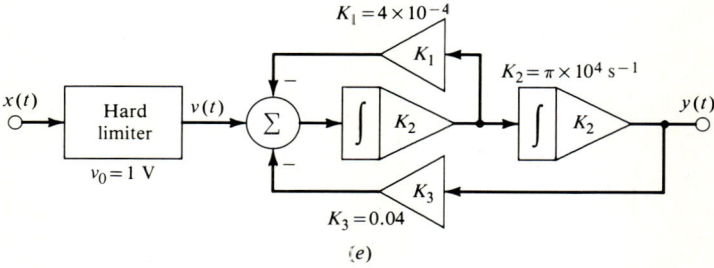

(e)

Figure P4-13

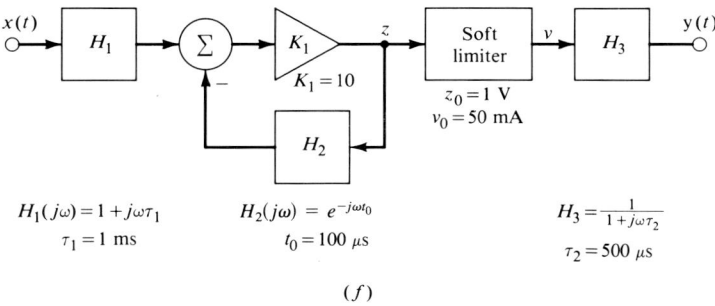

$H_1(j\omega) = 1 + j\omega\tau_1$ $H_2(j\omega) = e^{-j\omega t_0}$ $H_3 = \dfrac{1}{1+j\omega\tau_2}$
$\tau_1 = 1$ ms $t_0 = 100$ μs $\tau_2 = 500$ μs

(f)

Figure P4-13 *Continued*

4-14 Use Appendix B and the relations given in Prob. 4-10 to obtain Fourier coefficients for the signals of Fig. P4-14.

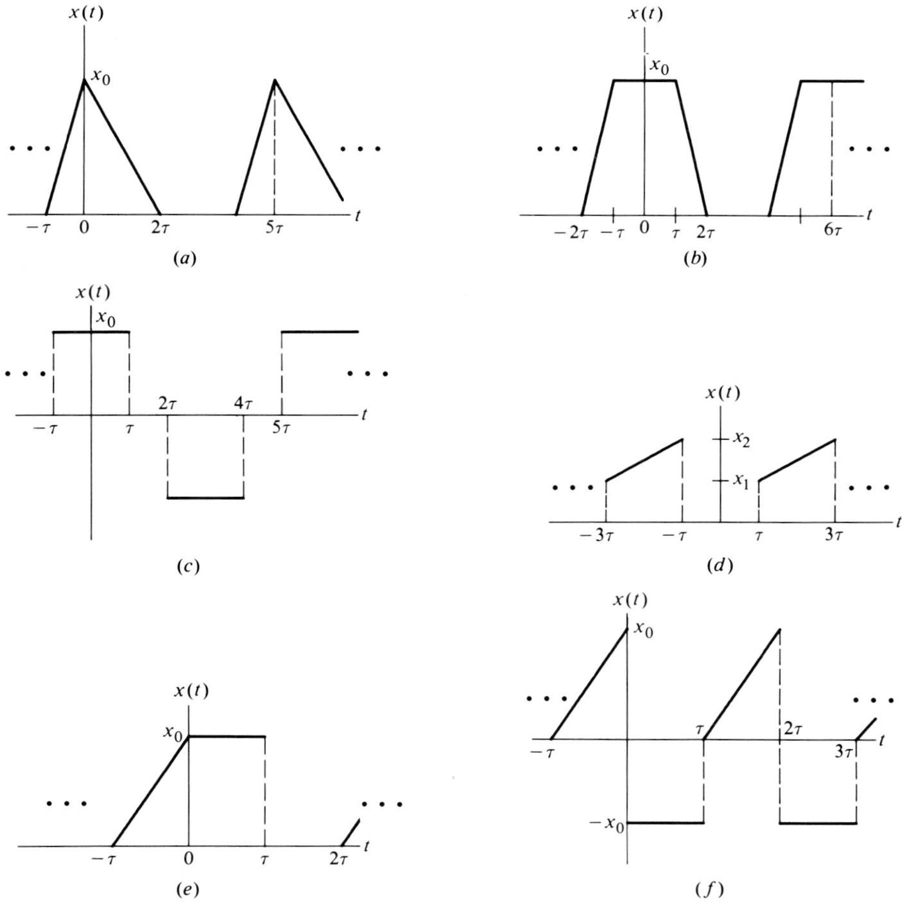

Figure P4-14

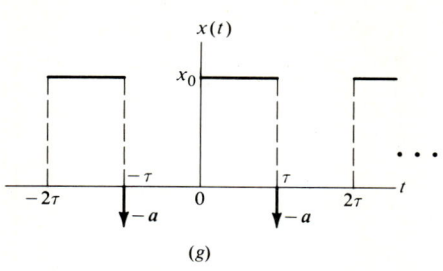
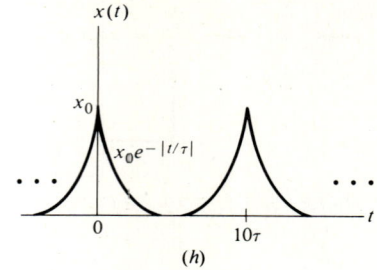

Figure P4-14 *Continued*

4-15 Figure P4-15 shows one period of each of four periodic signals. Give the values to which the Fourier series for each of these signals converges for $t = 0$, $t = 0^+$, $t = t_1^-$, $t = t_1$, $t = t_1^+$, $t = T^-$, $t = T$, and $t = T^+$.

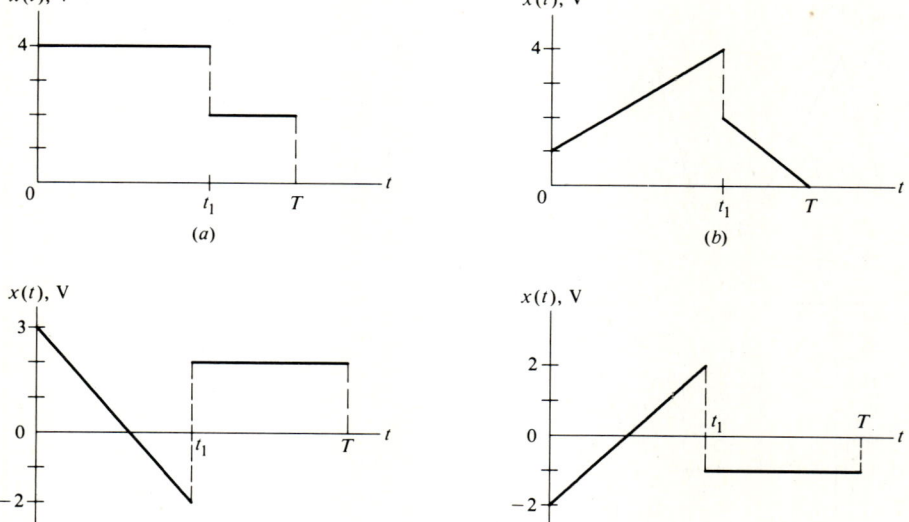

Figure P4-15

4-16 For each system of Fig. P4-16 calculate the third-harmonic distortion (percent) for input $x(t) = A \cos \omega_0 t$, where $A = 1$ mV and $f_0 = 1$ kHz.

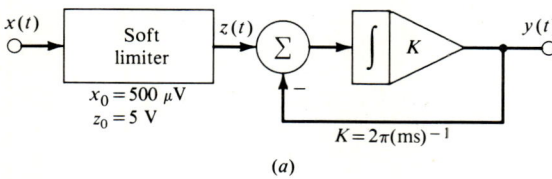

(a)

Figure P4-16

FOURIER SERIES, FOURIER INTEGRAL, AND FOURIER TRANSFORMATION 281

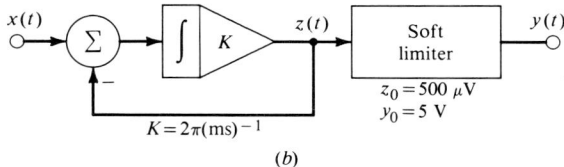

(b)

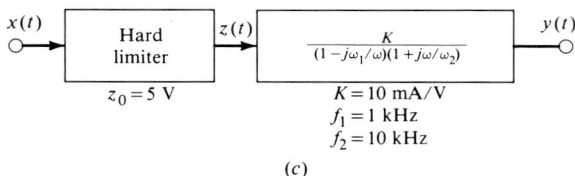

(c)

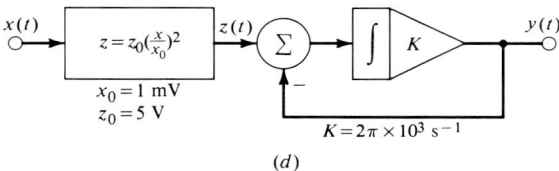

(d)

Figure P4-16 *Continued*

4-17 A rectangular pulse train having period T, amplitude x_0, and pulse duration $T/2$ is applied to an ideal low-pass filter having transfer function

$$H(j\omega) = \begin{cases} 0 & |\omega| > 2\pi/T \\ 1 & |\omega| \le 2\pi/T \end{cases}$$

Plot the output $y(t)$ as a function of time.

4-18 A square wave $x(t)$ having period T and peak-to-peak amplitude $2x_0$ is applied to a system that doubles the peak amplitude of the third harmonic and leaves all other harmonics unchanged. Plot the output $y(t)$ versus time. Give a block diagram for the system.

4-19 In the system of Fig. P4-19, the transfer function of the bandpass filter is

$$H(j\omega) = \frac{1}{1 + jQ[(\omega/3\omega_0)^2 - 1]}$$

where Q is large. Show that the system is a frequency tripler.

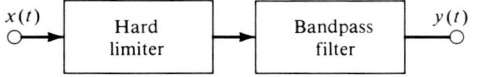

Figure P4-19

4-20 A rectangular pulse train $x(t)$ having period T and pulse duration $\tau = T/4$ is applied to an ideal comb filter that passes the even harmonics and stops the odd harmonics of $x(t)$. Plot the output as a function of time. Repeat for a comb filter that passes the odd harmonics and stops the even harmonics of $x(t)$.

4-21 Give the Fourier series in amplitude-phase form for the signal $x(t)$ having the two-sided spectrum of Fig. P4-21. What is the fundamental frequency of the series?

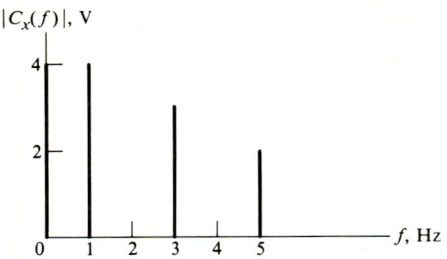
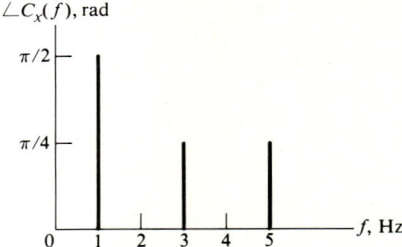

Figure P4-21

4-22 Show that the signal having the two-sided spectrum of Fig. P4-22 can be expressed either as a harmonic series having fundamental frequency ω_0 or as a harmonic series having fundamental frequency $\omega_0/2$. Tabulate the Fourier coefficients versus harmonic number n for both series.

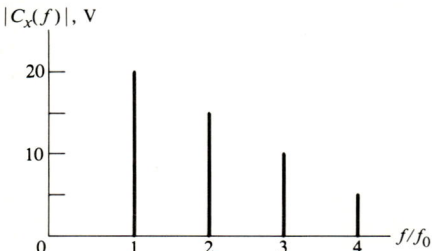
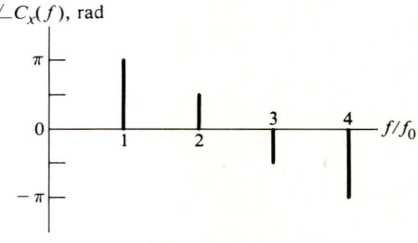

Figure P4-22

4-23 List the frequencies represented in the Fourier series for the signals of Fig. P4-23. *Hint:* The output of a sinusoidally excited static element contains only even (odd) harmonics of the input if the transfer characteristic of the element is even (odd) and contains all harmonics of the input if the transfer characteristic is neither even nor odd.

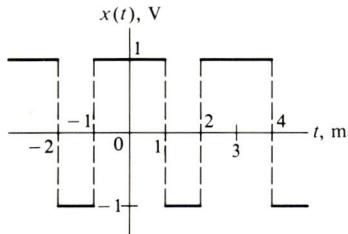
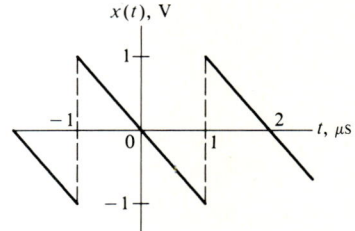

Figure P4-23

4-24 It can be shown that the magnitudes of the Fourier coefficients for a signal $x(t)$ decrease asymptotically with harmonic index n as n^{-k}, where k is the smallest integer for which $d^k x(t)/dt^k$ contains a delta function. Illustrate this rule using the triangular wave of Fig. P4-24.

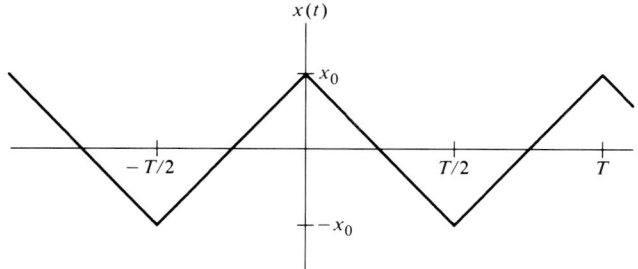

Figure P4-24

4-25 A signal $x(t)$ is to be approximated over an interval $t_0 \leq t < t_0 + T$ by a finite harmonic series

$$\hat{x}(t) = \sum_{n=-K}^{K} X_n e^{jn\omega_0 t}$$

where $\omega_0 = 2\pi/T$. Obtain an expression for the coefficients that give minimum mean-squared error ϵ^2 over the interval, where

$$\epsilon^2 = \frac{1}{T} \int_{t_0}^{t_0+T} [\hat{x}(t) - x(t)]^2 \, dt$$

by solving $\partial \epsilon^2 / \partial X_n = 0$ for X_n. Interpret the result in terms of the Fourier series for $x(t)$ over the interval.

4-26 A sinusoidal signal $x(t) = 5 \cos \omega_0 t$ is applied to a static element having the transfer characteristic of Fig. P4-26.

(a) Plot the output $y(t)$ versus time t.

(b) Obtain an expression for Fourier coefficients for the output.

(c) Calculate the peak amplitudes and the initial phases of the first three nonzero terms of the Fourier series for the output.

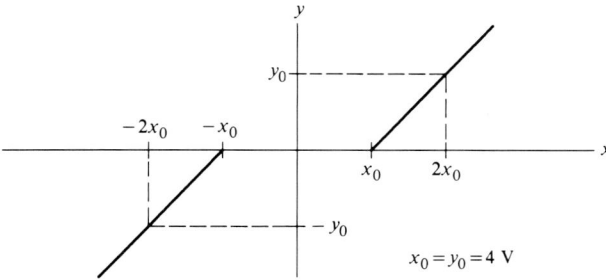

Figure P4-26

4-27 A Fourier series

$$\hat{x}(t) = \sum_{n=-\infty}^{\infty} X_n e^{j2\pi nt/T}$$

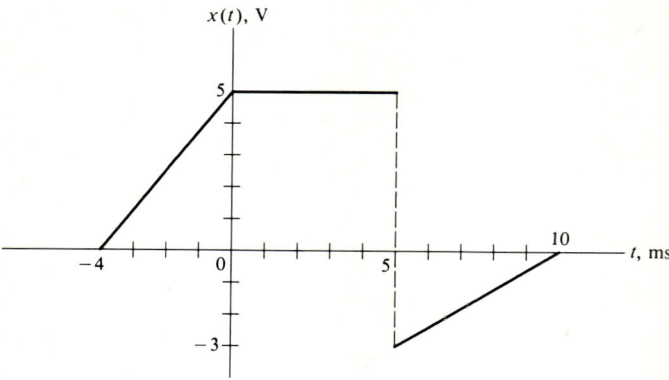

Figure P4-27

is obtained for the nonperiodic signal $x(t)$ of Fig. P4-27 using (4-5). Plot two periods of $\hat{x}(t)$ versus time t for:

(a) $t_0 = 0$ and $T = 4$ ms   (b) $t_0 = -4$ ms and $T = 6$ ms
(c) $t_0 = -4$ ms and $T = 2$ ms   (d) $t_0 = -8$ ms and $T = 8$ ms

4-28 A Fourier series $\hat{x}(t)$ can be used to obtain the output $y(t)$ of a linear stationary system for a nonperiodic input $x(t)$, as illustrated in Example 4-5. For each of the following inputs and systems give a condition on the period of $\hat{x}(t)$ under which a period of the periodic output $\hat{y}(t)$ for input $\hat{x}(t)$ is a good approximation to the actual (nonperiodic) output $y(t)$.

(a) $x(t) = x_0 e^{-t/\tau} u(t)$ $h(t) = \dfrac{1}{\tau} e^{-t/\tau} u(t)$

(b) $x(t) = a\delta(t)$ $\tau \dfrac{dy}{dt} + y = x$

(c) $x(t) = x_0 \dfrac{t}{\tau_0} r\left(\dfrac{t}{\tau_0}\right)$ $h(t) = \dfrac{1}{\tau_1} r\left(\dfrac{t}{\tau_1}\right)$

(d) $x(t) = x_0 r\left(\dfrac{t}{\tau}\right)$ $h(t) = \omega_0 e^{-r\tau_1}(\sin \omega_0 t) u(t)$

4-29 Use (4-11) to obtain the spectral density of each of the signals of Fig. P4-29.

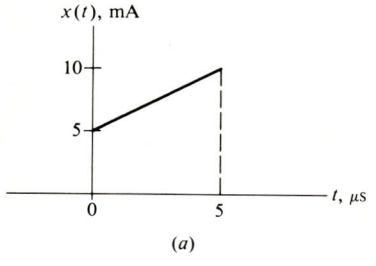

(a)

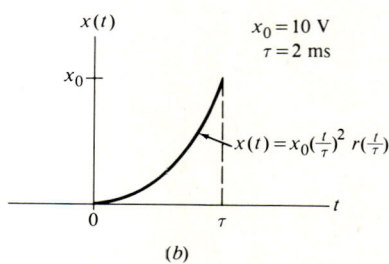

(b)

Figure P4-29

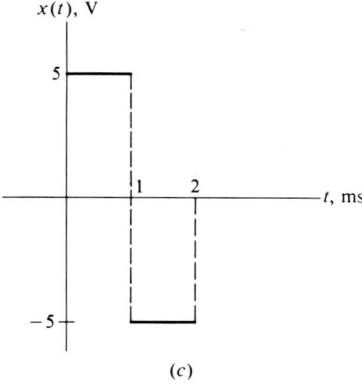
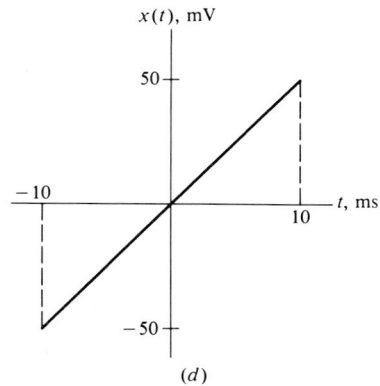

Figure P4-29 Continued

4-30 Use the result of Example 4-20 to obtain the spectral density of the signal $x(t) = x_0 \, \text{sgn} \, t$, where sgn α is the *sign function* defined by

$$\text{sgn} \, \alpha = \begin{cases} -1 & \alpha < 0 \\ 0 & \alpha = 0 \\ 1 & \alpha > 0 \end{cases}$$

4-31 Obtain the spectral density of $x(t) = A \cos(\omega_0 t + \theta)$.

4-32 Show that

$$\delta(2\pi f) = \frac{1}{2\pi} \delta(f)$$

(a) by using the definition

$$\delta(\alpha) = \lim_{\Delta\alpha \to 0} \left[\frac{1}{\Delta\alpha} r\left(\frac{\alpha}{\Delta\alpha}\right) \right]$$

and (b) by using the definition

$$\delta(\alpha) = \int_{-\infty}^{\infty} e^{-j\alpha t} \, dt$$

4-33 Derive (4-12) from (4-10) and (4-11).

4-34 Plot both the spectrum (Fourier coefficients) and the spectral density (the Fourier transform) versus frequency f for the rectangular pulse train $x(t)$ of Fig. P4-34. Describe similarities and differences.

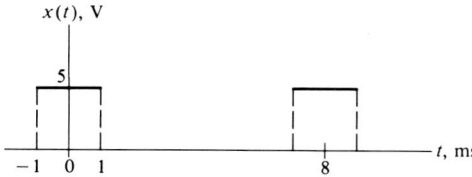

Figure P4-34

4-35 A periodic signal $\tilde{e}(t)$ is formed as the periodic extension of a nonperiodic pulse $e(t)$, where $e(t) = 0$ for $|t| > T/2$ and T is the period of $\tilde{e}(t)$. Obtain an expression for the Fourier coefficients of $\tilde{e}(t)$ in terms of the Fourier transform of $e(t)$.

4-36 Let $E(j\omega) = \mathcal{F}\{e(t)\}$. Show that

(a) $\mathcal{F}\{e(\eta t)\} = \dfrac{1}{|\eta|} E\left(\dfrac{j\omega}{\eta}\right)$

(b) $\mathcal{F}\left\{\dfrac{de(t)}{dt}\right\} = j\omega E(j\omega)$

(c) $\mathcal{F}\{e(t - t_0)\} = E(j\omega)e^{-j\omega t_0}$

(d) For $e_1(t)$ and $e_2(t)$ real, $\mathcal{F}\left\{\displaystyle\int_{-\infty}^{\infty} e_1(\lambda)e_2(t - \lambda)\,d\lambda\right\} = E_1(j\omega)E_2(j\omega)$

(e) $\mathcal{F}\{e_1(t)e_2(t)\} = \dfrac{1}{2\pi}\displaystyle\int_{-\infty}^{\infty} E_1(j\lambda)E_2(j\omega - j\lambda)\,d\lambda$

(f) $\mathcal{F}^{-1}\left\{\dfrac{dE(j\omega)}{d\omega}\right\} = -jte(t)$

4-37 Show that the frequency-translation property of the Fourier transformation is a special case of the property given in part (e) of Prob. 4-36.

4-38 Show that the integral of the spectral density of a signal $x(t)$ over all frequency equals $x(0)$.

4-39 Let $E(j\omega) = \mathcal{F}\{e(t)\}$. Show that:

(a) $e(-t) = e(t)$ implies $\text{Im } E(j\omega) = 0$ and vice versa.
(b) $e(-t) = -e(t)$ implies $\text{Re } E(j\omega) = 0$ and vice versa.
(c) $\mathcal{F}\{e(-t)\} = E^*(j\omega)$.

4-40 Show that

$$\mathcal{F}\left\{\dfrac{d^n e(t)}{dt^n}\right\} = (j\omega)^n \mathcal{F}\{e(t)\}$$

4-41 Use the Fourier transformation to obtain the impulse response of each system of Fig. P4-41.

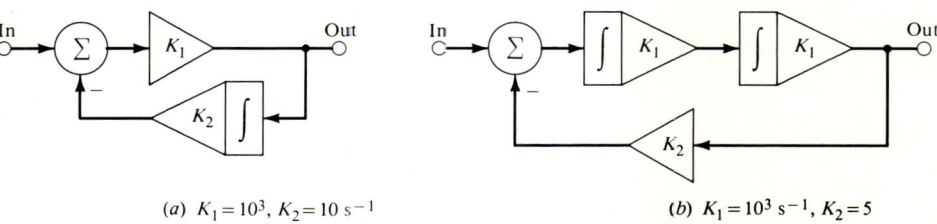

(a) $K_1 = 10^3$, $K_2 = 10\text{ s}^{-1}$ (b) $K_1 = 10^3 \text{ s}^{-1}$, $K_2 = 5$

Figure P4-41

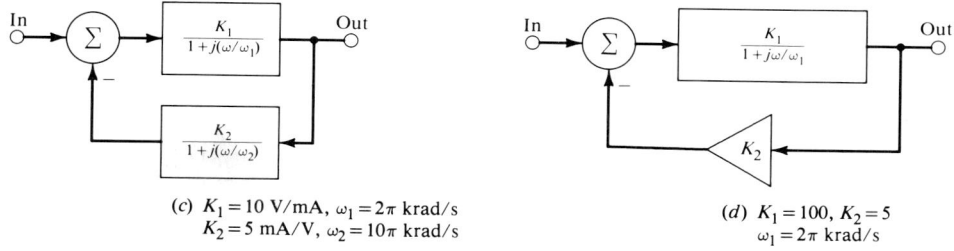

(c) $K_1 = 10$ V/mA, $\omega_1 = 2\pi$ krad/s
$K_2 = 5$ mA/V, $\omega_2 = 10\pi$ krad/s

(d) $K_1 = 100$, $K_2 = 5$
$\omega_1 = 2\pi$ krad/s

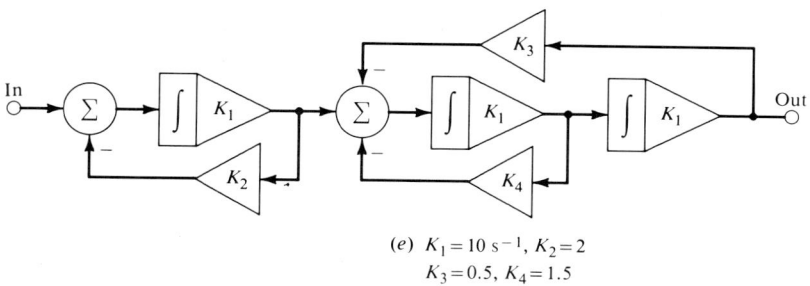

(e) $K_1 = 10$ s^{-1}, $K_2 = 2$
$K_3 = 0.5$, $K_4 = 1.5$

Figure P4-41 *Continued*

4-42 Show that Rayleigh's theorem holds for a complex function $x(t)$ by showing that

$$\int_{-\infty}^{\infty} |x(t)|^2 \, dt = \frac{1}{2\pi} \int_{-\infty}^{\infty} |X(j\omega)|^2 \, d\omega$$

where $|x(t)|^2 = x(t)x^*(t)$ and $X(j\omega) = \mathcal{F}\{x(t)\}$.

4-43 Show that Parseval's theorem holds for a complex periodic function $x(t)$ by showing that

$$\frac{1}{T} \int_{t_0}^{t_0+T} |x(t)|^2 \, dt = \sum_{n=-\infty}^{\infty} |X_n|^2$$

where $|x(t)|^2 = x(t)x^*(t)$ and $X_0, X_{\pm 1}, X_{\pm 2}, \ldots$ are Fourier coefficients for $x(t)$.

4-44 A generalization of Rayleigh's theorem, called *Plancherel's theorem*, is expressed by the relation

$$\int_{-\infty}^{\infty} x(t)y^*(t) \, dt = \frac{1}{2\pi} \int_{-\infty}^{\infty} X(j\omega)Y^*(j\omega) \, d\omega$$

where $X(j\omega) = \mathcal{F}\{x(t)\}$ and $Y(j\omega) = \mathcal{F}\{y(t)\}$. Derive this relation from the Fourier-transformation relations (4-19) and (4-20).

4-45 Obtain the energy density spectrum $E_x(f)$ for a sinusoidal pulse $x(t) = x_0 r(t/\tau) \cos \omega_0 t$, where $\omega_0 = 1000\pi/\tau$. Sketch a graph of $E_x(f)$ versus frequency f. Calculate the fraction of the normalized energy of $x(t)$ contributed by components in the band $f_0 - 1/\tau < f < f_0 + 1/\tau$. You may use

$$\int_0^{\pi} \text{sa}^2 \alpha \, d\alpha = 1.418$$

288 INTRODUCTION TO SYSTEM ANALYSIS

4-46 Obtain the energy density spectrum $E_x(f)$ for an exponential pulse $x(t) = x_0 e^{-t/\tau} u(t)$, where $\tau > 0$. Obtain an expression for the fraction γ of the normalized energy of $x(t)$ of components in a band $0 \leq f < W$ as a function of W. Then express W as a function of γ and determine the values of W for which $\gamma = 0.50$, $\gamma = 0.90$, and $\gamma = 0.99$. Interpret these calculations in terms of bandwidth of $x(t)$.

4-47 Obtain the power density spectrum of a rectangular pulse train

$$x(t) = \sum_{n=-\infty}^{\infty} x_0 r\left(\frac{t + nT}{\tau}\right) \quad \text{where } \tau = \frac{T}{8}$$

Obtain an expression for the normalized power of components in a band $0 \leq f < W$, where W is an integral multiple of $1/T$. Calculate the normalized power of components in the band $0 \leq f < 1/\tau$ and express the result as a fraction of the (total) normalized power of $x(t)$. Comment on the significance of the result.

4-48 An exponential pulse $x(t) = 5e^{-t/\tau} u(t)$ V is applied to an ideal low-pass filter having gain

$$\Gamma(\omega) = \begin{cases} 1 \text{ V/V} & |\omega| < \dfrac{1}{\tau} \\ 0 & |\omega| \geq \dfrac{1}{\tau} \end{cases}$$

calculate the normalized energy of the output $y(t)$. Express the result as a fraction of the normalized energy of the input $x(t)$ and comment on the physical significance of the result.

4-49 The signal $x(t) = 10 \text{ sa } 2\pi W t$ V, where $W = 1$ kHz, is applied to an ideal low-pass filter having unit dc gain. Express the normalized energy of the output as a function of the bandwidth of the filter.

4-50 For each system of Fig. P4-50, $K_1 = 1 \text{ s}^{-1}$. Determine all values of K_2 for which the system is stable.

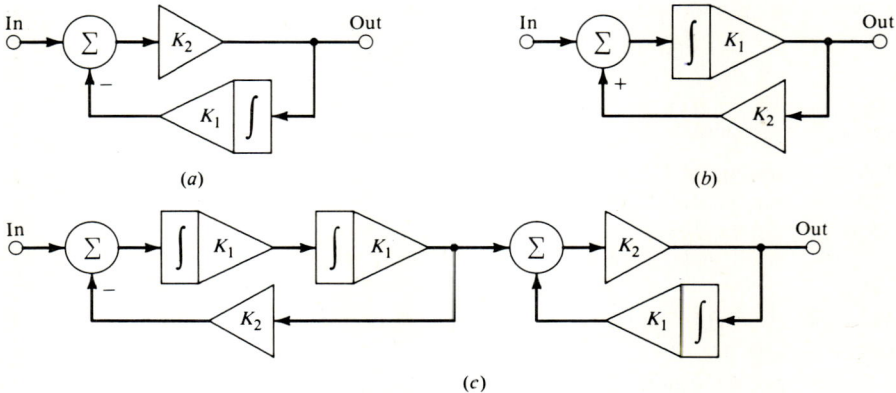

Figure P4-50

4-51 Show that for $p > 0$ the transfer function

$$H(j\omega) = \frac{p}{j\omega + p}$$

represents either a realizable stable system or a nonrealizable unstable system.

4-52 The spectral density of the output of a proposed system is given by $Y(j\omega) = X^*(j\omega)$, where $X(j\omega)$ is the spectral density of an input. Show that the system is nonrealizable.

4-53 The spectral density of the output of a proposed system is given by $Y(j\omega) = X(j\omega/2)$, where $X(j\omega)$ is the spectral density of an input. Show that the system is nonrealizable.

4-54 Draw a block diagram consisting of realizable elementary systems for each transfer function:

(a) $H(j\omega) = \dfrac{K_1 K_2}{(j\omega + p_1)(j\omega + p_2)}$

(b) $H(j\omega) = \dfrac{K(1 + j\omega/\omega_1)}{1 + j\omega/\omega_2}$

(c) $H(j\omega) = \dfrac{j\omega\omega_0}{-\omega^2 + 2j\omega\omega_0 + \omega_0^2}$

(d) $H(j\omega) = \dfrac{j(\omega/\omega_1)e^{-j\omega t_0}}{1 + j\omega/\omega_2}$

(e) $H(j\omega) = \dfrac{K}{j\omega(1 + j\omega/\omega_1)}$

4-55 Can the Paley-Wiener condition be used to show that a distortionless system is nonrealizable? Show how or explain why not.

4-56 Can the Paley-Wiener condition be used to show that an ideal integrator is nonrealizable? An ideal differentiator? Show how or explain why not.

4-57 With reference to Fig. 4-34, explain why the transfer function

$$H(j\omega) = \dfrac{1}{\sqrt{1 + (\omega/\omega_0)^2}}$$

is nonrealizable.

4-58 With reference to Fig. 4-37, explain why the transfer function $H(j\omega) = e^{-j\tan^{-1}(\omega/\omega_0)}$ is nonrealizable.

4-59 Give a condition on n and m under which the Paley-Wiener condition can be applied to a rational transfer function of the form

$$H(j\omega) = \dfrac{b_m(j\omega)^m + b_{n-1}(j\omega)^{m-1} + \cdots + b_0}{(j\omega)^n + a_{n-1}(j\omega)^{n-1} + \cdots + a_0}$$

4-60 Use the Paley-Wiener condition to show that an ideal low-pass filter is nonrealizable.

4-61 Use the Paley-Wiener condition to show that an ideal bandpass filter is nonrealizable

4-62 For each transfer function given below, obtain a condition on τ under which the system provides high-fidelity transmission for a rectangular pulse having duration τ. Sketch a graph of the output $y(t)$ versus time t for a value of τ satisfying the condition obtained.

(a) $H(j\omega) = \dfrac{1 + j\omega/\omega_1}{(1 + j\omega/\omega_0)(1 + j\omega/\omega_2)}$ V/V

where $f_0 = 1$ kHz, $f_1 = 5$ kHz, and $f_2 = 10$ kHz.

(b) $H(j\omega) = \dfrac{1 + j\omega/\omega_1}{1 + j\omega/\omega_2}$ V/V

where $f_1 = 10$ Hz and $f_2 = 100$ Hz.

(c) $H(j\omega) = \dfrac{\omega_0(1 + j\omega/\omega_1)}{j\omega(1 + j\omega/\omega_2)}$ V/V

where $f_0 = 1$ kHz, $f_1 = 10$ kHz, and $f_2 = 100$ kHz.

(d) $H(j\omega) = \dfrac{\omega_0^2}{(1 + j\omega/\omega_1)(\omega_0^2 + 2j\zeta\omega\omega_c - \omega^2)}$ V/V

where $\zeta = 0.4$, $f_0 = 10$ kHz, and $f_1 = 100$ kHz.

4-63 Figure P4-63 shows a graph of gain versus frequency for a stable second-order system. Find the rise time, the peak time, the percent overshoot, the steady-state amplitude, the ring frequency, and the settling time. Sketch a graph of the step response of the system versus time.

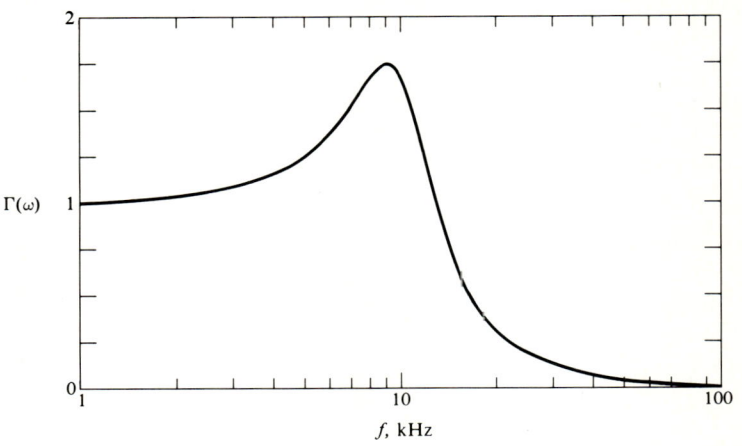

Figure P4-63

4-64 Figure P4-64 shows a graph of step response versus time for a second-order system. Give the dc gain, the peak gain, the damping ratio and the undamped natural frequency of the system. Sketch a graph of the gain of the system versus frequency.

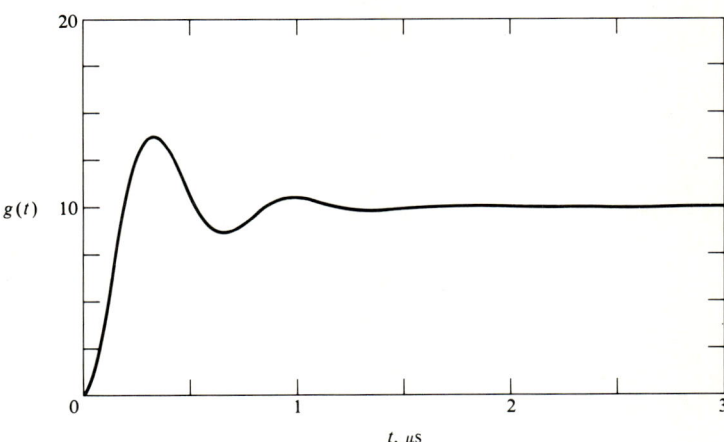

Figure P4-64

4-65 Use the Paley-Wiener condition to show that a Butterworth filter having gain

$$|H(j\omega)| = \dfrac{|K|}{\sqrt{1 + (\omega/\omega_0)^{2n}}}$$

is realizable for $n > 0$. Why is the Paley-Wiener condition inapplicable to this gain for $n = 0$?

4-66 Show that the system of Fig. P4-66 is a Butterworth filter. Determine the order n and the cutoff frequency f_c.

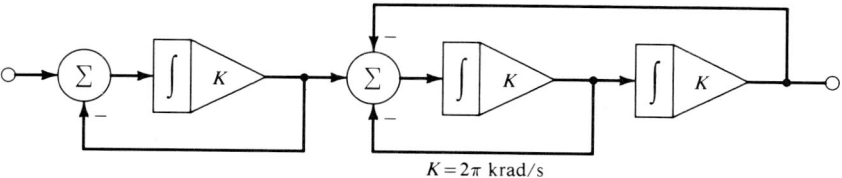

Figure P4-66

4-67 Determine the order n and the cutoff frequency f_c for a Butterworth filter meeting each set of specifications below (see Fig. 4-51):
  (a) $f_1 = 3$ kHz, $\Gamma_1 = 0.8$, $f_2 = 4$ kHz, $\Gamma_2 = 0.1$.
  (b) $f_1 = 1$ MHz, $\Gamma_1 = 0.9$, $f_2 = 1.5$ MHz, $\Gamma_2 = 0.2$.
  (c) $f_1 = 100$ Hz, $\Gamma_1 = 0.95$, $f_2 = 1$ kHz, $\Gamma_2 = 0.05$.

4-68 A fourth-order Butterworth filter is to meet the following specifications (see Fig. 4-51): $f_1 = 1$ kHz, $\Gamma_1 = 0.9$, and $\Gamma_2 = 0.1$. Determine the minimum obtainable value for f_2.

4-69 Refer to Fig. 4-51. For $f_1 = 1$ kHz, $\Gamma_1 = 0.9$, and $f_2 = 2$ kHz, plot order n and cutoff frequency f_c versus maximum stopband gain Γ_2. Explain the practical implications of the result.

4-70 A second-order Butterworth filter has cutoff frequency $f_c = 1$ kHz. Obtain the step response of the filter. Sketch a graph of the output for input $x(t) = x_0 r(t/\tau)$ for (a) $\tau = 10$ ms, (b) $\tau = 1$ ms, and (c) $\tau = 100$ μs. Explain the practical implications of the results.

4-71 Gain of a low-pass filter often is specified in *decibels* (dB), where $\Gamma(\omega)$ dB = $20 \log[\Gamma(\omega)/\Gamma(0)]$. Obtain the transfer function of a Butterworth filter meeting the following specifications (see Fig. 4-51): $f_1 = 1$ kHz, $\Gamma_1 = -0.5$ dB, $f_2 = 5$ kHz, $\Gamma_2 = -30$ dB.

4-72 The characteristic roots of a low-pass *Chebyshev filter*† lie on an ellipse in a complex plane. Figure P4-72 gives the characteristic roots for a third-order Chebyshev filter. Obtain the transfer

†After the Russian mathematician P. L. Chebyshev (1821–1894).

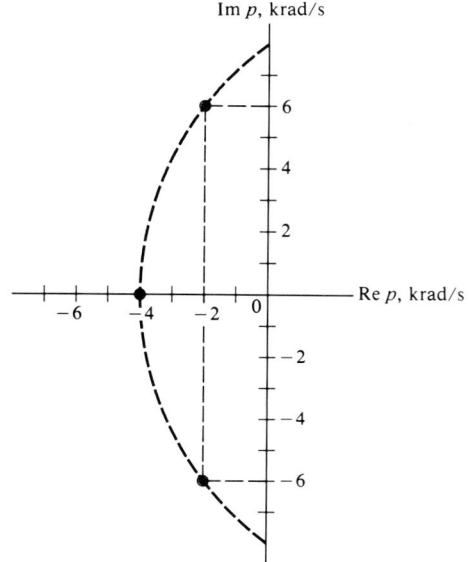

Figure P4-72

function of the filter. Plot the gain and the phase shift versus frequency of (a) the Chebyshev filter and (b) a comparable Butterworth filter. How is the Chebyshev filter superior to the Butterworth filter? Inferior?

4-73 Figure P4-73 shows a simple (idealized) frequency-division multiplex system used for simultaneous transmission of two signals $x_1(t)$ and $x_2(t)$ on a single cable. Both signals have bandwidth W. The carrier frequency f_c equals $2W$. Draw a block diagram for a system that recovers $x_1(t)$ and $x_2(t)$ (individually) from the transmitted signal $y(t)$. Assume that a replica of the carrier is available to the system.

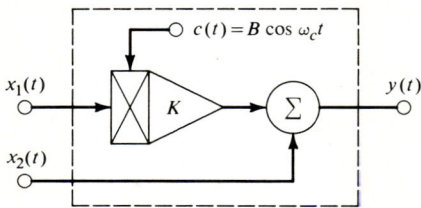

Figure P4-73

4-74 Determine the effects of using a local oscillator to demodulate a single-sideband signal, where the oscillator output differs in frequency and initial phase from the carrier used for transmission. In this context, explain why SSB modulation might be better than DSB modulation in certain phase-insensitive applications.

4-75 To thwart eavesdroppers, signals often are scrambled before transmission.
 (a) Draw a block diagram for a system that scrambles a signal $x(t)$ as shown in Fig. P4-75.
 (b) Draw a block diagram for a system that unscrambles the scrambled signal.

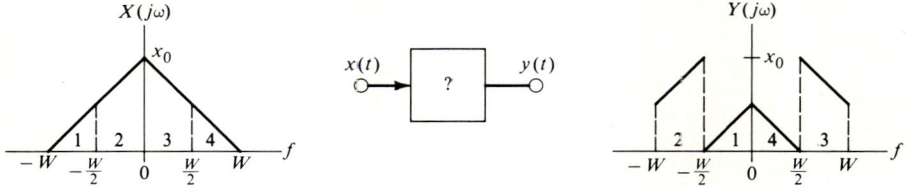

Figure P4-75

4-76 Two signals $x_1(t)$ and $x_2(t)$ can be multiplexed as shown in Fig. P4-76, where

$$y(t) = ax_1(t) \cos \omega_c t + bx_2(t) \sin \omega_c t$$

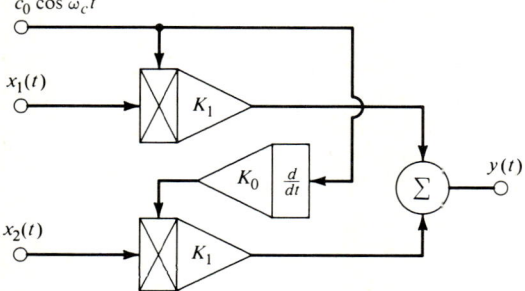

Figure P4-76

Draw a block diagram for a system that recovers $x_1(t)$ and $x_2(t)$ from $y(t)$. Assume that $x_1(t)$ and $x_2(t)$ have the same bandwidth W, that $f_c > W$, and that a replica of the carrier $c_0 \cos \omega_c t$ is available to the system.

4-77 Three signals representing speech, each having bandwidth $W = 3$ kHz, are to be multiplexed on a link having passband 10 kHz $\leq f \leq 25$ kHz. Draw a block diagram for a multiplex-demultiplex system for this application. Show how the carrier is supplied to the demultiplexer.

4-78 The servomechanism of Fig. 4-63 is to position a load having moment of inertia $J = 0.05$ Nm/(rad/s^2) and coefficient of viscous friction $F_L = 0.20$ Nm/(rad/s). Overshoot must not exceed 5 percent, and rise time must not exceed 200 ms. Determine values for $K_e K_m$ and added mechanical damping F_0 (if needed).

4-79 The servomechanism of Fig. 4-68 is to position a load having moment of inertia $J = 0.20$ Nm/(rad/s^2) and coefficient of viscous friction $F_L = 0.75$ Nm/(rad/s). The supply voltage is $V_0 = 28$ V. Overshoot must not exceed 15 percent, and rise time must not exceed 1 s. Determine values for K_m and K_t.

4-80 Figure P4-80 shows a schematic diagram for an armature-controlled dc motor driving a load having moment of inertia J_L and coefficient of viscous friction F_L. The input is a voltage $v(t)$, and the output is angular position $\theta(t)$. The motor produces torque $T(t)$ given by $T = K_T i_a$, where i_a is armature current. Back emf $v_b = K_T d\theta/dt$ is developed across the armature. Draw a block diagram for the system and obtain the transfer function of the system.

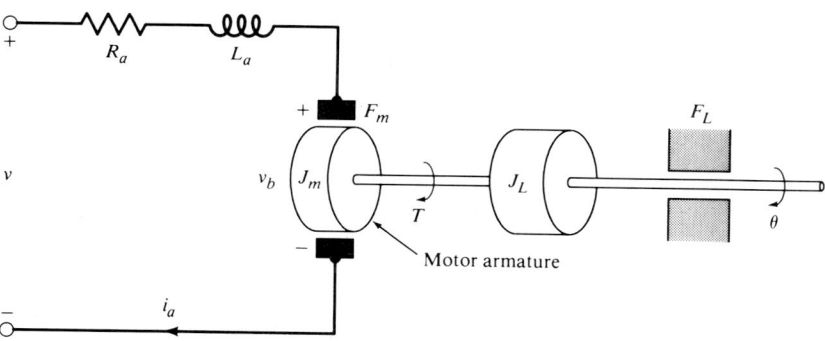

Figure P4-80

4-81 The servomechanism of Fig. 4-68 is to meet the following specifications: steady-state error $e_{ss} = 0$, overshoot PO ≤ 10 percent, and rise time $t_R \leq 500$ ms. The total moment of inertia (load and motor armature) is $J = 0.50$ Nm/(rad/s^2), and the total coefficient of viscous friction is $F = 2.00$ Nm/(rad/s). The constant $K_m = 0.30$ Nm/V and $\theta_i(t) = u(t)$ rad.
 (a) Determine values for the constant K_t and the supply voltage V_0.
 (b) Find the time for which the angular speed $\Omega = d\theta_0/dt$ is maximum.

4-82 Obtain a block diagram and the transfer function for the system of Fig. P4-82.

4-83 Figure P4-83 shows a schematic diagram for a servomechanism, where $J = 0.15$ Nm/(rad/s^2), $F = 1.00$ Nm/(rad/s), $K_e = 1.00$ V/rad, and $K_m = 0.30$ Nm/V.
 (a) Determine values of amplifier gain K_a and tachometer gain K_t for which overshoot does not exceed 15 percent and rise time does not exceed 500 ms. Can these specifications be met using mechanical damping instead of tachometer feedback?

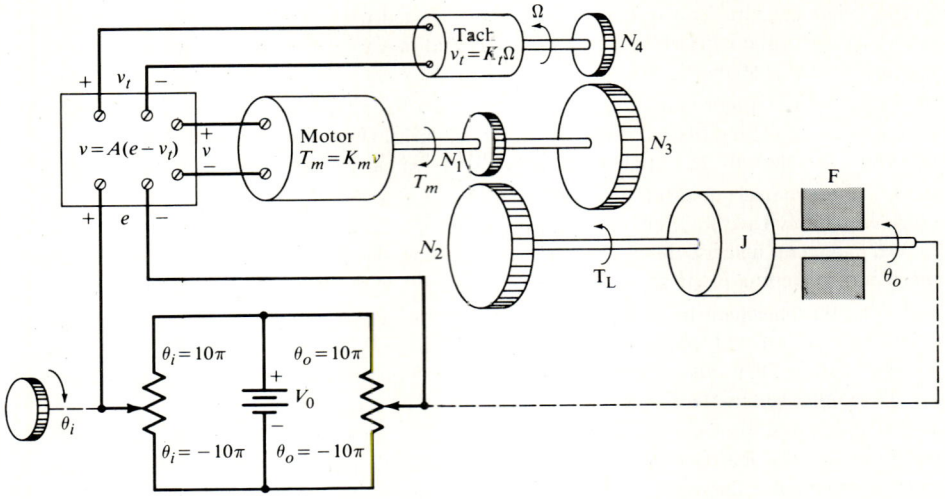

Figure P4-82

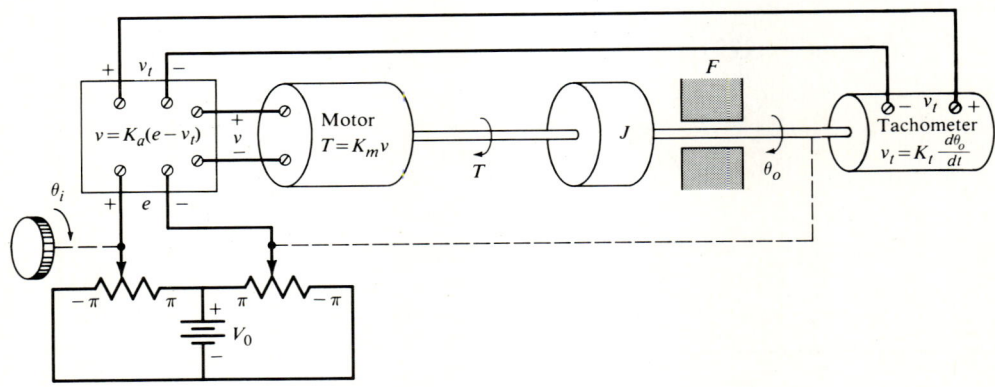

Figure P4-83

(b) For gain K_a obtained in part (a), calculate the sensitivities of damping ratio and undamped natural frequency to variations in K_a. Does tachometer feedback increase or decrease these sensitivities?

CHAPTER FIVE

LAPLACE TRANSFORMATION

The Fourier transformation is one of a handful of transformations that are widely used in signal and system analysis. In this chapter we describe the Laplace transformation,† which is mathematically similar to the Fourier transformation. Indeed, any practical problem that can be solved using the Laplace transformation can also be solved using the Fourier transformation and vice versa. Nonetheless, certain kinds of problems are more easily solved using the Laplace transformation. Also, the Laplace transformation provides additional insight into certain kinds of problems. Consequently, the Laplace transformation is preferred in certain applications.

In Sec. 5-1 we give the definition and the essential properties of the Laplace transformation. In Sec. 5-2 we show how the Laplace transformation is used to obtain an output of a linear stationary system. In Sec. 5-3 we describe how realizability, stability, and other important properties of a system are reflected in a Laplace-transform description of the system. In Sec. 5-4 we introduce the one-sided Laplace transformation, which is particularly well suited for solving initial-value problems.

5-1 DEFINITION AND PROPERTIES OF THE LAPLACE TRANSFORMATION

The *Laplace transform* of a function $e(t)$ is denoted by $E(s)$ and defined by

$$E(s) = \int_{-\infty}^{\infty} e(t) e^{-st} \, dt \tag{5-1}$$

To simplify notation the operator $\mathscr{L}$ is used to denote Laplace transformation; thus

†After the French mathematician Pierre Simon de Laplace (1749–1827).

$$E(s) = \mathcal{L}\{e(t)\} \tag{5-2}$$

denotes the operation on the right side of (5-1).

A function $e(t)$ whose Laplace transform is $E(s)$ is an *inverse Laplace transform* of $E(s)$. We denote this relation by

$$e(t) = \mathcal{L}^{-1}\{E(s)\} \tag{5-3}$$

Methods for obtaining inverse Laplace transforms are described in the sequel.

Example 5-1 The Laplace transform of a delta function is

$$\mathcal{L}\{\delta(t)\} = \int_{-\infty}^{\infty} \delta(t) e^{-st} \, dt = e^0 \int_{-\infty}^{\infty} \delta(t) \, dt = 1$$

The inverse relation is

$$\mathcal{L}^{-1}\{1\} = \delta(t)$$

The unit of a Laplace transform $E(s) = \mathcal{L}\{e(t)\}$ is the unit of $te(t)$; for example, if $e(t)$ is voltage and time t is expressed in seconds, the SI unit of $E(s)$ is volt-seconds. The independent variable s of a Laplace transform $E(s)$ is a complex quantity having the dimension of t^{-1} called *complex frequency*. The real part of s is denoted by σ, and the imaginary part of s is denoted by ω; thus

$$s = \sigma + j\omega \tag{5-4}$$

Values of complex frequency s can be associated with points in a complex plane called the *s plane* (Fig. 5-1). The s plane is an important graphical tool in the theory and application of the Laplace transformation.

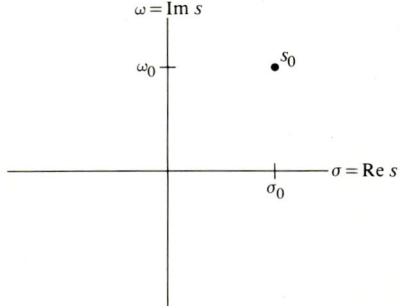

Figure 5-1 An s plane. A point (σ_0, ω_0) represents complex frequency $s_0 = \sigma_0 + j\omega_0$.

5-1A Convergence and Uniqueness

In general, the integral in (5-1) converges only for certain values of s, that is, only over a certain region in an s plane. The values of s for which the integral in (5-1) converges constitute the *region of convergence* for a Laplace transform $E(s)$.

Example 5-2 Determine the Laplace transform and associated region of convergence for the function

$$e_1(t) = e^{pt}u(t)$$

where p is real.

SOLUTION From (5-1),

$$E_1(s) = \int_{-\infty}^{\infty} e^{pt}u(t)e^{-st}\,dt = \int_0^{\infty} e^{-(s-p)t}\,dt = -\frac{1}{s-p}\left\{\lim_{t\to\infty} e^{-(s-p)t} - 1\right\}$$

The limit exists only if $\text{Re}(s - p) > 0$; that is, for $\text{Re}\,s > p$, in which case

$$\lim_{t\to\infty} e^{-(s-p)t} = 0 \qquad \text{Re}\,s > p$$

and

$$E_1(s) = \frac{1}{s-p}$$

Figure 5-2 shows a graph of the function $e_1(t)$ and the region of convergence for $\mathcal{L}\{e_1(t)\} = E_1(s)$.

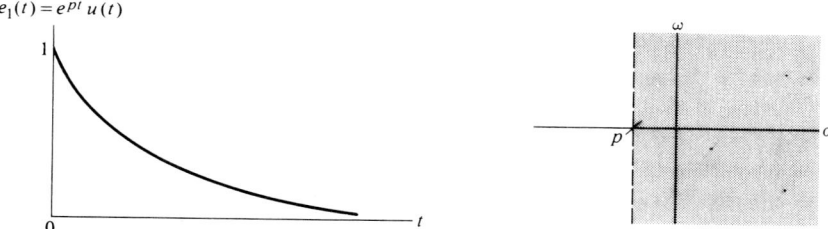

Figure 5-2 Graph of the function $e_1(t) = e^{pt}u(t)$ and the region of convergence $\text{Re}\,s > p$ for $\mathcal{L}\{e_1(t)\}$.

Example 5-3 Determine the Laplace transform and associated region of convergence for the function

$$e_2(t) = -e^{pt}u(-t)$$

where p is real.

SOLUTION From (5-1),

$$E_2(s) = -\int_{-\infty}^0 e^{-(s-p)t}\,dt = \frac{1}{s-p}\left(1 - \lim_{t\to-\infty} e^{-(s-p)t}\right)$$

The limit exists only if $\text{Re}(s - p) < 0$; that is, if $\text{Re}\,s < p$, in which case

$$\lim_{t\to-\infty} e^{-(s-p)t} = 0 \qquad \text{Re}\,s < p$$

and

$$E_2(s) = \frac{1}{s-p}$$

Figure 5-3 shows a graph of $e_2(t)$ and the region of convergence associated with $\mathcal{L}\{e_2(t)\} = E_2(s)$.

Examples 5-2 and 5-3 show that the functions

$$e_1(t) = e^{pt}u(t) \qquad e_2(t) = -e^{pt}u(-t)$$

298 INTRODUCTION TO SYSTEM ANALYSIS

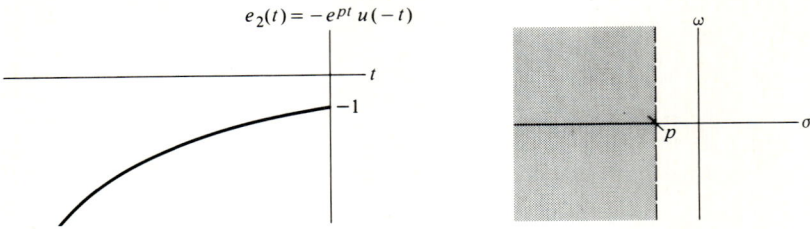

Figure 5-3 Graph of the function $e_2(t) = -e^{pt}u(-t)$ and the region of convergence Re $s < p$ for $\mathscr{L}\{e_2(t)\}$.

have the same Laplace transform, given by

$$E(s) = \frac{1}{s - p}$$

This illustrates the fact that (in general) a region of convergence must be specified if a Laplace transform is to have a unique inverse. For example, if Re $s > p$ is specified as the region of convergence for $E(s)$ above, then

$$\mathscr{L}^{-1}\{E(s)\} = e^{pt}u(t)$$

whereas if Re $s < p$ is specified as the region of convergence, then

$$\mathscr{L}^{-1}\{E(s)\} = -e^{pt}u(-t)$$

Thus, in general, a Laplace transform consists of a function of s and an associated region of convergence; however, in most practical applications, specifying a region of convergence for a Laplace transform $E(s)$ can be replaced by specifying an equivalent and more convenient condition on the associated inverse transform $e(t)$. We describe this condition below.

5-1B Right-Sided Inverse Laplace Transforms

A *right-sided function* of time t is one that is nonzero only for t greater than some fixed time t_0, as illustrated in Fig. 5-4a. A *left-sided function* of time is one that is nonzero only for t less than some fixed time t_0 (Fig. 5-4b). A *two-sided function* of time is one that is neither right- nor left-sided (Fig. 5-4c). A function having finite duration, e.g., a rectangular pulse, is both right- and left-sided.

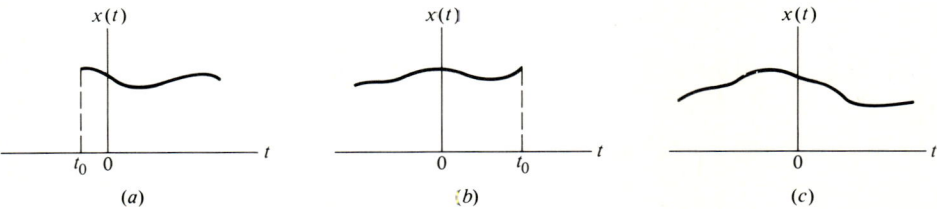

Figure 5-4 An illustration of the definitions of (a) a right-sided function, (b) a left-sided function, and (c) a two-sided function.

From Fig. 5-2, specifying the right-sided region Re $s > p$ as the region of convergence for $E(s) = 1/(s - p)$ is equivalent to specifying the right-sided function $e_1(t) = e^{pt}u(t)$ as the inverse Laplace transform of $E(s)$. From Fig. 5-3, specifying the left-sided region Re $s < p$ as the region of convergence for $E(s)$ is equivalent to specifying the left-sided function $e_2(t) = -e^{pt}u(-t)$ as the inverse Laplace transform of $E(s)$. In general, an inverse Laplace transform may be a right-, left-, or two-sided function, depending on the form of the transform and on the associated region of convergence. It can be shown that the right-sided inverse of a transform is unique (LePage, pp. 318–322). Therefore, specifying a region of convergence for a transform is unnecessary if the inverse transform is known to be right-sided. This is the case, for example, if the inverse transform is the output of a realizable system for a right-sided input. In the remainder of this chapter, with one exception (see Sec. 5-3D), we limit our discussion to right-sided signals. We adopt the convention that $\mathcal{L}^{-1}\{E(s)\}$ denotes the right-sided inverse Laplace transform of $E(s)$; for example,

$$\mathcal{L}^{-1}\left\{\frac{1}{s-p}\right\} = e^{pt}u(t)$$

Where it is necessary to consider other than the right-sided inverse of a transform $E(s)$, we put both the transform and its region of convergence under the operator $\mathcal{L}^{-1}$; for example,

$$\mathcal{L}^{-1}\left\{\frac{1}{s-p}; \quad \text{Re } s < p\right\} = -e^{pt}u(-t)$$

Using (5-1), we can construct a table of Laplace transforms. Because a right-sided inverse Laplace transform is unique, we can use the table to obtain right-sided inverse Laplace transforms. Table D-1 contains all Laplace-transform pairs needed in this book.

5-1C Relation to the Fourier Transformation

Taking $s = \sigma + j\omega$ in the definition (5-1) of the Laplace transform of a function $e(t)$ gives

$$\mathcal{L}\{e(t)\} = \int_{-\infty}^{\infty} e(t)e^{-\sigma t}e^{-j\omega t}\,dt$$

Comparing this relation with the defining relation (4-19) for the Fourier transform of $e(t)$ shows that

$$\mathcal{L}\{e(t)\} = \mathcal{F}\{e(t)e^{-\sigma t}\} \tag{5-5a}$$

This relation holds for all values of $s = \sigma + j\omega$ in the region of convergence for $\mathcal{L}\{e(t)\}$. If the region of convergence includes the line $\sigma = 0$ ($s = j\omega$) then

$$\mathcal{F}\{e(t)\} = \mathcal{L}\{e(t)\}\bigg|_{\sigma=0} = \mathcal{L}\{e(t)\}\bigg|_{s=j\omega} \tag{5-5b}$$

i.e., if the region of convergence for the Laplace transform of $e(t)$ includes the ω axis, the Fourier transform of $e(t)$ is the Laplace transform of $e(t)$ with s replaced by $j\omega$ and the Laplace transform of $e(t)$ is the Fourier transform of $e(t)$ with $j\omega$ replaced by s.

Example 5-4 Consider the function $e(t) = e^{pt}u(t)$, where p is real. From Example 5-2, the Laplace transform of $e(t)$ is

$$\mathcal{L}\{e(t)\} = \mathcal{L}\{e^{pt}u(t)\} = \frac{1}{s - p}$$

with associated region of convergence Re $s > p$ ($\sigma > p$). For $p < 0$, the region of convergence includes the ω axis (Fig. 5-2), in which case

$$\mathcal{F}\{e(t)\} = \mathcal{L}\{e(t)\}\Big|_{s=j\omega} = \frac{1}{j\omega - p}$$

For $p \geq 0$, the region of convergence excludes the ω axis (Fig. 5-3), in which case

$$\mathcal{F}\{e(t)\} \neq \mathcal{L}\{e(t)\}\Big|_{s=j\omega}$$

For example, for $p = 0$, $e(t) = u(t)$ and (see Table D-1)

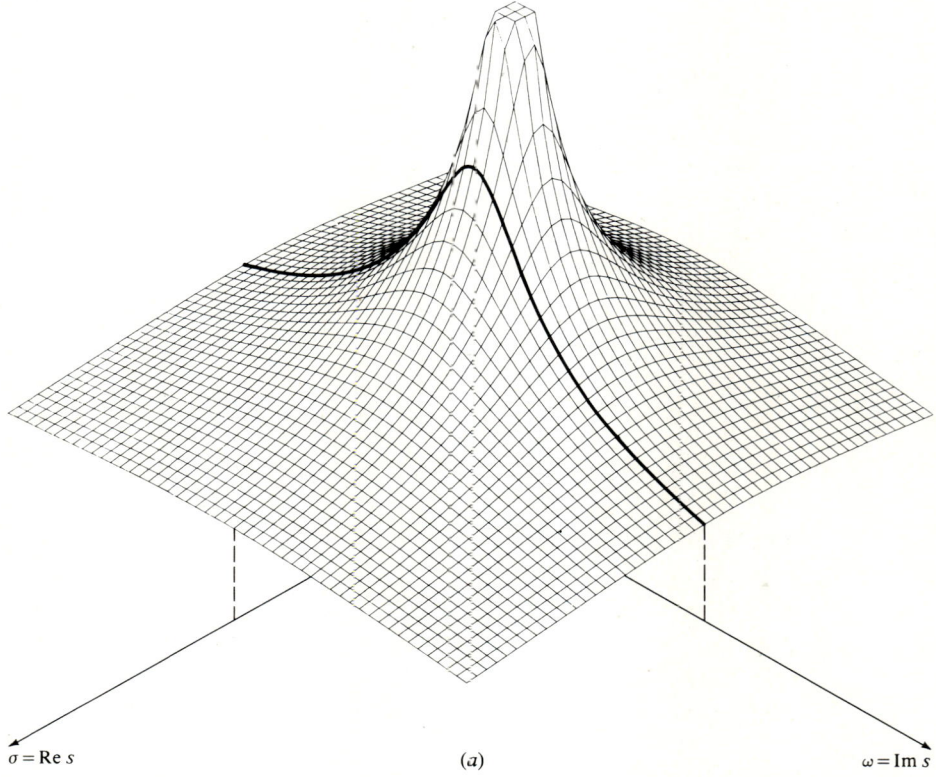

$\sigma = $ Re s (a) $\omega = $ Im s

Figure 5-5 Graphs of $|\mathcal{L}\{e(t)\}|$ versus $\sigma = $ Re s and $\omega = $ Im s, where $e(t) = e^{pt}u(t)$.

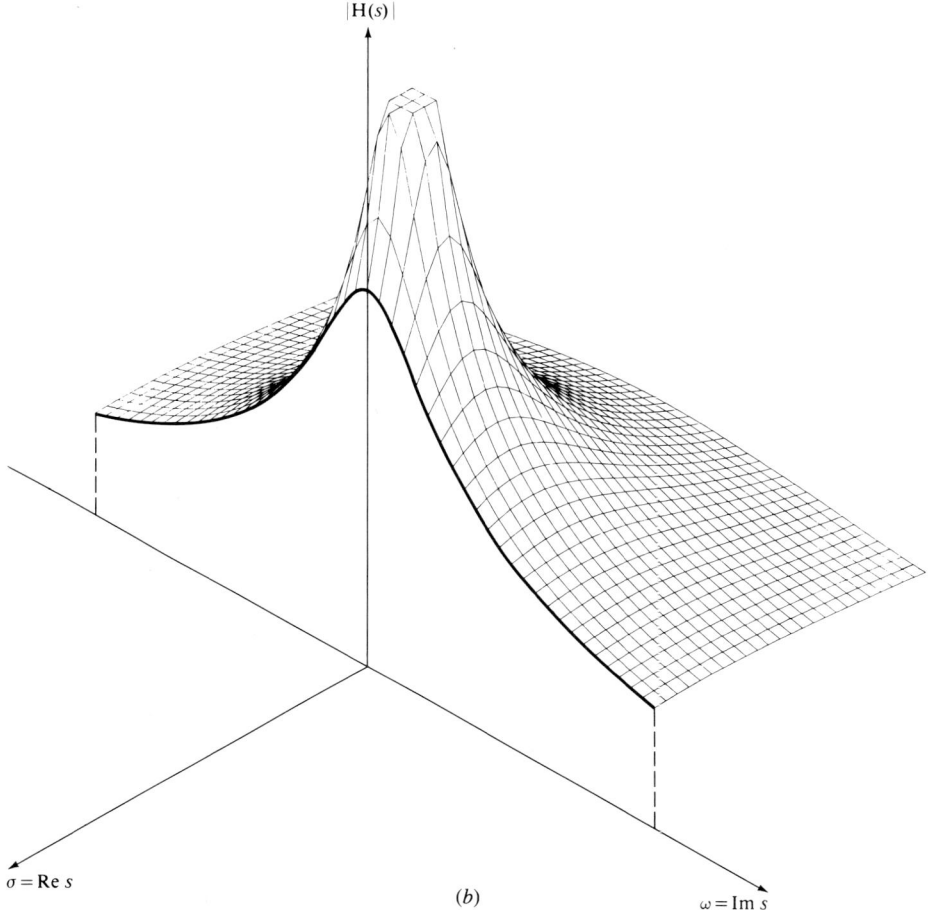

Figure 5-5 *Continued*

$$\mathcal{L}\{e(t)\} = \frac{1}{s}$$

whereas (strictly)

$$\mathcal{F}\{e(t)\} = \pi\delta(\omega) + \frac{1}{j\omega}$$

The Fourier transform of a function $e(t)$ is a complex function of an imaginary variable $j\omega$ or, equivalently, of a real variable ω. A Fourier transform can be plotted by plotting its magnitude and its angle as curves over a frequency (ω or f) axis. The Laplace transform of a function $e(t)$ is a complex function of a complex variable s. A Laplace transform can be plotted by plotting its magnitude and its angle as *surfaces* over an s plane. If the region of convergence of a transform $E(s)$ includes the ω axis, then [from Eq. (5-5)]

$$\mathcal{F}\{e(t)\} = \mathcal{L}\{e(t)\}\Big|_{s=j\omega}$$

This means that the graphs of $|\mathcal{F}\{e(t)\}|$ and $\angle\mathcal{F}\{e(t)\}$ can be obtained by taking thin slices of $|\mathcal{L}\{e(t)\}|$ and $\angle\mathcal{L}\{e(t)\}$, respectively, along the ω axis. Figure 5-5 illustrates this for the function $e(t) = e^{pt}u(t)$, where p is real and negative. The Laplace transform of $e(t)$ is

$$\mathcal{L}\{e(t)\} = \frac{1}{s-p}$$

where the associated region of convergence is Re $s > p$ ($\sigma > p$). Because $p < 0$, the region of convergence includes the ω axis (where $\sigma = 0$). It follows that

$$\mathcal{F}\{e(t)\} = \mathcal{L}\{e(t)\}\Big|_{s=j\omega} = \frac{1}{j\omega - p}$$

Figure 5-5a shows a graph of the surface defined by

$$|\mathcal{L}\{e(t)\}| = \frac{1}{|s-p|} = \frac{1}{|\sigma + j\omega - p|} = \frac{1}{\sqrt{(\sigma-p)^2 + \omega^2}}$$

Figure 5-5b shows the left half of the surface of Fig. 5-5a. The front face (for $\sigma = 0$) of the surface of Fig. 5-5b is a graph of

$$|\mathcal{F}\{e(t)\}| = \frac{1}{\sqrt{p^2 + \omega^2}}$$

because the region of convergence for $\mathcal{L}\{e(t)\}$ includes the ω axis.

5-1D Linearity

Laplace transformation and inverse Laplace transformation are *linear operations*. This means that

$$\mathcal{L}\{a_1 e_1(t) + a_2 e_2(t) + \cdots\} = a_1 \mathcal{L}\{e_1(t)\} + a_2 \mathcal{L}\{e_2(t)\} + \cdots \quad (5\text{-}6)$$

and

$$\mathcal{L}^{-1}\{a_1 E_1(s) + a_2 E_2(s) + \cdots\} = a_1 \mathcal{L}^{-1}\{E_1(s)\} + a_2 \mathcal{L}^{-1}\{E_2(s)\} + \cdots \quad (5\text{-}7)$$

Equation (5-6) follows from (5-1) because integration is a linear operation. Equation (5-7) follows from (5-6) because the inverse of a linear operation is a linear operation.

Example 5-5 Obtain the Laplace transform of the signal

$$x(t) = \mathbf{a}\delta(t) + x_0 u(t)$$

SOLUTION From (5-6),
$$\mathcal{L}\{x(t)\} = \mathbf{a}\mathcal{L}\{\delta(t)\} + x_0\mathcal{L}\{u(t)\}$$
From Table D-1, $\mathcal{L}\{\delta(t)\} = 1$ and $\mathcal{L}\{u(t)\} = 1/s$; thus
$$\mathcal{L}\{x(t)\} = X(s) = \mathbf{a} + x_0\frac{1}{s}$$

Example 5-6 Obtain the inverse Laplace transform of
$$X(s) = \frac{x_1}{s - p_1} + \frac{x_2}{s - p_2}$$

SOLUTION From (5-7),
$$\mathcal{L}^{-1}\{X(s)\} = x_1\mathcal{L}^{-1}\left\{\frac{1}{s - p_1}\right\} + x_2\mathcal{L}^{-1}\left\{\frac{1}{s - p_2}\right\}$$
From Table D-1, $\mathcal{L}^{-1}\{1/(s - p)\} = e^{pt}u(t)$; thus
$$x(t) = \mathcal{L}^{-1}\{X(s)\} = (x_1 e^{p_1 t} + x_2 e^{p_2 t})u(t)$$

Example 5-7 Use Euler's identity and the relation
$$\mathcal{L}\{e^{pt}u(t)\} = \frac{1}{s - p}$$
to obtain the Laplace transform of the signal
$$x(t) = A(\cos\omega_0 t)u(t)$$

SOLUTION Using Euler's identity $2\cos\alpha \equiv e^{j\alpha} + e^{-j\alpha}$ gives
$$x(t) = \tfrac{1}{2}Ae^{j\omega_0 t}u(t) + \tfrac{1}{2}Ae^{-j\omega_0 t}u(t)$$
Using (5-7) (with $p = \pm j\omega_0$) and the relation given above yields
$$X(s) = \mathcal{L}\{x(t)\} = \frac{A/2}{s - j\omega_0} + \frac{A/2}{s + j\omega_0} = \frac{As}{s^2 + \omega_0^2}$$

5-1E Operational Properties

Table D-2 gives the most important operational properties of the Laplace transformation. They are identical to the corresponding operational properties of the Fourier transformation except that s replaces $j\omega$. Below, we give a proof of the integration property of the Laplace transformation. Proofs of the other operational properties of the Laplace transformation, which are virtually identical to those of corresponding properties of the Fourier transformation, are left as exercises.

We wish to show that
$$\mathcal{L}\left\{\int_{-\infty}^{t} e(\alpha)\,d\alpha\right\} = \frac{1}{s}\mathcal{L}\{e(t)\} \qquad (5\text{-}8)$$

From the definition (5-1) of the Laplace transform,

$$\mathcal{L}\left\{\int_{-\infty}^{t} e(\alpha)\, d\alpha\right\} = \int_{-\infty}^{\infty}\int_{-\infty}^{t} e(\alpha)\, d\alpha\, e^{-st}\, dt$$

Partial integration gives†

$$\mathcal{L}\left\{\int_{-\infty}^{t} e(\alpha)\, d\alpha\right\} = -\frac{1}{s}\lim_{t\to\infty}\left[e^{-st}\int_{-\infty}^{t} e(\alpha)\, d\alpha\right] + \frac{1}{s}\lim_{t\to -\infty}\left[e^{-st}\int_{-\infty}^{t} e(\alpha)\, d\alpha\right]$$

$$+ \frac{1}{s}\int_{-\infty}^{\infty} e(t)e^{-st}\, dt$$

The first term on the right can be dropped because for any function $e(t)$ of interest there is a region of convergence where

$$\lim_{t\to\infty}\left[e^{-st}\int_{-\infty}^{t} e(\alpha)\, d\alpha\right] = 0$$

The second term can also be dropped because we assume that $e(t)$ is right-sided. This means that there is a finite value of t for which

$$\int_{-\infty}^{t} e(\alpha)\, d\alpha = 0$$

We are left with

$$\mathcal{L}\left\{\int_{-\infty}^{t} e(\alpha)\, d\alpha\right\} = \frac{1}{s}\int_{-\infty}^{\infty} e(t)e^{-st}\, dt$$

The integral on the right is the Laplace transform of $e(t)$. Equation (5-8) follows.

Example 5-8 In this example we use the integration property to obtain the Laplace transform of the function $e(t) = tu(t)$. We note that

$$tu(t) = \int_{-\infty}^{t} u(\alpha)\, d\alpha$$

The integration property (5-8) gives

$$\mathcal{L}\{tu(t)\} = \frac{1}{s}\mathcal{L}\{u(t)\}$$

From Table D-1, $\mathcal{L}\{u(t)\} = 1/s$; thus

$$\mathcal{L}\{tu(t)\} = \frac{1}{s^2}$$

The following examples illustrate other operational properties of the Laplace transformation.

†Here we use $\int u\, dv = uv - \int v\, du$ with $u = \int_{-\infty}^{t} e(\alpha)\, d\alpha$ and $dv = e^{-st}\, dt$ (see Thomas and Finney pp. 363–368).

Example 5-9 Obtain the Laplace transform of $f(t) = tu(t - t_0)$.

SOLUTION We can write
$$f(t) = (t - t_0)u(t - t_0) + t_0 u(t - t_0) = e(t - t_0)$$
where
$$e(t) = tu(t) + t_0 u(t)$$
From Table D-1, $\mathcal{L}\{tu(t)\} = 1/s^2$ and $\mathcal{L}\{u(t)\} = 1/s$. Using the linearity property of the Laplace transformation gives
$$\mathcal{L}\{e(t)\} = \frac{1}{s^2} + \frac{t_0}{s} = \frac{t_0 s + 1}{s^2}$$
Using the delay property $\mathcal{L}\{e(t - t_0)\} = e^{-st_0}\mathcal{L}\{e(t)\}$ gives
$$\mathcal{L}\{f(t)\} = \mathcal{L}\{e(t - t_0)\} = \frac{e^{-st_0}(t_0 s + 1)}{s^2}$$

Example 5-10 Use the translation property of the Laplace transformation to obtain the Laplace transform of
$$f(t) = e^{pt}\cos(\omega_0 t - \theta)u(t)$$

SOLUTION From the translation property, $\mathcal{L}\{e(t)e^{pt}\} = E(s - p)$, where $E(s) = \mathcal{L}\{e(t)\}$, thus
$$\mathcal{L}\{f(t)\} = \mathcal{L}\{\cos(\omega_0 t - \theta)u(t)\}\Big|_{s \to s-p}$$
From Table D-1,
$$\mathcal{L}\{\cos(\omega_0 t - \theta)u(t)\} = \frac{s\cos\theta + \omega_0 \sin\theta}{s^2 + \omega_0^2}$$
Therefore,
$$\mathcal{L}\{f(t)\} = \frac{(s - p)\cos\theta + \omega_0 \sin\theta}{(s - p)^2 + \omega_0^2}$$

5-2 APPLICATION TO LINEAR STATIONARY SYSTEMS

In this section we describe how the Laplace transformation is used to obtain an output of a linear stationary system. The procedures are essentially identical to those described in Sec. 4-3 for the Fourier transformation except that s replaces $j\omega$ as the transform variable.

5-2A System Function

The output $y(t)$ of a linear stationary system for input $x(t)$ is given by
$$y(t) = \int_{-\infty}^{\infty} x(\lambda)h(t - \lambda)\, d\lambda$$

where $h(t)$ is the impulse response of the system. Taking the Laplace transform of this relation gives (see Table D-2)

$$Y(s) = H(s)X(s) \quad \text{where } \begin{aligned} Y(s) &= \mathcal{L}\{y(t)\} \\ X(s) &= \mathcal{L}\{x(t)\} \end{aligned} \quad (5\text{-}9)$$

and
$$H(s) = \mathcal{L}\{h(t)\} \quad (5\text{-}10)$$

Note that (5-9) is identical to the corresponding relation

$$Y(j\omega) = H(j\omega)X(j\omega)$$

using Fourier transforms, except that s replaces $j\omega$. The function $H(s)$ given by (5-10) is called the *system function* of a system having impulse response $h(t)$. The system function $H(s)$ and the transfer function $H(j\omega)$ of a system are analogous quantities. Indeed, in most practical applications the transfer function $H(j\omega)$ of a system equals the system function $H(s)$ of the system with s replaced by $j\omega$, as the notation suggests. Furthermore, in most practical applications the Fourier transform (the spectral density) $X(j\omega)$ of a signal $x(t)$ equals the Laplace transform $X(s)$ of the signal with s replaced by $j\omega$, as the notation suggests. There are exceptions, however. As shown above, the Fourier transform of a function equals the Laplace transform of the function with s replaced by $j\omega$ only if the region of convergence for the Laplace transform of the function includes the ω axis. Thus, although we use the same symbol (H) for both a transfer function and a system function, there are cases where the transfer function $H(j\omega)$ of a system is not equal to the system function $H(s)$ of the system with s replaced by $j\omega$. Similar remarks apply to the Laplace transform $X(s)$ and the Fourier transform $X(j\omega)$ of a signal $x(t)$. Prudence requires us to think of the Laplace transform of a function and the Fourier transform of the function as different functions unless we can show that the region of convergence for the Laplace transform includes the ω axis.

We use (5-9) to obtain the output of a linear stationary system for input $x(t)$ as follows:

1. Obtain the Laplace transform $X(s)$ of the input $x(t)$ using Table D-1 or (5-1).
2. Obtain the system function $H(s)$ of the system using (5-9) and Table D-1 or (5-1).
3. Obtain the output $y(t)$ by taking the inverse Laplace transform of the product $H(s)X(s)$; that is,

$$y(t) = \mathcal{L}^{-1}\{H(s)X(s)\}$$

The following example illustrates this procedure.

Example 5-11 Obtain the output of a linear stationary system having impulse response

$$h(t) = \frac{1}{\tau} e^{-t/\tau} u(t)$$

and input $x(t) = x_0 u(t)$.

SOLUTION From Table D-1, the Laplace transform $X(s)$ of the input and the system function $H(s)$ (the Laplace transform of the impulse response) are

$$X(s) = \frac{x_0}{s} \qquad H(s) = \frac{1/\tau}{s + 1/\tau}$$

From (5-9), the Laplace transform $Y(s)$ of the output $y(t)$ is

$$Y(s) = H(s)X(s) = \frac{x_0/\tau}{s(s + 1/\tau)}$$

Table D-1 gives

$$y(t) = \mathcal{L}^{-1}\{Y(s)\} = x_0(1 - e^{-t/\tau})u(t)$$

Equation (5-10) provides the most direct route to the system function of a system described only by its impulse response. In practice, systems to be analyzed are described by differential equations or block diagrams more often than they are described by impulse responses. In Secs. 5-2B and 5-2C we show how to obtain the system function of a system described by a differential equation or a block diagram.

5-2B Systems Described by Differential Equations

The system function of a system described by the differential equation

$$\frac{d^n y}{dt^n} + a_{n-1}\frac{d^{n-1}y}{dt^{n-1}} + \cdots + a_0 y = b_m \frac{d^m x}{dt^m} + b_{m-1}\frac{d^{m-1}x}{dt^{m-1}} + \cdots + b_0 x \qquad (5\text{-}11)$$

can be obtained using the differentiation property

$$\mathcal{L}\left\{\frac{d^n e(t)}{dt^n}\right\} = s^n \mathcal{L}\{e(t)\} = s^n E(s) \qquad (5\text{-}12)$$

Taking the Laplace transform of (5-11) and using (5-12) gives

$$s^n Y(s) + a_{n-1}s^{n-1}Y(s) + \cdots + a_0 Y(s) = b_m s^m X(s)$$
$$+ b_{m-1}s^{m-1}X(s) + \cdots + b_0 X(s)$$

where $X(s) = \mathcal{L}\{x(t)\}$ and $Y(s) = \mathcal{L}\{y(t)\}$. Solving for $Y(s)$ gives

$$Y(s) = \frac{b_m s^m + b_{m-1}s^{m-1} + \cdots + b_0}{s^n + a_{n-1}s^{n-1} + \cdots + a_0} X(s) \qquad (5\text{-}13)$$

Comparing this relation with (5-9) shows that the system function $H(s)$ for a system described by (5-11) is

$$H(s) = \frac{b_m s^m + b_{m-1}s^{m-1} + \cdots + b_0}{s^n + a_{n-1}s^{n-1} + \cdots + a_0} \qquad (5\text{-}14)$$

Example 5-12 The input $x(t)$ and corresponding output $y(t)$ of a certain system are related by the differential equation

$$\frac{dy}{dt} + \frac{1}{\tau}y = \frac{1}{\tau}x$$

Obtain the output $y(t)$ for input $x(t) = x_0 u(t)$.

SOLUTION From Table D-1, the Laplace transform of the input is

$$X(s) = x_0 \mathcal{L}\{u(t)\} = \frac{x_0}{s}$$

From (5-14), the system function of the system is

$$H(s) = \frac{1/\tau}{s + 1/\tau}$$

From (5-9), the output is

$$y(t) = \mathcal{L}^{-1}\{H(s)X(s)\} = \mathcal{L}^{-1}\left\{\frac{x_0/\tau}{s(s + 1/\tau)}\right\}$$

Table D-1 gives

$$y(t) = x_0(1 - e^{-t/\tau})u(t)$$

5-2C Systems Described by Block Diagrams

A linear stationary system having system function $H(s)$ can be represented by the s-domain block diagram of Fig. 5-6, where $X(s)$ is the Laplace transform of an input and $Y(s)$ is the Laplace transform of the corresponding output. For example, the system of Example 5-12 can be represented by the block diagram of Fig. 5-7.

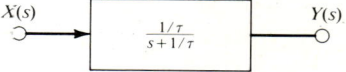

Figure 5-6 Block diagram for a linear stationary system having system function $H(s)$.

Figure 5-7 Block diagram for the system of Example 5-12.

Example 5-13 The transfer characteristic of an integrator having input $x(t)$ and output $y(t)$ is

$$y(t) = K \int_{-\infty}^{t} x(\alpha)\, d\alpha$$

Taking the Laplace transform of this relation gives (see Table D-2)

$$Y(s) = \frac{K}{s} X(s)$$

This shows that the system function of an integrator is

$$H(s) = \frac{K}{s}$$

An integrator is represented by the block diagram of Fig. 5-8.

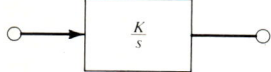

Figure 5-8 Block diagram for an integrator.

Table 5-1 Transform-domain symbols for elementary systems

Operation	Time-domain symbol	Transform-domain symbol
Proportion	$x(t) \to [K] \to y(t)$	$X(s) \to [K] \to Y(s)$
Delay	$x(t) \to [t - t_0] \to y(t)$	$X(s) \to [e^{-st_0}] \to Y(s)$
Differentiation	$x(t) \to [\frac{d}{dt} K] \to y(t)$	$X(s) \to [Ks] \to Y(s)$
Integration	$x(t) \to [\int K] \to y(t)$	$X(s) \to [\frac{K}{s}] \to Y(s)$
Addition	$x_1(t), x_2(t) \to (\Sigma) \to y(t)$	$X_1(s), X_2(s) \to (\Sigma) \to Y(s)$

Table 5-1 gives s-domain block diagrams for the five elementary linear stationary systems. They are identical to the corresponding frequency-domain block diagrams of Table 3-1 except that s replaces $j\omega$.

We can obtain the system function of a system described by an s-domain block diagram using block-diagram reduction. Tables 5-2 and 5-3 give rules for simplifying s-domain block diagrams. These rules are identical to their frequency-domain counterparts, and their derivations are identical to those of the corresponding rules for reducing frequency-domain block diagrams, except that s replaces $j\omega$.

Example 5-14 Obtain the output $y(t)$ of the system of Fig. 5-9a for input $x(t) = x_0 u(t)$.

SOLUTION With reference to Table 5-1, we transform the time-domain block diagram of Fig. 5-9a to the s-domain block diagram of Fig. 5-9b. With reference to Tables 5-2 and 5-3, we reduce the s-domain block diagram of Fig. 5-9b to a single block (Figs. 5-9c and d). The system function of the system is

$$H(s) = \frac{K_0 s + K_1}{s + K_1 K_2}$$

From Table D-1, the Laplace transform of the input is

$$X(s) = \mathcal{L}\{x(t)\} = \frac{x_0}{s}$$

310 INTRODUCTION TO SYSTEM ANALYSIS

Table 5-2 Rules for reducing series, parallel, and feedback connections

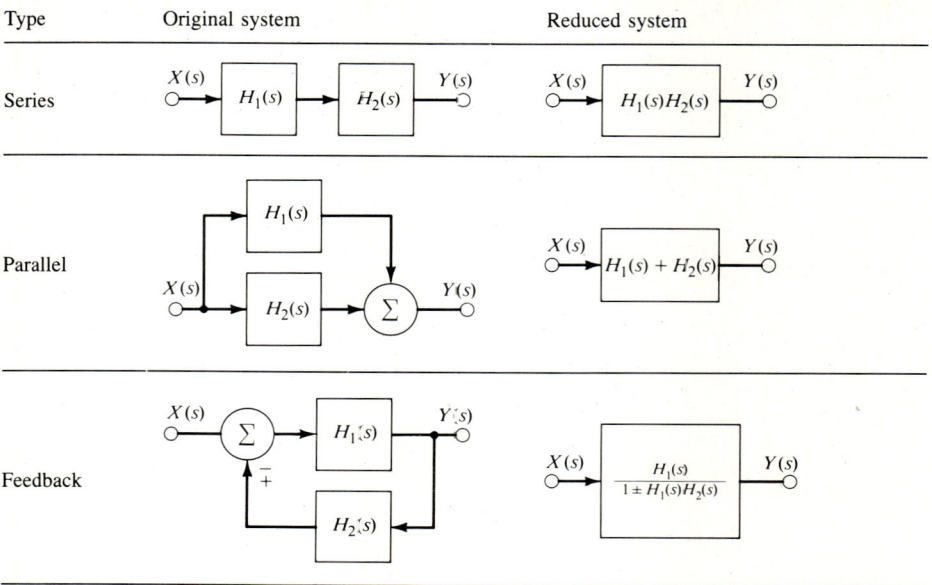

From (5-9), the Laplace transform of the output is

$$Y(s) = H(s)X(s) = \frac{K_0 x_0}{s + K_1 K_2} + \frac{K_1 x_0}{s(s + K_1 K_2)}$$

From Table D-1,

$$y(t) = \mathcal{L}^{-1}\{Y(s)\} = K_0 x_0 e^{-K_1 K_2 t} u(t) + \frac{x_0}{K_2}(1 - e^{-K_1 K_2 t})u(t)$$

$$= \frac{x_0}{K_2}[1 - (1 - K_0 K_2)e^{-K_1 K_2 t}]u(t)$$

We can summarize the development thus far as follows. The output $y(t)$ of a linear stationary system for input $x(t)$ is given by

$$y(t) = \mathcal{L}^{-1}\{H(s)\mathcal{L}\{x(t)\}\} \qquad (5\text{-}15)$$

where $H(s)$ is the system function of the system. A system function can be obtained from an impulse response using (5-10), from a differential equation using (5-14), or from a block diagram using block-diagram reduction.

Because of the great variety of inputs and transfer functions encountered in practice, it is impossible to tabulate all transforms that may arise from the product $H(s)\mathcal{L}\{x(t)\}$ in (5-15). In Sec. 5-2D we show how most transforms encountered in practice can be expressed as sums of terms appearing in Table D-1.

Table 5-3 Rules for moving an addition element or a pickoff from one side of a block to the other

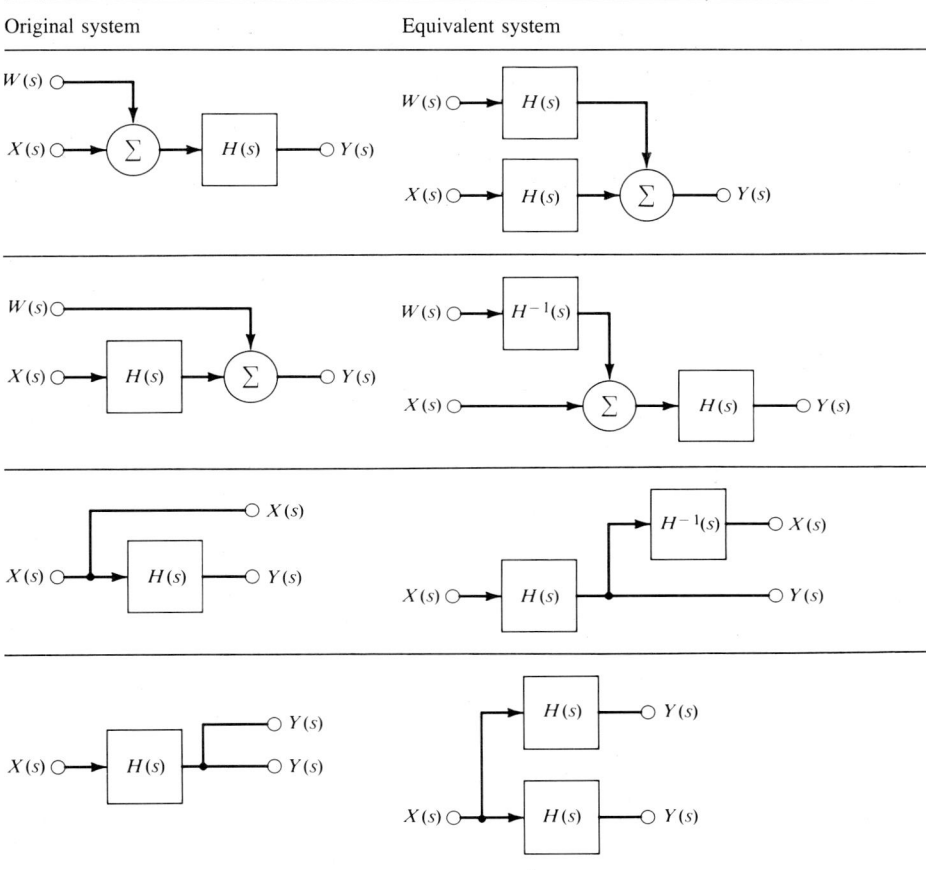

5-2D Partial-Fraction Expansion

We describe a procedure for finding the inverse Laplace transform of a rational function of the form

$$E(s) = \frac{N(s)}{D(s)} \quad \text{where } \begin{aligned} N(s) &= c_m s^m + c_{m-1} s^{m-1} + \cdots + c_0 \\ D(s) &= s^n + d_{n-1} s^{n-1} + \cdots + d_0 \end{aligned} \quad (5\text{-}16)$$

The procedure is based on the following assumptions:

1. The coefficient of s^n in the denominator $D(s)$ is unity. This assumption causes no loss of generality because we can always divide both numerator and denominator by the coefficient of the highest power of s appearing in the denominator.

312 INTRODUCTION TO SYSTEM ANALYSIS

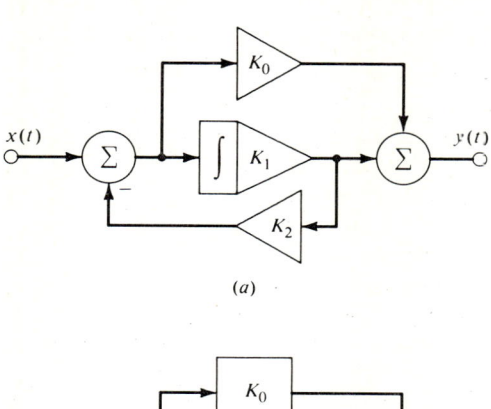

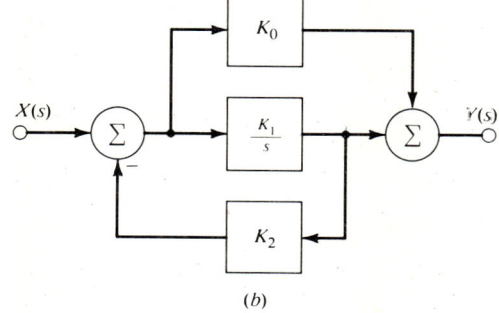

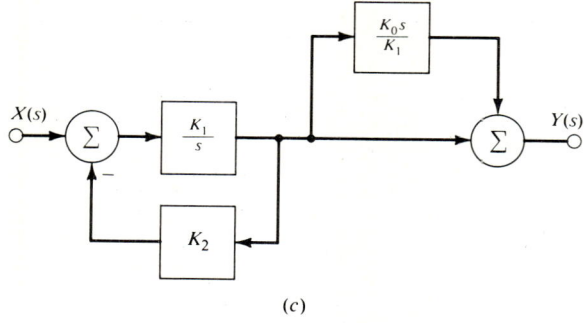

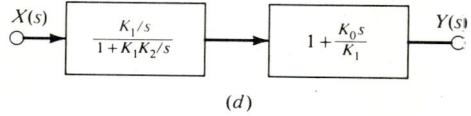

Figure 5-9 System of Example 5-14: (a) as originally given, (b) after transformation to the s domain, (c) and (d) in reduced form.

2. The degree m of the numerator does not exceed the degree n of the denominator. Cases where $m > n$ are no more difficult, but they almost never occur in practical applications.
3. The coefficients c_i ($i = 0, 1, \ldots, m$) and d_k ($k = 0, 1, \ldots, n - 1$) are all real. This is equivalent to assuming that $E(s) = N(s)/D(s)$ is the Laplace transform of a real function, e.g., an output of a physical system.

Under these assumptions, the general form of transforms considered below is

$$E(s) = \frac{N(s)}{D(s)} = \frac{c_n s^n + c_{n-1} s^{n-1} + \cdots + c_0}{s^n + d_{n-1} s^{n-1} + \cdots + d_0} \qquad (5\text{-}17)$$

Such a function can be expanded in partial fractions as follows:

1. Obtain the n roots of the denominator $D(s)$. In general, this must be done numerically if $n > 4$. In practice, the roots of $D(s)$ usually are obtained numerically for $n > 2$ because the algebraic formulas for the roots of third-degree and fourth-degree polynomials are cumbersome.
2. The first term of the partial-fraction expansion is c_n (the coefficient of s^n in the numerator). This term is zero if the degree of the numerator $N(s)$ is less than the degree of the denominator $D(s)$. For each root p of the denominator $D(s)$ include m additional terms

$$\frac{B_1}{s-p} + \frac{B_2}{(s-p)^2} + \cdots + \frac{B_m}{(s-p)^m}$$

where m is the multiplicity of p.
3. Determine the coefficients of the expansion, for example, $B_1, B_2, \ldots, B_m$ above, such that the expansion is identically equal to the function $E(s)$, that is, such that the expansion and $E(s)$ are equal for all values of s.

A partial-fraction expansion obtained according to this procedure consists of terms of the form $B/(s-p)^k$. Consequently, the inverse Laplace transform of the expansion (of the original function) can be found using Table D-1 and the linearity property of the Laplace transformation.

Example 5-15 Obtain the output $y(t)$ of the system of Fig. 5-10 for input $x(t) = x_0 u(t)$.

SOLUTION From (5-15),

$$y(t) = \mathcal{L}^{-1}\{H(s)\mathcal{L}\{x(t)\}\}$$

where (from Table D-1)

$$\mathcal{L}\{x(t)\} = \frac{x_0}{s}$$

and (by block-diagram reduction)

$$H(s) = \frac{K^2}{(s+K)^2}$$

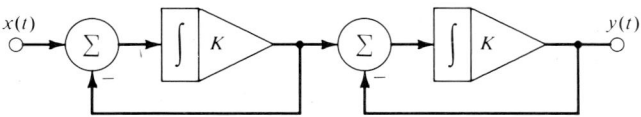

Figure 5-10 System of Example 5-15.

The Laplace transform of the output is

$$Y(s) = H(s)X(s) = \frac{K^2 x_0}{s(s + K)^2}$$

We expand $Y(s)$ in partial fractions as follows:

Step 1 The roots of the denominator are $p_1 = 0$ (multiplicity 1) and $p_2 = -K$ (multiplicity 2).

Step 2 The partial-fraction expansion is

$$\frac{K^2 x_0}{s(s + K)^2} = \frac{B_0}{s} + \frac{B_1}{s + K} + \frac{B_2}{(s + K)^2}$$

Step 3 Clearing fractions gives

$$K^2 x_0 = B_0(s + K)^2 + B_1 s(s + K) + B_2 s$$
$$= (B_0 + B_1)s^2 + (2B_0 K + B_1 K + B_2)s + B_0 K^2$$

Equating coefficients of like powers of s gives

$$B_0 + B_1 = 0$$
$$2KB_0 + KB_1 + B_2 = 0$$
$$K^2 B_0 = K^2 x_0$$

Solving these equations for B_0, B_1, and B_2 gives

$$B_0 = x_0 \qquad B_1 = -x_0 \qquad B_2 = -Kx_0$$

Thus

$$Y(s) = \frac{x_0}{s} - \frac{x_0}{s + K} - \frac{Kx_0}{(s + K)^2}$$

From Table D-1,

$$y(t) = x_0 u(t) - x_0 e^{-Kt} u(t) - Kx_c t e^{-Kt} u(t) = x_0[1 - (1 + Kt)e^{-Kt}] u(t)$$

In Example 5-15 the coefficients B_0, B_1, B_2 are determined by clearing fractions, writing both sides of the resulting expression as polynomials in s, and equating coefficients of like powers of s. Although this method always works, it is often unnecessarily tedious. A convenient shortcut for finding coefficients of a partial-fraction expansion was first formalized by Heaviside.†

Consider a rational function $E(s) = N(s)/D(s)$ whose denominator has *distinct* roots $p_1, p_2, \ldots, p_n$. The partial-fraction expansion for such a function has the form

$$\frac{N(s)}{D(s)} = \frac{N(s)}{(s - p_1)(s - p_2) \cdots (s - p_n)}$$

$$= c_n + \frac{B_1}{s - p_1} + \frac{B_2}{s - p_2} + \cdots + \frac{B_n}{s - p_n} \qquad (5\text{-}18)$$

†Oliver Heaviside (1850–1925), an English physicist who contributed extensively to electrical engineering.

where c_n is the coefficient of s^n in $N(s)$ and $p_i \neq p_k$ for $i \neq k$. Multiplying (5-18) by $s - p_1$ gives

$$\frac{N(s)}{(s - p_2)(s - p_3) \cdots (s - p_n)} = c_n(s - p_1) + B_1 + \frac{B_2(s - p_1)}{s - p_2}$$
$$+ \frac{B_3(s - p_1)}{s - p_3} + \cdots + \frac{B_n(s - p_1)}{s - p_n}$$

Taking $s = p_1$ in this expression yields

$$B_1 = \frac{N(p_1)}{(p_1 - p_2)(p_1 - p_3) \cdots (p_1 - p_n)}$$

Thus, to determine B_1 we simply strike the factor $s - p_1$ from the denominator of $E(s)$ and calculate the value of the resulting expression for $s = p_1$. The same procedure can be used to obtain $B_2, B_3, \ldots, B_n$. In general,

$$B_k = \left. \frac{(s - p_k)N(s)}{D(s)} \right|_{s \to p_k} \qquad k = 1, 2, \ldots, n \qquad (5\text{-}19)$$

Example 5-16 Obtain the inverse Laplace transform of

$$Y(s) = \frac{K(s + s_1)^2}{s^2 + qs + r}$$

where $K = 10$ Vs, $s_1 = 2$ s^{-1}, $q = 4$ s^{-1}, and $r = 3$ s^{-2}.

SOLUTION The roots of the denominator are $p_1 = -1$ s^{-1} and $p_2 = -3$ s^{-1}. The partial-fraction expansion for $Y(s)$ is

$$\frac{K(s + s_1)^2}{(s - p_1)(s - p_2)} = K + \frac{B_1}{s - p_1} + \frac{B_2}{s - p_2}$$

where, from (5-19),

$$B_1 = \left. \frac{(s - p_1)K(s + s_1)^2}{(s - p_1)(s - p_2)} \right|_{s \to p_1} = \frac{K(p_1 + s_1)^2}{p_1 - p_2} = 5 \text{ V}$$

$$B_2 = \left. \frac{(s - p_2)K(s + s_1)^2}{(s - p_1)(s - p_2)} \right|_{s \to p_2} = \frac{K(p_2 + s_1)^2}{p_2 - p_1} = -5 \text{ V}$$

From Table D-1,

$$y(t) = \mathcal{L}^{-1}\left\{ K + \frac{B_1}{s - p_1} + \frac{B_2}{s - p_2} \right\} = K\delta(t) + (B_1 e^{p_1 t} + B_2 e^{p_2 t})u(t)$$

where $K = 10$ Vs, $B_1 = -B_2 = 5$ V, $p_1 = -1$ s^{-1}, and $p_2 = -3$ s^{-1}.

Equation (5-19) is one of Heaviside's formulas for the coefficients of a partial-fraction expansion. Heaviside obtained other formulas for cases where a denominator has repeated roots (see Van Valkenburg, p. 186), but those formulas are cumbersome. When a denominator has repeated roots, it usually is easier to obtain partial-fraction coefficients by the direct method illustrated in Example 5-15, by repeated application of (5-19), or by a combination of these methods.

Example 5-17 To obtain a partial-fraction expansion for

$$E(s) = \frac{K}{s(s-p)^2}$$

by repeated application of (5-19) we factor $E(s)$ as

$$E(s) = \frac{K}{s-p}\left[\frac{1}{s(s-p)}\right]$$

Using (5-19) to expand the factor in the brackets gives

$$E(s) = \frac{K}{s-p}\left(-\frac{1/p}{s} + \frac{1/p}{s-p}\right)$$

which simplifies to

$$E(s) = -\frac{K/p}{s(s-p)} + \frac{K/p}{(s-p)^2}$$

Using (5-19) again to expand the first term on the right gives

$$E(s) = -\frac{K}{p}\left(-\frac{1/p}{s} + \frac{1/p}{s-p}\right) + \frac{K/p}{(s-p)^2}$$

$$= \frac{K}{p^2}\left[\frac{1}{s} - \frac{1}{s-p} + \frac{p}{(s-p)^2}\right]$$

Example 5-18 Obtain a partial-fraction expansion for

$$E(s) = \frac{K(s+s_0)^3}{s^2(s-p)}$$

where $K = 10.0$, $s_0 = 1.0$ s^{-1}, and $p = -2.0$ s^{-1}.

SOLUTION The partial-fraction expansion is

$$E(s) = \frac{K(s+s_0)^3}{s^2(s-p)} = K + \frac{B_1}{s^2} + \frac{B_2}{s} + \frac{B_3}{s-p}$$

Clearing fractions gives

$$K(s+s_0)^3 = Ks^2(s-p) + B_1(s-p) + B_2s(s-p) + B_3s^2$$

For $s = 0$, this yields $Ks_0^3 = B_1(-p)$, whence

$$B_1 = -\frac{Ks_0^3}{p} = 5 \text{ s}^{-1}$$

Similarly, taking $s = p$ gives $K(p + s_0)^3 = B_3 p^2$, whence

$$B_3 = \frac{K(p+s_0)^3}{p^2} = -2.5 \text{ s}^{-1}$$

Finally, taking $s = 2p$ gives

$$K(2p + s_0)^3 = K(2p)^2(2p - p) + B_1(2p - p) + B_2(2p)(2p - p) + B_3(2p)^2$$
$$= 4Kp^3 + B_1p + 2B_2p^2 + 4B_3p^2$$

which yields

$$B_2 = \frac{K(2p + s_0)^3 - 4Kp^3 - B_1p - 4B_3p^2}{2p^2} = 12.5 \text{ s}^{-1}$$

Example 5-19 Obtain the inverse Laplace transform of

$$Y(s) = \frac{K(s - s_0)}{s^2(s - p)^2}$$

where $K = 5$ V/s^2, $s_0 = -2$ s^{-1}, and $p = -1$ s^{-1}.

SOLUTION The expansion is

$$\frac{K(s - s_0)}{s^2(s - p)^2} = \frac{B_1}{s^2} + \frac{B_2}{s} + \frac{B_3}{(s - p)^2} + \frac{B_4}{s - p}$$

Clearing fractions gives

$$K(s - s_0) = B_1(s - p)^2 + B_2s(s - p)^2 + B_3s^2 + B_4s^2(s - p)$$

Taking $s = 0$ in this relation gives $-Ks_0 = B_1p^2$, whence

$$B_1 = \frac{-Ks_0}{p^2} = 10 \text{ V/s}$$

Similarly, taking $s = p$ gives $K(p - s_0) = B_3p^2$, whence

$$B_3 = \frac{K(p - s_0)}{p^2} = 5 \text{ V/s}$$

Taking $s = 2p$ gives

$$K(2p - s_0) = B_1(2p - p)^2 + B_2(2p)(2p - p)^2 + B_3(2p)^2 + B_4(2p)^2(2p - p)$$

and taking $s = 3p$ gives

$$K(3p - s_0) = B_1(3p - p)^2 + B_2(3p)(3p - p)^2 + B_3(3p)^2 + B_4(3p)^2(3p - p)$$

The last two relations yield two equations in the two unknowns B_2, B_4; thus

$$2p^3B_2 + 4p^3B_4 = -30 \text{ V/s}^3 \qquad 12p^3B_2 + 18p^3B_4 = -90 \text{ V/s}^3$$

Solving these equations for B_2 and B_4 yields

$$B_2 = -15 \text{ V} \qquad B_4 = 15 \text{ V}$$

The inverse Laplace transform of $Y(s)$ is

$$Y(s) = \mathcal{L}^{-1}\left\{\frac{B_1}{s^2} + \frac{B_2}{s} + \frac{B_3}{(s - p)^2} + \frac{B_4}{s - p}\right\}$$

$$= [B_1t + B_2 + (B_3t + B_4)e^{pt}]u(t)$$

where $B_1 = 10$ V/s, $B_2 = -B_4 = -15$ V, $B_3 = 5$ V/s, and $p = -1$ s^{-1}.

In the examples above the roots of the denominator are real. The same procedures can be used when some roots of the denominator are complex, but it is usually easier to combine the terms (in a partial-fraction expansion) arising from a conjugate pair of roots p, p^* as

$$\frac{C}{s-p} + \frac{C^*}{s-p^*} = \frac{As+B}{s^2 - 2(\operatorname{Re} p)s + |p|^2} \qquad \text{where } A = 2\operatorname{Re} C$$
$$B = -2\operatorname{Re} Cp^*$$

Example 5-20 This example illustrates two methods for obtaining a partial-fraction expansion of a function whose denominator has a conjugate pair of complex roots. We seek the inverse Laplace transform of the function

$$Y(s) = \frac{K(s-s_0)^2}{s(s^2+qs+r)}$$

where $K = 25.0$ V, $s_0 = 2.00 \text{ s}^{-1}$, $q = 6.00 \text{ s}^{-1}$, and $r = 25.0 \text{ s}^{-2}$.

SOLUTION 1 The roots of the denominator are 0, p, and p^*, where

$$p = \frac{-q + \sqrt{q^2 - 4r}}{2} = -3.00 + 4.00j \text{ s}^{-1}$$

The expansion can be written

$$Y(s) = \frac{B_0}{s} + \frac{B_1}{s-p} + \frac{B_1^*}{s-p^*}$$

The coefficients of $1/(s-p)$ and $1/(s-p^*)$ are conjugates because the coefficients of the numerator of $Y(s)$ are real. From (5-19),

$$B_0 = \left.\frac{K(s-s_0)^2}{s^2+qs+r}\right|_{s=0} = \frac{Ks_0^2}{r} = 4.00 \text{ V}$$

$$B_1 = \left.\frac{K(s-s_0)^2}{s(s-p^*)}\right|_{s=p} = \frac{K(p-s_0)^2}{p(p-p^*)} = \frac{(25.0)(-5.00+4.00j)^2}{(-3.00+4.00j)(8.00j)} = 25.6e^{1.15j} \text{ V}$$

The inverse Laplace transform of $Y(s)$ is

$$y(t) = (B_0 + B_1 e^{pt} + B_1^* e^{p^*t})u(t) = [B_0 + 2\operatorname{Re}(B_1 e^{pt})]u(t)$$
$$= [4.00 + 51.2e^{\sigma t}\cos(\omega t + 1.15)]u(t) \quad \text{V}$$

where $\sigma = \operatorname{Re} p = -3.00 \text{ s}^{-1}$ and $\omega = \operatorname{Im} p = 4.00$ rad/s.

SOLUTION 2 The partial-fraction expansion for $Y(s)$ can be written

$$Y(s) = \frac{K(s-s_0)^2}{s(s^2+qs+r)} = \frac{B_0}{s} + \frac{As+B}{s^2+qs+r}$$

Clearing fractions gives

$$K(s^2 - 2s_0 s + s_0^2) = B_0(s^2 + qs + r) + s(As + B)$$

Equating coefficients of like powers of s gives

$$B_0 + A = K \qquad B_0 q + B = -2Ks_0 \qquad B_0 r = Ks_0^2$$

Solving these equations for B_0, A, and B gives

$$B_0 = 4.00 \text{ V} \qquad A = 21.0 \text{ V} \qquad B = -124 \text{ V/s}$$

The inverse Laplace transform of $Y(s)$ is (see Table D-1)

$$y(t) = \mathscr{L}^{-1}\left\{\frac{B_0}{s}\right\} + \mathscr{L}^{-1}\left\{\frac{As+B}{s^2+qs+r}\right\}$$

$$= [4.00 + 51.2e^{\sigma t} \cos(\omega t + 1.15)] u(t) \quad \text{V}$$

as before, where $\sigma = -3.00 \text{ s}^{-1}$ and $\omega = 4.00 \text{ rad/s}$.

We conclude this section with a comprehensive example illustrating relations and procedures presented above.

Example 5-21 Obtain the output $y(t)$ of the system of Fig. 5-11 for input $x(t) = x_0 r(t/\tau)$, where $x_0 = 5$ V and $\tau = 30$ μs.

SOLUTION We obtain the output $y(t)$ by first finding the step response $g(t)$ and subsequently using superposition; thus

$$y(t) = x_0 [g(t) - g(t - \tau)]$$

The Laplace transform of the unit step $u(t)$ is

$$U(s) = \frac{1}{s}$$

The system function of the system is (by block-diagram reduction)

$$H(s) = \frac{K^3}{(s + K)(s^2 + Ks + K^2)}$$

where $K = 10^6 \text{ s}^{-1}$. The Laplace transform $G(s)$ of the step response $g(t)$ is

$$G(s) = \frac{K^3}{s(s + K)(s^2 + Ks + K^2)}$$

The roots of the quadratic factor $s^2 + Ks + K^2$ are

$$p_1 = \frac{K(-1 + \sqrt{3} j)}{2} \qquad p_2 = \frac{K(-1 - \sqrt{3} j)}{2} = p_1^*$$

We write the partial-fraction expansion for $G(s)$ as

$$G(s) = \frac{B_0}{s} + \frac{B_1}{s + K} + \frac{Cs + D}{s^2 + Ks + K^2}$$

The coefficients B_0, B_1 can be obtained from Heaviside's formula (5-19); thus

$$B_0 = \frac{sK^3}{s(s + K)(s^2 + Ks + K^2)} \bigg|_{s \to 0} = 1$$

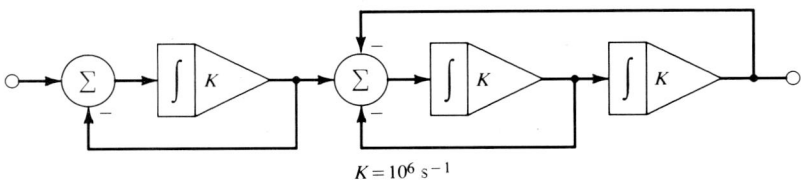

$K = 10^6 \text{ s}^{-1}$

Figure 5-11 System of Example 5-21.

$$B_1 = \frac{(s+K)K^3}{s(s+K)(s^2+Ks+K^2)}\bigg|_{s \to -K} = -1$$

Substituting these values for B_0, B_1 in the expansion above and clearing fractions gives

$$K^3 = (s+K)(s^2+Ks+K^2) - s(s^2+Ks+K^2) + (Cs+D)s(s+K)$$
$$= Cs^3 + (K+KC+D)s^2 + (K^2+KD)s + K^3$$

Equating coefficients of like powers of s gives

$$C = 0 \qquad D = -K$$

The partial-fraction expansion for the transform of the step response is

$$G(s) = \frac{1}{s} - \frac{1}{s+K} - \frac{K}{s^2+Ks+K^2}$$

The step response is

$$g(t) = \mathcal{L}^{-1}\{G(s)\} = \left[1 - e^{-Kt} - \frac{2}{\sqrt{3}}e^{-Kt/2}\sin\left(\frac{\sqrt{3}}{2}Kt\right)\right]u(t)$$

where $K = 10^6$ s^{-1}. The output $y(t)$ for input $x(t) = x_0 r(t/\tau)$ is given by

$$y(t) = x_0[g(t) - g(t-\tau)]$$

where $x_0 = 5$ V and $\tau = 30$ μs.

5-3 INTERPRETATION OF A SYSTEM FUNCTION

In this section we describe how important properties of a system, e.g., stability, are determined from the system function of the system. Also, we present some of the vocabulary of s-plane (Laplace-transform) analysis of systems. We consider only systems having rational system functions of the form $H(s) = N(s)/c(s)$, where $N(s)$ and $c(s)$ are polynomials in s and the degree of $N(s)$ does not exceed the degree of $c(s)$.

5-3A Poles and Zeros

The *zeros* of a Laplace transform $F(s)$ are the values of s for which $F(s) = 0$. The *poles* of $F(s)$ are the values of s for which $F(s) = \infty$ [for which $1/F(s) = 0$]. A rational system function can be expressed in factored form as

$$H(s) = \frac{K(s-z_1)(s-z_2)\cdots(s-z_m)}{(s-p_1)(s-p_2)\cdots(s-p_n)} \tag{5-20}$$

The zeros of $H(s)$ are $z_1, z_2, \ldots, z_m$ and the poles are $p_1, p_2, \ldots, p_n$. The poles of a system function are the characteristic roots for the system.

It is a common and useful practice to describe a system function graphically by a *pole-zero plot*, which shows the poles and the zeros of the system function on an s plane. Conventionally, poles are shown as crosses and zeros are shown as circles. In this book the multiplicity of a pole or zero is written beside the corresponding cross or circle. The quantity K in (5-20) cannot be determined from a pole-zero plot for $H(s)$.

If the value of K is significant, it can be given separately; but for many purposes the value of K is unimportant; e.g., the value of K has no bearing on realizability, stability, or fidelity of a system described by (5-20).

Example 5-22 Obtain the system function for the system described by the pole-zero plot of Fig. 5-12.

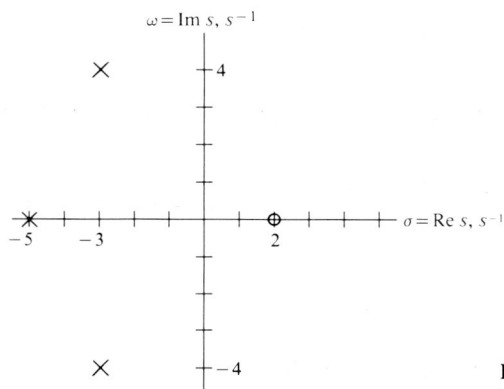

Figure 5-12 Pole-zero plot for the system of Example 5-22.

SOLUTION The system function has one zero $z_1 = 2$ s^{-1} and three poles $p_1 = -5$ s^{-1}, $p_2 = p_3^* = -3 + 4j$ s^{-1}. The system function is

$$H(s) = \frac{K(s - z_1)}{(s - p_1)(s - p_2)(s - p_2^*)} = \frac{K(s - z_1)}{(s - p_1)(s^2 + qs + r)}$$

where $q = -2 \operatorname{Re} p_2 = 6$ s^{-1} and $r = |p_2|^2 = 25$ s^{-2}. The value of K cannot be determined from the information given.

5-3B Impulse Response, Natural Modes, and Natural Frequencies

From (5-10), the impulse response of a system having system function $H(s)$ is

$$h(t) = \mathcal{L}^{-1}\{H(s)\} \tag{5-21}$$

Expanding $H(s)$ in partial fractions shows that (in general) $h(t)$ contains a term of the form $e^{pt}u(t)$ for each real pole p of $H(s)$ and a term of the form $e^{\sigma t}(\cos \omega t)u(t)$ for each pair $p, p^* = \sigma \pm j\omega$ of complex poles of $H(s)$. Such terms are called *natural modes* of the system.† The poles of a system function also are called the *natural frequencies* of the associated system. A repeated pole or a pair of complex-conjugate poles gives rise to only one natural mode and one natural frequency.

Example 5-23 Give the natural modes and natural frequencies of a system having system function

$$H(s) = \frac{K(s - z_1)(s - z_2)}{s(s - p_1)^2(s^2 + qs + r)}$$

†If $H(s)$ has as many zeros as poles, the impulse response also contains a delta function, but the delta function is not called a natural mode.

where $z_1 = 1\ \text{s}^{-1}$, $z_2 = -4\ \text{s}^{-1}$, $p_1 = -2\ \text{s}^{-1}$, $q = 2\ \text{s}^{-1}$, and $r = 2\ \text{s}^{-2}$.

SOLUTION The natural frequencies are 0, p_1, and $p_2 = -1 + j\ \text{s}^{-1}$ (usually, the upper half-plane pole of a conjugate pair is given as the natural frequency associated with the pair). The natural modes are $e^0 = 1$, $e^{p_1 t}$, and $e^{\sigma t} \cos \omega t$, where $\sigma = \text{Re}\ p_2$ and $\omega = \text{Im}\ p_2$.

Figure 5-13 illustrates the relation between the s-plane location of a pole and the form of the associated natural mode. The magnitude of a natural mode associated with a pole p grows (generally increases with time) if $\text{Re}\ p > 0$, decays (generally decreases with time) if $\text{Re}\ p < 0$, and neither grows nor decays if $\text{Re}\ p = 0$. The rate at which a natural mode grows or decays is proportional to the distance of the associated pole from the ω axis. The sign of a natural mode associated with a complex pole pair alternates if the poles are far enough from the σ axis. The rate at which the natural mode alternates in sign (rings) is proportional to the distance of either pole from the σ axis.

A partial-fraction expansion for the transform

$$Y(s) = H(s)X(s)$$

of the output of a system having system function $H(s)$ and input $x(t)$ contains terms arising from the poles of $H(s)$ and terms arising from the poles of $X(s)$. In general, all poles of $H(s)$ and all poles of $X(s)$ are represented in the expansion, and, in general, all natural modes of a system are represented in an output of the system. In principle, it is possible for a zero of $X(s)$ to cancel a pole of $H(s)$, but such cancellation is imperfect because poles and zeros can be specified (or determined) with only limited accuracy. If a zero of $X(s)$ almost cancels a pole of $H(s)$, the coefficient of the associated term in a partial-fraction expansion of $Y(s) = H(s)X(s)$ (the coefficient of the associated natural mode) is relatively small. If the real part of the almost-canceled pole is negative, the associated natural mode can be ignored for all practical purposes because it starts out small and decays. In such a case we say that the input $x(t)$ *suppresses* (or *does not excite*) the natural mode. If the real part of the partly canceled pole is positive, the associated natural mode grows exponentially. It eventually exceeds all bounds no matter how small its coefficient. Consequently, a pole having positive real part cannot be suppressed (in real life) by an input having a like zero.

5-3C Stability

A linear stationary system having impulse response $h(t)$ is stable if and only if

$$\int_{-\infty}^{\infty} |h(t)|\, dt < \infty$$

This implies that a system is stable if and only if all natural modes of the system decay. Therefore, a linear stationary system is stable if and only if all its poles lie in the left half of an s plane (have negative real parts).

Example 5-24 Find all values of K_1 for which the system of Fig. 5-14 is stable. Assume $K_2 > 0$.

SOLUTION By block-diagram reduction we find that the system function of the system is

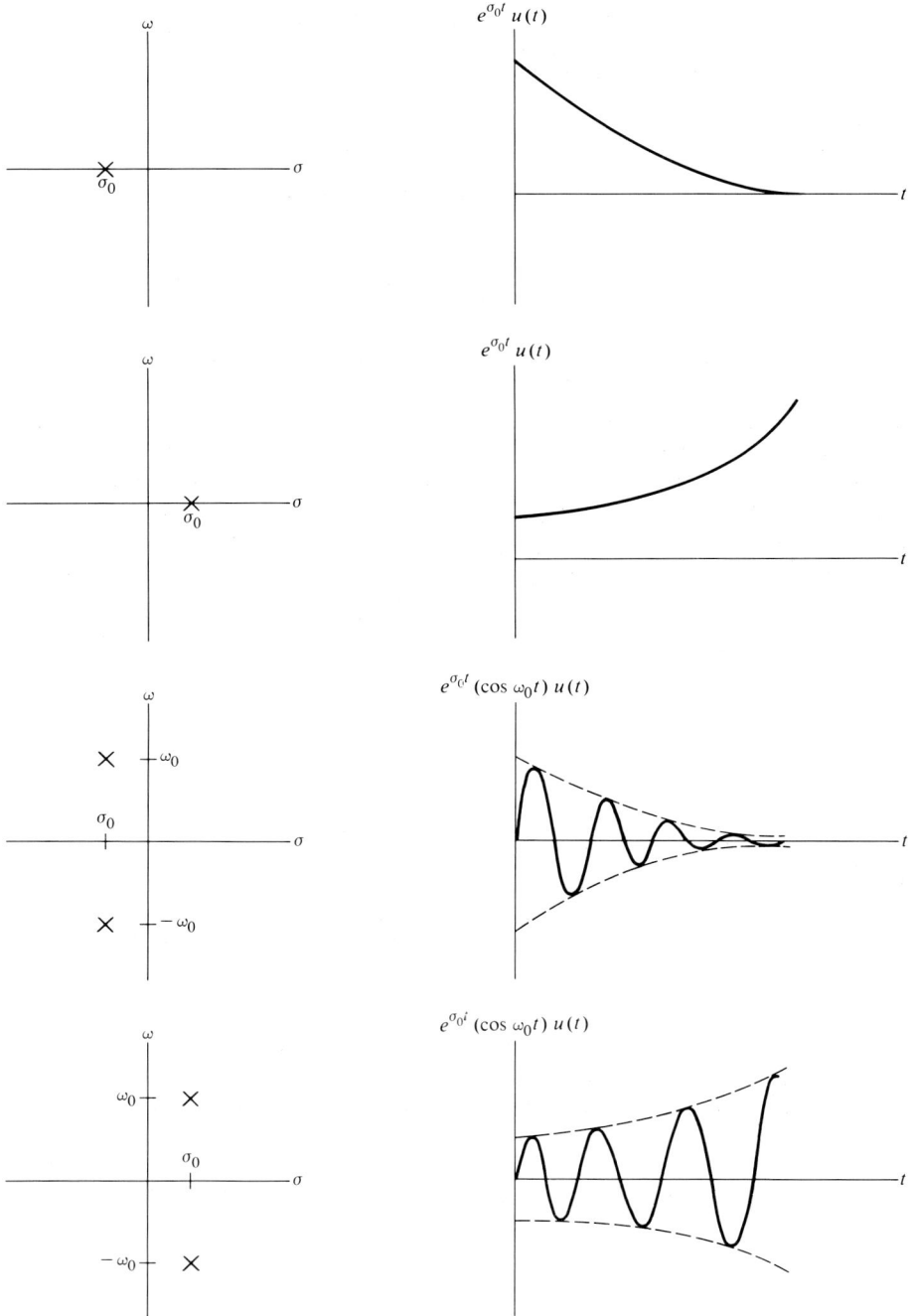

Figure 5-13 Relation between the waveform of a natural mode and the s-plane location of the associated pole(s).

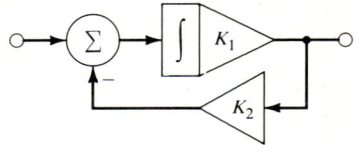

Figure 5-14 System of Example 5-24.

$$H(s) = \frac{K_1}{s + K_1 K_2}$$

The system function has one real pole $p = -K_1 K_2$. Because $K_2 > 0$, the pole p lies in the left half plane (the system is stable) for $K_1 > 0$.

Example 5-25 Find all values of K_1 for which the system of Fig. 5-15 is stable. Assume $K_2 > 0$.

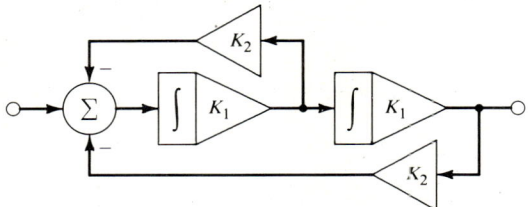

Figure 5-15 System of Example 5-25.

SOLUTION The system function is

$$H(s) = \frac{K_1^2}{s^2 + K_1 K_2 s + K_1^2 K_2}$$

The poles of $H(s)$ are

$$p_1, p_2 = \tfrac{1}{2} K_1 K_2 (-1 \pm \sqrt{1 - 4K_2^{-1}})$$

The poles are real if $K_2 \geq 4$ and complex if $0 < K_2 < 4$. We consider these cases separately.

If $K_2 > 4$, then $0 < \sqrt{1 - 4K_2^{-1}} < 1$ and $-1 \pm \sqrt{1 - 4K_2^{-1}} < 0$

Therefore both poles lie in the left half plane if $K_2 > 4$ and $K_1 > 0$.

If $0 < K_2 < 4$, the poles are a complex-conjugate pair, having real part $-K_1 K_2/2$. Therefore both poles lie in the left half plane if $0 < K_2 < 4$ and $K_1 > 0$.

We have shown that both poles lie in the left half plane for $K_1 > 0$, whether $0 < K_2 < 4$ or $K_2 > 4$. Therefore the system is stable for $K_1 > 0$.

Example 5-26 In the system of Fig. 5-16, $K_1 = 10 \text{ s}^{-1}$ and $p = 2 \text{ s}^{-1}$. Find all values of K_2 for which the system is stable.

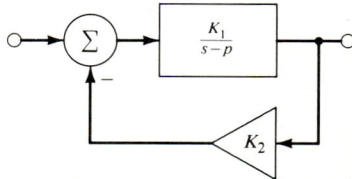

Figure 5-16 System of Example 5-26.

SOLUTION The system function of the system is

$$H(s) = \frac{K_1}{s - p + K_1 K_2} = \frac{K_1}{s - p_1}$$

where $p_1 = p - K_1 K_2$. The pole p_1 lies in the left half plane if $K_1 K_2 > p$. This gives $K_2 > p/K_1$ or $K_2 > 0.2$. This example illustrates the fact that negative feedback can stabilize an unstable system.

Example 5-27 In the system of Fig. 5-17, $p_1 = z_1 = 1 \text{ s}^{-1}$ and $p_2 = -1 \text{ s}^{-1}$. Is the system stable?

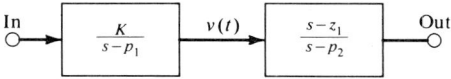

Figure 5-17 System of Example 5-27.

SOLUTION Evidently the system function of the system is

$$H(s) = \frac{K(s - z_1)}{(s - p_1)(s - p_2)} \stackrel{?}{=} \frac{K}{s - p_2}$$

because the zero at $s = z_1$ cancels the pole at $s = p_1$. If this is so, then the system is stable because $p_2 < 0$. However, if the subsystems (the individual blocks in Fig. 5-17) represent physical systems, the system is unstable because neither the zero nor the pole can be specified with sufficient accuracy to guarantee perfect cancellation. We can be certain that $z_1 = p_1 + \Delta p$, $\Delta p \neq 0$. Therefore the system function is actually

$$H(s) = \frac{K(s - p_1 - \Delta p)}{(s - p_1)(s - p_2)}$$

The system is unstable because the pole p_1 lies in the right half plane. Even if it were possible to make z_1 equal to p_1, the system (the linear model) represented by the first block would be unstable, the signal (the signal model) $v(t)$ would be unbounded, and the linear model of Fig. 5-17 would be invalid. This example illustrates the fact that it is practically impossible to cancel a right-half-plane pole by introducing a like zero.

5-3D Realizability

An impulse response $h(t)$ is realizable if and only if

$$h(t) = 0 \quad t < 0$$

Because the impulse response of a system is the inverse Laplace transform of the system function of the system, a system function is realizable if and only if

$$\mathcal{L}^{-1}\{H(s)\} = 0 \quad t < 0 \tag{5-22}$$

In assuming that $\mathcal{L}^{-1}$ means right-sided inverse Laplace transform we have implicitly assumed that all rational system functions are realizable; i.e., expanding a rational system function $H(s)$ in partial fractions and using Table D-1 to obtain $\mathcal{L}^{-1}\{H(s)\}$ always gives a realizable impulse response. To examine the realizability of a system function we must temporarily abandon our assumption that all inverse Laplace transforms are right-sided. To do so without giving up uniqueness we must associate a region of convergence with a system function. Depending on the region of convergence associated with a system function, the inverse transform of the system function (the impulse response) can be right-sided (realizable), left-sided (nonrealizable), or two-sided (nonrealizable).

To introduce the s-plane condition for realizability we consider the realizable impulse response $h_r(t)$ and the nonrealizable impulse response $h_{nr}(t)$, given by

$$h_r(t) = pe^{pt}u(t) \qquad h_{nr}(t) = -pe^{pt}u(-t)$$

From the definition (5-1) of the Laplace transform, the system functions and associated regions of convergence for these impulse responses are

$$H_r(s) = \frac{p}{s-p} \qquad \text{Re } s > \text{Re } p \qquad H_{nr}(s) = \frac{p}{s-p} \qquad \text{Re } s < \text{Re } p$$

These results show that a single-pole term of the form $1/(s-p)$ is realizable if the associated region of convergence lies to the right of the pole p and nonrealizable if the region of convergence lies to the left of the pole p. To generalize this result we use the fact that a rational system function $H(s)$ can be expanded in partial fractions, where each pole p of $H(s)$ gives rise to a term of the form $B/(s-p)$. Because such a term is realizable if and only if the associated region of convergence lies to the right of p, and because the region of convergence for $H(s)$ is the intersection of the regions of convergence for the terms in the partial-fraction expansion of $H(s)$, we find that *a rational system function $H(s)$ is realizable if and only if the region of convergence for $H(s)$ lies to the right of the rightmost pole of $H(s)$.*

Example 5-28 From a purely mathematical attack on a system design problem it is found that the optimum system function is

$$H_0(s) = \frac{K(p_1 - p_2)}{(s - p_1)(s - p_2)}$$

where $K = -15 \text{ s}^{-1}$, $p_1 = -2 \text{ s}^{-1}$, $p_2 = 1 \text{ s}^{-1}$, and the region of convergence is $p_1 < \text{Re } s < p_2$, as shown in Fig. 5-18. The system function is nonrealizable because the region of convergence lies to the left of the rightmost pole. Obtain a realizable system function $H(s)$ that approximates $H_0(s)$.

SOLUTION Expanding $H_0(s)$ in partial fractions gives

$$H_0(s) = \frac{K}{s - p_1} - \frac{K}{s - p_2}$$

The first term is realizable because the region of convergence is to the right of p_1. The second term is nonrealizable because the region of convergence is to the left of p_2. Therefore we identify the first term as the realizable part of $H_0(s)$, and we take

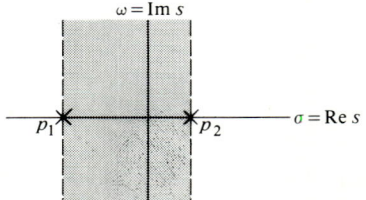

Figure 5-18 Region of convergence for the system function $H_0(s)$ of Example 5-28.

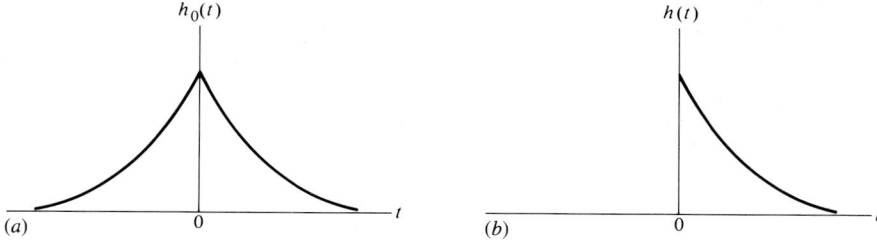

Figure 5-19 Impulse responses corresponding to (a) the optimum (but nonrealizable) system function $H_0(s)$ of Example 5-28 and (b) a realizable approximation obtained by discarding the nonrealizable part of $H_0(s)$.

$$H(s) = \frac{K}{s - p_1} \qquad \text{Re } s > p_1$$

as a realizable approximation to $H_0(s)$.

Figure 5-19 illustrates this approximation procedure in the time domain. Dropping the nonrealizable terms from a partial-fraction expansion of a system function corresponds to dropping the nonrealizable part (for $t < 0$) of the associated impulse response. The impulse response corresponding to the optimum nonrealizable system function $H_0(s)$ is

$$h_0(t) = Ke^{p_1 t}u(t) + Ke^{p_2 t}u(-t)$$

whereas the impulse response corresponding to the realizable approximation $H(s)$ is

$$h(t) = Ke^{p_1 t}u(t)$$

Comparing Figs. 5-19a and b, we see that we cannot expect an output of the realizable approximation to look much like the output of the optimum nonrealizable system for the same input. If the application can tolerate delay, an approach that is better than simply discarding the nonrealizable part of $h_0(t)$ is to introduce enough delay to move the significant part of $h_0(t)$ to the right of $t = 0$, as shown in Fig. 5-20a, and subsequently discard only the nonrealizable tail that lies to the left of $t = 0$, as shown in Fig. 5-20b. An output of a realizable approximation obtained in this manner will look like the output of the optimum nonrealizable system but delayed by delay time t_0. Unfortunately, some applications, e.g., fast-acting automatic control systems, cannot tolerate delay. Sometimes the approximation illustrated in Fig. 5-19 is the best we can do.

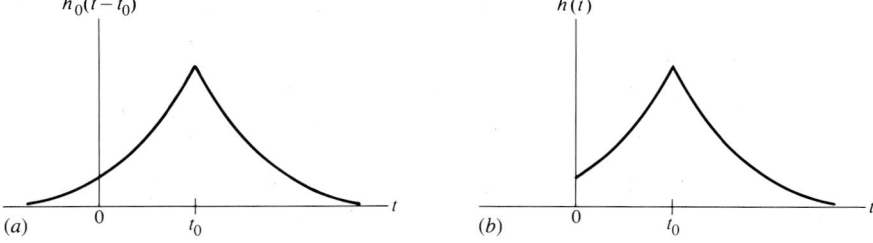

Figure 5-20 Using delay to obtain a realizable approximation to a nonrealizable system function.

5-3E Unforced Response and Forced Response

The output $y(t)$ of a linear stationary system having system function $H(s)$ and input $x(t)$ is given by

$$y(t) = \mathcal{L}^{-1}\{H(s)X(s)\} \tag{5-23}$$

where $X(s) = \mathcal{L}\{x(t)\}$. Expanding $H(s)X(s)$ in partial fractions shows that $y(t)$ consists (in general) of terms arising from the poles of $H(s)$ and terms arising from the poles of $X(s)$. The terms arising from poles of $H(s)$ (the natural modes of the system) are present no matter what input is applied and constitute the *unforced response* of the system. Terms not arising from the poles of $H(s)$ constitute the *forced response* of the system.

Example 5-29 Obtain the output $y(t)$ of a system having system function

$$H(s) = \frac{Kp^2}{s(s-p)}$$

and input $x(t) = x_0 u(t)$. Identify the unforced response and the forced response.

SOLUTION From (5-23), the output is

$$y(t) = \mathcal{L}^{-1}\left\{\frac{Kp^2 x_0}{s^2(s-p)}\right\} = Kx_0 \mathcal{L}^{-1}\left\{-\frac{p}{s^2} - \frac{1}{s} + \frac{1}{s-p}\right\}$$

$$= Kx_0(-pt - 1 + e^{pt})u(t)$$

The unforced response is

$$y_{uf}(t) = Kx_0(-1 + e^{pt})u(t)$$

because this component of $y(t)$ arises from the poles of the system function $H(s)$. The forced response is

$$y_f(t) = -Kx_0 pt u(t)$$

because this component does not arise from the poles of $H(s)$.

If a pole of the transform $X(s)$ of an input $x(t)$ equals a pole p of a system function $H(s)$, we say that the input $x(t)$ *forces the system at the pole* p; for example, in Example 5-29, the input $x(t)$ forces the system at the pole $p = 0$. Whether or not an input forces a system at a pole is noteworthy for mainly two reasons: (1) An output of a system forced at a pole usually is much larger than an output of the same system not forced at a pole. In fact, if a system is forced at a pole lying on the ω axis, the output is unbounded (see Example 5-29). Consequently, a system that is approximately linear under ordinary circumstances may become nonlinear (may perform poorly) if it is forced at a pole. (2) If a system is not forced at a pole, it is relatively easy to obtain the forced response of the system for an exponential input (or any other input, because any signal of interest can be described by a sum of exponentials). The *forced response* of a system having system function $H(s)$ and input

$$x(t) = x_0 e^{pt} u(t)$$

where p is not a pole of $H(s)$, is given by

$$y_f(t) = H(p) x_0 e^{pt} u(t) \tag{5-24}$$

To show this, we expand the Laplace transform of the output for input $x(t)$ above in partial fractions. This gives

$$Y(s) = H(s)X(s) = H(s)\frac{x_0}{s-p}$$

$$= \frac{B}{s-p} + \text{unforced terms arising from poles of } H(s)$$

where, from Heaviside's formula (5-19),

$$B = \left.\frac{(s-p)x_0 H(s)}{s-p}\right|_{s \to p} = H(p)x_0$$

because p is not a pole of $H(s)$. Therefore the forced response is

$$\mathcal{L}^{-1}\left\{\frac{H(p)x_0}{s-p}\right\} = H(p)x_0 e^{pt} u(t)$$

as was to be shown.

Test inputs for which we are often interested mainly in the forced response of a system are steps and "sinusoidal steps." From (5-24) (with $p = 0$), the forced response of a system having system function $H(s)$ and step input $x(t) = x_0 u(t)$ is given by

$$y_f(t) = H(0) x_0 u(t) \tag{5-25}$$

provided $H(s)$ has no pole at $s = 0$. The quantity $H(0)$ is called the *dc gain* of the system because the steady-state amplitude of the output for input $x(t) = x_0 u(t)$ is $H(0)x_0$.

A sinusoidal step $x(t) = A(\cos \omega_0 t) u(t)$ can be expressed (using Euler's identity) as

$$x(t) = \tfrac{1}{2} A e^{j\omega_0 t} u(t) + \tfrac{1}{2} A e^{-j\omega_0 t} u(t)$$

From (5-24) and superposition, the forced response of a system having system function $H(s)$ and sinusoidal-step input $x(t)$ above is given by

$$y_f(t) = H(j\omega_0)\frac{A}{2} e^{j\omega_0 t} u(t) + H(-j\omega_0)\frac{A}{2} e^{-j\omega_0 t} u(t)$$

provided $H(s)$ has no pole at $s = \pm j\omega_0$; that is, provided $x(t)$ does not force the system at a pole. This simplifies to

$$y_f(t) = A \operatorname{Re}[H(j\omega_0) e^{j\omega_0 t}] u(t) \tag{5-26}$$

330 INTRODUCTION TO SYSTEM ANALYSIS

Example 5-30 Obtain the forced response of a system having system function

$$H(s) = \frac{Kr}{s^2 + qs + r}$$

and input

$$x(t) = x_0 u(t) + x_0(\cos \omega_0 t) u(t)$$

where $K = 1$ V/V, $q = 6$ s^{-1}, $r = 25$ s^{-2}, $x_0 = 5$ V, and $\omega_0 = 4$ rad/s.

SOLUTION The poles of $H(s)$ are $p_1, p_2 = -3 \pm 4j$ s^{-1}. Since the input does not force the system at any of its poles, we can use (5-25) and (5-26) to obtain the forced response. By superposition, the forced response is

$$y_f(t) = H(0) x_0 u(t) + x_0 \, \text{Re}[H(j\omega_0) e^{j\omega_0 t}] u(t)$$

where

$$H(j\omega_0) = \frac{Kr}{-\omega_0^2 + jq\omega_0 + r} = \frac{25}{9 + 24j} = 0.98 e^{-1.21j} \quad \text{V/V}$$

Thus

$$y_f(t) = H(0) x_0 u(t) - x_0 \, \text{Re}(0.98 e^{-1.21j} e^{j\omega_0 t}) u(t)$$
$$= x_0[1 + 0.98 \cos(\omega_0 t - 1.21)] u(t)$$

5-3F Transient Response and Steady-State Response

The transient components of the output

$$y(t) = \mathcal{L}^{-1}\{H(s)X(s)\}$$

of a system having system function $H(s)$ and input $x(t)$ arise from left-half-plane poles of $Y(s) = H(s)X(s)$ (see Fig. 5-13). The steady-state components of the output arise from poles of $Y(s)$ on the ω axis and in the right half plane. In general (no cancellation), the poles of $Y(s)$ are the poles of $X(s)$ and the poles of $H(s)$; thus both a system and its input contribute both transient and steady-state components to an output, i.e., both the transient and the steady-state response have both forced and unforced components.

Example 5-31 Give the form of the output $y(t)$ of a system having system function

$$H(s) = \frac{Kp_1^2}{s(s - p_1)}$$

and input

$$x(t) = x_0 e^{p_2 t} u(t) + x_0(\cos \omega_0 t) u(t)$$

where $p_1 < p_2 < 0$. Identify the forced, unforced, transient, and steady-state components of the output.

SOLUTION The output has the form

$$y(t) = B_0 u(t) + B_1 e^{p_1 t} u(t) + B_2 e^{p_2 t} u(t) + B_3 \cos(\omega_0 t + \theta) u(t)$$

The first two terms are unforced because they arise from poles of $H(s)$. The last two terms are forced. The second and third terms are transient because they decay, i.e., because the poles p_1, p_2 lie in the left half plane. The first and last terms are steady-state because they do not decay, i.e., because the associated poles lie on the ω axis. We have

Unforced response $y_{uf}(t) = (B_0 + B_1 e^{p_1 t}) u(t)$

Forced response $y_f(t) = [B_2 e^{p_2 t} + B_3 \cos(\omega_0 t + \theta)] u(t)$

Transient response $y_t(t) = (B_1 e^{p_1 t} + B_2 e^{p_2 t}) u(t)$

Steady-state response $y_{ss}(t) = B_0 + B_3 \cos(\omega_0 t + \theta)$

Example 5-31 shows that, in general, the transient and unforced components of an output are different, as are the steady-state and forced components of an output; however, in most applications where it is useful to separate an output into steady-state and transient components, systems are stable (have only left-half-plane poles) and inputs of interest are steps and sinusoids (have poles only on the ω axis). In such cases, the steady-state and forced components of an output are identical, as are the transient and unforced components of the output. *For a stable system having a step or sinusoidal input,*

$$\text{Transient response} = \text{unforced response}$$

and $$\text{Steady-state response} = \text{forced response}$$

Therefore the steady-state output of a stable system having system function $H(s)$ and step input $x(t) = x_0 u(t)$ is given by [see (5-25)]

$$y_{ss}(t) = y_f(t) \Big|_{t>0} = H(0) x_0 \qquad (5\text{-}27)$$

and the steady-state output of a stable system having system function $H(s)$ and sinusoidal input $x(t) = x_0(\cos \omega_0 t) u(t)$ is given by [see (5-26)]

$$y_{ss}(t) = y_f(t) \Big|_{t>0} = x_0 \operatorname{Re}[H(j\omega_0) e^{j\omega_0 t}]$$

$$= |H(j\omega_0)| x_0 \cos[\omega_0 t + \angle H(j\omega_0)] \qquad (5\text{-}28)$$

Note that (5-27) is a special case of (5-28) (for $\omega_0 = 0$).

Example 5-32 Obtain the steady-state output of a system having system function

$$H(s) = \frac{Kp}{s - p}$$

and input

$$x(t) = x_0(\cos \omega_0 t) u(t)$$

where $K = -5$ V/V, $p = -2$ s^{-1}, $x_0 = 5$ V, and $\omega_0 = 2$ rad/s.

SOLUTION Since the system is stable and the input is a sinusoid, we can use (5-28), where

$$H(j\omega_0) = \frac{Kp}{j\omega_0 - p} = \frac{10}{2 + 2j} = 3.54e^{-0.79j} \text{ V/V}$$

whence $|H(j\omega_0)| = 3.54$ V/V, $\angle H(j\omega_0) = -0.79$ rad, and

$$y_{ss}(t) = 3.54x_0 \cos(\omega_0 t - 0.79)$$

5-3G Transfer Function and Frequency Response

Comparing (5-28) with (3-40) shows that the transfer function of a stable system equals the system function of the system with s replaced by $j\omega$. This also follows from the fact that the region of convergence of the system function of a stable system includes the ω axis. Neither an unstable nor a conditionally stable system has a transfer function because the steady-state output of such a system for step or sinusoidal input contains terms other than those arising from the input, i.e., terms arising from poles on the ω axis or in the right half plane. Nevertheless, we often treat a conditionally stable system as a barely stable system by assuming that poles on the ω axis actually lie just to the left of it. For example, the system function $H(s) = K/s$ of an ideal integrator has a pole on the ω axis (at $s = 0$). We assume that the system function of a real-life integrator is $H(s) = K/(s - \epsilon)$, where the pole ϵ may be very small and negative but not zero. Therefore the transfer function of a real-life integrator is

$$H(j\omega) = \frac{K}{j\omega - \epsilon} \approx \frac{K}{j\omega} \qquad \omega \neq 0$$

Although this transfer function can give an incorrect output for an input having a dc component, this rarely causes a problem in practice because we rarely apply a signal having a dc component to an integrator. We can use the Laplace transformation instead of the Fourier transformation when a signal having a dc component is applied to an integrator. (Alternatively, we can model the dc component as a very long rectangular pulse and use the Fourier transformation, but this is cumbersome.)

Example 5-33 Obtain the steady-state output of a system having system function

$$H(s) = \frac{Kp}{s - p}$$

and input $x(t) = x_0 \cos \omega_0 t$, where $K = -5$ V/V, $p = -4$ s^{-1}, $x_0 = 5$ V, and $\omega_0 = 3$ rad/s.

SOLUTION The system is stable because its pole p lies in the left half plane. Therefore the transfer function of the system is

$$H(j\omega) = H(s)\Big|_{s=j\omega} = \frac{Kp}{j\omega - p}$$

The gain $\Gamma(\omega)$ and the phase shift $\phi(\omega)$ of the system are

$$\Gamma(\omega) = |H(j\omega)| = \frac{|Kp|}{\sqrt{\omega^2 + p^2}}$$

$$\phi(\omega) = \angle H(j\omega) = -\text{Tan}^{-1}(\omega, -p) \qquad \text{rad}$$

The steady-state output for input $x(t)$ above is

$$y_{ss}(t) = \Gamma(\omega_0)x_0 \cos[\omega_0 t + \phi(\omega_0)] = y_0 \cos(\omega_0 t + \psi)$$

where $y_0 = \Gamma(\omega_0)x_0 = 20$ V and $\psi = \phi(\omega_0) = -0.64$ rad.

Figure 5-21 illustrates how poles and zeros of a stable system affect the shape of the gain of the system. The three-dimensional graphs represent the magnitudes $|H(s)|$ of system functions plotted versus $\sigma = \text{Re } s$ and $\omega = \text{Im } s$. The poles and zeros of the system functions are given by the associated pole-zero diagrams. The graphs of $|H(s)|$ are truncated along the ω axis, so the face of each three-dimensional graph is a graph of gain versus frequency. A pole (zero) near the ω axis produces a peak (valley) in the associated graph of gain versus frequency. Moving a pole (zero) toward the ω axis increases the height of the associated peak (the depth of the associated valley).

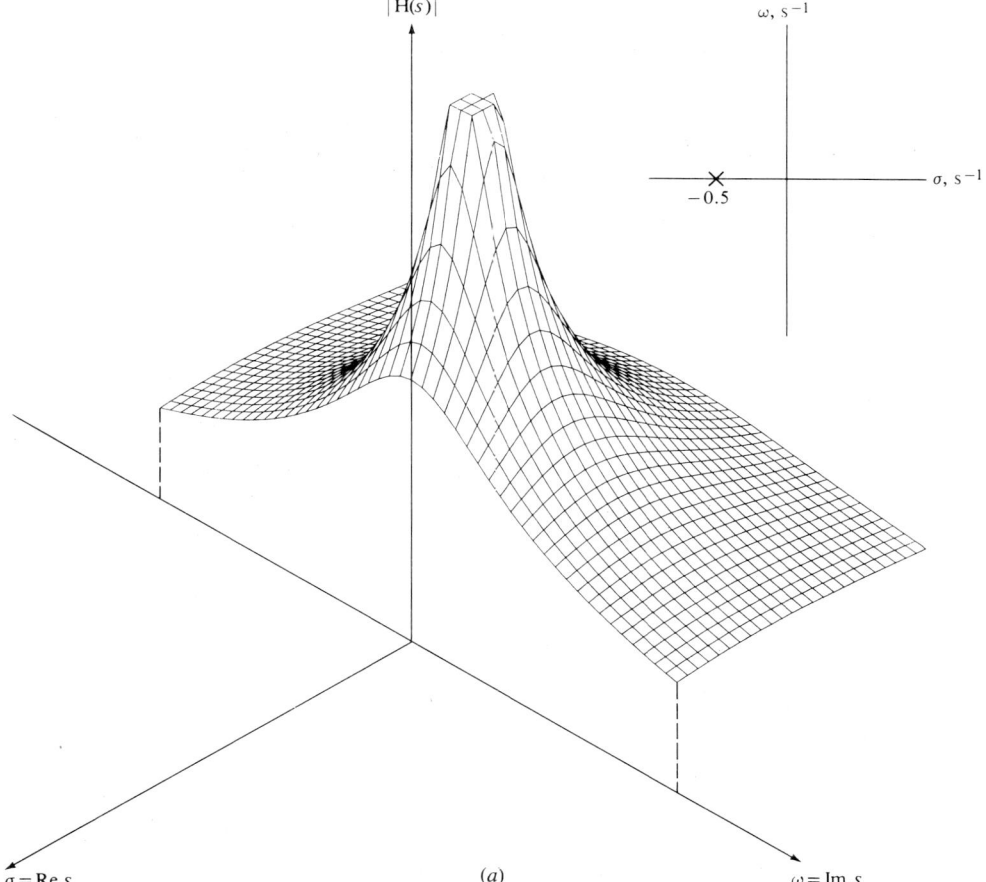

Figure 5-21 How s-plane location of poles and zeros of a system function $H(s)$ affects gain $\Gamma(\omega)$.

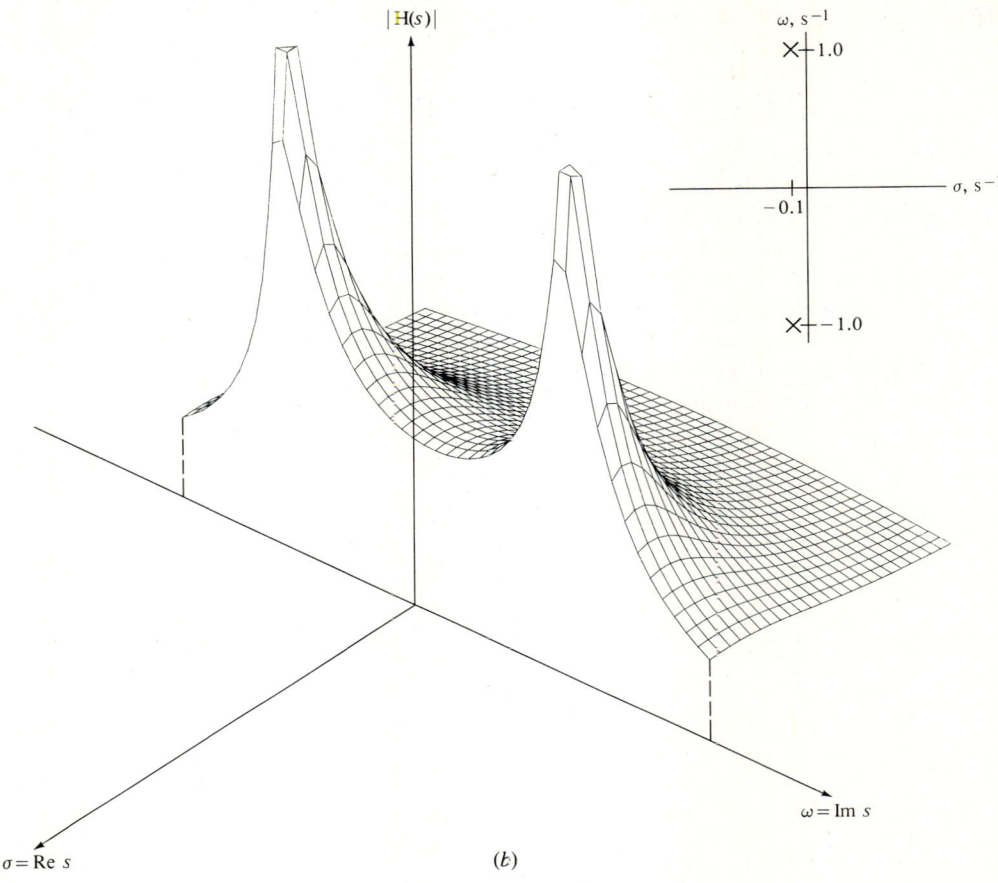

(b)

Figure 5-21 *Continued*

Figure 5-21a is drawn for a first-order system having system function

$$H(s) = \frac{Kp}{s - p}$$

Moving the pole p closer to the ω axis makes $\Gamma(\omega)$ more peaked near $\omega = 0$ and reduces the bandwidth of the system. Moving the pole farther from the ω axis makes $\Gamma(\omega)$ less peaked near ω and increases the bandwidth of the system. The system of Fig. 5-21b is a very lightly damped second-order system having poles p, $p^* = -0.1 \pm j$ s^{-1}. The gain is sharply peaked near $\omega = \operatorname{Im} p = 1$ rad/s. The system of Fig. 5-21c is a moderately damped second-order system having poles p, $p^* = -0.5 \pm 0.5j$ s^{-1} and one real zero $z = 0$. The gain is less sharply peaked than in Fig. 5-21b because the poles are farther from the ω axis. The dc gain of the system is zero because the system function has a zero at $s = 0$.

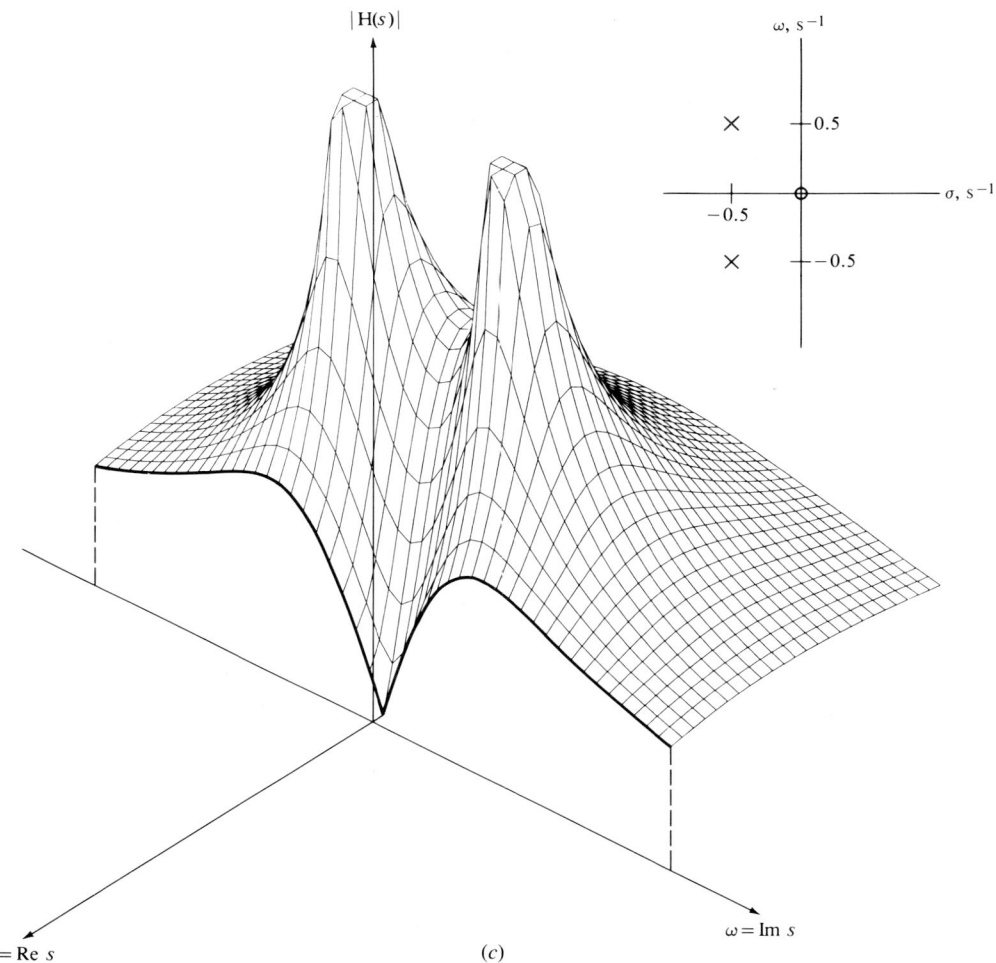

Figure 5-21 *Continued*

5-3H Minimum-Phase and Non-Minimum-Phase Systems

A system having only left-half-plane zeros is called a *minimum-phase system*.† A system having one or more right-half-plane zeros is called a *non-minimum-phase system*. This terminology arises because a system having right-half-plane zeros introduces more phase shift than a system having only left-half-plane zeros. This is illustrated by Fig. 5-22. Figure 5-22a shows graphs of gain and phase shift versus frequency for a minimum-phase system having one pole $p = -2 \text{ s}^{-1}$ and one zero $z = -4 \text{ s}^{-1}$. Figure 5-22b shows graphs of gain and phase shift versus

---

†More precisely, a minimum-phase system is one having only left-half-plane zeros *and poles;* however, the terminology is useful mainly in reference to stable systems.

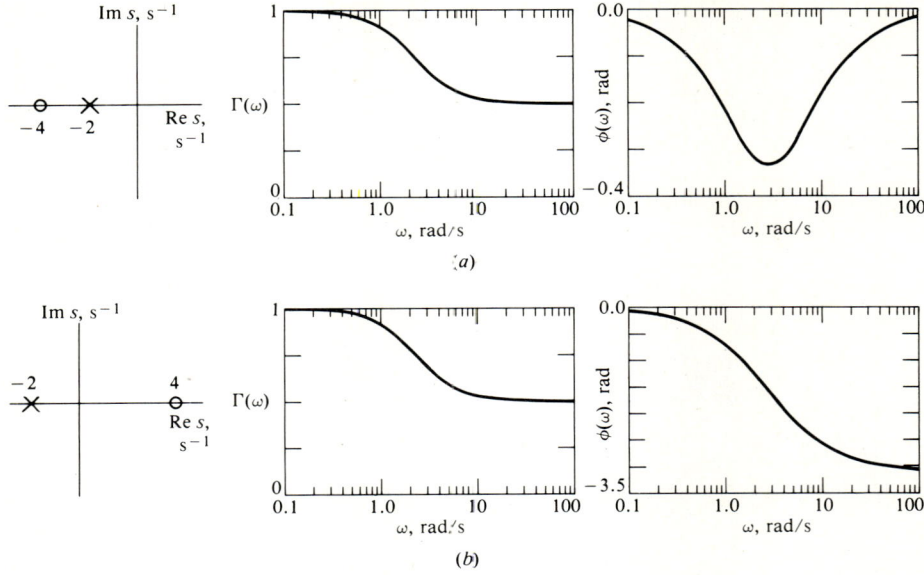

Figure 5-22 Graphs of gain and phase shift for (a) a first-order minimum-phase system and (b) a comparable non-minimum-phase system.

frequency for a comparable non-minimum-phase system having one pole $p = -2$ s^{-1} and one zero $z = 4$ s^{-1}. The gains of the systems are identical, but the maximum phase shift (magnitude) of the non-minimum-phase system is almost 10 times that of the minimum-phase system.

Whether a system is minimum-phase or non-minimum-phase is of interest for mainly three reasons:

1. A non-minimum-phase system generally introduces more phase distortion than a comparable minimum-phase system. This is illustrated by Fig. 5-23, which shows pulse responses of the systems of Fig. 5-22. The dashed lines represent the ideal (distortionless) pulse response $x_0 r(t/\tau)$, where $\tau = 1.5$ s. Figure 5-22a shows the pulse response of the minimum-phase system of Fig. 5-23a, and Fig. 5-23b shows the pulse response of the non-minimum-phase system of Fig. 5-22b. Note that the maximum instantaneous distortion (magnitude) introduced by the non-minimum-phase system is 3 times that introduced by the minimum-phase system.
2. The second reason has to do with the fact that the reciprocal of a non-minimum-phase system function is unstable. In certain applications in signal processing and automatic control it is desirable to realize a system whose system function is (at least in part) the reciprocal of a system function representing some fixed (unchangeable) physical system or process. For example, in designing an automatic control system we might wish to replace the system function of an uncooperative plant (system to be controlled) with one that suits our purpose. This entails connecting in series with the plant a compensator whose system function is in part the reciprocal of the plant system function. This fails for a non-minimum-

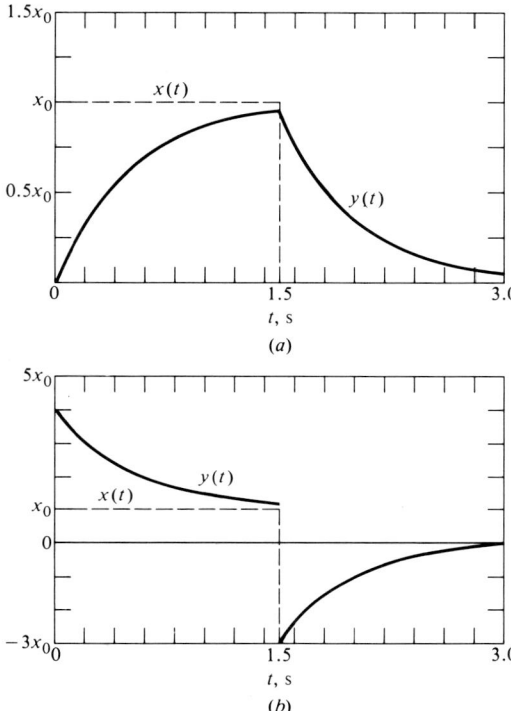

Figure 5-23 Pulse response of (*a*) the minimum-phase system of Fig. 5-22*a* and (*b*) the non-minimum-phase system of Fig. 5-22*b*.

phase plant because in that case the compensator is unstable; i.e., a right-half-plane zero in the plant system function gives rise to a right-half-plane pole in the compensator system function.

3. The system function of a stable minimum-phase system can be determined from the gain alone of the system. This is not true of a non-minimum-phase system. Proof of this assertion and a meaningful discussion of its consequences involve concepts beyond the scope of this book.

5-3I All-Pass Delay Equalizers

A filter designed to meet specifications on gain alone, e.g., a Butterworth filter, often has a poor delay characteristic (see Fig. 4-53). The resulting delay distortion is tolerable in some applications, e.g., some systems for voice communication, but in most applications delay distortion is at least as harmful as amplitude distortion. In most applications where a filter is designed from specifications on gain alone, some means for equalizing delay (linearizing phase shift) of the filter is necessary. Equalizing delay of a filter usually entails connecting in series with the filter an all-pass system whose delay complements the delay of the filter in such a way that overall delay (delay of the filter-equalizer combination) is almost independent of frequency (Fig. 5-24). The gain of the series combination is determined by the gain of the filter alone (as desired) because the gain of an all-pass delay equalizer is independent of frequency.

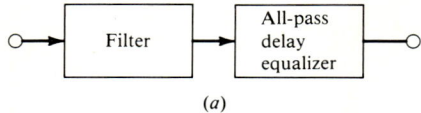

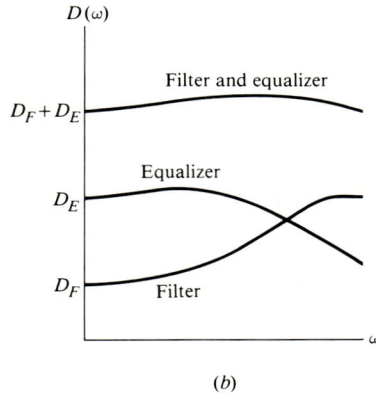

Figure 5-24 Delay equalization using an all-pass delay equalizer.

The system function of an *all-pass delay equalizer* has the form

$$H(s) = \frac{(s + p_1)(s + p_2)\cdots(s + p_n)}{(s - p_1)(s - p_2)\cdots(s - p_n)}$$

where the poles $p_1, p_2, \ldots, p_n$ lie in the left half plane and the zeros $-p_1, -p_2, \ldots, -p_n$ lie in the right half plane. An all-pass delay equalizer is a stable non-minimum-phase system. The system is all-pass because for each real pole p there is a zero $z = -p$ such that $|j\omega + p|/|j\omega - p| = 1$ and for each complex pole p there is a zero $z = -p^*$ such that $|j\omega + p^*|/|j\omega - p| = 1$. An all-pass delay equalizer can be realized as a series connection of first-order all-pass systems and second-order all-pass systems, where the first-order systems have real poles (and matching zeros) and the second-order systems have complex poles (and matching zeros). The system function of a first-order all-pass system has the form

$$H(s) = \frac{s + p}{s - p} \qquad (5\text{-}29)$$

where p is real and negative. The system function of a second-order all-pass system has the form

$$H(s) = \frac{(s - p)(s + p^*)}{(s - p)(s - p^*)} \qquad (5\text{-}30)$$

We describe first- and second-order all-pass systems separately below. Subsequently, we show (by example) how such systems are used to equalize the delay of a particular filter.

The transfer function of a first-order all-pass system has the form

$$H(j\omega) = \frac{j\omega + p}{j\omega - p}$$

The gain of the system is unity for all frequencies. The phase shift $\phi(\omega)$ and the delay $D(\omega)$ of the system are

$$\phi(\omega) = \tan^{-1}\frac{\omega}{p} - \tan^{-1}\left(-\frac{\omega}{p}\right) = 2\tan^{-1}\frac{\omega}{p}$$

$$D(\omega) = -\frac{d\phi}{d\omega} = -\frac{2}{p}\frac{1}{1+(\omega/p)^2}$$

Delay $D(\omega)$ is positive because p is negative. It is convenient to express normalized delay $|p|D(\omega)$ as a function of normalized frequency $\omega/|p|$; thus

$$|p|D(\omega) = \frac{2}{1+\lambda^2} \quad \text{where } \lambda = \frac{\omega}{|p|} \quad (5\text{-}31)$$

Figure 5-25 shows a graph of normalized delay $|p|D(\omega)$ versus normalized frequency λ. Normalized delay decreases monotonically (with increasing frequency) from $|p|D(0) = 2$ to $|p|D(\infty) = 0$. Actual delay $D(\omega)$ also decreases monotonically from $D(0) = -2/p$ to $D(\infty) = 0$. Moving the pole p (and the zero $z = -p$) farther from the ω axis decreases delay at all frequencies.

The transfer function of a second-order all-pass system has the form

$$H(j\omega) = \frac{(j\omega + p)(j\omega + p^*)}{(j\omega - p)(j\omega - p^*)} = \frac{|p|^2 - \omega^2 + 2(\text{Re } p)j\omega}{|p|^2 - \omega^2 - 2(\text{Re } p)j\omega}$$

An equalizer having a transfer function of this form is called an *all-pass biquad*.† It is convenient to express the transfer function $H(j\omega)$ as

$$H(j\omega) = \frac{1 - \lambda^2 - 2qj\lambda}{1 - \lambda^2 + 2qj\lambda} \quad (5\text{-}32a)$$

where normalized frequency λ and the parameter q are defined by

†Biquad is short for biquadratic, which refers to the fact that the transfer function is a ratio of two quadratic functions of frequency.

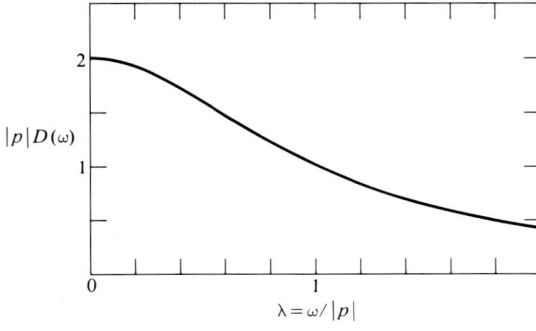

Figure 5-25 Graph of normalized delay versus normalized frequency for a first-order all-pass delay equalizer.

$$\lambda = \frac{\omega}{|p|} \qquad q = -\frac{\text{Re } p}{|p|} \qquad (5\text{-}32b)$$

The denominator of $H(j\omega)$ is the conjugate of the numerator. Therefore the gain $\Gamma(\omega)$ is unity, and the phase shift is

$$\phi(\omega) = 2 \tan^{-1} \frac{-2q\lambda}{1 - \lambda^2} \qquad \lambda = \frac{\omega}{|p|}$$

The delay is

$$D(\omega) = -\frac{d\phi(\omega)}{d\omega} = -\frac{d\phi(\omega)}{d\lambda} \frac{d\lambda}{d\omega}$$

where $d\lambda/d\omega = 1/|p|$. Figure 5-26 shows graphs of normalized delay $|p|D(\omega)$ versus

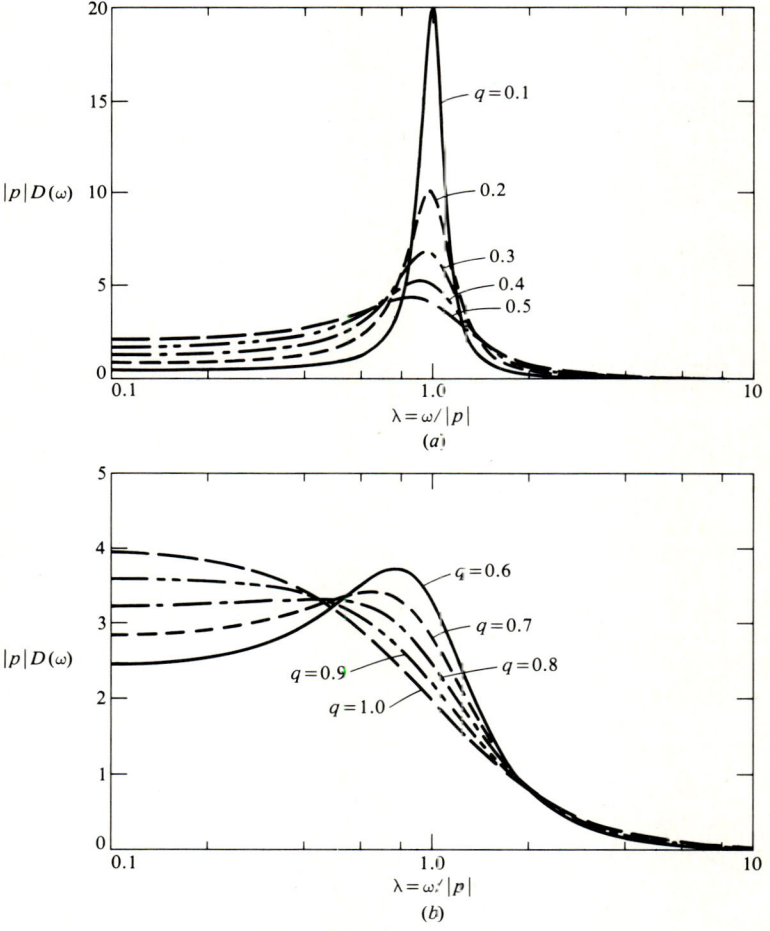

Figure 5-26 Graphs of normalized delay versus normalized frequency for a second-order all-pass delay equalizer.

normalized frequency $\lambda = \omega/|p|$, where

$$|p|D(\omega) = -\frac{d\phi(\omega)}{d\lambda} = \frac{4q(1 + \lambda^2)}{(1 - \lambda^2)^2 + 4(q\lambda)^2} \tag{5-33}$$

For $q < 0.9$, delay has a maximum near $\lambda = 1$ ($\omega = |p|$). For $q > 0.9$, delay decreases monotonically with increasing frequency, the transition (crossover) from $D(0)$ to $D(\infty)$ occurring largely in the vicinity of $\lambda = 1$ ($\omega = |p|$). Moving the poles radially away from (toward) the origin increases (decreases) the frequency corresponding to maximum delay (for $q < 0.9$) or to delay crossover (for $q > 0.9$). For $q < 0.9$, moving the pole p circumferentially clockwise (counterclockwise) increases (decreases) the maximum delay.

Example 5-34 In this example we use a cut-and-try method to design an equalized Butterworth filter meeting the following specifications:

Passband edge $f_1 = 1.0$ kHz
Stopband edge $f_2 = 2.0$ kHz
dc gain $H(0) = 1.0$

Minimum passband gain $\Gamma_1 = 0.80$
Maximum stopband gain $\Gamma_2 = 0.10$
Passband delay deviation $\Delta D = 0.05 D(0)$

Figure 5-27 illustrates these specifications. The delay specification illustrated by Fig. 5-27b means that delay in the passband must not deviate from $D(0)$ by more than 5 percent.

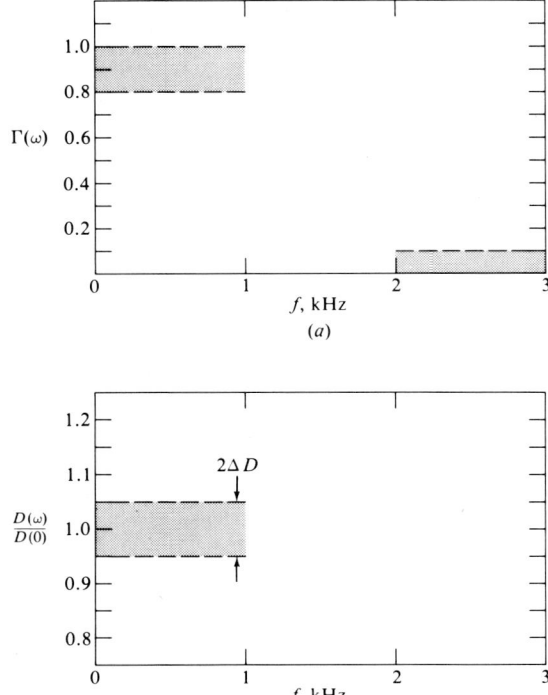

Figure 5-27 Specifications for the filter of Example 5-34.

First, we design a Butterworth filter from specifications on gain alone. The design procedure described in Sec. 4-5 yields a fourth-order Butterworth filter having poles

$$p_0, p_3 = -2.65 \pm 6.39j \text{ krad/s} \qquad p_1, p_2 = -6.39 \pm 2.65j \text{ krad/s}$$

The system function for the Butterworth filter is

$$H(s) = \frac{\omega_c^4}{(s^2 + as + \omega_c^2)(s^2 + bs + \omega_c^2)}$$

where $\omega_c = 6.92$ krad/s, $a = 5.30$ krad/s, and $b = 12.78$ krad/s. Figure 5-28 shows

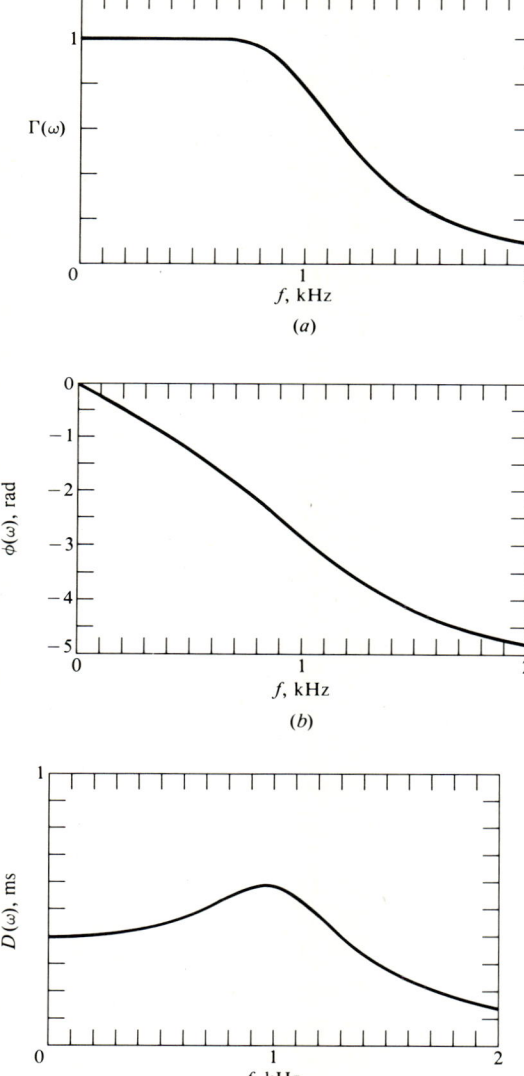

Figure 5-28 Graphs of gain, phase shift, and delay for the Butterworth filter of Example 5-34.

graphs of gain $\Gamma(\omega)$, phase shift $\phi(\omega)$, and delay $D(\omega)$ of the filter. The gain of the filter is given by

$$\Gamma(\omega) = \frac{1}{\sqrt{1 + (\omega/\omega_c)^8}}$$

the phase shift of the filter is given by

$$\phi(\omega) = -\sum_{k=0}^{3} \tan^{-1} \frac{\omega - \text{Im } p_k}{-\text{Re } p_k}$$

and the delay of the filter is given by

$$D(\omega) = -\sum_{k=0}^{3} \frac{\text{Re } p_k}{(\text{Re } p_k)^2 + (\omega - \text{Im } p_k)^2}$$

The bandwidth of the filter is $f_1 = 1$ kHz. In the passband, delay $D(\omega)$ ranges from $D(0) = 390$ μs to $D(2\pi f_1) = 580$ μs. The delay deviation of the filter is almost 50 percent of $D(0)$. Equalization is necessary to meet the specification $\Delta D \leq 5$ percent.

Figure 5-29 illustrates cut-and-try design of an all-pass delay equalizer. Initial choices for p and q in (5-33) are made by inspecting the graph of Fig. 5-26 and the phase shift of the filter. The next step is simply to try various values for p and q in (5-33). The specification on delay can be met using a single all-pass biquad in series with the Butterworth filter described above. For $|p| = 6.00$ s^{-1} and $q = 0.90$, overall delay in the passband ranges from $D(0) = 990$ μs to $D(2\pi f_1) = 940$ μs. The maximum deviation from $D(0)$ is less than 5 percent of $D(0)$, as specified. The system function of the equalizer has poles p_e, $p_e^* = -5.40 \pm 2.62j$ krad/s (and zeros $z_e = -p_e$, $z_e^* = -p_e^*$). The system function of the delay equalizer is

$$H(s) = \frac{s^2 - \gamma s + |p|^2}{s^2 + \gamma s + |p|^2}$$

where $\gamma = 10.80$ krad/s and $|p| = 6.00$ krad/s.

Figure 5-30 illustrates how the response of the filter is improved by delay equalization. Figure 5.30a shows the response of the Butterworth filter alone to a triangular pulse $x(t)$ having duration $\tau = 1$ ms. Since the bandwidth of the pulse is $W = 1/2\tau = 500$ Hz, the spectral density of the pulse is confined largely to the passband of the filter. However, delay in the passband deviates significantly (by about 50 percent) from $D(0)$, and gain in the passband deviates somewhat (by 20 percent) from $\Gamma(0)$. Consequently, the waveform of the output exhibits both delay distortion (overshoot, undershoot, and asymmetry) and amplitude distortion (rounding and smearing). Figure 5-30b shows the response of the equalized filter to the same pulse $x(t)$. The output exhibits as much amplitude distortion (rounding and smearing) as before because the delay equalizer has no effect on overall gain, but the output of the equalized filter exhibits less delay distortion than the output of the filter alone. The output is nearly symmetrical about its midpoint, and the overshoots and undershoots are much smaller than before.[†]

Figure 5-30 also illustrates the fact that $D(0)$ is (approximately) actual delay if $D(\omega)$ is approximately independent of frequency over the band occupied by an input. From Fig. 5-29, $D(0) \approx 400$ μs for the Butterworth filter alone, and $D(0) \approx 1$ ms for the equalized filter. In Fig. 5-30a the output lags behind the input by about 400 μs, and in

---

[†]Controlling overshoots and undershoots and maintaining symmetry is especially important in data-transmission systems, where successive pulses represent digits of binary numbers. In this application excessive overshoot or undershoot or excessive asymmetry causes interference between successive pulses, which can lead to a binary 1 being decoded as a binary 0 or vice versa.

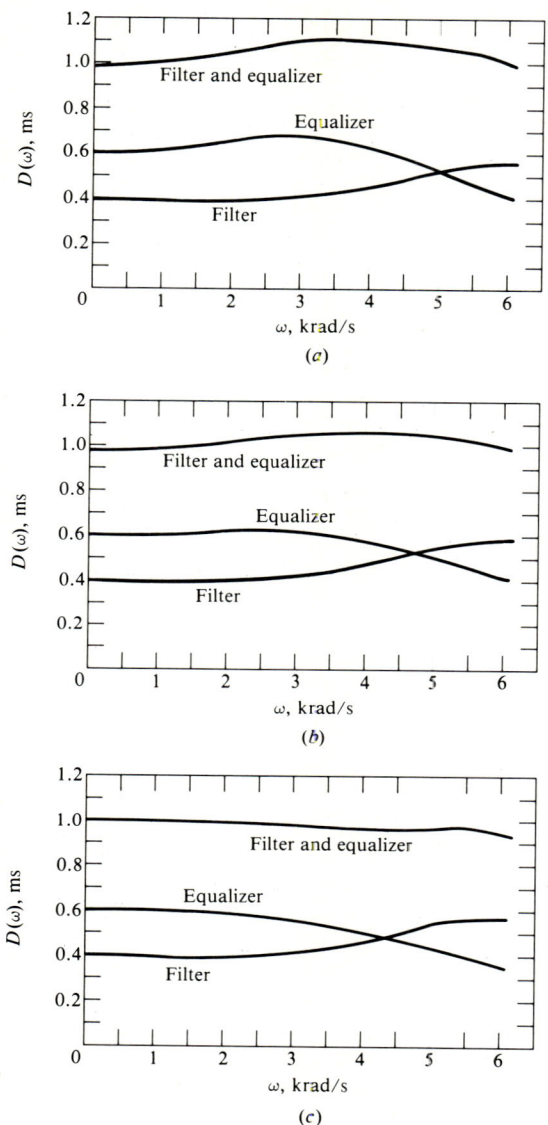

Figure 5-29 Cut-and-try design of an all-pass biquadratic delay equalizer [see (5-32) and (5-33)] for the Butterworth filter of Example 5-34: (a) first cut, where $|p| = 5.00$ krad/s and $q = 0.75$; (b) second cut, where $|p| = 5.50$ krad/s and $q = 0.80$; (c) third and final cut, where $|p| = 6.00$ krad/s and $q = 0.90$.

Fig. 5-30b the output lags behind the input by about 1 ms, in agreement with values obtained from Fig. 5-29. Finally, Fig. 5-30 illustrates the fact that equalization introduces considerable additional delay. The delay introduced by the equalized filter is about twice that introduced by the Butterworth filter alone.†

†The second-order equalizer can introduce more delay than the fourth-order filter because the equalizer is a non-minimum-phase system whereas the filter is a minimum-phase system.

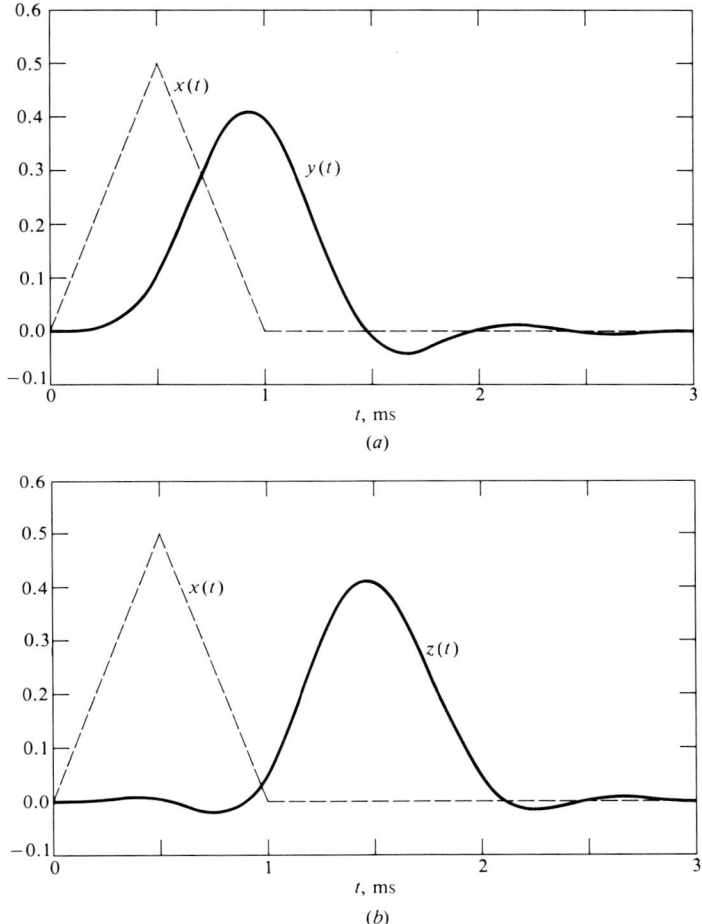

Figure 5-30 Improved response obtained through delay equalization: (*a*) response of the Butterworth filter alone to a triangular pulse; (*b*) response of the equalized filter to the same pulse.

We can decrease amplitude distortion introduced by the filter (for inputs confined to the passband) by using a higher-order Butterworth filter and additional delay equalization, but this would increase overall delay. In principle, we can make amplitude distortion and delay distortion (in the passband) arbitrarily small, but we pay for decreasing distortion by increasing overall delay. In any practical application there is a point where additional delay becomes intolerable or decreasing distortion further is not worth the additional cost. Also, increasing the order of a delay equalizer greatly increases the difficulty of determining the parameters p, q of each biquad. In practice, designing delay equalizers usually entails use of a computer.

5-3J Dominant-Pole Approximations

If one or two of the poles of a system largely determine the response of the system to any input of interest, the poles in question are called the *dominant poles* of the system.

Describing certain poles of a system as dominant implies that the system is adequately described by a system function having only those poles. We refer to such a system function as a *dominant-pole approximation*. Using a dominant-pole approximation to approximate a higher-order system function is the s-plane equivalent of using a low-order transfer function to approximate a higher-order transfer function (see Sec. 4-4E). In practice, the most important dominant-pole approximations have a single real pole or a single pair of complex poles.

Example 5-35 To illustrate dominant-pole approximation we consider a system whose system function (assumed exact) is

$$H(s) = \frac{p_1 p_2}{(s - p_1)(s - p_2)}$$

where p_1 and p_2 are real with $p_1 < 0$ and $p_2 = 10 \, p_1$, as shown by the pole-zero plot of Fig. 5-31.

The step response of the system is

$$g(t) = \mathcal{L}^{-1}\left\{\frac{H(s)}{s}\right\} = (1 + B_1 e^{p_1 t} + B_2 e^{p_2 t}) u(t)$$

where
$$B_1 = \frac{p_2}{p_1 - p_2} = -\frac{10}{9} \qquad B_2 = \frac{p_1}{p_2 - p_1} = \frac{1}{9}$$

The initial amplitude of the term arising from the pole p_1 is 10 times larger (in magnitude) than the initial amplitude of the term arising from the pole p_2. The time constant of the term arising from the pole p_1 is also 10 times larger than the time constant of the term arising from the pole p_2. This means that the term arising from the pole p_1 starts out 10 times larger and lasts 10 times longer than the term arising from the pole p_2. In other words, the pole p_1 dominates the transient component of the step response.

We can obtain a dominant-pole approximation $\hat{H}(s)$ for the system function $H(s)$ by dropping the weak pole p_2 and by forcing $\hat{H}(0) = H(0)$ (to make the dc gain of the dominant-pole approximation equal that of the actual system). This gives

$$\hat{H}(s) = \frac{-p_1}{s - p_1}$$

The step response of the dominant-pole approximation is

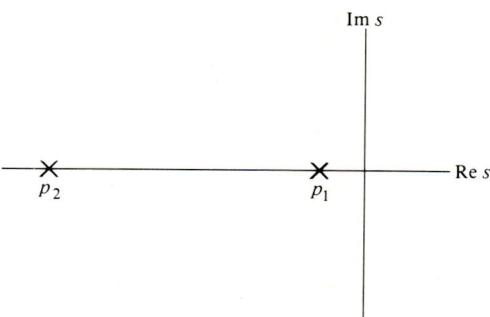

Figure 5-31 Pole-zero plot for the system of Example 5-35.

$$\hat{g}(t) = \mathcal{L}^{-1}\left\{\frac{\hat{H}(s)}{s}\right\} = (1 - e^{p_1 t})u(t)$$

Figure 5-32 shows the actual step response $g(t)$ and the dominant-pole step response $\hat{g}(t)$ for $0 \leq t \leq -2/p_1$. The approximation is quite good over this interval. For $t > -3/p_1$, the approximation is practically indistinguishable from the actual step response.

Figure 5-33 shows gain and phase shift of the actual system and the dominant-pole approximation. For $\omega \ll |p_2|$, the gain and phase shift of the dominant-pole approximation

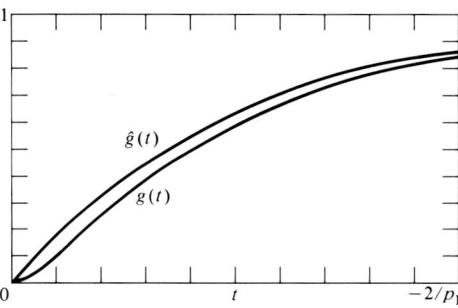

Figure 5-32 Actual step response $g(t)$ and approximate (dominant-pole) step response $\hat{g}(t)$ for the system of Example 5-35.

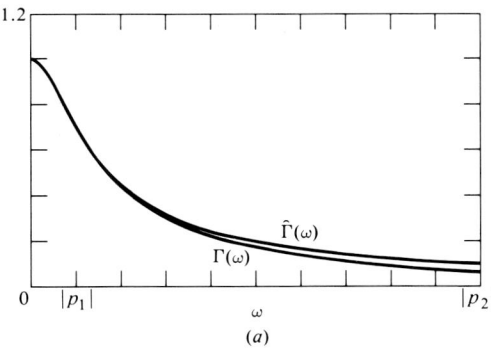

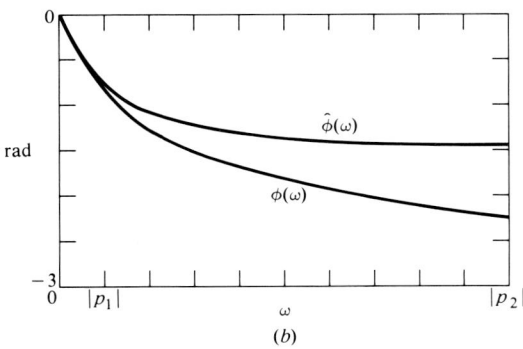

Figure 5-33 (a) Actual gain and phase shift and (b) approximate (dominant-pole) gain and phase shift for the system of Example 5-35.

is practically indistinguishable from those of the actual system. The dominant pole dominates the low-frequency response of the system.

The dominant poles of an all-pole system (a system having no finite zeros) are the poles nearest the origin in a pole-zero plot for the system. This is because both the initial amplitude and the time constant of a natural mode increase as the associated pole is moved toward the origin. Consequently, we can often identify the dominant poles of an all-pole system by inspecting a pole-zero plot for the system. For a system having one or more finite zeros, which poles (if any) dominate depends on the locations of the zeros as well as the locations of the poles because a nearby zero can weaken an otherwise strong pole. Thus, a weak pole (one far from the origin) can be made dominant by placing zeros near the stronger poles (those nearer the origin), and some strong poles can be made dominant by placing zeros near other strong poles.

Example 5-36 Consider the system function

$$H(s) = \frac{K(s - z)}{(s - p_1)(s - p_2)}$$

where $K = -p_1 p_2/z$ (for unit dc gain), $p_1 = -1 \text{ s}^{-1}$, $p_2 = -10 \text{ s}^{-1}$, and $z = -1.1 \text{ s}^{-1}$. Figure 5-34 shows a pole-zero plot for $H(s)$. Note that the zero z is very close to the pole p_1.

The step response of a system described by $H(s)$ above is

$$g(t) = (1 - 0.1e^{p_1 t} - 0.9e^{p_2 t})u(t)$$

The pole p_2 dominates the step response even though p_1 is much closer to the origin than p_2 because the pole p_1 is weakened by the nearby zero z.

To obtain a dominant-pole approximation $\hat{H}(s)$ for the system function $H(s)$ above we assume that the zero z cancels the pole p_1; thus

$$\hat{H}(s) = \frac{-p_2}{s - p_2}$$

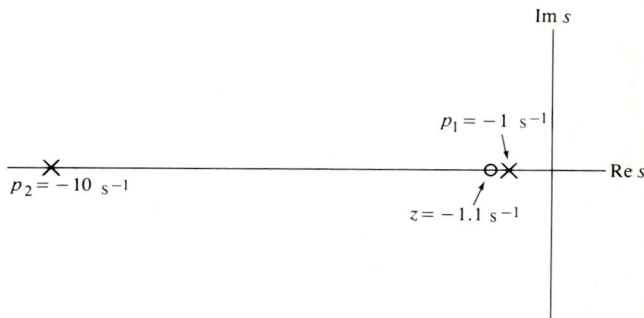

Figure 5-34 Pole-zero plot for the system of Example 5-36.

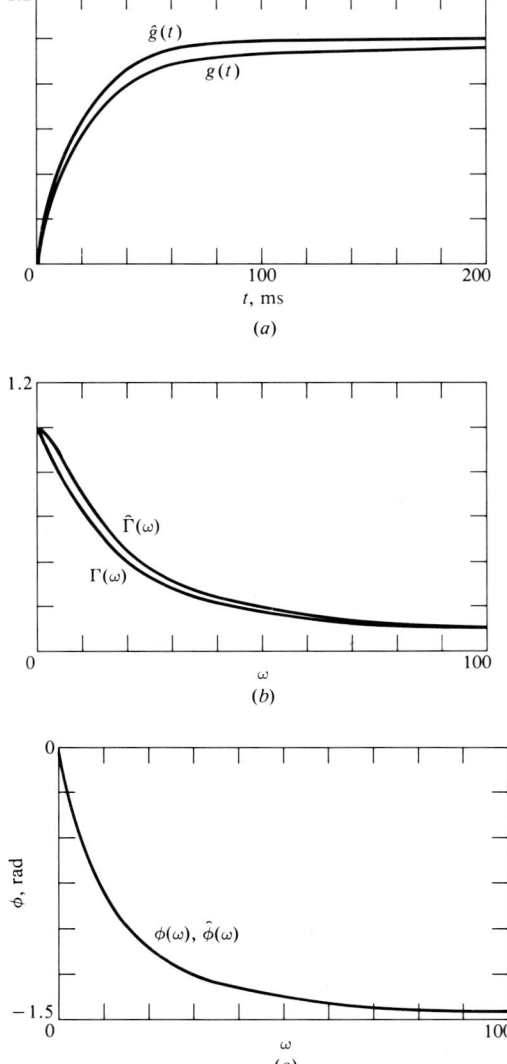

Figure 5-35 Actual step response $g(t)$ and approximate (dominant-pole) step response $\hat{g}(t)$ for the system of Example 5-36.

The step response of the dominant-pole approximation is

$$\hat{g}(t) = (1 - e^{p_2 t})u(t)$$

Figure 5-35a shows graphs of the actual step response $g(t)$ and the dominant-pole step response $\hat{g}(t)$ versus time. Figures 5-35b and c show graphs of actual and approximate (dominant-pole) gain and phase shift versus frequency. These graphs indicate that the dominant-pole system function is a reasonably good approximation to the actual system function. The dominant pole p_2 largely determines both rise time and bandwidth of the system.

Caution is called for in obtaining a dominant-pole approximation for a non-minimum-phase system. A right-half-plane zero near the mirror image of a left-half-plane pole strongly affects the strength of the pole, even though the zero may be far from the pole itself. For example, consider the system function

$$H(s) = \frac{p_2(s + p_1)}{(s - p_1)(s - p_2)}$$

where $p_1 = -1 \text{ s}^{-1}$ and $p_2 = -10 \text{ s}^{-1}$. The pole p_1 has no effect on gain $\Gamma(\omega) = |H(j\omega)|$ because

$$\frac{|j\omega + p_1|}{|j\omega - p_1|} = 1$$

for all frequencies. The bandwidth of the system is determined by the pole p_2, even though p_1 is much closer to the origin than p_2. On the other hand, the pole p_1 strongly influences the phase shift

$$\phi(\omega) = \angle H(j\omega) = 2 \text{ Tan}^{-1}(\omega, p_1) - \text{Tan}^{-1}(\omega, -p_2)$$

and the step response

$$g(t) = (1 - 2.22e^{p_1 t} + 1.22e^{p_2 t})u(t)$$

We cannot obtain a dominant-pole approximation for this system function because neither pole is dominant.

Dominant-pole approximations are useful because they simplify analysis and design, often allowing pencil-and-paper (or even mental) analysis where computer-aided analysis would otherwise be necessary.

5-3K Root-Locus Plots

A root-locus plot shows (on an s plane) paths followed by poles and zeros of a system function as some parameter of the system (usually a gain) is varied.

Example 5-37 Obtain a root-locus plot having parameter K for the system of Fig. 5-36, where

$$H_1(s) = \frac{s - z}{s - p_1} \quad \text{with } p_1 < 0$$

SOLUTION The system function of the system is

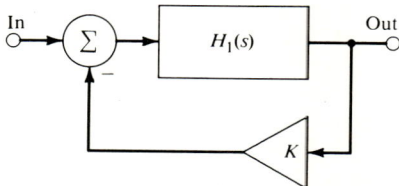

Figure 5-36 System of Example 5-37.

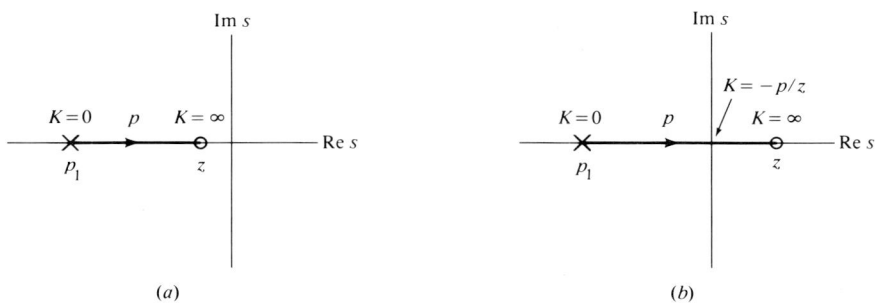

Figure 5-37 Root-locus plots (with parameter K) for the system of Example 5-37 where (a) $z < 0$ and (b) $z > 0$.

$$H(s) = \frac{K_1(s-z)}{s-p} \quad \text{where } K_1 = 1/(1+K)$$
$$p = (p_1 + Kz)/(1+K)$$

Figure 5-37a shows a root-locus plot for $z < 0$, where $H_1(s)$ is minimum phase. As K is increased from zero toward infinity, the pole p moves from $s = p_1$ along the σ axis toward $s = z$. For $z < 0$, the system is stable for $K > 0$ because the pole p lies in the left half plane for $K > 0$. Figure 5-37b shows a root-locus plot for $z > 0$, where $H_1(s)$ is nonminimum phase. As K is increased from zero toward infinity, the pole p moves from $s = p_1$ toward $s = z$ as before, but in this case the zero z lies in the right half plane and the system becomes unstable when the pole p crosses the ω axis. Since the pole p lies on the ω axis for $K = -p_1/z$, the system is unstable for $K > -p_1/z$.

Example 5-38 Obtain a root-locus plot having parameter ζ for the system function

$$H(s) = \frac{\omega_0^2}{s^2 + 2\zeta\omega_0 s + \omega_0^2} \quad \text{where } \zeta > 0$$
$$\omega_0 > 0$$

SOLUTION The poles of $H(s)$ are given by

$$p_1, p_2 = \begin{cases} \omega_0(-\zeta \pm \sqrt{\zeta^2 - 1}) & \zeta \geq 1 \\ \omega_0(-\zeta \pm j\sqrt{1 - \zeta^2}) & \zeta \leq 1 \end{cases}$$

In the discussion that follows we associate the positive sign with p_1.

Figure 5-38 shows a root-locus plot having parameter ζ for this system. For $\zeta = 0$, $p_1 = -p_2 = j\omega_0$; that is, both poles lie on the ω axis. As ζ is increased from zero to 1, the poles move to the σ axis on a circle having radius ω_0 (for $0 < \zeta < 1$, $|p_1| = |p_2| = \omega_0$), where p_1 moves counterclockwise and p_2 moves clockwise. For $\zeta = 1$, $p_1 = p_2 = -\omega_0$. As ζ is increased from zero toward infinity, p_1 moves to the right on the σ axis toward the origin and p_2 moves to the left on the σ axis toward $s = -\infty$. The root-locus plot shows that the system is stable for $\zeta > 0$ (if $\omega_0 > 0$). The root-locus plot also provides a graphical illustration of conditions for undamped response ($\zeta = 0$, imaginary poles), underdamped response ($0 < \zeta < 1$, complex poles), critically damped response ($\zeta = 1$, repeated real poles), and overdamped response ($\zeta > 1$, distinct real poles).

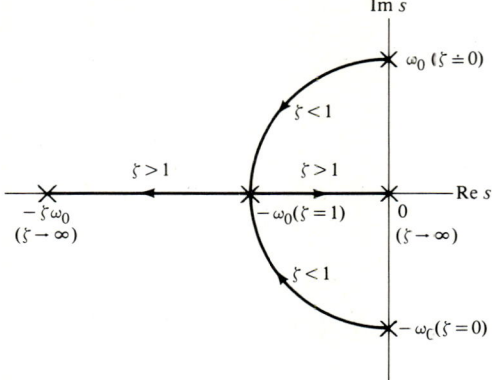

Figure 5-38 Root-locus plot with parameter ζ for the system of Example 5-38.

Root-locus plots are used most often in analysis and design of automatic control-systems. Most textbooks on that subject describe methods for sketching root-locus plots quickly and accurately (see D'Azzo and Houpis, chap. 7).

5-3L Servomechanisms

To conclude this section we apply concepts introduced above to analysis of the dc servomechanism of Fig. 5-39, where

$V_0 = 5.00$ V $K_T = 0.80$ Nm/A $K_1 = 6.28$ V/V

$J = 0.20$ Nm/(rad/s^2) $R = 0.50$ Ω $F = 1.00$ Nm/(rad/s)

$L = 10.00$ mH

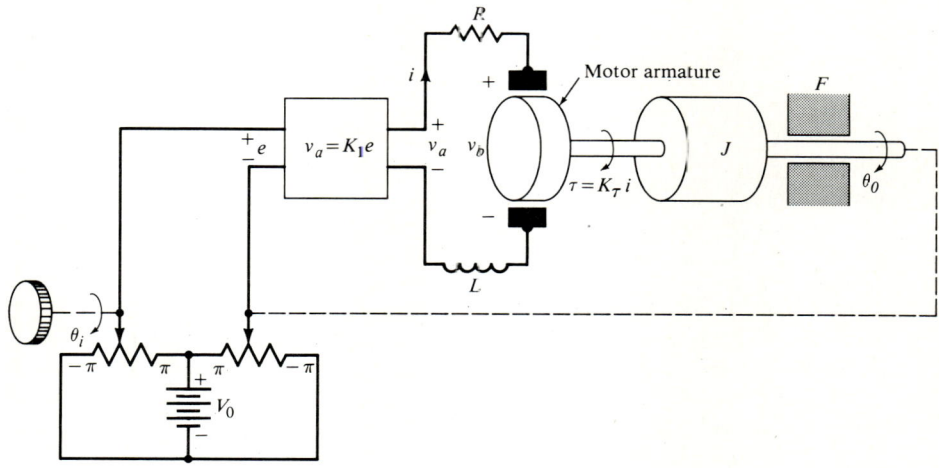

Figure 5-39 A dc servomechanism.

LAPLACE TRANSFORMATION 353

We seek a value for the amplifier gain K_1 for which the step response of the servomechanism exhibits only slight overshoot and little or no ringing (hunting).

To obtain the system function of the system we proceed as in Sec. 4-5. We first obtain the system function and a block diagram for each subsystem (error detector, amplifier, motor, and load). Then we connect these blocks together to obtain a block diagram for the system, and use block-diagram reduction to obtain the overall system function.

As shown in Sec. 4-5E, the transfer characteristic of the error-detecting circuit is

$$e(t) = K_e[\theta_i(t) - \theta_o(t)]$$

where $K_e = V_0/2\pi = 0.80$ V/rad. Taking the Laplace transform of this relation gives

$$E(s) = K_e[\Theta_i(s) - \Theta_o(s)]$$

Figure 5-40 shows a block diagram for the error-detecting circuit.

The transfer characteristic for the amplifier is

$$v_a(t) = K_1 e(t)$$

Taking the Laplace transform of this relation gives

$$V_a(s) = K_1 E(s)$$

Figure 5-41 shows a block diagram for the amplifier.

Figure 5-42a shows a schematic diagram for the armature-controlled dc servomotor. The torque produced by the motor is given by

$$\tau(t) = K_T i(t)$$

where $i(t)$ is armature current. The armature current $i(t)$ is related to applied voltage $v_a(t)$ and back emf v_b using Kirchhoff's voltage law; thus

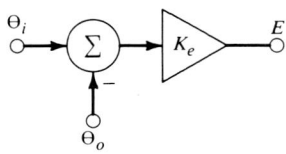

Figure 5-40 Block diagram for the error-detecting circuit.

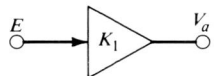

Figure 5-41 Block diagram for the amplifier.

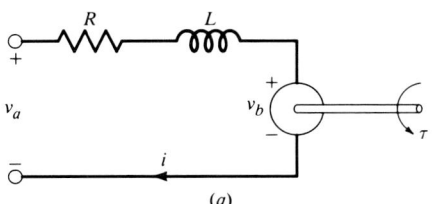

(a)

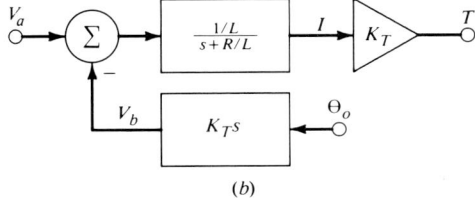

(b)

Figure 5-42 Armature-controlled dc servomotor: (a) schematic diagram; (b) block diagram.

$$L\frac{di}{dt} + Ri = v_a - v_b$$

where L is armature inductance and R is armature resistance. The back emf $v_b(t)$ is given by

$$v_b(t) = K_T \frac{d\theta_o}{dt}$$

(see Fitzgerald, Kingsley, and Uman, p. 203). Taking the Laplace transforms of the last three relations gives

$$T(s) = K_T I(s) \qquad I(s) = \frac{V_a(s) - V_b(s)}{sL + R} \qquad V_b(s) = K_T s \Theta_o(s)$$

where

$$T(s) = \mathcal{L}\{\tau(t)\} \qquad I(s) = \mathcal{L}\{i(t)\} \qquad V_a(s) = \mathcal{L}\{v_a(t)\} \qquad V_b(s) = \mathcal{L}\{v_b(t)\}$$

From these relations we obtain the block diagram of Fig. 5-42b.

Angular position of the load is related to applied torque using Newton's law for angular motion. This gives

$$J\frac{d^2\theta_o}{dt^2} + F\frac{d\theta_o}{dt} = \tau$$

For simplicity, we assume that J and F account for the load, the shaft, and the armature. Taking the Laplace transform of this relation and solving for $\Theta_o(s) = \mathcal{L}\{\theta_o(t)\}$ yields

$$\Theta_o(s) = \frac{T(s)}{s(sJ + F)}$$

Figure 5-43 shows a block diagram for the load.

We obtain a block diagram for the servomechanism by connecting the block diagrams obtained above, as shown in Fig. 5-44a. Then we obtain the transfer function of the servomechanism using block-diagram reduction. The result is

$$H(s) = \frac{c}{s^3 + as^2 + bs + c} \tag{5-34a}$$

where $\qquad a = \dfrac{R}{L} + \dfrac{F}{J} \qquad b = \dfrac{FR + K_T^2}{JL} \qquad c = \dfrac{K_T K_e K_1}{JL} \tag{5-34b}$

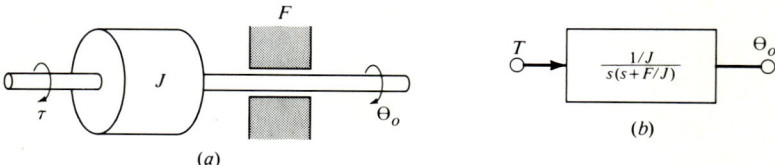

Figure 5-43 Block diagram for the load on the dc servomechanism of Fig. 5-39.

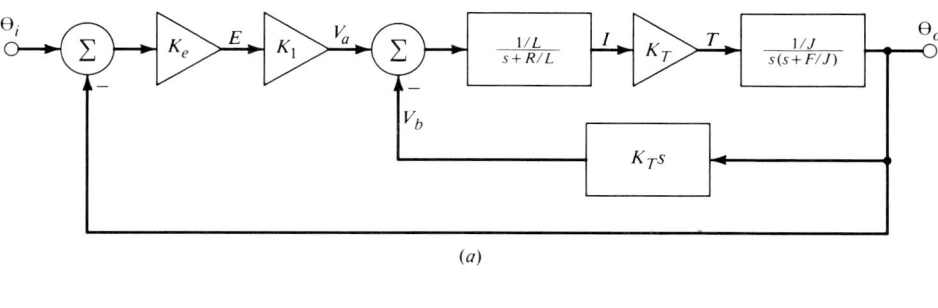

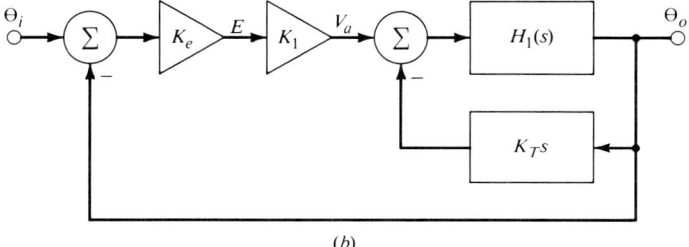

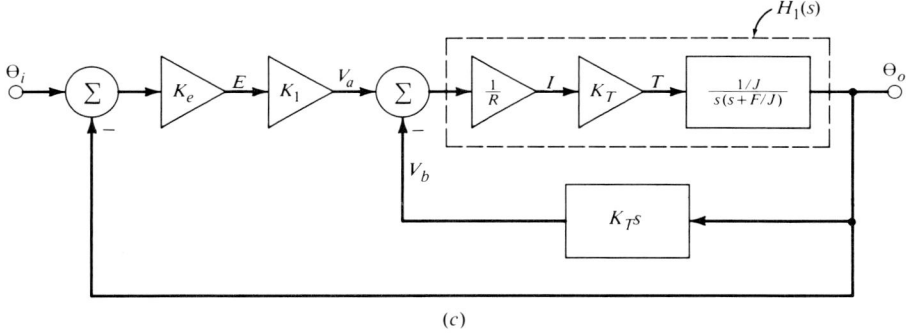

Figure 5-44 Block diagram for the dc servomechanism of Fig. 5-39.

The servomechanism of Fig. 5-44a is a third-order system; it can be simplified by a dominant-pole approximation. The system function for the forward path in the inner loop is

$$H_1(s) = \frac{K_T/JL}{s(s - p_L)(s - p_a)} = \frac{K_T/FR}{s(1 - s/p_L)(1 - s/p_a)}$$

where $p_L = -F/J = -5 \text{ s}^{-1}$ and $p_a = -R/L = -50 \text{ s}^{-1}$. Since the pole p_a is 10-times farther from the origin than the pole p_L, the pole p_L and the pole at the origin are the dominant poles of $H_1(s)$. Neglecting the weak pole p_a (without changing the low-frequency gain) gives the dominant-pole approximation

$$H_1(s) = \frac{K_T/FR}{s(1 - s/p_L)} = \frac{K_T/RJ}{s(s - p_L)}$$

Making this approximation is equivalent to ignoring armature inductance, which in turn is equivalent to assuming that the electrical time constant for the armature circuit is negligible relative to the mechanical time constant for the armature, the shaft, and the load. Thus the approximation assumes that speed of response, e.g., rise time of step response, is limited by mechanical inertia and not by the inductance of the armature winding.

Using the dominant-pole approximation, as illustrated in Figs. 5-44b and c gives the (approximate) overall system function

$$\hat{H}(s) = \frac{\omega_0^2}{s^2 - 2\zeta\omega_0 s + \omega_0^2}$$

where undamped natural frequency ω_0 and damping ratio ζ are given by

$$\omega_0^2 = \frac{K_1 K_e K_T}{RJ} \qquad \zeta = \frac{K_T^2 + FR}{2RJ\omega_0}$$

Choosing $\zeta = 0.71$ gives slight overshoot with no ringing (see Fig. 4-47). The corresponding undamped natural frequency is

$$\omega_0 = \frac{K_T^2 + FR}{2\zeta JR} = \frac{(0.80)^2 + (1.00)(0.50)}{(2)(0.71)(0.20)(0.50)} = 8.03 \text{ rad/s}$$

and the corresponding amplifier gain is

$$K_1 = \frac{JR\omega_0^2}{K_e K_T} = \frac{(0.20)(0.50)(8.06)^2}{(0.80)(0.80)} = 10.2 \text{ V/V}$$

Figure 5-45 shows the approximate (dominant-pole) step response $\hat{g}(t)$ and the actual step response $g(t)$ for the original third-order system. The dominant-pole approximation is quite good: it only slightly underestimates overshoot and rise time of the step response.

The dominant-pole approximation used above works well because the pole $p_a = -R/L$ is much farther from the origin than the pole $p_L = -F/J$. The quality of the approximation depends greatly on the dominance of the poles at $s = 0$ and $s = -F/J$. In less clear-cut cases, dominant-pole approximations should be used with caution.

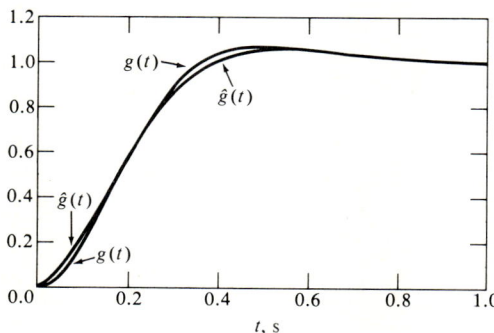

Figure 5-45 Actual step response $g(t)$ and approximate (dominant-pole) step response $\hat{g}(t)$ for the dc servomechanisms of Figs. 5-39 and 5-44.

A danger of using a dominant-pole approximation for system design is that the approximation can conceal a cause of unstable operation. For example, the weak pole p_a causes the actual (third-order) system considered above to be unstable for large amplifier gain. This is shown by the root-locus plot of Fig. 5-46a, where the complex poles of the third-order system lie in the right half plane for K_1 larger than about 100. On the other hand, the second-order dominant-pole approximation is stable for all values of K_1, as shown by the root-locus plot of Fig. 5-46b. Fortunately, in most practical applications, unstable operation concealed by a dominant-pole approximation is inconsequential if the approximation indicates that the system performs satisfactorily. For example, the dominant-pole approximation used above indicates that choosing $K_1 \approx 10$ gives a satisfactory step response. This gain is much smaller than the value $K_1 \approx 100$ for which the actual system is unstable.

Heretofore, we have described performance of servomechanisms in terms of step response alone, the idea being that a servomechanism should respond quickly to a sudden change of desired position and subsequently should maintain the new position without error. Many servomechanisms must be able to follow an input (desired output) that changes continually. For example, a servomechanism driving an automatic fire-control system must follow the trajectory of a target being tracked. Thus, in practice, responses of a servomechanism to other inputs (in addition to a step input) often are of interest.

An input $\theta_i(t)$ to a servomechanism can be expressed using Maclaurin's series as

$$\theta_i(t) = (d_0 + d_1 t + d_2 t^2 + \cdots)u(t)$$

Step response of a servomechanism describes the ability of the servomechanism to follow the first term $d_0 u(t)$. Responses to a ramp input $d_1 t u(t)$, a parabolic input $d_2 t^2 u(t)$, and so on, describe the ability of the servomechanism to follow a continually

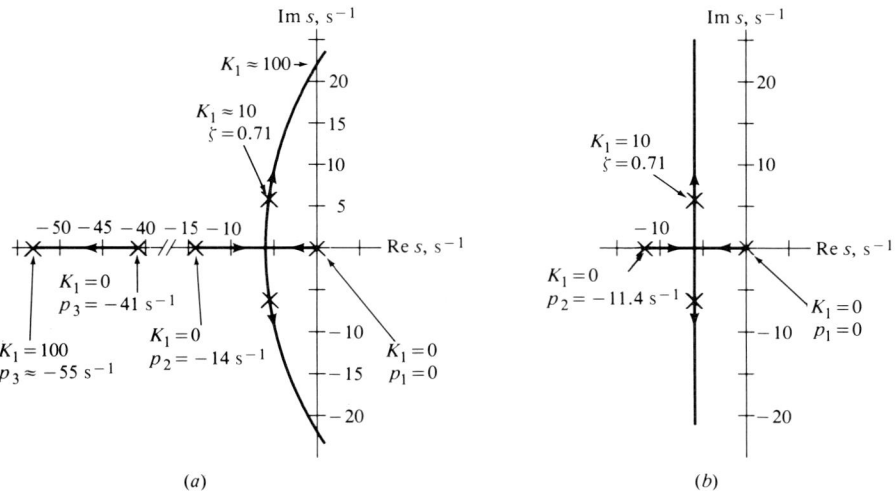

Figure 5-46 Root-locus plots with parameter K_1 for the servomechanism of Fig. 5-39: (a) actual (third-order) system; (b) dominant-pole (second-order) approximation.

changing input. Consequently, a servomechanism designer often is interested in *ramp response*, defined as the output for input $d_1 t u(t)$ with $d_1 = 1$ rad/s, and (less often) in responses to higher-order inputs (parabolic, cubic, etc.).

The *unit ramp response* $k(t)$ of a system having system function $H(s)$ is given by $k(t) = \mathcal{L}^{-1}\{H(s)/s^2\}$. For the servomechanism considered above, this gives [see (5-34)]

$$k(t) = \mathcal{L}^{-1}\left\{\frac{c}{s^2(s^3 + as^2 + bs + c)}\right\}$$

where $a = 55$ s^{-1}, $b = 570$ s^{-2}, and $c = 3200$ s^{-3}. The roots of the cubic factor in the denominator can be found numerically. The results are $p_0 = -43.60$ s^{-1} and $p_1, p_1^* = -5.70 \pm 6.40j$ s^{-1}. The transform above can be expressed in factored form as

$$K(s) = \frac{c}{s^2(s - p_0)(s^2 + 2\zeta\omega_0 s + \omega_0^2)}$$

where $\omega_0 = |p_1|^2 = 8.55$ s^{-1} and $\zeta = -(\mathrm{Re}\, p_1)/\omega_0 = 0.67$. Expanding this function in partial fractions gives

$$K(s) = \frac{B_0}{s^2} + \frac{B_1}{s} + \frac{B_2}{s - p_0} + \frac{B_3 s + B_4}{s^2 + 2\zeta\omega_0 s + \omega_0^2}$$

where $B_0 = 1$, $B_1 = -0.18$ s, $B_2 = 10^{-3}$ s, $B_3 = 0.18$ s, and $B_4 = 1.08$. Taking the inverse Laplace transform gives

$$k(t) = [B_0 t + B_1 + B_2 e^{p_0 t} - r e^{-\zeta \omega_0 t} \cos(\omega_1 - \theta)] u(t)$$

In practice, a quantity of interest is the error

$$e(t) = \omega_i[t u(t) - k(t)]$$

which indicates how well a servomechanism is able to follow a ramp input $\theta_i(t) = \omega_i t u(t)$. If the magnitude of $e(t)$ increases with time, the servomechanism cannot follow a ramp input. Figure 5-48 shows a graph of $e(t)$ versus time for $\omega_i = 1$ rad/s. The error has steady-state value $e_{ss} = 0.18$ rad. This indicates that the

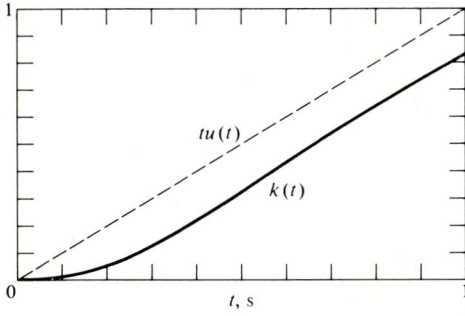

Figure 5-47 Unit ramp response of the servomechanism of Fig. 5-39.

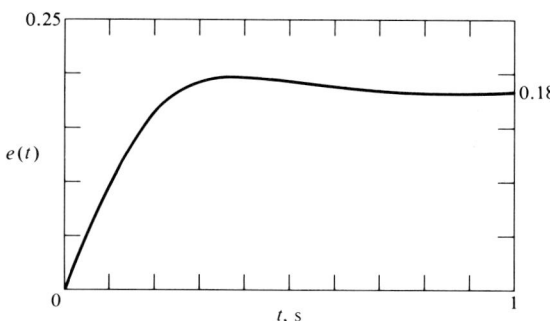

Figure 5-48 Instantaneous error (for unit ramp input) in the dc servomechanism of Fig. 5-39.

servomechanism can follow a ramp input, though the actual output (in steady state) lags the input by 18 percent.

5-4 ONE-SIDED LAPLACE TRANSFORMATION

The Laplace transformation described above is called the *two-sided Laplace transformation* because the limits of integration in (5-1) require that the function $e(t)$ be defined for $-\infty < t < \infty$. In this section we describe the one-sided Laplace transformation, which can be applied to functions defined only for $t \geq 0$. The one-sided Laplace transformation is especially useful for solving initial-value problems, where the condition (or state) of a system is specified at some initial time (usually, $t = 0$), an input is specified from that time on, and the output is to be determined from that time on.

5-4A Definition and Properties

The *one-sided Laplace transform of a function* $e(t)$ is denoted by $\tilde{E}(s)$ and defined by

$$\tilde{E}(s) = \int_0^\infty e(t) e^{-st}\, dt \tag{5-35}$$

In this book we use the operator $\mathcal{L}_1$ to denote one-sided Laplace transformation; thus

$$\tilde{E}(s) = \mathcal{L}_1\{e(t)\} \tag{5-36}$$

is shorthand for (5-35).

Example 5-39 The one-sided Laplace transform of the delta function $\delta(t)$ is

$$\tilde{\Delta}(s) = \mathcal{L}_1\{\delta(t)\} = \int_0^\infty \delta(t) e^{-st}\, dt = \int_0^\infty \delta(t) e^0\, dt = \int_0^\infty du(t) = u(\infty) - u(0) = 1$$

It is often claimed that using the one-sided Laplace transformation is equivalent to assuming that signals of interest are zero for $t < 0$. That is incorrect. The one-sided Laplace transformation assumes nothing whatever about $e(t)$ for $t < 0$. The definition

(5-35) simply ignores values of $e(t)$ for $t < 0$, so that any two (or more) signals that are identical for $t \geq 0$ have the same one-sided Laplace transform. This means that the one-sided Laplace transformation is a many-to-one transformation and that an inverse one-sided Laplace transform is defined uniquely only for $t \geq 0$. Thus, if $e(t)$ is a function having one-sided Laplace transform $\tilde{E}(s)$, we write

$$\mathcal{L}_1^{-1}\{\tilde{E}(s)\} = e(t) \qquad t \geq 0 \qquad (5\text{-}37)$$

where $\mathcal{L}_1^{-1}$ denotes inverse one-sided Laplace transformation. The function $e(t)$ thus obtained is undefined for $t < 0$. It is incorrect to write $\mathcal{L}_1^{-1}\{\tilde{E}(s)\} = e(t)u(t)$ because this implies that the inverse one-sided Laplace transform is defined (is equal to zero) for $t < 0$.

The *region of convergence* for a one-sided Laplace transform $\mathcal{L}_1\{e(t)\}$ comprises all values of s for which the integral on the right side of (5-35) exists.

Example 5-40 Obtain the one-sided Laplace transform and associated region of convergence for the function

$$e(t) = e^{pt}$$

SOLUTION From the definition (5-35),

$$\tilde{E}(s) = \int_0^\infty e^{pt} e^{-st}\, dt = \int_0^\infty e^{-(s-p)t}\, dt = \frac{1}{s-p}(1 - \lim_{t \to \infty} e^{-(s-p)t})$$

The limit exists only for Re $s >$ Re p, in which case it equals zero; thus

$$\tilde{E}(s) = \frac{1}{s-p}$$

The associated region of convergence is Re $s >$ Re p.

The region of convergence for a one-sided Laplace transform is of less interest than that for a two-sided Laplace transform because inverse one-sided Laplace transforms are defined only for $t \geq 0$; that is, choosing between left-sided, two-sided, and right-sided inverse transforms is not an issue. In applications considered in this book we can ignore regions of convergence for inverse one-sided Laplace transforms.

5-4B Linearity

One-sided Laplace transformation and inverse one-sided Laplace transformation, like their two-sided counterparts, are linear operations. This means that

$$\mathcal{L}_1\{a_1 e_1(t) + a_2 e_2(t) + \cdots\} = a_1 \mathcal{L}_1\{e_1(t)\} + a_2 \mathcal{L}_1\{e_2(t)\} + \cdots \qquad (5\text{-}38)$$

and

$$\mathcal{L}_1^{-1}\{a_1 \tilde{E}_1(s) + a_2 \tilde{E}(s) + \cdots\} = a_1 \mathcal{L}_1^{-1}\{\tilde{E}_1(s)\} + a_2 \mathcal{L}_1^{-1}\{\tilde{E}_2(s)\} + \cdots$$
$$(5\text{-}39)$$

Example 5-41 Obtain the inverse one-sided Laplace transform of

$$\tilde{E}(s) = 4 + \frac{6p}{s-p}$$

SOLUTION From (5-39),

$$\mathscr{L}_1^{-1}\{\tilde{E}(s)\} = 4\mathscr{L}_1^{-1}\{1\} + 6p\mathscr{L}_1^{-1}\left\{\frac{1}{s-p}\right\}$$

Using the results of Examples 5-39 and 5-40 gives

$$e(t) = 4\delta(t) + 6pe^{pt} \qquad t \geq 0$$

5-4C Relation to the Two-Sided Laplace Transform

Comparing the definition (5-35) of the one-sided Laplace transform with the definition (5-1) of the two-sided Laplace transform shows that

$$\mathscr{L}_1\{e(t)\} = \mathscr{L}\{e(t)u(t)\} \qquad (5\text{-}40a)$$

provided $e(t)$ contains no delta function at $t = 0$, that is, contains no delta function of the form $\mathbf{a}\delta(t)$. If $e(t)$ contains a delta function $\mathbf{a}\delta(t)$, then

$$\mathscr{L}_1\{e(t)\} = \mathscr{L}\{e(t)u(t)\} + \mathbf{a} \qquad (5\text{-}40b)$$

Equation (5-40) and Table D-1 can be used to obtain one-sided Laplace transforms.

Example 5-42 Obtain the one-sided Laplace transform of

$$e(t) = 2\cos\omega_0 t + 3\tau\delta(t - \tau) \qquad \text{where } \tau > 0$$

SOLUTION Since the function $e(t)$ contains no delta function at $t = 0$, we can use (5-40a). This gives

$$\mathscr{L}_1\{e(t)\} = 2\mathscr{L}_1\{\cos\omega_0 t\} + 3\tau\mathscr{L}_1\{\delta(t - \tau)\}$$
$$= 2\mathscr{L}\{\cos\omega_0 t)u(t)\} + 3\tau\mathscr{L}\{\delta(t - \tau)u(t)\}$$
$$= \frac{2s}{s^2 + \omega_0^2} + 3\tau e^{-s\tau}$$

Example 5-43 Obtain the one-sided Laplace transform of

$$e(t) = 5\delta(t) + pe^{pt}$$

SOLUTION We use (5-40b) because $e(t)$ contains a delta function at $t = 0$; thus

$$\mathscr{L}_1\{e(t)\} = 5 + \mathscr{L}\{pe^{pt}u(t)\} = 5 + \frac{p}{s-p}$$

The definition (5-35) of the one-sided Laplace transform shows that if $e(t) = 0$ for $t < 0$, then $\mathscr{L}_1\{e(t)\} = \mathscr{L}\{e(t)\}$ or $\tilde{E}(s) = E(s)$. It follows that we can use Table D-1 to obtain the inverse one-sided Laplace transform of any function $\tilde{E}(s)$ for which

$$\mathscr{L}^{-1}\{\tilde{E}(s)\} = 0 \qquad t < 0$$

i.e., if the inverse *two-sided* Laplace transform of a *one-sided* Laplace transform $\tilde{E}(s)$ equals zero for $t < 0$, then

$$\mathcal{L}_1^{-1}\{\tilde{E}(s)\} = \mathcal{L}^{-1}\{\tilde{E}(s)\} \qquad t \geq 0$$

Example 5-44 Obtain the inverse one-sided Laplace transform of

$$\tilde{E}(s) = \frac{p}{s(s-p)}$$

SOLUTION From Table D-1, the inverse two-sided Laplace transform of the function $\tilde{E}(s)$ above is

$$\mathcal{L}^{-1}\left\{\frac{p}{s(s-p)}\right\} = (-1 + e^{pt})u(t)$$

This function equals zero for $t < 0$. It follows that

$$\mathcal{L}_1^{-1}\left\{\frac{p}{s(s-p)}\right\} = -1 + e^{pt} \qquad t \geq 0$$

Example 5-45 Obtain the inverse one-sided Laplace transform of

$$\tilde{E}(s) = \frac{1}{s} e^{-st_0}$$

SOLUTION From Table D-1,

$$\mathcal{L}^{-1}\{\tilde{E}(s)\} = u(t - t_0)$$

If $t_0 > 0$, then $u(t - t_0)$ equals zero for $t < 0$ and the inverse one-sided Laplace transform of $\tilde{E}(s)$ equals the inverse two-sided Laplace transform of $\tilde{E}(s)$ for $t > 0$; thus, for $t_0 > 0$,

$$\mathcal{L}_1^{-1}\{\tilde{E}(s)\} = u(t - t_0) \qquad t \geq 0$$

If $t_0 < 0$, then $u(t - t_0)$ is not zero for (all) $t < 0$. In this case, we cannot obtain the inverse one-sided Laplace transform of $\tilde{E}(s)$ from Table D-1. In fact, if $t_0 < 0$, the function $\tilde{E}(s)$ above is not a valid one-sided Laplace transform.

5-4D Operational Properties

Several operational properties of the one-sided Laplace transformation are different from corresponding operational properties of the two-sided Laplace transformation. The differences account for the usefulness of the one-sided Laplace transformation for solving initial-value problems.

For problems considered in this book we need only two operational properties of the one-sided Laplace transformation: the *differentiation property*

$$\mathcal{L}_1\left\{\frac{de(t)}{dt}\right\} = s\mathcal{L}\{e(t)\} - e(0) \tag{5-41}$$

and the *integration property*

$$\mathcal{L}_1\left\{\int_{-\infty}^{t} e(\alpha)\, d\alpha\right\} = \frac{1}{s}\mathcal{L}_1\{e(t)\} + \frac{1}{s}\int_{-\infty}^{0} e(\alpha)\, d\alpha \qquad (5\text{-}42)$$

The differentiation property expressed by (5-41) assumes that $e(t)$ contains no delta function and no factor $u(t)$. The integration property expressed by (5-42) assumes that the integral on the left exists for all t. For example, (5-42) is inapplicable to $e(t) = e_0$ (a constant).

Equation (5-41) can be derived as follows. Consider

$$\frac{d}{dt}[e(t)u(t)] = \frac{de(t)}{dt}u(t) + e(t)\frac{du(t)}{dt} = \frac{de(t)}{dt}u(t) + e(0)\delta(t)$$

Taking the *two-sided* Laplace transform of this relation yields

$$\mathcal{L}\left\{\frac{d}{dt}[e(t)u(t)]\right\} = \mathcal{L}\left\{\frac{de(t)}{dt}u(t)\right\} + e(0)\mathcal{L}\{\delta(t)\}$$

or

$$s\mathcal{L}\{e(t)u(t)\} = \mathcal{L}\left\{\frac{de(t)}{dt}u(t)\right\} + e(0)$$

But, from (5-40a),

$$\mathcal{L}\{e(t)u(t)\} = \mathcal{L}_1\{e(t)\} \qquad \mathcal{L}\left\{\frac{de(t)}{dt}u(t)\right\} = \mathcal{L}_1\left\{\frac{de(t)}{dt}\right\}$$

provided $e(t)$ contains no delta function at $t = 0$ and $e(t)$ contains no factor $u(t)$. Combining the last three relations gives

$$s\mathcal{L}_1\{e(t)\} = \mathcal{L}_1\left\{\frac{de(t)}{dt}\right\} + e(0)$$

Equation (5-41) follows.

Equation (5-42) can be derived from (5-41) as follows. Let

$$\int_{-\infty}^{t} e(\alpha)\, d\alpha = f(t)$$

then

$$\frac{df(t)}{dt} = e(t)$$

From (5-41),

$$\mathcal{L}_1\left\{\frac{df(t)}{dt}\right\} = s\mathcal{L}_1\{f(t)\} - f(0)$$

or

$$\mathcal{L}_1\{e(t)\} = s\mathcal{L}_1\left\{\int_{-\infty}^{t} e(\alpha)\, d\alpha\right\} - \int_{-\infty}^{0} e(\alpha)\, d\alpha$$

Equation (5-42) follows.

364 INTRODUCTION TO SYSTEM ANALYSIS

Formulas for one-sided Laplace transforms of derivatives having order greater than 1 can be obtained by repeated application of (5-41); e.g.,

$$\mathcal{L}_1\left\{\frac{d^2 e(t)}{dt^2}\right\} = \mathcal{L}_1\left\{\frac{d}{dt}\left[\frac{de(t)}{dt}\right]\right\} = s\mathcal{L}_1\left\{\frac{de(t)}{dt}\right\} - \left.\frac{de(t)}{dt}\right|_{t=0}$$

$$= s[s\mathcal{L}_1\{e(t)\} - e(0)] - \left.\frac{de(t)}{dt}\right|_{t=0}$$

$$= s^2\mathcal{L}_1\{e(t)\} - se(0) - \left.\frac{de(t)}{dt}\right|_{t=0}$$

In general,

$$\mathcal{L}_1\left\{\frac{d^n e(t)}{dt^n}\right\} = s^n \mathcal{L}_1\{e(t)\} - s^{n-1} e(0) - s^{n-2} \left.\frac{de(t)}{dt}\right|_{t=0} - \cdots$$

$$- s \left.\frac{d^{n-2} e(t)}{dt^{n-2}}\right|_{t=0} - \left.\frac{d^{n-1} e(t)}{dt^{n-1}}\right|_{t=0} \quad (5\text{-}43)$$

Equation (5-43) assumes that $e(t)$ and its first $n - 1$ derivatives are continuous at $t = 0$.

5-4E Application to Initial-Value Problems

The fact that initial values appear explicitly in (5-41) to (5-43) accounts largely for the usefulness of the one-sided Laplace transformation in solving initial-value problems, where a system having some initial state is subjected to an input defined only for $t \geq 0$.

Example 5-46 In the circuit of Fig. 5-49 the voltage across the capacitor at $t = 0$ is $y(0) = y_0$, and the input (voltage) $x(t)$ is given for $t \geq 0$ by $x(t) = x_0$. Find the output (voltage) $y(t)$ for $t \geq 0$.

SOLUTION Writing Kirchhoff's current law for the node joining the resistor to the capacitor gives

$$C\frac{dy}{dt} + \frac{1}{R}(y - x) = 0$$

which can be written

$$\frac{dy}{dt} + \omega_0 y = \omega_0 x$$

where $\omega_0 = 1/RC$. Taking the one-sided Laplace transform of this equation gives

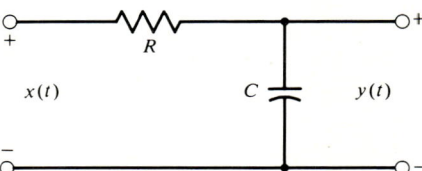

Figure 5-49 Circuit of Example 5-45.

$$s\mathcal{L}_1\{y(t)\} - y(0) + \omega_0\mathcal{L}_1\{y(t)\} = \omega_0\mathcal{L}_1\{x(t)\}$$

where we have used (5-38) and (5-41). This can be written

$$s\tilde{Y}(s) - y_0 + \omega_0\tilde{Y}(s) = \omega_0\tilde{X}(s)$$

where
$$\tilde{Y}(s) = \mathcal{L}_1\{y(t)\} \qquad \tilde{X}(s) = \mathcal{L}_1\{x(t)\} = \frac{x_0}{s}$$

Solving for $\tilde{Y}(s)$ yields

$$\tilde{Y}(s) = \frac{\omega_0\tilde{X}(s) + y_0}{s + \omega_0} = \frac{\omega_0 x_0}{s(s + \omega_0)} + \frac{y_0}{s + \omega_0}$$

Taking the inverse one-sided Laplace transform gives

$$y(t) = x_0 - (x_0 - y_0)e^{-\omega_0 t} \qquad t \geq 0$$

where we have used (5-40a) and Table D-1.

In practice, initial-value problems usually arise from treating nonlinear or nonstationary systems as piecewise-linear or piecewise-stationary systems, as illustrated in the following examples.

Example 5-47 In the circuit of Fig. 5-50, $R_1 = R_2 = 1$ kΩ, $C = 1$ μF, and $x(t) = V_0 \cos \omega_0 t$ with $V_0 = 5$ V and $\omega_0 = 1$ krad/s. The switch is open for $t < 0$ and closed for $t \geq 0$. Find the output (voltage) $y(t)$ for $-\infty < t < \infty$.

SOLUTION Writing Kirchhoff's voltage law for the node joining the resistors to the capacitor gives

$$\frac{dy}{dt} + \omega_1 y = \omega_1 V_0 \cos \omega_0 t \qquad t < 0$$

and
$$\frac{dy}{dt} + \omega_2 y = \omega_2 V_0 \cos \omega_0 t \qquad t \geq 0$$

where
$$\omega_1 = \frac{1}{(R_1 + R_2)C} = 500 \text{ s}^{-1} \qquad \omega_2 = \frac{1}{R_2 C} = 1000 \text{ s}^{-1}$$

Each differential equation above describes a linear stationary system. Together, they describe the circuit of Fig. 5-50 for all time. The circuit of Fig. 5-50 is a linear piecewise-stationary system.

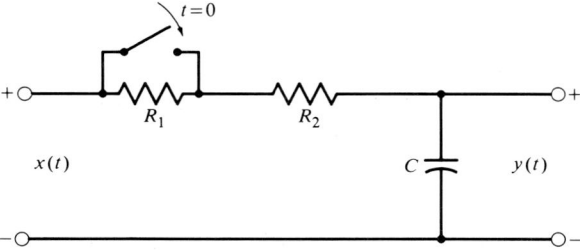

Figure 5-50 Circuit of Example 5-46.

We can obtain the output $y(t)$ for $t < 0$ by solving the first of the two differential equations above as if it were valid for all time because the output $y(t)$ cannot anticipate the closing of the switch (the output is causal). We use methods described in Chap. 3. The transfer function of the system (for $t < 0$) is

$$H(j\omega) = \frac{\omega_1}{j\omega + \omega_1} = \frac{1}{1 + j\omega/\omega_1}$$

For $\omega_1 = \omega_0$, the gain and phase shift are

$$\Gamma(\omega_0) = |H(j\omega_0)| = 0.89 \text{ V/V} \qquad \phi(\omega) = \angle H(j\omega_0) = -0.46 \text{ rad}$$

The output for $t < 0$ is

$$y(t) = \Gamma(\omega_0)x_0 \cos[\omega_0 t + \phi(\omega_0)] = 4.47 \cos(\omega_0 t - 0.46) \qquad \text{V}$$

To obtain the output for $t \geq 0$ we must solve the second differential equation above, which holds only for $t \geq 0$. Taking the one-sided Laplace transform of the equation yields

$$s\tilde{Y}(s) - y(0) + \omega_2 \tilde{Y}(s) = \omega_2 V_0 \mathcal{L}_1\{\cos \omega_0 t\}$$

From (5-40a) and Table D-1,

$$\mathcal{L}_1\{\cos \omega_0 t\} = \mathcal{L}\{(\cos \omega_0 t)u(t)\} = \frac{s}{s^2 + \omega_0^2}$$

Thus

$$\tilde{Y}(s) = \frac{\omega_2 V_0 s}{(s + \omega_2)(s^2 + \omega_0^2)} + \frac{y(0)}{s + \omega_2}$$

Expanding the first term on the right in partial fractions gives

$$\tilde{Y}(s) = \frac{V_0(s + \omega_2)}{2(s^2 + \omega_0^2)} + \frac{2y(0) - V_0}{2(s + \omega_2)}$$

From Table D-1,

$$y(t) = \mathcal{L}_1^{-1}\{\tilde{Y}(s)\} = 0.71 V_0 \cos(\omega_0 t - 0.79) + [y(0) - 0.50 V_0]e^{-\omega_2 t} \qquad t \geq 0$$

The initial value $y(0)$ is obtained from the solution for $t < 0$ given above using the fact that the voltage across a capacitor is continuous; thus

$$y(0) = y(0^-) = 4.47 \cos(-0.46) = 4.00 \text{ V}$$

and $\qquad y(t) = 3.54 \cos(\omega_0 t - 0.79) + 1.50 e^{-\omega_2 t} \qquad \text{V} \qquad t \geq 0$

Collecting results obtained above, we have

$$y(t) = \begin{cases} 4.47 \cos(\omega_0 t - 0.46) & \text{V} & t < 0 \\ 3.54 \cos(\omega_0 t - 0.79) + 1.50 e^{-\omega_2 t} & \text{V} & t \geq 0 \end{cases}$$

where $\omega_0 = \omega_2 = 1$ krad/s. Figure 5.51 shows a graph of the output $y(t)$.

Example 5-48 In this example we obtain the output $\theta_o(t)$ (rad) of the relay servomechanism of Fig. 5-52 for input $\theta_i(t) = \pi u(t)$ rad, where $E_0 = 12.50$ V, $K = 0.50$ A/V, $K_m = 0.10$ Nm/V (we assume that the armature circuit of the motor is purely resistive), $J = 0.10$ Nm/(rad/s²), and $F = 0.50$ Nm/(rad/s). Figure 5-53 shows the transfer characteristic of the relay, where $i_0 = 150$ mA and $v_0 = 12.5$ V.

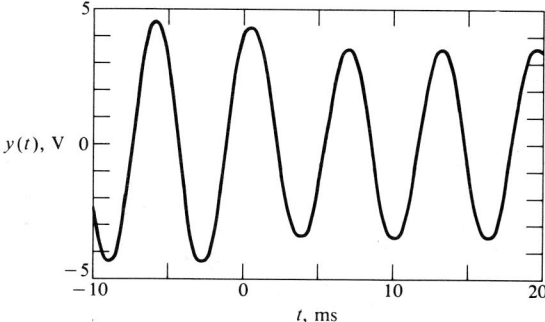

Figure 5-51 Output of the circuit of Example 5-50.

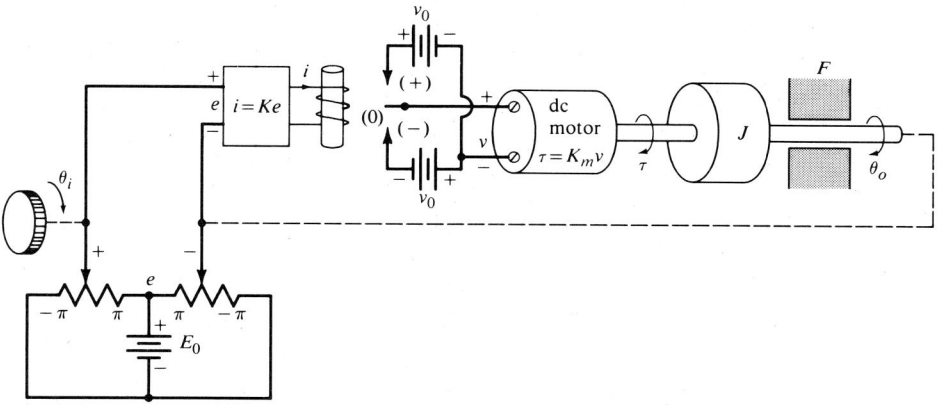

Figure 5-52 Relay servomechanism of Example 5-47.

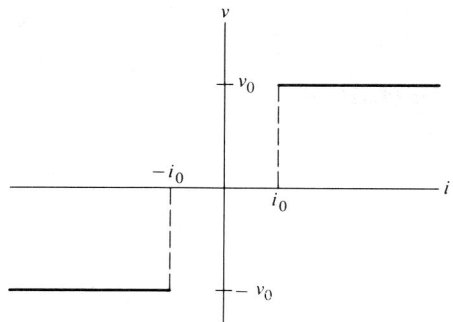

Figure 5-53 Transfer characteristic of the relay of Fig. 5-52.

We obtain the transfer characteristics of the error detector, the amplifier, the motor, and the load as described in Sec. 5-3. This gives the block diagram of Fig. 5-54, where $K_1 = KE_0/2\pi = 1.00$ A/rad, $K_2 = K_m/J = 1.00$ rad/(Vs^2), and $p = -F/J = -5.00$ s^{-1}.

368 INTRODUCTION TO SYSTEM ANALYSIS

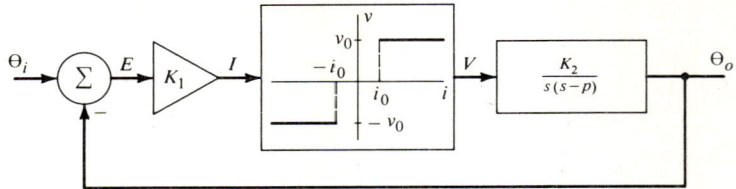

Figure 5-54 Block diagram for the relay servomechanism of Fig. 5-52.

To make the quantitative solution easier to follow we first give a qualitative description of a typical response of the system for step input $\theta_i(t) = \pi u(t)$ rad. We assume $\theta_o(t) = 0$ for $t \leq 0$. Figure 5-55 shows qualitative graphs of $\theta_i(t)$, $i(t)$, $v(t)$, and $\theta_o(t)$.

At $t = 0$ the input $\theta_i(t)$ suddenly changes from zero to π rad. In response, the current $i(t)$ suddenly goes from zero to

$$i(0^+) = K_1[\theta_i(0^+) - \theta_o(0^+)]$$

where we are ignoring inductance of the relay armature. The output $\theta_o(t)$ is continuous because the angular velocity of the load must be finite; thus $\theta_o(0^+) = 0$ and $i(0^+) = K_1\pi$. Assuming $K_1\pi > i_0$ gives $v(0^+) = v_0$. Consequently, at $t = 0^+$, torque $\tau(0^+) = K_m v_0$ is applied to the load, causing the load to accelerate. As $\theta_o(t)$ increases, the error $\theta_i(t) - \theta_o(t)$ and the current $i(t) = K_1[\theta_i(t) - \theta_o(t)]$ decrease. At time t_1, when the current has decreased to i_0, the relay opens, the armature voltage goes to zero, the torque applied to the load goes to zero, and the load begins to coast. The load coasts through the *deadband* defined by $-i_0 < i < i_0$ or

$$\theta_i - \frac{i_0}{K_1} < \theta_o < \theta_i + \frac{i_0}{K_1}$$

At time t_2, when θ_o first exceeds $\theta_i + i_0/K_1$, the current i becomes less than $-i_0$, the voltage v goes to $-v_0$, and the torque applied to the load goes to $-K_m v_0$. This torque reversal causes the load to slow down, stop for an instant, and turn back toward the deadband. At time t_3, when the load reenters the deadband (and when θ_o again equals $\theta_i + i_0/K_1$), the current i goes to zero, the voltage v goes to zero, the applied torque τ goes to zero, and the load begins to coast. If the load has sufficient momentum to coast through the deadband, the voltage v will again be switched by the relay to $+v_0$ and the load will be driven back toward the deadband. The graphs of Fig. 5-55 are drawn for a case where the load has insufficient momentum (at $t = t_3$) to reach the bottom of the deadband. The load coasts to a stop somewhere in the deadband.

The relay servomechanism of Fig. 5-52 is nonlinear because the relay is nonlinear; however, the servomechanism is piecewise-linear and stationary because each segment of an output $\theta_o(t)$ (e.g., for $0 < t < t_1$ in Fig. 5-55) can be determined by solving the differential equation

$$\frac{d^2\theta_o}{dt^2} - p\frac{d\theta_o}{dt^2} = K_2 v$$

where v equals $-v_0$, 0, or $+v_0$ and θ_o satisfies initial values established by the previous segment of θ_o. The initial values are obtained using the fact that both θ_o and its first derivative $d\theta_o/dt$ are continuous, as is necessary because neither velocity nor acceleration of the load can be infinite. Below, we determine the output $\theta_o(t)$ of the servomechanism for input

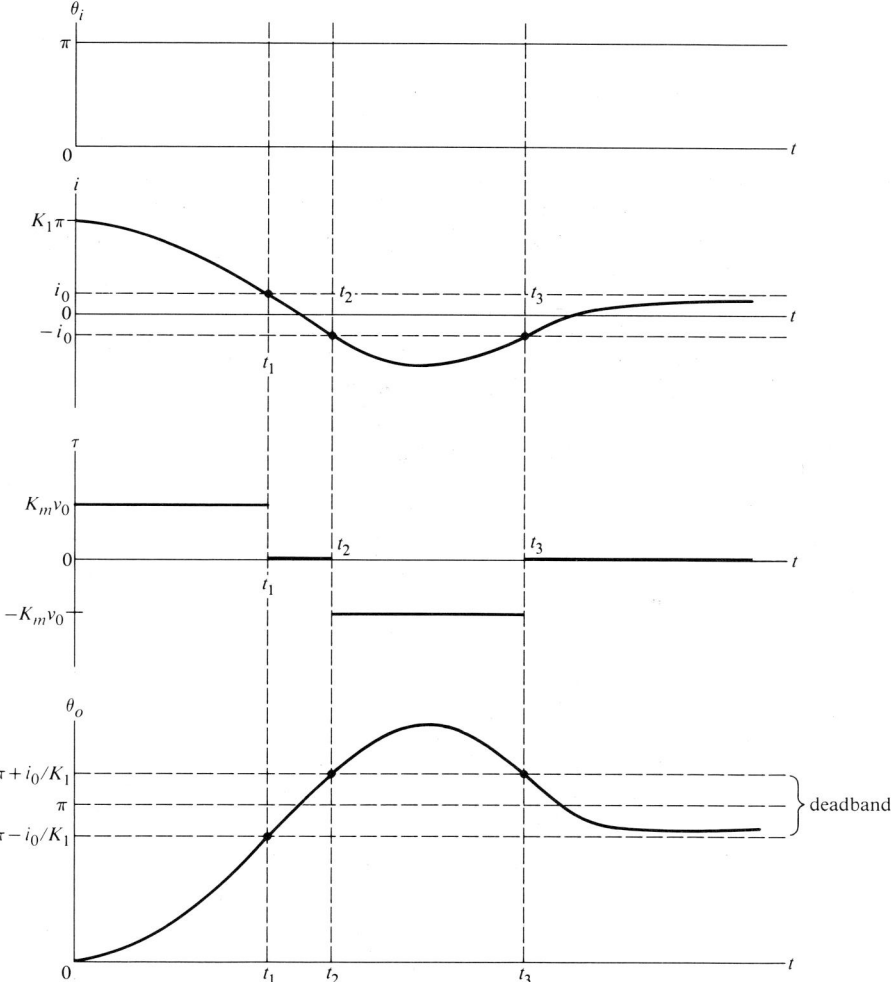

Figure 5-55 Qualitative graphs of signals in the relay servomechanism of Fig. 5-52.

$\theta_i(t) = \pi u(t)$ rad, where the calculations are guided by the qualitative graphs of Fig. 5-55. For convenience we use the dot notation for differentiation with respect to time, where

$$\dot{\theta}_o(t) = \frac{d\theta_o(t)}{dt} \qquad \ddot{\theta}_o(t) = \frac{d^2\theta_o(t)}{dt^2}$$

For $0 \leq t < t_1$, the voltage $v(t)$ equals v_0, and the output is obtained by solving the differential equation above with $v = v_0$; thus

$$\ddot{\theta}_o - p\dot{\theta}_o = K_2 v_0 \qquad 0 \leq t < t_1$$

where t_1 is the time (as yet unknown) at which θ_o first exceeds $\pi - i_0/K_1 = 2.99$ rad (the time at which the load first enters the deadband). Taking the one-sided Laplace transform of this equation gives

$$s^2\tilde{\Theta}_o(s) - s\theta_o(0) - \dot{\theta}_o(0) - p[s\tilde{\Theta}_o(s) - \theta_o(0)] = \frac{1}{s}K_2v_0$$

whence
$$\tilde{\Theta}_o(s) = \frac{K_1v_0}{s^2(s-p)} + \frac{\dot{\theta}_o(0)}{s(s-p)} + \frac{\theta_o(0)}{s}$$

For $t = 0$, the system is at rest. Therefore $\theta_o(0) = 0$, $\dot{\theta}_o(0) = 0$, and

$$\theta_o(t) = \mathcal{L}_1^{-1}\left\{\frac{K_2v_0}{s^2(s-p)}\right\} = \frac{K_2v_0}{p^2}(-pt - 1 + e^{pt})$$

$$= 0.50(-pt - 1 + e^{pt}) \text{ rad} \qquad 0 \le t < t_1$$

The time t_1 (the time at which the load first enters the deadband) is obtained by solving

$$\theta_o(t_1) = 0.50(-pt_1 - 1 + e^{pt_1}) = 2.99$$

This simplifies to

$$-pt_1 + e^{pt_1} = 6.98$$

Solving this equation numerically gives $pt_1 \approx -6.98$ and

$$t_1 = 1.40 \text{ s}$$

Next, we determine $\theta_o(t)$ for $t_1 \le t < t_2$, when the load is coasting through the deadband and where t_2 is the time at which $\theta_o(t)$ reaches the top of the deadband. For $t_1 \le t < t_2$, the differential equation above becomes

$$\ddot{\theta}_o - p\dot{\theta}_o = 0 \qquad t_1 \le t < t_2$$

because $v(t) = 0$ while $\theta_o(t)$ is in the deadband $2.99 < \theta_o < 3.29$ rad. Solving this equation using the one-sided Laplace transform is simplified by the substitution

$$\theta_o(t) = \phi(t - t_1)$$

which (in effect) translates the beginning of the segment of interest to the time origin; thus

$$\ddot{\phi} - p\dot{\phi} = 0 \qquad 0 \le t < t_2 - t_1$$

Taking the one-sided Laplace transform of this equation gives

$$s^2\tilde{\Phi}(s) - s\phi(0) - \dot{\phi}(0) - p[s\tilde{\Phi}(s) - \phi(0)] = 0$$

whence
$$\tilde{\Phi}(s) = \frac{\phi(0)}{s} + \frac{\dot{\phi}(0)}{s(s-p)}$$

The initial values $\phi(0)$, $\dot{\phi}(0)$ are obtained using the fact that $\theta_o(t)$ and $\dot{\theta}_o(t)$ are continuous; thus

$$\phi(0) = \theta_o(t_1) = 2.99 \text{ rad}$$

$$\dot{\phi}(0) = \dot{\theta}_o(t_1) = -\frac{K_2v_0}{p}(1 - e^{pt_1}) = 2.50 \text{ rad/s}$$

Taking the inverse one-sided Laplace transform of $\tilde{\Phi}(s)$ gives

$$\phi(t) = \phi(0) - \frac{\dot{\phi}(0)}{p}(1 - e^{pt}) = 3.49 - 0.50e^{pt} \qquad \text{rad} \qquad 0 \le t < t_2 - t_1$$

or $\quad \theta_o(t) = \phi(t - t_1) = 3.49 - 0.50e^{p(t-t_1)} \quad$ rad $\quad t_1 \leq t < t_2$

The time t_2 is the time at which θ_o reaches the top of the deadband (see Fig. 5-55); that is,

$$\theta_o(t_2) = 3.49 - 0.50e^{p(t_2-t_1)} = 3.29 \text{ rad}$$

This gives

$$t_2 = \frac{1}{p} \ln \frac{3.49 - 3.29}{0.50} + t_1 = 1.58 \text{ s}$$

When θ_o reaches the top of the deadband (at $t = t_2$), the voltage v goes to $-v_0$ and the differential equation relating $\theta_o(t)$ to $v(t)$ becomes

$$\ddot{\theta}_o - p\dot{\theta}_o = -K_2 v_0 \quad t_2 \leq t < t_3$$

where t_3 is the time at which θ_o reenters the deadband. Solving this equation is simplified by the substitution

$$\theta_o(t) = \phi(t - t_2)$$

thus $\quad \ddot{\phi} - p\dot{\phi} = -K_2 v_0 \quad 0 \leq t < t_3 - t_2$

Taking the one-sided Laplace transform of this equation and solving for $\tilde{\Phi}(s)$ gives

$$\tilde{\Phi}(s) = -\frac{K_2 v_0}{s^2(s-p)} + \frac{\dot{\phi}(0)}{s(s-p)} + \frac{\phi(0)}{s}$$

The initial values $\phi(0)$, $\dot{\phi}(0)$ are obtained using the fact that $\theta_o(t)$ and $\dot{\theta}_o(t)$ are continuous at $t = 0$; thus

$$\phi(0) = \theta_o(t_2) = 3.29 \text{ rad}$$
$$\dot{\phi}(0) = \dot{\theta}_o(t_2) = -0.50pe^{p(t_2-t_1)} = 1.02 \text{ rad/s}$$

and

$$\phi(t) = -\frac{K_2 v_0}{p^2}(-pt - 1 + e^{pt}) - \frac{\dot{\phi}(0)}{p}(1 - e^{pt}) + \phi(0)$$

$$= 0.50pt + 3.99 - 0.70e^{pt} \quad \text{rad} \quad 0 \leq t < t_3 - t_2$$

This gives

$$\theta_o(t) = \phi(t - t_2) = 0.50p(t - t_2) + 3.99 - 0.70e^{p(t-t_2)} \quad \text{rad} \quad t_2 \leq t < t_3$$

The time t_3 at which θ_o reenters the deadband is obtained by solving

$$\theta_o(t_3) = 0.50P(t_3 - t_2) + 3.99 - 0.70e^{p(t_3-t_2)} = 3.29 \text{ rad}$$

This simplifies to

$$1 - e^\alpha + 0.71\alpha = 0 \quad \text{where } \alpha = p(t_3 - t_2)$$

Solving this relation (numerically) yields† $\alpha = -0.72$ and

$$t_3 = t_2 - \frac{0.72}{p} = 1.72 \text{ s}$$

†Note that there are two solutions. One is $\alpha = 0$, corresponding to the time at which θ_o first exceeds 3.29 rad. The other (and the one sought here) is the time at which θ_o returns to 3.29 rad.

For the next interval,

$$\ddot{\theta}_o - p\dot{\theta}_o = 0 \qquad t_3 \leq t < t_4$$

where t_4 is the time at which θ_o reaches the bottom of the deadband. Proceeding as above, we let $\theta_o(t) = \phi(t - t_3)$; thus

$$\ddot{\phi} - p\dot{\phi} = 0$$

Taking the one-sided Laplace transform of this equation and solving for $\tilde{\Phi}(s)$ gives

$$\tilde{\Phi}(s) = \frac{\phi(0)}{s} + \frac{\dot{\phi}(0)}{s(s-p)}$$

where $\quad \phi(0) = \theta_o(t_3) = 3.29$ rad $\qquad \dot{\phi}(0) = \dot{\theta}_o(t_3) = -0.76$ rad/s

This yields

$$\phi(t) = \phi(0) - \frac{\dot{\phi}(0)}{p}(1 - e^{pt}) = 3.14 + 0.15e^{pt} \quad \text{rad} \qquad 0 \leq t < t_4 - t_3$$

whence $\quad \theta_o(t) = 3.14 + 0.15e^{p(t-t_3)} \quad$ rad $\qquad t_3 \leq t < t_4$

This function approaches 3.14 rad as t approaches infinity, which means that θ_o remains in the deadband for $t > t_3$. Therefore there are no more switching times, i.e., $t_4 = \infty$. Collecting the results obtained above, we have

$$\theta_o(t) = \begin{cases} 0.50(-pt - 1 + e^{pt}) & \text{rad} & 0 \leq t < t_1 \\ 3.49 - 0.50e^{p(t-t_1)} & \text{rad} & t_1 \leq t < t_2 \\ 0.50p(t - t_2) + 3.99 - 0.70e^{p(t-t_2)} & \text{rad} & t_2 \leq t < t_3 \\ 3.14 + 0.15e^{p(t-t_3)} & \text{rad} & t_3 \leq t < \infty \end{cases}$$

where $p = -5$ s^{-1}, $t_1 = 1.40$ s, $t_2 = 1.58$ s, and $t_3 = 1.72$ s.

Figures 5-56 to 5-59 show graphs of the output $\theta_o(t)$ and the voltage $v(t)$ for several values of the constant $K_2 = K_m/J$. Figure 5-56 shows the output for $K_2 = 1$ rad/(Vs^2), which is the value used in the analysis above. Figures 5-57 to 5-59 show the output for $K_2 = 2$, 4, and 8 rad/(Vs^2), respectively. As K_2 (the forward gain) is increased, rise time decreases and the response becomes more oscillatory. This behavior of the relay servomechanism of Fig. 5-52 is analogous to the behavior of a linear servomechanism.

SUMMARY

The (two-sided) *Laplace transform* of a function $e(t)$ is denoted by $E(s)$ or $\mathcal{L}\{e(t)\}$ and defined by

$$E(s) = \mathcal{L}\{e(t)\} = \int_{-\infty}^{\infty} e(t)e^{-st}\, dt$$

A function $e(t)$ whose Laplace transform is $E(s)$ is called an *inverse Laplace transform* of $E(s)$. This relation is denoted by

$$e(t) = \mathcal{L}^{-1}\{E(s)\}$$

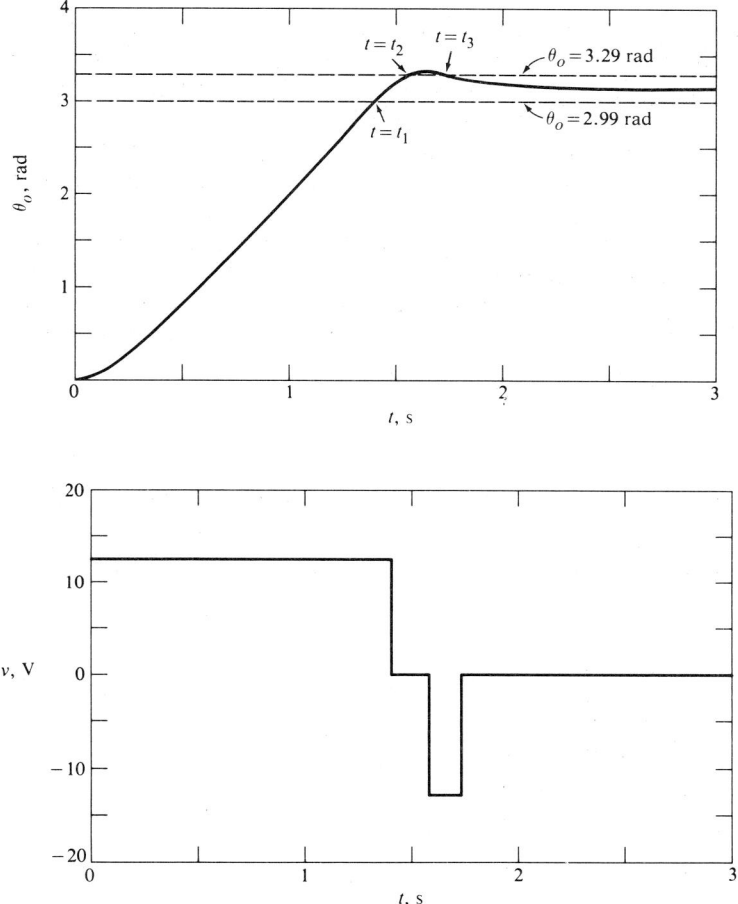

Figure 5-56 Output of the relay servomechanism of Fig. 5-52 for input $\theta_i(t) = \pi u(t)$ rad and $K_2 = 1$ rad/(Vs^2).

The independent variable s of a Laplace transform is a complex quantity called *complex frequency*. The real part of s is denoted by σ, and the imaginary part of s is denoted by ω. Values of s can be represented by points in a σ-ω plane called an *s plane*.

Values of s for which the integral above converges constitute the *region of convergence* for the Laplace transform $E(s)$. An inverse transform can be right-, left-, or two-sided, depending on the associated region of convergence. Except when examining realizability, we assume that all inverse Laplace transforms are right-sided. This makes specifying a region of convergence unnecessary because a right-sided inverse transform is unique.

Comparing the definition of the Laplace transform of a function $e(t)$ with the definition of the Fourier transform of the function shows that

$$\mathscr{L}\{e(t)\} = \mathscr{F}\{e(t)e^{-\sigma t}\}$$

374 INTRODUCTION TO SYSTEM ANALYSIS

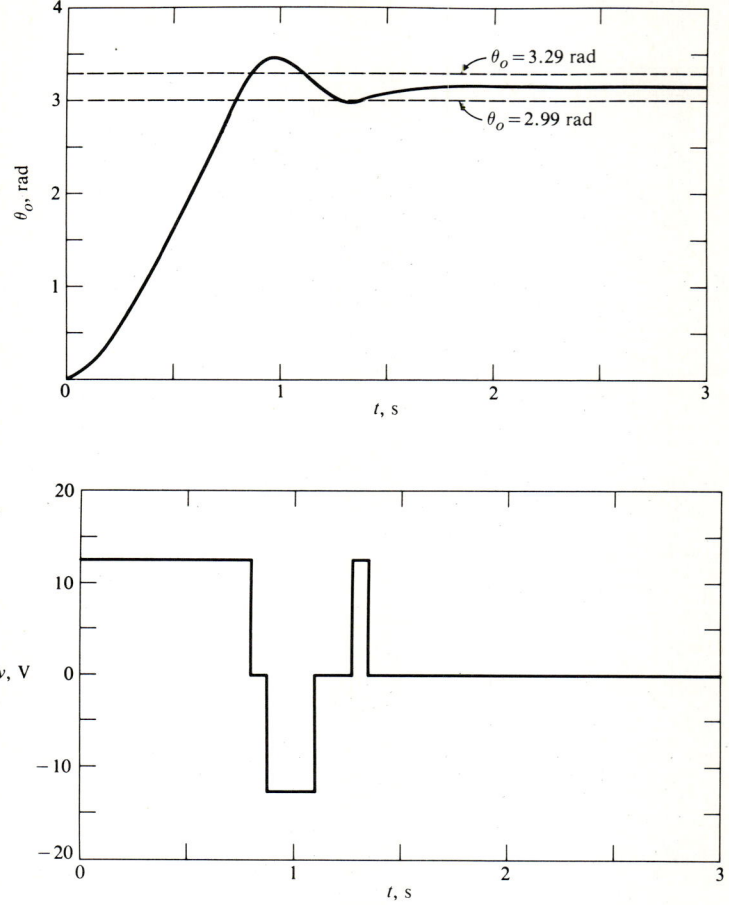

Figure 5-57 Output of the relay servomechanism of Fig. 5-52 for input $\theta_i(t) = \pi u(t)$ rad and $K_2 = 2$ rad/(Vs²).

If the Laplace transform $\mathcal{L}\{e(t)\}$ exists for $\sigma = 0$, that is, if the region of convergence includes the ω axis, then the Fourier transform $\mathcal{F}\{e(t)\}$ can be obtained by taking $s = j\omega$ in the Laplace transform $\mathcal{L}\{e(t)\}$. Otherwise the Laplace transform $E(s) = \mathcal{L}\{e(t)\}$ and the Fourier transform $E(j\omega) = \mathcal{F}\{e(t)\}$ are different functions, even though the symbol E is used for both.

The Laplace transformation, like the Fourier transformation, is linear. The operational properties of the Laplace transformation are essentially identical to corresponding operational properties of the Fourier transform except that s replaces $j\omega$ as the transform variable. In system analysis the most important operational properties of the Laplace transformation are the *convolution property*

$$\mathcal{L}\left\{\int_{-\infty}^{\infty} e_1(\lambda)e_2(t-\lambda)\,d\lambda\right\} = E_1(s)E_2(s)$$

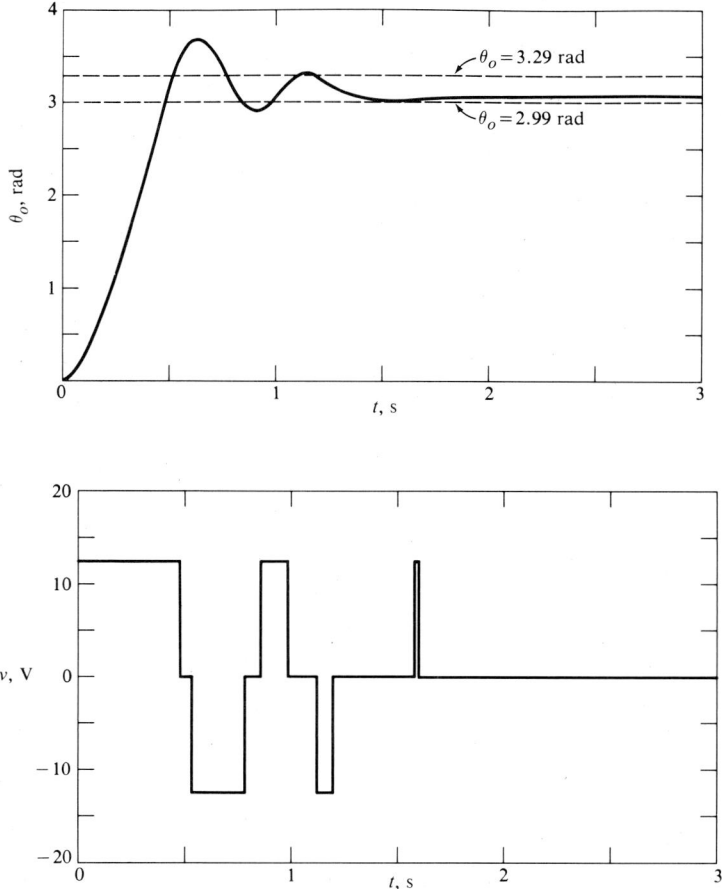

Figure 5-58 Output of the relay servomechanism of Fig. 5-52 for input $\theta_i(t) = \pi u(t)$ rad and $K_2 = 4$ rad/(Vs²).

and the *differentiation property*

$$\mathcal{L}\left\{\frac{d^n e(t)}{dt^n}\right\} = s^n \mathcal{L}\{e(t)\}$$

The convolution property shows that the output $y(t)$ of a linear stationary system having impulse response $h(t)$ and input $x(t)$ is given by

$$y(t) = \mathcal{L}^{-1}\{H(s)\mathcal{L}\{x(t)\}\}$$

where $H(s) = \mathcal{L}\{h(t)\}$ is called the *system function* of the system. This relation is a concise formulation of the following procedure for obtaining an output of a linear stationary system:

1. Obtain the Laplace transform $X(s) = \mathcal{L}\{x(t)\}$ of the input $x(t)$.

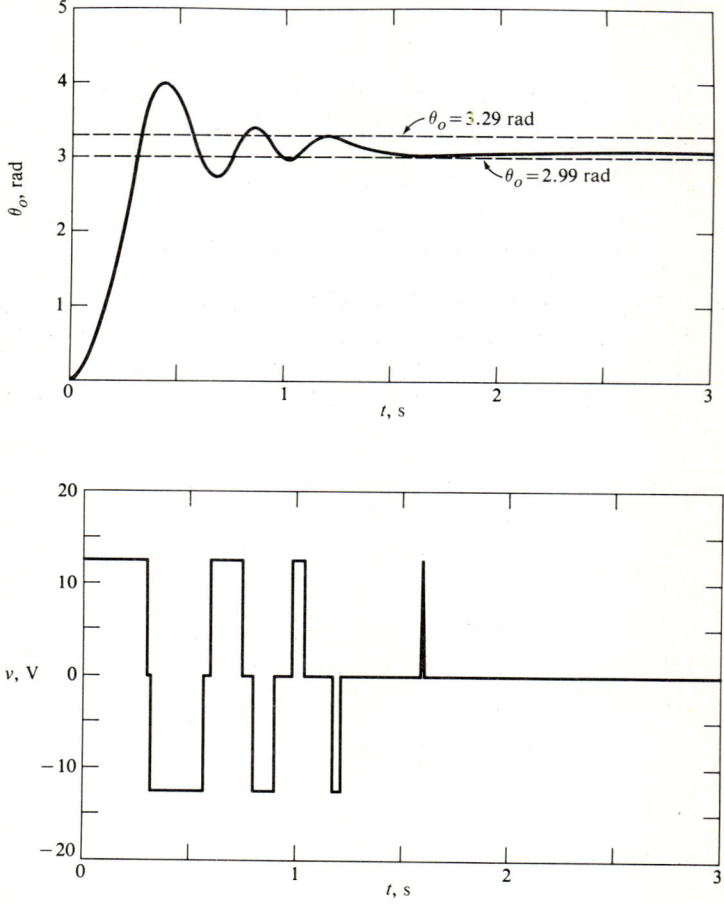

Figure 5-59 Output of the relay servomechanism of Fig. 5-52 for input $\theta_i(t) = \pi u(t)$ rad and $K_2 = 8$ rad/(Vs²).

2. Obtain the system function $H(s)$ of the system from the impulse response $h(t)$ of the system using $H(s) = \mathscr{L}\{h(t)\}$, from a differential equation describing the system using the differentiation property above, or from a block diagram describing the system using block-diagram reduction.
3. Obtain the output $y(t)$ from the inverse Laplace transform $y(t) = \mathscr{L}^{-1}\{H(s)X(s)\}$.

The differentiation property enables us to obtain the system function of a linear stationary system from a differential equation describing the system without first obtaining the impulse response of the system.

The *zeros* of a system function $H(s)$ are the values of s for which $H(s) = 0$. The *poles* of a system function are the values of s for which $H(s) = \infty$. The poles of a system function are the characteristic roots of the associated system. A system function is determined (except for a scale factor) by its poles and zeros. It is often useful to represent a system function graphically by a *pole-zero plot* showing the poles (as crosses) and the zeros (as circles) on an s plane. A pole-zero plot quickly conveys many important properties of a system.

A system function $H(s)$ is *realizable* if the associated region of convergence lies to the right of the rightmost pole of $H(s)$. A system described by system function $H(s)$ is *stable* if all poles of $H(s)$ lie in the left half plane. It is *conditionally stable* if it has no right-half-plane poles and has nonrepeated poles on the ω axis. It is *unstable* if it has a pole in the right half plane or a repeated pole on the ω axis.

The impulse response $h(t)$ and step response $g(t)$ of a system having system function $H(s)$ are given by

$$h(t) = \mathcal{L}^{-1}\{H(s)\} \qquad g(t) = \mathcal{L}^{-1}\left\{\frac{1}{s}H(s)\right\}$$

Expanding the transform $Y(s) = H(s)X(s)$ of an output in partial fractions shows that (in general) $y(t) = \mathcal{L}^{-1}\{Y(s)\}$ contains terms arising from poles of $H(s)$ as well as terms arising from poles of $X(s)$. Terms arising from poles of $H(s)$, called *natural modes*, consitute the *unforced response* of a system. Terms not arising from poles of $H(s)$ constitute the *forced response*. Terms arising from left-half-plane poles constitute the *transient response*. Terms arising from poles on the ω axis and in the right half plane constitute the *steady-state response*. For a stable system and a step or sinusoidal input, forced response equals steady-state response and unforced response equals transient response. The steady-state output $y_{ss}(t)$ of a stable system having step input $x(t) = x_0 u(t)$ is given by

$$y_{ss}(t) = H(0)x_0 u(t)$$

The steady-state output $y_{ss}(t)$ of a stable system which has sinusoidal input $x(t) = x_0(\cos \omega_0 t)u(t)$ is given by

$$y_{ss}(t) = |H(j\omega_0)|x_0 \cos[\omega_0 t + \angle H(j\omega_0)]$$

A high-order system function often has a few (one or two) *dominant poles*, meaning that the system function can be approximated for many purposes by a system function having only those poles. Such an approximation is called a *dominant-pole approximation*. In most cases where a dominant-pole approximation is useful the dominant poles of a system function are those nearest the origin of a pole-zero plot for the system function. When applicable, dominant-pole approximation greatly simplifies analysis and design of high-order systems.

A stable system whose system function has no right-half-plane zeros is called a *minimum-phase system*. A system whose system function has a right-half-plane zero is called a *non-minimum-phase system*. Generally, a non-minimum-phase system introduces much more phase shift and more phase distortion (delay distortion) than a comparable minimum-phase system. Also, the reciprocal of a minimum-phase system function describes a stable system, whereas the reciprocal of a non-minimum-phase system function describes an unstable system.

The *one-sided Laplace transform* of a function $e(t)$ is denoted by $\mathcal{L}_1\{e(t)\} = \tilde{E}(s)$ and is defined by

$$\tilde{E}(s) = \int_0^\infty e(t)e^{-st}\,dt$$

The one-sided Laplace transform of a function $e(t)$ ignores values of the function for $t < 0$. Consequently, the one-sided Laplace transformation is especially well suited for so-called initial-value problems, where inputs are defined only for $t \geq 0$ and outputs are determined only for $t \geq 0$. Initial-value problems usually arise in analysis of systems that are piecewise linear and stationary.

A one-sided Laplace transform $\tilde{E}(s) = \mathcal{L}_1\{e(t)\}$ is expressed in terms of the associated two-sided Laplace transform $E(s) = \mathcal{L}\{e(t)\}$ by the relation

$$\mathcal{L}_1\{e(t)\} = \mathcal{L}\{e(t)u(t)\} + \mathbf{a}$$

where $\mathbf{a}$ is the strength of a delta function $\mathbf{a}\delta(t)$ appearing in $e(t)$. If $e(t)$ contains no delta function at $t = 0$, this relation still applies but with $\mathbf{a} = 0$. We can use this relation and a table of two-sided Laplace transforms to obtain one-sided transforms. If the inverse two-sided Laplace transform of a function $\tilde{E}(s)$ equals zero for $t < 0$, the inverse two-sided transform and the inverse one-sided transform are identical for $t \geq 0$. Therefore, we can use a table of two-sided transforms to obtain one-sided inverse transforms.

The most important operational property of the one-sided Laplace transformation is the *differentiation property*

$$\mathcal{L}_1\left\{\frac{d^n e(t)}{dt^n}\right\} = s^n \mathcal{L}_1\{e(t)\} - s^{n-1}e(0) - s^{n-2}\left.\frac{de(t)}{dt}\right|_{t=0} - \cdots$$
$$- s\left.\frac{d^{n-2}e(t)}{dt^{n-2}}\right|_{t=0} - \left.\frac{d^{n-1}e(t)}{dt^{n-1}}\right|_{t=0}$$

Because the one-sided Laplace transform of the nth derivative of a function contains initial values of $e(t)$ and its first $n - 1$ derivatives, taking the one-sided Laplace transform of an nth-order differential equation automatically introduces the initial values necessary to obtain a unique solution (response) for $t \geq 0$. The two-sided Laplace transform can be used to solve initial-value problems, but an input must be defined (artificially) for $t < 0$ such that the specified initial values exist at $t = 0$. This approach to initial-value problems is cumbersome.

The Laplace transformation is mathematically similar to the Fourier transformation. Indeed, any practical problem that can be solved using the Laplace transformation can be solved using the Fourier transformation; however, certain problems are much easier to solve using the Laplace transformation. For example, finding the output of a system for ramp input $x(t) = x_0(t/\tau)u(t)$ is easier using the Laplace transformation. Using the Fourier transformation is more difficult because it requires the ramp input to be approximated by an absolutely integrable function such as $x(t) = x_0(t/\tau_1)r(t/\tau_1)$, where τ_1 is large but finite.

Sometimes a choice between Laplace-transform (s-plane) and Fourier-transform (frequency-domain) methods is based on computational considerations, as noted above; but usually the choice is based on which description (s-plane or frequency-domain) best illuminates performance of a system at hand. Thus, s-plane analysis is often used for analysis of automatic control systems because stability and transient response are depicted more clearly by an s-plane description, e.g., a pole-zero plot or root-locus diagram, than by plots of gain and phase shift versus frequency. On the other hand, frequency-domain methods generally are preferred for analysis of communication systems because fidelity is depicted more clearly by plots of gain and phase shift than by a pole-zero plot.

PROBLEMS

5-1 Use (5-1) to obtain the Laplace transform of each function below. Give the associated region of convergence.

(a) $e(t) = r\left(\dfrac{t}{\tau}\right)$ (b) $e(t) = u(t - t_0)$ (c) $e(t) = (\cos \omega_0 t) u(t)$

(d) $e(t) = (\sin \omega_0 t) u(t)$ (e) $e(t) = \delta(t - t_0)$ (f) $e(t) = tu(t)$

(g) $e(t) = tr\left(\dfrac{t}{\tau}\right)$ (h) $e(t) = e^{-t/\tau} r\left(\dfrac{t}{\tau}\right)$ (i) $e(t) = te^{-t/\tau} u(t)$

(j) $e(t) = e^{-t/\tau}(\cos \omega_0 t) u(t)$

5-2 Obtain the Laplace transform and associated region of convergence for the function $e(t) = e^{p|t|}$.

5-3 Use (5-1) and

$$\lim_{\tau \to \infty} r\left(\dfrac{t - \tau}{2\tau}\right) = 1 \quad -\infty < t < \infty$$

to show that $\mathcal{L}\{1\} = 2\pi \delta(s/j)$. What is the associated region of convergence?

5-4 Show that $\mathcal{L}\{u(t)\} = \mathcal{L}\{-u(-t)\} = 1/s$. How can we tell whether $\mathcal{L}^{-1}\{1/s\} = u(t)$ or $\mathcal{L}^{-1}\{1/s\} = -u(-t)$?

5-5 Using Table D-1, obtain the right-sided inverse Laplace transform of each of the following transforms:

(a) $X(s) = \dfrac{K}{s^4}$ (b) $X(s) = \dfrac{K}{s - p} + \dfrac{K^*}{s - p^*}$ $p = \sigma_0 + j\omega_0$

(c) $X(s) = \dfrac{K}{s(s - p)}$ (d) $X(s) = \dfrac{c}{(s - p)^2} + \dfrac{c^*}{(s - p^*)^2}$ $p = \sigma_0 + j\omega_0$ $c = re^{-j\theta}$

(e) $X(s) = \dfrac{K}{(s - p)^3}$

5-6 Show that if $\mathcal{L}\{e(t)\} = E(s)$, then:

(a) $\mathcal{L}\{e(t)e^{pt}\} = E(s - p)$ (b) $\mathcal{L}\{e(t - t_0)\} = E(s)e^{-st_0}$

(c) $\mathcal{L}\{te(t)\} = -\dfrac{dE(s)}{ds}$ (d) $\mathcal{L}\{e(t) \cos \omega_0 t\} = \tfrac{1}{2}E(s - j\omega_0) + \tfrac{1}{2}E(s + j\omega_0)$

5-7 Use Tables D-1 and D-2 to obtain the Laplace transform of each signal below:

(a) $x(t) = x_0 e^{pt} u(t - t_0)$

(b) $x(t) = x_0 r\left(\dfrac{t - t_0}{\tau}\right)$

(c) $x(t) = x_0 \dfrac{t}{\tau}(\cos \omega_0 t) u(t)$

(d) $x(t) = K \displaystyle\int_{-\infty}^{t} x_0[1 - e^{-\alpha/\tau}] u(\alpha)\, d\alpha$

(e) $x(t) = K \dfrac{d}{dt}\left[x_0 e^{-t/\tau} \cos\left(\omega_0 t - \dfrac{\pi}{4}\right) u(t)\right]$

(f) $x(t) = x_0 \left(\dfrac{t}{\tau}\right)^n (\cos \omega_0 t) u(t)$

(g) $x(t) = x_0 \dfrac{t - t_0}{\tau} e^{-t/\tau} u(t)$

(h) $x(t) = x_0 (\cosh pt) u(t)$

5-8 Use Tables D-1 and D-2 to obtain the inverse Laplace transform of each function below:

(a) $F(s) = \dfrac{1 - e^{-st_0}}{s^2}$

(b) $F(s) = \dfrac{1}{(s - p_1)(s - p_2)} \qquad p_1 \ne p_2$

(c) $F(s) = \dfrac{s}{s - p}$

(d) $F(s) = \dfrac{e^{-st_0}}{(s - p)^2}$

(e) $F(s) = \dfrac{se^{-st_0}}{s - p}$

(f) $F(s) = \dfrac{1}{s^2(s - p)}$

(g) $F(s) = \dfrac{s}{(s - p)^2} + \dfrac{s}{(s - p^*)^2} \qquad p = \sigma_0 + j\omega_0$

5-9 Use the Laplace transformation to obtain

$$e(t) = \int_{-\infty}^{\infty} f_1(\lambda) f_2(t - \lambda)\, d\lambda$$

for each pair of functions below:

(a) $f_1(t) = e^{-t/\tau} u(t)$
$f_2(t) = u(t)$

(b) $f_1(t) = e^{p_1 t} u(t)$
$f_2(t) = e^{p_2 t} u(t) \qquad p_1 \ne p_2$

(c) $f_1(t) = \delta(t - t_0)$
$f_2(t) = r\left(\dfrac{t}{\tau}\right)$

(d) $f_1(t) = r\left(\dfrac{t}{\tau}\right)$
$f_2(t) = \cos \dfrac{2\pi t}{\tau} u(t)$

(e) $f_1(t) = f_2(t) = r\left(\dfrac{t}{\tau}\right)$

5-10 Use the Laplace transformation to obtain the impulse response $h(t)$ and the step response $g(t)$ of the systems described by the differential equations below, where input $x = x(t)$ is voltage in volts and output $y = y(t)$ is current in amperes. Sketch graphs of $h(t)$ and $g(t)$ versus time t.

(a) $\dfrac{dy}{dt} + \omega_0 y = K\omega_0 x, \qquad \omega_0 = 1 \text{ krad/s}, \qquad$ and $\quad K = 0.1 \text{ A/V}$.

(b) $\dfrac{dy}{dt} + \omega_0 y = K\dfrac{dx}{dt}$, $\quad \omega_0 = 1 \text{ krad/s},\quad$ and $\quad K = 1 \text{ A/V}$.

(c) $\dfrac{d^2y}{dt^2} + a_1\dfrac{dy}{dt} + a_0 y = b_0 x$, $\quad a_1 = 4 \text{ s}^{-1},\quad a_0 = 2 \text{ s}^{-2},\quad$ and $\quad b_0 = 2 \text{ A/(Vs}^2)$.

(d) $\dfrac{d^2y}{dt^2} + a_1\dfrac{dy}{dt} = b_0 x$, $\quad a_1 = 10^3 \text{ s}^{-1},\quad$ and $\quad b_0 = 10^2 \text{ A/(Vs}^2)$.

5-11 Use the Laplace transformation to obtain the impulse response and the step response of each system of Fig. P5-11.

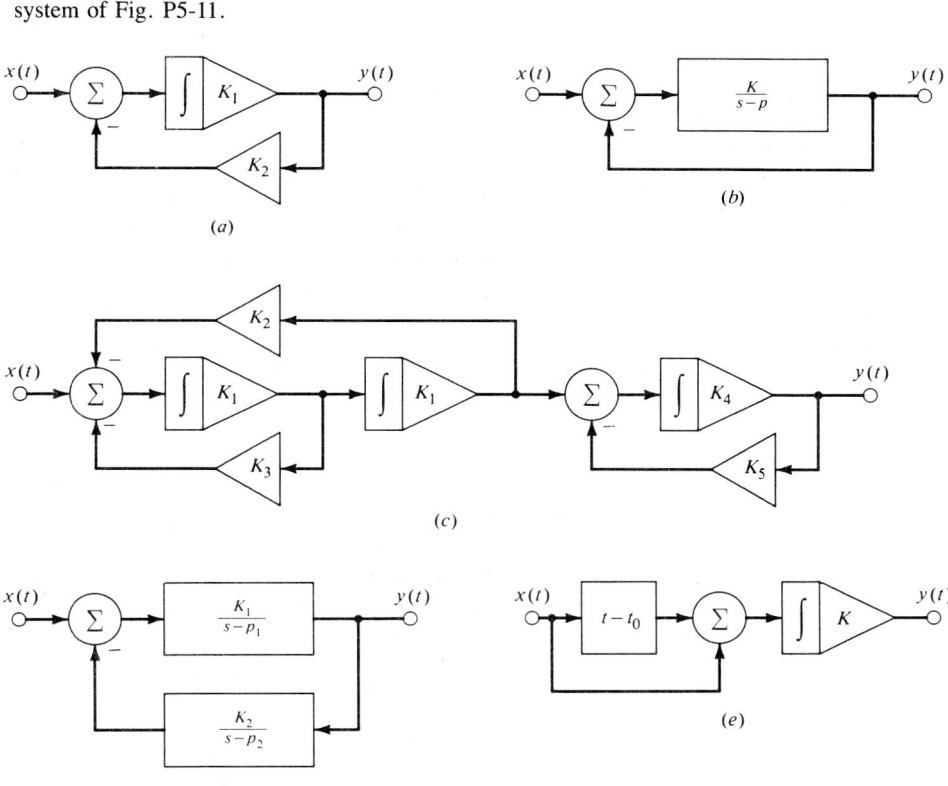

Figure P5-11 (a) $K_1 = 10 \text{ s}^{-1}$, $K_2 = 5$; (b) $K = 10 \text{ s}^{-1}$, $p = 5 \text{ s}^{-1}$; (c) $K_1 = 1 \text{ s}^{-1}$, $K_2 = 6.25$, $K_3 = 3$, $K_4 = 10 \text{ s}^{-1}$, $K_5 = 0.2$; (d) $K_1 = 6 \text{ s}^{-1}$, $K_2 = 1 \text{ s}^{-1}$, $p_1 = -2 \text{ s}^{-1}$, $p_2 = -3 \text{ s}^{-1}$; (e) $t_0 = 628$ ms, $K = 10 \text{ s}^{-1}$.

5-12 Use a partial-fraction expansion and Table D-1 to obtain the inverse Laplace transform of each function below:

(a) $X(s) = \dfrac{x_0 s}{(s - p_1)(s - p_2)}$

where $x_0 = 10 \text{ V}$, $p_1 = -1 \text{ s}^{-1}$, and $p_2 = -2 \text{ s}^{-1}$.

(b) $X(s) = \dfrac{Kx_0 s(s-z)}{(s-p_1)(s-p_2)}$

where $K = 1$ s, $x_0 = 10$ mA, $z = -5$ s^{-1}, $p_1 = -1$ s^{-1}, and $p_2 = -2$ s^{-1}.

(c) $X(s) = \dfrac{x_0 s(s-z)}{(s-p_1)(s-p_2)(s-p_3)}$

where $x_0 = 50$ mA, $z = -5$ s^{-1}, $p_1 = -10$ s^{-1}, $p_2 = -1$ s^{-1}, and $p_3 = -4$ s^{-1}.

(d) $X(s) = \dfrac{Kx_0}{s^2(s-p)}$

where $K = 100$ s^{-2}, $x_0 = 5$ V, and $p = -10$ s^{-1}.

(e) $X(s) = \dfrac{Kx_0(s-z)}{s^2(s-p)}$

where $K = 1$ s^{-1}, $x_0 = 10$ mA, $z = 1$ s^{-1}, and $p = -10$ s^{-1}.

(f) $X(s) = \dfrac{x_0 s(s-z)}{(s-p_1)(s-p_2)^2}$

where $x_0 = 25$ V, $z = -2$ s^{-1}, $p_1 = -1$ s^{-1}, and $p_2 = -5$ s^{-1}.

(g) $X(s) = \dfrac{x_0(s-z)^3}{s^3(s-p)}$

where $x_0 = 5$ V, $z = 1$ s^{-1}, and $p = -1$ s^{-1}.

(h) $X(s) = \dfrac{x_0 \omega_0^2}{s(s^2 + 2\zeta\omega_0 s + \omega_0^2)}$

where $x_0 = 10$ mA, $\omega_0 = 1$ krad/s, and $\zeta = 0.2$.

(i) $X(s) = \dfrac{x_0(s-z_1)(s-z_2)}{(s-p_1)(s^2+qs+r)}$

where $x_0 = 500$ mV, $z_1 = -1$ s^{-1}, $z_2 = -10$ s^{-1}, $p_1 = -5$ s^{-1}, $q = 2$ s^{-1}, and $r = 4$ s^{-2}.

5-13 Use the Laplace transformation to obtain the output $y(t)$ of a system having impulse response $h(t) = (1/\tau)r(t/\tau)$ and input $x(t) = x_0 r(t/\tau)$, where $\tau = 1$ ms and $x_0 = 5$ V.

5-14 Use the Laplace transformation to obtain the output $y(t)$ of a system having impulse response

$$h(t) = K\dfrac{t}{\tau}e^{-t/\tau}u(t)$$

and input

$$x(t) = x_0(\cos\omega_0 t)u(t)$$

where $K = 50$ mA/(Vs), $\tau = 100$ μs, $x_0 = 10$ V, and $\omega_0 = 10$ krad/s.

5-15 Use the Laplace transformation to obtain the output $y(t)$ for input $x(t) = x_0 r(t/\tau)$ of a system described by the differential equation

$$\dfrac{d^2 y}{dt^2} + a_1 \dfrac{dy}{dt} + a_0 y = b_1 \dfrac{dx}{dt} + b_0 x$$

where $a_1 = 3 \times 10^3 \text{ s}^{-1}$, $a_0 = 2 \times 10^6 \text{ s}^{-2}$, $b_1 = 10^4 \text{ mA/(Vs)}$, $b_0 = 10^7 \text{ mA/(Vs}^2)$, $x_0 = 5$ V, and $\tau = 1$ ms.

5-16 Use the Laplace transformation to obtain the output $y(t)$ for input $x(t) = x_0 u(t)$ of a system described by the differential equation

$$\frac{d^2 y}{dt^2} + a_1 \frac{dy}{dt} = b_1 \frac{dx}{dt}$$

where $a_1 = 10^3 \text{ s}^{-1}$, $b_1 = 10^3 \text{ s}^{-1}$, and $x_0 = 5$ V.

5-17 Obtain the output $y(t)$ of the system of Fig. P5-11a for input $x(t) = x_0 e^{-t/\tau} u(t)$, where $x_0 = 5$ V and $\tau = 50$ ms.

5-18 Obtain the output $y(t)$ of the system of Fig. P5-11b for input $x(t) = x_0 r(t/\tau)$, where $x_0 = 5$ V and $\tau = 50$ ms.

5-19 Obtain the output $y(t)$ of the system of Fig. P5-11c for input $x(t) = x_0 (\cos \omega_1 t) u(t)$, where $x_0 = 5$ V and $\omega_1 = 2.5$ rad/s.

5-20 Obtain the output $y(t)$ of the system of Fig. P5-11d for input $x(t) = x_0 r(t/\tau)$, where $x_0 = 5$ V and $\tau = 2$ ms.

5-21 For each system of Fig. P5-11 draw a pole-zero plot, list the natural modes and the natural frequencies, and determine whether the system is stable.

5-22 Repeat Prob. 5-21 for the system of Prob. 5-16.

5-23 Show that a system function of the form

$$H(s) = H_1(s) e^{-s t_0}$$

where $H_1(s)$ is a rational function, is realizable if and only if $t_0 \geq 0$.

5-24 Describe two ways to obtain a realizable approximation to the nonrealizable impulse response

$$h(t) = \frac{1}{\tau} \left[e^{t/\tau} u(-t) + r\left(\frac{t}{\tau}\right) \right] \quad \text{where } \tau > 0$$

5-25 Show that a rational system function having a pole in the right half of an s plane but an unspecified region of convergence describes either a nonrealizable system or an unstable system.

5-26 For each system function $H(s)$ and input transform $X(s)$ below, give the forms of the forced response, the unforced response, the transient response, and the steady-state response. All poles and zeros are real.

(a) $H(s) = \dfrac{Ks}{(s - p_1)(s - p_2)}$ $\quad X(s) = \dfrac{x_0}{s} + \dfrac{x_0}{s - p_1}$ $\quad$ with $p_1 < p_2 < 0$

(b) $H(s) = \dfrac{K}{s(s - p_1)}$ $\quad X(s) = \dfrac{x_0}{s - p_2}$ $\quad$ with $p_1 < p_2 < 0$

(c) $H(s) = \dfrac{K(s - z)}{s(s - p)}$ $\quad X(s) = \dfrac{x_0}{s} + a$ $\quad$ with $p < 0$

(d) $H(s) = \dfrac{Ks(s - z)^2}{(s - p_1)^2 (s - p_2)}$ $\quad X(s) = \dfrac{x_0}{s} + \dfrac{x_0}{s - 2p_1}$ $\quad$ with $p_1 < p_2 < 2p_2 < 0$

5-27 Obtain the steady-state output of each system of Fig. P5-11 for input $x(t) = x_0 (1 + \cos \omega_0 t) u(t)$, with $\omega_0 = 5$ rad/s.

384 INTRODUCTION TO SYSTEM ANALYSIS

5-28 For each pole-zero plot of Fig. P5-28 give the associated system function (within a constant factor), sketch graphs of the natural modes versus time, sketch graphs of gain and phase shift versus frequency, give a differential equation describing the system, and give a block diagram for the system.

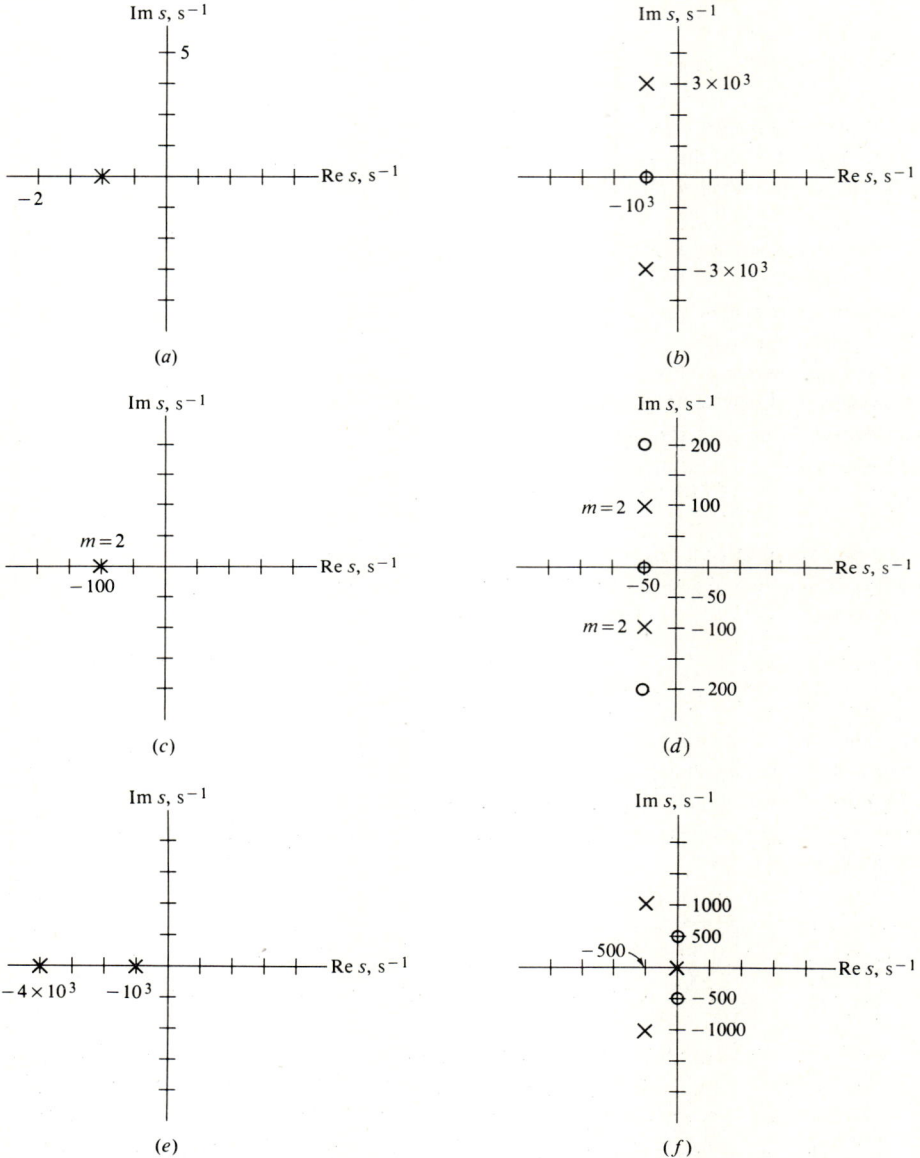

Figure P5-28

5-29 Use the Laplace transformation to show that a feedforward connection of stable linear stationary systems is stable.

5-30 Obtain the output of the system of Fig. P5-30 for input $x(t) = x_0(\cos \omega_0 t)u(t)$. Identify the forced, unforced, transient, and steady-state components of the output. Under what conditions can the steady-state output be obtained using the frequency-domain relation $\mathscr{F}\{y_{ss}(t)\} = H(j\omega_0)\mathscr{F}\{x_{ss}(t)\}$?

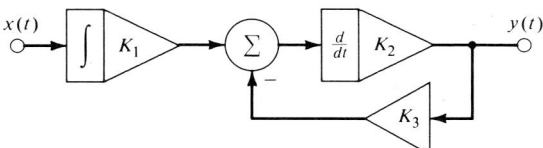

Figure P5-30

5-31 A second-order system has system function

$$H(s) = \frac{K\omega_0^2}{s^2 + 2\zeta\omega_0 s + \omega_0^2}$$

Express damping ratio ζ and undamped natural frequency ω_0 in terms of the poles p_1, p_2 of $H(s)$. On a cartesian plane whose horizontal axis is damping ratio and whose vertical axis is undamped natural frequency, show all values of ζ and ω_0 for which the system is stable. Sketch root-locus plots having parameters ζ and ω_0 for $-\infty < \zeta < \infty$ and $-\infty < \omega_0 < \infty$.

5-32 Explain why the phase shift of a system cannot be obtained unambiguously from a pole-zero plot for the system. Is the ambiguity significant?

5-33 Draw pole-zero plots for systems described as follows:
   (a) Five-pole, low-pass, stable, minimum-phase
   (b) Three-pole, low-pass, stable, non-minimum-phase
   (c) First-order, high-pass, stable, minimum-phase
   (d) Fourth-order, low-pass, stable, minimum-phase, dominant-pole pair at $s = -1 \pm j$ rad/s
   (e) Six-pole, all-pass, stable
   (f) All-pass, stable, biquadratic
   (g) Third-order, non-minimum-phase, stable, dominant pole at $s = -4$ krad/s
   (h) Bandpass, sixth-order, center frequency $= 10$ kHz, bandwidth $= 1$ kHz
   (i) Second-order, all-pole, overdamped
   (j) Second-order, all-pole, underdamped
   (k) Second-order, all-pole, damping ratio $= 0.707$, undamped natural frequency $= 10$ krad/s.

5-34 With reference to a pole-zero plot, explain why the system function

$$H(s) = \frac{s}{s - p} \qquad p < 0$$

describes a high-pass system.

5-35 With reference to a pole-zero plot and a partial-fraction expansion, explain why the pole p_1 is dominant in the system function

$$H(s) = \frac{K}{(s - p_1)(s - 10p_1)}$$

5-36 Describe (using sketches) how adding a zero at $s = -2$ s^{-1} changes gain and step response of a system represented by the pole-zero plot of Fig. P5-36.

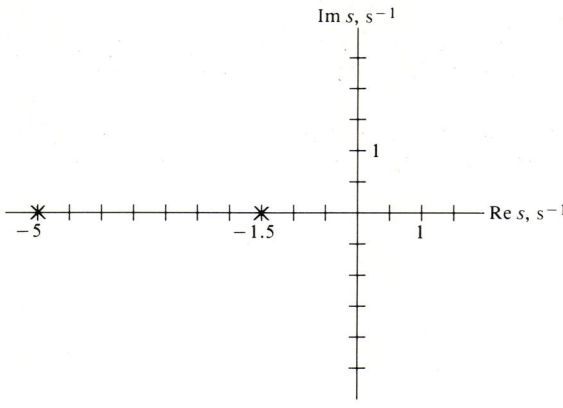

Figure P5-36

5-37 Describe (using sketches) how adding zeros at $-1 \pm j$ s^{-1} changes gain and step response of a system represented by the pole-zero plot of Fig. P5-37.

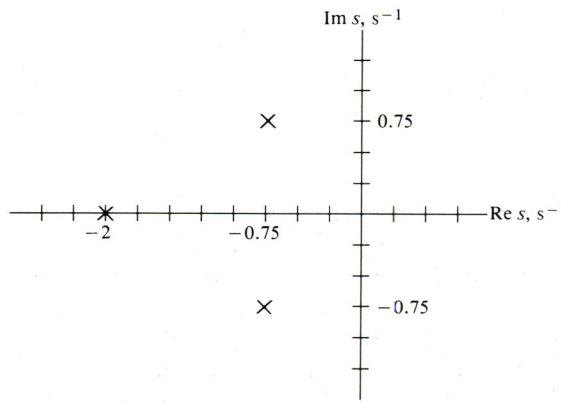

Figure P5-37

5-38 A series connection of all-pass biquads [see (5-32)] is to provide delay $t_0 \approx 100$ μs for any signal occupying the band $0 \le f \le 10$ kHz. How many biquads are necessary if $p = -40\pi$ krad/s and $q = 0.8$ for each one? Why are these reasonable values for p and q?

5-39 The parameters of an all-pass biquad described by (5-32) are $p = -10$ krad/s and $q = 0.6$. Obtain (approximately) the output of the biquad for input $x(t) = x_0 r(t/\tau)$ with $x_0 = 5$ V and $\tau = 500$ ms.

5-40 Figure P5-40 shows a pole-zero plot for a system having dc gain $H(0) = 1$. Obtain (approximately) the output of the system for input $x(t) = x_0 r(t/\tau)$ with $x_0 = 5$ V and $\tau = 1$ ms.

5-41 The system function of a certain system is

$$H(s) = \frac{K}{(s - p_1)(s - p_2)}$$

where $p_1 = -5$ s^{-1} and $p_2 = -100$ s^{-1}. The input is a rectangular pulse having duration τ. For what values of τ can the pole p_2 be ignored in obtaining the output?

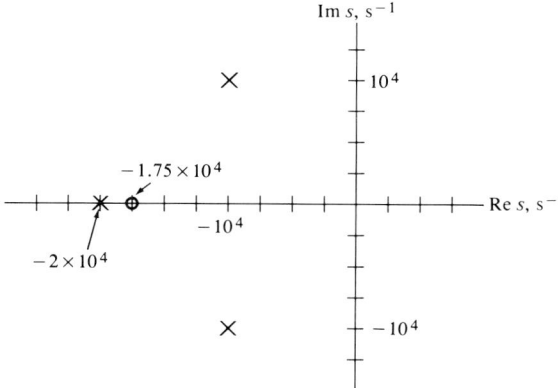

Figure P5-40

5-42 Obtain a dominant-pole approximation for each system function below. Give a condition on bandwidth of an input for which the approximation is valid.

(a) $H(s) = \dfrac{K}{s(s-p)}$

where $p = -10 \text{ s}^{-1}$.

(b) $H(s) = \dfrac{K(s-z)}{(s-p_1)(s-p_2)}$

where $z = -50 \text{ s}^{-1}$, $p_1 = -1 \text{ s}^{-1}$, and $p_2 = -55 \text{ s}^{-1}$.

(c) $H(s) = \dfrac{K(s-z)}{(s-p_1)(s-p_2)}$

where $z = -10 \text{ s}^{-1}$, $p_1 = -9 \text{ s}^{-1}$, and $p_2 = -100 \text{ s}^{-1}$.

(d) $H(s) = \dfrac{-\omega_0^2 p}{(s-p)(s^2 + 2\zeta\omega_0 s + \omega_0^2)}$

where $p = -5 \text{ s}^{-1}$, $\zeta = 0.5$, and $\omega_0 = 100 \text{ rad/s}$.

5-43 The system function of an all-pass biquad is

$$H(s) = \dfrac{s^2 - \gamma s + |p|^2}{s^2 + \gamma s + |p|^2}$$

(a) Write the system function as

$$H(s) = \dfrac{s^2 - 2\zeta\omega_0 s + \omega_0^2}{s^2 + 2\zeta\omega_0 s + \omega_0^2}$$

and express ζ and ω_0 in terms of p and γ.

(b) Using $H(s)$ obtained in part (a), obtain expressions for the step response of the biquad for $\zeta > 1$, $\zeta = 1$, and $\zeta < 1$.

(c) Find the maximum amplitude g_p of the step response as a function of ζ and ω_0. For what values of ζ does the step response exhibit overshoot?

(d) Let $|p| = 1$ s^{-1}. Plot percent overshoot PO versus damping ratio ζ for $0 \le \zeta \le 1$, where PO $= 100(g_p - 1)$.

(e) Describe the ways in which the biquad is different from and similar to an all-pole second-order system.

5-44 A second-order low-pass Butterworth filter has cutoff frequency $\omega_c = 1$ rad/s. Use a cut-and-try method to design an all-pass delay equalizer for the filter. The order of the equalizer should not exceed 2, and the delay for the equalized filter should be almost independent of frequency over the band $0 \le \omega \le \omega_c$.

5-45 Figure P5-45 shows a schematic diagram of a speed-control system. The input impedance of the amplifier is effectively infinite.

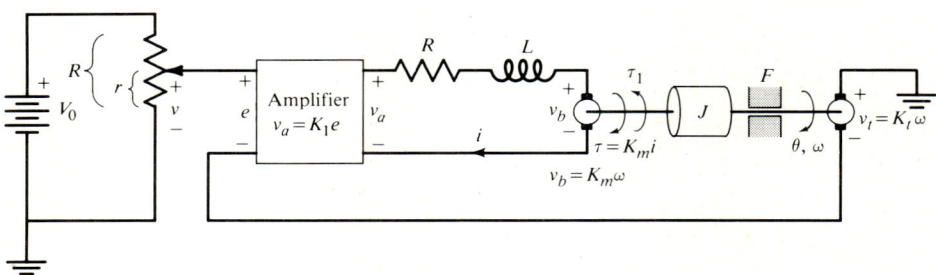

Figure P5-45

(a) Draw a block diagram for the system and obtain the system function of the system (input = voltage v, output = angular speed ω). Write the system function in the form

$$H(s) = \frac{H(0)\omega_0^2}{s^2 + 2\zeta\omega_0 s + \omega_0^2}$$

and express $H(0)$, ζ, and ω_0 in terms of the parameters of the system. Verify that the expressions are dimensionally correct.

(b) Let $V_0 = 10$ V, $K_1 = 5$ V/V, $K_m = 0.40$ Nm/A, $R = 0.50$ Ω, $L = 5.00$ mH, $J = 0.01$ Nm/(rad/s^2), $F = 0.25$ Nm/(rad/s), and $K_t = 0.25$ V/(rad/s). Find the peak time t_p and the percent overshoot PO $= 100[g(t_p) - H(0)]$ of the step response $g(t)$.

(c) Let the parameters have the values given in part (b). Find the full-scale steady-state speed (the steady-state speed for $v = V_0$).

(d) How should steady-state error be defined for the system? Assume that the potentiometer has linear taper and that $r = 0$ (fully counterclockwise) corresponds to $\omega = 0$ and $r = R$ (fully clockwise) corresponds to $\omega = H(0)V_0$. What is the steady-state error for $v = 0.5V_0$?

(e) Suppose the system is in steady state, with $v = 0.5V_0$. At $t = 0$, a load torque $\tau_1 = 1$ Nm is applied suddenly to the shaft in opposition to torque delivered by the motor. Determine the speed $\omega(t)$ for $t > 0$. Find the new steady-state speed. Find the steady-state error under the load [the desired speed is $0.5 H(0)V_0$]. What parameters should be changed to reduce steady-state error under load?

5-46 Obtain the one-sided Laplace transform of each of the following functions:

(a) $e(t) = r\left(\dfrac{t}{\tau}\right)$ (b) $e(t) = \tau\delta(t) + 5e^{-t/\tau}$ (c) $e(t) = t^2$

(d) $e(t) = \cos \omega_0 t$ (e) $e(t) = te^{pt}$ $p < 0$

5-47 Obtain the inverse one-sided Laplace transform of each of the following functions:

(a) $\tilde{E}(s) = \dfrac{\omega_0 s}{(s - p_1)(s - p_2)} \qquad p_1 < p_2$

(b) $\tilde{E}(s) = \dfrac{(s - z_1)(s - z_2)}{(s - p_1)(s - p_2)} \qquad z_1 < p_1 < z_2 < p_2$

(c) $\tilde{H}(s) = \dfrac{qr}{s(s^2 + qs + r)} \qquad \begin{array}{l} q = 2 \text{ s}^{-1} \\ r = 2 \text{ s}^{-2} \end{array}$

(d) $\tilde{H}(s) = \dfrac{e^{-st_0}}{s - p} \qquad \begin{array}{l} p < 0 \\ t_0 > 0 \end{array}$

5-48 Under what conditions is the relation

$$\mathcal{F}\{e(t)\} = \mathcal{L}_1\{e(t)\}\Big|_{s=j\omega}$$

valid?

5-49 Can we determine whether a system is realizable or not from the one-sided Laplace transform of the impulse response of the system? Justify your answer.

5-50 Use (5-42) to obtain an expression for the one-sided Laplace transform of

$$f(t) = \int_{-\infty}^{t} \int_{-\infty}^{\alpha} e(\beta) \, d\beta \, d\alpha \qquad \text{where } e(t) = 0 \text{ for } t < 0$$

5-51 The convolution property of the one-sided Laplace transformation is expressed by the relation

$$\mathcal{L}_1\left\{ \int_{-\infty}^{\infty} e(\lambda) f(t - \lambda) \, d\lambda \right\} = \tilde{E}(s)\tilde{F}(s)$$

Give conditions on $e(t)$ and $f(t)$ under which this relation holds.

5-52 The delay property of the one-sided Laplace transformation is expressed by the relation

$$\mathcal{L}_1\{e(t - t_0)\} = e^{-st_0}\tilde{E}(s)$$

Give conditions on $e(t)$ and t_0 under which this relation holds.

5-53 In the circuit of Fig. P5-53 the switch is open for $t \leq 0$ and closed for $t > 0$. Obtain an expression for the voltage $y(t)$.

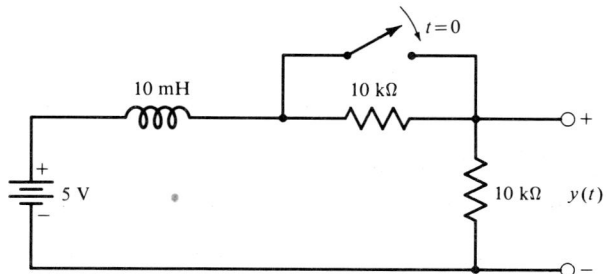

Figure P5-53

5-54 In the circuit of Fig. P5-54 the switch is closed for $t \leq 0$ and open for $t > 0$. Obtain an expression for the voltage $y(t)$.

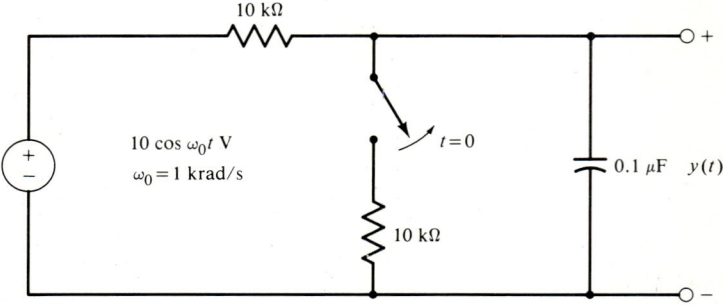

Figure P5-54

5-55 In the circuit of Fig. P5-55 the switch is closed for $t \leq 0$, open for $0 < t \leq t_1$, and closed for $t > t_1$, where $t_1 = 10$ ms. Obtain an expression for the voltage $y(t)$.

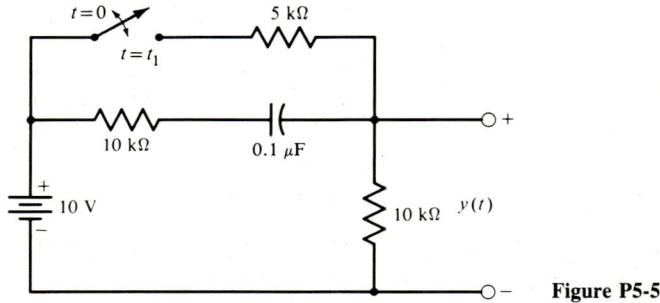

Figure P5-55

5-56 Refer to the relay servomechanism of Fig. 5-52. For $t < 0$, the system is at rest with $\dot{\theta}_o = 0$. Describe (a) effects on step response of varying (individually) K, K_m, i_0, F and (b) how the output for a step input varies with amplitude of the input.

5-57 In the servomechanism of Fig. 5-52 it is assumed that the relay closes (pulls in) for $|i| > i_0$ and opens (releases) for $|i| < i_0$. A real-life relay exhibits hysteresis, such that the magnitude of the pull-in current exceeds the magnitude of the release current. A real-life relay is modeled more accurately by the transfer characteristic of Fig. P5-57 than by that of Fig. 5-53. Use the transfer characteristic of Fig. P5-57 with pull-in current $i_1 = 180$ mA and release current $i_2 = 120$ mA to describe the relay in the servomechanism of Fig. 5-52 and determine the first four segments of the output $\theta_o(t)$ for input $\theta_i(t) = \pi u(t)$ rad. Use the values given in Example 5-47 for all other parameters.

5-58 For the relay servomechanism of Fig. 5-52 describe how increasing amplifier gain K affects steady-state error, rise time, and overshoot of step response.

5-59 Steady-state error of the relay servomechanism of Fig. 5-52 can be decreased by increasing amplifier gain K. Does this mean that K should be as large as possible? Why not?

5-60 For the relay servomechanism of Fig. 5-52 decreasing the width $2i_0$ of the relay deadband decreases steady-state error. Explain why it is inadvisable to make the deadband as small as possible.

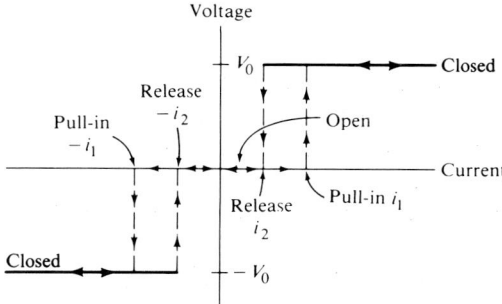

Figure P5-57

5-61 Increasing the voltage v_0 in the relay servomechanism of Fig. 5-52 has the same effect as increasing the motor torque constant K_m. Describe (qualitatively) how increasing v_0 affects overall performance (step response) of the servomechanism.

5-62 For the relay servomechanism of Fig. 5-52, $E_0 = 2.00$ V, $K = 3.14$ A/V, $i_0 = 150$ mA, $v_0 = 12.50$ V, $K_m = 0.40$ Nm/V, $J = 0$, and $F = 0.50$ Nm/(rad/s). Obtain the output $\theta_o(t)$ for input $\theta_i(t) = \pi u(t)$ rad. Describe how and why step response of this servomechanism differs from that of the linear dc servomechanism of Fig. 4-63 for the same load.

CHAPTER
SIX

DIGITAL SIGNALS AND SYSTEMS

Many signal-processing operations can be done digitally, as illustrated in Fig. 6-1. The AD (for analog to digital) digitizes regularly spaced samples of an analog input, e.g., voltage, $x(t)$. The output of the AD is a sequence of numbers B_0, $B_{\pm 1}$, $B_{\pm 2}$, ... representing values of the analog input at the sampling instants. The digital processor, e.g., a microcomputer, processes these numbers sequentially, according to some algorithm, and produces another sequence of numbers C_0, $C_{\pm 1}$, $C_{\pm 2}$, ... The DA (for digital to analog) constructs by interpolation (or otherwise) an analog output $y(t)$ from these numbers. The clock is a periodic signal that synchronizes operation of the AD, the digital processor, and the DA.

Digital signal processing is superior to conventional (analog) signal processing in many applications. Digital systems can be programmable, such that operations are determined by software rather than hardware. Consequently, they offer flexibility of purpose and standardization of hardware not easily obtained using analog systems. Also, digital systems are absolutely repeatable, so that one can build any number of systems having identical input-output properties. Further, digital systems generally are less sensitive to parameter variations and environmental disturbances than analog systems. Consequently, digital systems often require little or no in-field adjustment (tweaking). Finally, digital systems often are smaller and cheaper than comparable analog systems.

This is the first of three chapters devoted to digital signals and systems. In this chapter we analyze analog-to-digital and digital-to-analog conversion, we introduce block-diagram descriptions for digital systems (or digital signal-processing algorithms), and we discuss realizability, stability, fidelity, and other important concepts as they apply to digital systems.

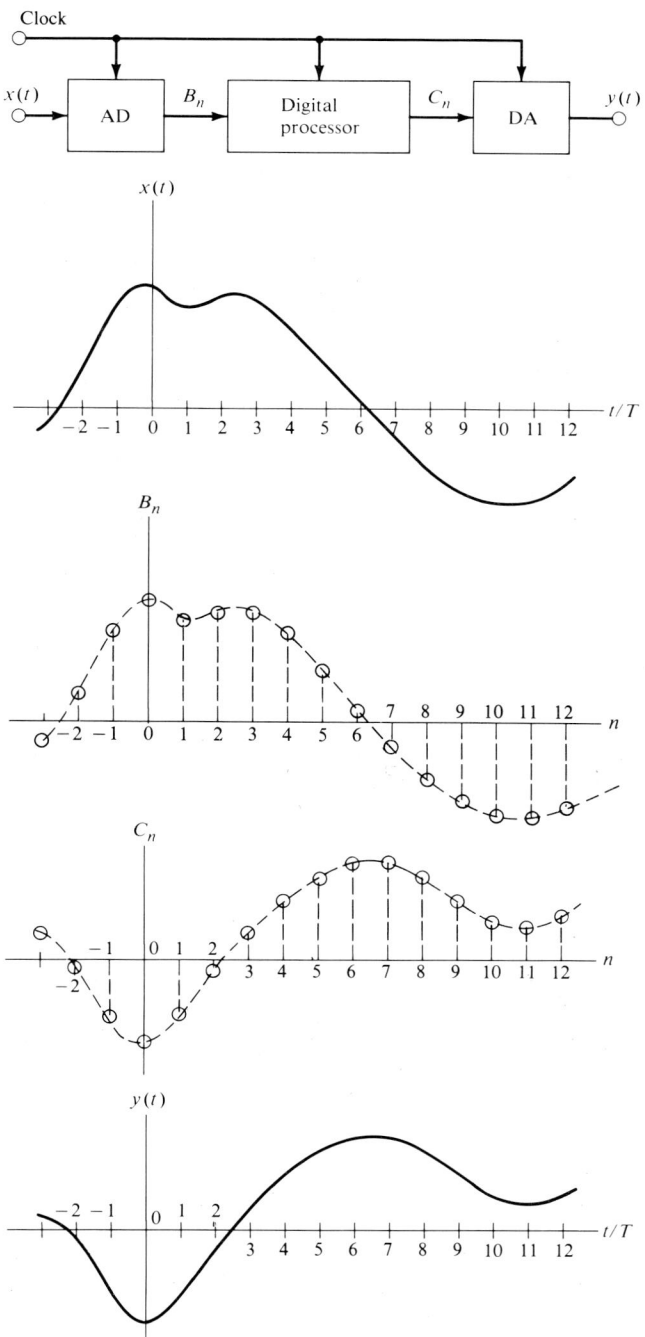

Figure 6-1 System for digital signal processing.

394 INTRODUCTION TO SYSTEM ANALYSIS

6-1 ANALOG-TO-DIGITAL AND DIGITAL-TO-ANALOG CONVERSION

In almost all applications the objective of digital signal processing is to emulate analog signal processing; e.g., the objective of digital filter design is to define a digital processor such that the system of Fig. 6-1 has some specified overall transfer function. Generally such emulation requires that a digital signal be a faithful representation of its analog counterpart. In any particular application this places constraints on sampling rate and resolution used in analog-to-digital (AD) and digital-to-analog (DA) conversion. In this section we analyze AD and DA conversion and develop guidelines for specifying sampling rate and resolution. We start by describing typical systems for AD and DA conversion in order to provide background for subsequent modeling and analysis.

6-1A Systems for AD and DA Conversion

Figure 6-2 shows a simplified block diagram for an electronic system for AD conversion, consisting of a *sample and hold* (S&H) and an *analog-to-digital converter*

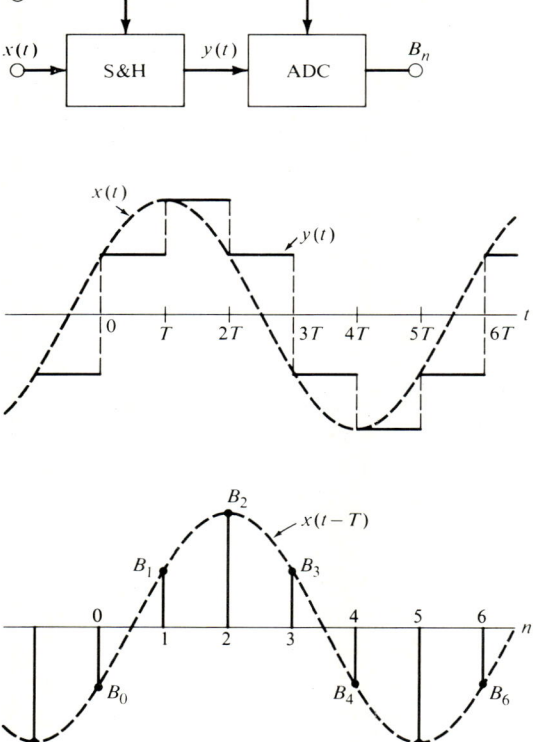

Figure 6-2 System for analog-to-digital conversion.

(ADC). The S&H samples an analog input $x(t)$, and the ADC digitizes the samples. These operations are synchronized by a *clock* (a periodic signal). On command from the clock (at $t = 0, \pm T, \pm 2T, \ldots$), the S&H takes a new sample of the analog input $x(t)$ and begins holding the value of that sample at the input of the ADC; this holding operation is necessary because the ADC requires time to digitize a sample. At the same time, the ADC passes the previously digitized sample to a digital processor (not shown) and begins digitizing the new sample. Note that AD conversion introduces delay T equal to one sampling interval because an ADC cannot digitize a sample instantaneously.

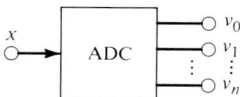

Figure 6-3 Analog-to-digital converter.

Figure 6-3 shows a block diagram for an ADC. The ADC outputs are voltages $v_0, v_1, \ldots, v_{n-1}$, which may take on one of two values, for example, $v_i = 0$ or $v_i = 5$ V. These n voltages represent n bits (binary digits) $b_0, b_1, \ldots, b_{n-1}$ of a binary code for the value of an input voltage x; for example, $b_i = 0$ if $v_i = 0$ and $b_i = 1$ if $v_i = 5$ V. For simplicity we assume that a signed binary code is used, where b_0 is a sign bit ($b_0 = 0$ if $x > 0$ and $b_0 = 1$ if $x < 0$) and the other bits $b_1, b_2, \ldots, b_{n-1}$ have place values $2^0, 2^1, \ldots, 2^{n-2}$, respectively; i.e., the outputs of the ADC represent a number B given by

$$B = \pm(b_{n-1}b_{n-2}\cdots b_1)_2 = \pm(b_1 + 2b_2 + \cdots + 2^{n-2}b_{n-1}) \tag{6-1a}$$

where the plus sign is used if $b_0 = 0$ and the minus sign is used if $b_0 = 1$, The *full-scale output* is

$$B_{FS} = 2^{n-1} - 1 \tag{6-1b}$$

The *full-scale input* is the smallest input that gives full-scale output. An input x whose magnitude exceeds the full-scale input of an ADC is represented by B_{FS} if $x > 0$ and by $-B_{FS}$ if $x < 0$. The *step size* is

$$x_0 = \frac{x_{FS}}{B_{FS}} \tag{6-1c}$$

where x_{FS} is the full-scale input.

Example 6-1 Obtain the transfer characteristic for a 4-bit ADC whose full-scale input is 2.1 V.

SOLUTION From (6-1b), the full-scale output is $B_{FS} = 2^3 - 1 = 7$. From (6-1c), the step size is $x_0 = 2.1/7 = 0.3$ V. Figure 6-4 shows the transfer characteristic for this ADC.

Figure 6-5 shows a simplified block diagram for an electronic system for DA conversion. The system consists of a *digital-to-analog converter* (DAC) and an S&H.

396 INTRODUCTION TO SYSTEM ANALYSIS

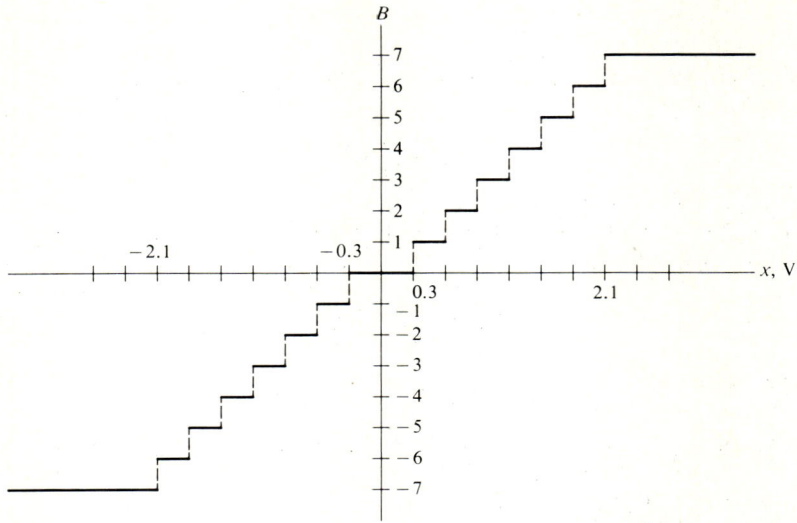

Figure 6-4 Transfer characteristic of a 4-bit ADC having full-scale output $v_{FS} = 2.1$ V.

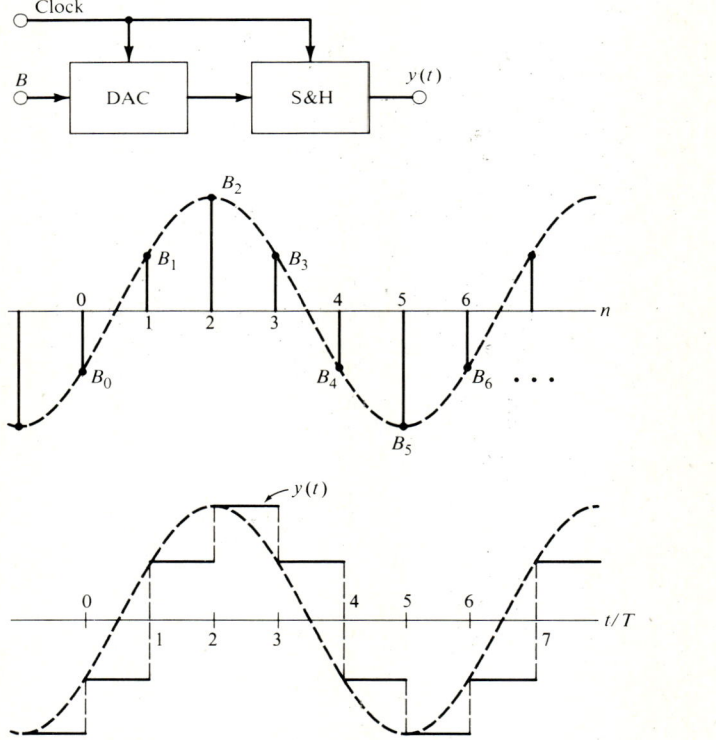

Figure 6-5 System for digital-to-analog conversion.

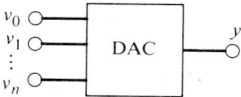

Figure 6-6 Digital-to-analog converter.

The DAC converts a digital input into a voltage. The S&H holds the voltage produced by the DAC until the next sampling instant. The clock synchronizes these operations. On command from the clock, the S&H samples the DAC output and begins holding that voltage at the S&H output. At the same time, the DAC begins decoding a new digital input.

Figure 6-6 shows a block diagram for a DAC. The inputs are voltages $v_0, v_1, \ldots, v_{n-1}$, which represent n bits of a signed binary number B as described above. The output is given by

$$y = By_0 \tag{6-2a}$$

where B is given by (6-1a) and y_0 is the *step size* of the DAC. The step size has the unit of the output. The *full-scale output* for the DAC is

$$y_{FS} = B_{FS} y_0 \tag{6-2b}$$

where B_{FS} is the full-scale input given by (6-1b).

Example 6-2 Obtain the transfer characteristic of a 4-bit DAC having step size $y_0 = 125$ mV.

SOLUTION From (6-1b) and (6-2b), the full-scale output is

$$y_{FS} = 7y_0 = 875 \text{ mV}$$

Figure 6-7 shows the transfer characteristic of the DAC.

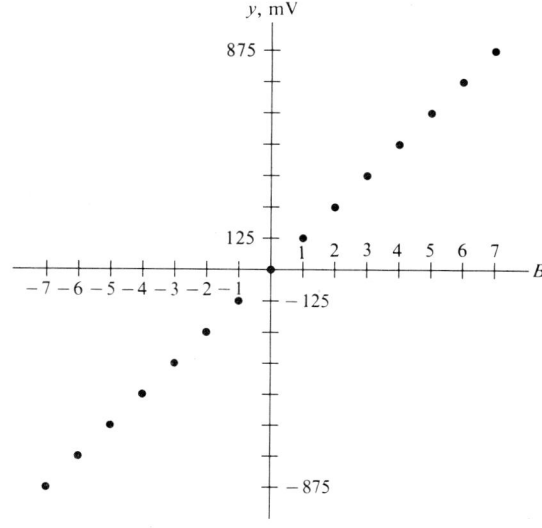

Figure 6-7 Transfer characteristic of a 4-bit DAC having step size $y_0 = 125$ mV.

398 INTRODUCTION TO SYSTEM ANALYSIS

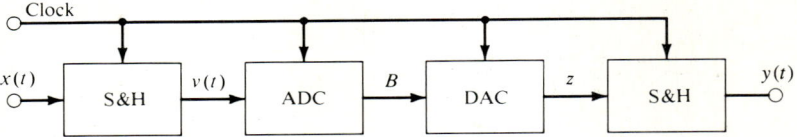

Figure 6-8 System considered in Example 6-3.

Example 6-3 In this example we describe operation of the system of Fig. 6-8. The ADC has full-scale input $x_{FS} = 2.1$ V and 4-bit output. The DAC has 4-bit input and step size $q = 125$ mV. The sampling rate (the clock frequency) is $F = 8$ kHz. Samples are taken at $t = 0, T, 2T, \ldots$, where $T = 1/F = 125$ μs. The analog input is a voltage given by $x(t) = 2 \cos \omega_0 t$ V, where $f_0 = 1$ kHz.

Figure 6-9 illustrates operation of this system. At $t = 0$ the first S&H samples the input and begins holding the sample value $x(0) = 2$ V at the input of the ADC. At the same time the second S&H samples the output of the DAC and begins holding that value (assumed to be zero) at the system output. During $0 < t < T$, the ADC produces the 4-bit binary representation 0110 for $x(0) = 2$ V (see Fig. 6-5), and the DAC determines the corresponding voltage $z = 750$ mV (see Fig. 6-7). At $t = T$ the first S&H samples the input and begins holding the sample value $x(T) = \cos \omega_0 T = 1.41$ V at the input of the ADC. At the same time the second S&H samples the output of the DAC and begins holding that value (750 mV) at the system output. During $T < t < 2T$ the ADC produces the 4-bit binary representation 0100 for $x(T) = 1.14$ V, and the DAC produces the corresponding voltage $z = 500$ mV. At $t = 2T$ the first S&H samples the input and begins holding the sample value $x(2T) = 0$ V at the input of the ADC. At the same time the second S&H samples the output of the DAC and begins holding that value (500 mV) at the system output. During $2T < t < 3T$ the ADC produces the 4-bit binary representation 0000 for $x(2T) = 0$, and the DAC produces the corresponding voltage $z = 0$ V. The reader should follow the operation of this system through a few more sampling periods.

6-1B Definition of Error Introduced by AD Conversion

Intuitively we can see that AD conversion error is decreased if samples are taken closer together or represented more precisely. On this basis, we would always use the largest possible sampling rate and code length; on the other hand, difficulty and cost of AD conversion, DA conversion, and digital processing increase with sampling rate and code length. In practice, we wish to use whatever sampling rate and code length are necessary, but we wish to avoid using unnecessarily large sampling rates and code lengths. To choose sampling rate and code length intelligently we need a quantitative measure of AD conversion error; i.e., we need a quantitative measure of the difference between an analog signal and its digital representation.

We cannot compare an analog signal and its digital representation directly. An analog signal is a dimensioned quantity, e.g., a voltage, whereas a digital signal is a sequence of numbers (is dimensionless). Further, an analog signal is defined at every instant, whereas a digital signal is defined only at sampling instants. When we say that a digital signal provides a faithful representation of an analog signal, we mean that the analog signal can be recovered from the digital signal with little error. Thus *AD conversion error* is defined as distortion introduced by the system of Fig. 6-10. Below,

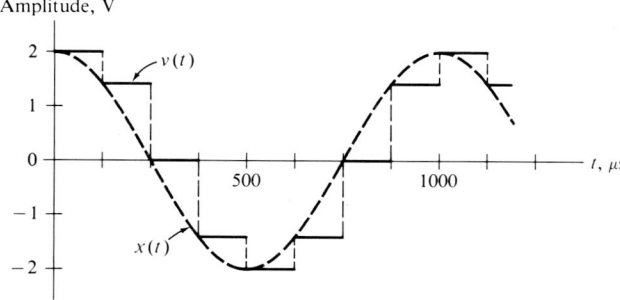
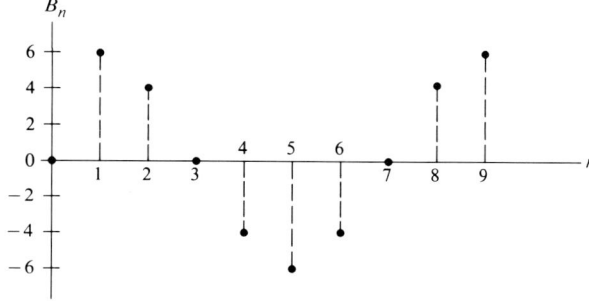
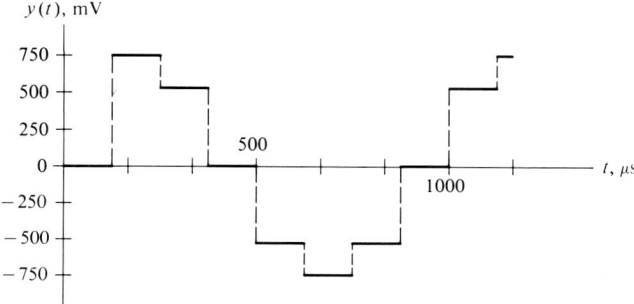

Figure 6-9 Signals illustrating operation of the system of Fig. 6-8.

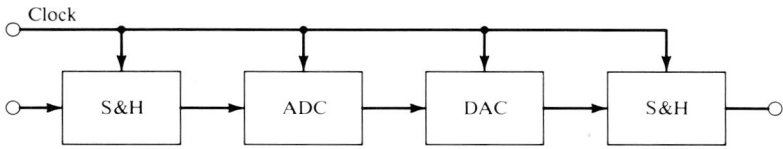

Figure 6-10 Block diagram illustrating the definition of error introduced by AD conversion.

400 INTRODUCTION TO SYSTEM ANALYSIS

we introduce mathematical models for the components of this system and use those models to analyze AD conversion error.

6-1C Quantizing Error

The ADC-DAC series combination in the system of Fig. 6-10 is modeled by the system of Fig. 6-11. The delay element accounts for the time (one sampling interval) required by the ADC and DAC to digitize and "undigitize" a sample (see Example 6-3). The quantizer in the model of Fig. 6-11 is a nonlinear static element whose transfer characteristic is shown in Fig. 6-12. An output x for input v is given by

$$x = \begin{cases} -x_{FS} = -B_{FS}x_0 & v < -v_{FS} \\ x_0 \text{ sgn } v \text{ int}\left|\dfrac{v}{v_0}\right| & -v_{FS} < v < v_{FS} \\ x_{FS} = B_{FS}x_0 & v > v_{FS} \end{cases} \qquad (6\text{-}3)$$

where v_0 = step size of the ADC
v_{FS} = full-scale input of the ADC
B_{FS} = full-scale output of the ADC
x_0 = step size of the DAC
x_{FS} = full-scale output of the DAC

$$\text{sgn } \alpha = \begin{cases} -1 & \alpha < 0 \\ 1 & \alpha \geq 0 \end{cases}$$

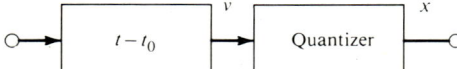

Figure 6-11 Model for the system of Fig. 6-10.

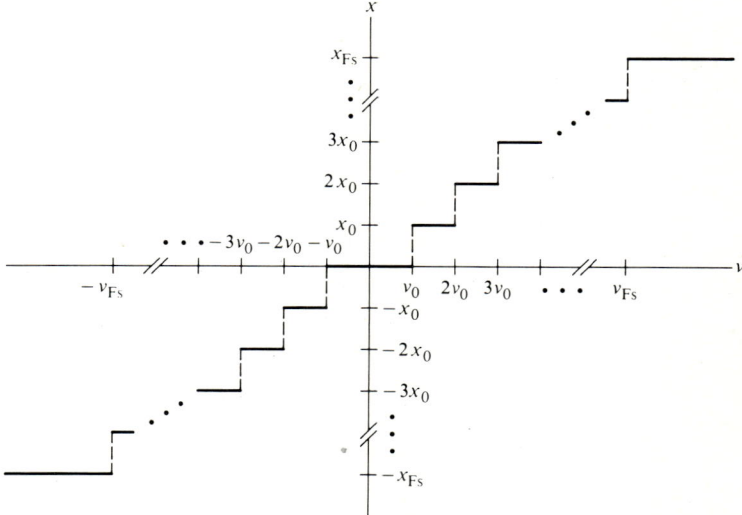

Figure 6-12 Transfer characteristic for a quantizer.

and int α denotes the largest integer less than α, for example, int $2.3 = 2$ and int$(-2.3) = -3$.

Example 6-4 An ADC and a DAC are connected in series, as in Fig. 6-10. Sampling rate is 10 kHz, full-scale input of the ADC is 1.5 V, step size of the DAC is 20 mA, and code length is 5 bits. Obtain a model for this system as shown in Fig. 6-11.

SOLUTION The delay introduced by the ADC-DAC series combination is one sampling interval; thus the delay time in the model of Fig. 6-11 is $T = 1/F = 100 \ \mu\text{s}$. The parameters of the quantizer are determined as follows. From (6-1), the full-scale output of the ADC is $B_{FS} = 2^{5-1} - 1 = 15$ and the step size of the ADC is $v_0 = 1.5/15 = 0.1$ V; from (6-2), the full-scale output of the DAC is $x_{FS} = (15)(20) = 300$ mA. Figure 6-13 shows the transfer characteristic of the quantizer (for $v > 0$).

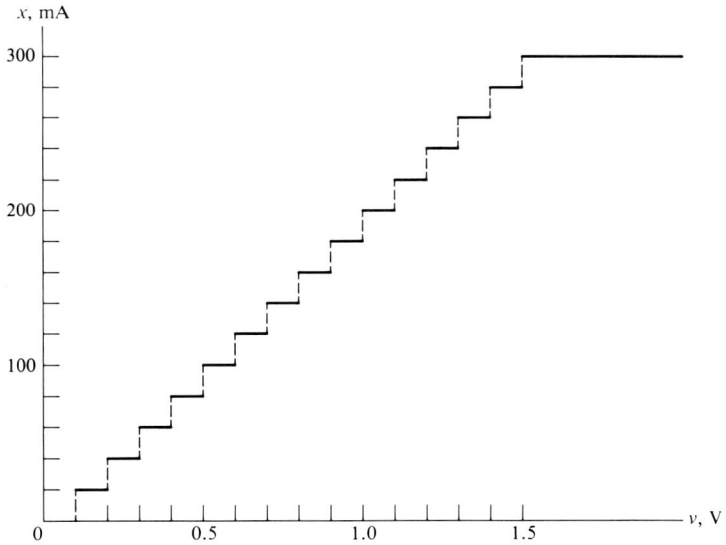

Figure 6-13 Transfer characteristic for the quantizer of Example 6-4.

It is convenient to express the quantizer transfer characteristic of (6-3) in the form

$$x = \frac{x_0}{v_0} \begin{cases} -B_{FS} v_0 & v < -v_{FS} \\ v_0 \ \text{sgn} \ v \ \text{int} \left| \frac{v}{v_0} \right| & -v_{FS} < v < v_{FS} \\ B_{FS} v_0 & v > v_{FS} \end{cases} \qquad (6\text{-}4)$$

This allows the quantizer of Fig. 6-12 to be represented by the series connection of a proportion element having gain $K = x_0/v_0$ and a *unit-gain quantizer* whose full-scale input and full-scale output are equal (Fig. 6-14). Figure 6-15 shows the transfer characteristic for a unit-gain quantizer.

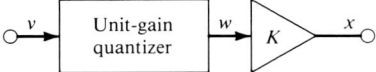

Figure 6-14 Separation of a quantizer into a proportion element and a unit-gain quantizer.

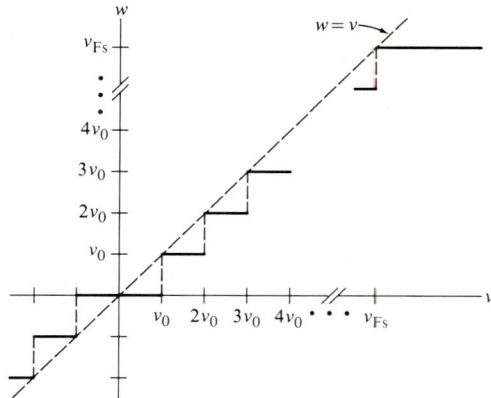

Figure 6-15 Transfer characteristic for a unit-gain quantizer.

The development above shows that the ADC-DAC series combination in the system of Fig. 6-10 can be modeled as shown in Fig. 6-16. The delay time T is one sampling interval. The transfer characteristic for the unit-gain quantizer is given by (see Fig. 6-15)

$$w = \begin{cases} -B_{FS}v_0 & v < -v_{FS} \\ v_0 \operatorname{sgn} v \operatorname{int} \left| \dfrac{v}{v_0} \right| & -v_{FS} < v < v_{FS} \\ B_{FS}v_0 & v > v_{FS} \end{cases} \quad (6\text{-}5)$$

where v_{FS} = full-scale input for the ADC
B_{FS} = full-scale output for the ADC
$v_0 = v_{FS}/B_{FS}$ = step size for the ADC

The gain of the proportion element is x_0/v_0, where x_0 is the step size for the DAC. In examining error (distortion) introduced by this system (by the ADC and DAC in the system of Fig. 6-10) we can ignore the delay element and the proportion element, because they provide distortionless transmission for any input. The quantizer is the only source of distortion in the system of Fig. 6-16.

The transfer characteristic for the unit-gain quantizer in the system of Fig. 6-16 is a staircase approximation to the (distortionless) transfer characteristic $w = v$, represented by the dashed line in Fig. 6-15. The quantizer introduces distortion (error) because its transfer characteristic departs from this relation. For input v, the distortion is the difference between the desired output given by $w = v$ and the actual output given by (6-5). This error has two components: *overload error* (peak clipping), which occurs when the magnitude of an input exceeds the full-scale input, and *quantizing error*, which occurs when the magnitude of an input is smaller than the full-scale input. In practice, overload error is avoided by proper scaling; say by attenuating an input; e.g.,

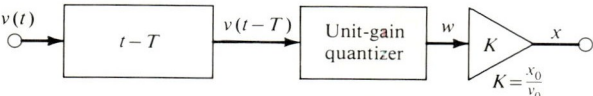

Figure 6-16 Model for the ADC-DAC series combination in the system of Fig. 6-10.

if a signal having peak amplitude 5 V is to be digitized using an ADC having full-scale input 2 V, the signal would be scaled by the factor $2/5 = 0.4$ before AD conversion. Consequently, overload error occurs only rarely in a well-designed system for AD conversion. Quantizing error is the dominant error introduced by digitizing a signal. In practice, specifications on quantizing error are often given in terms of ratio of peak signal to peak error expressed in decibels (dB) as

$$S/Q = 20 \log\left(\frac{S}{Q}\right) \qquad (6\text{-}6)$$

where S/Q = *signal-to-quantizing-error ratio*, in decibels
 S = peak amplitude of input
 Q = peak amplitude of quantizing error

Acceptable values for S/Q are different in different applications; e.g., digital representation of telephone-quality speech requires $S/Q \approx 40$ dB, and digital representation of studio-quality TV video requires $S/Q \approx 55$ dB.

In a well-designed system, where inputs are scaled to avoid overload error, peak amplitude S of an input equals full-scale input v_{FS} for the quantizer; thus

$$S = v_{FS} = B_{FS}v_0 = (2^{n-1} - 1)v_0 \qquad (6\text{-}7a)$$

where n is number of bits per sample (including the sign bit) and v_0 is step size for the quantizer. In most applications code length n is large enough for

$$S \approx 2^{n-1}v_0 \qquad (6\text{-}7b)$$

to be a good approximation. From Fig. 6-15 we see that the maximum magnitude of quantizing error is $0.5v_0$, where v_0 is step size; consequently,

$$Q = 0.5v_0 \qquad (6\text{-}8)$$

Using (6-7b) and (6-8) in (6-6) gives

$$S/Q = 20 \log 2^n = 6n \quad \text{dB} \qquad (6\text{-}9)$$

This shows that the ratio of peak signal to peak error equals $6n$ dB, where n is code length, provided $2^{n-1} - 1 \approx 2^{n-1}$ is a good approximation (in practice, this requires only $n \geq 4$). Equation (6-9) indicates that S/Q is a function of code length only. This is true only if overload is avoided (by proper scaling). Also, (6-8) holds only if code length is reasonably large ($n \geq 4$); otherwise, the relation between S/Q and code length is more complicated. Both of these conditions are met in almost all practical applications.

Example 6-5 Digital transmission of telephone-quality speech requires $S/Q \approx 40$ dB. From (6-8), the code length required in this application is $n = 7$ (code length must be an integer). In practice, an 8-bit code generally is used in this application. This provides insurance against unexpected interference, occasional overload, and variations of system parameters. An 8-bit code is also convenient from the viewpoint of a digital system designer.

In signal-processing applications, the matter of specifying code length (number of bits) is often dominated by a requirement for a certain *dynamic range*, that is, by a requirement to detect a small signal in the presence of a large one. For example, suppose that a voltage to be digitized has two components: a signal component having peak amplitude A and a noise component having peak amplitude B, where $A \ll B$. If we choose code length on the basis of B alone, we may find that the smaller signal is swamped by quantizing error. In such cases, full-scale input is determined by signal plus noise, and step size is determined by signal alone; together, full-scale input and step size determine code length.

Example 6-6 A voltage to be digitized has two components: a noise voltage whose peak amplitude is 5.00 V and a signal voltage whose peak amplitude is 500 mV. It is necessary that S/Q for the signal voltage exceed 15 dB after digitizing. Determine the code length required.

SOLUTION The peak amplitude of the input (signal plus noise) determines the full-scale input $v_{FS} = 5.50$ V. The specified S/Q for the small signal voltage determines step size v_0 according to (6-6), where S is the peak amplitude for the small signal and Q is peak quantizing error given by (6-8); thus,

$$S/Q = 15 = 20 \log \frac{S}{Q}$$

where $S = 0.50$ V and $Q = 0.50v_0$. Solving for step size v_0 gives

$$v_0 = \frac{2S}{10^{0.75}} = 180 \text{ mV}$$

The number of positive output levels given by $N = 2^{n-1} - 1$ must equal or exceed v_{FS}/v_0; this gives

$$2^{n-1} - 1 \geq \frac{5.50}{0.18}$$

from which we find $n = 6$. If the large noise voltage were absent, the required value for S/Q could be obtained using fewer bits ($6n \geq 15$ gives $n = 3$).

6-1D Sampling Error

Next, we examine errors introduced by sampling and boxcar (sample-and-hold) reconstruction used in the system of Fig. 6-10. In this analysis sampling and boxcar reconstruction are modeled as shown in Fig. 6-17, where a signal is multiplied by a periodic impulse train $d(t)$, given by

$$d(t) = T \sum_{n=-\infty}^{\infty} \delta(t - nT) \qquad (6\text{-}10)$$

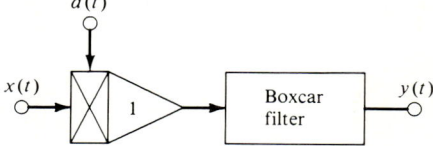

Figure 6-17 Model for sampling and boxcar (S&H) reconstruction.

where T is sampling interval. The sampled signal is given by

$$y(t) = x(t)d(t) = Tx(t) \sum_{n=-\infty}^{\infty} \delta(t - nT) \qquad (6\text{-}11a)$$

which can also be written

$$y(t) = T \sum_{n=-\infty}^{\infty} x(nT)\delta(t - nT) \qquad (6\text{-}11b)$$

because the delta functions are nonzero only at the sampling instants. Equation (6-11) shows that the sampled signal $y(t)$ is a train of impulses whose strengths (areas) are proportional to values of $x(t)$ at the corresponding sampling instants.

The *boxcar filter* in the model of Fig. 6-17 is a linear stationary system having impulse response

$$h(t) = \frac{1}{T} r\left(\frac{t}{T}\right)$$

It follows that input $Tx(nT)\delta(t)$ produces output $x(nT)r(t/T)$. By superposition and time invariance, the output for input $y(t)$ given by (6-11) is a boxcar approximation to the sampler input $x(t)$, as illustrated in Fig. 6-18.

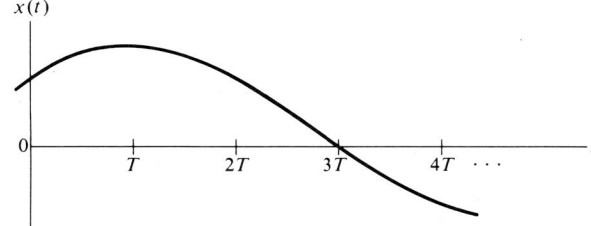

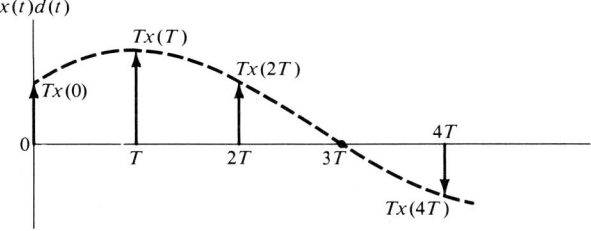

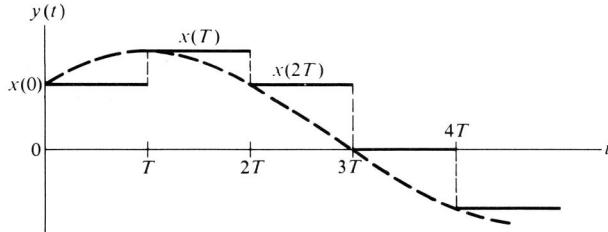

Figure 6-18 Boxcar reconstruction for an ideally sampled signal.

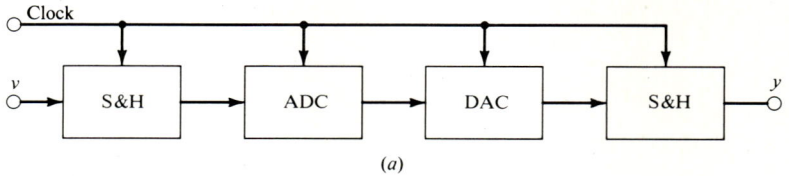

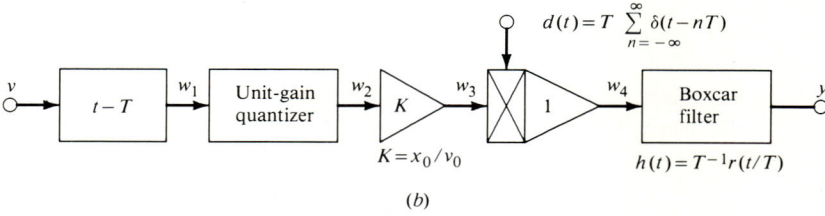

Figure 6-19 System of Example 6-7: (a) block diagram; (b) model.

Example 6-7 (Compare with Example 6-3). Model the system of Fig. 6-19a as shown in Fig. 6-19b. The sampling rate is $F = 8$ kHz, and the input is a sinusoidal voltage given by

$$v(t) = 2 \cos \omega_0 t \quad \text{V}$$

where $f_0 = 1$ kHz. The full-scale input for the ADC is $v_{FS} = 2.1$ V, the code length is $n = 4$ bits, and the step size for the DAC is $x_0 = 125$ mV.

SOLUTION The transfer characteristic for the quantizer in the system of Fig. 6-19b is given by (6-5), where $B_{FS} = 2^3 - 1 = 7$ and $v_0 = v_{FS}/7 = 0.3$ V. The gain of the proportion element is $K = x_0/v_0 = 0.125/0.300 = 0.417$ V/V, and the delay time (one sampling interval) is $T = 1/F = 125$ μs.

Figure 6-20 shows the signals that appear in this model. The output $z(t)$ of the model is identical to the output $z(t)$ of the system of Example 6-3.

Sampling introduces error to the extent that samples of a signal fail to describe the signal between sampling instants. It is intuitively clear that a rapidly changing signal must be sampled more frequently than a slowly changing signal. Below, we express this idea quantitatively by obtaining a relation between bandwidth of a signal and sampling rate required for high-fidelity reconstruction of the signal from its samples.

In the ideal sampler of Fig. 6-21 the period of the impulse train $d(t)$ is the sampling interval T. *Sampling rate* F is the reciprocal of sampling interval and sampling rate in radians per second is denoted by W; thus,

$$F = \frac{1}{T} \qquad W = 2\pi F = \frac{2\pi}{T} \qquad (6\text{-}12)$$

We wish to prove the following *sampling theorem:*

A strictly bandlimited signal can be recovered without error from samples of the signal provided the sampling rate exceeds twice the bandwidth of the signal.

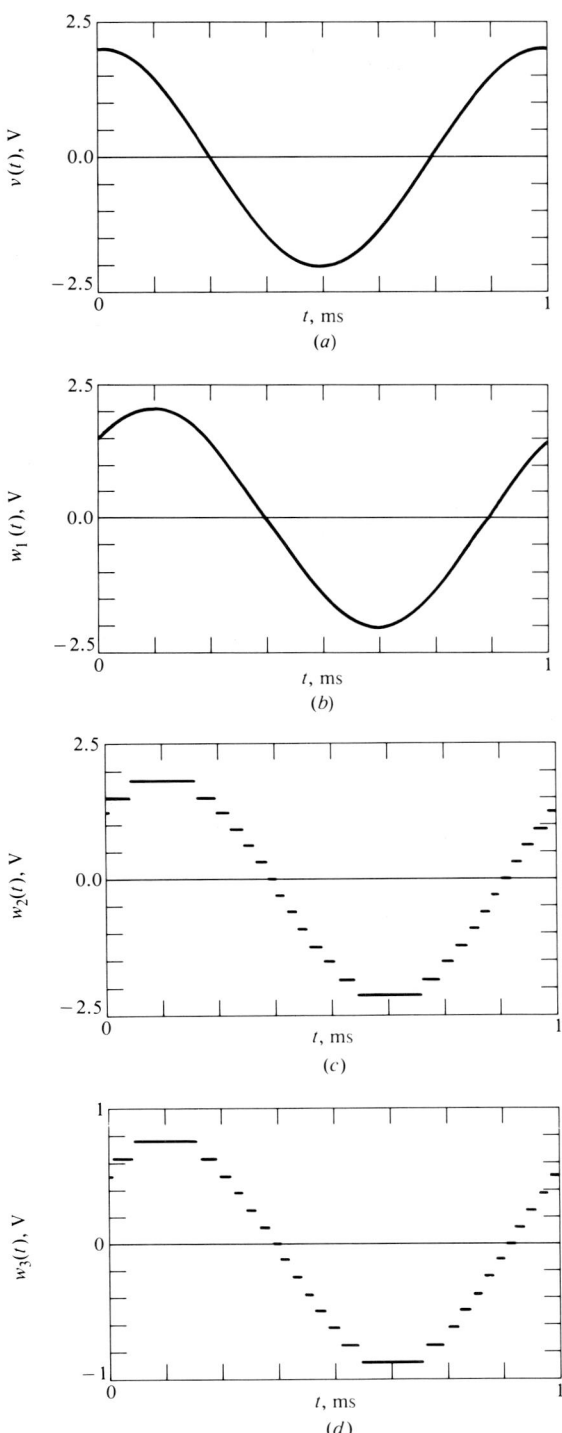

Figure 6-20 Signals appearing in the model of Fig. 6-19b.

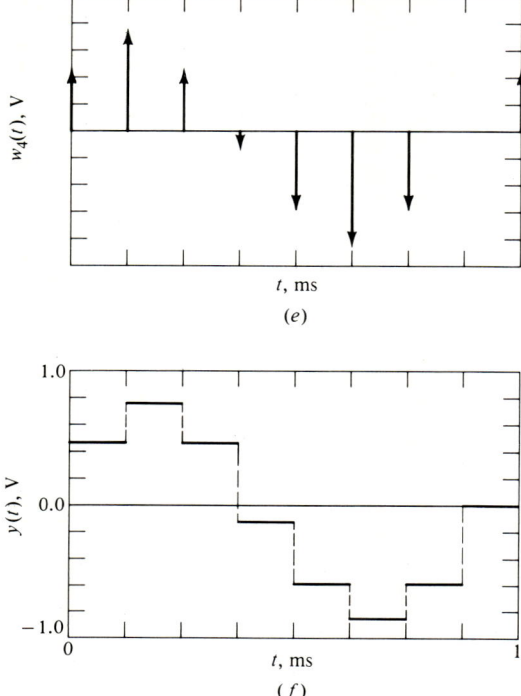

Figure 6-20 *Continued*

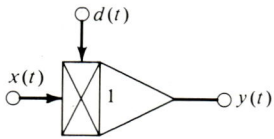

Figure 6-21 Ideal sampler.

Proving this theorem requires only inspection of the spectral density of a sampled signal.

To obtain an expression for the spectral density of a sampled signal $y(t)$ given by (6-11) we first determine the Fourier series for the impulse train $d(t)$ of (6-10). This gives

$$d(t) = \sum_{n=-\infty}^{\infty} D_n e^{jnWt}$$

where

$$D_n = \frac{1}{T} \int_{-T/2}^{T/2} T\delta(t) e^{-jnWt} \, dt = 1$$

Thus we have

$$d(t) = \sum_{n=-\infty}^{\infty} e^{jnWt}$$

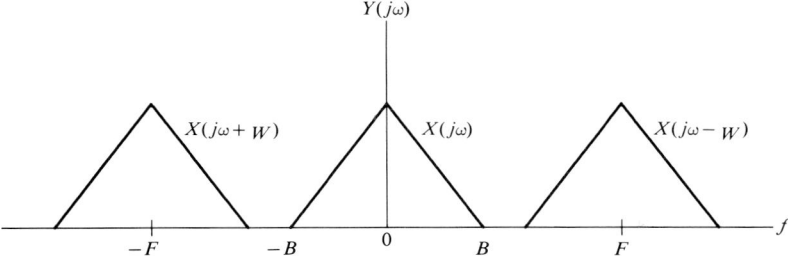

Figure 6-22 Spectral density for an ideally sampled signal.

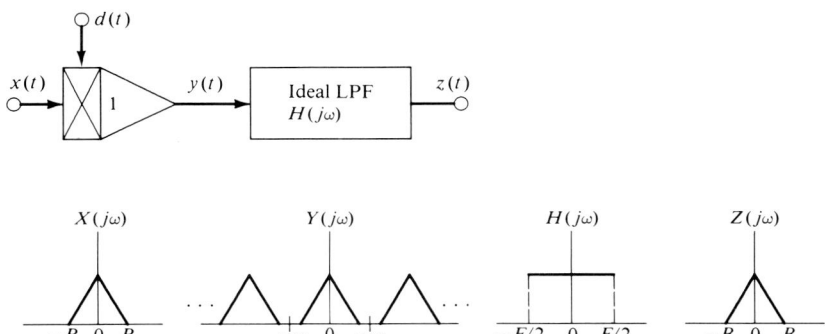

Figure 6-23 Recovery of a signal from samples of the signal by ideal low-pass filtering.

Using this expression in (6-11a) gives

$$y(t) = \sum_{n=-\infty}^{\infty} x(t) e^{jnWt}$$

From (4-23) and (4-29), the spectral density (the Fourier transform) of $y(t)$ above is given by

$$Y(j\omega) = \sum_{n=-\infty}^{\infty} X(j\omega - jnW) \tag{6-13}$$

This shows that the spectral density of a sampled signal is a superposition of translated replicas of the spectral density of the original signal. Figure 6-22 illustrates this result for a case where the original signal $x(t)$ is strictly bandlimited (to $f < B$) and the sampling rate exceeds twice the bandwidth of $x(t)$ ($F > 2B$). Under these conditions the original signal $x(t)$ can be recovered without error from the sampled signal $x(t)$ by ideal low-pass filtering, as illustrated in Fig. 6-23. This proves the sampling theorem above.

Example 6-8 A sinusoidal signal $x(t) = A \cos \omega_0 t$ is applied to the input of the system of Fig. 6-23. The sampling rate is 4 times the frequency of the sinusoid. Obtain the spectral density of the sampled signal $y(t)$ and the spectral density of the output $z(t)$.

410 INTRODUCTION TO SYSTEM ANALYSIS

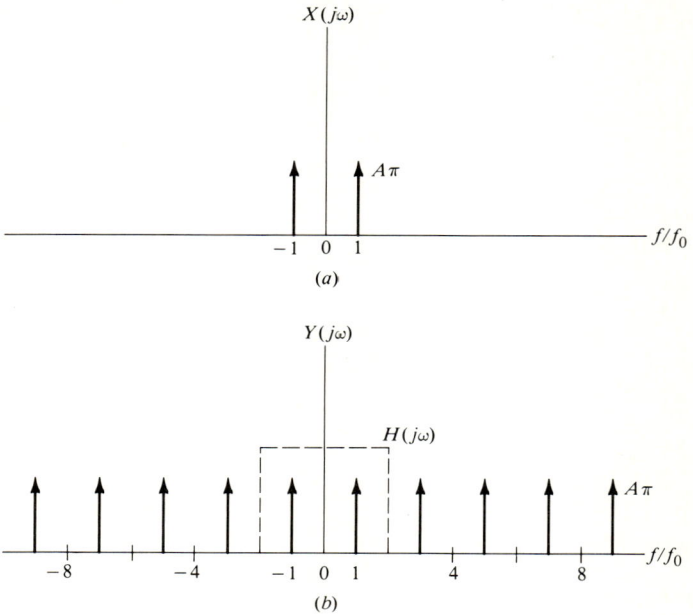

Figure 6-24 Ideal sampling and exact recovery of a sinusoidal signal (Example 6-8).

SOLUTION From Table C-1, the spectral density of the sinusoid $x(t)$ is

$$X(j\omega) = A\pi[\delta(\omega - \omega_0) + \delta(\omega + \omega_0)]$$

From (6-13), the spectral density of the sampled signal $y(t)$ is

$$Y(j\omega) = A\pi \sum_{n=-\infty}^{\infty} [\delta(\omega - \omega_0 - nW) + \delta(\omega + \omega_0 - nW)]$$

where $W = 4\omega_0$ is sampling rate in radians per second. Figure 6-24 shows the spectral densities of the sinusoidal input $x(t)$ and the sampler output $y(t)$.

The transfer function for the low-pass filter in the system of Fig. 6-23 is

$$H(j\omega) = \begin{cases} 1 & -\dfrac{W}{2} < \omega < \dfrac{W}{2} \\ 0 & \text{otherwise} \end{cases}$$

as indicated by the dashed line in Fig. 6-24b. The filter passes the components of $y(t)$ at $\omega = \pm\omega_0$ and stops all other components. The spectral density of the filter output is

$$Z(j\omega) = A\pi[\delta(\omega - \omega_0) + \delta(\omega + \omega_0)] = X(j\omega)$$

This shows that the input $x(t)$ is recovered without error from the sampled signal, as asserted by the sampling theorem above.

Neither ideal sampling nor error-free recovery can be accomplished by any physical system because no signal is strictly bandlimited and because the operations shown in Fig. 6-23 are impossible. Sampling introduces two kinds of error. One arises

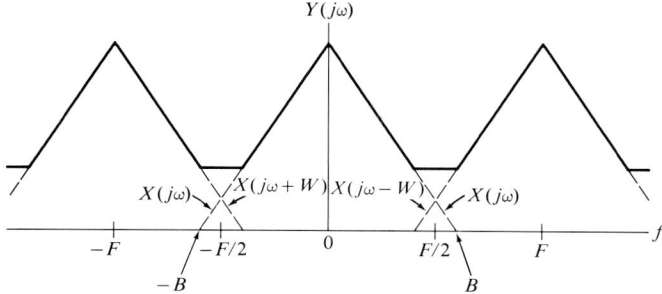

Figure 6-25 Spectral density for an undersampled signal.

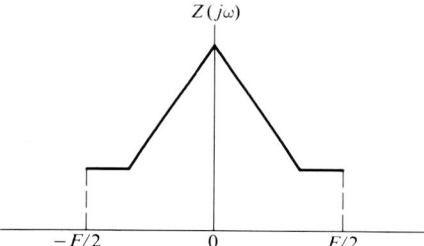

Figure 6-26 Illustration that an undersampled signal cannot be recovered from its samples.

from *undersampling,* where a signal is sampled at a rate less than twice the bandwidth of the signal. The other arises because neither ideal (instantaneous, perfectly periodic) sampling nor ideal low-pass filtering is possible.

Figure 6-25 shows the spectral density given by (6-13) for a case where a signal $x(t)$ is undersampled. Attempting to recover the signal $x(t)$ by low-pass filtering of the sampled signal $y(t)$, as in Fig. 6-23, yields the signal $z(t)$, whose spectral density $Z(j\omega)$ is shown in Fig. 6-26. Because $Z(j\omega) \neq X(j\omega)$, the original signal $x(t)$ is not recovered without error. Error introduced by undersampling is often called *foldover error* or *alias error*. The term foldover error is suggested by comparing $Z(j\omega)$ of Fig. 6-26 with $X(j\omega)$ of Fig. 6-22. The spectral density $Z(j\omega)$ looks like the original spectral density $X(j\omega)$ except that the tails of $X(j\omega)$ have been folded about $f = \pm F/2$ toward $f = 0$. The term *alias error* is suggested by the fact that components lying above $F/2$ appear under aliases (at other frequencies) in the reconstructed signal.

Example 6-9 This example illustrates alias error using a sinusoidal signal $x(t) = A\cos\omega_0 t$ as the input of the system of Fig. 6-23, where the sampling rate is $W = 1.5\omega_0$; that is, the sampling rate is less than twice the bandwidth of the input.

As shown in Example 6-8, the spectral density of the sampled signal is

$$Y(j\omega) = A\pi \sum_{n=-\infty}^{\infty} [\delta(\omega - \omega_0 - nW) + \delta(\omega + \omega_0 - nW)]$$

This spectral density is shown in Fig. 6-27, where the dashed line represents the transfer function for the ideal low-pass filter in the system of Fig. 6-23. The spectral density of the filter output is

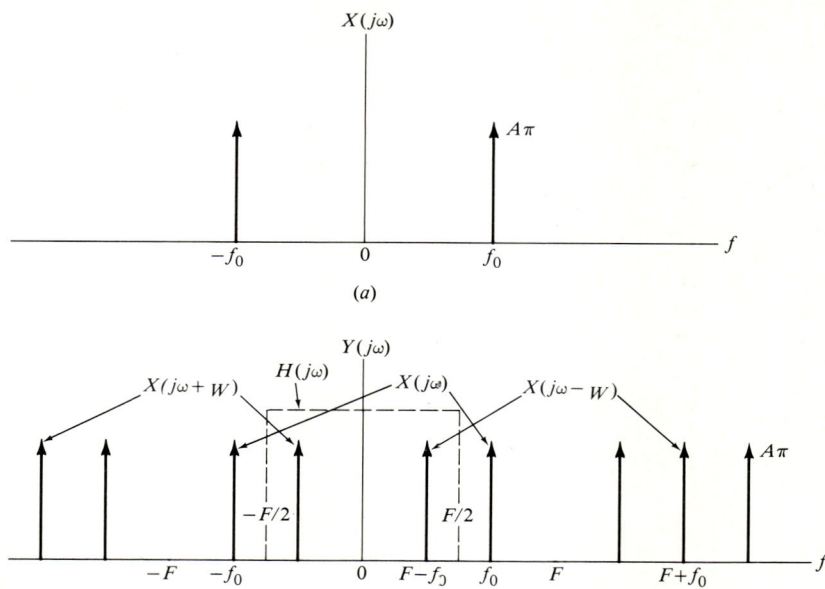

Figure 6-27 Spectral density of an undersampled sinusoid (Example 6-9).

$$Z(j\omega) = A\pi[\delta(\omega - 0.5\omega_0) + \delta(\omega + 0.5\omega_0)]$$

and the filter output is given by

$$z(t) = A \cos 0.5\omega_0 t$$

Although the recovered signal $z(t)$ is a sinusoid having the same peak amplitude as the input $x(t)$, the frequency of $z(t)$ is different from that of $x(t)$. Attempting to recover an undersampled sinusoid by low-pass filtering causes the sinusoid to appear under an alias. In general, attempting to recover an undersampled signal by low-pass filtering causes all components lying above half the sampling rate F to appear under aliases in the band $0 < f < F/2$.

Alias error introduced by undersampling a signal can be defined quantitatively in terms of the power density spectrum for the signal (Fig. 6-28a). If a signal $x(t)$ is sampled (ideally) and then reconstructed using an ideal low-pass filter, as in Fig. 6-23, the normalized average power of the alias error is given by

$$P_a = \int_{F/2}^{\infty} p_x(f) \, df \qquad (6\text{-}14a)$$

where $p_x(f)$ is the power density spectrum for the signal and F is sampling rate. The normalized average power of components passed by the system without alias error is given by

$$P_b = \int_{0}^{F/2} p_x(f) \, df \qquad (6\text{-}14b)$$

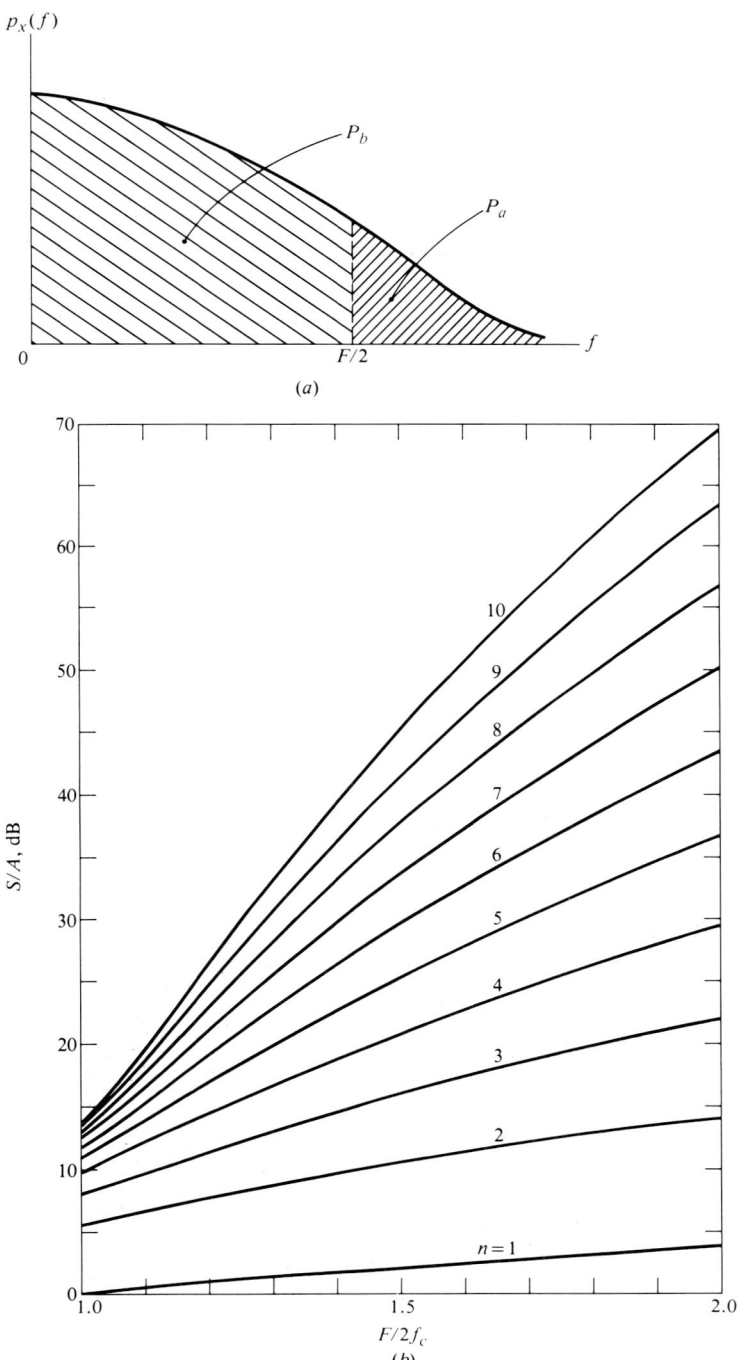

Figure 6-28 S/A ratio versus ratio of sampling rate to twice signal bandwidth for a signal having an nth-order Butterworth power density spectrum.

The signal-to-alias-error ratio (S/A) is expressed in decibels by

$$S/A = 10 \log \frac{P_b}{P_a} \tag{6-14c}$$

Example 6-10 A signal $x(t)$ has power density spectrum

$$p_x(f) = \frac{p_0}{1 + (f/f_c)^2}$$

This signal is sampled (ideally) with sampling rate $F = 4f_c$. The sampled signal is applied to an ideal low-pass filter having bandwidth $F/2$. Calculate the S/A ratio defined by (6-14).

SOLUTION From (6-14a), the normalized power of the alias error is

$$P_a = \int_{F/2}^{\infty} p_x(f)\,df = p_0 f_c \left(\frac{\pi}{2} - \tan^{-1} \frac{F}{2f_c} \right)$$

$$= p_0 f_c (0.5\pi - \tan^{-1} 2) = 0.47 p_0 f_c$$

From (6-14b), the normalized power of the nonalias component of the reconstructed signal is

$$P_b = \int_0^{F/2} p_x(f)\,df = p_0 f_c \tan^{-1} \frac{F}{2f_c}$$

$$= p_0 f_c \tan^{-1} 2 = 1.11 p_0 f_c$$

From (6-14c), the signal-to-error ratio is

$$S/A = 10 \log \frac{P_b}{P_a} = 10 \log \frac{1.11}{0.47} = 3.73 \text{ dB}$$

In practice, alias error is controlled by using a sharp-cutoff low-pass *anti-alias filter* before sampling and by using a sampling rate somewhat greater than twice the bandwidth of the anti-alias filter; e.g., in systems for digital transmission of telephone-quality speech, signals representing speech are often bandlimited to about 3 kHz (by an anti-alias filter) before being sampled at 8 kHz. The bandwidth of an anti-alias filter is determined by the bandwidth required for satisfactory transmission of signals to be digitized. Sampling rate depends on required S/A ratio and on how sharply the gain of the anti-alias filter decreases with frequency. As a rule, the S/A ratio should be at least as large as the S/Q ratio.

To give a rough idea of what is required to meet realistic specifications on alias error we assume that the power density spectrum of the output of an anti-alias filter can be approximated as

$$p_x(f) \approx \frac{p_0}{1 + (f/f_c)^{2n}}$$

This approximation is good for an nth-order Butterworth anti-alias filter and an input whose power density spectrum is almost independent of frequency over the passband ($0 \leq f \leq f_c$) of the filter. Figure 6-28b shows graphs of S/A ratio versus $F/2f_c$, where F is sampling rate and f_c is (loosely) signal bandwidth. These graphs illustrate the fact that alias error is decreased by using a higher-order (sharper-cutoff) anti-alias filter or by increasing sampling rate. These or similar graphs can be used to specify sampling rate and the order of an anti-alias filter, as shown in the following example.

Example 6-11 A signal representing speech is to be digitized. The S/Q ratio must be greater than 40 dB, and the bandwidth of the signal must exceed 3 kHz. Use the graphs of Fig. 6-28 to estimate the sampling rate and the order of an anti-alias filter.

SOLUTION Since S/A should be roughly the same as S/Q,

$$\text{S/A} \approx \text{S/Q} \approx 40 \text{ dB}$$

From Fig. 6-28 we see that several combinations of filter order and sampling rate meet the specifications, e.g.,

$$n = 10 \quad F = (1.40)(2f_c) = 8.4 \text{ kHz}$$

or

$$n = 6 \quad F = (1.85)(2f_c) = 11.1 \text{ kHz}$$

We can use a tenth-order filter and sample at about 8 kHz, or we can use a sixth-order filter and sample at about 11 kHz. Choosing between these alternatives is an economic problem.

Other errors associated with sampling are *jitter* and *aperture error,* which arise because the ideal operations described by Fig. 6-14 are only approximated in any physical system for AD conversion. Jitter refers to random fluctuation of actual sampling instants about their nominal values. Some jitter is unavoidable in practical systems because it is impossible to produce a perfectly periodic, noise-free clock signal. Aperture error arises from the impossibility of instantaneous sampling. Instantaneous sampling requires instantaneous measurement, which in turn requires instantaneous transfer (or conversion) of energy (requires infinite power). We ignore jitter and aperture error because detailed examination of these errors involves concepts beyond the scope of this book. We point out only that jitter and aperture error are controlled by ensuring that timing fluctuations and measurement time (aperture time) are small fractions of sampling interval.

6-1E Reconstruction Error

The sampling theorem given above shows that a strictly bandlimited signal can be recovered from samples of the signal by ideal low-pass filtering. Above, we describe error introduced because signals are not strictly bandlimited (alias error). Next, we describe error introduced by using S&H (boxcar) reconstruction to recover a signal from its samples (as is usually done in practice).

The S&H operation on the output of the DAC in the system of Fig. 6-10 is modeled by boxcar filtering of an ideally sampled signal (see Fig. 6-18). The boxcar filter has impulse response $h(t) = (1/T)r(t/T)$, where T is sampling interval. From (4-26) and Table C-1, the gain of the boxcar filter is

$$|H(j\omega)| = \left| \mathscr{F}\left\{ \left(\frac{1}{T} r\left(\frac{t}{T}\right) \right) \right\} \right| = \left| \text{sa} \frac{\omega T}{2} \right|$$

Figure 6-29 shows a graph of gain versus frequency for the boxcar filter. The filter is

416 INTRODUCTION TO SYSTEM ANALYSIS

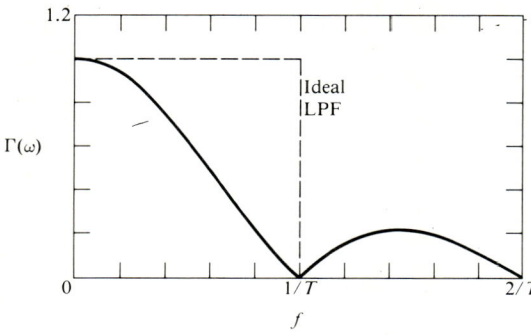

Figure 6-29 Gain versus frequency for a boxcar filter (T = sampling interval).

low-pass but it certainly is not ideal. Sample-and-hold reconstruction (boxcar filtering) alone is satisfactory in many applications but unsatisfactory in many others. When S&H reconstruction alone is unsatisfactory, fidelity can be improved considerably by equalizing gain and delay (Fig. 6-30). In the system passband (from zero to half the sampling rate) the gain of the equalizer is the reciprocal of the gain for the boxcar filter, and the delay of the equalizer is such that the overall delay is nearly linear. Outside the passband the gain of the equalizer drops sharply, so that spurious components produced by sampling, i.e., those in (6-13) for $n = \pm 1, \pm 2, \ldots$, are suppressed. With proper equalization S&H reconstruction can approach ideal low-pass filtering. Design of such equalizers is beyond the scope of this book.

6-1F Error-Free AD and DA Conversion

Unless otherwise noted, we henceforth assume that AD and DA conversion are *error-free* and *instantaneous*. As a practical matter this is equivalent to assuming that alias error, quantizing error, reconstruction error, and other errors associated with AD and

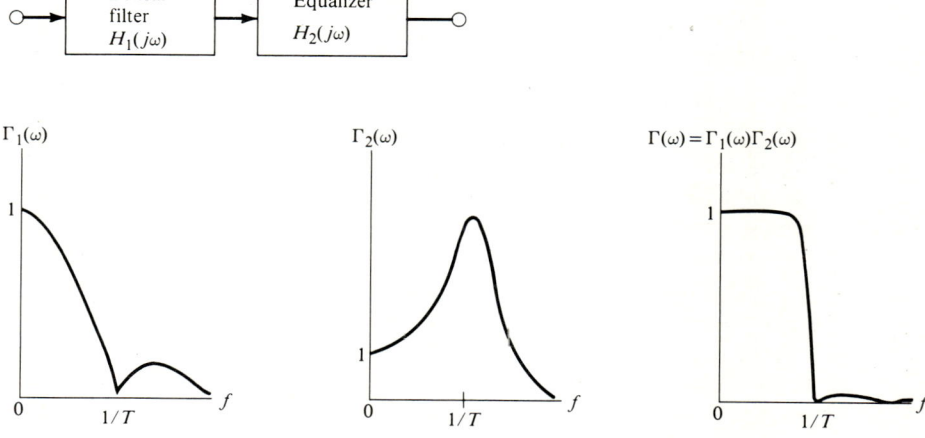

Figure 6-30 Gain equalization applied to a boxcar filter.

DA conversion are negligible and that delays introduced by AD and DA conversion are insignificant.

With this assumption, a digital signal $x_d(n)$ obtained by digitizing an analog signal $x_a(t)$ is given by

$$x_d(n) = A^{-1}x_a(nT) \tag{6-15}$$

and an analog signal $x_a(t)$ obtained by DA conversion from a digital signal $x_d(n)$ is given by

$$x_a(t) = Ax_d\left(\frac{t}{T}\right) \tag{6-16}$$

where T is sampling interval and A is a constant having the unit of $x_a(t)$. The magnitude of A is irrelevant, but for convenience we usually assume that $A = 1$ in the unit of $x_a(t)$; for example, where $x_a(t)$ is voltage in millivolts, we take $A = 1$ mV. In block diagrams error-free instantaneous AD and DA conversion are denoted by the symbols of Fig. 6-31.

Figure 6-31 Symbols for error-free AD and DA conversion.

Example 6-12 A voltage given by $x_a(t) = 10 \cos \omega_0 t$ V with $f_0 = 5$ kHz is digitized without error. The sampling rate is $F = 20$ kHz. The digital signal $x_d(n)$ is given by

$$x_d(n) = 10 \cos \omega_0 nT = 10 \cos \frac{2\pi f_0 n}{F} = 10 \cos \frac{n\pi}{2}$$

where we have used (6-15) with $A = 1$ V and $T = 1/F$. Figure 6-32 shows a graph of the analog signal versus time and a graph of the digital signal versus sample number.

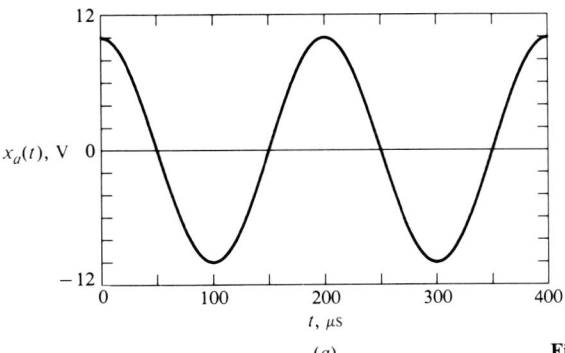

(a)

Figure 6-32 Signals of Example 6-12.

Example 6-13 A digital signal $y_d(n) = 20(1 - e^{-n/2})u(n)$ is applied to the input of an error-free DA converter whose output is current in milliamperes. The sampling rate is $F = 100$ kHz. The analog output is given by

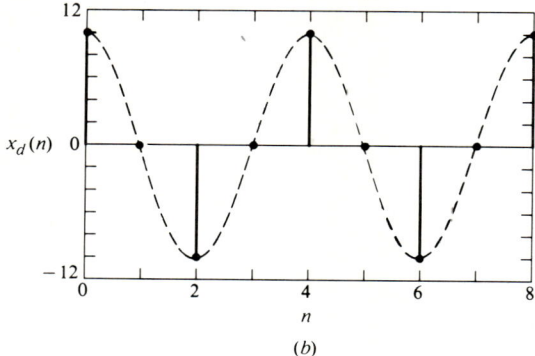

(b)

Figure 6-32 *Continued*

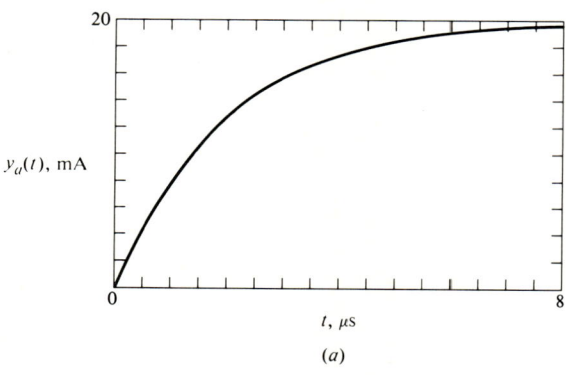

(a)

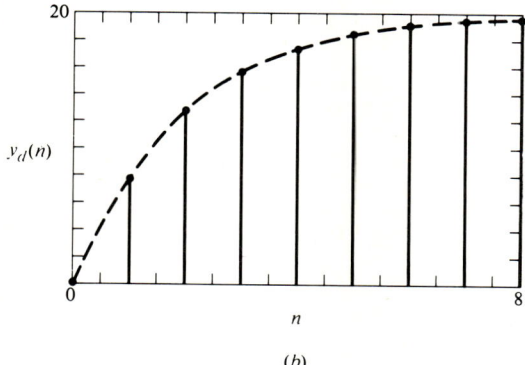

(b)

Figure 6-33 Signals of Example 6-13.

$$y_a(t) = 20(1 - e^{-0.5t/T})u\left(\frac{t}{T}\right) = 20(1 - e^{-0.5t/T})u(t) \quad \text{mA}$$

where we have used (6-16) with $A = 1$ mA and $T = 1/F = 10$ μs. We also have used the fact that $u(t/T) = u(t)$ (because $T > 0$). Figure 6-33 shows a graph of the digital signal versus sample number and a graph of the analog signal versus time.

6-1G Electromechanical AD and DA Converters

For concreteness, the discussion above focuses on electronic AD and DA conversion, where analog signals are voltages and currents. Nonetheless, the results apply with minor modifications to other kinds of AD and DA conversion processes, where analog signals may be other than electrical quantities. For example, angular position of a rotating shaft can be digitized using the numerical encoding disk shown in Fig. 6-34. The disk has four tracks. The outermost track is called the *window track*. The three inside tracks are *data tracks*. Along any radius the data tracks represent a 3-bit binary code for angular position, where a darkened area corresponds to binary 1 and a light area corresponds to binary 0. Angular position can be sensed electrically (using wipers), magnetically (where dark and light regions represent magnetized and non-magnetized material, respectively), or optically (where dark and light represent opaque and transparent, respectively). With any of these schemes ambiguous readings result if the data tracks are read along a radial boundary between adjacent data patterns. The window track eliminates these ambiguous readings by allowing a reading only if the window track is dark, i.e., only if a data value is centered in the reading window.

Numerical shaft encoders allow direct conversion of a mechanical signal (angular position) to a digital code. There are also systems for direct conversion of digital signals into mechanical signals. An example of such a system is a stepper motor whose armature moves by a fixed fraction of 2π rad when adjacent field windings are energized consecutively. Stepper motors are used to control head position in disk drives, to control workpiece position in numerically controlled machining tools, and in other applications.

Models for electromechanical AD and DA converters differ somewhat from the models for electronic AD and DA converters given in Figs. 6-31 and 6-32; e.g., electrical anti-alias filtering cannot be used in connection with numerical shaft encoding. Nonetheless, basic principles for design of electromechanical AD and DA converters are the same as those for design of electronic AD and DA converters. Sampling rate must exceed twice the bandwidth of an analog signal being digitized,

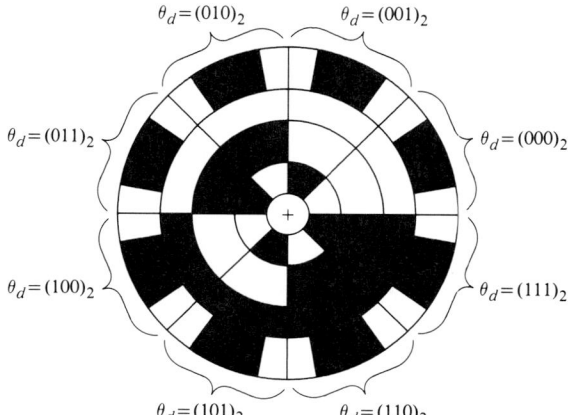

Figure 6-34 Digital encoding disk.

420 INTRODUCTION TO SYSTEM ANALYSIS

overload must be avoided, code length must be sufficient for the resolution desired, and reconstruction must be achieved by a process equivalent to low-pass filtering.

6-2 DIGITAL SIGNALS AND SYSTEMS

In this section we begin our study of digital systems (systems represented by the middle block in Fig. 6-1). We describe elementary digital signals and several elementary digital systems. We illustrate methods for finding outputs of simple systems for simple inputs and we illustrate how digital systems can be used to emulate analog systems.

6-2A Elementary Signals and Systems

A *digital signal* is a sequence of numbers representing consecutive samples of an analog signal. We denote a digital signal by a function of an integer variable n called *sample number;* for example, $x(n)$ denotes a digital signal $\ldots, x(-1), x(0), x(1), \ldots$. We refer to numbers constituting a digital signal as *samples;* e.g., we refer to $x(2)$ as the second sample of $x(n)$. A digital signal is represented graphically as illustrated in Fig. 6-35. Lines are often drawn from the coordinates (n, x) to the horizontal axis to make such graphs easier to read.

A *digital system* is a system, e.g., a computer, that accepts digital signals as inputs and produces other digital signals as outputs. A digital system can be represented by a block diagram as shown in Fig. 6-36, where $x_1(n), x_2(n), \ldots$ denote inputs and $y_1(n), y_2(n), \ldots$ denote outputs.

Five *elementary digital signals* often used as test inputs and as constituents of more complicated signals are the *unit step,* the *unit delta,* the *unit rectangular pulse,* the *unit sinusoid,* and the *unit exponential pulse,* defined in Table 6-1. These elementary signals are digital counterparts of the elementary analog signals defined in Sec. 1-2. Except for the unit delta, the elementary digital signals of Table 6-1 can be obtained

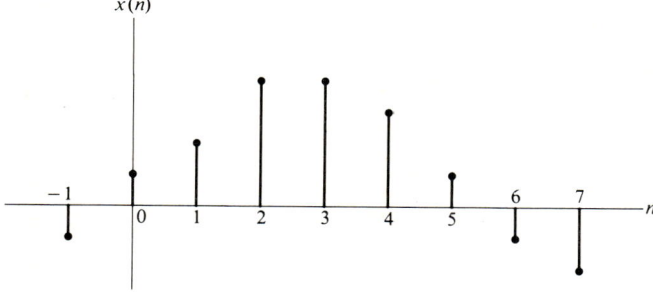

Figure 6-35 Graphical representation of a digital signal.

Figure 6-36 Block diagram for a digital system.

Table 6-1 Elementary digital signals

Name	Definition	Graph
Unit step	$u(n) = \begin{cases} 0 & n \leq 0 \\ 1 & n > 0 \end{cases}$	
Unit delta	$\Delta(n) = \begin{cases} 1 & n = 0 \\ 0 & n \neq 0 \end{cases}$	
Unit pulse	$r(\lambda n) = \begin{cases} 0 & n \leq 0 \\ 1 & 0 < n \leq \lambda^{-1} \\ 0 & n > \lambda^{-1} \end{cases}$	
Unit sinusoid	$\cos \lambda n$	
Unit exponential pulse	$e^{\lambda n} u(n)$	

from their analog counterparts by ideal AD conversion. The unit delta cannot be obtained by digitizing an impulse because an impulse $\delta(t)$ cannot be digitized.

Table 6-2 gives names, input-output relations, and symbols for five *elementary digital systems*. A *proportion element* (Fig. 6-37) multiplies a digital signal by a dimensionless scale factor. The scale factor is called gain, regardless of its magnitude. A *shift element* (Fig. 6-38) shifts a digital signal. A shift element having shift parameter k introduces delay kT, where T is sampling interval for the system in which the shift element appears. An *adder* (Fig. 6-39) adds two (or more) digital signals. Addition is algebraic, where a negative sign associated with an input is given next to the corresponding arrowhead. A *multiplier* (Fig. 6-40) multiplies one digital signal by another. Unlike analog multiplication, digital multiplication requires no dimensioning constant because all digital signals are dimensionless. The last entry in Table 6-2 is a generic description for a class of digital systems called *static* or *memoryless digital systems*, where N denotes a single-valued real function. Here the meaning of the term static is the same as for an analog system: a static digital system is one for which

Table 6-2 Elementary digital systems

Name	Transfer characteristic	Symbol
Proportion	$y(n) = Kx(n)$	$x(n) \to [K] \to y(n)$
Shift	$y(n) = x(n - k)$	$x(n) \to [n-k] \to y(n)$
Adder	$y(n) = x_1(n) + x_2(n)$	$x_1(n), x_2(n) \to [\Sigma] \to y(n)$
Multiplier	$y(n) = x_1(n)x_2(n)$	$x_1(n), x_2(n) \to [\times] \to y(n)$
Nonlinear static element	$y(n) = N[x(n)]$	$x(n) \to [y=N(x)] \to y(n)$

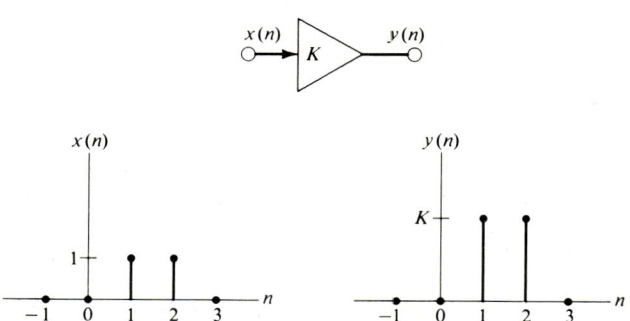

Figure 6-37 A proportion element.

the nth sample of an output depends only on the nth sample of the corresponding input; e.g., a proportion element is a static element. No dimensioning constants are needed in the transfer characteristic of a static digital system because all digital signals are dimensionless.

The following examples illustrate how overall transfer characteristics are obtained for simple series, parallel, and feedback connections of elementary digital systems.

DIGITAL SIGNALS AND SYSTEMS **423**

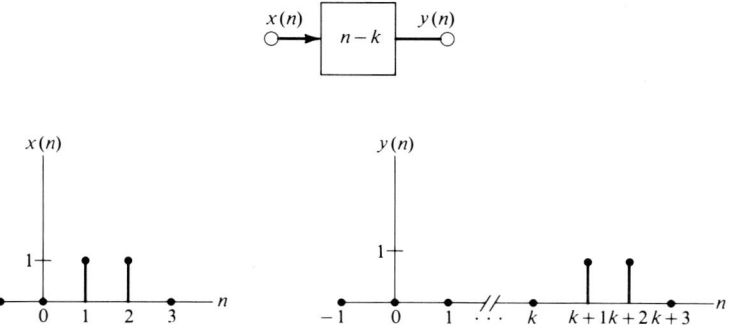

Figure 6-38 A shift element.

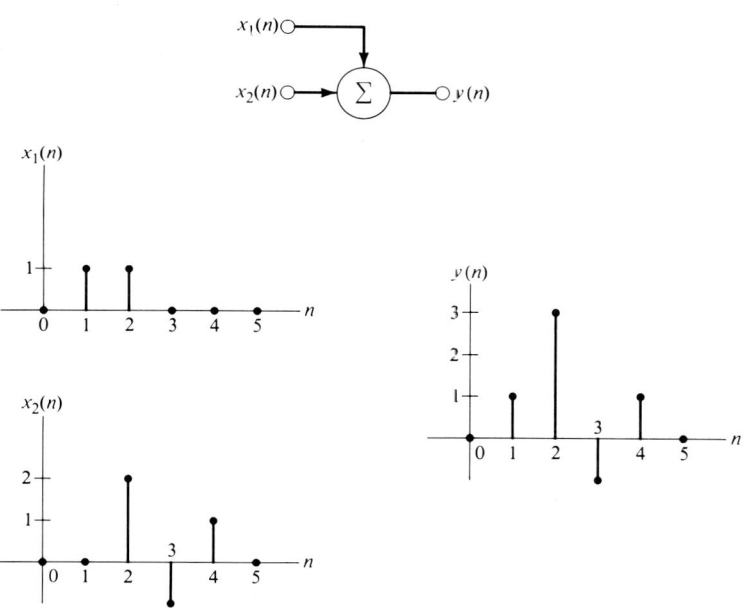

Figure 6-39 An adder.

Example 6-14 Obtain the output $y(n)$ of the system of Fig. 6-41 for input $x(n) = u(n)$.

SOLUTION From Fig. 6-41 (and Table 6-2),

$$y(n) = x(n) + 0.8x(n-1) + 0.4x(n-2) = u(n) + 0.8u(n-1) + 0.4u(n-2)$$

Figure 6-42 shows a graph of the output.

Example 6-15 Obtain the output of the system of Fig. 6-43 for input $x(n) = 10 \cos \omega_0 nT$, where $f_0 = 1$ kHz and $T = 200$ μs.

424 INTRODUCTION TO SYSTEM ANALYSIS

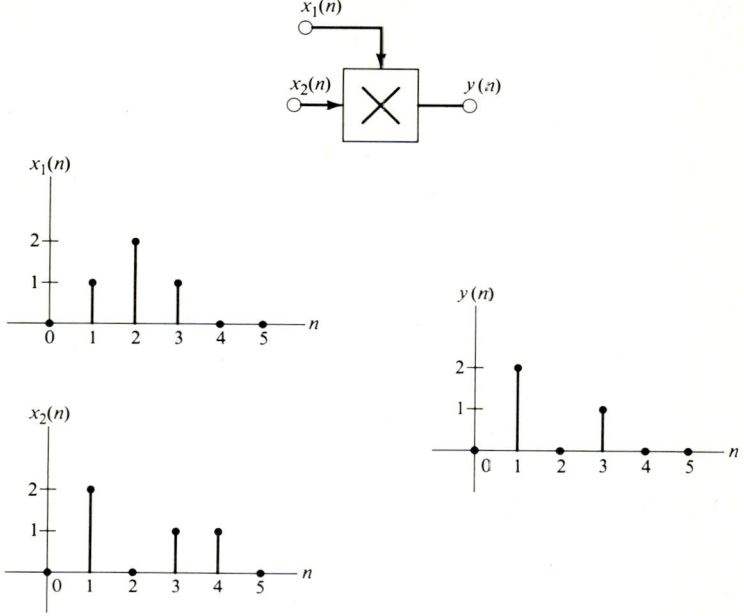

Figure 6-40 A multiplier.

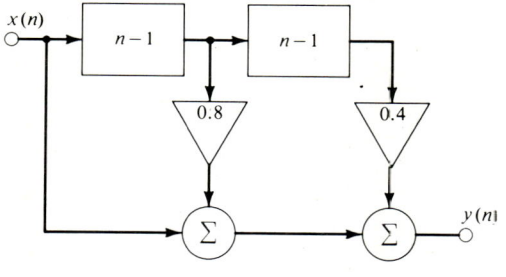

Figure 6-41 System of Example 6-14.

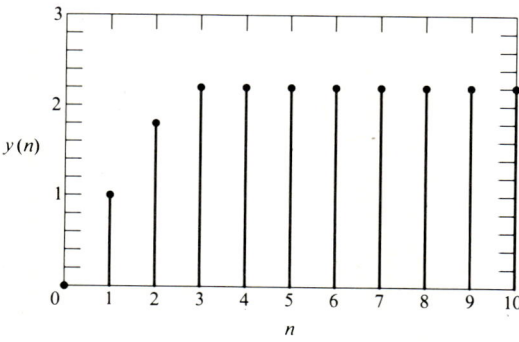

Figure 6-42 Output of the system of Fig. 6-41 for input $x(n) = u(n)$.

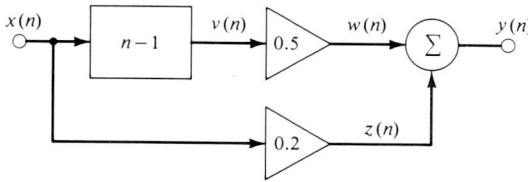

Figure 6-43 System considered in Example 6-15.

SOLUTION From Fig. 6-43 (and Table 6-2),

$$y(n) = w(n) + z(n) = 0.5v(n) + 0.2x(n) = 0.5x(n-1) + 0.2x(n)$$
$$= 5\cos[\omega_0(n-1)T] + 2\cos\omega_0 nT$$
$$= 5\cos(0.4\pi n - 0.4\pi) + 2\cos 0.4\pi n$$

Using the identity

$$\cos(\alpha - \beta) \equiv \cos\alpha\cos\beta + \sin\alpha\sin\beta$$

gives

$$\cos(0.4\pi n - 0.4\pi) = \cos 0.4\pi n \cos 0.4\pi + \sin 0.4\pi n \sin 0.4\pi$$
$$= 0.31\cos 0.4\pi n + 0.95\sin 0.4\pi n$$

whence
$$y(n) = 3.55\cos 0.4\pi n + 4.76\sin 0.4\pi n$$

Using the identity

$$a\cos\alpha + b\sin\alpha \equiv \sqrt{a^2 + b^2}\cos\left(\alpha - \tan^{-1}\frac{b}{a}\right)$$

gives

$$y(n) = 5.94\cos(0.4\pi n - 0.93)$$

Figure 6-44 shows a graph of the output.

Example 6-16 Obtain the output of the system of Fig. 6-45 for input $x(n) = \Delta(n)$.

SOLUTION We use a recursive procedure. From Fig. 6-45 (and Table 6-2),

$$y(n) = Kx(n-1) + Ky(n-1) \qquad (6\text{-}17)$$

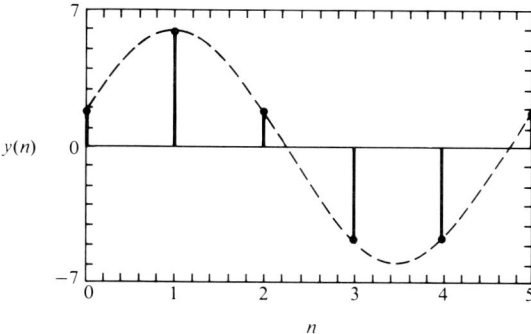

Figure 6-44 Output of the system of Fig. 6-43 for a sinusoidal input (Example 6-15).

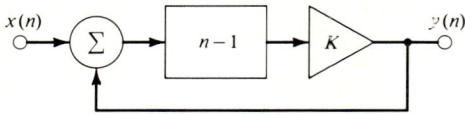

Figure 6-45 System of Example 6-16.

This gives

$$y(n-1) = Kx(n-2) + Ky(n-2) \qquad (6\text{-}18)$$
$$y(n-2) = Kx(n-3) + Ky(n-3) \qquad (6\text{-}19)$$

and so on. Using (6-18) to replace $y(n-1)$ in (6-17) gives

$$y(n) = Kx(n-1) + K[Kx(n-2) + Ky(n-2)]$$
$$= Kx(n-1) + K^2x(n-2) + K^2y(n-2) \qquad (6\text{-}20)$$

Using (6-19) to replace $y(n-2)$ in (6-20) gives

$$y(n) = Kx(n-1) + K^2x(n-2) + K^3x(n-3) + K^3y(n-4)$$

Continuing in this way, we find by induction that

$$y(n) = \sum_{m=1}^{\infty} K^m x(n-m)$$

For $x(n) = \Delta(n)$ this gives

$$y(n) = \sum_{m=1}^{\infty} K^m \Delta(n-m)$$

Figure 6-46 shows graphs of the output for four values of K.

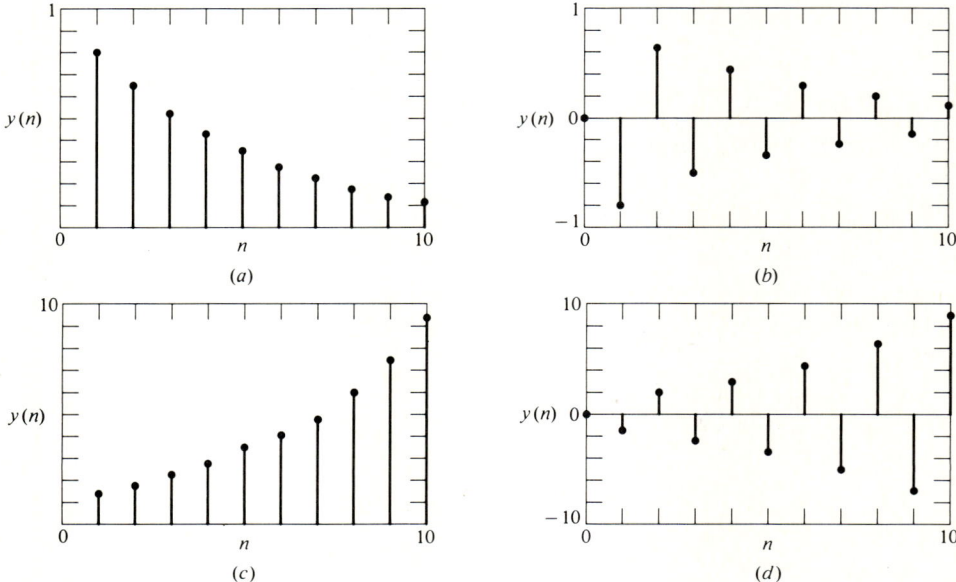

Figure 6-46 Outputs of the system of Fig. 6-45 for input $x(n) = \Delta(n)$: (a) $K = 0.8$, (b) $K = -0.8$, (c) $K = 1.25$, (d) $K = -1.25$.

Recall (Sec. 3-2) that an output of a nonlinear analog system generally has larger bandwidth than the corresponding input; e.g., for input $x(t) = A \cos \omega_0 t$, the output $y(t)$ of a system described by $y = y_0(x/x_0)^3$ contains a component having frequency $3\omega_0$. We must take this property of a nonlinear system into account when choosing sampling rate for a digital system that emulates a nonlinear analog system. Avoiding alias error in a nonlinear digital system requires that the sampling rate exceed twice the bandwidth of a signal produced by a comparable analog system.

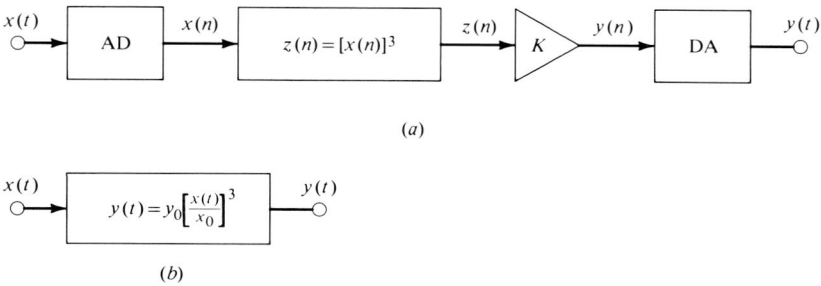

Figure 6-47 System of Example 6-17.

Example 6-17 The digital system of Fig. 6-47a is to emulate the analog system of Fig. 6-47b; i.e., when the inputs $x_a(t)$ and $x(t)$ are identical, the output $y_a(t)$ of the digital system should approximate the output $y(t)$ of the analog system. This requires that the sampling rate for the digital system exceed twice the bandwidth of the signal $y(t)$. From Fig. 6-47b, $y(t) = y_0[x(t)/x_0]^3$. A component of $x(t)$ having frequency f gives rise to a component of $y(t)$ having frequency $3f$, and so the bandwidth of $y(t)$ is roughly 3 times the bandwidth of $x(t)$. This means that the sampling rate required to provide a good representation of $y(t)$ is 3 times that required for $x(t)$. Therefore, the sampling rate for the nonlinear digital system of Fig. 6-47a must exceed 6 times the bandwidth of the input $x_a(t)$, whereas the sampling rate for a linear digital system must exceed only twice the bandwidth of an input.

6-2B Digital Integrators and Differentiators

Digital systems can be designed to emulate integration and differentiation. The emulations are only approximate because integration and differentiation entail limiting operations that can only be approximated in digital systems. In principle, error introduced by these approximations can be made as small as desired. In practice, the error has a minimum value imposed by cost and state-of-the-art limits on AD conversion, DA conversion, and processing speed. Below, we describe a simple digital integrator and a simple digital differentiator.

The transfer characteristic for an analog integrator is

$$y(t) = K \int_{-\infty}^{t} x(\alpha)\, d\alpha$$

where $x(t)$ is an input and $y(t)$ is the corresponding output. Figure 6-48 illustrates approximating this relation by

428 INTRODUCTION TO SYSTEM ANALYSIS

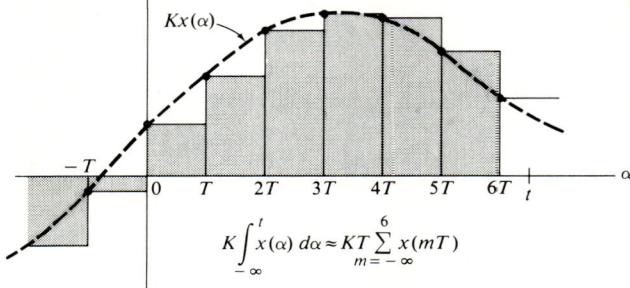

Figure 6-48 Numerical integration (rectangular rule). An integral at time t is approximated by the area of all completed rectangles to the left of t.

$$y(t) \approx KT \sum_{m=-\infty}^{n} x(mT)$$

where $n = \text{int}(t/T)$ is the largest integer less than t/T. Since the value of the scale factor KT is irrelevant in the present discussion, we let $KT = 1$ for simplicity; thus

$$y(t) \approx \sum_{m=-\infty}^{n} x(mT) \qquad (6\text{-}21)$$

With this result in mind, consider the system of Fig. 6-49.

Using the recursive procedure illustrated in Example 6-16, we find that the transfer characteristic for the digital system of Fig. 6-49 is

$$y_d(n) = \sum_{i=0}^{\infty} x_d(n-i)$$

Changing the variable of summation from i to $m = n - i$ gives

$$y_d(n) = \sum_{m=-\infty}^{n} x_d(m)$$

Using (6-15) yields

$$y_d(n) = A^{-1} \sum_{m=-\infty}^{n} x_a(mT)$$

where $A = 1$ in the unit of $x_a(t)$. Using (6-16) and interpreting t/T as $\text{int}(t/T) = n$ in the upper limit of summation gives

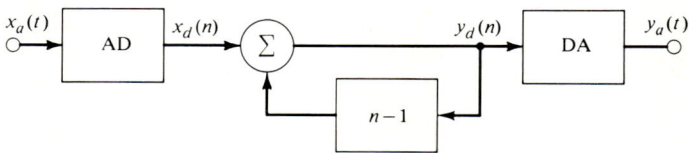

Figure 6-49 A simple digital integrator.

$$y_a(t) = B y_d\left(\frac{t}{T}\right) = BA^{-1} \sum_{m=-\infty}^{n} x_a(mT)$$

where $B = 1$ in the unit of $y_a(t)$. For simplicity, we assume that $B = A$, i.e., that x_a and y_a are expressed in the same unit. Then

$$y_a(t) = y_d\left(\frac{t}{T}\right) = \sum_{m=-\infty}^{n} x_a(mT)$$

Since this relation is identical to (6-21), the system of Fig. 6-49 approximates an analog integrator.

Example 6-18 A voltage given by

$$x_a(t) = x_0 r\left(\frac{t}{\tau}\right)$$

with $x_0 = 5$ V and $\tau = 10$ ms is applied to the digital integrator of Fig. 6-49. The sampling rate is $F = 1$ kHz. Obtain the analog output $y_a(t)$.

SOLUTION The sampling interval is $T = 1/F = 1$ ms. From (6-15), the output of the AD converter in Fig. 6-49 is

$$x_d(n) = 5r\left(\frac{nT}{\tau}\right) = 5r(0.1n)$$

From (6-21), the digital signal $y_d(n)$ in Fig. 6-49 is

$$y_d(n) = 5 \sum_{m=-\infty}^{n} r(0.1m)$$

We can change the lower limit of summation to 1 because $r(\alpha) = 0$ for $\alpha \leq 0$. From (6-16), the analog output is

$$y_a(t) = y_d\left(\frac{t}{T}\right) = 5 \sum_{m=1}^{n} r(0.1m)u(n)$$

where $n = \text{int}(t/T)$ in the upper limit of summation. Figure 6-50 shows a graph of the output.

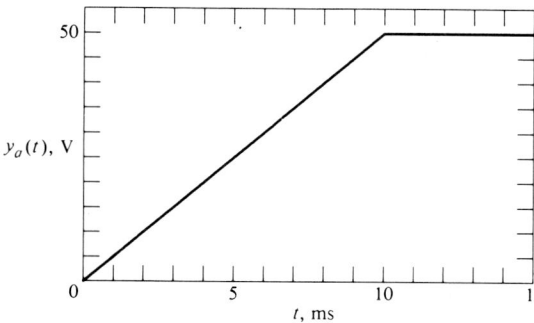

Figure 6-50 Output of the digital integrator of Fig. 6-49 for input $x(n) = 5\ r(t/\tau)$ V, with $\tau = 10$ ms.

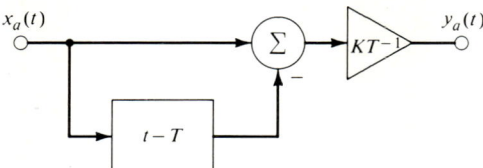

Figure 6-51 A simple analog differentiator.

We can approximate the transfer characteristic of an analog differentiator as

$$y_a(t) = K\frac{dx_a(t)}{dt} \approx \frac{K[x_a(t) - x_a(t - T)]}{T}$$

as illustrated in Fig. 6-51. Since the value of the scale factor K/T is irrelevant to the present discussion, we take $K/T = 1$ for simplicity; thus

$$y_a(t) \approx x_a(t) - x_a(t - T) \qquad (6\text{-}22)$$

With this approximation in mind, consider the system of Fig. 6-52. The output of the AD converter is

$$x_d(n) = A^{-1}x_a(nT)$$

where $A = 1$ in the unit of $x_a(t)$. The output of the digital system is

$$y_d(n) = x_d(nT) - x_d[(n - 1)T]$$

From (6-16), the analog output is

$$y_a(t) = By_d\left(\frac{t}{T}\right) = BA^{-1}[x_a(t) - x_a(t - T)]$$

where $B = 1$ in the unit of $y_a(t)$. We assume that $B = A$; then

$$y_a(t) = x_a(t) - x_a(t - T)$$

Since this relation is identical to (6-22), the system of Fig. 6-52 approximates an analog differentiator.

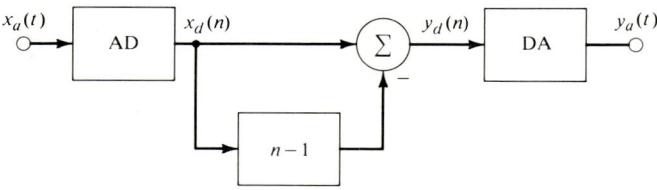

Figure 6-52 A simple digital differentiator.

Example 6-19 A voltage $x_a(t) = 2u(t)$ V is applied to the digital differentiator of Fig. 6-52. The sampling rate is $F = 50$ kHz. Obtain the output $y_a(t)$.

SOLUTION From (6-22),

$$y_a(t) = 2u(t) - 2u(t - T) \quad \text{V}$$

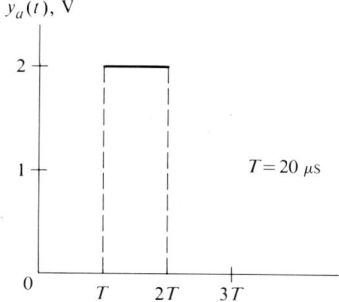

Figure 6-53 Output of the system of Fig. 6-52 for input $x(n) = u(n)$.

where $T = 1/F = 20 \mu$s. Figure 6-53 shows a graph of the output.

The discussion above is expanded in subsequent chapters, where we treat more complicated systems and more powerful methods for obtaining outputs.

6-3 FUNDAMENTAL CONCEPTS

Realizability, stability, fidelity, transient response, steady-state response, and sensitivity are fundamental concepts in analysis and design of systems—both digital and analog. In this section we describe how these concepts are applied to digital systems (systems having digital inputs and outputs). The treatment parallels that of Sec. 1-5 for analog systems.

6-3A Realizability

A transfer characteristic for a digital system *is realizable* if and only if every sample of an output is determined by the current sample and previous samples of the input; e.g., for input x and output y the transfer characteristic $y(n) = x(n) + 0.5x(n - 1)$ is realizable, whereas the transfer characteristic $y(n) = x(n) + 0.5x(n + 1)$ is nonrealizable.

> **Example 6-20** A purely mathematical attack on a design problem shows that it is desirable to build a digital compression system whose output $y(n)$ for input $x(n)$ is given by $y(n) = x(2n)$. Unfortunately, this transfer characteristic is nonrealizable because for all positive n, $2n > n$, which means that a current output sample depends on a future input sample.

In so-called *offline* (or non-real-time) signal processing, nonrealizable operations are made realizable for a limited class of inputs by incorporating sufficient delay. For example, in processing seismic data taken in exploring for oil, signals are recorded on magnetic tape and processed later on a digital computer. Here the delay is the time between recording and processing, and the limited class of inputs consists of inputs that will fit on the tape. Often, it is claimed that offline processing allows use of nonrealizable systems (or signal-processing operations) but that is not so. Offline processing does not attempt to make use of future data, i.e., data not yet recorded. It simply introduces enough delay to permit data recorded later to be used in extracting information from data recorded earlier. Always, apparent circumvention of realizability is achieved only by introducing delay.

Also it is often claimed that realizability is irrelevant if the independent variable (sample number) for a sequence represents some physical quantity other than time; e.g., in a seismic profile of density as a function of depth. This too is wrong because it is *how data are processed* (and not what the data represent) that matters in questions of realizability. Realizability is necessary whenever data are to be processed sequentially, some samples being made available to a processor before others. Whether sample number represents time or some other quantity (depth, shoe size, or whatever) is irrelevant.

Example 6-21 Find the smallest value of the shift k (which represents delay equal to k sampling intervals) for which the system of Fig. 6-54 has a causal (nonanticipatory) output for input $x(n) = r(0.1n)$.

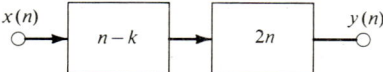

Figure 6-54 System of Example 6-21.

SOLUTION The output for input $x(n) = r(0.1n)$ is

$$y(n) = r[0.1(2n - k)] = r(0.2n - 0.1k)$$

Causality requires that the leading edge of the output occur no earlier than the leading edge of the input and that the trailing edge of the output occur no earlier than the trailing edge of the input. Recall that $r(\alpha) = 1$ for $0 < \alpha \leq 1$. The leading edge of the input occurs for the smallest integral value of n such that $0.1n > 0$; that is, for $n = 1$. The leading edge of the output occurs for the smallest integral value of n such that $0.2n - 0.1k > 0$; that is, for $n > k/2$. Therefore the leading edge of the output occurs no earlier than the leading edge of the input provided that $k \geq 0$. The trailing edge of the input occurs for the largest integer n such that $0.1n \leq 1$; that is, for $n = 10$. The trailing edge of the output occurs for the largest integral value of n such that $0.2n - 0.1k \leq 1$; that is, for the largest integer $n \leq 5 + 0.5k$. Requiring that the trailing edge of the output occur no earlier than the trailing edge of the input gives

$$5 + 0.5k \geq 10$$

or $k \geq 10$. Therefore the system gives a causal output for input $x(n) = r(0.1n)$ provided the shift k is not smaller than 10 (provided the delay is not less than 10 sampling intervals). This example illustrates that compression can be achieved for a limited class of inputs at cost of delay.

6-3B Stability

We use the bounded-input–bounded-output (BIBO) condition for stability for both digital and analog systems. This condition is based on the following definition. A digital signal $x(n)$ is *bounded* if there is a number M_x such that $|x(n)| < M_x$ for all n; for example, the signal $x(n) = e^{\lambda n}u(n)$ is bounded if $\lambda < 0$ (let $M_x = 1$) and is unbounded if $\lambda > 0$. A digital system is *stable* if it gives a bounded output for any bounded input, it is *conditionally stable* if it gives bounded outputs for some (but not all) nonzero bounded inputs, and it is *unstable* if it gives an unbounded output for any nonzero bounded input.

Example 6-22 Show that the system of Fig. 6-55 is conditionally stable.

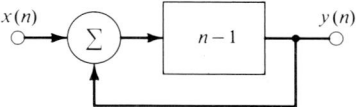

Figure 6-55 System of Example 6-22.

SOLUTION Using a recursive procedure (as in Example 6-16), we find that the output $y(n)$ for input $x(n)$ is given by

$$y(n) = \sum_{m=-\infty}^{n-1} x(m)$$

Figure 6-56 shows graphs of the outputs for input $x(n) = \Delta(n)$ (unit delta) and $x(n) = u(n)$ (unit step). For input $x(n) = \Delta(n)$ the output is bounded, whereas for input $x(n) = u(n)$ the output is unbounded. The system is conditionally stable because it gives a bounded output for one nonzero input (a delta) and an unbounded output for another (a step).

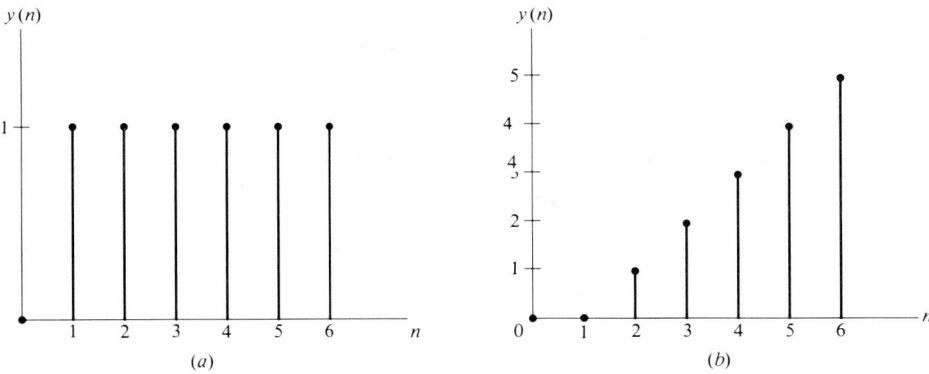

Figure 6-56 Output of the system of Fig. 6-55 for input (a) $x(n) = \Delta(n)$, (b) $x(n) = u(n)$.

To show that a system is stable (or unstable) we must show that the system gives a bounded (or unbounded) output for every bounded input. This is possible only for a limited class of systems. In practice, simulation is used extensively in stability studies.

6-3C Transient and Steady-State Response

Loosely, *transient response* is the part of a response that vanishes as sample number approaches infinity, and *steady-state response* is the part of the response that does not vanish as sample number approaches infinity.

Example 6-23 The response

$$y(n) = [(0.9)^n + \cos n\lambda] u(n)$$

has transient component $y_t(n) = (0.9)^n u(n)$ and steady-state component $y_{ss}(n) = \cos n\lambda$.

434 INTRODUCTION TO SYSTEM ANALYSIS

In practice, separating a response into transient and steady-state components is useful mainly where the steady-state component is a constant, a sinusoid, or (less often) a ramp and where the steady-state response can be determined independently, e.g., where a stable system having known dc gain is excited by a step. In such cases, transient response is obtained by subtracting the steady-state response from the complete response.

Example 6-24 In this example we examine the transient response of the system of Fig. 6-57 for a step input and for $0 < K < 1$.

From Fig. 6-57, the output of the system for input $x(n)$ is given by

$$y(n) = x(n-1) + Ky(n-1)$$

Using recursion to obtain the transfer characteristic of the system yields

$$y(n) = \sum_{m=1}^{\infty} K^{m-1} x(n-m)$$

For $x(n) = u(n)$, the transfer characteristic above gives

$$y(n) = \sum_{m=1}^{\infty} K^{m-1} u(n-m)$$

The step $u(n-m)$ is zero for $n \leq m$ and is unity for $n > m$. It follows that

$$y(n) = u(n-1) \sum_{m=1}^{n-1} K^{m-1}$$

Using the identity

$$\sum_{i=0}^{l-1} a^i = \frac{1 - A^l}{1 - A} \qquad \alpha \neq 1$$

gives

$$y(n) = \frac{1 - K^{n-1}}{1 - K} u(n-1) \qquad K \neq 1$$

Figure 6-58 shows a graph of the output $y(n)$ versus sample number n for $0 < K < 1$. The output rises monotonically from $y(0) = 0$ to the steady-state amplitude $y(\infty) = 1/(1-K)$. The sample number for which the output reaches a specified fraction α of the steady-state output is determined by solving the equation

$$\frac{1 - K^{n-1}}{1 - K} = \frac{\alpha}{1 - K}$$

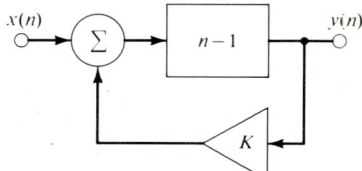

Figure 6-57 System of Example 6-24.

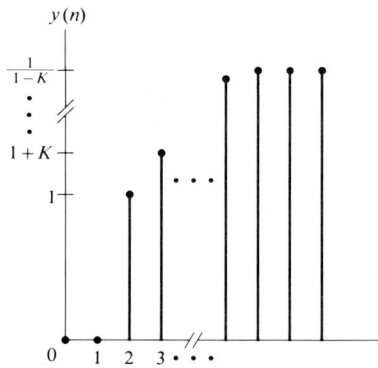

Figure 6-58 Output of the system of Fig. 6-57 for input $x(n) = u(n)$.

for n. The solution is

$$n = 1 + \frac{\ln(1 - \alpha)}{\ln K} \tag{6-23}$$

where n is a nonnegative integer. One reasonable choice for α in (6-23) is $\alpha = 0.9$. This corresponds to defining the duration of the transient response as the sample number n for which the output first exceeds 90 percent of its steady-state value. Figure 6-59 shows a graph of the relation given by (6-23) for $\alpha = 0.9$. Figure 6-60 shows output versus sample number for a step input; in Fig. 6-60a, $K = 0.5$, and in Fig. 6-60b, $K = 0.9$. For $K = 0.5$, the

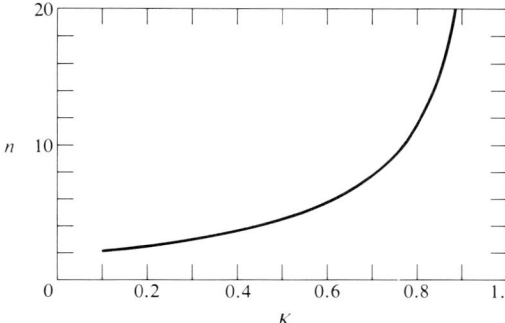

Figure 6-59 Duration of transient (rise time) for the system of Fig. 6-57 versus K for input $x(n) = u(n)$.

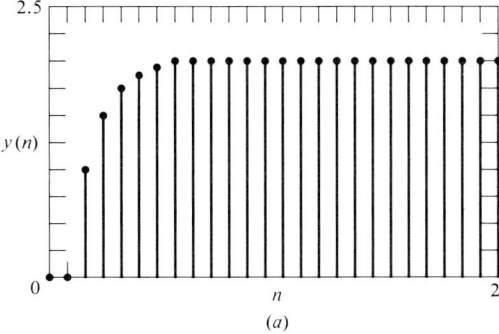

Figure 6-60 Output of the system of Fig. 6-57 for input $x(n) = u(n)$ and (a) $K = 0.5$, (b) $K = 0.9$.

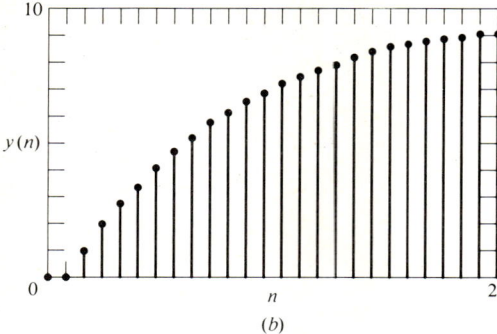

(b)

Figure 6-60 *Continued*

output exceeds 90 percent of its steady-state value after only five sampling intervals, whereas for $K = 0.9$, the output does not exceed 90 percent of the steady-state value until the twenty-third sampling instant. For $K > 0.5$, the duration of the transient (the rise time) increases sharply with K.

6-3D Fidelity

A digital system provides *distortionless transmission* if its output for input $x(n)$ is given by

$$y(n) = Kx(n - k)$$

where K and k are independent of n. A system whose transfer characteristic closely approximates the relation above for some inputs provides *high-fidelity transmission* for those inputs.

Example 6-25 In this example we describe conditions under which the system of Fig. 6-57 provides high-fidelity transmission for a rectangular pulse $r(\lambda n)$.

A rectangular pulse can be represented as

$$r(\lambda n) = u(n) - u(n - p)$$

where $p = \lambda^{-1}$. It follows from this relation and the transfer characteristic of the system (Example 6-24) that the output for input $r(\lambda n)$ is given by

$$y(n) = g(n) - g(n - p)$$

where $g(n)$ is the output for step input $u(n)$. This result implies that the system provides high-fidelity transmission for a pulse if the duration of the transient component of the step response $g(n)$ is much smaller than the duration of the pulse. From Example 6-24, the duration of the transient is given by

$$n = 1 + \frac{\ln(1 - \alpha)}{\ln K}$$

which is derived by defining duration of transient response as the sample number for which the step response reaches a fraction α of its steady-state value. The duration of a pulse $r(\lambda n)$ is $p = \lambda^{-1}$. Thus the requirement for high-fidelity transmission of a pulse by the system of Fig. 6-57 is

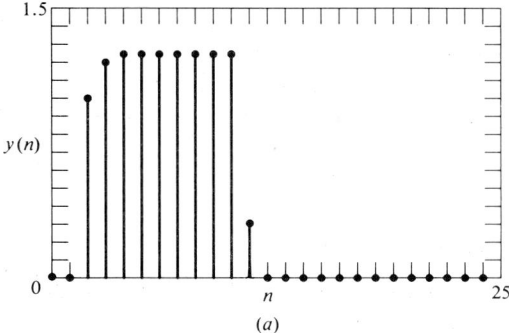

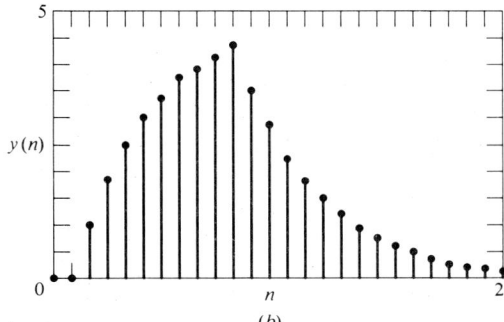

Figure 6-61 Output of the system of Fig. 6-57 for input $x(n) = r(n/10)$ and (a) $K = 0.2$, (b) $K = 0.8$.

$$\frac{\ln(1-\alpha)}{\ln K} \ll \lambda^{-1}$$

where $0 < K < 1$. Because $\ln K < 0$, this gives

$$\ln K \ll \lambda \ln(1-\alpha) \quad \text{or} \quad K \ll (1-\alpha)^\lambda$$

Thus, for $\alpha = 0.9$ and $\lambda = 0.1$ ($p = 10$), we require $K \ll 0.8$.

Figure 6-61 shows outputs for input $r(0.1n)$ for two values of K. In Fig. 6-61a, $K = 0.2$ ($K \ll 0.8$), and distortion is relatively small. In Fig. 6-61b, $K = 0.8$, and distortion is relatively large.

Instantaneous distortion introduced by a system is defined as shown in Fig. 6-62, where $x(n)$ is an appropriate test input, $y(n)$ is the actual output, $y'(n)$ is the desired (distortionless) output, and $v(n)$ is instantaneous distortion. Any measure of the size of instantaneous distortion is called a *distortion measure*. One widely used distortion measure is the *sum of squared errors* (SSE), given by

$$\text{SSE} = \sum_{n=-\infty}^{\infty} v^2(n) \qquad (6\text{-}24)$$

where $v(n)$ is instantaneous distortion.

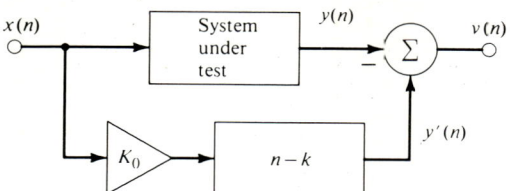

Figure 6-62 Definition of instantaneous distortion for a digital system.

Example 6-26 Calculate the SSE for the system of Fig. 6-57 for unit step input $x(n) = u(n)$ and $|K| < 1$.

SOLUTION We must first find the ideal response, i.e., values for gain K_0 and shift k introduced by a comparable distortionless system. From Example 6-24, the steady-state step response of the system of Fig. 6-57 is

$$y_{ss}(n) = \frac{1}{1-K}$$

which is also the gain K_0 of a comparable distortionless system. Again from Example 6-24, the input is zero for $n \leq 1$, and the output is zero for $n \leq 2$. Therefore the shift introduced by a comparable distortionless system is one sampling interval ($k = 1$). From Fig. 6-62, the instantaneous distortion is given by

$$v(n) = K_0 x(n-k) - y(n) = \frac{u(n-1)}{1-K} - y(n)$$

where $y(n)$ is the actual output, given by (Example 6-24)

$$y(n) = \frac{1 - K^{n-1}}{1-K} u(n-1)$$

Thus the instantaneous distortion is

$$v(n) = \frac{1 - (1 - K^{n-1})}{1-K} u(n-1)$$

and

$$\text{SSE} = \sum_{n=2}^{\infty} \frac{[1 - (1 - K^{n-1})]^2}{(1-K)^2} = \frac{1}{(1-K)^2} \sum_{n=2}^{\infty} K^{2n-2}$$

$$= \frac{K^2(1 + K^2 + K^4 + \cdots)}{(1-K)^2} = \frac{K^2}{(1-K)^2(1-K^2)}$$

Figure 6-63 shows a graph of SSE versus K. The SSE increases with K, as we expect (from Example 6-25).

6-3E Sensitivity

The definition and interpretation of sensitivity for digital systems is the same as for analog systems (Sec. 1-5E). The *sensitivity* of a performance measure ϕ to variations of a parameter p is defined by

$$S_\phi(p) = \frac{\partial \phi}{\partial p} \frac{p}{\phi} \qquad (6\text{-}25)$$

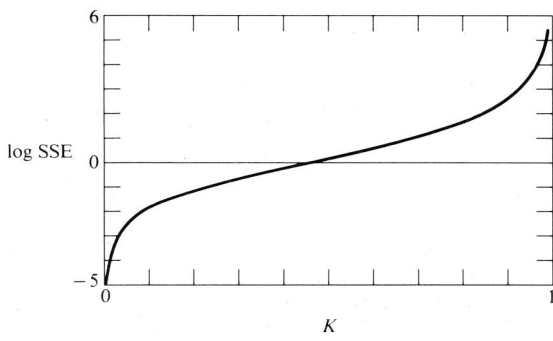

Figure 6-63 Sum of squared errors (SSE) for the system of Fig. 6-57 with input $x(n) = u(n)$.

Example 6-27 Calculate the sensitivity of the dc gain of the system of Fig. 6-64 to variations of the parameter K_1 for $K_1 = 5$ and $K_2 = 0.1$.

SOLUTION The dc gain of the system is the steady-state amplitude of the output for a unit step input. From Fig. 6-64,

$$y(n) = K_1 x(n-1) + K_1 K_2 y(n-1)$$

Using the usual recursive procedure to determine the transfer characteristic yields

$$y(n) = \sum_{m=1}^{\infty} K_1^m K_2^{m-1} x(n-m)$$

For $x(n) = u(n)$ this gives

$$y(n) = u(n-1) \sum_{m=1}^{n-1} K_1^m K_2^{m-1} = K_1 u(n-1) \sum_{p=0}^{n-2} (K_1 K_2)^p$$

$$= K_1 \frac{1 - (K_1 K_2)^{n-1}}{1 - K_1 K_2} u(n-1) \qquad K_1 K_2 \neq 1$$

where we have used the identity

$$\sum_{i=0}^{I-1} \alpha^i = \frac{1 - \alpha^I}{1 - \alpha} \qquad \alpha \neq 1$$

For $K_1 K_2 < 1$, the steady-state output (the dc gain) is

$$y_{ss} = \phi = \frac{K_1}{1 - K_1 K_2}$$

From (6-25), the sensitivity of dc gain to variations of the forward gain K_1 is given by

$$S_\phi(K_1) = \frac{1}{1 - K_1 K_2}$$

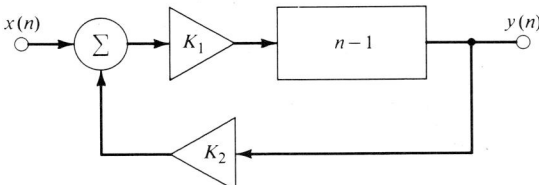

Figure 6-64 System of Example 6-27.

For $K_1 = 5$ and $K_2 = 0.1$, this gives

$$S_\phi(K_1) = 2$$

which means that a 1 percent change in K_1 produces a 2 percent change in dc gain. Proceeding as above, we find that the sensitivity of dc gain to variations in K_2 is given by

$$S_\phi(K_2) = \frac{K_1 K_2}{1 - K_1 K_2} = 1$$

which means that a 1 percent change in K_2 produces a 1 percent change in dc gain. For $K_1 = 5$ and $K_2 = 0.1$, dc gain is more sensitive to variations of the forward gain K_1 than to variations of the feedback gain K_2.

SUMMARY

Error introduced by AD conversion is defined as distortion introduced by the system of Fig. 6-10, where an analog signal is digitized and the digital signal is immediately "undigitized," without any intervening digital processing. The system of Fig. 6-10 can be modeled as shown in Fig. 6-15. The quantizer, the proportion element, and the delay element account for effects of digitizing. The ideal sampler accounts for effects of sampling, and the boxcar filter models S&H reconstruction. The proportion element and the delay element can be ignored in calculating distortion introduced by this model (by the system of Fig. 6-10).

Distortion introduced by AD conversion consists of *overload error, quantizing error, alias error, jitter, aperture error,* and *reconstruction error.* In a well-designed system an input is scaled to just fit within the full-scale limits of the ADC, thus avoiding overload error and making maximum use of available code length. The code length is specified to obtain the desired dynamic range and S/Q ratio. Alias error is limited by a sharp-cutoff low-pass anti-alias filter at the input and by sampling at a rate exceeding twice the bandwidth of the filter. Jitter and aperture error are minimized by using a stable, e.g., crystal-controlled, almost noise-free clock, by careful design of clock-detection circuitry, and by using very fast measurement (sampling) circuitry that introduces very little loading. Reconstruction error is reduced by equalizing gain and delay after DA conversion.

The output of an *error-free AD converter* for analog input $x_a(t)$ and sampling interval T is given by

$$x_d(n) = A^{-1} x_a(nT)$$

where $A = 1$ in the unit of $x_a(t)$. The output of an *error-free DA converter* for digital input $x_d(n)$ and sampling interval T is given by

$$x_a(t) = A x_d\left(\frac{t}{T}\right)$$

where $A = 1$ in the unit of $x_a(t)$. Both relations above assume that the analog signals involved are strictly bandlimited to $0 \leq f \leq F/2$, where $F = 1/T$ is sampling rate. Henceforth, unless otherwise noted, we assume that AD and DA conversion are error-free and instantaneous. This is equivalent to assuming that error and delay introduced by AD and DA conversion are insignificant. With this assumption, the AD and DA operations in the prototype system of Fig. 6-1 are transparent. The overall transfer characteristic for the system of Fig. 6-1 is determined by the digital processor alone.

A *digital signal* is a sequence of numbers representing samples of an analog signal. A digital signal is denoted by a function of an integer variable n called sample number. Five important elementary digital signals are the *unit step*, the *unit rectangular pulse*, the *unit delta*, the *unit sinusoid*, and the *unit exponential pulse*. These elementary digital signals are used both to describe more complicated digital signals and as test inputs in specifying performance of digital systems.

A *digital system* is one that accepts digital signals as inputs and produces digital signals as outputs. Important elementary digital systems are defined in Table 6-2. These elementary digital systems are connected in series, parallel, and feedback configurations to represent more complicated systems. A digital system can be designed to emulate virtually any analog system within technological limits on sampling rate, code length, and processing speed.

Definitions and interpretations of realizability, stability, fidelity, transient response, steady-state response, and sensitivity to parameter variations for digital systems are essentially the same as for analog systems. *Realizability* requires a causal output for any input. *Stability* requires a bounded output for any bounded input. *Fidelity* requires that the system introduce only delay and gain. *Transient response* is the part of a response that arises from a change in the character of an input. *Steady-state response* (for a step or sinusoidal input) is the part of a response that persists after the associated transient has decayed. *Sensitivity* indicates how strongly variations of a parameter affect a performance measure for a system. At least one of these concepts arises in virtually every practical application of digital systems. It is essential that the reader learn the definitions and implications of these important concepts.

PROBLEMS

6-1 Plot the transfer characteristic of a 4-bit ADC having full-scale input $v_{FS} = 5$ V.

6-2 Plot the transfer characteristic of an ADC having full-scale input $v_{FS} = 1.5$ V and full-scale output $B_{FS} = (01111)_2$.

6-3 The system of Fig. P6-3 has input $v(t) = 5 \cos \omega_0 t$ V, where $f_0 = 5$ kHz. The sampling rate is $F = 25$ kHz, the code length is $n = 6$ bits, and the full-scale input for the ADC is $v_{FS} = 4$ V. Determine the output B for $0 \leq t < 200$ μs, assuming that samples are taken for $t = 0, \pm T, \pm 2T, \ldots,$ where $T = 1/F$.

Figure P6-3

6-4 Plot the transfer characteristic of a 4-bit DAC having step size $y_0 = 50$ mA.

6-5 Plot the transfer characteristic of a DAC having full-scale input $B_{FS} = 15$ and full-scale output $y_{FS} = 3$ V.

6-6 The system of Fig. P6-6 has input $v(t) = 3 \cos \omega_0 t$ V, where $f_0 = 1$ kHz. The sampling rate is $F = 8$ kHz, the full-scale input for the ADC is 3 V, the code length is 5, and the step size

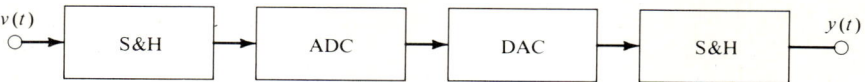

Figure P6-6

for the DAC is 2 mV. Determine the output $y(t)$ for $0 \le t < 500\ \mu s$, assuming that samples are taken for $t = 0, \pm T, \pm 2T, \ldots,$ where $T = 1/F$.

6-7 An ADC and DAC are connected in series as shown in Fig. P6-7. The sampling rate is 5 kHz, the full-scale input for the ADC is 5 V, the code length is 5, and the step size for the DAC is 250 mV. The system is to be modeled as shown in Fig. 6-11. Determine the parameters (delay time, etc.) of the model. Factor the transfer characteristic for the quantizer as shown in Fig. 6-14 and determine the gain of the proportion element and the transfer characteristic for the unit-gain quantizer.

Figure P6-7

6-8 Do the elements in the system of Fig. 6-14 commute? Justify your answer.

6-9 The system of Fig. P6-9 is to be used to measure error introduced by AD conversion. The full-scale input for the ADC is 5 V, the sampling rate is 20 kHz, the code length is 8, and the step size for the DAC is 25 mA.

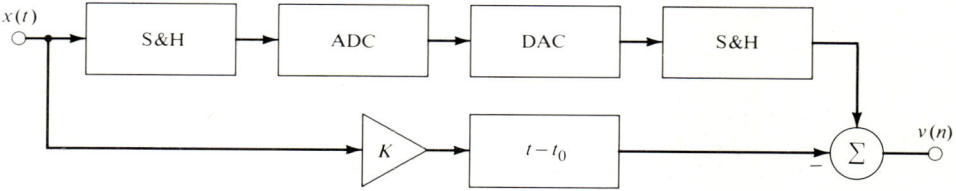

Figure P6-9

(a) What values should be used for the gain of the proportion element and the delay time of the delay element?

(b) For what input amplitudes is the error zero?

(c) For what input amplitudes is the error maximum (assuming no overload)?

(d) What is the maximum error (no overload)?

6-10 Determine the ratio of peak signal to peak quantizing error in decibels for the system of Prob. 6-7.

6-11 Digital representation of studio-quality TV video requires that the ratio of peak signal to peak quantizing error exceed 55 dB. What code length is required in this application?

6-12 A certain sonar system is to detect a sinusoidal signal in the presence of interference. The amplitude of the interference is 100 times larger than the amplitude of the sinusoid, and the signal-to-error ratio required for reliable detection of the sinusoid is 15 dB. The detection processing is to be done digitally. What code length is required?

6-13 Under certain circumstances a person having normal hearing can detect a sinusoidal tone in the presence of noise whose amplitude is about 1000 times larger than that of the tone. What code length would be required for a digital processor to duplicate this feat if detection requires that S/Q for the sinusoid exceed 10 dB?

6-14 A signal to be digitized has three components. Two are sinusoids having peak amplitudes 5 and 1 V; the other is noise (interference) having peak amplitude 10 V. The S/Q ratio for each sinusoid must exceed 20 dB. What code length is required?

6-15 The signal whose spectral density is shown in Fig. P6-15 is sampled using the system of Fig. 6-21. Plot the spectral density of the sampled signal for sampling rate (a) $F = 8$ kHz and (b) $F = 6$ kHz.

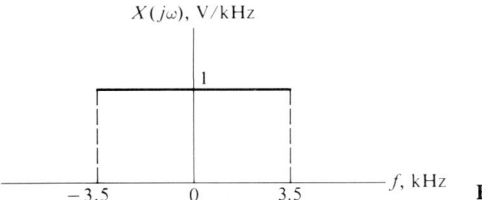

Figure P6-15

6-16 The signal $x(t)$ given by

$$x(t) = 9 + 6 \cos \omega_0 t + 3 \cos 3\omega_0 t \quad \text{V}$$

is ideally sampled using sampling rate F. The sampled signal is then applied to a filter having transfer function

$$H(j\omega) = \begin{cases} 1 & \dfrac{-F}{2} < f < \dfrac{F}{2} \\ 0 & \text{otherwise} \end{cases}$$

Determine the filter output for (a) $F = 8\omega_0$ and (b) $F = 4\omega_0$.

6-17 In a certain western film a stagecoach accelerates from rest. At first, the wheels of the coach rotate faster and faster in the proper direction, but then they appear to slow down, stop, and begin to rotate in the opposite direction. They rotate faster and faster in this direction, then appear to slow down, stop again, and begin rotating in the proper direction.

(a) Explain this phenomenon in terms of ideal sampling and alias error.
(b) What plays the role of the reconstruction filter in a model of this phenomenon?
(c) How fast is the coach going the third time the wheels appear to stop if the frame rate (for camera and projector) is 24 frames per second, the wheel diameter is 1 m, and each wheel has 12 spokes?

6-18 The signal of Example 6-10 is sampled ideally. The sampled signal is applied to the input of an ideal low-pass filter having bandwidth $F/2$, where F denotes sampling rate. Plot the S/A ratio at the filter output versus the ratio $F/2f_c$. Compare your graph with the appropriate graph in Fig. 6-28.

6-19 A signal representing TV video is to be digitized. The S/Q ratio must exceed 55 dB, and the bandwidth of the signal must exceed 5 MHz. Use the graphs of Fig. 6-28 to estimate the order of the anti-alias filter and the sampling rate required in this application.

6-20 A voltage given by $v(t) = v_{FS} \cos \omega_0 t$ is sampled and reconstructed using the system of Fig. P6-20. Spurious components produced by sampling appear in the reconstructed signal because the boxcar filter is not an ideal low-pass filter.

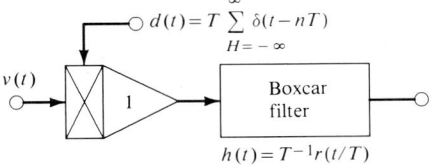

Figure P6-20

(a) Determine the ratio in decibels of the amplitude of the signal to that of the lowest frequency spurious component as a function of $F/2f_0$, where $F > 2f_0$ is sampling rate.

(b) What component added to the system would increase this ratio without requiring the sampling rate to be any larger than that necessary to avoid alias error?

6-21 The analog signals below are digitized without error as described in connection with Fig. 6-34. The sampling rate F is as specified. Plot the resulting digital signals.

(a) $x_a(t) = 20e^{-t/\tau}u(t)$ mA; $\tau = 10$ ms; $F = 1$ kHz.

(b) $x_a(t) = 500r(t/\tau)$ mV; $\tau = 5$ ms; $F = 2$ kHz.

(c) $x_a(t) = 15 \cos \omega_0 t$ V; $f_0 = 5$ kHz; $F = 40$ kHz.

(d) $x_a(t) = 10(1 - e^{-t/\tau})u(t)$ V; $\tau = 10$ ms; $F = 1$ kHz.

6-22 The digital signals below are converted without error into voltages, as described in connection with Fig. 6-36. The sampling rate F is specified. Plot the resulting analog signals.

(a) $x_d(n) = 25(\sin 0.1\pi n)u(n)$; $F = 5$ Hz.

(b) $x_d(n) = 500e^{-2n}u(n)$; $F = 500$ kHz.

(c) $x_d(n) = 10nr(0.1n)$; $F = 10$ kHz.

(d) $x_d(n) = 5n^{-1}u(n)$; $F = 5$ kHz.

6-23 Explain why an impulse $x_a(t) = \mathbf{a}\delta(t)$ cannot be digitized.

6-24 An ideal (error-free) DA converter can be modeled as shown in Fig. P6-24. The box labeled $\Delta(n) \to \delta(t)$ somehow converts a unit delta $\Delta(n)$ into an impulse $\mathbf{a}\delta(t)$ having strength $\mathbf{a} = 1$ Vs. The ideal low-pass filter has dc gain $1/F$ and bandwidth $F/2$, where F is sampling rate.

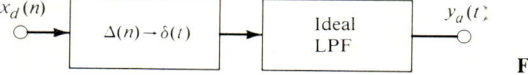

Figure P6-24

(a) Determine the analog output $y_a(t)$ of this system for input $x_d(n) = \Delta(n)$.

(b) Show that the analog output $y_a(t)$ for any digital input $x_d(n)$ is given at any sampling instant by (6-16) (with $K = 1$ V). *Hint:* A digital input can be represented as a linear combination of unit deltas; e.g.,

$$\cos 10n = \cdots + \Delta(n) + (\cos 10)\Delta(n-1) + (\cos 20)\Delta(n-2) + \cdots$$

Determine analog output for each delta; examine the sum of the outputs at a sampling instant.

6-25 In the system of Fig. P6-25 the analog input is given by $x_a(t) = 5u(t)$ V, and the sampling rate is 20 kHz. Plot the digital output $y_d(n)$ for $0 \le n \le 5$.

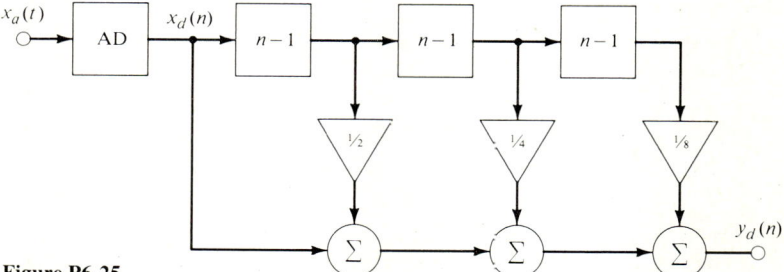

Figure P6-25

6-26 For each system of Fig. P6-26 plot the output for input $x_d(n) = \Delta(n)$.

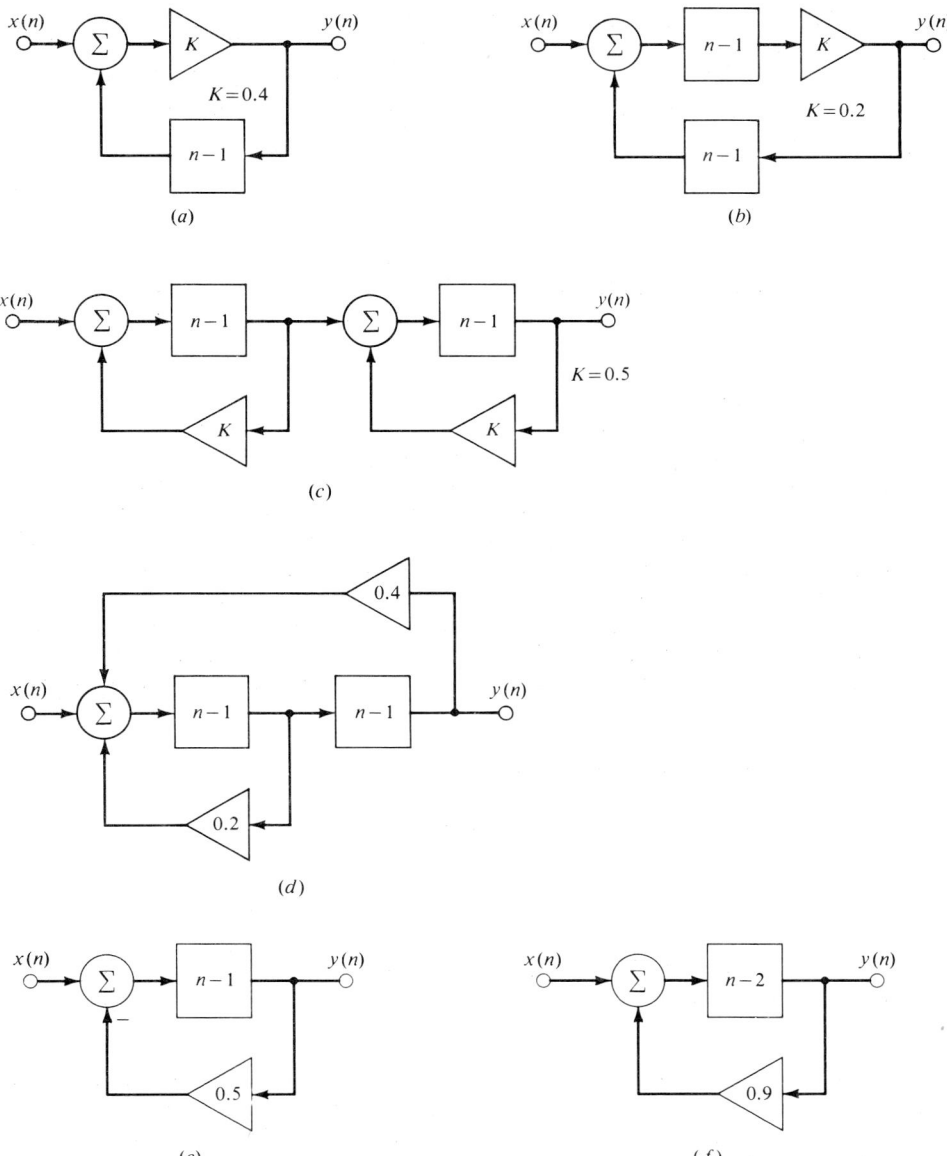

Figure P6-26

6-27 Plot the analog output of the digital integrator of Fig. 6-49 for input $x_a(t) = 5u(t)$ V and sampling rate $F = 1$ kHz.

6-28 Obtain the analog output of the digital differentiator of Fig. 6-52 for input $x_a(t) = 20(t/\tau)u(t)$ mA, where $\tau = 10$ ms and sampling rate $F = 10$ kHz.

6-29 The digital system of Fig. P6-29 is to emulate an analog square-law rectifier for an analog input whose bandwidth is 5 kHz. What sampling rate should be used?

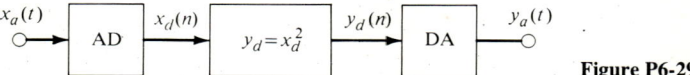

Figure P6-29

6-30 It is necessary to design a digital system that emulates an analog system whose transfer characteristic is

$$y_a = y_0 \ln\left(1 + \frac{x_a}{x_0}\right)$$

Assume that the input $x_a(t)$ is small enough to permit the transfer characteristic to be approximated using a third-degree polynomial in x_a/x_0. Draw a block diagram for the digital system and determine a suitable sampling rate.

6-31 It is necessary to design a digital system that emulates an analog system described by the differential equation

$$\frac{dy}{dt} + a_0 y = b_0 x$$

where a_0 and b_0 are known.

(a) Using a numerical approximation to the derivative dy/dt, determine a digital system (block diagram) that emulates this analog system and specify a suitable sampling rate in terms of a_0.

(b) Solving the above differential equation for dy/dt and integrating to obtain an expression for $y(t)$ gives

$$y(t) = \int_{-\infty}^{t} [b_0 x(\alpha) - a_0 y(\alpha)] \, d\alpha$$

Using this relation and a numerical approximation to the integral, obtain a digital system (block diagram) that emulates the analog system. Choose a suitable sampling rate.

6-32 Determine the smallest shift k for which the system of Fig. P6-32 gives a causal output for input $x(n) = r(0.2n)$.

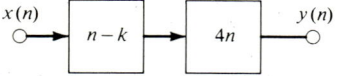

Figure P6-32

6-33 Determine the smallest shift k for which the system of Fig. P6-33 gives a causal output for input $x(n) = r(0.1n)$.

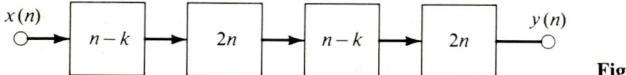

Figure P6-33

6-34 Determine the smallest shift k for which the system of Fig. P6-34 gives a causal output for input $x(n) = r(0.1n)$.

6-35 Can the system of Fig. P6-35 be made to give a causal output for input $x(n) = r(n/4)$ by introducing sufficient shift (delay) in the feedback path? Justify your answer.

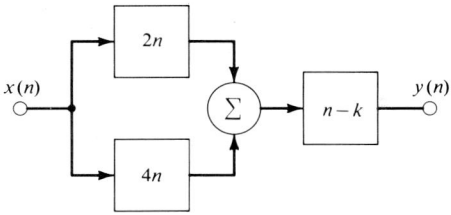

Figure P6-34

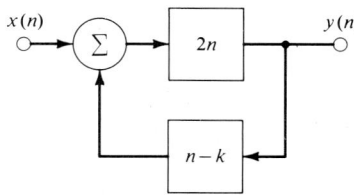

Figure P6-35

6-36 Determine the largest value of λ for which the system of Fig. P6-36 gives a causal output for input $x(n) = r(0.2n)$.

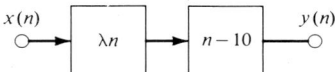

Figure P6-36

6-37 Show that the system of Fig. P6-37 is at least conditionally stable (is not unstable) for $-1 < K < 1$.

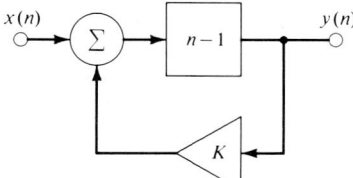

Figure P6-37

6-38 Show that the system of Fig. P6-38 is at least conditionally stable for $-1 < K < 1$.

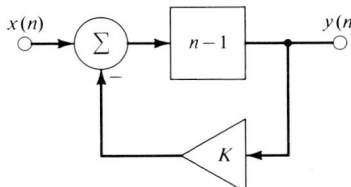

Figure P6-38

6-39 Show that the system of Fig. P6-39 is at least conditionally stable for $-1 < K_1 K_2 < 1$.

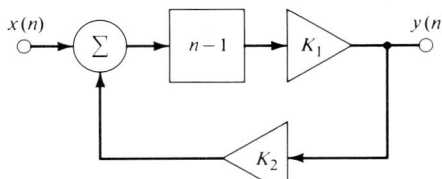

Figure P6-39

6-40 Show that the system of Fig. P6-40 is at least conditionally stable for $-1 < K_1 K_2 < 1$.

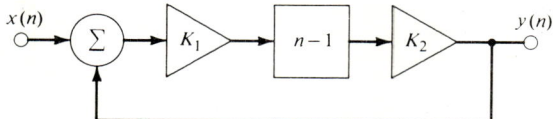

Figure P6-40

6-41 Determine the output of the system of Fig. P6-41 for input $x(n) = u(n)$ as a function of K and n. For what values of K is the output bounded? For $0 < K < 1$, determine the sample number for which the output first exceeds 90 percent of the steady-state output. Use this result to determine a condition on K under which the system provides high-fidelity transmission for input $x(n) = r(0.1n)$.

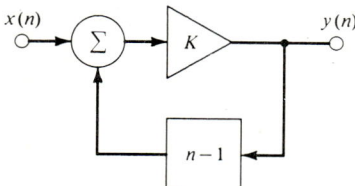

Figure P6-41

6-42 Repeat Prob. 6-41 for the system of Fig. P6-42.

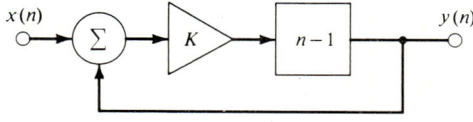

Figure P6-42

6-43 Determine a condition on λ under which the system of Fig. P6-42 provides high-fidelity transmission for input $x(n) = r(\lambda n)$.

6-44 Determine the steady-state output of the system of Fig. P6-38 for input $x(n) = u(n)$. Determine the sample number n_0 such that the output is within 5 percent of its steady-state value for all $n > n_0$. Use this result to obtain a condition on λ under which the system provides high-fidelity transmission for input $x(n) = r(\lambda n)$.

6-45 Determine the sensitivity of the dc gain of the system of Fig. P6-26a to variations of K for $K = 0.4$.

6-46 Determine the sensitivities of the dc gain of the system of Fig. P6-40 to variations of K_1 and K_2 for $K_1 = 0.1$ and $K_2 = 8$.

6-47 Determine the sensitivity of the dc gain of the system of Fig. P6-26b to variations of K for $K = 0.2$.

6-48 The system of Fig. P6-42 is to provide high-fidelity transmission for input $x(n) = r(0.1n)$. Determine the SSE for this input and show that it decreases with K.

6-49 Obtain an expression for the SSE for the system of Fig. P6-42 for input $x(n) = u(n)$. For $K = 0.5$, calculate the sensitivity of the SSE to variations of K.

CHAPTER
SEVEN

LINEAR SHIFT-INVARIANT DIGITAL SYSTEMS

Linear shift-invariant digital systems are the digital counterparts of linear stationary (time-invariant) analog systems. They obey superposition and a limited form of time invariance. Consequently, they are relatively easy to analyze and design and they constitute an important class of digital system.

The treatment of linear shift-invariant digital systems in this chapter parallels the treatment of linear stationary analog systems in Chap. 2. We show that a linear shift-invariant system is described completely by its response to a unit delta, in that the output for any other input can be determined if the output for input $\Delta(n)$ is known. This leads to a convolution sum which is analogous to the convolution integral described in Chap. 2. We describe how important properties such as realizability, stability, and fidelity are reflected in a delta response (the response to a unit delta), and we show how the delta response of a system can be obtained from a difference equation describing the system. We treat sinusoidally excited linear shift-invariant digital systems and introduce digital filtering, which is probably the single most important application of linear shift-invariant digital systems.

7-1 DEFINITION OF A LINEAR SHIFT-INVARIANT SYSTEM

In this section we describe superposition and shift invariance, which are the defining properties of a linear shift-invariant system. These properties are important because they provide the basis for virtually all methods for analysis and design of such systems.

7-1A Superposition

A *linear* digital system is a system that obeys the following *principle of superposition:*

If input $x_1(n)$ acting alone produces output $y_1(n)$ and input $x_2(n)$ acting alone produces output $y_2(n)$, then input

$$x(n) = a_1 x_1(n) + a_2 x_2(n)$$

produces output

$$y(n) = a_1 y_1(n) + a_2 y_2(n)$$

Example 7-1 Show that the system of Fig. 7-1 is linear.

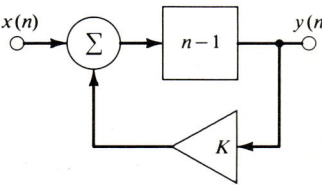

Figure 7-1 System of Example 7-1.

SOLUTION From Fig. 7-1,

$$y(n) = x(n-1) + Ky(n-1) \qquad (7\text{-}1)$$

For inputs $x_1(n)$ and $x_2(n)$ applied individually this gives

$$y_1(n) = x_1(n-1) + Ky_1(n-1) \qquad (7\text{-}2)$$

and

$$y_2(n) = x_2(n-1) + Ky_2(n-1) \qquad (7\text{-}3)$$

where $y_1(n)$ is the output for input $x_1(n)$ and $y_2(n)$ is the output for input $x_2(n)$. Adding a_1 times (7-2) to a_2 times (7-3) gives

$$a_1 y_1(n) + a_2 y_2(n) = a_1 x_1(n-1) + a_2 x_2(n-1) + K[a_1 y_1(n-1) + a_2 y_2(n-1)]$$

This relation is identical to (7-1) except that $x(n)$ is replaced by $a_1 x_1(n) + a_2 x_2(n)$ and $y(n)$ is replaced by $a_1 y_1(n) + a_2 y_2(n)$. This shows that the output for input $a_1 x_1(n) + a_2 x_2(n)$ is $a_1 y_1(n) + a_2 y_2(n)$, where $y_1(n)$ is the output for input $x_1(n)$ and $y_2(n)$ is the output for input $x_2(n)$. The system is linear because it obeys superposition.

Example 7-2 Show that the system of Fig. 7-2 is nonlinear.

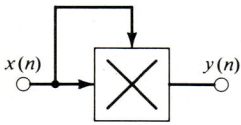

Figure 7-2 System of Example 7-2.

SOLUTION The output $y(n)$ for input $x(n)$ is given by

$$y(n) = x^2(n)$$

Taking $x(n) = a_1 x_1(n) + a_2 x_2(n)$ in this relation gives

$$y(n) = a_1^2 x_1^2(n) + a_2^2 x_2^2(n) + 2a_1 a_2 x_1(n) x_2(n)$$

whereas if the system obeyed superposition, the output would be

$$y'(n) = a_1 x_1^2(n) + a_2 x_2^2(n)$$

The system is nonlinear because it fails to obey superposition.

Superposition can be extended by induction to any number of inputs. The output of a linear system for input

$$x(n) = a_1 x_1(n) + a_2 x_2(n) + a_3 x_3(n) + \cdots$$

is given by

$$y(n) = a_1 y_1(n) + a_2 y_2(n) + a_3 y_3(n) + \cdots$$

where $y_i(n)$ is the output for input $x_i(n)$ $(i = 1, 2, \ldots)$.

It can be shown that any system consisting entirely of linear subsystems is a linear system. Because proportion, shift, compression, and addition elements are linear, any system that contains only those elements is linear. We can usually tell whether a system is linear by inspecting a block diagram for the system.

7-1B Shift Invariance

A *shift-invariant* digital system is one that obeys the following *principle of shift invariance:*

If input $x(n)$ produces output $y(n)$, then input $x(n - k)$ produces output $y(n - k)$, where k is an arbitrary integer.

This implies that a shift-invariant system commutes with a shift element.

Example 7-3 Show that the system of Fig. 7-3 is shift-invariant.

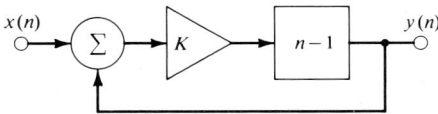

Figure 7-3 System of Example 7-3.

SOLUTION The output $y(n)$ of the system for input $x(n)$ is given by

$$y(n) = \sum_{m=1}^{\infty} K^m x(n - m)$$

It follows that the output for input $x(n - k)$ is

$$y'(n) = \sum_{m=1}^{\infty} K^m x(n - k - m) = y(n - k)$$

Therefore the system is shift-invariant.

Example 7-4 Show that the system of Fig. 7-4 is not shift-invariant.

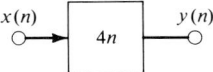

Figure 7-4 System of Example 7-4.

SOLUTION The output $y(n)$ of the system for input $x(n)$ is given by

$$y(n) = x(4n)$$

and the output $y'(n)$ for input $x(n - k)$ is given by

$$y'(n) = x(4n - k) \neq y(n - k)$$

Therefore the system is not shift-invariant.

It can be shown that a system consisting entirely of shift-invariant subsystems is shift-invariant. Because proportion, shift, and addition elements are shift-invariant, a system that contains only these elements is shift-invariant. We can usually tell whether a system is shift-invariant by inspecting a block diagram for the system.

A multiplication element is linear, shift-invariant, or neither, depending on whether the multiplication element represents a time-varying parameter, a nonlinear operation, or both. If only one input to a multiplier is derived from an input to the system (the other being generated internally and independently of the input), the multiplier is linear but not shift-invariant. If both inputs to a multiplier are derived entirely from an input to the system, the multiplier is shift-invariant but nonlinear. If the inputs to a multiplier are derived from both an input to the system and an internally generated signal, the multiplier is neither linear nor shift-invariant. For example, the system of Fig. 7-5a is linear but not shift-invariant, that of Fig. 7-5b is shift-invariant but nonlinear, and that of Fig. 7-5c is neither linear nor shift-invariant.

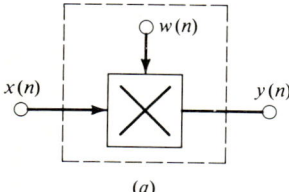

(a)

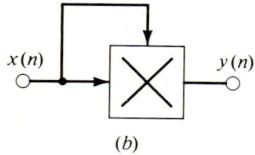

(b)

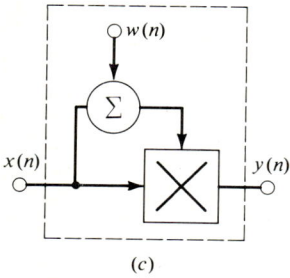

(c)

Figure 7-5 A multiplication element can be (a) linear, (b) shift-invariant, or (c) neither.

7-1C Linear Shift-Invariant Systems

A *linear shift-invariant system* is one that obeys both superposition and shift invariance: if input $x_m(n)$ acting alone produces output $y_m(n)$, then input

$$x(n) = a_1 x_1(n - k_1) + a_2 x_2(n - k_2) + \cdots$$

produces output

$$y(n) = a_1 y_1(n - k_1) + a_2 y_2(n - k_2) + \cdots$$

Example 7-5 Obtain the output of the system of Fig. 7-3 for input $x(n) = r(0.2n)$.

SOLUTION The input can be represented as the sum of two step functions; thus

$$x(n) = r(0.2n) = u(n) - u(n - 5)$$

The system is linear and shift-invariant. By superposition and shift invariance, the output is

$$y(n) = g(n) - g(n - 5)$$

where $g(n)$ is the output for input $u(n)$, given by

$$g(n) = \sum_{m=1}^{\infty} K^m u(n - m)$$

Figure 7-6 shows graphs of $g(n)$ and $y(n)$ for $K = 0.6$.

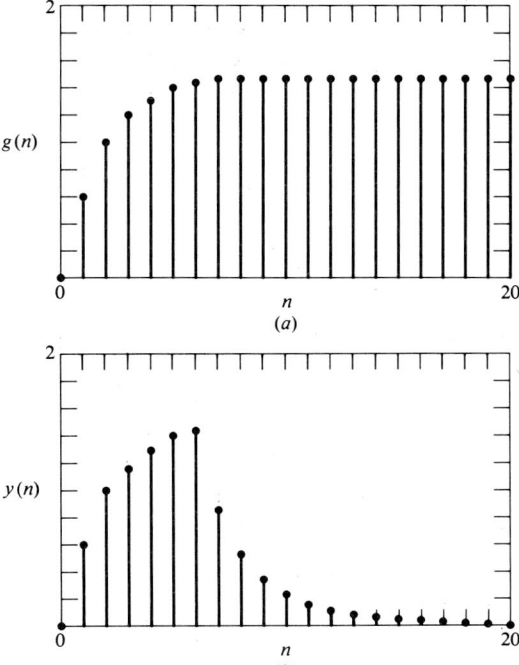

Figure 7-6 Output of the system of Fig. 7-3 for (a) step input $u(n)$, (b) pulse input $r(0.2n) = r(n/5)$.

7-2 CONVOLUTION

In this section we present a convolution sum which gives the output of a linear shift-invariant system for any input in terms of the output for a unit delta input. This convolution sum is the digital counterpart of the convolution integral treated in Sec. 2-2. It shows that any linear shift-invariant system can be described by a transfer characteristic and that all important properties of a linear shift-invariant system can be determined from the delta response of the system.

7-2A Delta Response

The *delta response* of a digital system is the output of the system for input $\Delta(n)$ (a unit delta). The delta response of a digital system is analogous to the impulse response of an analog system. The delta response of a system is denoted throughout this book by the symbol $h(n)$ (with subscripts, if necessary).

Example 7-6 Obtain the delta response of the system of Fig. 7-7.

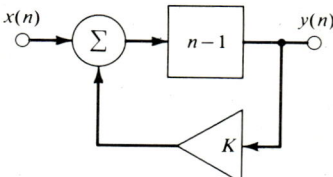

Figure 7-7 System of Example 7-6.

SOLUTION From Fig. 7-7, an output $y(n)$ is related to the corresponding input $x(n)$ by

$$y(n) = x(n-1) + Ky(n-1)$$

Using the usual recursive procedure to obtain the transfer characteristic gives

$$y(n) = \sum_{m=1}^{\infty} K^{m-1} x(n-m)$$

For $x(n) = \Delta(n)$, this relation gives the delta response

$$h(n) = \sum_{m=1}^{\infty} K^{m-1} \Delta(n-m)$$

Because $h(n) = 0$ for $n < 1$ and $\Delta(n-m) = 0$ for $m \neq n$, the delta response simplifies to

$$h(n) = K^{n-1} u(n)$$

Figure 7-8 shows graphs of the delta response for various values of the parameter K.

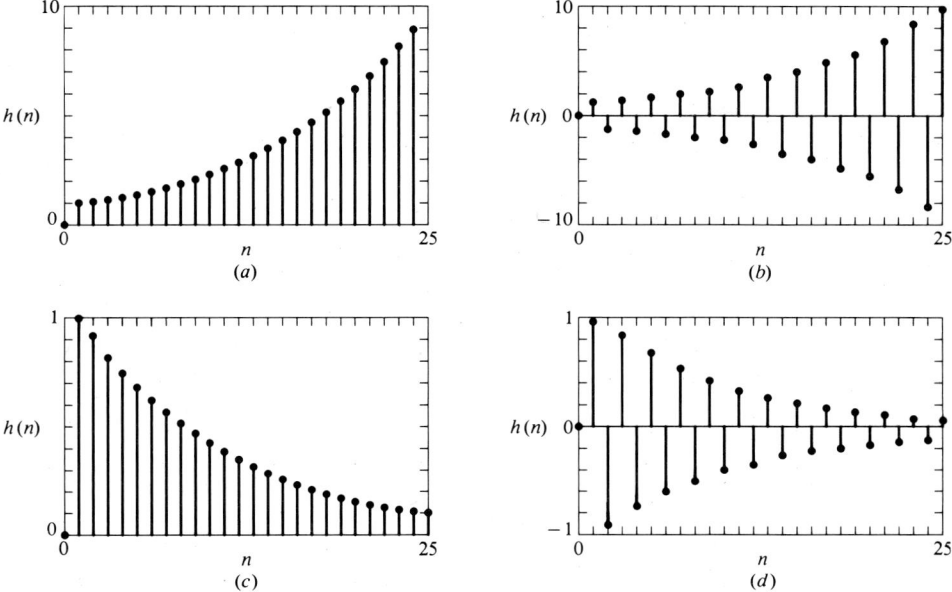

Figure 7-8 Delta response of the system of Fig. 7-7: (a) $k = 1.1$, (b) $k = -1.1$, (c) $k = 0.9$, (d) $k = -0.9$.

7-2B Convolution Sum

Figure 7-9 illustrates that any digital signal $x(n)$ can be represented as a superposition of scaled and shifted unit deltas; thus

$$x(n) = \sum_{k=-\infty}^{\infty} x(k)\Delta(n - k) \qquad (7\text{-}4)$$

A sum of this form is called a *superposition sum*.

For a linear shift-invariant system, representing an input as a sum of scaled and shifted deltas allows the corresponding output to be expressed as a sum of scaled and shifted replicas of the delta response for the system. Representing an input $x(n)$ by a superposition sum gives

$$x(n) = \sum_{k=-\infty}^{\infty} x(k)\Delta(n - k)$$

Let $h(n)$ denote the delta response of a linear shift-invariant system. By shift invariance, the output of the system for input $\Delta(n - k)$ is $h(n - k)$. By superposition, the output $y(n)$ for input $x(n)$ above is given by

$$y(n) = \sum_{k=-\infty}^{\infty} x(k)h(n - k) \qquad (7\text{-}5)$$

Thus the output $y(n)$ is represented as a sum of scaled and shifted replicas of the delta response of the system. A sum of this form is called a *convolution sum*. If we know

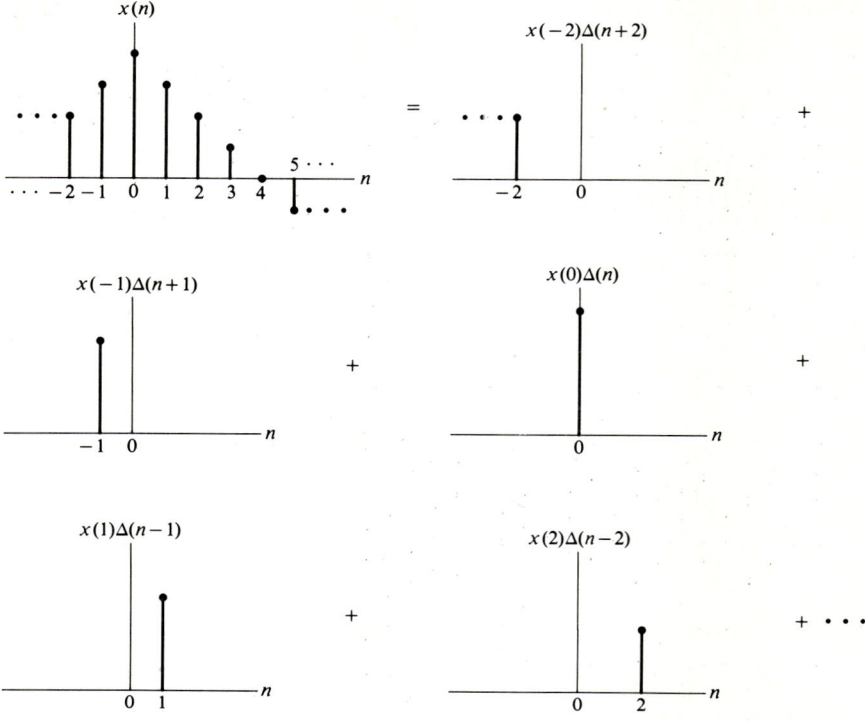

Figure 7-9 A digital signal can be represented as a superposition of scaled and shifted unit deltas.

the delta response of a system, we can use convolution to obtain the output of the system for any other input.

Example 7-7 Obtain the output of the system of Fig. 7-7 for input
$$x(n) = \lambda^{n-1} u(n)$$

SOLUTION From Example 7-6, the delta response of the system is
$$h(n) = K^{n-1} u(n)$$

From (7-5), the output for input $x(n)$ is given by
$$y(n) = \sum_{k=-\infty}^{\infty} \lambda^{k-1} u(k) K^{n-k-1} u(n-k)$$

Replacing $u(n-k)$ by zero for $k > n-1$ and by unity for $k \leq n-1$ gives
$$y(n) = \sum_{k=-\infty}^{n-1} \lambda^{k-1} u(k) K^{n-k-1}$$

Replacing $u(k)$ by zero for $k < 1$ and by unity for $k \geq 1$ gives
$$y(n) = u(n-1) \sum_{k=1}^{n-1} \lambda^{k-1} K^{n-k-1}$$

where the step $u(n-1)$ is necessary because replacing the step $u(k)$ by unity implies that the sum is over only positive values of k. This relation can be written

$$y(n) = u(n-1)\lambda^{-1}K^{n-1}\sum_{k=1}^{n-1}\left(\frac{\lambda}{K}\right)^k = u(n-1)K^{n-2}\sum_{i=0}^{n-2}\left(\frac{\lambda}{K}\right)^i$$

Using the identity

$$\sum_{i=0}^{I-1}\alpha^i = \begin{cases}\dfrac{1-\alpha^I}{1-\alpha} & \alpha \neq 1 \\ I & \alpha = 1\end{cases}$$

gives

$$y(n) = \begin{cases}\dfrac{(K^{n-1} - \lambda^{n-1})}{(K-\lambda)}u(n-1) & \lambda \neq K \\ (n-1)K^{n-2}u(n-1) & \lambda = K\end{cases}$$

Figure 7-10 shows graphs of the output for various values of K and λ.

7-2C Graphical Interpretation of Convolution

A convolution sum

$$y(n) = \sum_{k=-\infty}^{\infty} x(k)h(n-k)$$

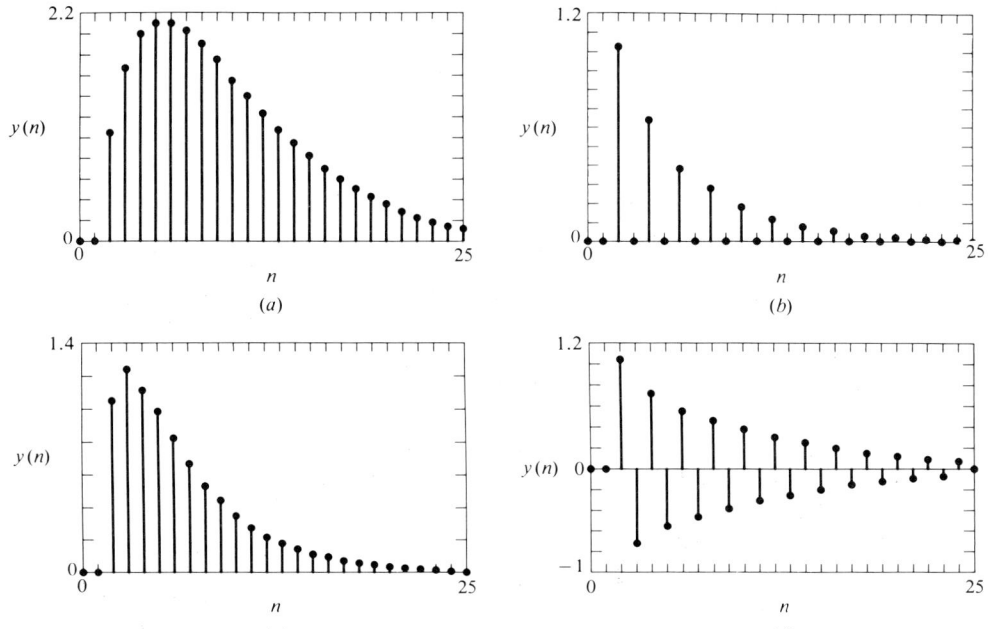

Figure 7-10 Output of the system of Fig. 7-7 for input $x(n) = \lambda^{n-1}u(n)$: (a) $k = \lambda = 0.8$; (b) $k = -0.8$, $\lambda = 0.8$; (c) $k = 0.4$, $\lambda = 0.8$; (d) $k = -0.9$, $\lambda = 0.2$.

458 INTRODUCTION TO SYSTEM ANALYSIS

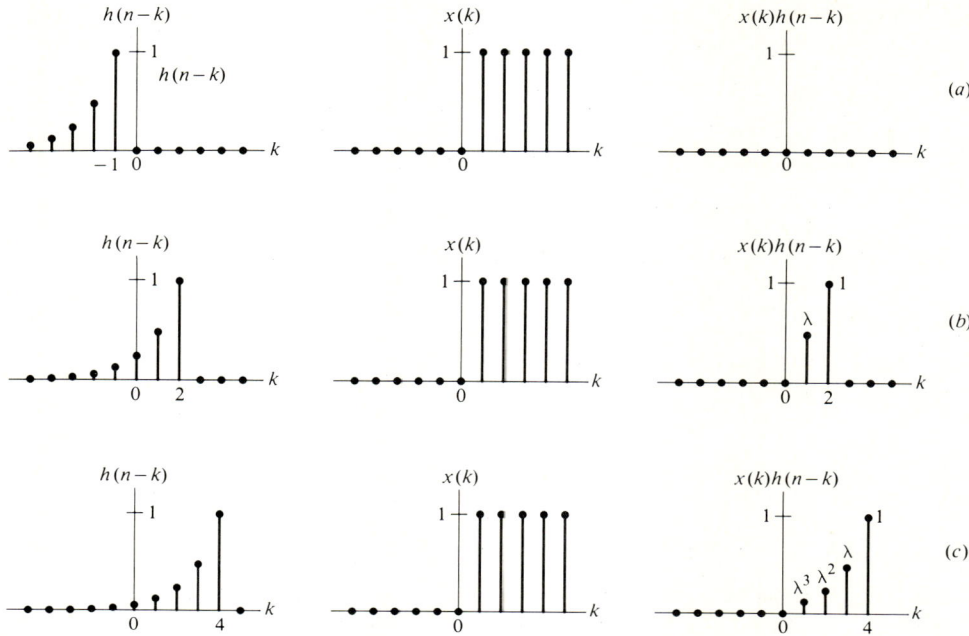

Figure 7-11 Graphical representation of a convolution operation: (a) $n = 0$, (b) $n = 3$, (c) $n = 5$.

can be represented graphically as illustrated in Fig. 7-11, where

$$h(k) = \lambda^{k-1} u(k) \qquad x(k) = u(k)$$

with $\lambda = 0.5$. For any particular sample number n, the output $y(n)$ is the sum of the product of the input and a reversed and shifted (by n) replica of the delta response. The graph of Fig. 7-11 can be viewed as three frames (taken at $n = 0$, $n = 3$, and $n = 5$) from a motion picture of a convolution operation. The signal $x(k)$ does not change from one frame to the next because $x(k)$ is independent of sample number n; however, the reversed delta response $h(n - k)$ moves to the right along the k axis. For $n = 0$ the functions $x(k)$ and $h(n - k)$ do not overlap, the product $x(k)h(n - k)$ is zero for all k, and the output is zero. For $n = 3$, about half of the reversed delta response $h(n - k)$ overlaps the input $x(k)$. The corresponding output is

$$y(3) = \sum_{k=-\infty}^{\infty} u(k) \lambda^{2-k} u(3 - k) = \lambda + 1 = 1.50$$

For $n = 5$, the output is

$$y(5) = \sum_{k=-\infty}^{\infty} u(k) \lambda^{4-k} u(5 - k) = \lambda^3 + \lambda^2 + \lambda + 1 = 1.88$$

which is 94 percent of the steady-state output $y(\infty) = 2$.

Graphical representation of convolution helps make the delta response of a linear shift-invariant system a meaningful description of the system. For example, we can infer several things from Fig. 7-11:

1. For any sample number n, the output $y(n)$ depends only on earlier values of the input. This shows that the system is realizable (at least for a step input).
2. The contribution to $y(n)$ of an input sample $x(n_0)$ decreases rapidly with $n - n_0$. This shows that the system has a limited memory of past inputs.
3. For $n > 5$, the output has almost reached its steady-state value. This illustrates that the duration of the transient is equal to the duration of the delta response.
4. The output is bounded. This shows that the system is at least conditionally stable.

Interpretation of a delta response is treated further in Sec. 7-3.

7-2D Block Diagrams and Commutativity

Any linear shift-invariant system, no matter how many elements it contains and whether it is feedforward or feedback, can be represented by a transfer characteristic (a convolution sum). Therefore, a linear shift-invariant system can be represented by a single block, as in Fig. 7-12. By convention, only the delta response is written in the block. It is understood that the transfer characteristic is given by (7-5).

Methods for reducing series, parallel, and feedback connections of linear shift-invariant systems to single blocks are given in Sec. 7-5F. Here we wish to show only that linear shift-invariant systems in series commute. Figure 7-13a shows a series connection of two linear shift-invariant systems. Figure 7-13b shows the same two systems connected in the reverse order. We wish to show that these two series connections have the same overall transfer characteristic.

From (7-5), the output of the first (leftmost) system in Fig. 7-13a for input $x(n)$ is given by

$$y(n) = \sum_{k=-\infty}^{\infty} x(k) h_1(n - k)$$

and the output of the second system is given by

$$z(n) = \sum_{p=-\infty}^{\infty} y(p) h_2(n - p)$$

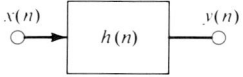

Figure 7-12 Symbol for a linear shift-invariant system having delta response $h(n)$.

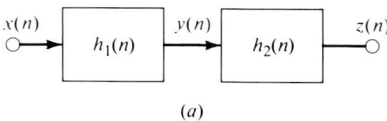

(a)

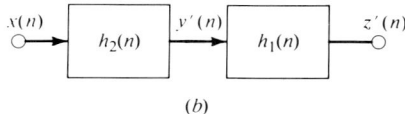

(b)

Figure 7-13 Linear shift-invariant systems commute. Systems (a) and (b) have the same overall transfer characteristic.

Combining these relations gives

$$z(n) = \sum_{p=-\infty}^{\infty} \sum_{k=-\infty}^{\infty} x(k) h_1(p-k) h_2(n-p)$$

$$= \sum_{k=-\infty}^{\infty} x(k) \sum_{p=-\infty}^{\infty} h_1(p-k) h_2(n-p) \qquad (7\text{-}6a)$$

Similarly, we find that the output of the system of Fig. 7-13b is given by

$$z'(n) = \sum_{k=-\infty}^{\infty} x(k) \sum_{p=-\infty}^{\infty} h_2(p-k) h_1(n-p) \qquad (7\text{-}6b)$$

To show that the systems of Fig. 7-13a and b have the same transfer characteristic we must show that $z(n)$ and $z'(n)$ above are identical, which in turn requires that we show that the inner sums (over p) in (7-6a) and (7-6b) are identical. In the inner sum in (7-6a) we use the change of variable $p - k = m$. This gives

$$\sum_{p=-\infty}^{\infty} h_1(p-k) h_2(n-p) = \sum_{m=-\infty}^{\infty} h_1(m) h_2(n-m-k)$$

In the inner sum in (7-6b) we use the change of variable $n - p = m$. This gives

$$\sum_{p=-\infty}^{\infty} h_2(p-k) h_1(n-p) = \sum_{m=-\infty}^{\infty} h_2(n-m-k) h_1(m)$$

Comparing the right sides of the last two relations shows that the inner sums in (7-6a) and (7-6b) are identical; therefore $z(n) = z'(n)$ in Fig. 7-13, and two linear shift-invariant systems in series commute. It follows by induction that the transfer characteristic of a series connection of any number of linear shift-invariant systems is independent of the order in which the systems are connected. This property of linear shift-invariant systems often simplifies analysis and design of such systems.

7-2E Step Response

The *step response* of a linear shift-invariant system is the output of the system for unit step input $u(n)$. In this book a step response is denoted by the symbol $g(n)$ (with subscripts if necessary). Step response is important in its own right, e.g., as a description of performance, and because of its relation to delta response. A unit delta can be represented as a sum of two steps; thus

$$\Delta(n) = u(n-1) - u(n) \qquad (7\text{-}7)$$

It follows by superposition and shift invariance that the delta response of a linear shift-invariant system is given by

$$h(n) = g(n+1) - g(n) \qquad (7\text{-}8)$$

where $g(n)$ is the step response of the system. This relation is the digital counterpart of the relation

$$h(t) = \frac{dg(t)}{dt}$$

between impulse response $h(t)$ and step response $g(t)$ of a linear stationary analog system. Equation (7-8) can also be written

$$g(n) = h(n-1) + g(n-1)$$

whence

$$g(n) = \sum_{m=1}^{\infty} h(n-m)$$

Using the change of variable $n - m = p$ gives

$$g(n) = \sum_{p=-\infty}^{n-1} h(p) \tag{7-9}$$

This relation describes a digital integrator (see Fig. 6-49 and surrounding discussion); thus (7-9) is the digital counterpart of the relation

$$g(t) = \int_{-\infty}^{t} h(\alpha)\, d\alpha$$

for a linear stationary analog system.

Example 7-8 Obtain the delta response and the step response of the system of Fig. 7-14.

SOLUTION From Fig. 7-14, an output $y(n)$ and input $x(n)$ are related by

$$y(n) = x(n-1) + 0.8 y(n-1)$$

whence

$$y(n) = \sum_{m=1}^{\infty} (0.8)^{m-1} x(n-m)$$

The delta response is

$$h(n) = \sum_{m=1}^{\infty} (0.8)^{m-1} \Delta(n-m) = (0.8)^{n-1} u(n)$$

and the step response is

$$g(n) = \sum_{m=1}^{\infty} (0.8)^{m-1} u(n-m)$$

We can also obtain the step response from the delta response using (7-9); thus

$$g(n) = \sum_{p=-\infty}^{n-1} (0.8)^{p-1} u(p) = u(n-1) \sum_{p=1}^{n-1} (0.8)^{p-1}$$

$$= u(n-1) \sum_{i=0}^{n-2} (0.8)^i = \frac{1 - (0.8)^{n-1}}{1 - 0.8} u(n-1)$$

$$= 5[1 - (0.8)^{n-1}] u(n-1)$$

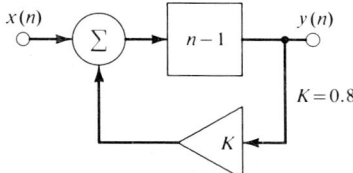

Figure 7-14 System of Example 7-8.

462 INTRODUCTION TO SYSTEM ANALYSIS

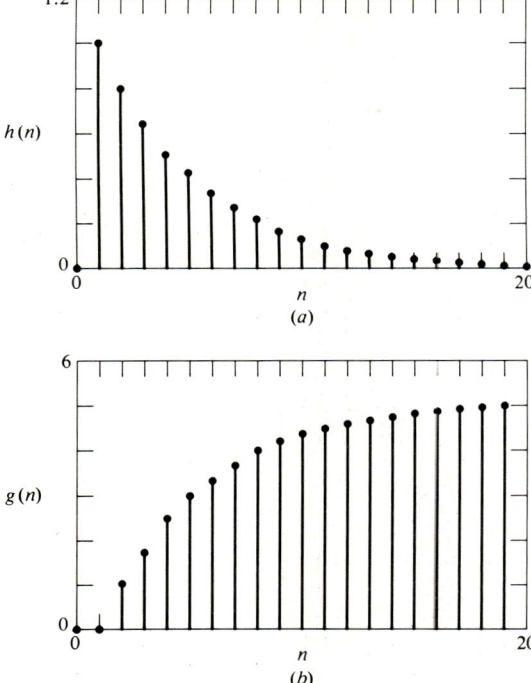

Figure 7-15 (*a*) Delta response and (*b*) step response for the system of Fig. 7-14.

Figure 7-15 shows graphs of the delta response $h(n)$ and the step response $g(n)$.

7-2F Response to a Rectangular Pulse

A rectangular pulse can be written as a sum of two steps; thus

$$r(\lambda n) = r\left(\frac{n}{k}\right) = u(n) - u(n - k)$$

where k is the largest integer value of n for which $\lambda n \leq 1$. It follows that the output of a linear shift-invariant system for input $r(n/k)$ is given by

$$f(n) = g(n) - g(n - k)$$

where $g(n)$ is the step response of the system.

Example 7-9 Determine the output of the system of Example 7-8 for input $r(0.3n)$.

SOLUTION The largest integer for which $0.3n \leq 1$ is $k = 3$; thus the input can be expressed as

$$r(0.3n) = r\left(\frac{n}{3}\right) = u(n) - u(n - 3)$$

The corresponding output is

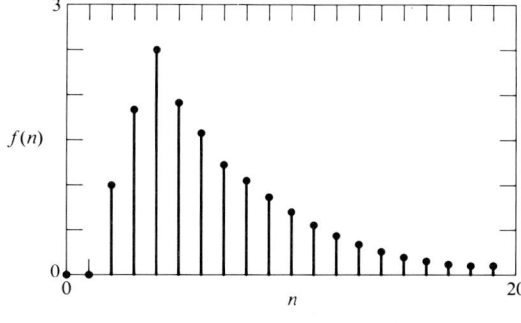

Figure 7-16 Output of the system of Fig. 7-8 for input $r(0.3n)$.

$$f(n) = 5\{[1 - (0.8)^{n-1}]u(n-1) - [1 - (0.8)^{n-4}]u(n-4)\}$$

Figure 7-16 shows a graph of the output $f(n)$.

7-2G Comparison of Shift Invariance and Time Invariance

A time-invariant (stationary) analog system is one whose output for $x(t - t_0)$ is $y(t - t_0)$, where $y(t)$ is the output for input $x(t)$. With this in mind, consider the system of Fig. 7-17, where the AD and DA converters are error-free and the digital system is linear and shift-invariant. Below, we show that this system is approximately a linear stationary analog system.

From (6-15), the digital signal $x_d(n)$ is given by

$$x_d(n) = A^{-1}x_a(nT)$$

where T is sampling interval and $A = 1$ in the unit of $x_a(t)$. From (6-17), the analog signal $y_a(t)$ is given by

$$y_a(t) = By_d\left(\frac{t}{T}\right)$$

where $B = 1$ in the unit of $y_a(t)$. Because the digital system is shift-invariant, its output for input $x_d(n - k)$ is $y_d(n - k)$, where $y_d(n)$ is the output for input $x_d(n)$. This implies that the analog output for analog input $x_a(t - kT)$ is $y_a(t - kT)$; that is, if an input is delayed by an integral number of sampling intervals, the corresponding output is delayed by the same amount. Therefore the system of Fig. 7-17 is time-invariant provided the delay time is an integral multiple of the sampling interval. For most practical purposes, the system of Fig. 7-17 is time-invariant provided only that the sampling interval is a small fraction of any other time of interest. Generally, this is also necessary for satisfactory AD conversion, so that most practical systems of the form shown in Fig. 7-17 can be treated as though they were truly time-invariant.

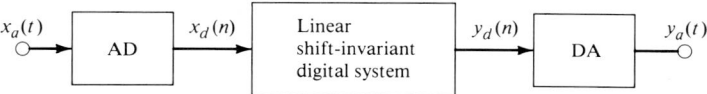

Figure 7-17 A system that is approximately linear and stationary if the AD and DA converters are error-free.

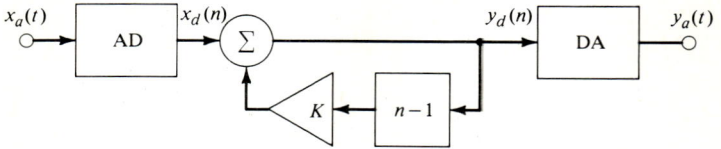

Figure 7-18 A system that emulates a first-order analog system.

We conclude this section by illustrating the relation (from a signal-processing point of view) between linear shift-invariant digital systems and linear stationary analog systems. Specifically, we wish to show that the system of Fig. 7-18 emulates a first-order linear stationary system. We assume that the AD and DA converters are error-free. The digital processor is a linear shift-invariant digital system. Therefore the overall system is linear and approximately time-invariant, as described above.

We seek the step response and the impulse response of the system of Fig. 7-18. To obtain the step response, we determine the output for input $x_a(t) = x_0 u(t)$. From (6-15), the output of the error-free AD converter is

$$x_d(n) = A^{-1} x_0 u(nT)$$

where $A = 1$ in the unit of $x(t)$ and T is the sampling interval for the system. Because T is positive, $u(nT) = u(n)$. To simplify notation, we define $\beta = A^{-1} x_0$. Thus

$$x_d(n) = \beta u(n)$$

The output of the digital system for input $x_d(n)$ is given by

$$y_d(n) = \sum_{m=0}^{\infty} K^m x_d(n-m) = \sum_{m=0}^{\infty} K^m \beta u(n-m) = \beta u(n-1) \sum_{m=0}^{n-1} K^m$$

Using the identity

$$\sum_{i=0}^{I-1} \alpha^i = \begin{cases} \dfrac{1-\alpha^I}{1-\alpha} & \alpha \neq 1 \\ I & \alpha = 1 \end{cases}$$

gives

$$y_d(n) = \begin{cases} \dfrac{\beta(1 - K^n)}{1 - K} u(n-1) & K \neq 1 \\ \beta n u(n-1) & K = 1 \end{cases}$$

We assume $0 < K < 1$. From (6-16), the output of the error-free DA converter is

$$y_a(t) = B y_d\left(\frac{t}{T}\right) = B\beta D(1 - K^{t/T}) u\left(\frac{t}{T} - 1\right)$$

where $D = 1/(1 - K)$ and $B = 1$ in the unit of $y_a(t)$. Using $\beta = x_0/A$ gives

$$y_a(t) = \frac{BD}{A} x_0 (1 - K^{t/T}) u(t - T)$$

By definition, the step response of the system is

$$g(t) = C(1 - K^{t/T})u(t - T)$$

where $C = B/[A(1 - K)]$. The step response can be written in a familiar form by noting that $K^{t/T} = e^{(\ln K)t/T}$; thus

$$g(t) = C(1 - e^{-t/\tau})u(t - T)$$

where $\tau = -T/(\ln K)$. In most applications, sampling interval T is a small fraction of any other time of interest, e.g., the time constant τ; thus $u(t - T) \approx u(t)$, and the step response can be approximated as

$$g(t) \approx C(1 - e^{-t/\tau})u(t)$$

The impulse response of the system is

$$h(t) = \frac{dg(t)}{dt} \approx \frac{C}{\tau} e^{-t/\tau} u(t)$$

These results show that the system of Fig. 7-18 emulates a first-order linear stationary analog system described by the block diagram of Fig. 7-19 (see Sec. 2-4C).

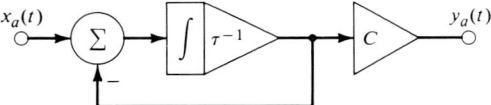

Figure 7-19 First-order analog system emulated by the system of Fig. 7-18.

7-3 INTERPRETATION OF DELTA RESPONSE

The transfer characteristic of a linear shift-invariant system is determined by the delta response of the system. Consequently, many important properties of a linear shift-invariant system can be deduced from the delta response of the system. In this section we show how realizability, stability, and fidelity are reflected in the delta response of a linear shift-invariant system.

7-3A Realizability

A linear shift-invariant system *is realizable* if and only if its delta response $h(n)$ is zero for $n < 0$. This follows from the convolution relation, which gives the output $y(n)$ for input $x(n)$ as

$$y(n) = \sum_{m=-\infty}^{\infty} x(m)h(n - m)$$

If $h(n) = 0$ for $n < 0$, then $h(n - m) = 0$ for $m > n$ and

$$y(n) = \sum_{m=-\infty}^{n} x(m)h(n - m)$$

The only values of the input $x(m)$ used in finding $y(n)$ are those for which $m \leq n$; thus $y(n)$ is causal (the system is realizable) if $h(n) = 0$ for $n < 0$. On the other hand, if

466 INTRODUCTION TO SYSTEM ANALYSIS

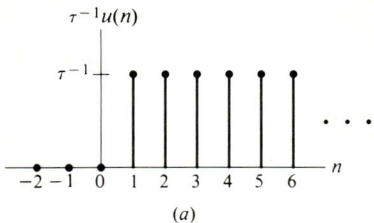
(a)

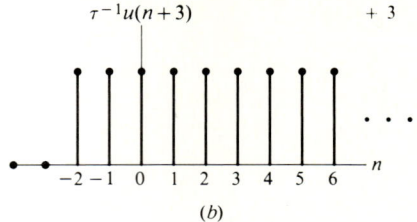
(b)

Figure 7-20 Delta response for (a) a realizable system and (b) a nonrealizable system.

$h(n) \neq 0$ for some $n < 0$, the nth sample of an output depends on at least one later sample of the corresponding input. In that case, the output is noncausal and the system is nonrealizable.

Example 7-10 Figure 7-20 shows a graph of a delta response given by

$$h(n) = \frac{1}{\tau} u(n - m)$$

This delta response describes a realizable system if $m \geq 0$ (Fig. 7-20a) and a nonrealizable system if $m < 0$ (Fig. 7-20b).

The realizability condition $h(n) = 0$ for $n < 0$ applies only to a linear shift-invariant system. It is inapplicable to a nonlinear or non-shift-invariant system because such a system is not described completely by its response to a delta.

Example 7-11 The output of a compression element for input $\Delta(n)$ is $\Delta(\lambda n)$, where λ is the compression ratio. For $\lambda > 0$, this becomes $\Delta(\lambda n) = \Delta(n)$; thus the "delta response" of a compression element is causal. Nonetheless, a compression element is nonrealizable. The condition $h(n) = 0$ for $n < 0$ is insufficient in this case because a compression element is not shift-invariant (see Example 7-4).

7-3B Stability

A linear shift-invariant system is *stable* if and only if its delta response $h(n)$ is absolutely summable; i.e., if and only if

$$\sum_{n=-\infty}^{\infty} |h(n)| < \infty \qquad (7\text{-}10)$$

Otherwise, the system is either conditionally stable or unstable.

To show that (7-10) is a sufficient condition for stability we must show that (7-10) implies that any bounded input produces a bounded output. Using convolution, the output $y(n)$ for input $x(n)$ can be expressed as

$$y(n) = \sum_{m=-\infty}^{\infty} x(m) h(n - m)$$

The triangle inequality gives

$$|y(n)| \le \sum_{m=-\infty}^{\infty} |x(m)||h(n-m)|$$

Consequently, the condition $|x(n)| \le M_x$, where M_x is a positive constant, implies that

$$|y(n)| \le M_x \sum_{m=-\infty}^{\infty} |h(n-m)| \quad \text{or} \quad |y(n)| \le M_x \sum_{m=-\infty}^{\infty} |h(m)|$$

This shows that $y(n)$ is bounded if $x(n)$ is bounded and (7-10) holds. Therefore, (7-10) is a sufficient condition for stability.

To show that (7-10) is a necessary condition for stability we need only find one system for which (7-10) fails to hold and for which one bounded input produces an unbounded output. We consider a system having delta response $h(n) = u(n)$. For this system the sum

$$\sum_{m=-\infty}^{\infty} |h(m)| = \sum_{m=-\infty}^{\infty} |u(m)| = \sum_{m=1}^{\infty} 1$$

does not exist; that is, (7-10) fails to hold. Furthermore, the output $y(n)$ for input $x(n) = u(n)$, given by

$$y(n) = \sum_{m=-\infty}^{\infty} u(m)u(n-m) = u(n-1)\sum_{m=1}^{n-1} 1 = (n-1)u(n-1)$$

is unbounded. Therefore, (7-10) is a necessary condition for stability.

Example 7-12 The delta response for a certain linear shift-invariant system is

$$h(n) = K^n u(n)$$

Find all values of K for which the system is stable.

SOLUTION We seek the values of K for which the sum

$$\sum_{m=-\infty}^{\infty} |h(m)| = \sum_{m=-\infty}^{\infty} |K^m u(m)| = \sum_{m=1}^{\infty} |K^m|$$

is finite. This requires

$$\lim_{n \to \infty} \left(\sum_{m=1}^{n} |K^m| \right) = \lim_{n \to \infty} |K(1-K)^{-1}(1-K^n)| \le \infty$$

which in turn requires $|K| < 1$. Therefore, the system is stable for $-1 < K < 1$.

Example 7-13 Find all values of K for which the system of Fig. 7-21 is stable.

SOLUTION The output $y(n)$ for input $x(n)$ is given by

$$y(n) = \sum_{p=0}^{M} K^p x(n-p)$$

whence

$$|y(n)| \le \sum_{p=0}^{M} |K^p||x(n-p)|$$

468 INTRODUCTION TO SYSTEM ANALYSIS

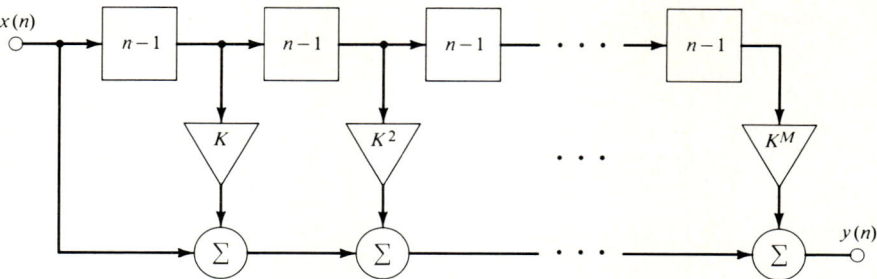

Figure 7-21 System of Example 7-13.

For $|x(n)| \leq M_x$, this gives

$$|y(n)| \leq M_x \sum_{p=0}^{M} |K^p|$$

which shows that a bounded input gives a bounded output if K is finite. Therefore the system is stable for all finite values of K.

7-3C Fidelity

A *distortionless system* is a linear shift-invariant system whose output $y(n)$ for input $x(n)$ is given by

$$y(n) = Kx(n - n_0)$$

where K and n_0 are independent of n. The corresponding delta response is

$$h(n) = K\Delta(n - n_0) \qquad (7\text{-}11)$$

The parameters K and n_0 are arbitrary because gain and delay do not introduce distortion no matter how much they may degrade other measures of performance.

Equation (7-11) shows that the delta response of a distortionless system is a delta. The closer the delta response of a system approximates a delta the smaller the distortion introduced by the system. Loosely speaking, a linear shift-invariant system provides high-fidelity transmission for an input if the delta response $h(n)$ of the system is much narrower than the narrowest significant feature of the input and

$$\sum_{n=-\infty}^{\infty} h(n) \neq 0$$

(This requires that the dc gain of the system be nonzero.)

Example 7-14 Determine a condition on k under which the system of Fig. 7-22 provides high-fidelity transmission for input

$$x(n) = r\left(\frac{n}{k}\right)$$

SOLUTION The delta response of the system is

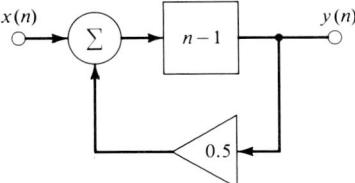

Figure 7-22 System of Example 7-14.

$$h(n) = (0.5)^{n-1} u(n)$$

We require that the duration of the input be much greater than the duration of the delta response. The duration of the input is k sampling intervals. The duration of the delta response above can be defined as the sample number for which the delta response equals a small fraction (say 1 percent) of its initial value $h(1) = 1$. From $(0.5)^{n-1} \le (0.01)h(1)$ we find $n = 8$.

It follows that the system provides high-fidelity transmission for a rectangular pulse $r(n/k)$ if the duration of the pulse is much greater than eight sampling intervals; i.e., provided $k \gg 8$. Figure 7-23 shows outputs of the system for inputs $r(n/8)$ and $r(n/40)$. For input $r(n/8)$, considerable distortion is evident. For input $r(n/40)$, the output looks much like the input.

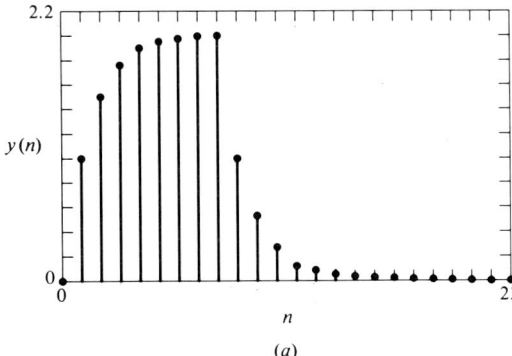

(a)

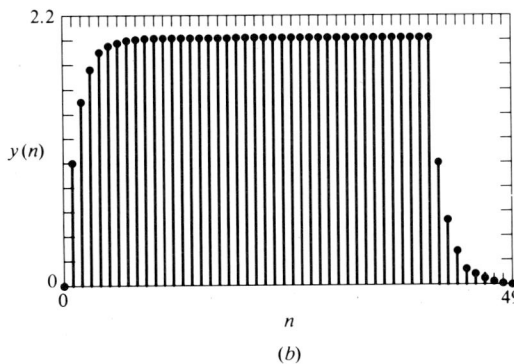

(b)

Figure 7-23 Output of the system of Fig. 7-22 for input $r(n/k)$ with (a) $k=8$ and (b) $k=40$.

470 INTRODUCTION TO SYSTEM ANALYSIS

Above, we describe how the delta response of a linear shift-invariant system can be used to obtain an output of the system and how important properties of the system can be determined from the delta response. Next we show how to obtain the delta response of a linear shift-invariant system described by a difference equation.

7-4 SYSTEMS DESCRIBED BY DIFFERENCE EQUATIONS

An input $x(n)$ and corresponding output $y(n)$ for a linear shift-invariant system are related by a difference equation of the form

$$y(n) + a_1 y(n-1) + \cdots + a_k y(n-k)$$
$$= b_0 x(n) + b_1 x(n-1) + \cdots + b_m x(n-m) \quad (7\text{-}12)$$

The integer k is called the *order* of the equation (the order of the system). Systems described by (7-12) are called *recursive* if $k \neq 0$ and are called *nonrecursive* if $k = 0$ (and m is finite). The coefficients a_i ($i = 1, 2, \ldots, k$) and b_i ($i = 0, 1, \ldots, m$) are determined by parameters of a system described by (7-12). In (7-12) the coefficient of $y(n)$ is equal to 1. This convention can always be enforced by dividing a difference equation by the coefficient of $y(n)$.

In this section we present a standard block diagram representing a system described by (7-12), we show how (7-12) can be used to determine an output numerically (by recursion), and we describe a method for finding a closed-form expression for the delta response of a system described by (7-12).

7-4A Block Diagrams

A system described by (7-12) with $k = m$ can be represented by the block diagram of Fig. 7-24. If $k < m$, then $a_i = 0$ for $i > k$. If $k > m$, then $b_i = 0$ for $i > m$. For a

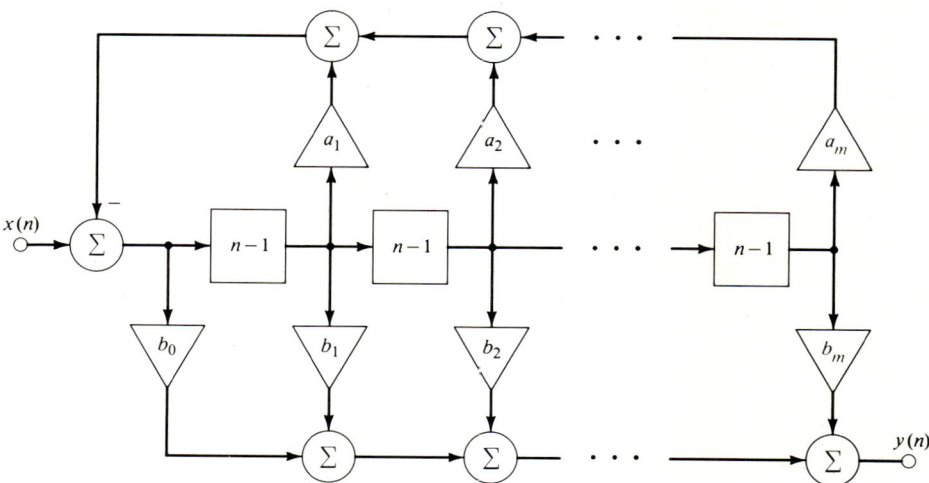

Figure 7-24 Block diagram representing a system described by (7-12).

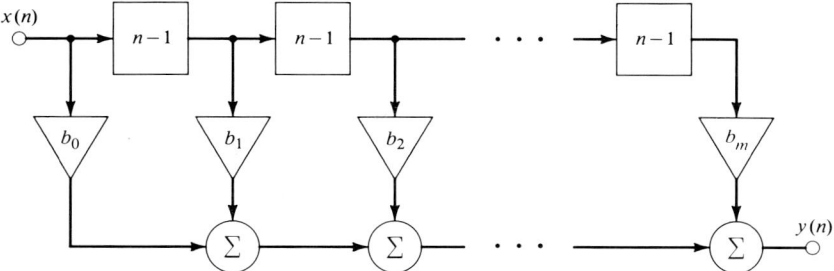

Figure 7-25 Block diagram representing a nonrecursive system.

nonrecursive system, $a_i = 0$ for all i, and the block diagram of Fig. 7-24 simplifies to that of Fig. 7-25. We show below that a second-order recursive system described by (7-12) is represented by the block diagram of Fig. 7-24. Generalization to systems of arbitrary order is straightforward.

Figure 7-26 shows the block diagram of Fig. 7-24 for $k = m = 2$. To obtain a difference equation describing this system we first write expressions for the outputs of the addition elements. From Fig. 7-26,

$$y(n) = b_0 e(n) + b_1 e(n-1) + b_2 e(n-2) \qquad (7\text{-}13)$$

and
$$e(n) = x(n) - a_1 e(n-1) - a_2 e(n-2) \qquad (7\text{-}14)$$

Using (7-14) to replace $e(n)$, $e(n-1)$, and $e(n-2)$ in (7-13) yields

$$\begin{aligned} y(n) = &\, b_0[x(n) - a_1 e(n-1) - a_2 e(n-2)] \\ &+ b_1[x(n-1) - a_1 e(n-2) - a_2 e(n-3)] \\ &+ b_2[x(n-2) - a_1 e(n-3) - a_2 e(n-4)] \end{aligned}$$

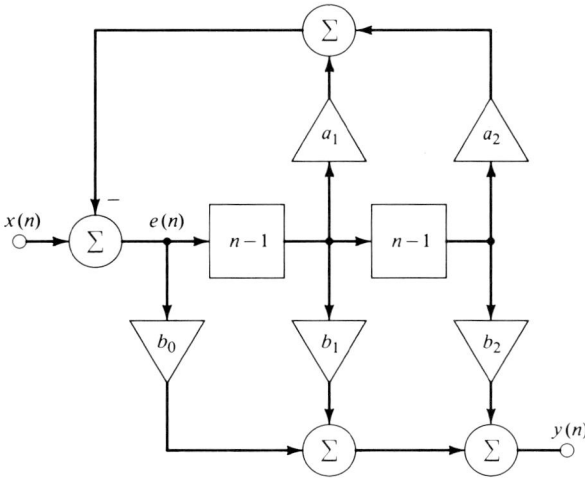

Figure 7-26 Block diagram for a second-order recursive system.

472 INTRODUCTION TO SYSTEM ANALYSIS

Rearranging terms gives

$$y(n) = b_0 x(n) + b_1 x(n-1) + b_2 x(n-2)$$
$$- a_1[b_0 e(n-1) + b_1 e(n-2) + b_2 e(n-3)]$$
$$- a_2[b_0 e(n-2) + b_1 e(n-3) + b_2 e(n-4)]$$

From (7-13), the quantities in brackets equal $y(n-1)$ and $y(n-2)$, respectively; thus the last relation above simplifies to

$$y(n) + a_1 y(n-1) + a_2 y(n-2) = b_0 x(n) + b_1 x(n-1) + b_2 x(n-2)$$

which has the form of (7-12) with $k = m = 2$. This shows that the block diagram of Fig. 7-26 represents a recursive second-order system. Generalization to the block diagram of Fig. 7-24 is straightforward.

7-4B Numerical Recursion

An output of a system can be obtained numerically by repeated use of a difference equation describing the system. The procedure is illustrated in the following example.

Example 7-15 Obtain the first 10 samples of the output $y(n)$ of the system of Fig. 7-27 for input $x(n) = u(n)$.

SOLUTION Comparing the block diagram of Fig. 7-27 with that of Fig. 7-24 shows that the system is described by the difference equation

$$y(n) + 0.9y(n-1) + 0.2y(n-2) = x(n) + 0.8x(n-1) + 0.6x(n-2)$$

We assume that the system is realizable. It follows that the output $y(n)$ for input $x(n) = u(n)$ equals zero for $n < 1$. To obtain the output $y(n)$ for $n \geq 1$ we use the difference equation above for $n = 1, 2, 3, \ldots$; thus

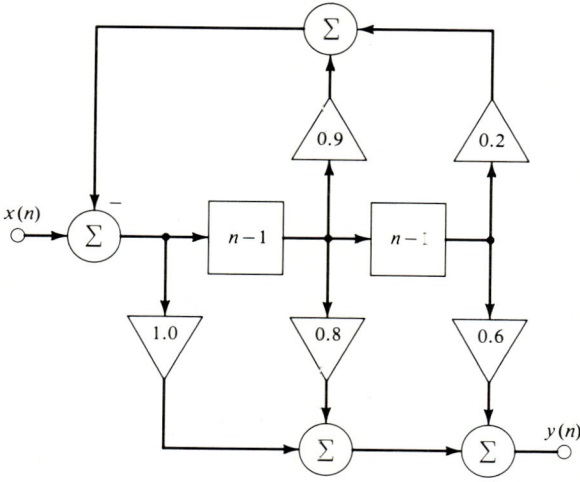

Figure 7-27 System of Example 7-15.

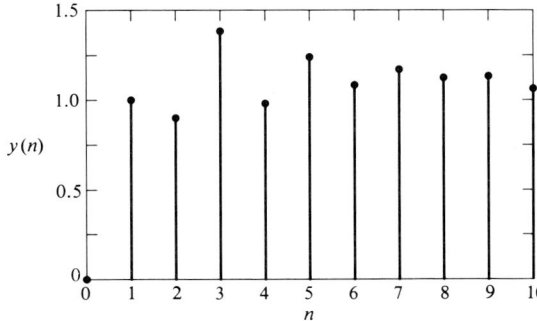

Figure 7-28 Output of the system of Fig. 7-27 for input $x(n) = u(n)$.

$$y(1) = x(1) + 0.8x(0) + 0.6x(-1) - 0.9y(0) - 0.2y(-1)$$
$$= 1.0 + (0.8)(0) + (0.6)(0) - (0.9)(0) - (0.2)(0) = 1.0$$
$$y(2) = x(2) + 0.8x(1) + 0.6x(0) - 0.9y(1) - 0.2y(0)$$
$$= 1.0 + (0.8)(1.0) + (0.6)(0) - (0.9)(1.0) - (0.2)(0) = 0.9$$
$$y(3) = x(3) + 0.8x(2) + 0.6x(1) - 0.9y(2) - 0.2y(1)$$
$$= 1.0 + (0.8)(1.0) + (0.6)(1.0) - (0.9)(0.9) - (0.2)(1.0) = 1.4$$

and so on. Figure 7-28 shows a graph of the output $y(n)$ for $0 \le n \le 10$.

Finding an output by numerical recursion is useful when a closed-form expression is not needed, e.g., in examining fidelity for a specific input. Numerical recursion is also useful for checking closed-form expressions for correctness and for obtaining initial values for a closed-form solution to (7-12).

7-4C Delta Response and Convolution

To obtain a closed-form expression for the delta response of a linear shift-invariant system we solve the associated difference equation for the output for input $\Delta(n)$. This task is simplified by separating the system into feedforward (nonrecursive) and feedback (recursive) subsystems, as follows. We can write (7-12) as two equations:

$$v(n) = b_0 x(n) + b_1 x(n-1) + b_2 x(n-2) + \cdots + b_m x(n-m) \quad (7\text{-}15)$$

which describes a nonrecursive (feedforward) system with input $x(n)$ and output $v(n)$, and

$$y(n) + a_1 y(n-1) + a_2 y(n-2) + \cdots + a_k y(n-k) = v(n) \quad (7\text{-}16)$$

which describes a recursive (feedback) system with input $v(n)$ and output $y(n)$. Equation (7-12) describes a series connection of these systems, as shown in Fig. 7-29a. The systems described by (7-15) and (7-16) commute because they are linear and shift-invariant. Therefore the systems of Fig. 7-29a can be reordered as shown in Fig. 7-29b. The system of Fig. 7-29b is described by the difference equations

$$w(n) + a_1 w(n-1) + a_2 w(n-2) + \cdots + a_k w(n-k) = x(n) \quad (7\text{-}17)$$

474 INTRODUCTION TO SYSTEM ANALYSIS

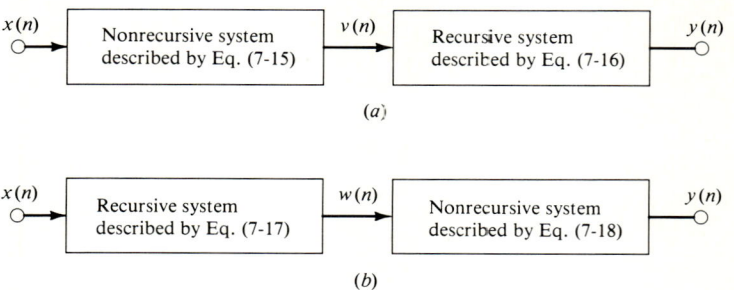

Figure 7-29 Separation of a linear shift-invariant system into feedforward (nonrecursive) and feedback (recursive) subsystems.

$$y(n) = b_0 w(n) + b_1 w(n-1) + b_2 w(n-2) + \cdots + b_m w(n-m) \qquad (7\text{-}18)$$

We determine the delta response of a system described by (7-12) as follows. First, we determine the delta response $h'(n)$ of the recursive system described by (7-17) by solving the difference equation

$$h'(n) + a_1 h'(n-1) + \cdots + a_k h'(n-k) = \Delta(n) \qquad (7\text{-}19)$$

Then we determine the delta response $h(n)$ of the system as a whole using (7-18)

$$h(n) = b_0 h'(n) + b_1 h'(n-1) + \cdots + b_m h'(n-m) \qquad (7\text{-}20)$$

We assume that the system described by (7-17) is realizable. This means that the delta response $h'(n)$ equals zero for $n < 0$. We also use the fact that the effect of the input $\Delta(n)$ is to establish an initial condition from which the system relaxes for $n > 0$; that is, for $n > 0$, the input $\Delta(n)$ equals zero and the system is unforced. This means that for $n > 0$ the delta response $h'(n)$ satisfies the so-called *homogeneous equation*

$$h'(n) + a_1 h'(n-1) + \cdots + a_k h'(n-k) = 0 \qquad (7\text{-}21)$$

It follows that the delta response of the recursive subsystem described by (7-17) is obtained by finding the complete solution $h'(n)$ of (7-21) subject to the conditions

$$h'(-1) = h'(-2) = \cdots = 0 \qquad h'(0) = \Delta(0) = 1 \qquad (7\text{-}22)$$

It can be shown that the complete solution of (7-21) consists of k linearly independent terms at least one of which has the form cz^n, where c and z are constants. Substituting cz^n for $h'(n)$ in (7-21) gives

$$cz^n(1 + a_1 z^{-1} + \cdots + a_k z^{-k}) = 0$$

For a nontrivial solution, it is necessary that $c \neq 0$ and $z \neq 0$. The equation above simplifies to

$$z^k + a_1 z^{k-1} + \cdots + a_k = 0 \qquad (7\text{-}23)$$

Equation (7-23) is called the *characteristic equation* for a system described by (7-12). The left side of (7-23) is called the *characteristic polynomial* for the system. The roots of the characteristic polynomial (the solutions of the characteristic equation) are called

LINEAR SHIFT-INVARIANT DIGITAL SYSTEMS **475**

the *characteristic roots* for the system. Equation (7-23) has k roots (a system of order k has k characteristic roots). We denote them by $p_1, p_2, \ldots, p_k$. Every function of the form cp^n, where p is a characteristic root and c is an arbitrary constant, is a solution of (7-21). The complete solution of (7-21) is a function consisting of k linearly independent terms, as follows:

For each nonrepeated characteristic root p, the solution contains a term of the form

$$cp^n$$

For each k-fold-repeated root p, the solution contains a term of the form

$$(c_0 + c_1 n + c_2 n^2 + \cdots + c_{k-1} n^{k-1}) p^n$$

For example, suppose the characteristic equation for a system can be written in factored form as

$$(z - p_1)(z - p_2)(z - p_3)^2 = 0$$

where p_1, p_2, p_3 are all different. In this case there are two nonrepeated characteristic roots (p_1 and p_2) and one twofold-repeated root p_3, and the complete solution of (7-21) is

$$h'(n) = c_1 p_1^n + c_2 p_2^n + (c_3 + c_4 n) p_3^n$$

The solution consists of $k = 4$ linearly independent terms, as required.

So far as (7-21) is concerned, the constants $c_1, c_2, \ldots, c_k$ that appear in a complete solution are arbitrary. To determine these k constants we require k relations. We can obtain these relations from (7-21) using recursion and the conditions on $h'(n)$ given by (7-22); thus

$$h'(0) = \Delta(0) - a_1 h'(-1) - a_2 h'(-2) - \cdots - a_k h'(-k) = 1$$
$$h'(1) = \Delta(1) - a_1 h'(0) - a_2 h'(-1) - \cdots - a_k h'(-k+1) = -a_1$$
$$h'(2) = \Delta(2) - a_1 h'(1) - a_2 h'(0) - \cdots - a_k h'(-k+2) = a_1^2 - a_2$$

and so on, until k relations are obtained. These k relations give k simultaneous equations which can be solved for the k constants $c_1, c_2, \ldots, c_k$ that appear in the solution to (7-21).

Example 7-16 Determine the delta response for the system of Fig. 7-30.

SOLUTION Comparing the block diagram of Fig. 7-30 with that of Fig. 7-24 shows that the system of Fig. 7-30 is described by the difference equation

$$y(n) + 1.20 y(n-1) + 0.32 y(n-2) = x(n) + 0.50 x(n-1)$$

The characteristic equation for the system is

$$z^2 + 1.20 z + 0.32 = 0$$

The characteristic roots are

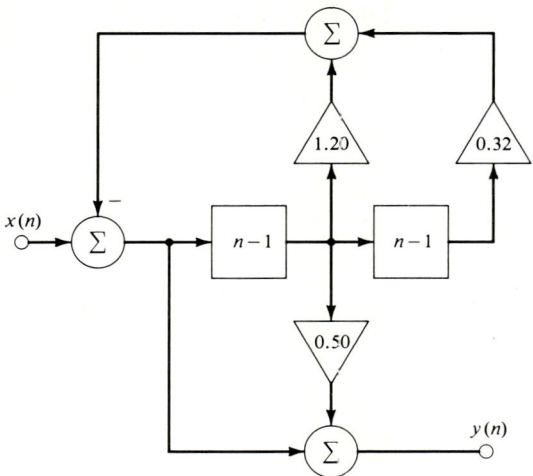

Figure 7-30 System of Example 7-16.

$$p_1 = -0.40 \qquad p_2 = -0.80$$

The delta response of the recursive subsystem is

$$h'(n) = c_1 p_1^n + c_2 p_2^n$$

To determine the constants c_1, c_2 we need two conditions on $h'(n)$. We obtain them from (7-22) using recursion and the difference equation

$$h'(n) + 1.20 h'(n-1) + 0.32 h'(n-2) = \Delta(n)$$

which describes the recursive subsystem; thus

$$h'(0) = 1.00$$
$$h'(1) = -1.20 h'(0) = -1.20$$

From these conditions and the expression given above for $h'(n)$ we obtain the two simultaneous equations

$$c_1 + c_2 = 1.00$$
$$c_1 p_1 + c_2 p_2 = -1.20$$

where $p_1 = -0.40$ and $p_2 = -0.80$. Solving these equations gives

$$c_1 = -1.00 \qquad c_2 = 2.00$$

Therefore, the delta response of the recursive subsystem is

$$\begin{aligned}
h'(n) &= -p_1^n + 2.00 p_2^n & n \geq 0 \\
&= -(-0.40)^n + 2.00(-0.80)^n & n \geq 0 \\
&= [-(-0.40)^n + 2.00(-0.80)^n] u(n+1)
\end{aligned}$$

We can check this result by comparing the value of $h'(2)$ given by the expression above with that obtained by recursion from the difference equation describing the recursive subsystem. The expression above for $h'(n)$ gives

$$h'(2) = -(-0.4)^2 + 2(-0.8)^2 = 1.12$$

The difference equation describing the recursive subsystem gives

$$h'(2) = -1.20h'(1) - 0.32h'(0) = -1.20(-1.20) - (0.32)(1.00) = 1.12$$

The fact that these two calculations give the same value for $h'(2)$ indicates that the expression above for $h'(n)$ is correct.

The delta response $h(n)$ of the system as a whole is the output of the nonrecursive subsystem for input $h'(n)$ obtained above:

$$h(n) = h'(n) + 0.50h'(n-1)$$
$$= [-(-0.40)^n + 2.00(-0.80)^n]u(n+1)$$
$$+ 0.50[-(-0.40)^{n-1} + 2.00(-0.80)^{n-1}]u(n)$$
$$= \begin{cases} 1.00 & n = 0 \\ 0.25(-0.40)^n + 0.75(-0.80)^n & n > 0 \end{cases}$$
$$= [0.25(-0.40)^n + 0.75(-0.80)^n]u(n+1)$$

Figure 7-31 shows a graph of the delta response $h(n)$ versus sample number n.

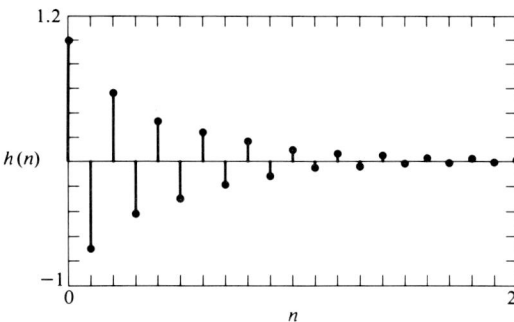

Figure 7-31 Delta response of the system of Fig. 7-30.

Once we have obtained the delta response of a system, we can obtain the output for any other input by convolution. This avoids solving the difference equation repeatedly when outputs for several different inputs are to be obtained. More important, a delta response is often easier to interpret (in terms of stability, fidelity, transient response, and other important properties) than the associated difference equation.

Example 7-17 Obtain the output $y(n)$ of the system of Example 7-16 for input

$$x(n) = 5.00(0.60)^n u(n)$$

SOLUTION From Example 7-16, the delta response of the system is

$$h(n) = [0.25(-0.40)^n + 0.75(-0.80)^n]u(n+1)$$

By convolution,

$$y(n) = \sum_{k=-\infty}^{\infty} x(k)h(n-k)$$
$$= \sum_{k=-\infty}^{\infty} 5.00(0.60)^k u(k)[0.25(-0.40)^{n-k} + 0.75(-0.80)^{n-k}]u(n-k+1)$$

$$= 5.00u(n) \sum_{k=1}^{n} [0.25(0.60)^k(-0.40)^{n-k} + 0.75(0.60)^k(-0.80)^{n-k}]$$

$$= 1.25(-0.40)^n u(n) \sum_{k=1}^{n} \left(-\frac{0.60}{0.40}\right)^k + 3.75(-0.80)^n u(n) \sum_{k=1}^{n} \left(-\frac{0.60}{0.80}\right)^k$$

$$= (1.25)(-1.50)(-0.40)^n u(n) \sum_{k=0}^{n-1} (-1.50)^k$$

$$+ (3.75)(-0.75)(-0.80)^n u(n) \sum_{k=0}^{n-1} (-0.75)^k$$

$$= -0.75(-0.40)^n[1 - (-1.50)^n]u(n) - 1.61(-0.80)^n[1 - (-0.75)^n]u(n)$$

This simplifies to

$$y(n) = [0.75(-0.40)^n - 1.61(-0.80)^n + 2.36(0.60)^n]u(n)$$

To check this result we calculate $y(1)$ using the expression above and again using the difference equation describing the system. The expression above gives

$$y(1) = -(0.75)(-0.40) - (1.61)(-0.80) + (2.36)(0.60) = 3.00$$

The difference equation gives

$$y(1) = x(1) + 0.50x(0) - 1.20y(0) - 0.32y(-1)$$
$$= (5.00)(0.60) + (0.50)(0) - (1.20)(0) - (0.32)(0) = 3.00$$

The fact that these calculations give the same value for $y(n)$ indicates that the expression above for $y(n)$ is correct.

7-4D Realizability

A system is *realizable* if the nth sample of an output depends only on the nth and earlier samples of the corresponding input.

Example 7-18 A system whose input $x(n)$ and output $y(n)$ are related by

$$y(n) + 0.4y(n - 1) = 1.2x(n + 1) - 0.5x(n)$$

is nonrealizable because the output at the nth sampling instant depends on a future sample of the input.

Example 7-19 A system whose input $x(n)$ and output $y(n)$ are related by

$$y(n + 4) - 2y(n + 2) + 4y(n - 1) = 5x(n) + 3x(n - 3)$$

is realizable because the most recent output $y(n + 4)$ depends only on earlier samples $x(n)$, $x(n - 3)$ of an input. This becomes clear on writing the difference equation above as

$$y(p) - 2y(p - 2) + 4y(p - 5) = 5x(p - 4) + 3x(p - 7)$$

where we have used the change of variable (shift of time origin) $p = n + 4$.

7-4E Stability

A linear shift-invariant system is *stable* if and only if its delta response $h(n)$ is absolutely summable; i.e., if and only if

$$\sum_{n=-\infty}^{\infty} |h(n)| < \infty$$

The delta response of a system consists of linearly independent terms of the form

$$(c_0 + c_1 n + c_2 n^2 + \cdots + c_{i-1} n^{i-1}) p^n$$

where p is a characteristic root having multiplicity i. It follows that the delta response is absolutely summable if and only if each such term is absolutely summable. A term of the form above is absolutely summable if and only if $|p| < 1$. Therefore, a linear shift-invariant system is stable if and only if all characteristic roots for the system have magnitudes smaller than unity.

Example 7-20 Determine whether a system having input $x(n)$ and output $y(n)$ related by the difference equation

$$y(n) - 1.5y(n-1) - y(n-2) = x(n) - 4.0x(n-1)$$

is stable.

SOLUTION The characteristic equation for the system is

$$z^2 - 1.5z - 1.0 = 0$$

and the characteristic roots are

$$p_1 = -2.0 \qquad p_2 = 0.5$$

The system is not stable because one characteristic root (p_1) has magnitude greater than unity.

The delta response of a nonrecursive system has finite duration and finite magnitude. Therefore the delta response of a nonrecursive system is absolutely summable, and a nonrecursive system, i.e., a system described by (7-12) with $k = 0$, is stable.

7-5 RESPONSE TO SINUSOIDAL EXCITATION

In this section we define the transfer function of a linear shift-invariant system and describe how the transfer function of a system is used to obtain the output of the system for an input consisting of one or more sinusoidal components.

7-5A Representations of Sinusoidal Signals

From (6-15), error-free AD conversion of a sinusoidal analog signal $x_a(t)$ produces a sinusoidal digital signal $x_d(n)$; thus if

$$x_a(t) = x_0 \cos(\omega t + \theta)$$

then

$$x_d(n) = C^{-1} x_0 \cos(n\omega T + \theta)$$

where $C = 1$ in the unit of x_0 and T is sampling interval. To simplify the notation we let $A = C^{-1} x_0$ and define

$$\lambda = \omega T = \frac{2\pi f}{F} \qquad (7\text{-}24)$$

where F is sampling rate. The sinusoidal digital signal $x_d(n)$ is then expressed as

$$x_d(n) = A \cos(\lambda n + \theta) \qquad (7\text{-}25)$$

The quantity λ defined by (7-24) is called *digital frequency*. Note that it is a dimensionless quantity. Note also that $\lambda = \pi$ corresponds to $f = F/2$. In AD conversion using sampling rate F, an anti-alias filter generally is used to eliminate components having frequencies greater than $F/2$. Also, only components having frequencies below $F/2$ appear in the output of an (ideal) DA converter. Consequently, in analysis of a digital system we are interested primarily in digital frequencies in the range $0 \leq \lambda \leq \pi$ corresponding to actual frequencies (analog frequencies) in the range $0 \leq f < F/2$, where F is sampling rate.

A sinusoidal signal $x_d(n)$ expressed in amplitude-phase form as in (7-25) can be expressed in exponential form as

$$x_d(n) = X e^{j\lambda n} + X^* e^{-j\lambda n} \qquad (7\text{-}26)$$

where X is called *complex amplitude*. The exponential form of (7-26) is often more convenient than the amplitude-phase form of (7-25) for analysis of linear shift-invariant systems. To obtain the exponential form of a sinusoidal signal from the amplitude-phase form of the signal (to obtain X from A and θ) we use Euler's identity

$$\cos \alpha \equiv \tfrac{1}{2} e^{j\alpha} + \tfrac{1}{2} e^{-j\alpha}$$

Applying this identity to the right side of (7-25) gives the right side of (7-26) with

$$X = \frac{A}{2} e^{j\theta} \qquad (7\text{-}27)$$

Equation (7-27) gives complex amplitude X in polar form. From (7-27), peak amplitude A and initial phase θ of a sinusoidal signal expressed in exponential form are given by

$$A = 2|X| \qquad \theta = \angle X \qquad (7\text{-}28)$$

respectively, where X is the complex amplitude of the signal.

Example 7-21 The sinusoidal signal

$$x(n) = 10 \cos\left(\lambda n - \frac{\pi}{4}\right)$$

is expressed in exponential form as

$$x(n) = X e^{j\lambda n} + X^* e^{-j\lambda n} \qquad \text{where } X = 5 e^{-j\pi/4}$$

Example 7-22 The sinusoidal signal

$$x(n) = 4 e^{j\pi/3} e^{j\lambda n} + 4 e^{-j\pi/3} e^{-j\lambda n}$$

is expressed in amplitude-phase form as

$$x(n) = 8 \cos\left(\lambda n + \frac{\pi}{3}\right)$$

7-5B Transfer Function of a Linear Shift-Invariant System

By convolution, the output $y(n)$ of a linear shift-invariant system for input $x(n)$ is given by

$$y(n) = \sum_{k=-\infty}^{\infty} x(k)h(n-k)$$

where $h(n)$ is the delta response of the system. An alternative form of the convolution sum is more convenient for present purposes. Using the change of variable $m = n - k$ in the sum above gives

$$y(n) = \sum_{m=-\infty}^{\infty} x(n-m)h(m) \qquad (7\text{-}29)$$

We wish to derive an expression for the output $y(n)$ of a system for input $x(n) = Xe^{j\lambda n}$. From (7-29), the output is given by

$$y(n) = \sum_{m=-\infty}^{\infty} Xe^{j\lambda(n-m)}h(m)$$

Using the law of exponents to factor the exponential gives

$$y(n) = Xe^{j\lambda n} \sum_{m=-\infty}^{\infty} h(m)e^{-j\lambda m}$$

The sum is independent of sample number n. It is a function of digital frequency λ. We denote the sum by $H(j\lambda)$; thus

$$H(j\lambda) = \sum_{m=-\infty}^{\infty} h(m)e^{-j\lambda m} \qquad (7\text{-}30)$$

With this definition, the previous expression can be written

$$y(n) = Ye^{j\lambda n} \quad \text{where } Y = H(j\lambda)X \qquad (7\text{-}31)$$

The function $H(j\lambda)$ defined by (7-30) is called the *transfer function* of a linear shift-invariant system having delta response $h(n)$. Equation (7-31) shows that the output of a linear shift-invariant system for exponential input $e^{j\lambda n}$ is an exponential whose frequency equals that of the input and whose complex amplitude equals the complex amplitude of the input multiplied by the transfer function for the system.

The delta response $h(n)$ of a linear shift-invariant system consists of terms of the form $n^k p^n$, where p is a characteristic root for the system. For the sum in (7-30) to exist, all such terms in $h(n)$ must approach zero as n approaches infinity. This requires that all characteristic roots for the system have magnitudes smaller than unity. Therefore, for a linear shift-invariant system to have a transfer function the system must be stable.

We can use (7-26), (7-31), and superposition to obtain the output of a linear shift-invariant system for sinusoidal input $x(n) = A \cos(\lambda n + \theta)$; thus

$$x(n) = Xe^{j\lambda n} + X^* e^{-j\lambda n}$$

where $X = \frac{1}{2}Ae^{j\theta}$. From (7-31) and superposition, the output is

$$y(n) = H(j\lambda)Xe^{j\lambda n} + H(-j\lambda)X^*e^{-j\lambda n}$$

This can be written

$$y(n) = Ye^{j\lambda n} + Y^*e^{-j\lambda n} \quad \text{where } Y = H(j\lambda)X \quad (7\text{-}32)$$

The output is expressed in amplitude-phase form as

$$y(n) = B\cos(\lambda n + \psi) \quad \text{where } B = 2|Y| \quad (7\text{-}33)$$
$$\psi = \angle Y$$

We can summarize the development above by giving a procedure for finding the output of a sinusoidally excited, stable, linear, shift-invariant system:

1. Obtain the complex amplitude of the input using

$$X = \frac{A}{2}e^{j\theta}$$

where A and θ are the peak amplitude and initial phase, respectively, of the input.

2. Obtain the transfer function of the system using

$$H(j\lambda) = \sum_{m=-\infty}^{\infty} h(m)e^{-j\lambda m}$$

where $h(n)$ is the delta response for the system.

3. Obtain the complex amplitude Y of the output using

$$Y = H(j\lambda_0)X$$

where λ_0 is the digital frequency of the input.

4. Obtain the peak amplitude B and the initial phase ψ of the output using

$$B = 2|Y| \quad \psi = \angle Y$$

and express the output $y(n)$ in amplitude-phase form as

$$y(n) = B\cos(\lambda_0 n + \psi)$$

Example 7-23 Use the procedure above to obtain the output of the system of Fig. 7-32 for input

$$x(n) = 10\cos\left(100n + \frac{\pi}{4}\right)$$

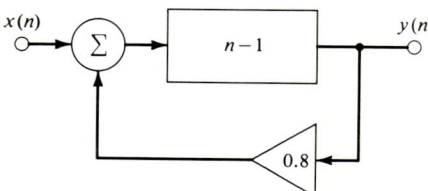

Figure 7-32 System of Example 7-23.

LINEAR SHIFT-INVARIANT DIGITAL SYSTEMS **483**

SOLUTION *Step 1* The complex amplitude of the input is
$$X = 5e^{j\pi/4}$$

Step 2 The delta response of the system is
$$h(n) = (0.8)^{n-1} u(n)$$
From (7-30), the transfer function of the system is
$$H(j\lambda) = \sum_{m=-\infty}^{\infty} (0.8)^{m-1} u(m) e^{-j\lambda m} = (0.8)^{-1} \sum_{m=1}^{\infty} (0.8 e^{-j\lambda})^m$$
$$= e^{-j\lambda} \sum_{m=0}^{\infty} (0.8 e^{-j\lambda})^m = \frac{e^{-j\lambda}}{1 - 0.8 e^{-j\lambda}} = \frac{1}{e^{j\lambda} - 0.8}$$

Step 3 The complex amplitude of the output is
$$Y = H(j100)X = (1.91 e^{1.84j})(5 e^{j\pi/4}) = 9.53 e^{2.62j}$$

Step 4 The peak amplitude and the initial phase of the output are
$$B = (2)(9.53) = 19.1 \qquad \psi = 2.62 \text{ rad}$$
The output is expressed in amplitude-phase form as
$$y(n) = 19.1 \cos(100n + 2.62)$$

7-5C Gain and Phase Shift

The development above shows that the output $y(n)$ of a stable linear shift-invariant system for input
$$x(n) = A \cos(\lambda_0 n + \theta)$$
is given by
$$y(n) = |H(j\lambda_0)| A \cos[\lambda_0 n + \theta + \angle H(j\lambda_0)] \tag{7-34}$$
where $H(j\lambda)$ is the transfer function for the system. To obtain the peak amplitude of the output we scale the peak amplitude of the input by $|H(j\lambda_0)|$. To obtain the initial phase of the output we shift the initial phase of the input by $\angle H(j\lambda_0)$. We can regard a sinusoidally excited stable linear shift-invariant system as one that introduces a frequency-dependent gain and a frequency-dependent phase shift. We denote the *gain* by $\Gamma(\lambda)$ and the *phase shift* by $\phi(\lambda)$, where
$$\Gamma(\lambda) = |H(j\lambda)| \qquad \phi(\lambda) = \angle H(j\lambda) \tag{7-35}$$
With these definitions (7-34) becomes
$$y(n) = \Gamma(\lambda_0) A \cos[\lambda_0 n + \theta + \phi(\lambda_0)] \tag{7-36}$$

Example 7-24 The transfer function for a certain linear shift-invariant system is
$$H(j\lambda) = \frac{1 - e^{-j\lambda}}{1 + 0.5 e^{-j\lambda}}$$

Obtain the output for input

$$x(n) = 5\cos\left(10^4 n - \frac{\pi}{3}\right)$$

SOLUTION First we determine the gain and phase shift. For $\lambda = 10^4$, the transfer function is expressed in polar form as

$$H(j10^4) = \frac{1 - e^{-j10^4}}{1 + 0.5e^{-j10^4}} = \frac{1.98 e^{-j0.16}}{0.55 e^{-j0.28}} = 3.60 e^{j0.12}$$

Therefore, for $\lambda = 10^4$, the gain of the system is $\Gamma(10^4) = 3.60$ and the phase shift of the system is $\phi(10^4) = 0.12$ rad. From (7-36), the output for input $x(n)$ given above is

$$y(n) = (3.60)(5.00)\cos\left(10^4 n - \frac{\pi}{3} + 0.12\right)$$

$$= 18.00\cos(10^4 n - 0.93)$$

7-5D Constant (dc) Signals

It is often convenient to regard a constant (dc) digital signal as a sinusoidal digital signal whose digital frequency is zero. Taking $\lambda = 0$ in (7-25) gives the amplitude-phase form for a dc signal $x(n) = x_0$; thus

$$x_0 = A\cos\theta$$

In order to abide by the convention that peak amplitude is nonnegative we define

$$A = |x_0| \qquad \theta = \angle x_0 \qquad (7\text{-}37)$$

For example, the signal $x(n) = 5$ has peak amplitude $A = 5$ and initial phase $\theta = 0$, whereas the signal $x(n) = -5$ has peak amplitude $A = 5$ and initial phase $\theta = \pi$ rad.

The exponential form for a dc signal is the signal itself; that is, the complex amplitude of a dc signal x_0 is

$$X = x_0 = A e^{j\theta} = A\cos\theta \qquad (7\text{-}38)$$

Note that (7-37) and (7-38) differ by a factor of 2 from (7-27) and (7-28).

An advantage of regarding a dc signal as a zero-frequency sinusoid is that we can use the transfer function (with $\lambda = 0$) for a system to determine the output of the system for a constant input. The output of a linear shift-invariant system for input $x(n) = x_0$ is given by

$$y(n) = H(0)x_0 = y_0 \qquad (7\text{-}39)$$

where $H(j\lambda)$ is the transfer function of the system. The quantity $H(0)$ is called the *dc gain* of a system having transfer function $H(j\lambda)$.

Example 7-25 Obtain the output of the system of Example 7-23 for input $x(n) = -12$.

SOLUTION From Example 7-23, the transfer function of the system is

$$H(j\lambda) = \frac{1}{e^{j\lambda} - 0.8}$$

The dc gain of the system is

$$H(0) = \frac{1}{e^0 - 0.8} = 5$$

From (7-39), the output for input $x(n) = -12$ is

$$y(n) = H(0)(-12) = -60$$

Equation (7-30) provides the most direct route to the transfer function of a system described by (only) an impulse response. Using (7-30) for a system described by a difference equation or a block diagram is tedious because it requires that we first obtain the impulse response of the system. Next, we show how the transfer function of a system can be derived directly from a difference equation or a block diagram describing the system.

7-5E Systems Described by Difference Equations

A system described by a difference equation of the form

$$y(n) + a_1 y(n-1) + \cdots + a_k y(n-k)$$
$$= b_0 x(n) + b_1 x(n-1) + \cdots + b_m x(n-m)$$

is a linear shift-invariant system. We know from (7-31) that the output of such a system for input $e^{j\lambda n}$ is

$$y(n) = H(j\lambda)e^{j\lambda n}$$

where $H(j\lambda)$ is the transfer function for the system; thus we know that the difference equation above is satisfied for

$$x(n) = e^{j\lambda n} \qquad y(n) = H(j\lambda)e^{j\lambda n}$$

Substituting these expressions for $x(n)$ and $y(n)$ into the difference equation above and solving for $H(j\lambda)$ gives

$$H(j\lambda) = \frac{b_0 + b_1 e^{-j\lambda} + \cdots + b_m e^{-j\lambda m}}{1 + a_1 e^{-j\lambda} + \cdots + a_k e^{-j\lambda k}} \qquad (7\text{-}40)$$

Example 7-26 The input $x(n)$ and output $y(n)$ of a certain linear shift-invariant system are related by

$$y(n) - 1.20 y(n-1) - 0.37 y(n-2) = x(n) + 0.50 x(n-1)$$

Find the output for input

$$x(n) = 5.00 \cos 50n$$

SOLUTION First we find whether the system is stable (otherwise, it cannot be described by a transfer function). The characteristic equation for the system is

$$z^2 - 1.20z - 0.32 = 0$$

The characteristic roots are $p_1 = 0.80$ and $p_2 = 0.40$. Since the magnitudes of both characteristic roots are less than unity, the system is stable. From (7-40), the transfer function for the system is

$$H(j\lambda) = \frac{1.00 + 0.50e^{-j\lambda}}{1.00 - 1.20e^{-j\lambda} - 0.32e^{-2j\lambda}}$$

For $\lambda = 50$, this gives

$$H(j50) = 2.31e^{2.40j}$$

For $\lambda = 50$, the gain of the system is 2.31 and the phase shift of the system is 2.40 rad. The output for input $x(n)$ above is

$$y(n) = (2.31)(5.00)\cos(50n + 2.40) = 11.54\cos(50n + 2.40)$$

7-5F Block-Diagram Reduction

A system having transfer function $H(j\lambda)$ and input $Xe^{j\lambda n}$ can be represented as shown in Fig. 7-33, where

$$Y = H(j\lambda)X$$

A system consisting of several such subsystems can be reduced to a single block by block-diagram reduction. The transfer function for the system is the transfer function for the single block thus obtained.

The rules of block-diagram reduction for digital systems are identical to those for analog systems. For derivations of the rules and examples illustrating application of the rules, refer to the discussion of block-diagram reduction for analog systems in Sec. 3-1. The derivations and examples given there for analog systems apply to digital systems as well if ω is replaced by λ.

We can apply block-diagram reduction to a block diagram whose constituents (blocks) are described by transfer functions. We can convert a block diagram whose blocks are described by transfer characteristics into a form suitable for block-diagram reduction by obtaining the transfer function for each block appearing in the block diagram. The transfer function for a block described by a delta response can be obtained using (7-30). The transfer function for a block described by a difference equation can be obtained using (7-40). Figure 7-34 gives transfer functions and symbols for elementary linear shift-invariant systems. Derivations of these transfer functions follow.

The transfer characteristic for a proportion element having input $x(n)$, gain K, and output $y(n)$ is

$$y(n) = Kx(n)$$

The output for input $x(n) = Xe^{j\lambda n}$ is given by

$$y(n) = KXe^{j\lambda n}$$

This shows that the transfer function for a proportion element having gain K is

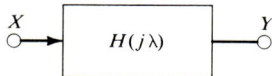

Figure 7-33 Symbol for a linear shift-invariant system having transfer function $H(j\lambda)$.

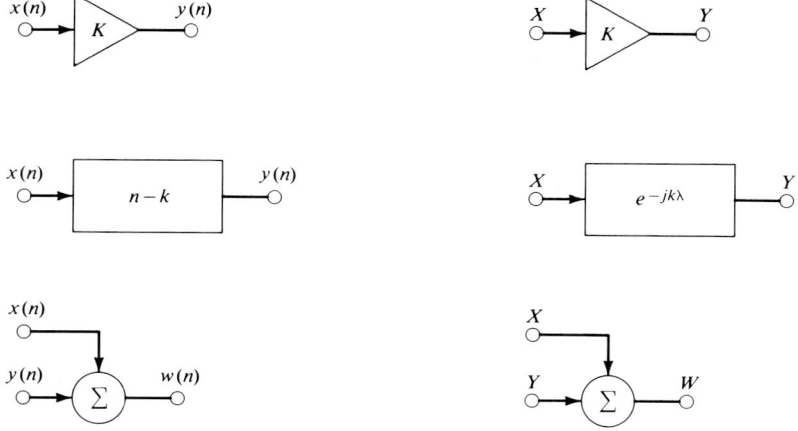

Figure 7-34 Transfer functions and block-diagram symbols for elementary linear shift-invariant systems.

$$H(j\lambda) = K$$

The transfer characteristic for a shift element having input $x(n)$, shift k, and output $y(n)$ is

$$y(n) = x(n - k)$$

The output for input $x(n) = Xe^{j\lambda n}$ is given by

$$y(n) = Xe^{j\lambda(n-k)} = e^{-j\lambda k}Xe^{j\lambda n}$$

This shows that the transfer function for a shift element having shift k is

$$H(j\lambda) = e^{-jk\lambda}$$

The output of an addition element having inputs

$$x(n) = Xe^{j\lambda n} \qquad y(n) = Ye^{j\lambda n}$$

is given by

$$w(n) = (X + Y)e^{j\lambda n} = We^{j\lambda n}$$

where $W = X + Y$. This shows that an addition element can be represented as shown in Fig. 7-34c provided all inputs have the same digital frequency.

Example 7-27 Obtain the transfer function of the system of Fig. 7-35.

SOLUTION Following the rules given in Fig. 7-34, we transform the block diagram of Fig. 7-35 into that of Fig. 7-36a. We then apply block-diagram reduction, as shown in Figs. 7-36b and c, where

$$H_1(j\lambda) = 0.1e^{-j\lambda} \qquad H_2(j\lambda) = \frac{e^{-j\lambda}}{1 + 0.5e^{-j\lambda}} \qquad H_3(j\lambda) = \frac{H_2(j\lambda)e^{-j\lambda}}{1 + 0.2H_2(j\lambda)e^{-j\lambda}}$$

488 INTRODUCTION TO SYSTEM ANALYSIS

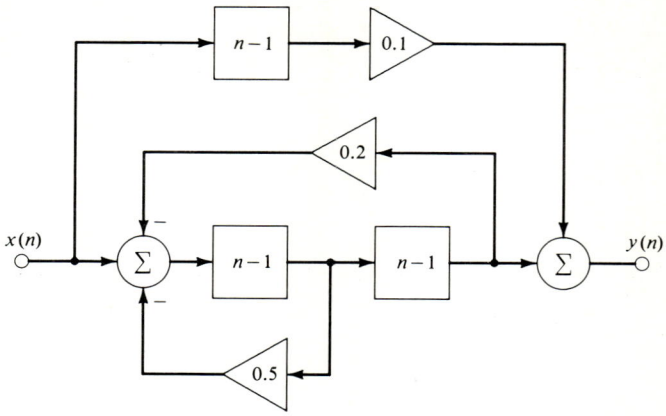

Figure 7-35 System of Example 7-27.

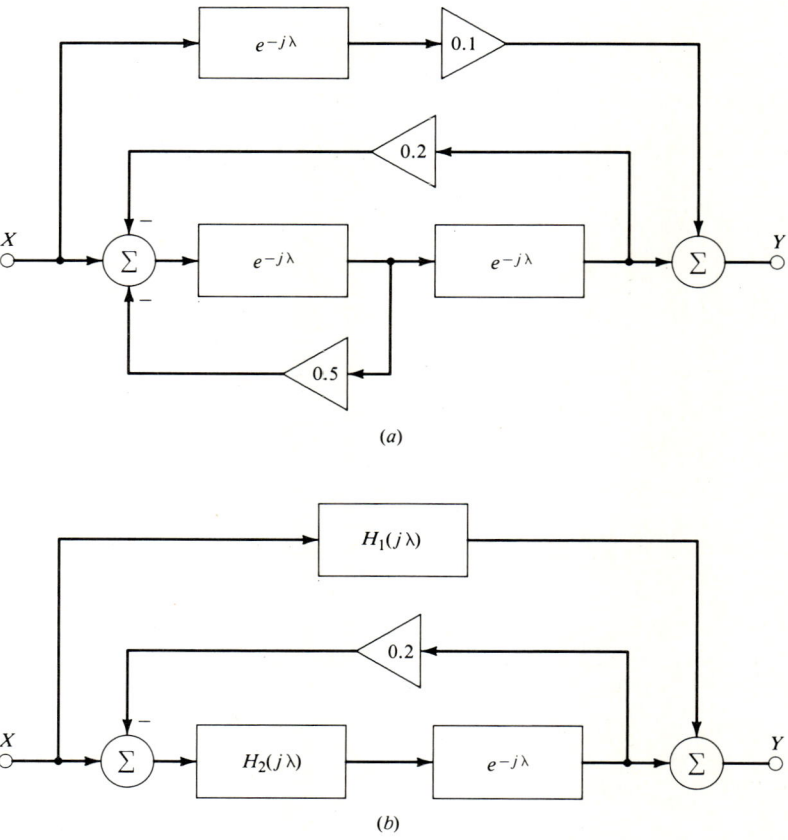

Figure 7-36 Application of block-diagram reduction to the system of Fig. 7-35.

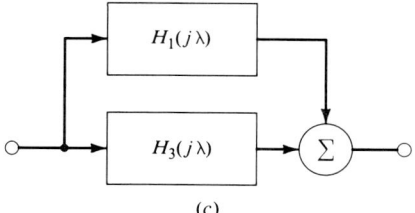

(c)   Figure 7-36 *Continued*

The transfer function of the system is

$$H(j\lambda) = H_1(j\lambda) + H_3(j\lambda) = \frac{0.10e^{-j\lambda} + 1.05e^{-j2\lambda} + 0.02e^{-j3\lambda}}{1.00 + 0.50e^{-j\lambda} + 0.20e^{-j2\lambda}}$$

7-5G Superposition

We use superposition to obtain the output of a linear shift-invariant system for an input consisting of two or more sinusoidal components. We use the methods described above to determine the output for each component considered individually. The complete output is the sum of the individual outputs thus obtained.

Example 7-28 The transfer function of a certain linear shift-invariant system is

$$H(j\lambda) = \frac{5}{1 + e^{-j\lambda}}$$

Obtain the output in amplitude-phase form for input

$$x(n) = 3 + 2\cos(10n) + \cos(20n + 0.79)$$

SOLUTION The dc gain of the system is

$$H(0) = \frac{5}{2} = 2.5$$

Therefore the output for the dc term of the input is $(2.5)(3.0) = 7.5$. The value of the transfer function for $\lambda = 10$ is

$$H(j10) = 8.81e^{-j1.28}$$

Therefore the gain and phase shift for $\lambda = 10$ are 8.81 and -1.28 rad, respectively, and the output for the term $2\cos(10n)$ of the input is $17.62\cos(10n - 1.28)$. The value of the transfer function for $\lambda = 20$ is

$$H(j20) = 2.98e^{j0.58}$$

Therefore the gain and phase shift for $\lambda = 20$ are 2.98 and 0.58 rad, respectively, and the output for the term $\cos(20n + 0.79)$ of the input is $2.98\cos(20n + 1.37)$. By superposition, the output for input

$$x(n) = 3 + 2\cos 10n + \cos(20n + 0.79)$$

is

$$y(n) = 7.5 + 17.62\cos(10n - 1.28) + 2.98\cos(20n + 1.37)$$

490 INTRODUCTION TO SYSTEM ANALYSIS

Above, we show how to obtain the output of a linear shift-invariant digital system for an input consisting of one or more sinusoidal components. We find that a linear shift-invariant digital system is described by a gain and a phase shift which are functions of digital frequency λ. Thus, we find that a linear shift-invariant digital system exhibits frequency selectivity, where the form of the frequency response is determined by the order and the coefficients of the difference equation describing the system. Next we show how frequency-selective properties of linear shift-invariant digital systems can be used for digital filtering of analog signals.

7-6 DIGITAL FILTERS

We wish to show that the system of Fig. 7-37 serves as an analog filter. Thus we seek the output of the system for sinusoidal input

$$x_a(t) = x_0 \cos \omega t$$

The output of the (error-free) AD converter for input $x_a(t)$ above is

$$x_d(n) = C^{-1}x_a(nT) = C^{-1}x_0 \cos \omega nT$$

where $C = 1$ in the unit of x_0 and T is sampling interval. We assume that the sampling rate $F = 1/T$ exceeds twice the frequency $f = \omega/2\pi$ of the analog input $x_a(t)$. For input $x_d(n)$ above, the output of the linear shift-invariant digital subsystem is

$$y_d(n) = \Gamma_d(\lambda)C^{-1}x_0 \cos[\lambda n + \phi_d(\lambda)]$$

where $\Gamma_d(\lambda)$ and $\phi_d(\lambda)$ are the gain the the phase shift, respectively, of the digital subsystem for digital frequency $\lambda = \omega T$. For input $y_d(n)$ above, the output of the (error-free) DA converter is

$$y_a(t) = Dy_d\left(\frac{t}{T}\right) = \Gamma_d(\omega T)DC^{-1}x_0 \cos[\omega t + \phi_d(\omega T)]$$

where $D = 1$ in the unit of $y_a(t)$. For present purposes we may assume that $x_a(t)$ and $y_a(t)$ are expressed in the same unit; thus $DC^{-1} = 1$. The expression above for $y_a(t)$ can be written

$$y_a(t) = \Gamma(\omega)x_0 \cos[\omega t + \phi(\omega)]$$

where $\qquad \Gamma(\omega) = \Gamma_d(\omega T) \quad$ and $\quad \phi(\omega) = \phi_d(\omega T) \qquad$ (7-41)

This shows that the system of Fig. 7-37 is (for error-free AD and DA conversion) a linear stationary system having gain $\Gamma(\omega)$ and phase shift $\phi(\omega)$ given by (7-41) for $f < F/2$. For $f > F/2$ the gain is zero (and phase shift is irrelevant) because the

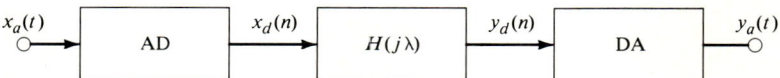

Figure 7-37 Block diagram for digital filtering of an analog signal.

error-free AD converter contains an ideal low-pass anti-alias filter with bandwidth $F/2$ and because the error-free DA converter acts as an ideal low-pass reconstruction filter with bandwidth $F/2$. The anti-alias filter eliminates components above $F/2$ from an analog input and the DA converter prevents components above $F/2$ (spurious components produced by sampling) from contributing to an output. Consequently, we are interested in the frequency response of the system over the band $0 \le f \le F/2$, corresponding to digital frequencies in the range $0 \le \lambda \le \pi$. The digital system can be designed so that the system of Fig. 7-37 has a specified (overall) frequency response for $0 \le f \le F/2$; for example, the system of Fig. 7-37 can be designed to achieve any one of the common filtering operations (low-pass, bandpass, high-pass, and bandstop) over that band. In this application, the system of Fig. 7-37 is called a *digital filter*.

To introduce digital filtering we analyze a first-order nonrecursive digital filter and compare the performance of the filter with that of a first-order analog filter.

7-6A Analysis of a First-Order Recursive Filter

Figure 7-38 shows a block diagram for a first-order recursive digital filter. The transfer function for the digital subsystem is

$$H_d(j\lambda) = \frac{1}{e^{j\lambda} - K}$$

If we assume error-free AD and DA conversion, the overall transfer function of the system is (for $0 \le f \le F/2$)

$$H(j\omega) = H_d(j\omega T) = \frac{1}{e^{j\omega T} - K}$$

For $f < F/2$, the gain of the system of Fig. 7-38 is given by

$$\Gamma(\omega) = |H(j\omega)| = \frac{1}{\sqrt{1 + K^2 - 2K \cos \omega T}}$$

For ideal AD and DA conversion, the gain is zero for $f > F/2$. In a practical system the gain for $f > F/2$ is nonzero because practical anti-alias and reconstruction filters are not ideal; however, in a well-designed system the gain for $f > F/2$ is not significantly different from zero.

The dc gain and the gain for $f = F/2$ are

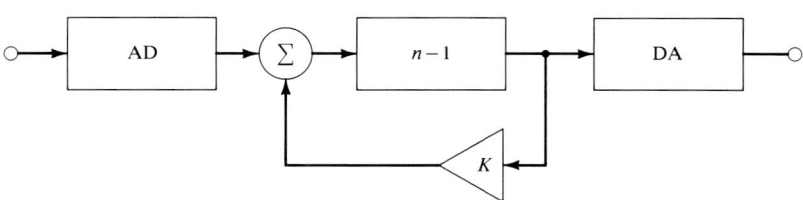

Figure 7-38 A first-order digital filter.

$$\Gamma(0) = \frac{1}{1-K} \qquad \Gamma(\pi F) = \frac{1}{1+K}$$

The parameter K determines the dc gain and the gain for $f = F/2$. Together the parameter K and the sampling rate F determine the (actual) bandwidth of the system. (Note that for the system to be stable, $|K|$ must be less than 1.) Figure 7-39 shows graphs of gain versus frequency for the system of Fig. 7-38. Figure 7-39a shows gain (normalized to unity for $f = 0$) versus frequency for several values of K in the range $0 < K < 1$. For $0 < K < 1$, the dc gain exceeds the gain for $f = F/2$ and the system of Fig. 7-38 is a low-pass filter. Figure 7-39b shows gain (normalized to unity for $f = F/2$) versus frequency for several values of K in the range $-1 < K < 0$. For $-1 < K < 0$, the gain for $f = F/2$ exceeds the dc gain and the system of Fig. 7-38 is a high-pass filter.†

The phase shift of the system of Fig. 7-38 is given by

$$\phi(\omega) = -\text{Tan}^{-1}(\sin \omega T, \cos \omega T - K)$$

Usually delay is of more interest than phase shift in filter design. Delay as a function of frequency for the system of Fig. 7-38 is given by

†Ordinarily the bandwidth of an analog input to the system is less than $F/2$. The system is a high-pass filter insofar as such an input is concerned, even though the gain is approximately zero for $f > F/2$.

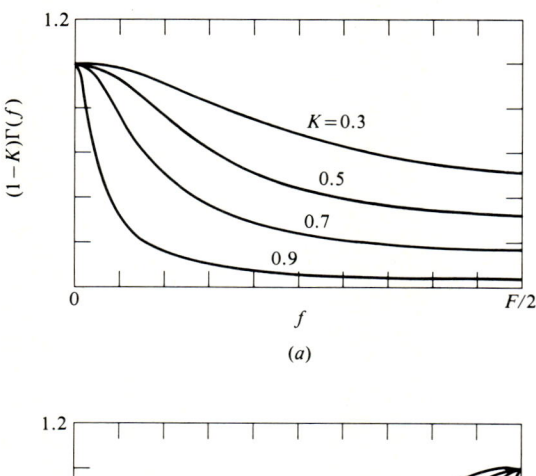

(a)

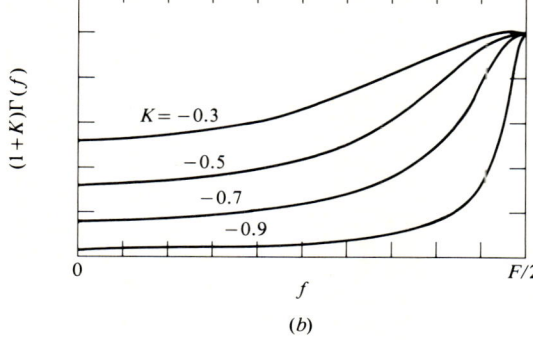

(b)

Figure 7-39 Gain versus frequency for the digital filter of Fig. 7-38: (a) low-pass filter ($K<0$), (b) high-pass filter ($K>0$).

$$D(\omega) = -\frac{d\phi(\omega)}{d\omega} = \frac{(1 - K\cos\omega T)T}{1 + K^2 - 2\cos\omega T}$$

The values of delay $D(\omega)$ for $f = 0$ and $f = F/2$ are given by

$$D(0) = \frac{T}{1 - K} \qquad D(\pi F) = \frac{T}{1 + K}$$

Figure 7-40 shows graphs of delay versus frequency for the system of Fig. 7-38. For $0 \le K \le 1$ (for a low-pass filter), delay is maximum for $f = 0$. For $-1 \le K \le 0$ (for a high-pass filter), delay is maximum for $f = F/2$. In either case, maximum delay is introduced in the passband, minimum delay is introduced in the stopband, and the transition from maximum to minimum delay is monotonic. Note that for $K = 0$ delay is independent of frequency and is equal to T (the sampling interval). This is because for $K = 0$ the digital subsystem in Fig. 7-38 simplifies to a shift element having shift $k = 1$ (one sampling interval). Note also that the delay function given above does not account for delay introduced by AD and DA conversion. In a practical system anti-alias filtering, low-pass reconstruction filtering, and associated equalization might well be the principal sources of delay.

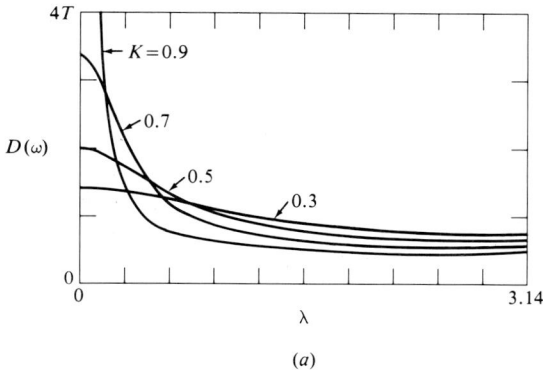

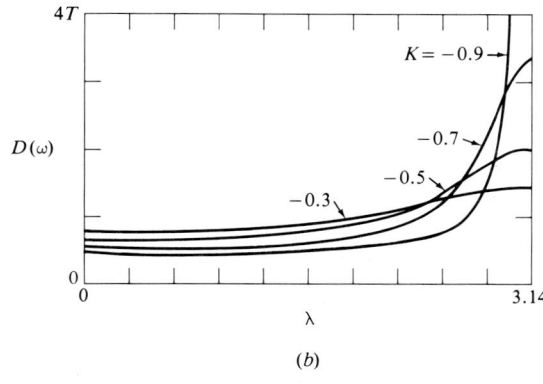

Figure 7-40 Delay versus frequency for the digital filter of Fig. 7-38: (a) $K > 0$, (b) $K < 0$.

7-6B Comparison of First-Order Analog and Digital Filters

It is instructive to compare the frequency response of the first-order digital filter of Fig. 7-38 with that of a first-order analog filter having transfer function

$$H(j\omega) = \frac{1}{1 + j\omega/\omega_0}$$

For definiteness we use sampling rate $F = 10f_0$. To facilitate comparison we normalize the gain of the digital filter such that the dc gain equals unity (equals the dc gain of the analog filter), and we choose the parameter K for the digital filter such that the bandwidth of the digital filter equals the bandwidth $f_0 = \omega_0/2\pi$ of the analog filter. Under these conditions the gain of the digital filter is given by

$$\Gamma_1(\omega) = \frac{1 - K}{\sqrt{1 + K^2 - 2K\cos(0.2\pi f/f_0)}}$$

where K is to be specified such that $\Gamma_1(\omega_0) = 1/\sqrt{2}$ (such that the bandwidth of the digital filter equals the bandwidth of the analog filter). This requirement yields the equation

$$2(1 - K)^2 = 1 + K^2 - 2K\cos 0.2\pi$$

which simplifies to

$$K^2 - (4 - 2C)K + 1 = 0$$

where $C = \cos 0.2\pi = 0.809$. Solving this quadratic equation for K yields $K = 1.838$ and $K = 0.544$. We choose $K = 0.544$ because for $K = 1.838$ the system is unstable.

The gain of the analog filter is given by

$$\Gamma_2(\omega) = \frac{1}{\sqrt{1 + (f/f_0)^2}}$$

For $K = 0.544$ and $F = 10f_0$ the gain of the digital filter is given by

$$\Gamma_1(\omega) = \frac{0.456}{\sqrt{1.296 - 1.088\cos(0.628 f/f_0)}}$$

Figure 7-41 shows graphs of the gains above versus frequency for $0 \leq f \leq 5f_0$ ($5f_0 = F/2$). The gains are essentially identical within the nominal passband $0 \leq f \leq f_0$. So far as gain is concerned, the first-order digital filter of Fig. 7-38 emulates a first-order analog filter.

The phase shift of the analog filter is given by

$$\phi_2(\omega) = -\tan^{-1}\frac{f}{f_0}$$

For $F = 10f_0$ and $K = 0.544$, the phase shift of the digital filter is given by

$$\phi_1(\omega) = -\text{Tan}^{-1}\left(\sin\frac{0.628f}{f_0}, \cos\frac{0.628f}{f_0} - 0.544\right)$$

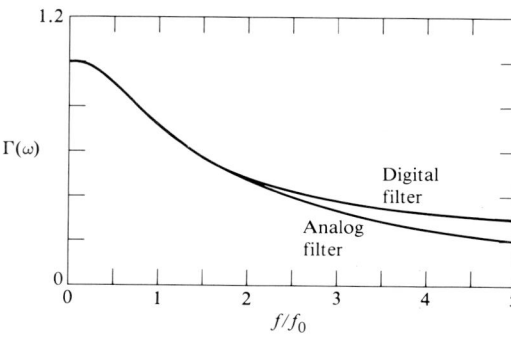

Figure 7-41 Graphs of gain versus frequency for a first-order digital and comparable first-order analog filter.

Figure 7-42 shows graphs of phase shift versus frequency for the systems being compared. The digital filter introduces significantly more phase shift than the analog filter. This is typical. In many applications the larger phase shift introduced by a digital filter (as opposed to a comparable analog filter) is inconsequential; however, it sometimes militates against using a digital filter in applications where phase shift is critical, e.g., in certain feedback control systems where phase shift has a destabilizing influence.

The delay of the analog filter is given by

$$D_2(\omega) = -\frac{d\phi_2(\omega)}{d\omega} = \frac{10T}{2\pi[1 + (f/f_0)^2]}$$

For $F = 10f_0$ and $K = 0.544$, delay of the digital filter is given by

$$D_1(\omega) = -\frac{d\phi_1(\omega)}{d\omega} = \frac{T[1 - 0.544 \cos(0.628 f/f_0)]}{1.296 - 1.088 \cos(0.628 f/f_0)}$$

where T is sampling interval. Figure 7-43 shows graphs of delay versus frequency for the two filters. The graphs have approximately the same shape, indicating that the digital filter and the analog filter introduce approximately the same delay distortion; however, delay introduced by the digital filter exceeds (by about 30 percent) delay introduced by the analog filter in the passband $0 \leq f \leq f_0$. If delay introduced by AD and DA conversion (and associated equalization) is included, delay introduced by a digital filter can be much greater than that introduced by a comparable analog filter. Additional delay introduced by a digital filter is insignificant in certain applications

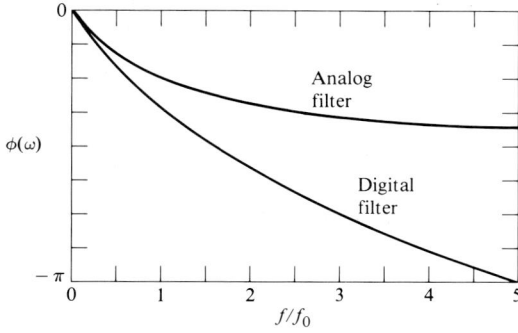

Figure 7-42 Graphs of phase shift versus frequency for a first-order digital and comparable first-order analog filter.

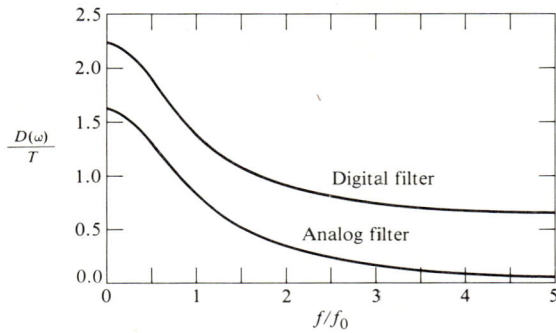

Figure 7-43 Graphs of delay versus frequency for a first-order digital and comparable first-order analog filter.

(e.g., where fidelity alone is important) but is significant in applications where overall delay is important (e.g., in some automatic control systems, where excessive delay prevents timely corrective action).

Using a digital filter often has one or more advantages over using a comparable analog filter. For low-frequency applications, where filter bandwidth is a few hundred hertz, a digital filter can be smaller and less expensive than an analog filter. This is because size and cost of components, e.g., capacitors, of analog filters increase as frequencies corresponding to edges of passbands (or stopbands) are decreased. Also, the transfer function of a digital filter is repeatable; i.e., it is possible to build any number of digital filters having identical transfer functions. On the other hand, no two analog filters can have exactly the same frequency response because no two analog components are identical. Further, the frequency response of a digital filter, unlike that of an analog filter, does not change with temperature and age (barring failure). Finally, the frequency response of a digital filter is relatively easy to change (by changing sampling rate and digital gains), whereas changing the frequency response of an analog filter may entail changing components or completely redesigning the filter. The flexibility of digital filters allows a degree of hardware standardization not readily obtainable with analog filters.

On the other hand, using an analog filter is advantageous or even necessary in many applications. In many moderate- to high-frequency applications analog filters are less complicated and less expensive than digital filters because AD and DA conversion are unnecessary, because passive (RLC) filters often are less expensive than comparable digital filters, and because digital filters require a source of power whereas passive analog filters do not. Analog filters may be preferred or required in applications where phase shift or delay is critical. Further, digital filtering is impractical or impossible in very high-frequency applications, where sufficiently fast and accurate AD and DA conversion is difficult or impossible. Finally, a recursive digital filter, being an active feedback system, can become unstable under certain conditions (e.g., when an abnormally large signal drives the filter into nonlinear operation) whereas a passive analog filter is stable.

7-6C Nonrecursive Digital Filters

Figure 7-44 shows a block diagram for a nonrecursive digital filter. The points at which samples $x_d(n)$, $x_d(n-1)$, ..., $x_d(n-M)$ of an input appear are called *taps*. The

LINEAR SHIFT-INVARIANT DIGITAL SYSTEMS **497**

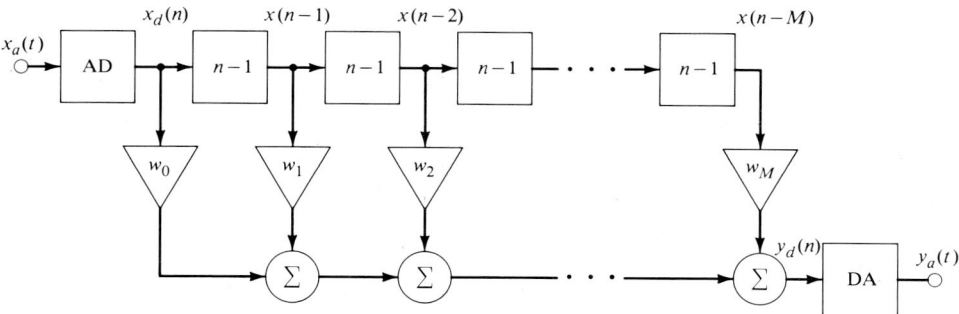

Figure 7-44 A nonrecursive digital filter having $M + 1$ taps.

gains $w_0, w_1, \ldots, w_M$ are called *tap weights*. The output of the digital subsystem for input $x_d(n)$ is given by

$$y_d(n) = \sum_{m=0}^{M} w_m x_d(n - m) \tag{7-42}$$

We wish to describe a method for determining tap weights such that the transfer function of a nonrecursive digital filter approximates a specified function of frequency. We start by obtaining an expression for the transfer function of a nonrecursive digital filter. For input $x_d(n) = e^{j\lambda n}$, (7-42) gives

$$y_d(n) = e^{j\lambda n} \sum_{m=0}^{M} w_m e^{-j\lambda m}$$

This shows that the transfer function for the digital subsystem in the system of Fig. 7-44 is

$$H_d(j\lambda) = \sum_{m=0}^{M} w_m e^{-j\lambda m} \tag{7-43}$$

Assuming error-free AD and DA conversion, the overall transfer function for the system of Fig. 7-44 for $0 \le f \le F/2$ is

$$H_a(j\omega) = H_d(j\omega T) = \sum_{m=0}^{M} w_m e^{-jm\omega T} \tag{7-44}$$

where T is the sampling interval for the system. For $f > F/2$, the transfer function is zero (because of the low-pass nature of the AD and DA converters).

One method of designing nonrecursive filters is based on the fact that the right side of (7-44) can be expressed (in part) as a finite Fourier series. To show this we use the change of variable $p = m - M/2$, where we assume that the number of shift elements M is even (the number of taps is odd). This gives

$$H_a(j\omega) = e^{-jM\omega T/2} \sum_{p=-M/2}^{M/2} w_{p+M/2} e^{-jp\omega T} \tag{7-45}$$

The factor $e^{-jM\omega T/2}$ is the transfer function of a delay element. Equation (7-45) shows that a nonrecursive filter can be represented as a series connection of a delay element having delay time $MT/2$ and a system having transfer function

$$H_M(j\omega) = \sum_{p=-M/2}^{M/2} w_{p+M/2} e^{-jp\omega T} \qquad (7\text{-}46)$$

The delay element represented by the factor $e^{-j\omega MT/2}$ introduces unit gain and linear phase shift $-\omega MT/2$ (frequency-independent delay $MT/2$). It introduces neither amplitude distortion nor phase (delay) distortion. The gain $\Gamma(\omega)$ of a nonrecursive filter is determined by the function $H_M(j\omega)$ alone; thus

$$\Gamma_a(\omega) = |H_M(j\omega)|$$

The phase shift introduced by the filter is given by

$$\phi_a(\omega) = \angle H_M(j\omega) - \frac{\omega MT}{2}$$

and the delay $D(\omega)$ introduced by the filter is given by

$$D_a(\omega) = D_M(\omega) + \frac{MT}{2}$$

where $D_M(\omega)$ is delay introduced by $H_M(j\omega)$. Except for a fixed delay, the essential properties of a nonrecursive filter are determined by the function $H_M(j\omega)$ alone. This means that a nonrecursive filter can be designed by specifying $H_M(j\omega)$.

In the right side of (7-46) we use the change of variable $k = -p$ and replace T by $1/F$. This gives

$$H_M(j\omega) = \sum_{k=-M/2}^{M/2} C_k e^{jk\omega/F} \qquad (7\text{-}47)$$

where we have defined

$$C_k = w_{M/2-k} \qquad (7\text{-}48)$$

Equation (7-47) gives the transfer function $H_M(j\omega)$ as a finite Fourier series. The period of the series is the sampling rate F. The coefficients of the series are related to the tap weights for the filter by (7-48). Letting $M/2 - k = m$ in (7-48) yields

$$w_m = C_{M/2-m} \qquad (7\text{-}49)$$

which gives the tap weights in terms of the Fourier coefficients for $H_M(j\omega)$.

The transfer function $H_M(j\omega)$ given by (7-47) is nonrealizable because it represents a system whose output $y(n)$ for input $x(n)$ is given by

$$y(n) = C_{-M/2} x\left(\frac{n-M}{2}\right) + \cdots + C_{-1} x(n-1) + C_0 x(n)$$

$$+ C_1 x(n+1) + \cdots + C_{M/2} x\left(\frac{n+M}{2}\right)$$

where a current output sample $y(n)$ depends on future input samples $x(n+1), \ldots, x(n+M/2)$. The factor $e^{-j\omega MT/2}$ appearing in (7-45) introduces just enough delay for the overall transfer function

LINEAR SHIFT-INVARIANT DIGITAL SYSTEMS **499**

$$H_a(j\omega) = H_M(j\omega)e^{-j\omega MT/2}$$

to be realizable.

The development above suggests a procedure for designing a nonrecursive digital filter whose transfer function approximates a specified transfer function $H(j\omega)$:

1. Determine the necessary sampling rate F based on the bandwidths of signals to be filtered.
2. Approximate the desired transfer function $H(j\omega)$ by a finite Fourier series over the band $0 \le f \le F/2$. The only restriction on the desired transfer function is that it have conjugate symmetry about $f = 0$; that is,

$$H(-j\omega) = H^*(j\omega)$$

This is necessary because the Fourier coefficients for $H(j\omega)$ must be real.
3. Determine the tap weights for the filter from the coefficients of the series using (7-48).

The following example illustrates the procedure.

Example 7-29 Design an 11-tap nonrecursive digital filter whose transfer function approximates the transfer function $H_0(j\omega)$ of Fig. 7-45, where $W = 2$ kHz. The bandwidth of inputs of interest is $B = 4$ kHz.

SOLUTION *Step 1* Determine sampling rate. Assuming error-free AD and DA conversion, the necessary sampling rate is

$$F = 2B = 8 \text{ kHz}$$

Step 2 Obtain a finite Fourier series for the desired transfer function $H_0(j\omega)$ over the band $-F/2 \le f \le F/2$, where $F/2 = B = 4$ kHz. This gives

$$H_M(j\omega) = \sum_{k=-M/2}^{M/2} C_k e^{jk\omega/F}$$

where $M = 10$ and (from Appendix B)

$$C_k = \frac{W}{F} \text{sa}^2 \frac{k\pi W}{F} = 0.25 \text{ sa}^2 \ 0.25k\pi$$

Step 3 Determine the tap weights for the filter from the coefficients for the series using (7-49) with $M = 10$; thus

$$w_m = C_{5-m} = 0.25 \text{ sa}^2[0.25(5-m)\pi]$$

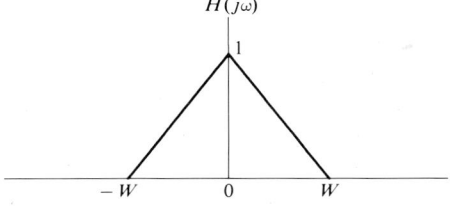

Figure 7-45 Specified transfer function for the nonrecursive filter of Example 7-29.

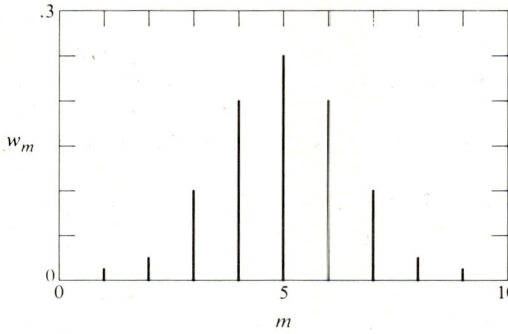

Figure 7-46 Tap weights for the nonrecursive filter of Example 7-29.

for $m = 0, 1, 2, \ldots, 10$. Figure 7-46 shows a graph of tap weight versus tap number m. The tap weights have even symmetry about the midpoint $m = 5$ because the desired transfer function $H_0(j\omega)$ is a real, even function of frequency.

The transfer function of the nonrecursive filter (assuming ideal AD and DA conversion) is

$$H_a(j\omega) = e^{-5j\omega T} H_M(j\omega)$$

The function $H_M(j\omega)$ is real because the desired transfer function approximated by $H_M(j\omega)$ is real; thus

$$H_M(j\omega) = \text{Re } H_M(j\omega) = \text{Re}\left(\sum_{k=-5}^{5} C_k e^{-j\omega k/F}\right) = C_0 + 2\sum_{k=1}^{5} C_k \cos\frac{\omega k}{F}$$

The gain of the filter is given by

$$\Gamma_a(\omega) = \left| C_0 + 2\sum_{k=1}^{5} C_k \cos\frac{\omega k}{F} \right|$$

Figure 7-47 shows graphs of actual gain given above and desired gain $|H_0(j\omega)|$ versus frequency. The accuracy with which the actual gain approximates the desired gain can be improved by increasing M (using more taps).

The phase shift of the filter is given by

$$\phi_a(\omega) = \angle H_M(j\omega) - 5\omega T$$

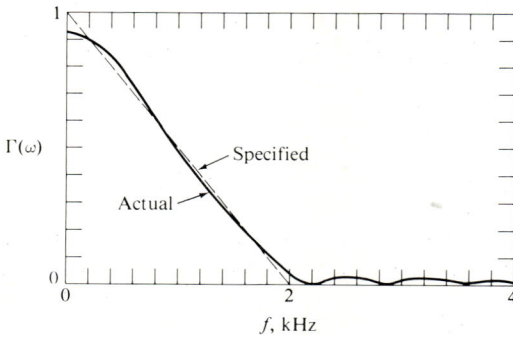

Figure 7-47 Actual gain of the nonrecursive filter of Example 7-29.

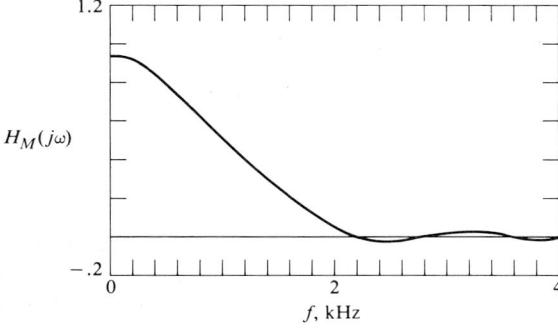

Figure 7-48 Graph of a finite Fourier series for the desired gain of the nonrecursive filter of Example 7-29.

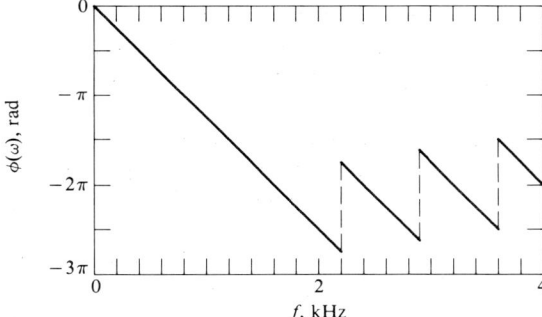

Figure 7-49 Actual phase shift of the nonrecursive filter of Example 7-29.

Figure 7-48 shows a graph of $H_M(j\omega)$ versus frequency f. The angle of $H_M(j\omega)$ equals zero for $|f| < W$ because $H_M(j\omega)$ is positive for $|f| < W$. Therefore, the phase shift in the passband is given by

$$\phi_a(\omega) = -5\omega T \qquad 0 < f < W$$

Figure 7-49 shows a graph of phase shift versus frequency. The discontinuities outside the passband arise from changes in the sign of $H_M(j\omega)$ (see Fig. 7-48). The phase shift is a linear function of frequency within the passband. The delay in the passband is

$$D_a(\omega) = -\frac{d\phi_a(\omega)}{d\omega} = 5T \qquad 0 < f < W$$

The delay in the passband is independent of frequency. This result illustrates a very important property of nonrecursive digital filters: the phase shift is (exactly) linear (and the delay is strictly frequency-independent) in the passband of a nonrecursive digital filter if the tap weights are symmetric about their midpoint, i.e., if the desired transfer function is a real, positive even function of frequency. This means that such a filter introduces no phase distortion (no delay distortion).

The Fourier-series method described above can be used to design bandpass, highpass, and bandstop filters as well as low-pass filters. It can also be used to design other kinds of filters, as illustrated in the next example.

Example 7-30 In this example we use the Fourier-series method described above to design a nonrecursive differentiating filter having effective bandwidth $B = 5$ kHz. This means that the filter should differentiate any input whose spectral density lies entirely in the band $0 \leq f \leq 5$ kHz. Assuming error-free AD and DA conversion, the minimum acceptable sampling rate is twice the desired bandwidth; thus $F = 10$ kHz.

The transfer function of an ideal differentiator has the form

$$H_0(j\omega) = j\omega\tau$$

where τ is independent of frequency. In this example $\tau = 1$ ms. Figure 7-50 shows a graph of the imaginary part of this transfer function versus frequency. The real part of the transfer function is zero. To obtain the transfer function of a nonrecursive differentiator having bandwidth B we obtain a finite Fourier series describing the function of Fig. 7-50 for $|f| < B$. Figure 7-51 shows a graph of the periodic sawtooth function actually represented by the series.

From Appendix B, the Fourier coefficients for the sawtooth function of Fig. 7-51 are given by

$$C_0 = 0$$

$$C_k = \frac{j(j2\pi\tau F)(-1)^k}{2k\pi} = \frac{\tau F(-1)^{k+1}}{k} \qquad k = \pm 1, \pm 2, \ldots$$

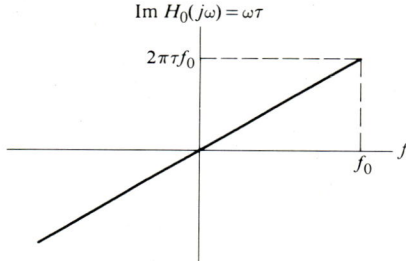

Figure 7-50 Transfer function for a differentiating filter.

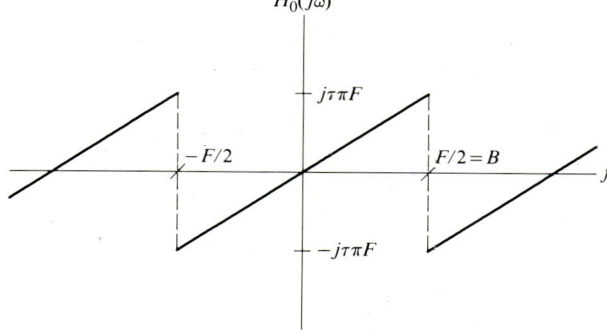

Figure 7-51 Periodic (sawtooth) function used to describe the transfer function of a differentiating filter over $-F/2 \leq f \leq F/2$.

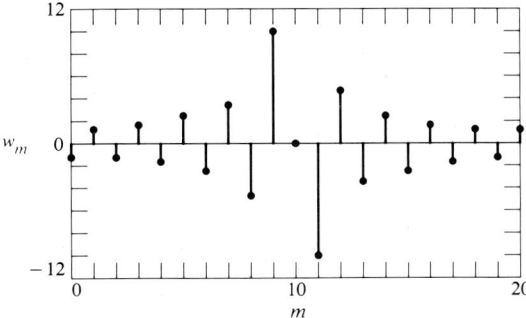

Figure 7-52 Tap weights for a 21-tap nonrecursive differentiating filter.

The coefficients have odd symmetry about $k = 0$ because the function $H_0(j\omega) = j\omega\tau$ is an imaginary, odd function of frequency. From (7-49), the tap weights are

$$w_{M/2} = C_0 = 0 \qquad w_m = \frac{\tau F(-1)^{M/2-m+1}}{M/2 - m} \qquad m \neq \frac{M}{2}$$

Figure 7-52 shows a graph of tap weight versus tap number m. The tap weights have odd symmetry about $m = M/2$.

For M even, the transfer function of an $(M + 1)$ − tap differentiator is

$$H_a(j\omega) = H_M(j\omega)e^{-j\omega MT/2} \qquad \text{where } H_M(j\omega) = \sum_{k=-M/2}^{M/2} C_k e^{j\omega k/F}$$

The function $H_M(j\omega)$ is an imaginary, odd function of frequency because it is a finite Fourier series for an imaginary, odd function of frequency. Since $H_M(j\omega)$ is imaginary,

$$H_M(j\omega) = j \operatorname{Im} H_M(j\omega) = 2j \sum_{k=1}^{M/2} C_k \sin \frac{\omega k}{F}$$

In examining performance of the nonrecursive differentiating filter we ignore the delay factor $e^{-j\omega MT/2}$. This factor introduces delay $MT/2$ but otherwise does not influence the performance of the differentiating filter. The gain of the filter is given by

$$\Gamma_a(\omega) = \left| 2 \sum_{k=1}^{M/2} C_k \sin \frac{\omega k}{F} \right|$$

If we ignore the delay factor $e^{-j\omega MT/2}$, the phase shift of the filter is given by

$$\phi_a(\omega) = \angle H_M(j\omega) = \frac{\pi}{2} \text{ rad} \qquad f < \frac{F}{2}$$

The phase shift equals $\pi/2$ rad for $f < F/2$ because $\operatorname{Im} H_M(j\omega) > 0$ and $\operatorname{Re} H_M(j\omega) = 0$ for $f < F/2$. Figure 7-53 shows graphs of gain for a 21-tap ($M = 20$) nonrecursive differentiator and for an ideal differentiator (for $\tau = 1$ ms). The ripples in the gain of the nonrecursive differentiator are Gibbs' oscillations. They arise from the jump discontinuities of the sawtooth function described by the finite Fourier series $H_M(j\omega)$ (see Sec. 4-1B). To decrease the amplitude of the Gibbs' oscillations we can (1) increase the sampling rate to move the discontinuity to higher frequencies, (2) eliminate the discontinuity by using a smoothed sawtooth function to approximate the desired transfer function, or (3) use more taps. In practice a combination of these three remedies is generally used.

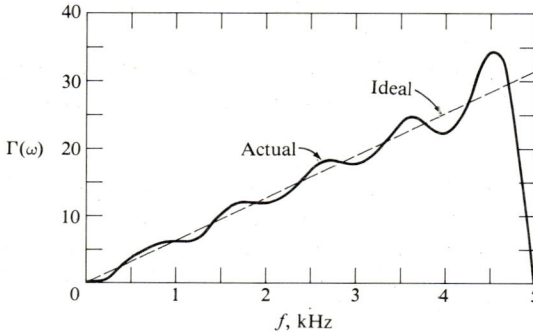

Figure 7-53 Gain of a 21-tap nonrecursive differentiating filter.

Nonrecursive digital filters offer several advantages over recursive digital filters:

1. A nonrecursive filter is unconditionally stable whereas a recursive filter can be driven to unstable operation under certain conditions. This makes nonrecursive filters more attractive than recursive filters in applications where unexpected instability is disastrous, e.g., in a control system on board a high-performance, highly automated aircraft.
2. A nonrecursive filter can have (exactly) linear phase shift (frequency-independent delay) in its passband, making it attractive where phase (delay) distortion is especially harmful, e.g., in data transmission.
3. Parameters of a nonrecursive filter (tap weights) are related to a specified transfer function more directly than the parameters of a recursive filter. This often makes it easier to design a nonrecursive filter than a comparable recursive filter, especially if the desired filter is other than a simple low-pass, high-pass, bandpass, or bandstop filter.
4. All nonrecursive digital filters have the same structure, whereas recursive digital filters are configured in various ways. Hardware and software for nonrecursive filters can therefore be standardized to a greater extent than for recursive filters.

On the other hand, nonrecursive digital filters have some disadvantages. A nonrecursive filter generally requires more hardware than a comparable recursive filter and requires more operations (multiplications and additions) per unit time than a recursive filter having a comparable gain characteristic. Recursive filters may therefore be preferred where cost and speed are more important than unconditional stability and (exactly) linear phase shift, e.g., in a music synthesizer, and in applications where a nonrecursive filter cannot operate at the necessary sampling rate unless an expensive parallel processor is used for the computations, e.g., in certain fast automatic control systems.

SUMMARY

A *linear shift-invariant system* is a system whose output $y(n)$ for input

$$x(n) = a_1 x_1(n - k_1) + a_2 x_2(n - k_2) + \cdots$$

is given by

$$y(n) = a_1 y_1(n - k_1) + a_2 y_2(n - k_2) + \cdots$$

where $y_i(n)$ is the output for input $x_i(n)$. Linear shift-invariant systems are important because they are used for a number of signal-processing operations and because they are relatively easy to analyze and design.

Any digital signal can be represented as a superposition of deltas; thus

$$x(n) = \sum_{m=-\infty}^{\infty} x(m) \Delta(n - m)$$

By using superposition and shift invariance the output of a linear shift-invariant system for input $x(n)$ above can be expressed as a *convolution sum*

$$y(n) = \sum_{m=-\infty}^{\infty} x(m) h(n - m) = \sum_{m=-\infty}^{\infty} x(n - m) h(m)$$

where $h(n)$ is the *delta response* of the system. Consequently, a linear shift-invariant system is completely described by its delta response, i.e., by its output for input $\Delta(n)$. This means that important properties of a linear shift-invariant system can be determined from the delta response of the system. A linear shift-invariant system is *realizable* if and only if its delta response equals zero for time $t < 0$. A linear shift-invariant system is *stable* if and only if its delta response is absolutely summable. A linear shift-invariant system provides *high-fidelity transmission* if its delta response approximates a delta (for inputs of interest). This means that the delta response must be much narrower than the narrowest significant feature of an input and that the sum of the delta response over all sample numbers must be nonzero.

A block diagram for a linear shift-invariant system contains only proportion, shift, and addition elements. Consequently, a linear shift-invariant system can be described by a *difference equation* of the form

$$y(n) + a_1 y(n - 1) + \cdots + a_k y(n - k)$$
$$= b_0 x(n) + b_1 x(n - 1) + \cdots + b_m x(n - m)$$

where k is the *order* of the system and $y(n)$ is the output of the system for input $x(n)$. The delta response of a system can be determined by solving the difference equation describing the system for the output for input $\Delta(n)$. A method for doing this is described in Sec. 7-4C. An important by-product of the method is that it shows that a linear shift-invariant system described by the difference equation above is stable if and only if all roots of the associated *characteristic equation*

$$z^k + a_1 z^{k-1} + \cdots + a_{k-1} z + a_k = 0$$

have magnitude smaller than 1. A simpler method for obtaining a delta response is described in Chap. 8.

A linear shift-invariant system is described in the digital frequency domain by its *transfer function* $H_d(j\lambda)$, defined as the ratio $y(n)/x(n)$, where $y(n)$ is the output of the system for input $x(n) = e^{j\lambda n}$. The transfer function for a linear shift-invariant system can be obtained from the delta response of the system using (7-30), from the coefficients of a difference equation describing the system using (7-40), or from a block diagram describing the system using block-diagram reduction. The magnitude of the transfer function for a system gives the gain of the system as a function of frequency, and the angle of the transfer function gives the phase shift of the system as a function of frequency. Using superposition and the transfer function for a system, we can express the output of the system for an input of the form

$$x(n) = \sum_{k=1}^{\infty} A_k \cos(\lambda_k n + \theta_k)$$

as

$$y(n) = \sum_{k=1}^{\infty} \Gamma_d(\lambda_k) A_k \cos[\lambda_k n + \theta_k + \phi_d(\lambda_k)]$$

where

$$\Gamma_d(\lambda) = |H_d(j\lambda)| \quad \text{and} \quad \phi_d(\lambda) = \angle H_d(j\lambda)$$

are the *gain* and the *phase shift*, respectively, of the system.

One of the most important applications of linear shift-invariant systems is digital filtering. A *digital filter* is a linear shift-invariant digital system connected to the outside (analog) world through AD and DA converters (Fig. 7-37). For ideal AD and DA conversion, a digital filter is (at its terminals) a linear stationary analog system having transfer function $H_a(j\omega) = H_d(j\omega T)$, where $H_d(j\lambda)$ is the transfer function of the linear shift-invariant digital system and T is the sampling rate of the system. Because of the low-pass nature of AD and DA conversion, the overall gain of a digital filter is approximately zero for frequencies greater than half the sampling frequency $F = 1/T$. Virtually any realizable filtering operation can be realized in the band $f < F/2$, subject only to constraints on cost and state-of-the-art limits on speed and accuracy of AD conversion, digital processing, and DA conversion.

PROBLEMS

7-1 Use the definitions of linearity and shift invariance to determine whether the systems of Fig. P7-1 are (a) linear and (b) shift-invariant.

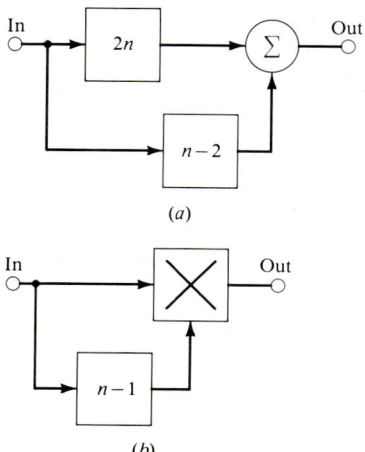

Figure P7-1

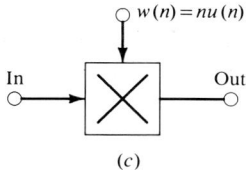

(c)

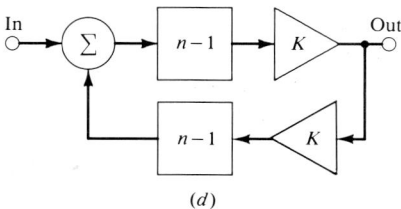

(d)

Figure P7-1 *Continued*

7-2 Obtain the ouput of each system of Fig. P7-2 for input $x(n) = r(0.2n)$.

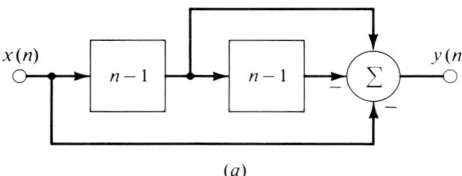

(a)

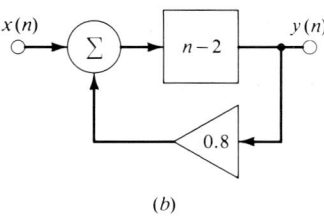

(b)

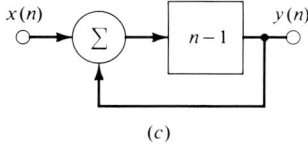

(c)

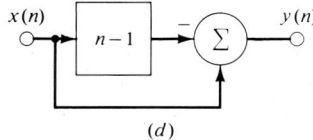

(d)

Figure P7-2

7-3 Show that

$$\sum_{k=-\infty}^{\infty} x(k)h(n-k) = \sum_{k=-\infty}^{\infty} x(n-k)h(k)$$

7-4 Obtain the delta response and the step response of each system of Fig. P7-4.

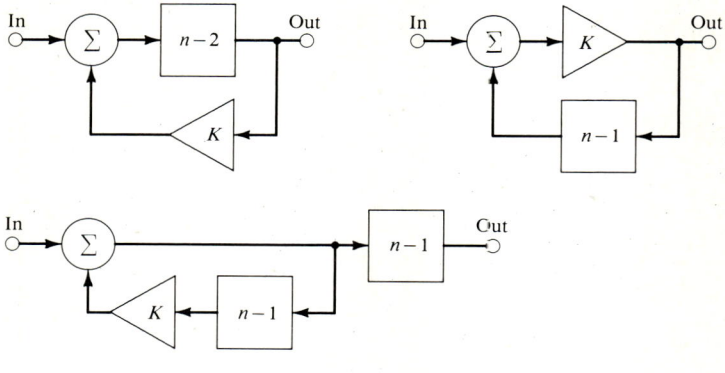

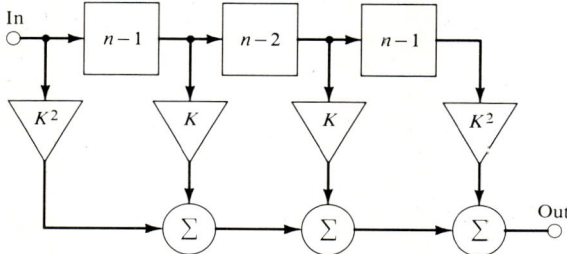

Figure P7-4

7-5 Find a value for K such that the systems of Fig. P7-5 have approximately the same delta response.

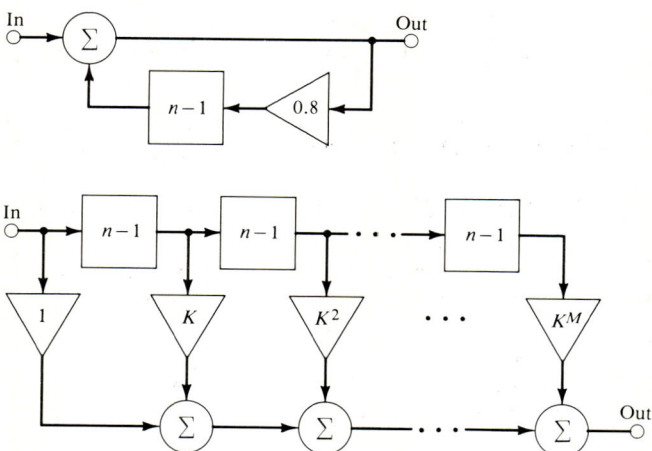

Figure P7-5

7-6 Obtain the step response of a system having delta response $h(n) = \lambda^n u(n)$. Check the result using (7-8).

7-7 Find the value of K for the system of Fig. P7-7a such that the step response of the system approximates the step response of the analog system of Fig. P7-7b.

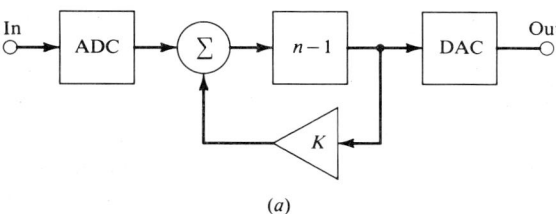

(a)

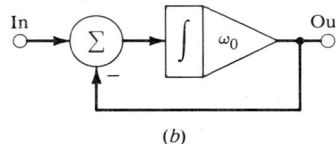

(b)

Figure P7-7

7-8 Find the smallest value of k for which the system of Fig. P7-8 is realizable.

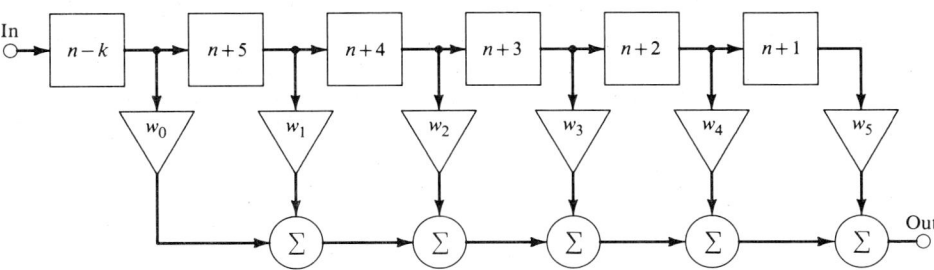

Figure P7-8

7-9 For each system of Fig. P7-9 find all values of all parameters for which the system is realizable and stable.

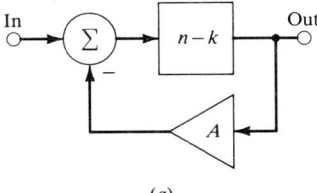

(a)

Figure P7-9

510 INTRODUCTION TO SYSTEM ANALYSIS

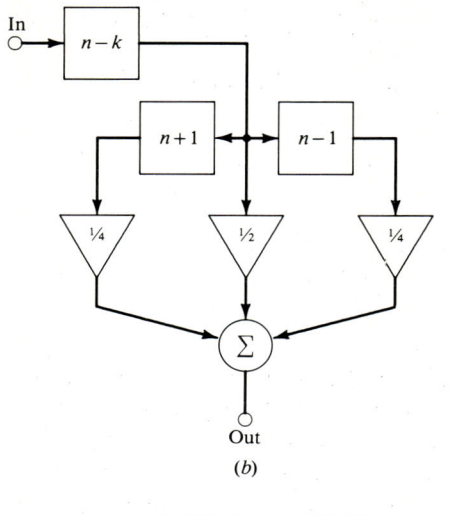

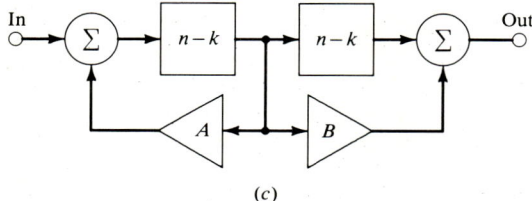

Figure P7-9 *Continued*

7-10 For each system of Fig. P7-10 obtain a condition on k under which the system provides high-fidelity transmission for input $x(n) = r(n/k)$.

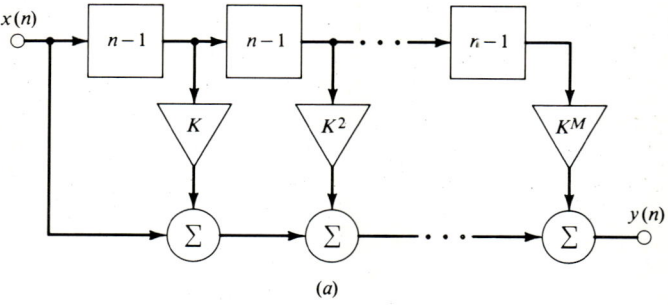

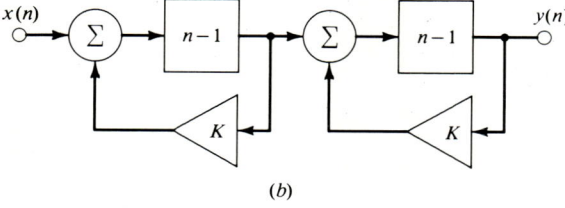

Figure P7-10

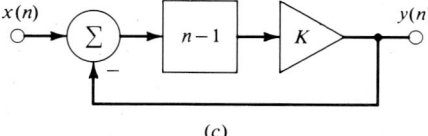

(c)   **Figure P7-10** *Continued*

7-11 For each system of Fig. P7-11 obtain a difference equation relating an output $y(n)$ to the corresponding input $x(n)$. Give the difference equation in the form of (7-12).

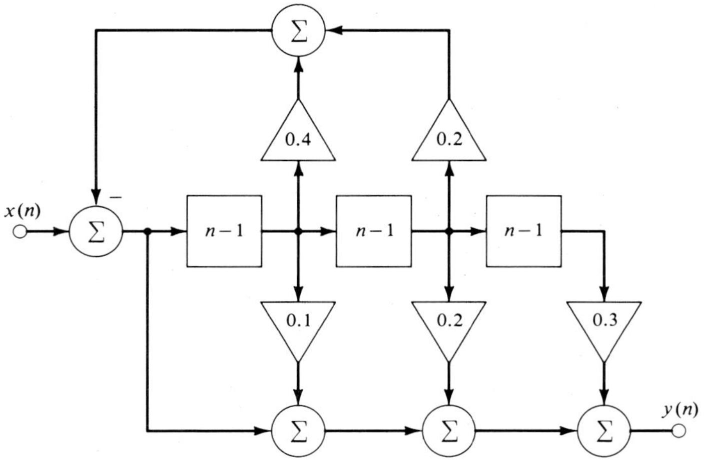

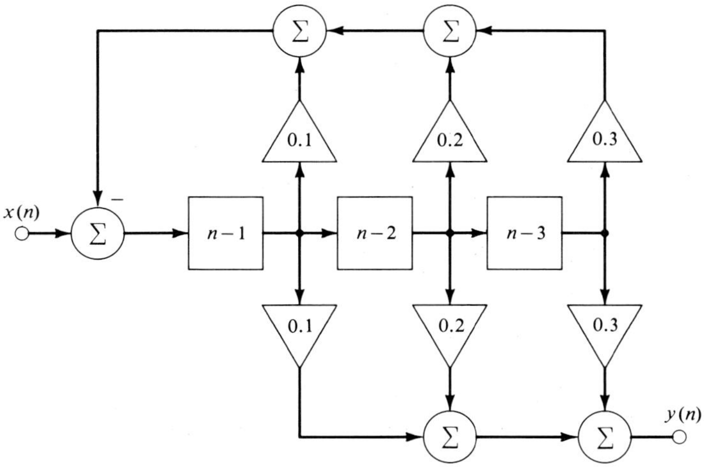

Figure P7-11

7-12 Draw a block diagram like that of Fig. 7-24 for each difference equation:

(a) $y(n) - 1.40y(n-1) + 0.48y(n-2)$
$$= 4.00x(n) + 3.00x(n-1) + 2.00x(n-2) + x(n-3)$$

(b) $y(n) + y(n-1) + y(n-2) = x(n) - x(n-1)$

(c) $y(n+2) - y(n-2) = x(n)$

(d) $y(n) = x(n+1) + x(n) + x(n-1)$

7-13 Use numerical recursion to obtain the first five samples of the step response for each realizable system of Prob. 7-12.

7-14 Obtain closed-form expressions for the delta responses and the step responses of systems described by the difference equations below, where $x(n)$ denotes an input and $y(n)$ denotes the corresponding output:

(a) $y(n) - 1.40y(n-1) + 0.48y(n-2) = x(n)$

(b) $y(n) + y(n-1) + 0.25y(n-2) = x(n) - x(n-1)$

(c) $y(n) + 1.5y(n-1) + 0.5y(n-2) = 0.5x(n) + x(n-1)$

(d) $y(n) - 1.3y(n-1) + 0.4y(n-2) = x(n) - 0.5x(n-1)$

(e) $y(n) + y(n-1) + y(n-2) = x(n-1)$

7-15 Obtain a difference equation and a block diagram describing a system having delta response:

(a) $h(n) = [2(0.6)^n - 3(0.8)^n]u(n+1)$

(b) $h(n) = [(-0.5)^{n-1} + 2(0.5)^{n-1}]u(n)$

(c) $h(n) = r(n/10)$

(d) $h(n) = [2n(0.4)^n + (0.4)^n - (0.5)^n]u(n)$

(e) $h(n) = [(0.8)^{n+1} + (0.6)^{n-1}]u(n+1)$

7-16 Delta responses $h(n)$ and inputs $x(n)$ for several linear shift-invariant systems are given below. Use convolution to determine the output $y(n)$ of each system.

(a) $h(n) = 5(0.4)^n u(n)$
$x(n) = (0.5)^{n-1} u(n)$

(b) $h(n) = (0.5)^{n-1} u(n)$
$x(n) = \cos 0.2n$

(c) $h(n) = (0.5)^n (\cos 0.1n) u(n)$
$x(n) = (0.5)^n u(n)$

(d) $h(n) = [2(0.8)^n - 3(-0.8)^n]u(n+1)$
$x(n) = (0.9)^{n-1} u(n+1)$

7-17 Obtain the delta response of a system whose input $x(n)$ and output $y(n)$ are related by the difference equation

$$y(n-1) + 1.2y(n-2) + 0.6y(n-3) = x(n)$$

Use the delta response to show that the system is nonrealizable and stable.

7-18 Show that the delta response of a linear shift-invariant system described by (7-12) with $m = 0$ is determined within a constant multiplier by the characteristic roots of the system.

7-19 The characteristic roots for a certain realizable second-order system described by (7-12) with $m = 1$ are $p_1 = 0.5$ and $p_2 = -0.5$. The delta response of the system satisfies $h(0) = 1.0$ and $h(1) = 2.0$. Obtain a difference equation describing the system.

7-20 Find all values of K for which the system of Fig. P7-20 is stable.

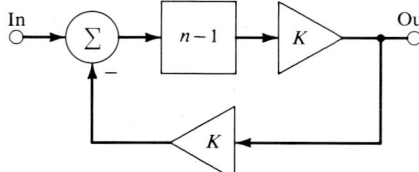

Figure P7-20

7-21 Find all values of K_1 and K_2 for which the system of Fig. P7-21 is stable.

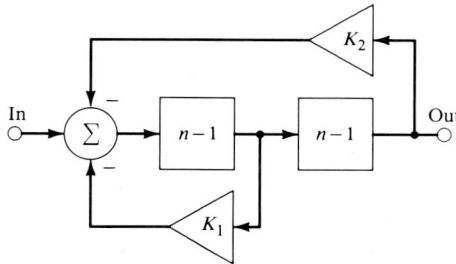

Figure P7-21

7-22 Express each signal in exponential form:

(a) $x(n) = 5 \cos(0.3n + 1.2)$

(b) $x(n) = 10 - 5 \cos(0.2n) + 2 \cos(0.4n - 1.5)$

(c) $x(n) = 2 + 2 \sin 0.2n$

7-23 Express each signal in amplitude-phase form:

(a) $x(n) = 2e^{j0.1n} + 2e^{-j0.1n}$

(b) $x(n) = (1 + j)e^{j0.2n} + (1 - j)e^{-j0.2n}$

(c) $x(n) = 3e^{-j}e^{j0.3n} + 3e^{j}e^{-j0.3n}$

(d) $x(n) = 5 + je^{j0.4n} - je^{-j0.4n}$

7-24 Obtain the transfer function (if it exists) for each system of Prob. 7-15. Use the transfer function to obtain the output for input

$$x(n) = 10 \cos(0.2n + 1.7)$$

7-25 Obtain the transfer function (if it exists) for each system of Prob. 7-14. Use the transfer function to obtain the output of each system for input

$$x(n) = 5 + 4 \cos 0.1n + 3 \cos\left(0.2n - \frac{\pi}{4}\right)$$

7-26 Obtain the output of each system of Fig. P7-26 for input

$$x(n) = 10 + 5 \cos 0.2n$$

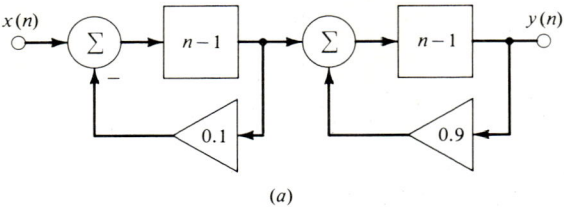

(a)

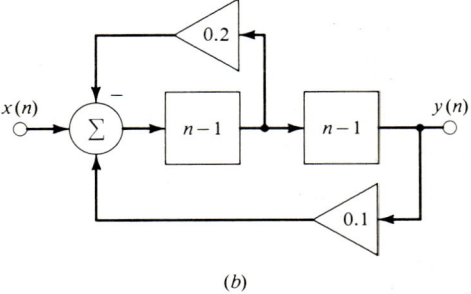

(b)

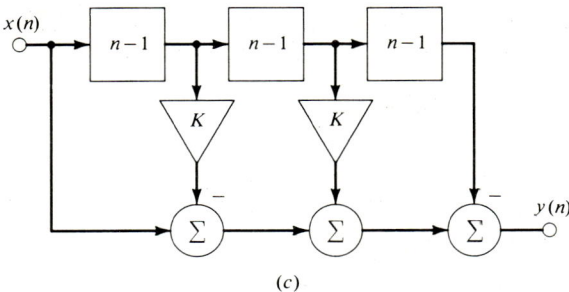

(c)

Figure P7-26 In part (c), $K = 2.96$.

7-27 For the first-order recursive digital filter of Fig. 7-38 the sampling rate is 10 kHz and the gain K equals 0.8. For what analog frequency is the overall gain equal to $1/\sqrt{2}$ times the dc gain? How does this frequency change if the sampling rate is doubled?

7-28 Modify the first-order recursive filter of Fig. 7-38 so that the dc gain of the filter equals unity for $0 < K < 1$. Determine the gain of the modified filter for digital frequency $\lambda = \pi$. Determine the delay introduced by the modified filter as a function of actual frequency ω.

7-29 For the first-order recursive filter of Fig. 7-38, the sampling rate is $F = 5$ kHz and $K = -0.8$. Give a block diagram for a comparable analog filter (one having the same bandwidth). Plot gain (normalized to unity maximum value) versus frequency and delay versus frequency for both filters. Discuss advantages and disadvantages of each filter (assume that the analog filter is a passive RC filter).

7-30 Obtain an expression for the tap weights of a nonrecursive digital filter whose transfer function approximates the function defined in Fig. P7-30. The bandwidth of inputs of interest is $B = 2$ kHz.

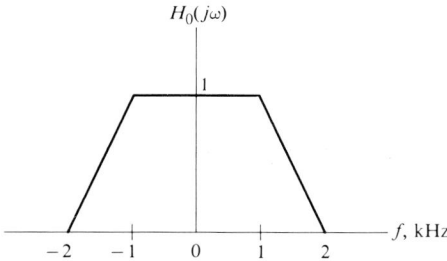

Figure P7-30

7-31 Obtain an expression for the tap weights of a nonrecursive digital filter whose transfer function approximates the function defined in Fig. P7-31. The bandwidth of inputs of interest is 5 kHz.

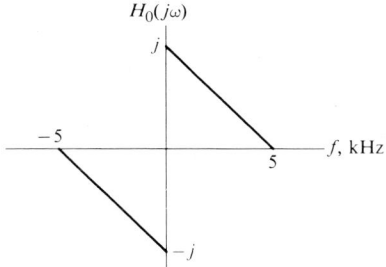

Figure P7-31

7-32 A three-tap nonrecursive digital filter has tap weights $w_0 = w_2 = 0.25$ and $w_1 = 0.5$. Plot the gain and phase shift of the filter versus digital frequency λ.

7-33 The simple differentiating filter described in Sec. 6–2B is a two-tap nonrecursive filter. Plot the gain and phase shift of the filter versus digital frequency. Compare the effective bandwidth of the two-tap filter and that of the 21-tap differentiating filter of Example 7-30 if:

(a) Both filters employ the same sampling rate.

(b) The sampling rate for the two-tap filter is 10 times the sampling rate for the 21-tap filter.

7-34 Obtain a relation between delta response and tap weights of a nonrecursive filter. This relation suggests that a nonrecursive filter can be designed by choosing tap weights equal to (or proportional to) samples $h(0), h(T), \ldots, h(MT)$ of a specified impulse response. Describe any advantages this approach may have over the Fourier series method.

7-35 A transfer function for a digital filter (including AD and DA converters) can be obtained either (a) by determining the ratio $y_a(t)/x_a(t)$, where $y_a(t)$ is the output for input $x_a(t) = x_0 e^{j\omega t}$, or (b) by using $H(j\omega) = \mathcal{F}\{h_a(t)\}$ and the relation $h_a(t) = dg_a(t)/dt$, where $g_a(t)$ is the step response for the filter. Use both methods to determine a transfer function for the first-order filter of Fig. 7-38. Explain why the methods give different transfer functions for the filter. Show that the transfer functions become identical if the sampling rate approaches infinity. *Hint:* Error-free AD and DA conversion is possible only for a strictly bandlimited signal. What is the bandwidth of a step?

7-36 In the text we describe a method for designing a nonrecursive filter having a specified gain and linear phase shift. Describe a method for designing a nonrecursive filter having frequency-independent gain and a specified phase shift (note that the phase shift must be an odd function of frequency). Where might such a filter be useful?

CHAPTER
EIGHT
z TRANSFORMATION

A transformation called the z transformation does for analysis and design of digital systems what the Laplace transformation does for analysis and design of analog systems. The z transform of a function $f(n)$ of an integer variable n is a function of a complex variable z. Whereas the variable s for a Laplace transform is interpreted as a differentiation operator, the variable z for a z transform is interpreted as a unit advance operator; i.e., multiplying the z transform of a digital signal $x(n)$ by z corresponds to advancing the signal by one sampling interval to obtain $x(n + 1)$. Conversely, whereas s^{-1} is interpreted as an integration operator, z^{-1} is interpreted as a unit delay operator. Taking the z transformation of a difference equation describing a linear shift-invariant system yields an algebraic equation, just as taking the Laplace transformation of a differential equation describing a linear stationary system yields an algebraic equation. The algebraic equation can be solved for the z transform of an output, and the output itself can be obtained by inverse z transformation. The whole procedure is quite similar — both conceptually and mechanically — to using the Laplace transformation for solving a differential equation for an output of a linear stationary analog system. Further, applying the z transformation to digital systems leads to defining a function called the system function whose role in analysis and design of digital systems parallels that of the system function for analog systems. The system function for a digital system is the z transform of the delta response for the system, just as the system function for an analog system is the Laplace transform of the impulse response for the system. The system function for a digital system can be described graphically using a pole-zero diagram on a z plane, just as the system function for an analog system can be described using a pole-zero diagram on an s plane. The pole-zero diagram (or the system function) for a digital system can be interpreted in terms of realizability, stability, transient response, and fidelity, just as the pole-zero diagram (or the system function) for an analog system can be interpreted in those terms.

  This chapter describes the z transformation and its application to analysis and design of digital systems. Although the presentation parallels that of the Laplace transformation in Chap. 5, it is more concise because many concepts and procedures associated with the z transformation are analogous to concepts and procedures discussed in Chap. 5.

8-1 DEFINITION AND FUNDAMENTAL PROPERTIES

In this section we give the defining relation for the z transform of a digital signal, we introduce z-plane representation of a z transform, we discuss uniqueness and convergence of z transforms, we explain the use of tables for obtaining z transforms and inverse z transforms, and we describe the most important operational properties of the z transformation.

8-1A Definition of the z Transformation

The *z transform* of a digital function $x(n)$ is denoted by $X(z)$ and defined by

$$X(z) = \sum_{n=-\infty}^{\infty} x(n)z^{-n} \qquad (8\text{-}1)$$

where the independent variable z is a dimensionless complex quantity. The operator $\mathscr{L}$ denotes z transformation; thus

$$X(z) = \mathscr{L}\{x(n)\} \qquad (8\text{-}2)$$

A signal $x(n)$ whose z transform is $X(z)$ is the *inverse z transform* of $X(z)$. We denote this relation by

$$x(n) = \mathscr{L}^{-1}\{X(z)\} \qquad (8\text{-}3)$$

Methods for obtaining inverse z transforms are described in Secs. 8-1G and 8-1H.

Example 8-1 The z transform of the unit delta $x(n) = \Delta(n)$ is

$$X(z) = \mathscr{L}\{\Delta(n)\} = \sum_{n=-\infty}^{\infty} \Delta(n)z^{-n}$$

By definition, $\Delta(0) = 1$ and $\Delta(n) = 0$ for $n \neq 0$; thus

$$\mathscr{L}\{\Delta(n)\} = 1$$

The inverse relation is

$$\mathscr{L}^{-1}\{1\} = \Delta(n)$$

8-1B z Plane

The independent variable z of a z transform is a complex variable. Values of z can be associated with points in a plane called the *z plane* (Fig. 8-1). The z plane is an important graphical tool in theory and application of the z transformation.

518 INTRODUCTION TO SYSTEM ANALYSIS

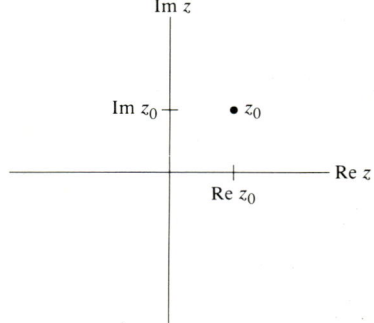

Figure 8-1 A z plane, where a particular value z_0 for z is represented by the point (Re z_0, Im z_0).

8-1C Uniqueness and Convergence

From (8-1), the z transform of a signal is fundamentally an infinite series. The coefficients of the series are in one-to-one correspondence with the samples of the signal; e.g., the coefficient of z^{-3} is $x(3)$. Therefore, a signal $x(n)$ and its z transform $X(z)$ constitute a unique pair so long as $X(z)$ is expressed as a series of the form

$$X(z) = \cdots + x(-1)z + x(0) + x(1)z^{-1} + \cdots$$

Example 8-2 The inverse z transform of

$$X(z) = 5z + 10 - z^{-1} + 2z^{-3}$$

is a signal $x(n)$ having sample values $x(-1) = 5$, $x(0) = 10$, $x(1) = -1$, $x(3) = 2$, and $x(n) = 0$ for all other n.

Manipulation of infinite series is cumbersome. For practical reasons, it usually is necessary to express a z transform in closed form. This requires that we obtain a closed-form expression for the series in (8-1), which in turn requires that the series converge for at least some values of z. The values of z for which the series converges constitute the *region of convergence* for the corresponding z transform.

Example 8-3 Obtain a closed-form expression and the associated region of convergence for the z transform of the right-sided exponential signal

$$x(n) = \lambda^n u(n)$$

SOLUTION From (8-1),

$$X(z) = \sum_{n=-\infty}^{\infty} \lambda^n z^{-n} u(n) = \sum_{n=1}^{\infty} (\lambda z^{-1})^n$$

$$= \lambda z^{-1}[1 + \lambda z^{-1} + (\lambda z^{-1})^2 + \cdots]$$

The quantity in brackets is a geometric series with ratio $r = \lambda z^{-1}$. For $|r| < 1$, the series converges to

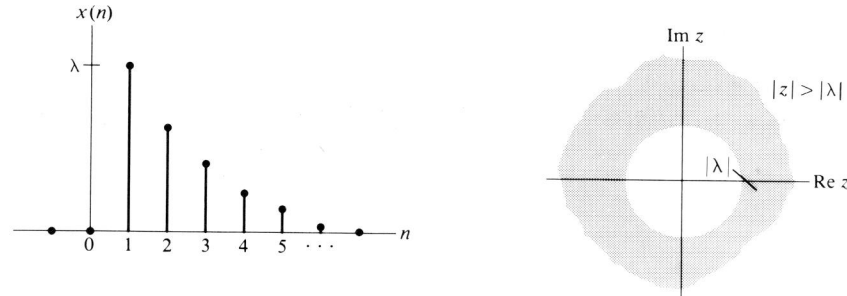

Figure 8-2 A right-sided exponential signal $x(n) = \lambda^n u(n)$ and the region of convergence for its z transform (see Example 8-3).

$$1 + r + r^2 + \cdots = \frac{1}{1-r} \qquad |r| < 1$$

The series diverges for $|r| \geq 1$. Therefore, for $|\lambda z^{-1}| < 1$ (for $|z| > \lambda$), the z transform above can be expressed in closed form as

$$X(z) = \frac{\lambda z^{-1}}{1 - \lambda z^{-1}} = \frac{\lambda}{z - \lambda} \qquad |z| > |\lambda|$$

The associated region of convergence is given by $|z| > |\lambda|$. Figure 8-2 shows a graph of the signal and a z-plane representation of the region of convergence for the z transform of the signal. The region of convergence is the exterior of a circle having radius $|\lambda|$.

Example 8-4 Obtain a closed-form expression and the associated region of convergence for the z transform of the left-sided exponential signal

$$y(n) = -\lambda^n u(1 - n)$$

SOLUTION From (8-1),

$$Y(z) = -\sum_{n=-\infty}^{\infty} \lambda^n z^{-n} u(1-n) = -\sum_{n=-\infty}^{-1} (\lambda z^{-1})^n$$

$$= -\sum_{n=1}^{\infty} (\lambda^{-1} z)^n = -[1 + \lambda^{-1} z + (\lambda^{-1} z)^2 + \cdots]$$

The quantity in brackets is a geometric series with ratio $r = \lambda^{-1} z$. For $|r| < 1$, the series converges to

$$1 + r + r^2 + \cdots = \frac{1}{1-r} \qquad |r| < 1$$

The series diverges for $|r| \geq 1$. Therefore, for $|\lambda^{-1} z| < 1$ (for $|z| < |\lambda|$), the z transform above can be expressed in closed form as

$$Y(z) = -\frac{1}{1 - \lambda^{-1} z} = \frac{\lambda}{z - \lambda} \qquad |z| < |\lambda|$$

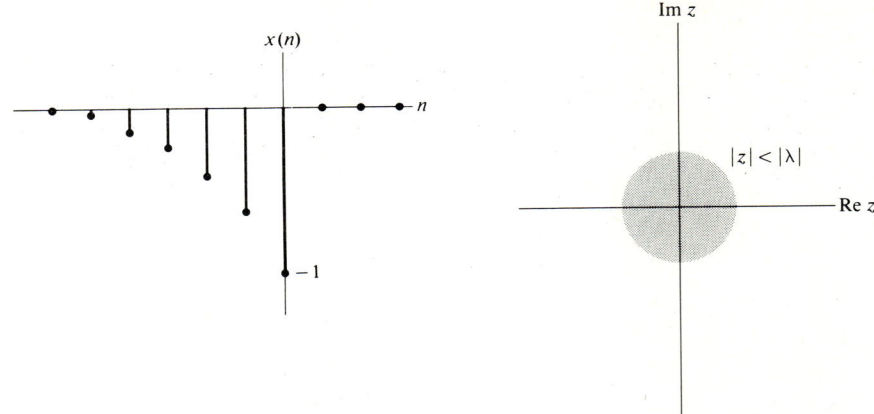

Figure 8-3 A left-sided exponential signal $x(n) = -\lambda^n u(1 - n)$ and the region of convergence for its z transform (see Example 8-4).

The associated region of convergence is given by $|z| < |\lambda|$. Figure 8-3 shows a graph of the signal and a z-plane representation of the region of convergence for the z transform of the signal. The region of convergence is the interior of a circle having radius $|\lambda|$.

Examples 8-3 and 8-4 show that the closed-form expression for the z transform of the right-sided signal $\lambda^n u(n)$ is identical to the closed-form expression for the z transform of the left-sided signal $-\lambda^n u(1 - n)$; that is,

$$\mathcal{Z}\{\lambda^n u(n)\} = \mathcal{Z}\{-\lambda^n u(1 - n)\} = \frac{\lambda}{z - \lambda}$$

This illustrates the fact that a closed-form expression for a z transform has a unique inverse only if a region of convergence for the transform is specified. For example, if the region of convergence for the transform above is specified as $|z| > |\lambda|$, the inverse transform is

$$\mathcal{Z}^{-1}\left\{\frac{\lambda}{z - \lambda}\right\} = \lambda^n u(n)$$

whereas if the region of convergence is $|z| < |\lambda|$, the inverse transform is

$$\mathcal{Z}^{-1}\left\{\frac{\lambda}{z - \lambda}\right\} = -\lambda^n u(1 - n)$$

Together the regions $|z| < |\lambda|$ and $|z| > |\lambda|$ cover the entire z plane. Consequently, the signals $\lambda^n u(n)$ and $-\lambda^n u(1 - n)$ are the only possible inverse transforms of the function $\lambda/(z - \lambda)$.

Example 8-5 Obtain the inverse z transform of

$$X(z) = \frac{\beta}{z - \beta} + \frac{\lambda}{z - \lambda}$$

with associated region of convergence $|\beta| < |z| < |\lambda|$.

SOLUTION We apply the expansion

$$\frac{1}{1-r} = \begin{cases} 1 + r + r^2 + r^3 + \cdots & |r| < 1 \\ -r^{1}(1 + r^{-1} + r^{-2} + \cdots) & |r| > 1 \end{cases}$$

to each term of the expression for $X(z)$ above. The first term is represented by

$$\frac{\beta}{z - \beta} = \begin{cases} \beta z^{-1}(1 + \beta z^{-1} + \beta^2 z^{-2} + \cdots) & |z| > |\beta| \\ -(1 + \beta^{-1}z + \beta^{-2}z^2 + \cdots) & |z| < |\beta| \end{cases}$$

where we have used $r = \beta z^{-1}$ in the series expansion above. The second term is represented by

$$\frac{\lambda}{z - \lambda} = \begin{cases} \lambda z^{-1}(1 + \lambda z^{-1} + \lambda^2 z^{-2} + \cdots) & |z| > |\lambda| \\ -(1 + \lambda^{-1}z + \lambda^{-2}z^2 + \cdots) & |z| < |\lambda| \end{cases}$$

where we have used $r = \lambda z^{-1}$ in the series expansion above. Thus two different series expansions are identified with each of the two terms in the closed-form expression above for $X(z)$. Consequently, four different series expansions are identified with $X(z)$. Only one of these, namely,

$$X(z) = \beta z^{-1}(1 + \beta z^{-1} + \beta^2 z^{-2} + \cdots) - (1 + \lambda^{-1}z + \lambda^{-2}z^2 + \cdots)$$

converges within the specified region of convergence. Therefore, for the specified region of convergence, the inverse transform of $X(z)$ is

$$x(n) = \mathscr{L}^{-1}\{X(z)\} = \beta^n u(n) - \lambda^{-n} u(1 - n)$$

Figure 8-4 shows a graph of the signal $x(n)$ and a z-plane description of the region of convergence for the z transform of the signal. The region of convergence is an annular region (a region bounded by concentric circles).

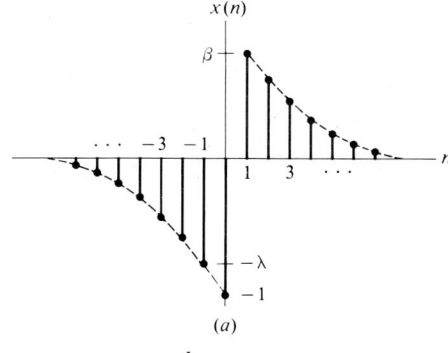

(a)

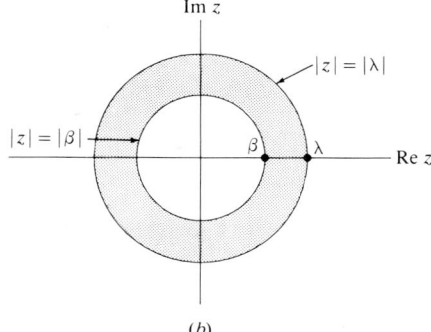

(b)

Figure 8-4 A two-sided exponential signal $x(n) = \beta^n u(n) - \lambda^{-n} u(1 - n)$ and the region of convergence for its z transform (see Example 8-5).

8-1D Right-Sided Inverse z Transform

Digital signals, like analog signals, are either *right-sided, left-sided,* or *two-sided.* We regard a signal having finite duration (e.g., a rectangular pulse) as a right-sided signal. Example 8-3 illustrates the fact that the region of convergence for the z transform of a right-sided signal is the exterior of a circle in the z plane. Example 8-4 illustrates the fact that the region of convergence for the z transform of a left-sided signal is the interior of a circle in the z plane. Example 8-5 illustrates the fact that the region of convergence for the z transform of a two-sided signal is the region between two concentric circles in the z plane. Examples 8-1 and 8-2 illustrate the fact that the z transform of a signal having finite duration converges for all z except (possibly) $z = 0$. The region of convergence can be described as the exterior of a circle having zero radius, as the interior of a circle having infinite radius, or as the region between a circle having zero radius and one having infinite radius.

The inverse transform of a function $X(z)$ can be a right-sided signal (including a finite-duration signal as a special case), a left-sided signal, or a two-sided signal, depending on the region of convergence for $X(z)$. The right-sided inverse transform of a function is unique; e.g., the function $\lambda/(z - \lambda)$ has only one right-sided inverse transform, namely, the signal $\lambda^n u(n)$. Except where otherwise noted, we limit application of the z transformation to right-sided signals. This makes inverse z transforms unique and eliminates the need to specify a region of convergence for a z transform. By $\mathcal{L}^{-1}\{X(z)\}$ we mean the *right-sided inverse transform* of $X(z)$; for example,

$$\mathcal{L}^{-1}\left\{\frac{\lambda}{z - \lambda}\right\} = \lambda^n u(n)$$

8-1E Table of z Transforms

We can use (8-1) to make a table of z transforms. Because a right-sided inverse z transform is unique, we can use the table to obtain right-sided inverse z transforms. Table E-1 contains all z-transform pairs needed in this book. Two entries in Table E-1 are derived in Examples 8-1 and 8-3 above. Two other entries are derived in the next two examples.

Example 8-6 Obtain the z transform of the signal

$$x(n) = n\lambda^n u(n)$$

SOLUTION From Example 8-3, we have $\mathcal{L}\{\lambda^n u(n)\} = \lambda/(z - \lambda)$, or

$$\sum_{n=-\infty}^{\infty} \lambda^n u(n) z^{-n} = \frac{\lambda}{z - \lambda}$$

Differentiating this relation once with respect to z and multiplying the result by $-z$ gives

$$\sum_{n=-\infty}^{\infty} n\lambda^n u(n) z^{-n} = \frac{\lambda z}{(z - \lambda)^2} = \mathcal{L}\{n\lambda^n u(n)\}$$

Entries 4 to 9 in Table E-1 are useful for finding z transforms but not for finding inverse z transforms, because the forms of the numerators of the transforms do not match those in partial-fraction expansions for more complicated transforms. Entries 10 to 15 in Table E-1 are included to facilitate inversion using partial-fraction expansion. Entry 10 is just entry 3 divided by λ. Entries 11 to 14 can be obtained by differentiating previous entries with respect to λ.

Example 8-7 From Table E-1,

$$\mathcal{L}\{\lambda^{n-1}u(n)\} = \sum_{n=-\infty}^{\infty} \lambda^{n-1}u(n)z^{-n} = \frac{1}{z-\lambda}$$

Differentiating the last two members once with respect to λ gives

$$\sum_{n=-\infty}^{\infty} (n-1)\lambda^{n-2}u(n)z^{-n} = \frac{1}{(z-\lambda)^2} = \mathcal{L}\{(n-1)\lambda^{n-2}u(n)\}$$

This is entry 11 in the table (where $c = \lambda$).

8-1F Linearity

Both z transformation and inverse z transformation are linear operations; i.e.,

$$\mathcal{L}\{a_1x_1(n) + a_2x_2(n) + \cdots\} = a_1\mathcal{L}\{x_1(n)\} + a_2\mathcal{L}\{x_2(n)\} + \cdots \quad (8\text{-}4)$$

and

$$\mathcal{L}^{-1}\{b_1X_1(z) + b_2X_2(z) + \cdots\} = b_1\mathcal{L}^{-1}\{X_1(z)\} + b_2\mathcal{L}^{-1}\{X_2(z)\} + \cdots \quad (8\text{-}5)$$

Equation (8-4) follows from (8-1) because summation is a linear operation. Equation (8-5) follows from (8-4) because the inverse of a linear operation is a linear operation.

Example 8-8 Obtain the z transform of the signal

$$x(n) = (\cos n\lambda)u(n)$$

SOLUTION Expressing the sinusoid in exponential form gives

$$x(n) = (0.5e^{jn\lambda} + 0.5e^{-jn\lambda})u(n)$$

which can be written

$$x(n) = [0.5\beta^n + 0.5(\beta^*)^n]u(n)$$

where $\beta = e^{j\lambda}$. From (8-4) and the result of Example 8-3,

$$\mathcal{L}\{x(n)\} = 0.5\mathcal{L}\{\beta^n u(n)\} + 0.5\mathcal{L}\{(\beta^*)^n u(n)\}$$

$$= \frac{0.5\beta}{z-\beta} + \frac{0.5\beta^*}{z-\beta^*} = \frac{0.5(\beta-\beta^*)z - \beta\beta^*}{z^2 - (\beta+\beta^*)z + \beta\beta^*}$$

$$= \frac{(\operatorname{Re}\beta)z - |\beta|^2}{z^2 - 2(\operatorname{Re}\beta)z + |\beta|^2} = \frac{(\cos\lambda)z - 1}{z^2 - 2(\cos\lambda)z + 1}$$

Note that this verifies entry 7 in Table E-1 for $c = 1$.

8-1G Partial-Fraction Expansion

Inverting a rational z transform of the form

$$F(z) = \frac{c_m z^m + c_{m-1} z^{m-1} + \cdots + c_0}{z^r + d_{n-1} z^{a-1} + \cdots + d_0}$$

is facilitated by expanding the function in partial fractions such that the terms of the expansion have the forms given in Table E-1. The procedure is exactly the same as for Laplace transforms (Sec. 5-2D) except that the independent variable is named z instead of s.

Example 8-9 Obtain the inverse z transform of

$$X(z) = \frac{2z^2}{(z-\alpha)(z-\beta)} \qquad \alpha \neq \beta$$

SOLUTION Expanding $X(z)$ in partial fractions gives

$$X(z) = 2 + \frac{A}{z-\alpha} + \frac{B}{z-\beta}$$

where

$$A = (z-\alpha)X(z)\bigg|_{z=\alpha} = \frac{2\alpha^2}{\alpha-\beta} \qquad B = (z-\beta)X(z)\bigg|_{z=\beta} = \frac{2\beta^2}{\beta-\alpha}$$

Using (8-5) and Table E-1 gives

$$x(n) = 2\mathscr{L}^{-1}\{1\} + \frac{2\alpha}{\alpha-\beta}\mathscr{L}^{-1}\left\{\frac{\alpha}{z-\alpha}\right\} + \frac{2\beta}{\beta-\alpha}\mathscr{L}^{-1}\left\{\frac{\beta}{z-\beta}\right\}$$

$$= 2\Delta(n) + \frac{2\alpha}{\alpha-\beta}\alpha^n u(n) + \frac{2\beta}{\beta-\alpha}\beta^n u(n)$$

$$= 2\Delta(n) + \frac{2}{\alpha-\beta}(\alpha^{n+1} - \beta^{n+1})u(n)$$

Example 8-10 Obtain the inverse z transform of

$$X(z) = \frac{z}{(z-0.4)(z-0.5)^2}$$

SOLUTION Expanding $X(z)$ in partial fractions gives

$$X(z) = \frac{1}{z-0.5}\cdot\frac{z}{(z-0.4)(z-0.5)} = \frac{1}{z-0.5}\left(\frac{-4}{z-0.4} + \frac{5}{z-0.5}\right)$$

$$= \frac{5}{(z-0.5)^2} - \frac{4}{(z-0.4)(z-0.5)} = \frac{5}{(z-0.5)^2} + \frac{40}{z-0.4} - \frac{40}{z-0.5}$$

Using (8-5) and Table E-1 gives

$$x(n) = \mathscr{L}^{-1}\{X(z)\}$$

$$= [5(n-1)(0.5)^{n-2} + 40(0.4)^{n-1} - 40(0.5)^{n-1}]u(n)$$

8-1H Numerical Inversion by Long Division

The coefficients of a z transform expressed as an infinite series are in one-to-one correspondence with the samples of the associated digital signal (see Example 8-2). We can use long division to obtain a limited number of terms of the infinite series for a rational z transform. This is useful for checking a closed-form expression for an inverse transform and for obtaining numerically the inverse of a transform difficult to invert by any other method.

Example 8-11 To illustrate the inversion of a rational z transform by long division we consider

$$X(z) = \frac{0.5}{z - 0.5}$$

The indicated division can be done in two ways, according to whether we arrange the terms of the denominator in order of increasing powers of z or in order of decreasing powers of z.

Case 1

$$
\begin{array}{r}
-1 - 2z - 4z^2 - \cdots + \\
0.5 - z \overline{\smash{)} -0.5 } \\
\underline{-0.5 + z} \\
-z \\
\underline{-z + 2z^2} \\
-2z^2
\end{array}
$$

or

Case 2

$$
\begin{array}{r}
0.5z^{-1} + 0.25z^{-2} + 0.125z^{-3} + \cdots \\
z - 0.5 \overline{\smash{)} 0.5 } \\
\underline{0.5 - 0.25z^{-1}} \\
0.25z^{-1} \\
\underline{0.25z^{-1} - 0.125z^{-2}} \\
0.125z^{-2}
\end{array}
$$

In case 1, where terms of the divisor are ordered according to increasing powers of z, the coefficients of the quotient [the series expansion for $X(z)$] are associated with positive powers of z, implying that the quotient is the transform of a left-sided signal [see (8-1)]. In case 2, where terms of the divisor are ordered according to decreasing powers of z, the coefficients of the quotient (series expansion) are associated with negative powers of z, implying that the quotient is the transform of a right-sided signal. Because we limit application of the z transformation to right-sided signals, we must always order terms of the divisor (of the denominator of a transform) according to decreasing powers of z before using long division to obtain an inverse transform. The right-sided inverse transform of $X(z)$ above is given by

$$\mathscr{Z}^{-1}\{X(z)\} = x(n) = \mathscr{Z}^{-1}\{0.5z^{-1} + 0.25z^{-2} + 0.125z^{-3} + \cdots\}$$

Thus the first three samples of $x(n)$ are $x(1) = 0.5$, $x(2) = 0.25$, and $x(3) = 0.125$ × [$x(n) = 0$ for $n < 1$].

Example 8-12 Obtain the first three nonzero samples of the inverse transform of

$$X(z) = \frac{1 - 2z^{-1}}{(z - 0.2)(z + 0.1)}$$

SOLUTION First, we express the divisor (denominator) as a polynomial in z. This gives

$$(z - 0.2)(z + 0.1) = z^2 - 0.1z - 0.02$$

where we have ordered the terms of the divisor according to decreasing powers of z. Next we separate $X(z)$ into a number of terms corresponding to the number of terms in the numerator as

$$X(z) = \frac{1}{z^2 - 0.1z - 0.02} - \frac{2z^{-1}}{z^2 - 0.1z - 0.02}$$

Using long division to expand the first term on the right gives

$$
\begin{array}{r}
z^{-2} + 0.1z^{-3} + 0.03z^{-4} + \cdots \\
z^2 - 0.1z - 0.02 \overline{\smash{\big)}\, 1} \\
\underline{1 - 0.1z^{-1} - 0.02z^{-2}} \\
0.1z^{-1} + 0.02z^{-2} \\
\underline{0.1z^{-1} - 0.01z^{-2} - 0.002z^{-3}} \\
0.03z^{-2} + 0.002z^{-3}
\end{array}
$$

The second term in the expression above for $X(z)$ is just $-2z^{-1}$ times the first; thus the first three terms of the second quotient are given by

$$-2z^{-1}(z^{-2} + 0.1z^{-3} + 0.03z^{-4}) = -2z^{-3} - 0.2z^{-4} - 0.06z^{-5}$$

It follows that the expansion for $X(z)$ (first three nonzero terms) is

$$X(z) = z^{-2} + (0.1 - 2)z^{-3} + (0.03 - 0.20)z^{-4}$$

and that the first three nonzero samples of the inverse transform are $x(2) = 1.0$, $x(3) = -1.9$, and $x(4) = -0.17$. Note that $x(n) = 0$ for $n < 2$ because the zeroth and first powers of z are absent from the quotient above.

8-1I Operational Properties

The two most important operational properties of the z transformation are expressed by the relations

$$\mathscr{L}\{x(n - k)\} = z^{-k}\mathscr{L}\{x(n)\} \tag{8-6}$$

$$\mathscr{L}\left\{\sum_{m=-\infty}^{\infty} x(m)h(n - m)\right\} = \mathscr{L}\{x(n)\}\mathscr{L}\{h(n)\} \tag{8-7}$$

Equation (8-6) shows that shifting a function by k samples corresponds to multiplying the z transform of the (unshifted) function by z^{-k}. This property is called the *shift property* of the z transformation. It is analogous to the differentiation property $\mathscr{L}\{d^n x(t)/dt^n\} = s^n \mathscr{L}\{x(t)\}$ of the Laplace transformation. In Sec. 8-2B we show how the shift property of the z transformation is used to transform a difference equation into an algebraic equation in the same way that the differentiation property of the Laplace transformation is used to transform a differential equation into an algebraic equation.

Equation (8-7) shows that the z transform of a convolution of two functions is the product of the z transforms of the functions. This property is analogous to the convolution property (5-9) of the Laplace transformation. It shows that the output $y(n)$ of a linear shift-invariant system for input $x(n)$ is given by

$$y(n) = \mathcal{L}^{-1}\{X(z)H(z)\} \tag{8-8}$$

where $X(z)$ is the z transform of the input and $H(z)$ is the z transform of the delta response of the system. The function $H(z)$ defined by

$$H(z) = \mathcal{L}\{h(n)\} \tag{8-9}$$

is called the *system function* for a linear shift-invariant system having delta response $h(n)$.

The shift property (8-6) can be derived from the definition of the z transformation as follows. From (8-1),

$$\mathcal{L}\{x(n-k)\} = \sum_{n=-\infty}^{\infty} x(n-k)z^{-n}$$

Introducing the change of variable $m = n - k$ gives

$$\mathcal{L}\{x(n-k)\} = \sum_{m=-\infty}^{\infty} x(m)z^{-m-k} = z^{-k} \sum_{m=-\infty}^{\infty} x(m)z^{-m}$$

The last sum above is the z transform of $x(n)$. Equation (8-6) follows.

The convolution property (8-7) of the z transformation can be derived as follows. Let

$$y(n) = \sum_{m=-\infty}^{\infty} x(m)h(n-m)$$

From (8-1),

$$\mathcal{L}\{y(n)\} = \sum_{n=-\infty}^{\infty} \sum_{m=-\infty}^{\infty} x(m)h(n-m)z^{-n}$$

Reversing the order of the summations gives

$$\mathcal{L}\{y(n)\} = \sum_{m=-\infty}^{\infty} x(m) \sum_{n=-\infty}^{\infty} h(n-m)z^{-n}$$

The inner sum is the z transform of $h(n-m)$. By the shift property (8-6),

$$\mathcal{L}\{y(n)\} = \sum_{m=-\infty}^{\infty} x(m)z^{-m}\mathcal{L}\{h(n)\}$$

The sum is the z transform of $x(n)$. Equation (8-7) follows.

In the next section we apply the z transformation to analysis of linear shift-invariant digital systems.

8-2 APPLICATION TO LINEAR SHIFT-INVARIANT SYSTEMS

In this section we show how the z transformation is used to obtain an output of a linear shift-invariant system described by a delta response, a difference equation, or a block diagram. We also describe two methods, impulse invariance and step invariance, for designing digital systems (e.g., digital filters) that emulate linear stationary analog systems.

The output $y(n)$ of a linear shift-invariant system for input $x(n)$ can be obtained using the z transformation as follows:

1. Obtain the z transform $X(z)$ of the input.
2. Obtain the system function $H(z)$.
3. Obtain the output $y(n)$ from

$$y(n) = \mathscr{L}^{-1}\{H(z)X(z)\}$$

For systems considered in this book, the transforms and inverse transforms can be obtained using Table E-1 (and partial-fraction expansion). The details of the second step of the procedure depend on whether a system is described by a delta response, a difference equation, or a block diagram.

8-2A Systems Described by Delta Responses

From (8-9), the system function for a system having delta response $h(n)$ is given by

$$H(z) = \mathscr{L}\{h(n)\}$$

For simple systems (systems considered in this book), the transform can be obtained using Table E-1.

Example 8-13 The delta response for a certain system is

$$h(n) = (0.5)^n u(n)$$

Obtain the output $y(n)$ for input

$$x(n) = u(n)$$

SOLUTION We follow the procedure given above.

Step 1 From Table E-1, the z transform of the input is

$$X(z) = \mathscr{L}\{x(n)\} = \mathscr{L}\{u(n)\} = \frac{1}{z-1}$$

Step 2 From (8-9) and Table E-1, the system function is

$$H(z) = \mathscr{L}\{h(n)\} = \mathscr{L}\{(0.5)^n u(n)\} = \frac{0.5}{z-0.5}$$

Step 3 From (8-8), the z transform of the output is given by

$$Y(z) = H(z)X(z) = \frac{0.5}{(z-1)(z-0.5)}$$

Expanding the right side in partial fractions gives

$$Y(z) = \frac{1}{z-1} - \frac{1}{z-0.5}$$

Using Table E-1 gives

$$y(n) = u(n) - (0.5)^{n-1}u(n) = [1 - (0.5)^{n-1}]u(n)$$

8-2B Systems Described by Difference Equations

A linear shift-invariant system having input $x(n)$ and output $y(n)$ can be described by a difference equation of the form

$$y(n) + a_1 y(n-1) + \cdots + a_k y(n-k)$$
$$= b_0 x(n) + b_1 x(n-1) + \cdots + b_m x(n-m)$$

Taking the z transform of this equation gives

$$(1 + a_1 z^{-1} + \cdots + a_k z^{-k})Y(z) = (b_0 + b_1 z^{-1} + \cdots + b_m z^{-m})X(z)$$

where we have used (8-4) and (8-6) (linearity and shift properties of the z transformation). This gives

$$Y(z) = H(z)X(z)$$

where
$$H(z) = \frac{b_0 + b_1 z^{-1} + \cdots + b_m z^{-m}}{1 + a_1 z^{-1} + \cdots + a_k z^{-k}} \qquad (8\text{-}10)$$

Equation (8-10) gives the system function for a system in terms of the coefficients of a difference equation describing the system.

Example 8-14 The output $y(n)$ of a certain system is related to an input $x(n)$ by the difference equation

$$y(n) + 0.1y(n-1) - 0.2y(n-2) = 2.0x(n) - x(n-1)$$

Obtain the output of the system for input

$$x(n) = (0.4)^n u(n)$$

SOLUTION We follow the procedure given at the beginning of this section. The z transform of the input is

$$X(z) = \mathscr{Z}\{(0.4)^n u(n)\} = \frac{0.4}{z - 0.4}$$

From (8-10), the system function for the system is

$$H(z) = \frac{2.0 - z^{-1}}{1 + 0.1z^{-1} - 0.2z^{-2}} = \frac{2.0z^2 - z}{z^2 + 0.1z - 0.2} = \frac{2.0z^2 - z}{(z - 0.4)(z + 0.5)}$$

The z transform of the output is

$$Y(z) = \frac{2.0z^2 - z}{(z - 0.4)^2(z + 0.5)}$$

530 INTRODUCTION TO SYSTEM ANALYSIS

Expanding the right side in partial fractions gives

$$Y(z) = -\frac{0.089}{(z-0.4)^2} + \frac{0.765}{z-0.4} + \frac{1.235}{z+0.5}$$

The output is

$$y(n) = \mathscr{L}^{-1}\{Y(z)\}$$
$$= -0.089(n-1)(0.4)^{n-2}u(n) + 0.765(0.4)^{n-1}u(n) + 1.235(-0.5)^{n-1}u(n)$$
$$= [(2.470 - 0.556n)(0.4)^n - 2.470(-0.5)^n]u(n)$$

8-2C Systems Described by Block Diagrams

A linear shift-invariant system having system function $H(z)$ can be represented by a z-domain block diagram as shown in Fig. 8-5, where $X(z)$ is the z transform of an input and $Y(z)$ is the z transform of the corresponding output. The block diagram of Fig. 8-5 denotes the relation

$$Y(z) = H(z)X(z)$$

Block diagrams for the three elementary linear shift-invariant systems are given in Table 8-1. The system functions for these elementary systems are derived by taking the z transforms of the associated transfer characteristics. For example, the transfer characteristic for a shift element having input $x(n)$ and output $y(n)$ is

$$y(n) = x(n-k)$$

where k is the shift. Taking the z transform of this relation gives

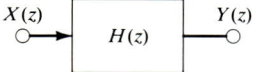

Figure 8-5 Symbol for a linear shift-invariant system having system function $H(z)$.

Table 8-1 Symbols for elementary linear shift-invariant digital systems

Element	Transfer characteristic	System function	Symbol
Proportion	$y(n) = Kx(n)$	$Y(z) = KX(z)$	X →[K]→ Y
Delay	$y(n) = x(n-k)$	$Y(z) = z^{-k}X(z)$	X →[z^{-k}]→ Y
Addition	$y(n) = w(n) + x(n)$	$Y(z) = W(z) + X(z)$	W, X →[Σ]→ Y

$$Y(z) = z^{-k}X(z)$$

where we have used the shift property of the z transformation. This shows that the system function for a shift element having shift k is

$$H(z) = z^{-k}$$

as given in Table 8-2.

Block diagrams representing digital systems in the z domain can be simplified in the same way as those representing analog systems in the s (Laplace) domain. Rules for simplifying series, parallel, and feedback connections of linear shift-invariant systems are given in Table 8-2. Rules for moving addition elements and pickoff points are given in Table 8-3. The rules in Tables 8-2 and 8-3 are identical to those given in Chap. 5 (for s-domain block diagrams) with s replaced by z. Further, derivations of the rules given in Tables 8-2 and 8-3 are identical to the derivations given in Chap. 5 with s replaced by z.

Example 8-15 Obtain the output of the system of Fig. 8-6 for input $x(n) = u(n)$.

SOLUTION **Step 1** The z transform of the input is

$$X(z) = \frac{1}{z-1}$$

Step 2 We use block-diagram reduction to obtain the system function of the system. Steps in the reduction are shown in Fig. 8-7. The system function is

Table 8-2 Rules for reducing series, parallel, and feedback connections

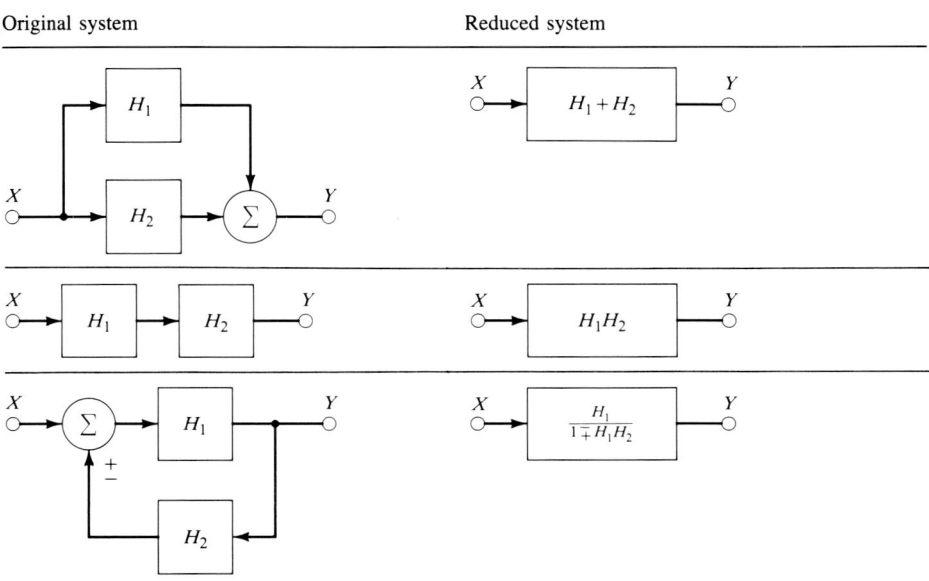

Table 8-3 Rules for moving addition elements and pickoff points from one side of a block to the other

Original system	Reduced system
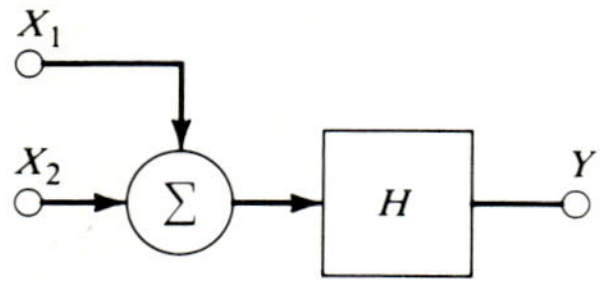 X_1, $X_2 \to \Sigma \to H \to Y$	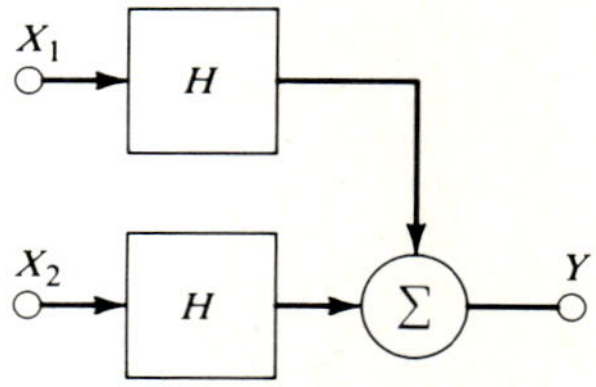 $X_1 \to H$; $X_2 \to H \to \Sigma \to Y$
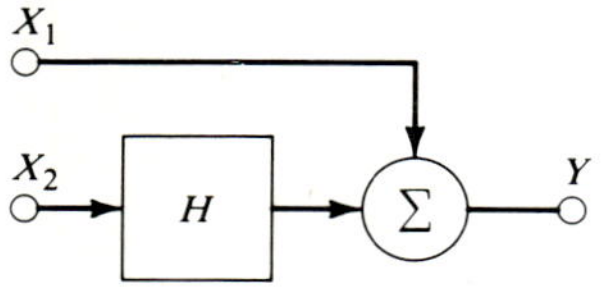 X_1, $X_2 \to H \to \Sigma \to Y$	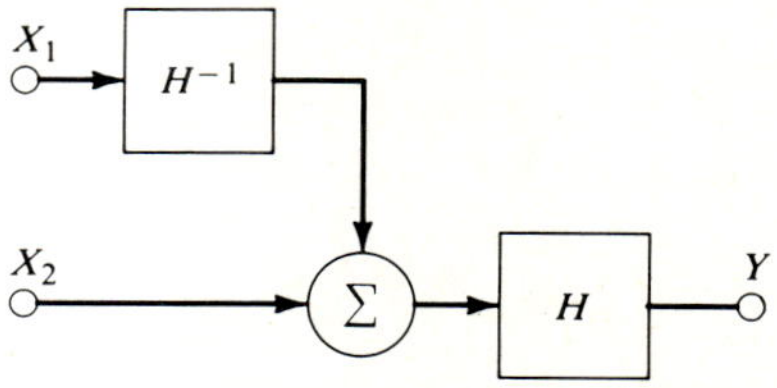 $X_1 \to H^{-1}$; $X_2 \to \Sigma \to H \to Y$
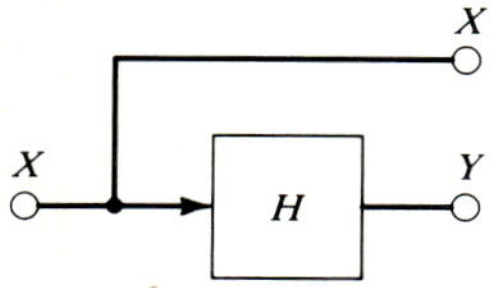 $X \to H \to Y$, with pickoff X	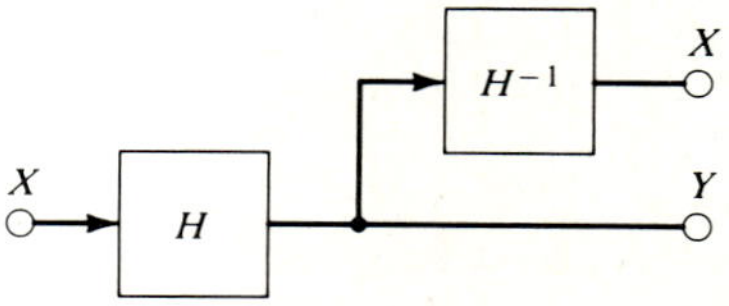 $X \to H \to Y$, with $H^{-1} \to X$ pickoff
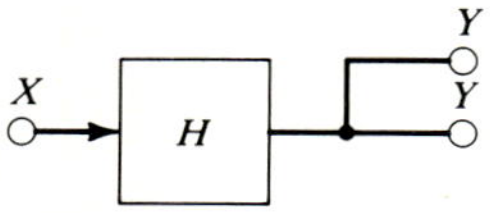 $X \to H \to Y$, pickoff Y	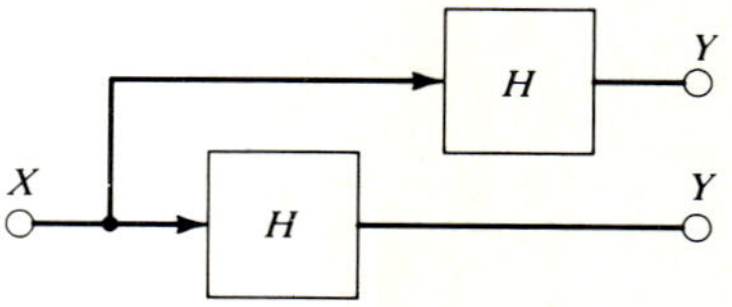 $X \to H \to Y$; $\to H \to Y$

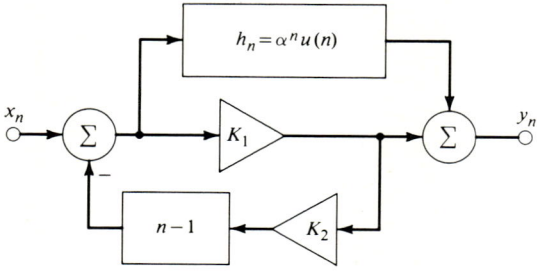

Figure 8-6 System of Example 8-15; $\alpha = 0.5$, $K_1 = -1.0$, and $K_2 = 0.8$.

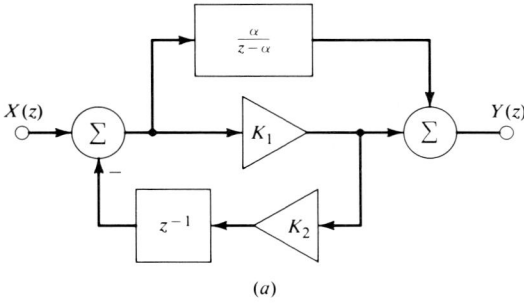

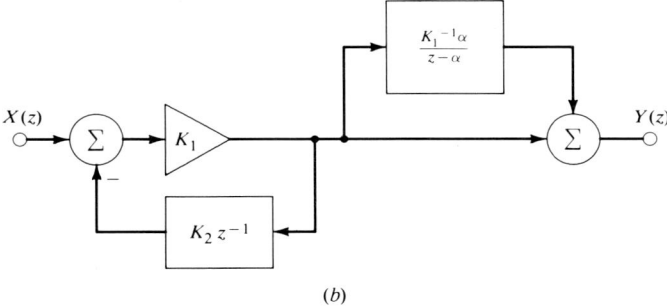

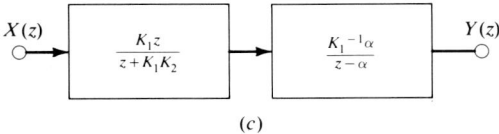

Figure 8-7 Block-diagram reduction applied to the system of Fig. 8-6.

$$H(z) = \frac{0.5z}{(z - 0.5)(z - 0.8)}$$

Step 3 The z transform of the output is

$$Y(z) = H(z)X(z) = \frac{0.5z}{(z - 1)(z - 0.5)(z - 0.8)}$$

Expanding the right side in partial fractions gives

$$Y(z) = \frac{5.00}{z - 1} + \frac{1.67}{z - 0.5} - \frac{6.67}{z - 0.8}$$

It follows that

$$y(n) = [5.00 + (1.67)(0.5)^{n-1} - (6.67)(0.8)^{n-1}]u(n)$$
$$= [5.00 + (3.33)(0.5)^n - (8.33)(0.8)^n]u(n)$$

8-2D Impulse Invariance

A digital system can be designed to emulate an analog system in several ways. One is to make the frequency-dependent gain of the digital system approximate that of the analog system (Sec. 7-6B). Another is to derive a difference equation describing the digital system from the differential equation describing the analog system using finite differences to approximate derivatives (see Prob. 6-31). Still another, described below, is based on a concept called *impulse invariance*.

Figure 8-8 illustrates the concept of impulse invariance. The idea is to design a digital system whose delta response is proportional to a sampled version of the impulse response of a parent analog system. The delta response of the impulse-invariant digital system is given by

$$h_d(n) = A^{-1} h_a(nT)$$

where $h_a(t)$ is the impulse response of the parent analog system, T is the sampling interval to be used, and A is a constant having the dimension of $h_a(t)$. Taking the z transform of this relation gives

$$H_d(z) = A^{-1} \mathscr{L}\{h_a(nT)\} \qquad (8\text{-}11)$$

where $H_d(z)$ is the system function for the impulse-invariant digital system.

Example 8-16 Obtain an impulse-invariant digital system for which the parent analog impulse response is

$$h_a(t) = \frac{1}{\tau} e^{-t/\tau} (\sin \omega_0 t) u(t)$$

with $\tau = 1$ ms and $\omega_0 = 2\pi$ krad/s ($f_0 = 1$ kHz). Use sampling rate $F = 20$ kHz.

SOLUTION The delta response of the impulse-invariant digital system is

$$h_d(n) = \frac{1}{A\tau} e^{-nT/\tau} (\sin \omega_0 nT) u(nT)$$

where $T = 1/F = 50$ μs is the sampling interval. Since the value of the constant $A\tau$ is irrelevant to the present discussion, we take $A\tau = 1$ for convenience; thus

$$h_d(n) = e^{-nT/\tau} (\sin \omega_0 nT) u(n) = c^n (\sin \lambda n) u(n)$$

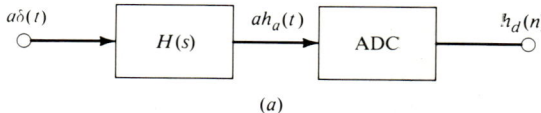

(a)

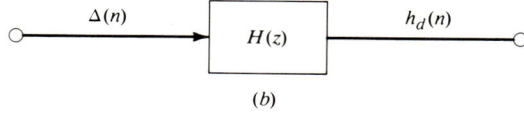

(b)

Figure 8-8 The concept of impulse invariance: (*a*) parent analog system, (*b*) impulse-invariant digital system.

where $c = e^{-T/\tau} = 0.951$ and $\lambda = \omega_0 T = 0.314$. Referring to Table E-1, we find that the z transform of $h_d(n)$ is

$$H(z) = \frac{c(\sin \lambda)z}{z^2 - 2c(\cos \lambda)z + c^2} = \frac{0.294z^{-1}}{1 - 1.809z^{-1} + 0.905z^{-2}}$$

This is the system function of the impulse-invariant digital system. Working backward from this system function [see (8-10)], we obtain the difference equation

$$y(n) - 1.809y(n-1) + 0.905y(n-2) = 0.294x(n-1)$$

describing the system.

8-2E Step Invariance

The idea behind impulse invariance can be generalized to other inputs. A step-invariant digital system is one whose step response is proportional to a sampled version of the step response of a parent analog system; i.e., the step response of a step-invariant digital system is given by

$$g_d(n) = A^{-1}g_a(nT)$$

where $g_a(t)$ is the step response of the parent analog system, T is sampling interval, and A is a constant having the unit of $g_a(t)$. In general, the impulse-invariant and step-invariant digital systems for a common parent analog system are different.

Taking the z transform of the relation above gives

$$\mathcal{Z}\{g_d(n)\} = A^{-1}\mathcal{Z}\{g_a(nT)\}$$

From (8-8),

$$\mathcal{Z}\{g_d(n)\} = \mathcal{Z}\{u(n)\}H_d(z)$$

where $H_d(z)$ is the system function for the step-invariant digital system. From Table E-1,

$$\mathcal{Z}\{u(n)\} = \frac{1}{z-1}$$

Combining the last three relations yields

$$H(z) = A^{-1}(z-1)\mathcal{Z}\{g_a(nT)\} \qquad (8\text{-}12)$$

Example 8-17 Obtain a step-invariant digital system from a parent analog system having step response

$$g_a(t) = (1 - e^{-t/\tau})u(t)$$

where $\tau = 5$ ms. Use sampling rate $F = 50$ kHz.

SOLUTION From (8-12),

$$H_d(z) = A^{-1}(z-1)\mathcal{Z}\{(1 - e^{-nT/\tau})u(nT)\} = (z-1)A^{-1}\mathcal{Z}\{(1 - c^n)u(n)\}$$

where $c = e^{-T/\tau} = 0.996$. Since the value of A is irrelevant to the present discussion, we take $A = 1$ for convenience. Taking the indicated transform gives

$$H_d(z) = (z-1)\left(\frac{1}{z-1} - \frac{c}{z-c}\right) = \frac{(1-c)z}{z-c} = \frac{1-c}{1-cz^{-1}}$$

The difference equation describing the step-invariant digital system is

$$y(n) - cy(n-1) = (1-c)x(n)$$

where $c = 0.996$.

In this section we define the system function of a linear shift-invariant digital system and illustrate using system functions for obtaining outputs of systems. We also show how to obtain the impulse- and step-invariant digital systems from a parent analog system. In the next section we show how to interpret a system function in terms of realizability, stability, and other important properties of digital systems.

8-3 INTERPRETING A SYSTEM FUNCTION

In this section we describe how the system function for a system reflects realizability, stability, fidelity, transient response, steady-state response, and frequency response of the system. The fact that these properties are reflected clearly in a system function accounts for the importance of system functions in analysis and design of linear shift-invariant systems. We begin by introducing some terminology used in describing system functions and associated systems.

8-3A Poles, Zeros, and Natural Modes

The *zeros* of a z transform $F(z)$ are the values of z for which $F(z) = 0$. The *poles* of a z transform are the values of z for which $F(z) = \infty$ [for which $1/F(z) = 0$]. The system function for a linear shift-invariant system is a rational function of z that can be expressed in factored form as

$$H(z) = \frac{(z-z_1)(z-z_2)\cdots(z-z_m)}{(z-p_1)(z-p_2)\cdots(z-p_k)} \tag{8-13}$$

The system function above has m zeros $z_1, z_2, \ldots, z_m$ and k poles $p_1, p_2, \ldots, p_k$. The poles of a system function are the characteristic roots for the system because the denominator of a system function expressed as in (8-13) is the characteristic polynomial for the system. Complex poles and zeros occur in conjugate pairs because both the numerator and the denominator of a realizable rational system function are polynomials in z having real coefficients.

A system function can be represented graphically by a *pole-zero diagram* which shows poles (as crosses) and zeros (as circles) on a z plane. Multiplicities of repeated poles or zeros are written next to the associated crosses or circles.

Example 8-18 Figure 8-9 shows a pole-zero plot for the system function

$$H(z) = \frac{K(z-1)^2}{(z-0.3)(z^2-z+0.5)}$$

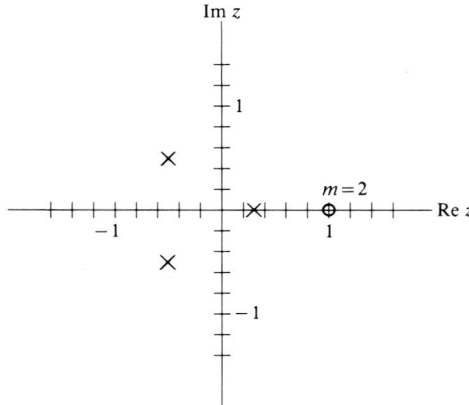

Figure 8-9 Pole-zero diagram for the system of Example 8-18.

The delta response of a system having system function $H(z)$ is given by

$$h(n) = \mathscr{Z}^{-1}\{\mathscr{Z}\{\Delta(n)\}H(z)\} = \mathscr{Z}^{-1}\{H(z)\}$$

because $\mathscr{Z}\{\Delta(n)\} = 1$. Expanding a system function in partial fractions and using the relation above shows that the associated delta response contains a term of the form $p^n u(n)$ for each distinct real pole p and a term of the form $c^n(\cos \lambda n)u(n)$ for each distinct pair $p_1, p_2 = ce^{\pm j\lambda}$ of complex poles. These terms are called the *natural modes* of the associated system.

Example 8-19 The poles of the system of Example 8-18 are $p_1 = 0.3$ and $p_2, p_3 = 0.707 e^{\pm j 3\pi/4}$. The associated natural modes are $(0.3)^n u(n)$ and $(0.707)^n \cos(3n\pi/4)u(n)$.

Figures 8-10 and 8-11 illustrate the relation between the z-plane location of a pole or pole pair and the form of the corresponding natural mode. The natural mode associated with a pole p grows if $|p| > 1$ and decays if $|p| < 1$. The natural mode associated with a nonrepeated pole having unit magnitude neither grows nor decays. The natural mode associated with a repeated pole having unit magnitude grows. Thus the form of the natural mode associated with a pole p depends strongly on the location of the pole relative to the *unit circle* described by $|z| = 1$. The natural mode associated with a pole near the origin decays more rapidly than one associated with a pole near (but inside) the unit circle. The natural mode associated with a pole far outside the unit circle grows more rapidly than one associated with a pole near (but outside) the unit circle. Also, the form of the natural mode associated with a real pole p depends on the sign of p; for example, the natural mode associated with a positive real pole is monotonic whereas the natural mode associated with a negative real pole is oscillatory.

8-3B Stability

A linear shift-invariant system is stable if and only if the delta response $h(n)$ for the system is absolutely summable, i.e., if and only if

$$\sum_{n=-\infty}^{\infty} |h(n)| < \infty$$

538 INTRODUCTION TO SYSTEM ANALYSIS

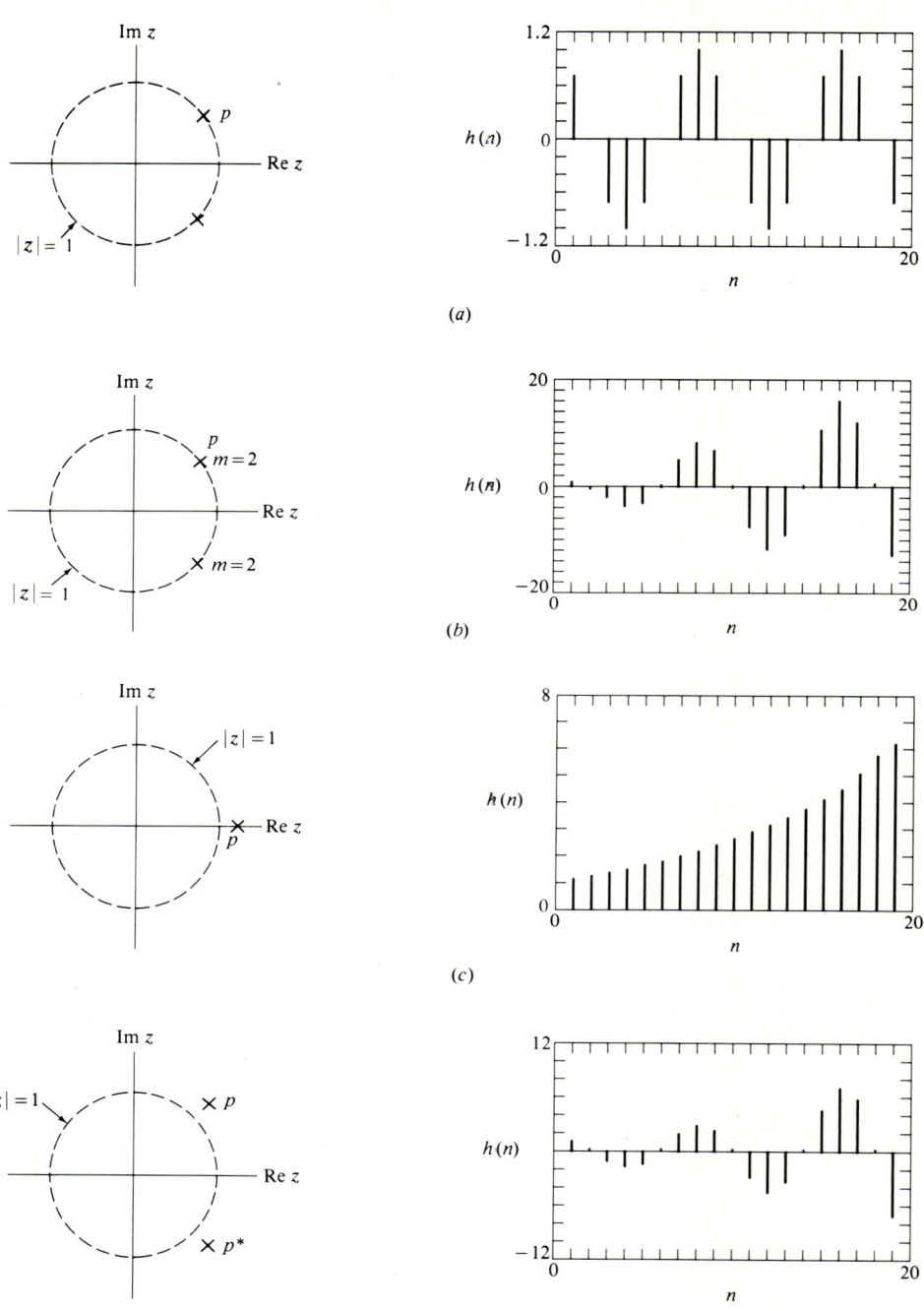

Figure 8-10 Relation between location of a pole and waveform of the associated natural mode: (a) $p = e^{j\pi/4}$, (b) $p_1 = p_2 = e^{j\pi/4}$, (c) $p = 1.1$, (d) $p = 0.8 + 0.8j$.

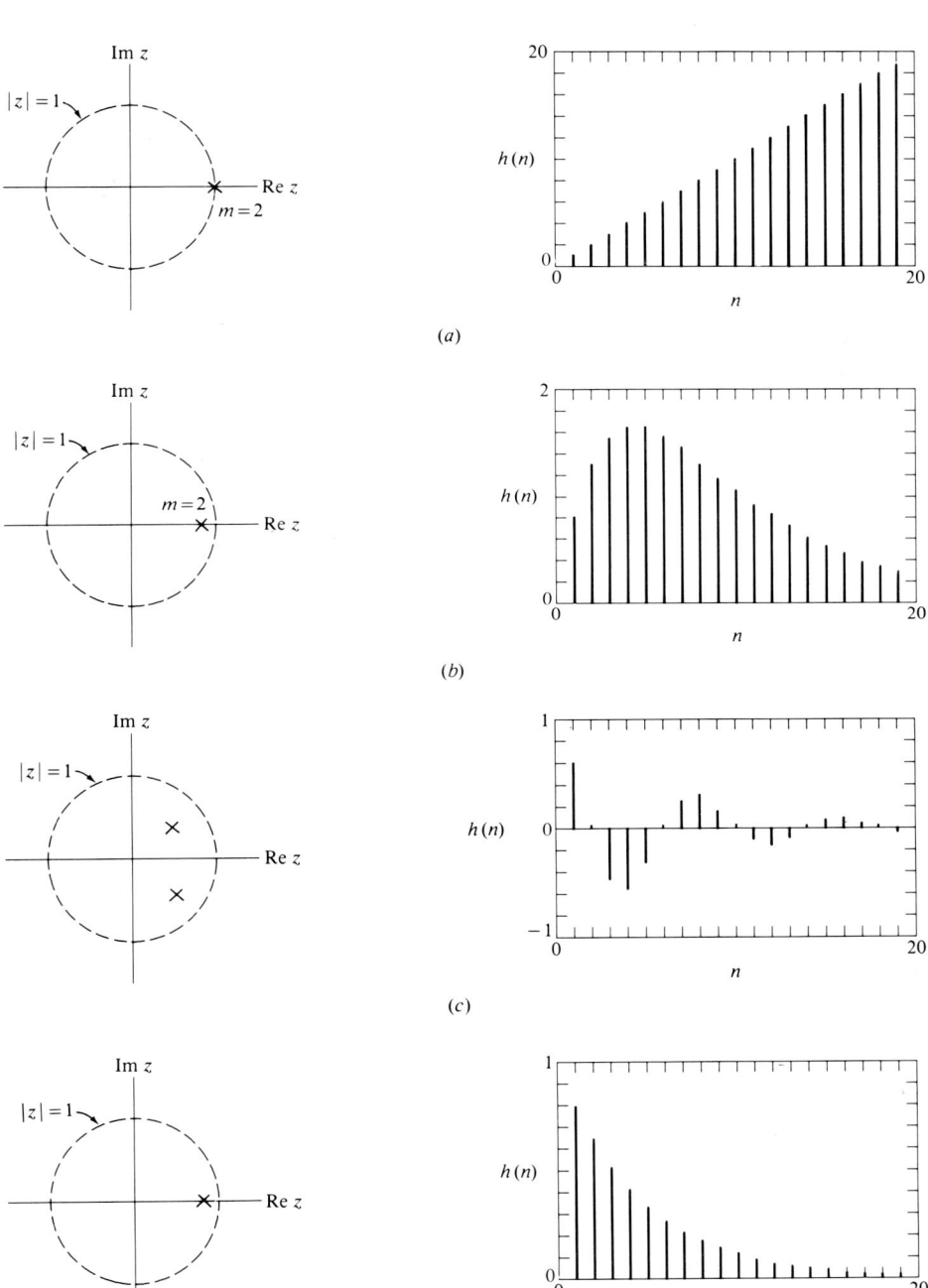

Figure 8-11 Relation between location of a pole and waveform of the associated natural mode: (a) $p_1 = p_2 = 1$, (b) $p_1 = p_2 = 0.8$, (c) $p = 0.6 + 0.6j$, (d) $p = 0.8$.

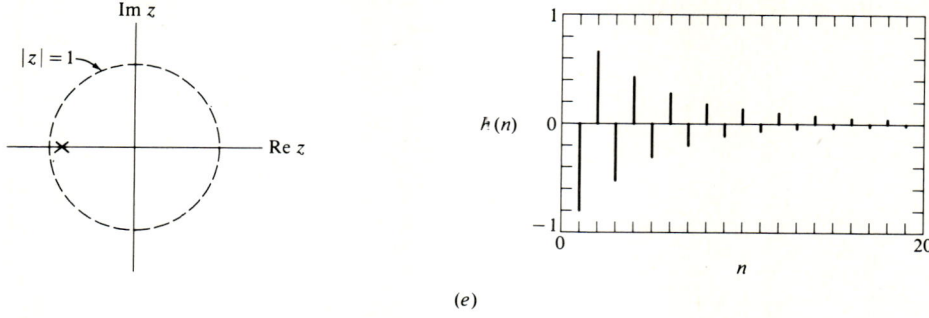

(e)

Figure 8-11 *Continued* (e) $p = -0.8$.

This requires that all natural modes of the system decay, which in turn requires that all poles of the system lie inside the unit circle described by $|z| = 1$.

Example 8-20 Find all values of K for which the system of Fig. 8-12 is stable.

SOLUTION Using block-diagram reduction, we find that the system function for the system is

$$H(z) = \frac{1}{z - K}$$

The system function has one real pole $p = K$. For the system to be stable this pole must lie inside the unit circle described by $|z| = 1$. This requires $|K| < 1$. Therefore the system is stable for $-1 < K < 1$.

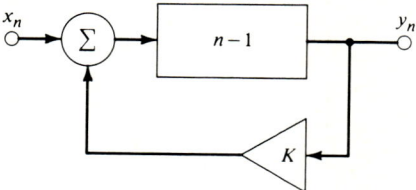

Figure 8-12 System of Example 8-20.

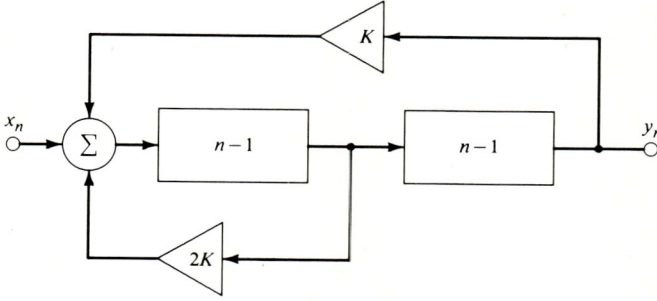

Figure 8-13 System of Example 8-21.

Example 8-21 Find all values of K for which the system of Fig. 8-13 is stable.

SOLUTION The system function for the system is

$$H(z) = \frac{1}{z^2 - 2Kz - K}$$

The poles of the system function are given by

$$p_1 = K + \sqrt{K^2 + K} \qquad p_2 = K - \sqrt{K^2 + K}$$

Note that the poles are complex for $-1 < K < 0$ and that

$$\lim_{K \to \infty} p_1 = \infty \qquad \lim_{K \to -\infty} p_1 = 0 \qquad \lim_{K \to \infty} p_2 = 0, \qquad \lim_{K \to -\infty} p_2 = -\infty$$

because $\sqrt{K^2 + K} \approx K$ for large K. Finding values of K for which the system is stable is facilitated by the root-locus diagrams of Fig. 8-14, which show the z-plane paths followed by the poles as K is increased from $-\infty$ to ∞.

Figure 8-14a shows a root-locus diagram for the pole p_1. For $K = -\infty$, the pole p_1 is at the origin. As K is increased, the pole p_1 moves to the left on the negative real axis until $K = -1$. As K is increased from -1, the pole p_1 becomes complex and moves clockwise on the curved path in the second quadrant until K reaches zero. For $K = 0$, the pole p_1 is again at the origin. As K is increased from zero, the pole p_1 moves to the right on the positive real axis, approaching infinity as K approaches infinity.

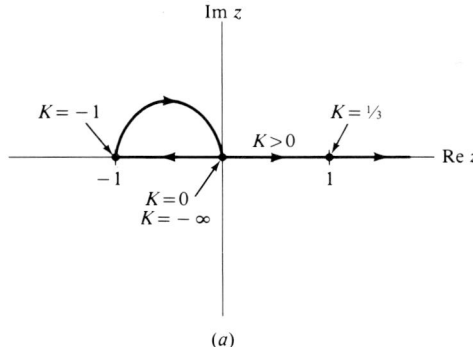

(a)

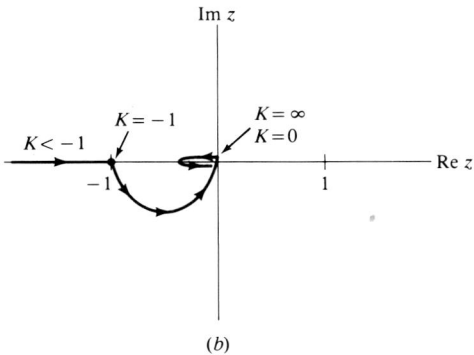

(b)

Figure 8-14 Root-locus diagrams (with parameter K) for the system of Example 8-22: (a) pole $p_1 = K + \sqrt{K^2 + K}$, (b) pole $p_2 = K - \sqrt{K^2 + K}$.

Figure 8-14b shows a root-locus diagram for the pole p_2. For $K = -\infty$, the pole p_2 is at $-\infty$. As K is increased, the pole moves to the right on the negative real axis until $K = -1$. As K is increased from -1, the pole p_2 moves counterclockwise on the curved path in the third quadrant and reaches the origin when $K = 0$. As K is increased from zero, the pole p_2 moves first to the left and then back to the right on the negative real axis, returning to the origin as K approaches infinity.

The root-locus diagrams of Fig. 8-14 show that the system is unstable for $K < -1$ because for $K < -1$ pole p_2 lies outside the unit circle. We need now consider only $K > -1$. First, we consider values $-1 < K < 0$, for which the poles are complex. In that case,

$$p_1, p_2 = K \pm j\sqrt{-K^2 - K}$$

and $|p_1| = |p_2| = |K|$. Therefore the system is stable for $-1 < K < 0$. Next, we find the value of $K > 0$ for which the pole p_1 crosses the unit circle. Setting $p_1 = 1$ gives

$$\sqrt{K^2 + K} = 1 - K$$

which yields $K = \frac{1}{3}$. For $0 < K < \frac{1}{3}$, the pole p_1 lies inside the unit circle. For $K > \frac{1}{3}$, the pole p_1 is outside the unit circle and the system is unstable. Finally, we determine whether the pole p_2 lies outside the unit circle for any $K > 0$. Setting $p_2 = -1$ gives

$$-\sqrt{K^2 + K} = -1 - K$$

which has no positive solution. Therefore, the pole p_2 lies inside the unit circle for all $K > 0$. We conclude from these calculations that the system is stable for $-1 < K < \frac{1}{3}$.

It is relatively easy to determine whether a system is stable if all parameters of the system are fixed. This requires only finding all characteristic roots. It is more difficult to find all values of a single parameter for which a system is stable. This requires that we find all characteristic roots as functions of the parameter. It is much more difficult to find all values of two or more parameters for which a system is stable. This requires using a computer in all but very simple cases.

8-3C Realizability

A rational system function of the form

$$H(z) = \frac{b_0 z^m + b_1 z^{m-1} + \cdots + b_m}{z^k + a_1 z^{k-1} + \cdots + a_k}$$

where $b_0 \neq 0$, is *realizable* if and only if the degree of the numerator does not exceed the degree of the denominator; i.e., the system function $H(z)$ above is realizable if and only if $m \leq k$. Otherwise, the associated delta response $h(n) = \mathcal{L}^{-1}\{H(z)\}$ is nonzero for some $n < 0$. This can be shown by long division. The first term in the quotient represented by $H(z)$ above is $b_0 z^{m-k}$. If $m > k$, then $m - k$ is positive. This implies that the associated delta response $h(n)$ is nonzero for $n = -(m - k) < 0$ [see (8-1)].

Example 8-22 Determine whether

$$H(z) = \frac{5z^2 - 2z}{5z + 4 + 3z^{-1}}$$

is a realizable system function.

SOLUTION We express $H(z)$ above as a ratio of polynomials in z by multiplying both numerator and denominator by z. This gives

$$H(z) = \frac{5z^3 - 2z^2}{5z^2 + 4z + 3}$$

The system function is nonrealizable because the degree of the numerator ($m = 3$) exceeds the degree of the denominator ($k = 2$).

The condition for realizability given above ($m < k$) assumes that the system function in question is the transform of a right-sided delta response, i.e., that the associated region of convergence is the *exterior* of a circle in a z plane. In general, both the condition given above and the region of convergence for a system function must be examined to determine whether a system function is realizable. For example, the system function $H(z) = 1/(z - c)$ having region of convergence $|z| < |c|$ is non-realizable (see Example 8-4) even though the degree of the denominator ($k = 1$) exceeds the degree of the numerator ($m = 0$).

In this book a system function is assumed to be the z transform of a right-sided delta response, and the condition $m < k$ given above is both necessary and sufficient.

8-3D Unforced Response and Forced Response

The output $y(n)$ of a linear shift-invariant system for input $x(n)$ is given by

$$y(n) = \mathscr{L}^{-1}\{H(z)X(z)\}$$

where $H(z)$ is the system function for the system and $X(z)$ is the z transform of the input. Expanding $H(z)X(z)$ in partial fractions allows the terms of $y(n)$ to be separated into two groups: terms arising from poles of $H(z)$ constitute the *unforced response*, and the remaining terms constitute the *forced response*.

Example 8-23 Identify the forced response and the unforced response for a system having system function

$$H(z) = \frac{9}{(z - 0.2)(z - 0.5)}$$

and input $x(n) = (0.2)^n u(n)$.

SOLUTION The z transform of the output is

$$Y(z) = \frac{(9)(0.2)}{(z - 0.2)^2(z - 0.5)}$$

Expanding $Y(z)$ in partial fractions gives

$$Y(z) = -\frac{6}{(z - 0.2)^2} - \frac{20}{z - 0.2} + \frac{20}{z - 0.5}$$

whence

$$y(n) = [-6(n-1)(0.2)^{n-2} - 20(0.2)^{n-1} + 20(0.5)^{n-1}]u(n)$$

The last two terms constitute the unforced response. Terms of this form arise from the poles of the system function no matter what the input. The first term is the forced response. A term of this form does not appear in $\mathscr{L}^{-1}\{H(z)\}$.

If a pole of an input to a system equals a pole of the system function for the system, we say that the input *forces* the system at that pole. For example, the input of Example 8-23 forces the system of Example 8-23 at the pole $p = 0.2$. Forcing a system at one of its poles gives a repeated pole in the z transform of the output. This tends to produce a response that has larger amplitude and longer duration than a response obtained when the system is not forced at one of its poles because a repeated pole gives rise to a term of the form nc^n whereas a simple pole gives rise to a term of the form c^n. This is especially true if a system is forced at a pole lying near the unit circle. Forcing a digital system at a pole can give rise to an unexpectedly large, long-lasting response which can saturate one or more arithmetic registers and cause the system to latch up. Avoiding this condition often requires either very conservative scaling, which leads to loss of precision on smaller signals, or some technique for automatically clearing saturated registers, which disables the system for a time and introduces artificial transients. This problem is especially acute in a system having a pole very near the unit circle.

8-3E Forced Response for Step or Sinusoidal Input

The forced response of a system having system function $H(z)$ and input $x(n) = c^n u(n)$, where c is not a pole of $H(z)$, is given by

$$y_f(n) = H(c)c^n u(n) \tag{8-14}$$

To show this we expand the z transform for the output in partial fractions:

$$Y(z) = H(z)X(z) = \frac{cH(z)}{z-c} = \frac{B}{z-c} + \text{terms arising from poles of } H(z)$$

where, since c is not a pole of $H(z)$,

$$B = (z-c)\frac{cH(z)}{z-c}\bigg|_{z \to c} = cH(c)$$

It follows that the forced response is given by

$$\mathscr{L}^{-1}\left\{\frac{cH(c)}{z-c}\right\} = H(c)c^n u(n)$$

as was to be shown.

To obtain the forced response of a system for a step input we use (8-14) with $c = 1$. This gives

$$y_f(n) = H(1)u(n) \tag{8-15}$$

provided $H(z)$ has no pole at $z = 1$. The quantity $H(1)$ is called the *dc gain* of a system having system function $H(z)$.

To obtain the forced response of a system for a sinusoidal input we use (8-14) and superposition, as follows. A sinusoidal input $x(n) = A(\cos \lambda)u(n)$ can be expressed as

$$A(\cos \lambda n)u(n) = 0.5A(e^{j\lambda n} + e^{-j\lambda n})u(n)$$
$$= 0.5A[c^n + (c*)^n]u(n) \quad \text{where } c = e^{j\lambda}$$

Using (8-14) and superposition, the corresponding forced response of a system having system function $H(z)$ is given by

$$y_f(n) = 0.5A[H(c)c^n + H(c*)(c*)^n]u(n)$$

Because $H(c*) = H*(c)$ and $(c*)^n = (c^n)*$, this expression for $y_f(n)$ reduces to

$$y_f(n) = A \operatorname{Re}[H(c)c^n]u(n) = A \operatorname{Re}[H(e^{j\lambda})e^{j\lambda n}]u(n) \qquad (8\text{-}16)$$

Example 8-24 The system function for a certain system is

$$H(z) = \frac{4}{z^2 - 0.7z + 0.1}$$

Obtain the forced response for input

$$x(n) = (5 + 2 \cos 0.2n)u(n)$$

SOLUTION The poles of the system function $H(z)$ are $p_1 = 0.5$ and $p_2 = 0.2$. Since neither term of the input forces the system at one of these poles, we can use (8-15) and (8-16) to obtain the forced response. The forced response for the term $5u(n)$ is

$$H(1)5u(n) = (10)(5)u(n) = 50u(n)$$

The forced response for the term $2(\cos 0.2n)u(n)$ is

$$2 \operatorname{Re}[H(e^{j0.2})e^{j0.2n}]u(n)$$

where

$$H(e^{j0.2}) = \frac{4}{e^{j0.4} - 0.7e^{j0.2} + 0.1} = 9.564e^{-j0.642}$$

Thus, the forced response for the sinusoidal term is

$$2 \operatorname{Re}(9.564e^{-j0.642}e^{j0.2n})u(n) = 19.129 \cos(0.2n - 0.642)$$

By superposition, the complete forced response is

$$y_f(n) = [50 + 19.129 \cos(0.2n - 0.642)]u(n)$$

8-3F Transient Response and Steady-State Response

An output $y(n)$ of a linear shift-invariant system is given by

$$y(n) = \mathcal{Z}^{-1}\{H(z)X(z)\}$$

where $X(z)$ is the z transform of the input and $H(z)$ is the system function for the system. The transient response is that part of $y(n)$ which decays; it arises from poles of $H(z)X(z)$ lying inside the unit circle described by $|z| = 1$. The steady-state response is what remains after the transient has disappeared; it arises from poles of $H(z)X(z)$ lying on or outside the unit circle. In general (no cancellation), the poles of $H(z)X(z)$ are the poles of $H(z)$ and the poles of $X(z)$; that is, both a system and its input contribute to both the transient and the steady-state components of an output. This means, in general, that both the transient response and the steady-state response have both forced and unforced components.

Example 8-25 Obtain the forced, unforced, transient, and steady-state responses for a system having system function

$$H(z) = \frac{1}{(z-1)(z+0.5)}$$

and input

$$x(n) = \left[(0.2)^n + \cos\frac{n\pi}{4}\right]u(n)$$

SOLUTION The z transform of the input is

$$X(z) = \frac{0.2}{z - 0.2} + \frac{\cos(\pi/4)z - 1}{z^2 - 2\cos(\pi/4)z + 1}$$

The poles of the system function are $p_1 = 1$ and $p_2 = -0.5$. The poles of $X(z)$ are $p_3 = 0.2$ and $p_4, p_5 = e^{\pm j\pi/4}$. Expanding $H(z)X(z)$ in partial fractions shows that the output has the form

$$y(n) = B_1 u(n) + B_2(-0.5)^n u(n) + B_3(0.2)^n u(n) + B_4 \cos\left(\frac{n\pi}{4} + \theta\right)u(n)$$

The first two terms constitute the unforced response because they arise from poles of $H(z)$. The last two terms constitute the forced response. The first and last terms constitute the steady-state response. The second and third terms constitute the transient response. Note that both the transient response and the steady-state response contain both forced and unforced components.

The example above illustrates the fact that, in general, transient response is not the same as unforced response and steady-state response is not the same as forced response. However, in almost every case where it is useful to consider transient response and steady-state response individually the system in question is stable (its poles lie inside the unit circle) and the input is either a step or a sinusoid (its transform has only simple poles on the unit circle). In such a case, the poles of the system contribute only transient components to the output, and the poles associated with the input contribute only steady-state components to the output; i.e., for a stable system with a step or a sinusoidal input, the transient response is the same as the unforced response, and the steady-state response is the same as the forced response. From (8-15), the steady-state response of a stable system for a step input $u(n)$ is given by

$$y_{ss}(n) = H(1)u(n) \tag{8-17}$$

where $H(z)$ is the system function for the system. From (8-16), the steady-state response of a stable system for a sinusoidal input $A(\cos \lambda n)u(n)$ is given by

$$y_{ss}(n) = A\ \text{Re}[H(e^{j\lambda})e^{j\lambda n}] \tag{8-18}$$

Example 8-26 The system function for a certain system is

$$H(z) = \frac{1}{z - 0.5}$$

Find the steady-state output for input

$$x(n) = (\cos 0.4n)u(n)$$

SOLUTION Since the system is stable, we can use (8-18). We have

$$H(e^{j0.4}) = \frac{1}{e^{j0.4} - 0.5} = 1.744 e^{-j0.746}$$

whence

$$y_{ss}(n) = \text{Re}(1.744 e^{-j0.746} e^{j0.4n}) = 1.744 \cos(0.4n - 0.746)$$

8-3G Transfer Function and Frequency Response

From (8-14), the forced response of a linear shift-invariant system for input $e^{j\lambda n}$ (let $c = e^{j\lambda}$) is given by

$$y_f(n) = H(e^{j\lambda}) e^{j\lambda n} u(n)$$

where $H(z)$ is the system function for the system. If the system is stable, the steady-state response for input $e^{j\lambda n}$ equals the forced response; thus

$$y_{ss}(n) = H(e^{j\lambda}) e^{j\lambda n}$$

where the step $u(n)$ is omitted because $u(n) = 1$ in steady state. This shows that the transfer function of a stable system is the system function for the system with z replaced by $e^{j\lambda}$; that is,

$$H_d(j\lambda) = H(e^{j\lambda}) \tag{8-19}$$

This relation gives the frequency response of a digital system as a function of digital frequency λ.

Example 8-27 Figure 8-15 shows a digital filter. The system function for the digital processor is

$$H(z) = \frac{z - 1}{z^2 - 1.273z + 0.81}$$

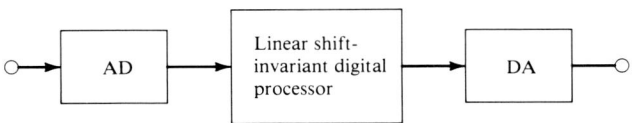

Figure 8-15 Digital filter of Example 8-27.

548 INTRODUCTION TO SYSTEM ANALYSIS

The sampling rate is $F = 10$ kHz, and the AD and DA converters are assumed to be error-free. Obtain the gain and phase shift of the digital filter as functions of (analog) frequency f.

SOLUTION The poles of the system function $H(z)$ are

$$p_1, p_2 = 0.9 e^{\pm j\pi/4}$$

Both poles lie inside the unit circle described by $|z| = 1$. The system is stable, and the transfer function for the digital processor is

$$H_d(j\lambda) = H(e^{j\lambda}) = \frac{e^{j\lambda} - 1}{e^{j2\lambda} - 1.273 e^{j\lambda} + 0.81}$$

Assuming error-free AD and DA conversion, the overall transfer function for the system is

$$H_a(j\omega) = H_d(j\omega T) = \frac{e^{j\omega T} - 1}{e^{j 2\omega T} - 1.273 e^{j\omega T} + 0.81}$$

This can be written

$$H_a(j\omega) = \frac{(C_1 - 1) + j S_1}{(C_2 - 1.273 C_1 - 0.81) + j(S_2 - 1.273 S_1)}$$

where $C_2 = \cos 2\omega T$, $S_2 = \sin 2\omega T$, $C_1 = \cos \omega T$, and $S_1 = \sin \omega T$. The gain of the system is given by

$$\Gamma_a(\omega) = \frac{\sqrt{(C_1 - 1)^2 + S_1^2}}{\sqrt{(C_2 - 1.273 C_1 + 0.81)^2 + (S_2 - 1.273 S_1)^2}}$$

and the phase shift of the system is given by

$$\phi_a(\omega) = \text{Tan}^{-1}(S_1, C_1 - 1) - \text{Tan}^{-1}(S_2 - 1.273 S_1, C_2 - 1.273 C_1 + 0.81)$$

Figure 8-16 shows graphs of gain and phase shift versus frequency f for $0 \leq f \leq F/2$. For error-free AD and DA conversion, the gain is zero (and the phase shift is irrelevant) for $f > F/2$.

Since the system function $H(z)$ has a zero at $z = 1$, the dc gain $H(1)$ equals zero. The gain has a sharp peak between $0.1F$ and $0.15F$ (between 1.0 and 1.5 kHz), indicating ringing (lightly damped oscillation) in the step response of the system. The step response can be obtained by recursion from a difference equation describing the system. The system function $H(z)$ shows that the digital processor is described by the difference equation

$$y(n) - 1.273 y(n - 1) + 0.81 y(n - 2) = x(n - 1) - x(n - 2)$$

where $y(n)$ is the output for input $x(n)$. It follows that the step response is given by

$$g(n) = u(n - 1) - u(n - 2) + 1.273 g(n - 1) - 0.81 g(n - 2)$$

where $g(n) = 0$ for $n \leq 0$. Figure 8-17 shows a graph of the step response versus sample number. The steady-state amplitude of the step response is zero because the dc gain of the system is zero. The ringing indicated by the peak in the graph of Fig. 8-16a is evident.

The magnitude of the first jump discontinuity (in the passband) of the phase shift equals π rad, indicating sign reversal (also called phase reversal) of a sinusoidal output as the frequency of the input is increased from a value slightly below the discontinuity to a value slightly above the discontinuity. The second jump discontinuity indicates the same thing, but it is in the stopband and is of less interest than the first.

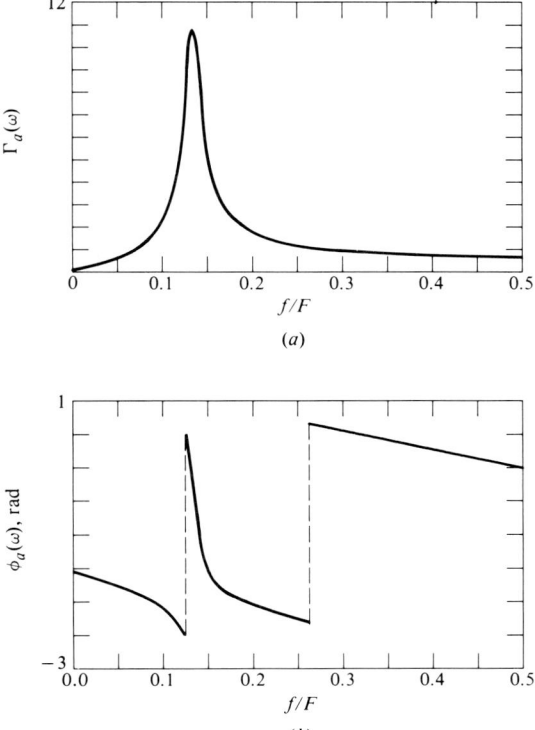

Figure 8-16 (a) Gain and (b) phase shift for the system of Fig. 8-15.

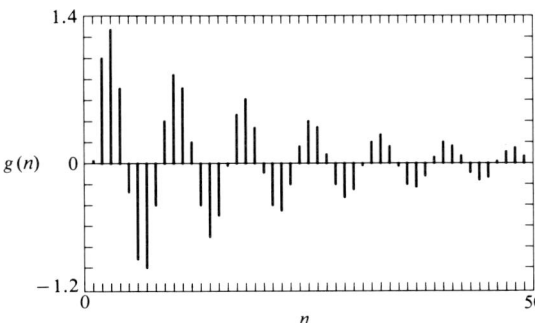

Figure 8-17 Step response for the system of Fig. 8-15.

To conclude this section we illustrate a cut-and-try method for digital-filter design. The method entails choosing the poles and zeros of a digital processor (as in Fig. 8-15), calculating frequency response, adjusting (and perhaps adding or deleting) poles and zeros, calculating frequency response again, and so on, until a satisfactory frequency response is obtained. Using the method effectively requires insight into how the poles and zeros of a system function determine the associated frequency response. Before illustrating the method we describe relations between a pole-zero plot and the associated graph of gain versus frequency.

The transfer function for the system of Fig. 8-15 is given by

$$H_a(j\omega) = H_d(j\omega T) = H(e^{j\omega T})$$

where $H(z)$ = the system function for the digital processor
$H_d(j\lambda)$ = transfer function for the digital processor
T = sampling interval employed by the system

The relation between the first and last members can be written

$$H_a(j2\pi f) = H(z)\Big|_{z=\exp(j2\pi f/F)}$$

where $F = 1/T$ is the sampling rate for the system. This shows that the frequency response of the system of Fig. 8-15 is determined by values of the system function $H(z)$ on the unit circle described by $z = e^{j2\pi f/F}$ in a z-plane description of the digital processor. In particular, the gain of the system is given by

$$\Gamma_a(2\pi f) = |H(e^{j2\pi f/F})|$$

As frequency f is increased from $f = 0$ to $f = F/2$, the point represented by $z = e^{j2\pi f/F}$ moves counterclockwise on the unit circle from $z = e^{j0} = 1$ to $z = e^{j\pi} = -1$. We can visualize the transfer function of the system of Fig. 8-15 (for $0 \leq f \leq F/2$) as a semicircular slice (from $z = 1$ to $z = -1$) of the system

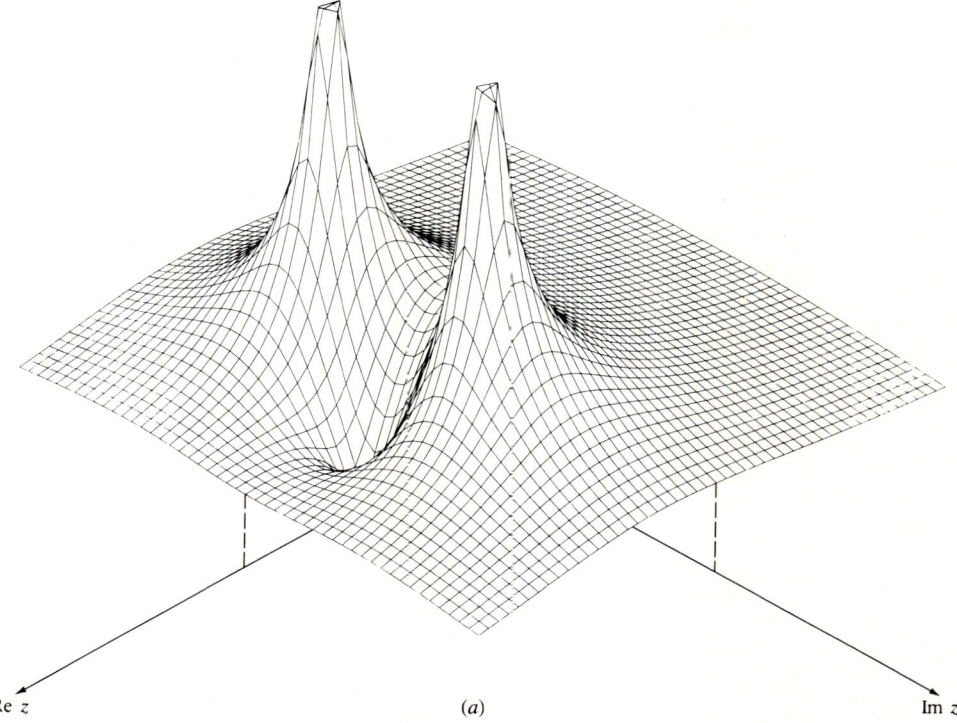

Figure 8-18 Graphs of $H(z)$ versus z for the system function of Example 8-27. In (b) the graph is truncated on the unit circle described by $|z| = 1$.

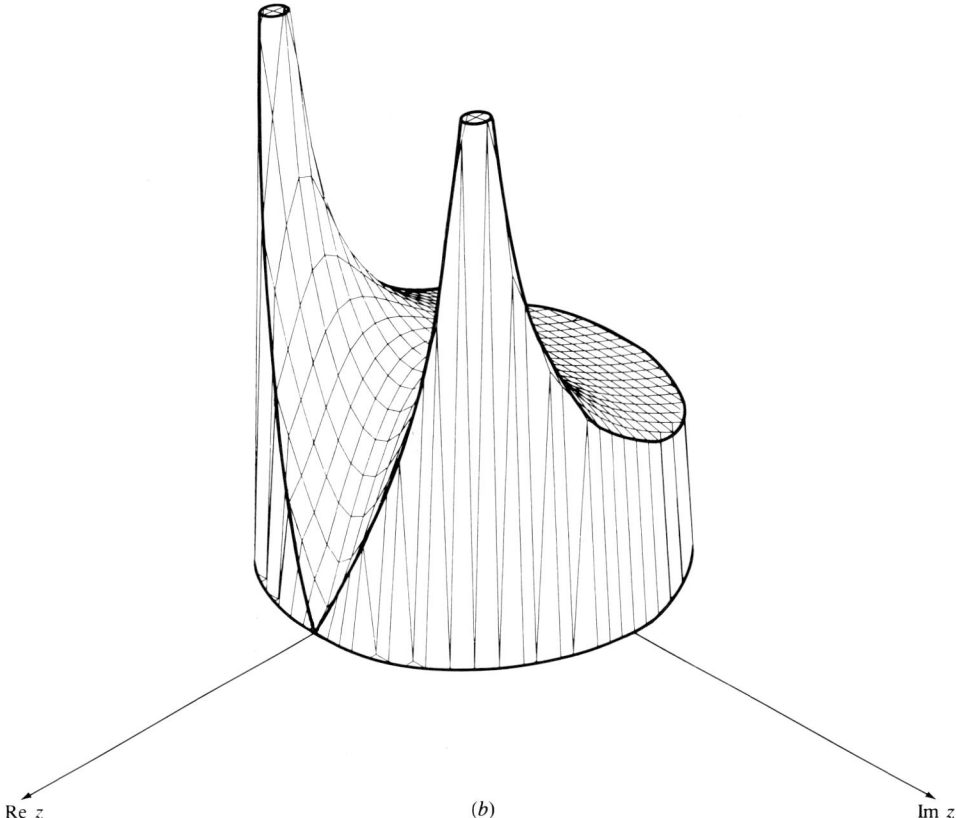

Re z   (b)   Im z

Figure 8-18 *Continued*

function for the digital processor. Figure 8-18 illustrates this for the system of Example 8-27, where

$$H(z) = \frac{z - 1}{z^2 + 1.273z + 0.81}$$

The system function $H(z)$ has a zero at $z = 1$ and poles at $z = 0.9e^{\pm j\pi/4}$. Figure 8-18a shows a graph of the magnitude of $H(z)$ versus z. Figure 8-18b shows the same graph truncated on the unit circle described by $|z| = 1$. The magnitude of $H(z)$ is small for values of z near the zero of $H(z)$ and is large for values of z near the poles of $H(z)$. If we take a thin vertical peeling along the unit circle from $z = 1$ counterclockwise to $z = -1$ and lay the peeling flat, we obtain the graph of gain versus frequency given in Fig. 8-16, where $z = 1$ corresponds to $f = 0$ and $z = -1$ corresponds to $f = F/2$.

Figures 8-19 and 8-20 further illustrate how gain is determined by z-plane locations of poles and zeros of the digital processor. Figure 8-19 shows pole-zero diagrams and corresponding graphs of gain versus frequency for a case where the digital processor has a single pair of complex-conjugate poles and no (finite) zeros. Figures 8-19a to d illustrate the relation between angular position of the poles and the frequency of

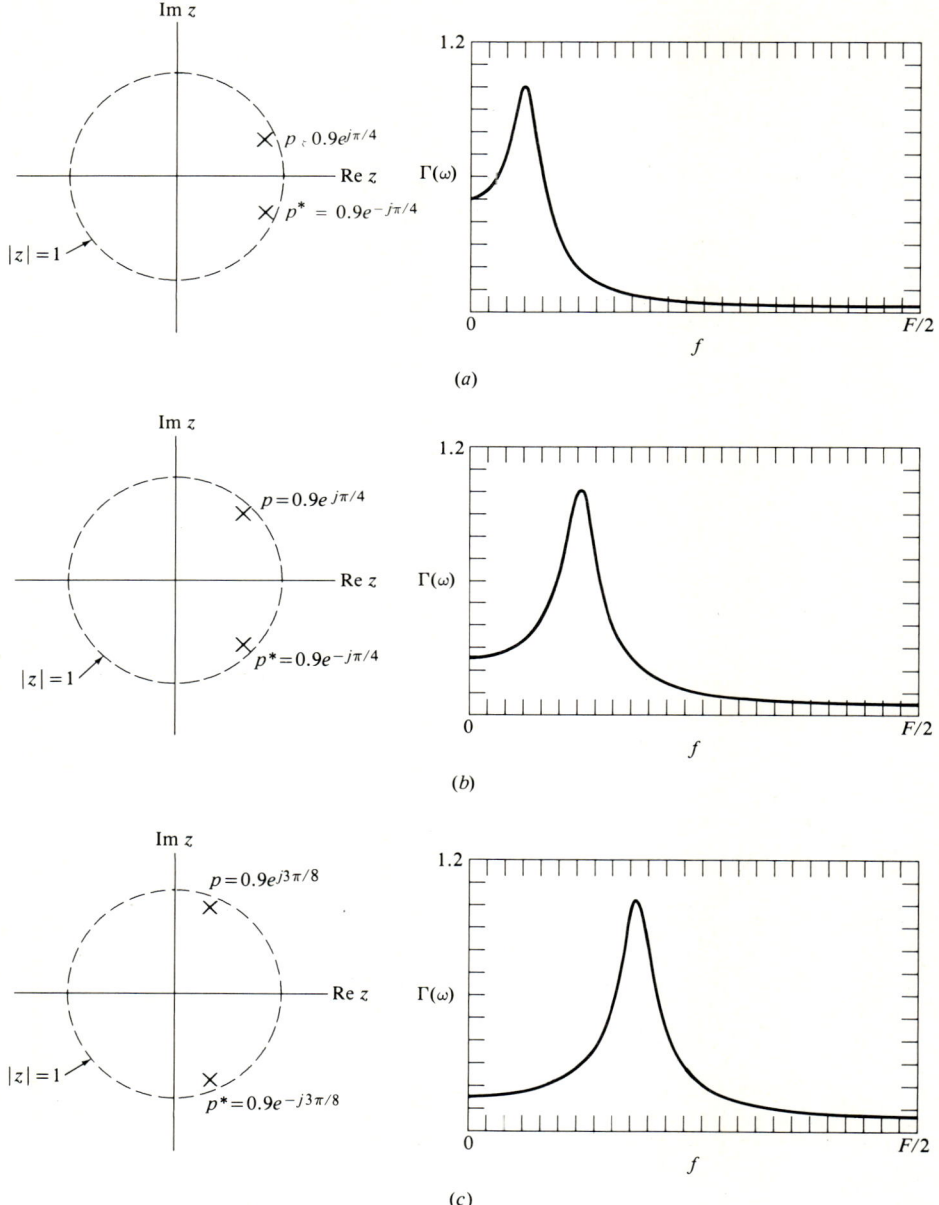

Figure 8-19 Influence of poles on gain.

maximum gain. Moving a pole counterclockwise moves the associated gain peak toward higher frequencies. Figures 8-19d to f illustrate the relation between radial position of the poles and the sharpness of the peak in overall frequency response. Moving a pole radially inward (outward) flattens (sharpens) the associated gain peak. Figure 8-20 illustrates how a zero on the unit circle affects gain. A zero on the unit

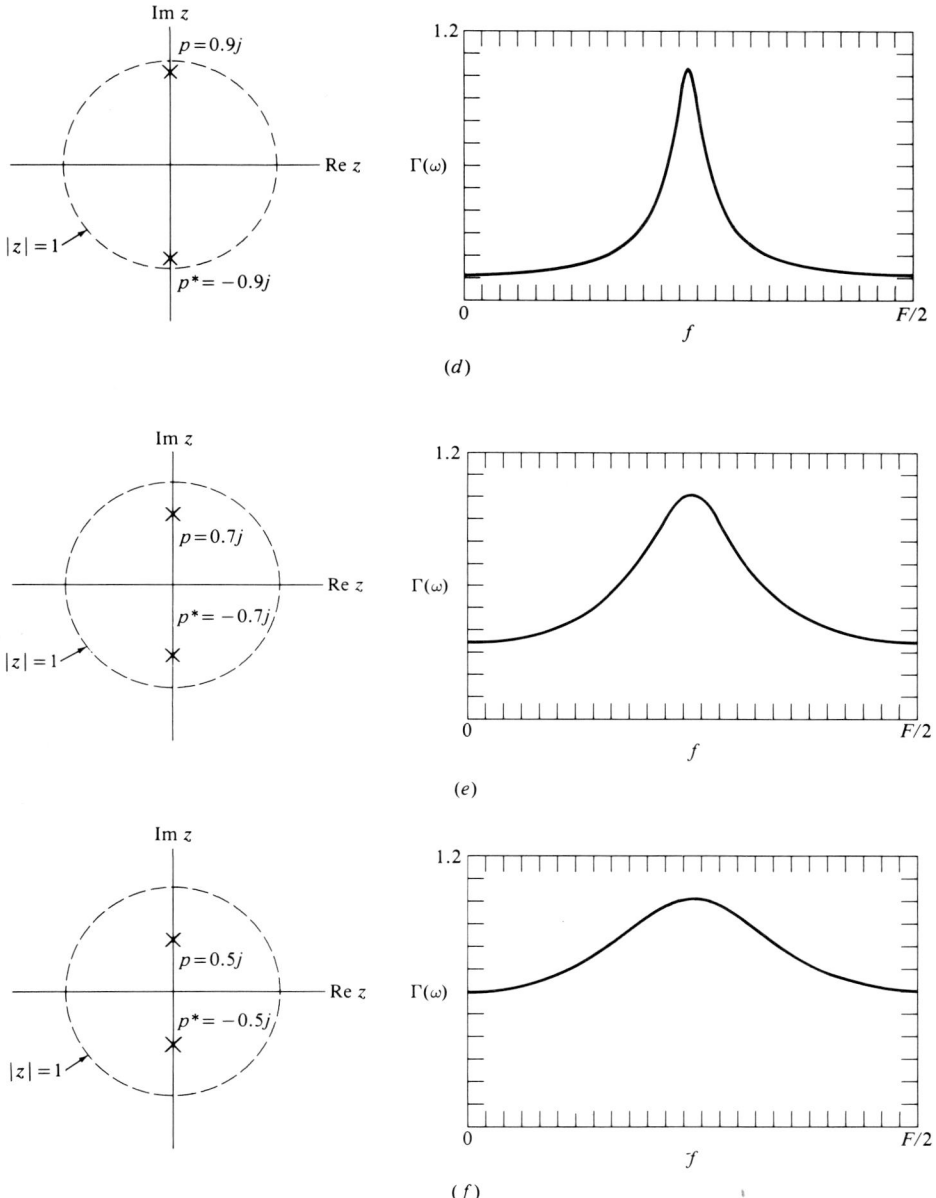

Figure 8-19 *Continued*

circle forces the gain to zero at the corresponding frequency; thus, the zero at $z = 1$ forces the dc gain to zero, and the zero at $z = -1$ forces the gain to zero for $f = F/2$.

Example 8-28 We wish to design a fourth-order (four-pole) bandpass digital filter having bandwidth $B = 3$ kHz, center frequency $f_c = 10$ kHz, and sampling rate $F = 50$ kHz.

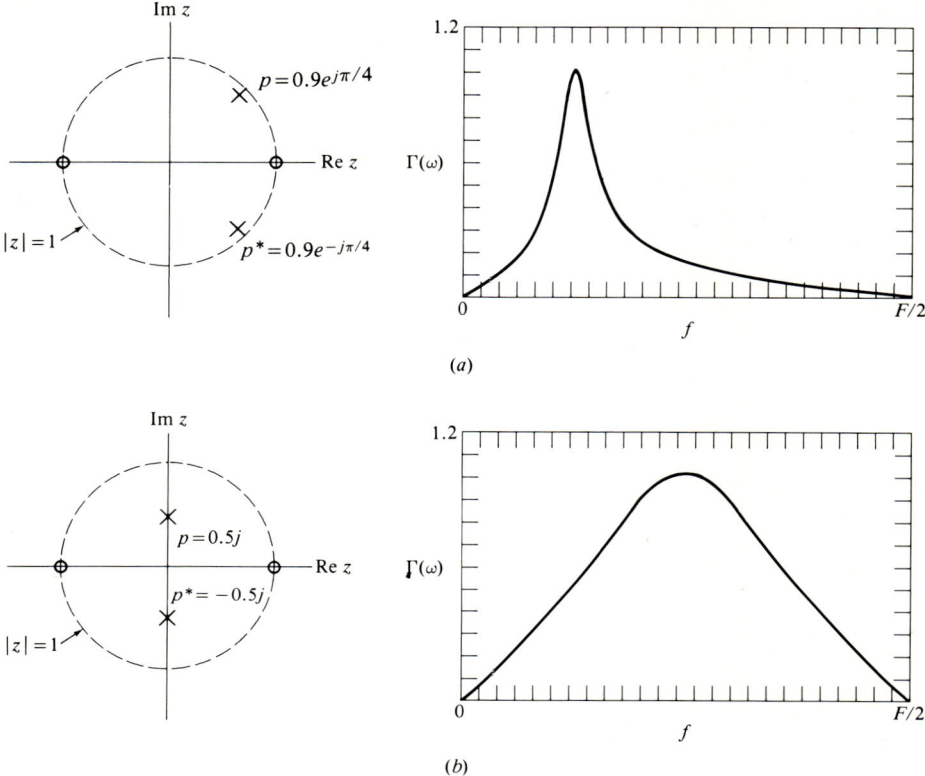

Figure 8-20 Influence of zeros on gain.

We begin by finding the arc on the unit circle (in the z plane) that corresponds to the specified passband on the frequency axis, as illustrated in Fig. 8-21. Angles corresponding to the upper and lower edges of the specified passband are

$$\frac{2\pi(f_c \pm B/2)}{F} = 1.257 \pm 0.188 \text{ rad}$$

These angles define the endpoints of the arc. The magnitude of the system function for the digital processor should be relatively large for values of z on or near the arc and relatively small elsewhere (on the upper semicircle). This suggests that two poles should be placed in the shaded region shown in Fig. 8-21a. The other two poles must lie at conjugate positions in the lower half of the plane.

Next, we choose first-cut values for the two upper-half-plane poles. To make the passband fairly sharp we should place the poles relatively near the unit circle. To account for the nonzero width of the peak associated with each pole we should place the poles inside (not on) the radial boundaries of the shaded region of Fig. 8-21a. To get started we choose

$$p_1 = 0.9e^{j1.131} \qquad p_2 = 0.9e^{j1.383}$$

as illustrated in Fig. 8-21b. The angle 1.131 rad corresponds to frequency 9 kHz, and the angle 1.383 rad corresponds to frequency 11 kHz. Because complex poles of a realizable

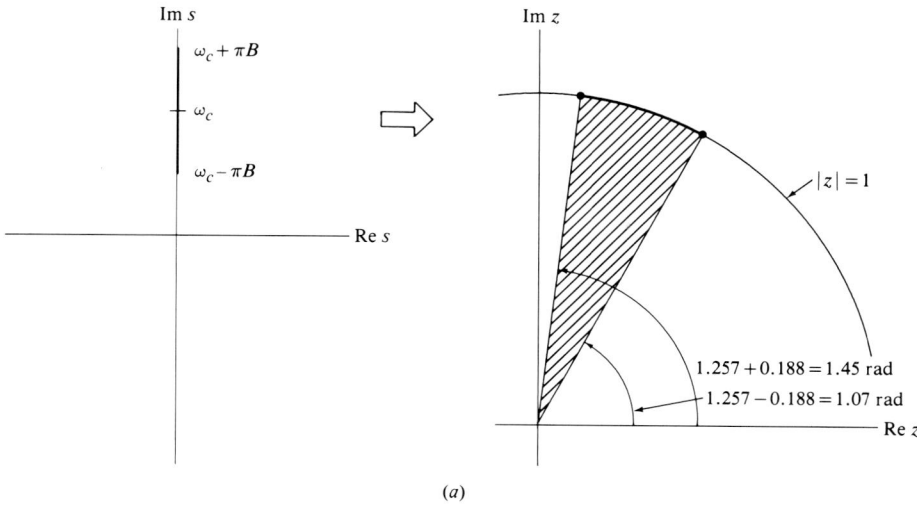

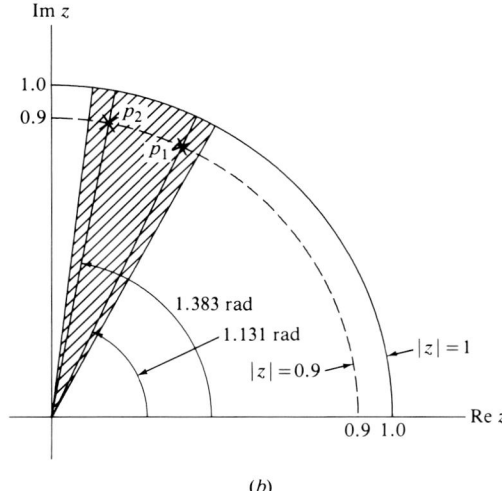

Figure 8-21 Cut-and-try design of a four-pole bandpass digital filter: (*a*) mapping the passband of the filter of Example 8-28 onto an arc on the unit circle in the z plane; (*b*) choosing the poles of the first-cut filter.

digital processor occur in conjugate pairs, we also have poles

$$p_3 = p_1^* = 0.9e^{-j1.131} \qquad p_4 = p_2^* = 0.9e^{-j1.383}$$

in the lower half of the z plane. Generally, we want a bandpass filter to have zero gain for $f = 0$ and $f = F/2$; otherwise, stopband gain tends to be too large. In order to force the gain to zero for $f = 0$ and $f = F/2$ we place zeros at $z = \pm 1$. Figure 8-22 shows a pole-zero diagram for the first cut. The corresponding (first-cut) system function for the digital processor is

$$H(z) = \frac{K(z-1)(z+1)}{(z-p_1)(z-p_2)(z-p_3)(z-p_4)}$$

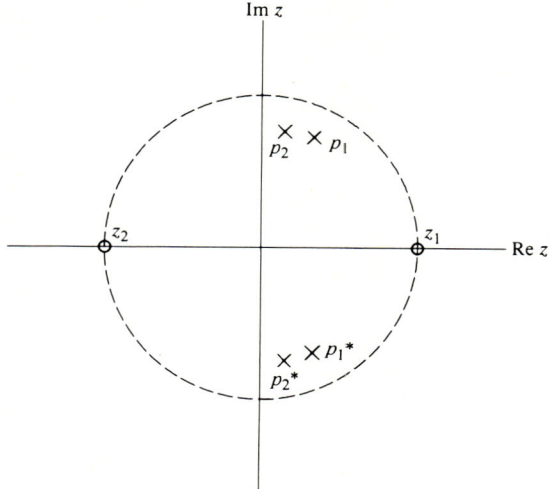

Figure 8-22 Pole-zero diagram for the first-cut filter of Example 8-28.

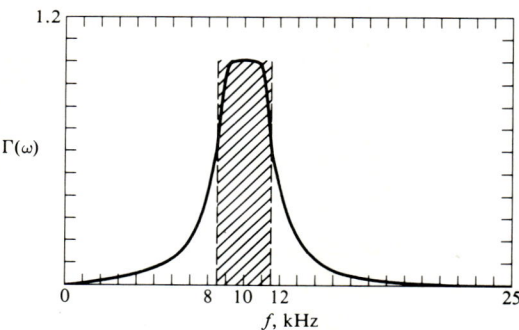

Figure 8-23 Gain versus frequency for the first-cut filter of Example 8-28.

where K is an arbitrary constant. This gives

$$H(z) = \frac{K(z^2 - 1)}{[z^2 - 2(\text{Re } p_1)z + |p_1^2|][z^2 - 2(\text{Re } p_2)z + |p_2^2|]}$$

$$= \frac{K(z^2 - 1)}{(z^2 - 0.766z + 0.810)(z^2 - 0.336z + 0.810)}$$

Figure 8-23 shows a graph of gain (normalized to unit maximum gain) for the first-cut system function given above. The shaded rectangle represents the gain of an ideal filter having the properties specified in the statement of the problem. The first-cut filter has the specified center frequency and approximately the specified bandwidth. The design could be improved (in a second cut), either by moving the poles slightly closer together (on a circle) or by moving the poles slightly outward (radially), or by a combination of these adjustments. The first cut described above is enough to illustrate the ideas involved. We describe a more scientific method for designing digital filters in the next section.

8-4 DESIGN OF RECURSIVE DIGITAL FILTERS BY BILINEAR SUBSTITUTION

In this section we describe a method for designing a recursive digital filter whose gain approximates a specified function of frequency. Since our purpose is to illustrate how concepts and procedures described above arise in a practical application, we do not attempt to treat digital-filter design in depth.

Methods for designing digital filters fall into two groups. One group consists of so-called *direct methods,* in which the system function for a digital filter is determined directly from specifications on frequency response. The cut-and-try method illustrated in the previous section is a direct method. The other group consists of so-called *indirect methods,* in which a digital filter is designed by first designing an analog filter and then using a substitution $s \to f(z)$ to obtain the system function for the digital filter. The method described in this section is an indirect method called *bilinear substitution.*

8-4A Bilinear Substitution

Designing a digital filter using bilinear substitution entails first designing an analog filter subject to certain specifications and then replacing s using

$$s = \frac{2F(z-1)}{z+1} \tag{8-20}$$

in the system function $H'(s)$ of the analog filter, where F is the sampling rate to be employed. Thus, the system function $H(z)$ for the digital filter is given by

$$H(z) = H'\left[\frac{2F(z-1)}{z+1}\right]$$

This is called a bilinear substitution because the quantity on the right side of (8-20) is a ratio of two linear functions of z. The bilinear substitution of (8-20) is often referred to as the bilinear z transformation, probably because it "transforms" an analog system function into a digital system function; however, it is not a transformation in the sense that the ordinary z transformation is a transformation; e.g., one cannot take the bilinear z transform of a digital signal. Consequently, we prefer the term bilinear substitution. Below, we give a heuristic derivation of the bilinear substitution (8-20) and describe some of its important properties.

An analog system function represents a differential equation relating two signals (an input and an output). Each appearance of the variable s in the analog system function represents differentiation of one signal or the other. We can transform an analog system function into a more or less equivalent digital system function by substituting an appropriate function of z (an appropriate digital operation) for the variable (differentiation operator) s in the analog system function. Many different substitutions are possible because there are many ways to define a digital equivalent of differentiation. The bilinear substitution expressed by (8-20) can be justified as follows. Because multiplication of the Laplace transform of a signal by s corresponds to differentiating the signal with respect to time, we consider

$$y_a(t) = \frac{dx_a(t)}{dt}$$

where $x_a(t)$ and $y_a(t)$ denote analog signals. Integrating this relation gives

$$\int_{-\infty}^{t} y_a(\alpha)\, d\alpha = x_a(t)$$

which can be written

$$\int_{-\infty}^{t-T} y_a(\alpha)\, d\alpha + \int_{t-T}^{t} y_a(\alpha)\, d\alpha = x_a(t)$$

or

$$x_a(t - T) + \int_{t-T}^{t} y_a(\alpha)\, d\alpha = x_a(t)$$

Using a trapezoidal approximation for the last integral gives

$$x_a(t - T) + 0.5T[y_a(t) + y_a(t - T)] = x_a(t)$$

where we assume that T is a suitably small sampling interval. The digital counterpart of this relation is

$$x_d(n - 1) + 0.5T[y_d(n) + y_d(n - 1)] = x_d(n)$$

Taking the z transform of the last relation and solving for $Y_d(z)$ yields

$$Y_d(z) = \frac{2(1 - z^{-1})}{T(1 + z^{-1})} X_d(z) = \frac{2F(z - 1)}{z + 1} X_d(z)$$

where $F = 1/T$. This result implies that multiplying the Laplace transform of an analog signal by s corresponds to multiplying the z transform of the associated digital signal by $2F(z - 1)/(z + 1)$. Consequently, using the substitution (8-20) to replace s in the system function of an analog system gives the system function of a digital system that is in some sense equivalent to the analog system. Note that other substitutions could be obtained by using other approximations (quadratic, cubic, etc.) for the integral in the derivation above.

The bilinear substitution constitutes a mapping from an s plane to a z plane. Rewriting (8-20) so that z is expressed as a function of s shows that a point s is mapped to a point z given by

$$z = \frac{2F + s}{2F - s} \tag{8-21}$$

For example, $s = 0$ maps to $z = 1$, and $s = j$ maps to $z = (2F + j)/(2F - j)$. One important property of this mapping is that points on the ω axis in an s plane are mapped onto the unit circle described by $|z| = 1$ in a z plane. To show this, let $s = j\omega$ (where ω is real) in (8-21); thus

$$|z| = \frac{|2F + j\omega|}{|2F - j\omega|} = \frac{\sqrt{(2F)^2 + \omega^2}}{\sqrt{(2F)^2 + \omega^2}} = 1$$

This means that the frequency response of a digital system obtained by bilinear substitution is in one-to-one correspondence with the frequency response of the parent analog system. This correspondence is described in greater detail below.

Another important property of the bilinear substitution is that points, e.g., poles, in the left half of an s plane are mapped onto the interior of the unit circle in a z plane. To show this, let $s = -\sigma + j\omega$, where σ and ω are real and $\sigma > 0$. Then

$$|z| = \frac{|2F - \sigma + j\omega|}{|2F + \sigma - j\omega|} = \frac{\sqrt{(2F - \sigma)^2 + \omega^2}}{\sqrt{(2F + \sigma)^2 + \omega^2}} < 1$$

This means that a digital system obtained by bilinear substitution is stable if the parent analog system is stable.

Since frequency response is of utmost importance in filter design, we wish to describe in more detail how points on the ω axis of an s plane are mapped to points on the unit circle in a z plane by bilinear substitution. We suppose that a digital system function $H(z)$ is obtained from an analog system function $H'(s)$ using the bilinear substitution (8-20). Then the system function for the digital system is given by

$$H(z) = H'\left[\frac{2F(z-1)}{z+1}\right]$$

and the transfer function of the digital system is given by

$$H(e^{j\omega T}) = H'\left[\frac{2F(e^{j\omega T} - 1)}{e^{j\omega T} + 1}\right]$$

To calculate gain (or phase shift) of the parent analog system for frequency f we replace s by $j2\pi f = j\omega$ in the system function of the analog system. To calculate gain (or phase shift) of the digital system for frequency ω we replace s by the quantity

$$j\omega' = \frac{2F(e^{j\omega T} - 1)}{e^{j\omega T} + 1}$$

in the system function for the analog system. The quantity $j\omega'$ defined above can be expressed as

$$j\omega' = \frac{2F[e^{j\omega T/2} - e^{-j\omega T/2}]}{e^{j\omega T/2} + e^{-j\omega T/2}} = \frac{2F[2j \sin(\omega T/2)]}{2 \cos(\omega T/2)} = j2F \tan \frac{\pi f}{F}$$

This shows that the frequency response (gain and phase shift) of the digital system for frequency f equals the frequency response of the analog system for frequency f' given by

$$f' = \frac{F}{\pi} \tan \frac{\pi f}{F} \tag{8-22}$$

In other words, the frequency response of the analog system for frequency f' is mapped (by the bilinear substitution) to frequency f given by

$$f = \frac{F}{\pi} \tan^{-1} \frac{\pi f'}{F} \tag{8-23}$$

560 INTRODUCTION TO SYSTEM ANALYSIS

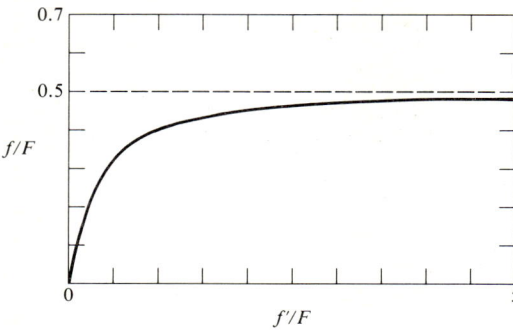

Figure 8-24 Graph of the compression relation (8-23) for bilinear substitution.

Figure 8-24 shows a graph of this relation. As f' ranges from zero to infinity, f ranges from zero to $F/2$; thus the entire frequency response (for $0 \leq f' \leq \infty$) of a parent analog system is squeezed into the band $0 \leq f \leq F/2$. This means that any particular point on a graph of gain (or phase shift) versus frequency for the parent analog system appears at a lower frequency given by (8-23) in a graph of gain (or phase shift) versus frequency for the digital system. Consequently, the frequency response of a digital system obtained by bilinear substitution is a compressed version of the frequency response of the parent analog system.

Example 8-29 This example illustrates bilinear substitution and its consequences. Figure 8-25 shows the systems considered in this example. The system function for the parent analog system is

$$H'(s) = \frac{\omega_0^2}{s^2 + 2\zeta\omega_0 s + \omega_0^2}$$

where $\zeta = 0.1$ and $\omega_0 = \pi$ krad/s. The system function for the digital processor in the system of Fig. 8-25 is obtained from $H'(s)$ above using the bilinear substitution

$$s = \frac{2F(z-1)}{z+1}$$

with sampling rate $F = 2$ kHz. This gives

$$H(z) = H'\left[\frac{2F(z-1)}{z+1}\right] = \frac{z^2 + 2z + 1}{az^2 + bz + c}$$

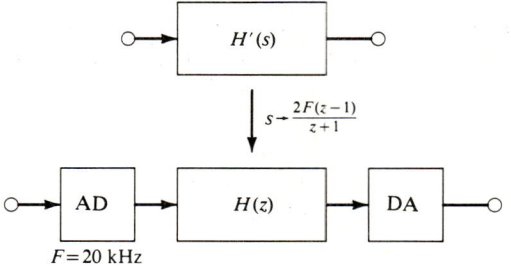

$F = 20$ kHz

Figure 8-25 Systems of Example 8-29.

where $a = (4F^2 + 4\zeta\omega_0 F + \omega_0^2)/\omega_0^2 = 2.88$
$b = (2\omega_0^2 - 8F^2)/\omega_0^2 = -1.24$
$c = (4F^2 - 4\zeta\omega_0 F + \omega_0^2)/\omega_0^2 = 2.37$

To illustrate compression caused by bilinear substitution we compare the gain of the digital system with that of the parent analog system. The gain of the latter is

$$\Gamma'(\omega) = |H'(j\omega)| = \left| \frac{\omega_0^2}{\omega_0^2 - \omega^2 + 2j\zeta\omega_0\omega} \right|$$

The gain of the digital system is

$$\Gamma_d(\omega) = |H_d(j\omega)| = |H(e^{j\omega T})| = \left| H'\left(2jF \tan \frac{\pi f}{F}\right) \right|$$

Figure 8-26 shows graphs of gain versus frequency for the parent analog system and the digital system. The effect of bilinear substitution is to compress the gain of the analog system so that it fits in the band $0 \le f \le F/2$. The compression is nonlinear (see Fig. 8-24), being greater for high frequencies than for low frequencies. For frequencies much smaller than the sampling rate, the gain of the digital system is nearly the same as that of the parent analog system. For larger frequencies, values of gain for the digital system appear at lower frequencies than corresponding values of gain for the parent analog system; e.g., maximum gain for the digital system occurs for $f \approx 420$ Hz whereas maximum gain for the parent

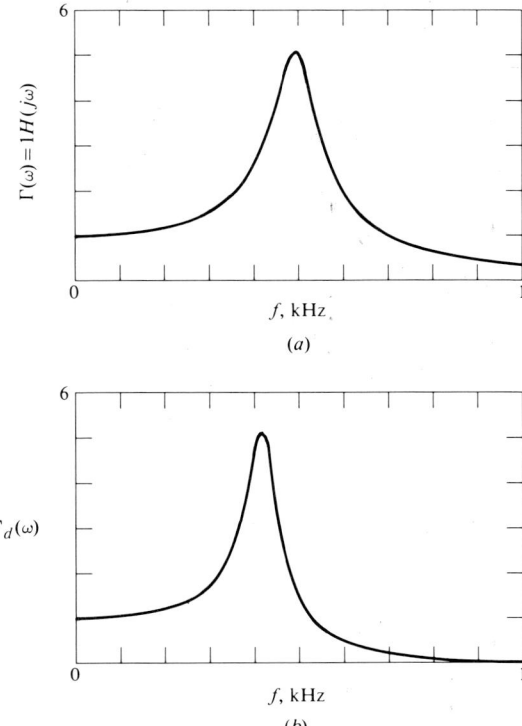

Figure 8-26 Graphs of gain versus frequency for (a) the parent analog system and (b) the digital system obtained by bilinear substitution in Example 8-29 (see Fig. 8-25).

analog system occurs for $f \approx 500$ hz. This is consistent with (8-23): the gain of the analog system for frequency $f' = 500$ Hz appears at the lower frequency f given by

$$f = \pi^{-1} F \tan^{-1} \frac{\pi f'}{F} = \pi^{-1}(2000) \tan^{-1} \frac{\pi(500)}{2000} = 424 \text{ Hz}$$

in the frequency response of the digital system.

8-4B Digital-Filter Design Using Bilinear Substitution

Bilinear substitution can be used to obtain the system function for a digital filter from the system function for an analog filter. Compression of frequency response that accompanies bilinear substitution can be corrected to some extent by expanding the specified frequency response before designing the parent analog system. The appropriate relation for expansion is given by (8-22). The following example illustrates design of a digital Butterworth low-pass filter using bilinear substitution.

Example 8-30 We wish to design a low-pass digital filter whose gain meets the specifications given in Fig. 8-27. The sampling rate for the filter is $F = 20$ kHz.

Anticipating compression of the frequency response that accompanies bilinear substitution, we first expand the specified frequency response using (8-22). The expanded passband- and stopband-edge frequencies are

$$f'_1 = \pi^{-1} F \tan \frac{\pi f_1}{F} = \pi^{-1}(20,000) \tan \frac{\pi(4000)}{20,000} = 4.625 \text{ kHz}$$

$$f'_2 = \pi^{-1} F \tan \frac{\pi f_2}{F} = \pi^{-1}(20,000) \tan \frac{\pi(6000)}{20,000} = 8.762 \text{ kHz}$$

Next we design a Butterworth low-pass filter using the expanded specifications of Fig. 8-28. Following the procedure described in Sec. 4-4, we find that the filter has order $n = 3$ and poles

$$p_1 = p_3^* = 16.100 + j27.886 \text{ krad/s} \qquad p_2 = -32.200 \text{ krad/s}$$

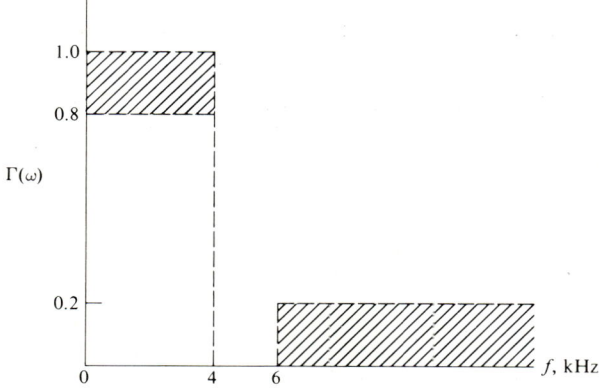

Figure 8-27 Specifications for the filter of Example 8-30.

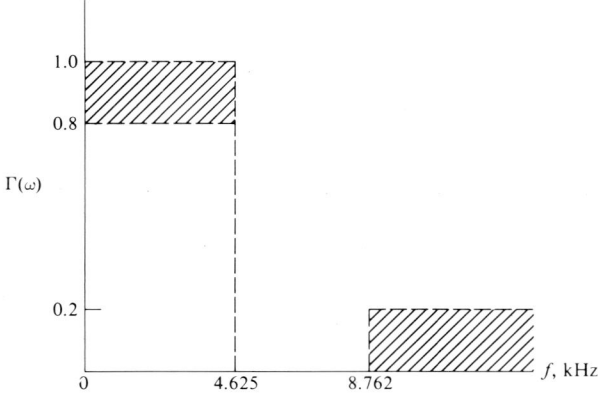

Figure 8-28 Expanded specifications for the filter of Example 8-30.

The system function for the parent analog filter is

$$H'(s) = \frac{-p_1 p_2 p_3}{(s - p_1)(s - p_2)(s - p_3)}$$

We use the bilinear substitution $s = 2F(z - 1)/(z + 1)$ ($F = 20$ kHz) to obtain the system function for the digital processor. This gives

$$H(z) = \frac{K(z + 1)^3}{(z - p_1')(z - p_2')(z - p_3')}$$

where

$$K = \frac{-p_1 p_2 p_3}{(2F - p_1)(2F - p_2)(2F - p_3)}$$

and

$$p_i' = \frac{2F + p_i}{2F - p_i} \quad \text{for } i = 1, 2, 3$$

The value given above for the constant K is irrelevant because we wish to force the dc gain of the processor to unity. We assume that the system function is

$$H(z) = \frac{K'(z + 1)^3}{(z - p_1')(z - p_2')(z - p_3')}$$

where K' is to be determined such that $H(1) = 1$.

The poles of the digital filter are

$$p_1' = 0.143 + j0.568 \qquad p_2' = 0.108 \qquad p_3' = 0.143 - j0.568$$

Using these values in the expression above gives

$$H(1) = \frac{K'(2)^3}{(1 - p_1')(1 - p_2')(1 - p_3')} = 8.482 K'$$

Solving $H(1) = 1$ for K' gives $K' = 0.118$; thus the system function for the digital processor is

$$H(z) = \frac{0.118(z^3 + 3z^2 + 3z + 1)}{z^3 - 0.394 z^2 + 0.374 z - 0.037}$$

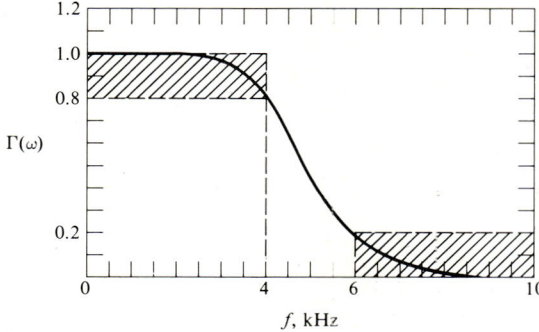

Figure 8-29 Graph of gain versus frequency for the digital filter designed in Example 8-30.

Figure 8-29 shows a graph of gain versus frequency for the digital filter. The digital processor can be implemented in at least three different forms (Fig. 8-30).

Direct form This form (Fig. 8-30a) results from obtaining the difference equation

$$y(n) = 0.118[x(n) + 3x(n-1) + 3x(n-2) + x(n-3)]$$
$$+ 0.394y(n-1) - 0.374y(n-2) + 0.037y(n-3)$$

directly from the system function $H(z)$ above.

Series (or cascade) form This form (Fig. 8-30b) results from factoring the system function as

$$H(z) = \frac{K'(z^2 + 2z + 1)}{z^2 - 2(\text{Re } p_1')z + |p_1'|^2} \frac{z+1}{z - p_2'}$$

$$= \frac{0.118(z^2 + 2z + 1)}{z^2 - 0.286z + 0.343} \frac{z+1}{z - 0.108}$$

This gives

$$w(n) = 0.118[x(n) + 2x(n-1) + x(n-2)] + 0.286w(n-1) - 0.343w(n-2)$$
$$y(n) = w(n) + w(n-1) + 0.108y(n-1)$$

Parallel form This form (Fig. 8-30c) results from expanding the system function in partial fractions as

$$H(z) = B_0 + \frac{B_1}{z - p_2'} + \frac{B_2 z + B_3}{z^2 - 2(\text{Re } p_1')z + |p_1'|^2}$$

$$= 0.118\left(1 + \frac{4.200}{z - 0.108} - \frac{0.805z - 3.744}{z^2 - 0.286z + 0.343}\right)$$

This gives

$$w_1(n) = 4.200x(n-1) + 0.108w_1(n-1)$$
$$w_2(n) = 0.805x(n) - 3.744x(n-1) + 0.286w_2(n-1) - 0.343w_2(n-2)$$
$$y(n) = 0.118[x(n) + w_1(n) - w_2(n)]$$

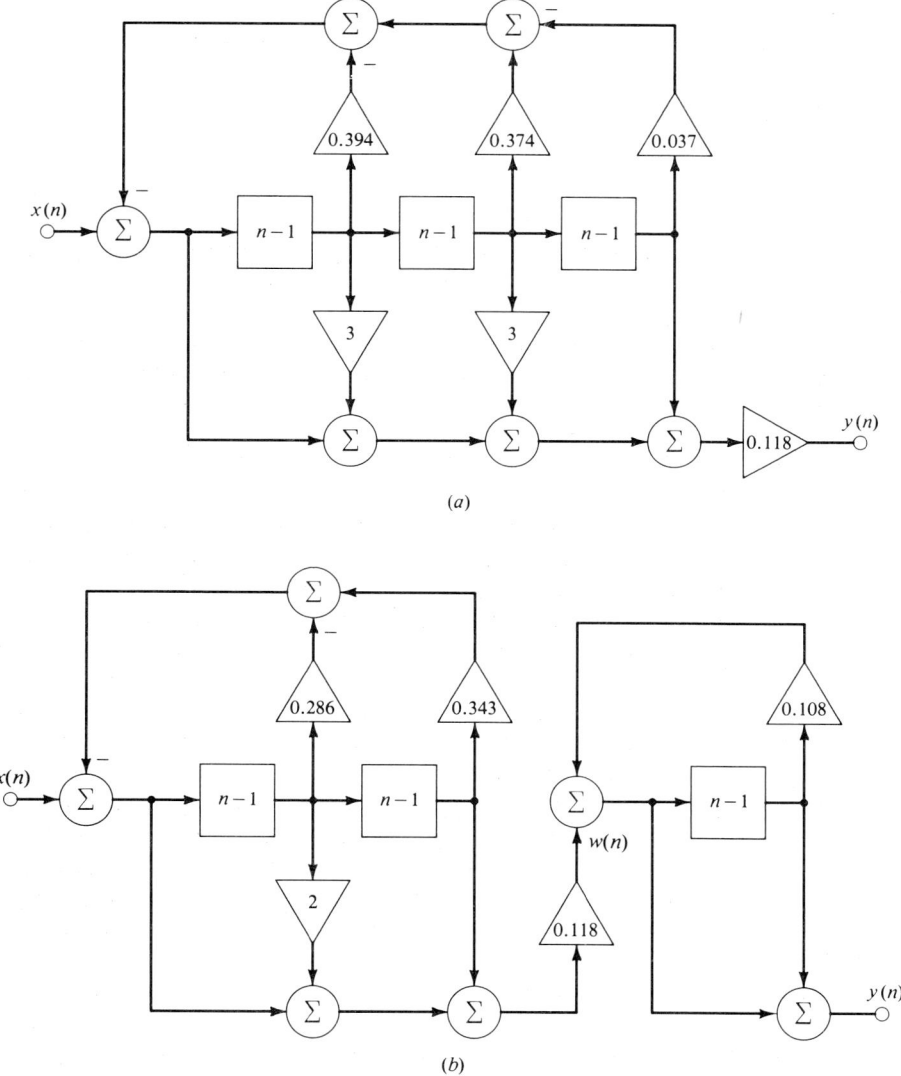

Figure 8-30 Realizations of the digital processor of Example 8-30: (*a*) direct, (*b*) series.

These three implementations are identical if system parameters and signals are specified with unlimited precision, but since unlimited precision is impossible in a real digital processor, the three implementations are not identical in a practical system. One of the problems of digital-filter design is choosing a direct, parallel, or series realization. A series realization like that of Fig. 8-30*b* usually is preferred (reasons why are given in books treating digital-filter design in depth), but there remains the problem of choosing the order of the subsystems; different orderings lead to different overall properties in a practical system.

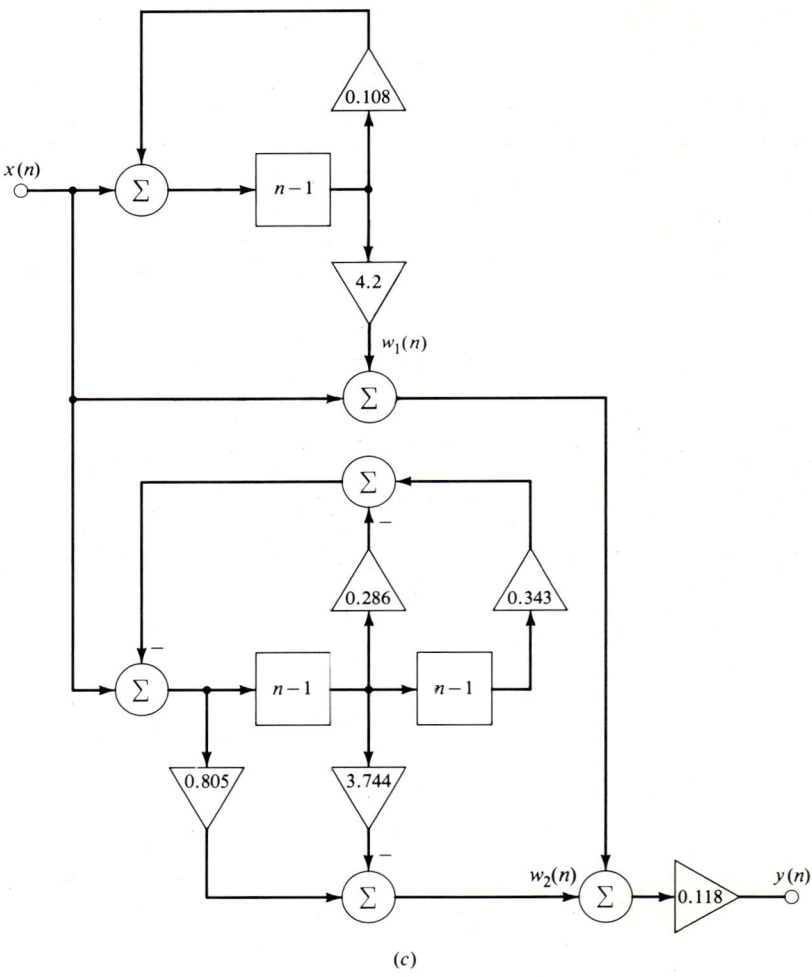

Figure 8-30 *Continued* (c) parallel.

It is often claimed that bilinear substitution is appropriate only for designing almost piecewise-constant (low-pass, bandpass, etc.) filters because of the nonlinear compression that accompanies bilinear substitution, i.e., because the compression changes the shapes of frequency-dependent parts of the frequency response of a parent analog system. In fact, bilinear substitution can be used whenever an analog system can be designed to fit properly expanded specifications if all significant parts of the specifications (or the desired frequency response) are expanded using (8-22) before applying the substitution.

Finally, we note that one effect of expanding original specifications (to account for compression accompanying bilinear substitution) is to relax them. In Example 8-30 the ratio of stopband edge to passband edge in the original specifications is

(6 kHz)/(4 kHz) = 1.5 whereas the same ratio for the expanded specifications is $8762/4625 \approx 2$. This means that an analog filter meeting the original specifications must have sharper cutoff than one meeting the expanded specifications, which may mean that an analog filter meeting the expanded specifications has lower order than one meeting the original specifications. In Example 8-30 a third-order Butterworth filter meets the expanded specifications whereas a fifth-order Butterworth filter is required to meet the original specifications. Reduction of filter order, though not always achieved, is one potential advantage of bilinear substitution over other methods for designing recursive digital filters.

SUMMARY

The *z transform* of a digital function $x(n)$ is denoted by $X(z)$ and defined by

$$X(z) = \sum_{n=-\infty}^{\infty} x(n)z^{-n}$$

The operator $\mathscr{L}$ denotes the operation on the right side of the relation above; thus

$$x(n) = \mathscr{L}^{-1}\{X(z)\}$$

Both z transformation and inverse z transformation are linear operations.

In general, an inverse z transform $\mathscr{L}^{-1}\{X(z)\}$ can be left-, right-, or two-sided, depending on the region of convergence associated with $X(z)$. The right-sided inverse transform is unique. In this book we apply the z transform only to right-sided functions, making it unnecessary to specify a region of convergence for a z transform.

The most important operational properties of the z transformation (for analysis and design of digital systems) are the *convolution property*, expressed by

$$\mathscr{L}\left\{\sum_{m=-\infty}^{\infty} x(m)h(n-m)\right\} = X(z)H(z)$$

and the *shift property*, expressed by

$$\mathscr{L}\{x(n-k)\} = z^{-k}X(z)$$

The convolution property shows that the z transform of an output $y(n)$ of a linear shift-invariant system for input $x(n)$ is given by

$$Y(z) = H(z)X(z)$$

where $H(z)$ is the z transform of the delta response of the system. This relation represents a powerful procedure for determining an output of a linear shift-invariant system, as it substitutes table-lookup operations and relatively simple algebraic operations for convolution. The shift property allows us to interpret the transform variable z^{-1} as a unit delay operator. This makes it easy to obtain the system function of a system from a difference equation describing the system and vice versa.

The function $H(z) = \mathcal{Z}\{h(n)\}$ is called the *system function* of a system having delta response $h(n)$. The system function of a linear shift-invariant system is a rational function of z. The values of z for which a system function $H(z)$ equals zero (the roots of the numerator polynomial) are called the *zeros* of $H(z)$. The values of z for which $1/H(z)$ equals zero (the roots of the denominator polynomial) are called the *poles* of $H(z)$. A system function is determined to within a constant multiplier by its poles and zeros. Consequently, all important properties of a linear shift-invariant system are determined by the poles and zeros of the system function of the system. A system function can be represented graphically by a *pole-zero plot*, which shows the poles (as crosses) and the zeros (as circles) of the system function on a z plane.

The inverse z transform of a system function is the delta response of the associated system. Expanding a system function in partial fractions shows that each pole p of the system function gives rise to a term of the form cp^n in the delta response of the system. These terms are called the *natural modes* of the system. A natural mode arising from a pole inside the unit circle defined by $|z| = 1$ decays. A natural mode arising from a pole outside the unit circle or a repeated pole on the unit circle grows. A natural mode arising from a nonrepeated pole on the unit circle neither grows nor decays. The rate at which a pole inside (outside) the unit circle decays (grows) increases with distance of the pole from the unit circle.

For the delta response of a system to be absolutely summable all natural modes of the system must decay. Consequently, a system is stable if and only if all poles of its system function lie inside the unit circle.

The unit circle described by $|z| = 1$ in a z plane corresponds to the ω (frequency) axis in an s plane. This allows cut-and-try design of a digital filter, guided by the fact that gain of a digital system is large where the unit circle passes near a pole and small where the unit circle passes near a zero.

The interior of the unit circle in a z plane corresponds to the left half of an s plane. This suggests that an analog system can be transformed into a more or less equivalent digital system by a substitution of the form $s \to f(z)$, where $f(z)$ is a function that maps the interior of the unit circle in a z plane to the left half of an s plane. Many such substitutions exist. The most useful one is the bilinear substitution described in Sec. 8-4. Bilinear substitution is often used for designing recursive digital filters because with proper use the substitution preserves significant properties of the parent analog filter, i.e., stability and passbands.

PROBLEMS

Unless otherwise specified, "inverse z transform" means "right-sided inverse z transform" in all problems below.

8-1 Give the first three nonzero terms of the z transforms (in series form) of the following digital signals:

(a) $x(n) = (5^{-n} + 2)u(n)$

(b) $x(n) = \Delta(n + 2) - \Delta(n + 1) + 2\Delta(n) + u(n + 2)$

(c) $x(n) = 5\left(\cos\dfrac{\pi n}{4}\right)u(n)$ (d) $x(n) = 2\left(\sin\dfrac{\pi n}{4}\right)u(n)$

(e) $x(n) = 3^n e^{-n/3} u(n)$ (f) $x(n) = 2^n e^{-0.2n}(\sin 0.4n)u(n)$

8-2 Give the two-sided inverse z transforms for $n = -1$, $n = 0$, and $n = 1$:

(a) $X(z) = z + 2 + 4z^{-1} - 5z^{-2}$ (b) $X(z) = (z - 1)(z^{-2} + 3z^{-1} + 2)$

(c) $X(z) = 1 - z^{-1} - z^{-2} - \cdots$ (d) $X(z) = \sum\limits_{n=-2}^{8} (0.8)^n z^{-n}$

8-3 Express the following z transforms in closed form. Give the region of convergence.

(a) $X(z) = z^{-1} + z^{-2} + z^{-3} + \cdots$ (b) $X(z) = z^{-1} - z^{-2} + z^{-3} - \cdots$

(c) $X(z) = z^{-2} + z^{-4} + z^{-6} + \cdots$ (d) $X(z) = z + z^2 + z^3 + \cdots$

(e) $X(z) = 2z^{-1} + 4z^{-2} + 8z^{-3} + \cdots$ (f) $X(z) = 3z - 9z^2 + 27z^3 - \cdots$

(g) $X(z) = z + z^{-1} + 0.50(z^2 + z^{-2}) + 0.25(z^3 + z^{-3}) + \cdots$

8-4 Use the definition (8-1) to obtain the z transforms of the following digital signals. Express the z transforms in closed form and give the regions of convergence.

(a) $x(n) = u(n)$ (b) $x(n) = r\left(\dfrac{n}{k}\right)$ $k > 0$

(c) $x(n) = e^{\lambda n} u(n + 1)$ (d) $x(n) = \sin[(n - k)\lambda] u(n)$

8-5 Find all possible (left-, right-, and two-sided) inverse z transforms and associated regions of convergence for each of the following functions:

(a) $X(z) = \dfrac{z}{z - 1}$ (b) $X(z) = \dfrac{z + 1}{z - 1}$ (c) $X(z) = \dfrac{2}{z + 1}$

(d) $X(z) = \dfrac{1}{z - 1} + \dfrac{1}{z + 1}$

8-6 Use Table E-1 (and partial-fraction expansion if necessary) to obtain the right-sided inverse z transforms of the following functions:

(a) $H(z) = \dfrac{z + 1}{(z - 0.2)(z + 0.5)}$ (b) $H(z) = \dfrac{z^{-1} + z}{(z + 1)(z - 0.4)(z - 0.5)}$

(c) $H(z) = \dfrac{z^2}{(z - 1)(z + 0.1)^2}$ (d) $H(z) = \dfrac{(z + 1)^2}{(z - 0.2)(z^2 - 0.60z + 0.25)}$

(e) $H(z) = \dfrac{16(z + 1)^2}{(2z - 1)(2z + 1)(4z - 1)}$ (f) $H(z) = \dfrac{(z + 1)(z + 0.2)}{(z - 1)(z - 0.2)}$

(g) $H(z) = \dfrac{4(z + 1)^3}{(z - 1)(z - 0.5)}$

8-7 Show that

(a) $\mathcal{L}\{n^2\lambda^n u(n)\} = \dfrac{\lambda z(z+\lambda)}{(z-\lambda)^3}$ (b) $\mathcal{L}\{\lambda^{-n} x(n)\} = X(\lambda z)$

(c) $\mathcal{L}\{c^n(\sin \lambda n) u(n)\} = \dfrac{cz \sin \lambda}{z^2 - 2c(\cos \lambda)z + c^2}$

8-8 Use long division to obtain the first three samples of the inverse z transform of each function below:

(a) $X(z) = \dfrac{1}{(z-1)(z+0.2)}$ (b) $Y(z) = \dfrac{z+1}{z^2 - 0.60z + 0.25}$

(c) $Y(z) = \dfrac{z^{-5} + 2z^{-3} + 1}{z^3 + z^2 + z + 1}$

8-9 Obtain the outputs for the following input $x(n)$ and delta-response $h(n)$ pairs:

(a) $x(n) = 5u(n)$
    $h(n) = 2u(n)$
(b) $x(n) = 2r(0.25n)$
    $h(n) = (0.5)^n u(n)$
(c) $x(n) = (0.9)^n u(n)$
    $h(n) = (0.5)^n u(n)$
(d) $x(n) = nu(n)$
    $h(n) = (0.8)^n u(n)$

8-10 Obtain the outputs $y(n)$ for inputs $x(n)$ of systems described by the following difference equations:

(a) $y(n) + 2y(n-1) = x(n-1)$
    $x(n) = (\sin 0.2n) u(n)$

(b) $y(n) + 0.20y(n-1) + 0.04y(n-2) = x(n-1) + x(n-2)$
    $x(n) = (0.8)^n u(n)$

(c) $y(n) - 0.60y(n-1) + 0.25y(n-2) = 2x(n) - x(n-1) + 2x(n-2)$
    $x(n) = 5r(0.1n)$

(d) $y(n) + y(n-1) = x(n) - x(n-1)$
    $x(n) = u(n)$

8-11 Give a block diagram in standard form (see Fig. 7-24) for each system of Prob. 8-10. Then use block-diagram reduction to find the system function for the system. Check the result using (8-10).

8-12 Find the delta response and the step response of each system of Fig. P8-12.

8-13 Obtain the impulse-invariant digital system for a parent analog system having impulse response

$$h(t) = \dfrac{t}{\tau^2} e^{-t/\tau} u(t)$$

where $\tau = 2$ ms.

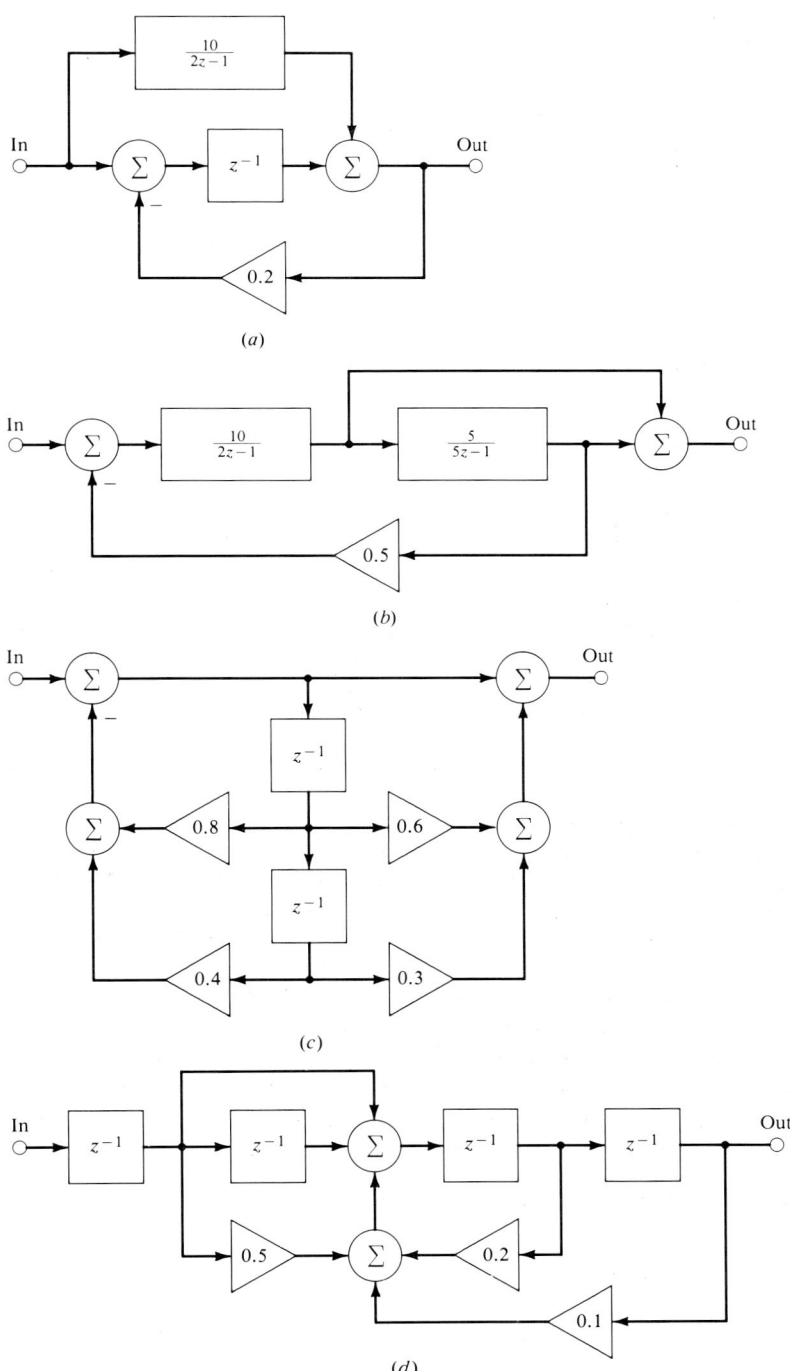

Figure P8-12

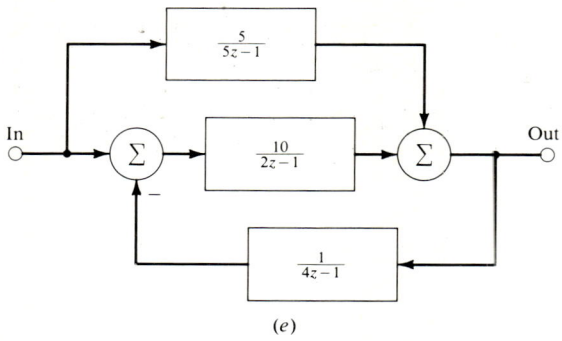

(e)

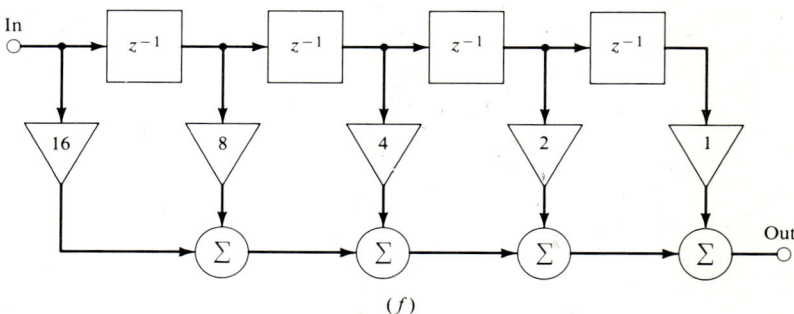

(f)

Figure P8-12 *Continued*

8-14 Obtain the impulse-invariant digital system for a parent analog system having system function

$$H(s) = \frac{\omega_0^2}{s^2 + 2\zeta\omega_0 s + \omega_0^2}$$

where $\zeta = 0.5$ and $\omega_0 = 2\pi$ krad/s.

8-15 Obtain the impulse-invariant digital system for a parent analog system whose input $x(t)$ and output $y(t)$ are related by the differential equation

$$\frac{d^2y}{dt^2} + a_1\frac{dy}{dt} + a_0 y = b_1\frac{dx}{dt} + b_0 x$$

where $a_1 = 3$ s^{-1}, $a_0 = 2$ s^{-2}, $b_1 = 10$ s^{-1}, and $b_0 = 5$ s^{-2}.

8-16 Obtain the impulse-invariant digital system for a parent analog system whose step response is given by

$$g(t) = (1 - e^{-t/\tau}\cos\omega t)u(t)$$

where $\tau = 1$ ms and $\omega = 1$ krad/s.

8-17 Obtain the step-invariant digital system for the parent analog system of:

(a) Prob. 8-13   (b) Prob. 8-14   (c) Prob. 8-15   (d) Prob. 8-16

8-18 Determine which of the following system functions are realizable and whether the realizable system functions represent stable systems.

(a) $H(z) = \dfrac{z^{-3}}{z - 0.2}$ (b) $H(z) = \dfrac{z^{-3} + z^{-2} + z^{-1} + 1}{z - 0.5}$

(c) $H(z) = \dfrac{0.5z^3}{(z - 0.5)^2}$ (d) $H(z) = \dfrac{10z^{-4}}{(1 - z^{-1})^2(1 - z^{-2})^2}$

(e) $H(z) = \dfrac{z^{-3}(z^5 + z^3 + z)}{(2z + 1)(4z - 1)(5z + 1)}$

8-19 Give a pole-zero diagram for each system function of Prob. 8-6. Give the form of each natural mode. Determine whether the system is stable; if so, predict the duration of the transient response for a step input.

8-20 Determine all values of K for which the system of Fig. P8-20 is stable.

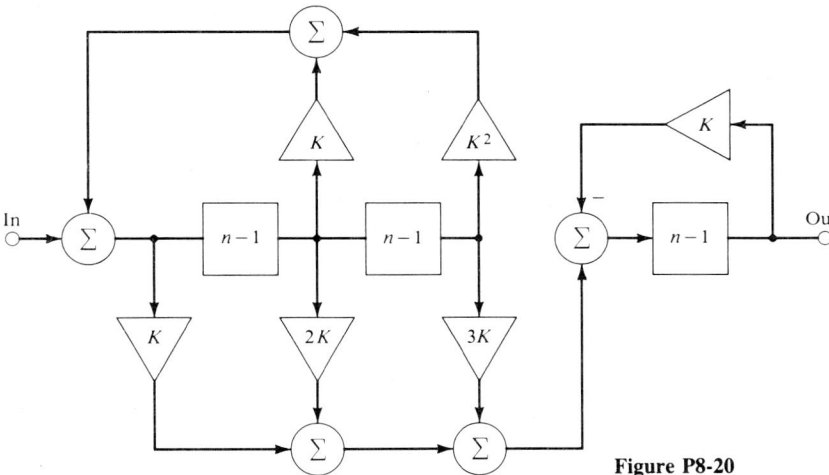

Figure P8-20

8-21 Identify forced, unforced, transient, and steady-state components of the step response of each system of Prob. 8-12.

8-22 Give the forms of the forced, unforced, transient, and steady-state components of the output of each system of Prob. 8-12 for input $x(n) = (0.5)^n(\cos 0.4n)u(n)$.

8-23 Obtain the forced response of each system of Prob. 8-12 for input $x(n) = [1 + (0.2)^n + \cos 0.4n]u(n)$.

8-24 Obtain the transfer function (as a function of digital frequency λ) for each stable system of Prob. 8-6. Use the transfer function to determine the steady-state output for input

$$x(n) = (1 + \cos 0.1n)u(n)$$

8-25 Figure P8-25 shows a system for digital signal processing, where

$$H(z) = \dfrac{z + 1}{z^2 - 1.344z + 0.950}$$

The AD and DA converters are error-free. Determine the output of this system for input

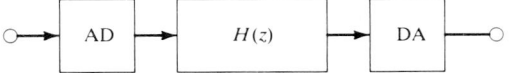

Figure P8-25

$$x_a(t) = 1 + \cos\frac{2\pi Ft}{8} + \cos\frac{2\pi Ft}{4} + \cos\frac{6\pi Ft}{8}$$

where F is sampling rate.

8-26 Plot gain and phase shift versus frequency in hertz for the system of Fig. P-25 if the sampling rate is $F = 10$ kHz and

$$H(z) = \frac{0.834(z-1)}{z - 0.667}$$

8-27 Figure P8-27 shows a pole-zero diagram for the digital processor in the system of Fig. P8-25. The sampling rate for the system is $F = 20$ kHz. Sketch a graph of gain versus frequency in hertz for the system.

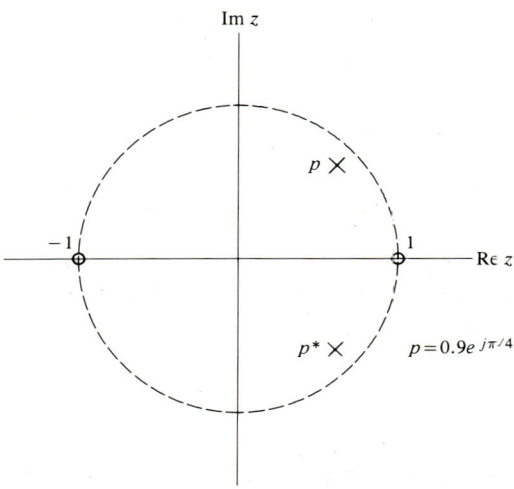

Figure P8-27

8-28 Repeat Prob. 8-27 for sampling rate $F = 10$ kHz and the pole-zero diagram of Fig. P8-28.

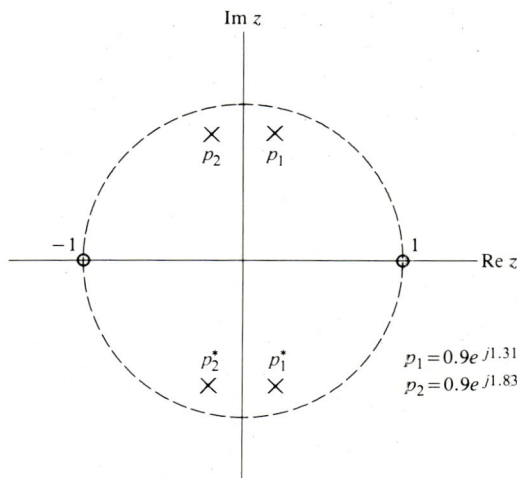

Figure P8-28

8-29 Use the cut-and-try method described in the text to design a three-pole low-pass digital filter having bandwidth 4 kHz and sampling rate 10 kHz. The gain should equal zero for $f = 5$ kHz. The gain in the passband should be nearly frequency-independent, and the edges of the passband should be as sharp as possible. Plot gain versus phase shift for one cut only and describe adjustments for the second cut.

8-30 Use bilinear substitution (after proper expansion) to design a Butterworth low-pass digital filter using sampling rate $F = 10$ kHz and meeting the specifications given in Fig. P8-30.

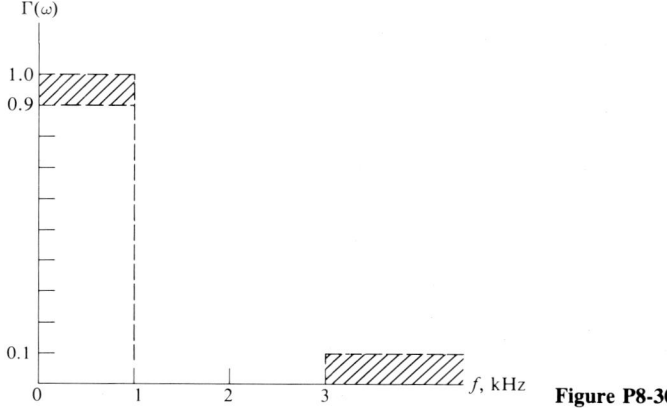

Figure P8-30

8-31 Repeat Prob. 8-30 for sampling rate $F = 40$ kHz and the specifications given in Fig. P8-31.

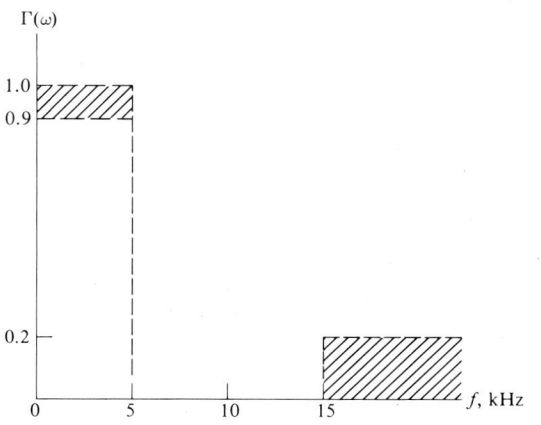

Figure P8-31

8-32 The derivative of an analog signal $x_a(t)$ can be approximated by $[x_a(t) - x_a(t - T)]/T$ provided T is sufficiently small. Show that this suggests the substitution $s \to 1 - z^{-1}$ as a means of obtaining an equivalent digital system from a parent analog system having system function $H'(s)$. Describe the substitution as a mapping from an s plane to a z plane. Under what condition(s) is the resulting digital system stable? How is the imaginary (frequency) axis mapped by this substitution? What are the disadvantages of this mapping for designing a digital filter?

CHAPTER
NINE
COMPUTER-AIDED ANALYSIS AND DESIGN

This chapter describes simulation and other methods for computer-aided analysis and design. These methods are very important for two reasons. One is that few practical systems are entirely amenable to closed-form or pencil-and-paper methods. The other is economy; it is wise to check a proposed design by a relatively inexpensive simulation before proceeding with costly construction of a prototype.

Section 9-1 provides an introduction to simulation using the simulation language CSMP (*C*ontinuous *S*ystem *M*odeling *P*rogram). Section 9-2 describes application of CSMP to linear stationary systems and also describes numerical convolution. Section 9-3 describes computer-aided frequency-domain analysis using a fast-Fourier-transformation algorithm for calculating Fourier coefficients, Fourier series, Fourier transforms, and Fourier integrals. Section 9-4 introduces interactive computer-aided analysis and design using programs for finding characteristic roots, obtaining partial-fraction expansions, and designing simple digital filters.

9-1 INTRODUCTION TO SIMULATION USING CSMP

Several languages have been developed to facilitate simulation of systems on digital computers. CSMP was selected for use in this book for three reasons: (1) it is relatively easy to learn, even by someone having no previous programming experience; (2) it is supported at almost all installations for scientific computing; and (3) it is applicable to a number of different kinds of problems. In this section we give a brief description of the CSMP language together with examples illustrating its use. The descriptive material should be read quickly and with the understanding that much of it will fall into place only with study of the examples that follow.

9-1A CSMP Statements

A CSMP program is a set of CSMP statements† typed on consecutive lines. A statement can be continued on up to eight additional lines using three periods (...) as the last three characters on each line to be continued. A CSMP statement can begin in any column, but it must end or be continued before column 73 (CSMP reads only columns 1 to 72). Blank lines are ignored by CSMP.

A statement beginning with an asterisk in column 1 is printed in a program listing but is otherwise ignored by CSMP. Such statements are used to insert comments in a CSMP program. It is good practice to use comments, blank lines, and indentation liberally to make programs as readable and self-explanatory as possible.

9-1B Reserved Words in CSMP

Certain words have special meanings in CSMP. These *reserved words* should not be used for any other purpose. A list of them is given in Table F-1. Meanings of these reserved words and their use in CSMP programs are explained and illustrated as they arise. Reserved words are printed in capitals in subsequent discussion and examples. One important reserved word is TIME, which denotes the independent variable in a simulation. This variable is kept up to date by CSMP as a simulation proceeds; its value should not be changed by any program statement.

9-1C Names, Constants, and Operators in CSMP

Signals, system parameters, and other quantities can be assigned names in a CSMP program. Acceptable names consist of one to six alphanumeric characters (letters‡ and numerals). The first character must be a letter (A to Z). Any convenient names can be used except the reserved words listed in Table F-1. Examples of acceptable names are VIN, IOUT, GAIN3, TAU2, and T0. Examples of unacceptable names are 2T (does not begin with a letter), V#3 (contains a character other than a letter or a number), THETAOUT (contains more than six characters), and DELAY (a reserved word). It is good practice to use names that suggest the meaning of the variable referred to, e.g., VIN for voltage at the input of a system. Except in a few special cases, CSMP assumes that all names refer to real (not integer) variables. Unlike FORTRAN, CSMP does not assume that names beginning with letters I to N identify integer variables.

Constants may be expressed using fixed-point notation (-20.58) or floating-point notation ($-0.2058E+02$). In either case, only seven digits are significant; for example, 1.23456789 is truncated by CSMP to 1.234567. This is a minor limitation, for system parameters rarely are specified with anything approaching seven-digit precision. Unless otherwise noted, decimal points should be typed when giving a numerical value (type GAIN = 50., not GAIN = 50).

†Almost all installations require certain job-control statements preceding and following a CSMP program. These statements can be obtained from a programming consultant.
‡Most versions of CSMP do not distinguish between lowercase and uppercase letters; e.g., vin and VIN refer to the same quantity in a CSMP program.

Since the user of CSMP must keep track of dimensions and units, we recommend that all dimensioned quantities used in a CSMP simulation be given in standard SI units *without prefixes;* e.g., frequency $f_0 = 100$ kHz should be specified as F0 = 1.E5 in a CSMP program. Following this rule eliminates many costly errors.

Symbols for operations and hierarchy of operations in CSMP are the same as in FORTRAN; thus A * * B means "raise A to the power B," A * B means "multiply A by B," A/B means "divide A by B," A + B means "Add A to B," A − B means "subtract B from A," and A = B means "assign the value of B to A." Exponentiations are performed from right to left. Other operations are performed from left to right. In any one statement exponentiations are done first, then multiplications and divisions, then additions and subtractions, and finally replacement; e.g., the statement A = B * * C/D * E + F causes CSMP to calculate $(B^C/D)E + F$ and assign the resulting value to A. Parentheses can be used to alter the normal order of operations; for example, A/(B * C) causes A to be divided by the product of B and C. Blanks between variable names and operations are ignored by CSMP. Often, inserting a few well-placed blanks makes a statement more readable; for example, A = B + C looks better than A=B+C.

9-1D Format of a CSMP Program

Figure 9-1 shows the skeleton of a typical CSMP program. The first statement begins with the reserved word TITLE, which causes a title for the simulation to be included on plots and other output produced by CSMP. The last two statements in a CSMP program must be the statements END and STOP to signify the end of a simulation. The main body of a typical CSMP program consists of an initial section, a dynamic section, and a terminal section, labeled by the reserved words INITIAL, DYNAMIC, and TERMINAL.

Statements in the initial section are executed once before actual simulation begins. They may specify initial values, assign values to fixed parameters, or perform calculations that are preliminary to an actual simulation.

Statements in the dynamic section cause CSMP to produce inputs and simulate components of a system. In effect, the dynamic section of a CSMP program is a verbal description of a system (block diagram) and its inputs. Each statement in a dynamic section is executed hundreds or thousands of times during a simulation as CSMP uses numerical algorithms to perform integration, differentiation, and other operations.

Statements in the terminal section specify duration of a simulation, output to be produced by CSMP, and other quantities that control execution of CSMP. A timer statement beginning with the reserved word TIMER should appear in the terminal section to specify the duration of a simulation using the reserved word FINTIM; e.g., the statement

$$\text{TIMER FINTIM} = 0.5$$

instructs CSMP to simulate a system (described in the dynamic section) for $0 \leq \text{TIME} \leq 0.5$. Other facilities of a timer statement are described in Sec. 9-1F.

COMPUTER-AIDED ANALYSIS AND DESIGN **579**

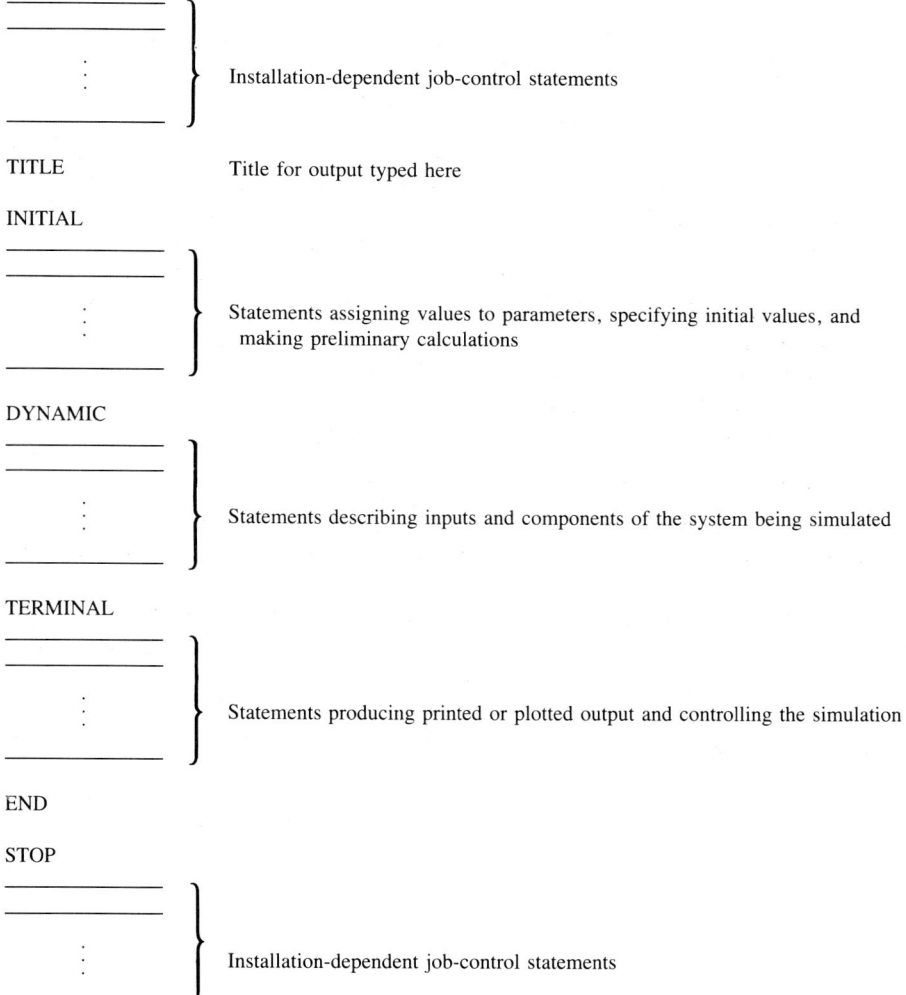

Figure 9-1 Skeleton of a typical CSMP program.

At least one output statement should appear in the terminal section of a CSMP program. Both plotted and printed output can be produced by CSMP; e.g., the statement PRTPLT Y instructs CSMP to produce a plot of a variable Y versus time and the statement PRINT Y instructs CSMP to print values of time and corresponding values of a variable Y. These output statements are described in greater detail below.

CSMP is a computer program that links various FORTRAN subroutines together. These subroutines actually execute the statements constituting a CSMP program. Statements in a dynamic section are executed in a loop. Before control is passed to the loop, TIME is set to zero. At the end of each pass through the loop, TIME is increased by a fraction of FINTIM. The loop terminates when TIME first exceeds FINTIM. During

580 INTRODUCTION TO SYSTEM ANALYSIS

execution of the loop, variables appearing in output statements are saved in arrays (CSMP sets up these arrays automatically, and a user need not worry about them). After the loop terminates, control passes to output statements in the terminal section.

Example 9-1 In this example we use CSMP to obtain a graph of the output of the system of Fig. 9-2 for input

$$x(t) = x_0(\cos \omega_0 t)u(t)$$

where $x_0 = 5$ V, $f_0 = 10$ kHz, and $K = 2$ mA/V. Figure 9-3 shows a CSMP program for this purpose.

The TITLE statement gives a short description of the simulation. Comments (statements beginning with an asterisk) describe variables and explain program logic. Blank lines and indentation improve readability.

The first three executable statements in the initial section define parameters associated with the system and input. The fourth converts frequency in hertz into frequency in

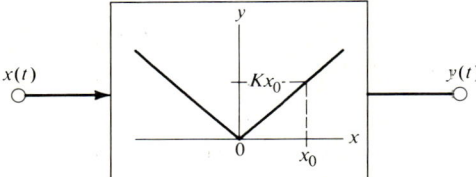

Figure 9-2 System of Example 9-1.

```
TITLE FULL-WAVE RECTIFIER WITH SINUSOIDAL INPUT

INITIAL

* X0 = PEAK AMPLITUDE (V) OF INPUT

* F0 = FREQUENCY (HZ) OF INPUT

     X0 = 5.
     F0 = 1.E4

* K = PARAMETER OF THE FULL WAVE RECTIFIER (A/V)

     K = 0.002

* W0 = ANGULAR FREQUENCY (RAD/S)

     W0 = 6.28 * F0

DYNAMIC

* X = INPUT (V), Y = OUTPUT (A)

     X = X0 * COS(W0 * TIME)
     Y = K * ABS(X)

TERMINAL

     TIMER FINTIM = 2.E-4
     PRTPLT Y

END
STOP
```

Figure 9-3 CSMP simulation for the system of Example 9-1.

radians per second in order to streamline generation of the input in the dynamic section, which follows.

The first statement in the dynamic section describes the input $x(t)$ (denoted by X) using the CSMP variable TIME. It is unnecessary to multiply the sinusoid by a step function because CSMP assumes that all simulations begin at $t = 0$. The second statement in the dynamic section uses a FORTRAN function† to describe the transfer characteristic

$$y(t) = K|x(t)|$$

of the full-wave rectifier in the system of Fig. 9-2. When the simulation is run, these two statements are executed repeatedly as TIME is increased from zero to FINTIM. Figure 9-4 shows a FORTRAN program segment that illustrates execution of the dynamic section.

The first statement in the terminal section is a timer statement that specifies FINTIM = 2.E−4. This is the duration of the simulation in seconds (because we use standard SI units). This value of FINTIM corresponds to two periods of the input $x(t)$, that is, $2/f_0 = 200$ μs. The statement PRTPLT Y causes CSMP to produce a plot of the output $y(t)$ versus time t, as shown in Fig. 9-5.‡ The plot begins at $t = 0$ and ends at $t = 200$ μs (as specified by FINTIM = 2.E−4). The interval between consecutive points is 2μs. This interval can be specified by the timer statement in a CSMP program, as described subsequently. If it is unspecified, as in this example, CSMP uses FINTIM/100 for the plotting time step. For FINTIM = 2.E−4 (200 μs) this gives $200/100 = 2$ μs, as noted above.

---

†CSMP accepts and interprets correctly all standard FORTRAN statements. The FORTRAN function ABS returns the absolute value of its argument.

‡A CSMP PRTPLT statement produces a so-called printer plot on the system line printer. Since such plots are not suitable for reproduction in a textbook, the plots produced by CSMP have been redrawn for this book.

```
            I = 1
            TIME = 0.
100         IF (TIME.GT.FINTIM) GO TO 200
            X = X0 * COS(W0 * TIME)
            Y = K * ABS(X)
            XPLOT(I) = X
            I = I + 1
            TIME = TIME + DELT
            GO TO 100
200         CONTINUE
```

Figure 9-4 FORTRAN program segment illustrating calculations performed in the dynamic section of the CSMP program of Fig. 9-3.

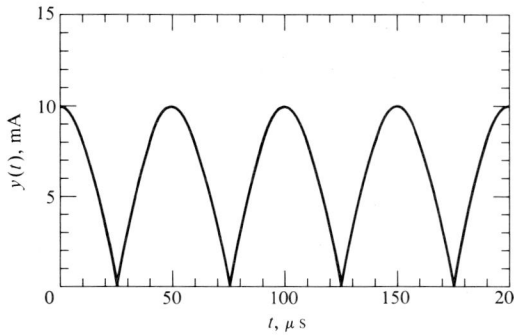

Figure 9-5 Output produced by the CSMP program of Fig. 9-3.

The discussion and example above give only a simplified overview of simulation using CSMP. Below, we describe facilities of CSMP in more detail.

9-1E Parameter Statement

A parameter statement begins with the reserved word PARAMETER. A parameter statement can be used in an initial section to define parameters of a simulation. A number of parameters can be defined by one parameter statement; e.g., the single statement

$$\text{PARAMETER X0} = 5., \text{F0} = 1.\text{E4}, \text{K} = 0.002$$

can replace the first three statements in the initial section of the program of Fig. 9-3. In addition, a parameter statement can cause an entire simulation to be repeated for as many as 50 values of a parameter; e.g., the statement

$$\text{PARAMETER X0} = 5., \text{F0} = 1.\text{E4}, \text{K} = (0.002, 0.005, 0.010)$$

in the initial section of a simulation causes the simulation to be run 3 times: once for $K = 0.002$, again for $K = 0.005$, and again for $K = 0.010$. Only one parameter can be varied in this way. This facility of CSMP should be used only in a proven program.

9-1F Timer Statement

In addition to controlling the duration of a simulation by specifying FINTIM (Sec. 9-1D), a timer statement can specify time increments for printed or plotted output using the reserved words PRDEL (for printed data) and OUTDEL (for plotted data); e.g., the statements

$$\text{TIMER FINTIM} = 0.100, \text{OUTDEL} = 0.001, \text{PRDEL} = 0.0005$$
PRTPLT Y
PRINT Z

in a terminal section instruct CSMP to plot the variable Y versus time at intervals of 1 ms and to print the variable Z at intervals of 500 μs.

If neither OUTDEL nor PRDEL is specified by a timer statement, CSMP uses OUTDEL = PRDEL = FINTIM/100. If one is specified but not the other, CSMP uses the specified value for both. If both OUTDEL and PRDEL are specified, the larger of the two must be an integral multiple of the smaller. If it is not, CSMP makes it so by adjusting the smaller before proceeding with the simulation.

Another important time step in a CSMP simulation is denoted by the reserved word DELT. This is the incremental time by which time (denoted by TIME) is increased on a pass through the dynamic section of a simulation; e.g., the statement

$$\text{TIMER FINTIM} = 0.100, \text{DELT} = 0.002$$

instructs CSMP to increase TIME by 0.002 (2 ms) after each pass through the dynamic section and to terminate the simulation when TIME first exceeds 0.100 (100 ms).

If DELT is unspecified, CSMP uses DELT = OUTDEL/16 or DELT = PRDEL/16, whichever is smaller. If DELT is specified and OUTDEL and PRDEL are unspecified, DELT is adjusted (if necessary) to be an integral submultiple of FINTIM/100. If none of these time steps is specified, CSMP uses

$$OUTDEL = PRDEL = FINTIM/100$$
$$DELT = OUTDEL/16$$

9-1G Elementary Signal Models in CSMP

All elementary signals except impulses can be modeled easily using CSMP. Many models can be invoked using built-in CSMP functions. Others can be modeled using FORTRAN statements and the independent variable TIME.

A step signal

$$x(t) = x_0 u(t - t_0)$$

is described in a CSMP simulation by the statement

$$X = X0 * STEP(T0)$$

where X0 is the amplitude of the step and T0 is the time at which the step switches from zero to X0. The independent variable TIME does not appear explicitly. For T0 = 0, this statement causes CSMP to produce a signal X defined by

$$X = \begin{cases} 0 & \text{TIME} < 0 \\ X0 & \text{TIME} \geq 0 \end{cases}$$

This definition is slightly different from that of (1-1), where a step is defined to be zero for $t \leq 0$. Ordinarily this difference is insignificant, but a user of CSMP should be aware of it when comparing mathematically derived signals with simulated signals; e.g., a simulated step response will be advanced slightly (by DELT) relative to the same response obtained mathematically.

A step $x(t) = x_0 u(t)$ that begins at $t = 0$ can be simulated using X = X0 * STEP(0.) or, because CSMP assumes that all simulations begin at $t = 0$, by the simpler (and faster) statement X = X0. The fact that the CSMP model for a step allows a predetermined delay (T0) to be specified simplifies modeling of a rectangular pulse; e.g., a signal

$$x(t) = x_0 r\left(\frac{t}{\tau}\right)$$

is produced in a CSMP simulation by the statement

$$X = X0 * (1. - STEP(TAU))$$

where X0 denotes x_0 and TAU denotes τ.

Other elementary signals can be produced using the independent variable TIME maintained internally by CSMP; e.g., a sinusoidal signal

$$x(t) = x_0 \cos \omega_0 t$$

is produced (for $t \geq 0$) by the statement

$$X = X0 * COS(W0 * TIME)$$

where X0 denotes peak amplitude x_0 and W0 denotes angular frequency ω_0 (both in standard SI units). As another example, an exponential pulse

$$x(t) = x_0 e^{-t/\tau} u(t)$$

is produced by the statement

$$X = X0 * EXP(-TIME/TAU)$$

where X0 denotes x_0 and TAU denotes τ. A factor STEP(0.) is unnecessary because CSMP assumes that a simulation begins at $t = 0$.

9-1H Elementary System Models in CSMP

Almost all elementary systems can be modeled under CSMP using either built-in CSMP functions or FORTRAN statements.† Below, we describe models for a few important elementary systems. Other models are described in subsequent sections as needed. Table F-2 gives all models used in this book.

9-1I Integration

The output $y(t)$ of an integrator for input $x(t)$ is given by

$$y(t) = K \int_{-\infty}^{t} x(\alpha)\, d\alpha$$

where K is a constant having the unit of $y(t)\,[x(t)t]^{-1}$. Realizing that a simulation under CSMP cannot begin at $t = -\infty$, we write this relation as

$$y(t) = K \int_{-\infty}^{0} x(\alpha)\, d\alpha + K \int_{0}^{t} x(\alpha)\, d\alpha$$

or

$$y(t) = y(0) + K \int_{0}^{t} x(\alpha)\, d\alpha$$

Using this expression requires that an input be specified only for positive times (provided the output is required only for positive times); effects of the input for negative times are accounted for by the initial value $y(0)$. This relation is described in a CSMP program by the statement

$$Y = INTGRL(Y0, K * X)$$

where X denotes $x(t)$, Y0 denotes $y(0)$, and Y denotes $y(t)$.

Seven different numerical methods for integration are available under standard CSMP: five are fixed-step methods and two are variable-step (adaptive) methods. Although a detailed description of these methods is beyond the scope of this book,

†Modeling a compression element is difficult.

rudimentary knowledge of fixed- and variable-step numerical integration is required for intelligent use of CSMP. Below, we illustrate fixed- and variable-step integration using the trapezoidal rule for numerical integration.

Figure 9-6 illustrates numerical integration using the trapezoidal rule and fixed step size $\Delta \alpha$. The integral is approximated at time $t = n\Delta\alpha$ as the area of all trapezoids to the left of $t = n\Delta\alpha$. The shaded areas represent error introduced at each step. This error is small where the integrand $x(\alpha)$ is almost linear and is large where the slope of $x(\alpha)$ changes significantly in one step. To keep error introduced at each step smaller than a specified value, step size must be small relative to the smallest interval over which the slope of the integrand changes significantly.

Figure 9-7 illustrates numerical integration using the trapezoidal rule and variable step size. Step size is small where the slope of the integrand $x(\alpha)$ changes rapidly and larger where the slope of $x(\alpha)$ changes slowly, so that error introduced at each step is approximately constant. In practice, variable-step integration is accomplished by various adaptive schemes, one of which is illustrated in Fig. 9-8. After a time step $\Delta\alpha$, incremental area to be added to that already accumulated is calculated twice, once using one trapezoid of width $\Delta\alpha$ and again using two trapezoids of width $\Delta\alpha/2$. If these two calculations of incremental area give nearly identical results, the integral is updated and

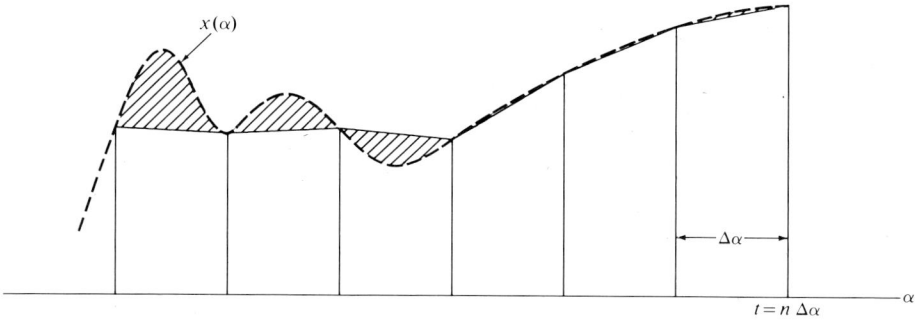

Figure 9-6 Fixed-step integration.

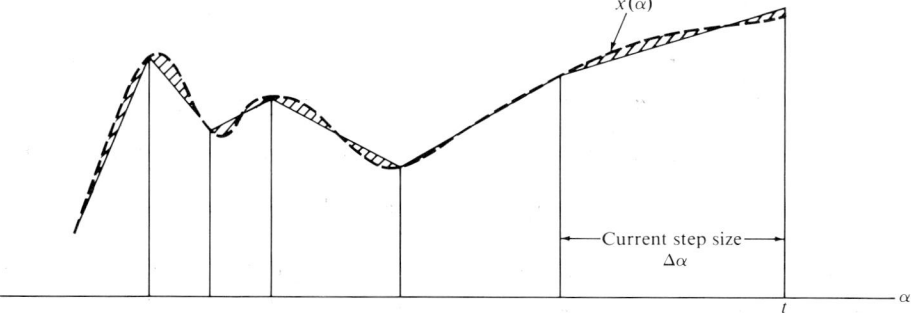

Figure 9-7 Variable-step integration.

586 INTRODUCTION TO SYSTEM ANALYSIS

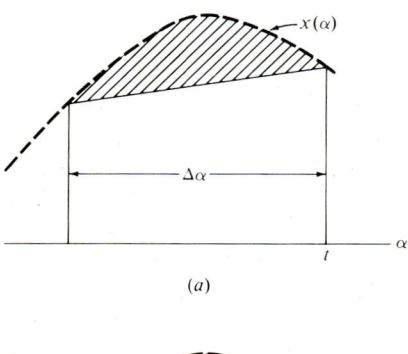

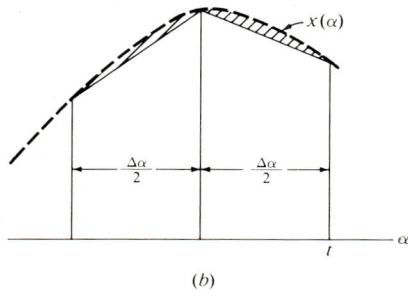

Figure 9-8 Interval halving method for adjusting step size in variable-step integration.

step size $\Delta\alpha$ is set to twice its previous value for the next step. If these two calculations give significantly different incremental areas, step size is halved again and incremental area is calculated using four trapezoids. This result is then compared with that calculated before using two trapezoids, and so on. This process continues until consecutive calculations yield sufficiently similar incremental areas or until roundoff error begins to offset any improvement gained by reducing step size further.

The minimum step size used in variable-step integration methods can be specified in a timer statement using the reserved word DELMIN; for example, as in

$$\text{TIMER FINTIM} = 0.200, \text{DELT} = 0.002, \text{DELMIN} = 2.\text{E}-5$$

where the minimum step size is specified as one one-hundredth of the nominal step size DELT. If DELMIN is unspecified, CSMP uses DELMIN = FINTIM $* 1.\text{E}-7$. Specifying DELMIN can avoid excessively time-consuming (and costly) calculations.

The integration method can be specified by a METHOD statement in a terminal section. If no method is specified, i.e., if no method statement appears in the terminal section, CSMP uses a variable-step Runge-Kutta method (see Cheny and Kincaid, pp. 184–197). In this book we use either this default method or a fixed-step Runge-Kutta method invoked by METHOD RKSFX. As a rule, variable-step methods are more accurate and more efficient than fixed-step methods. We use the fixed-step method only when necessary, e.g., when simulating a system containing a delay element, as described in Sec. 9-1J.

Example 9-2 This example illustrates simulation of an integration element using the fixed-step Runge-Kutta method. The integrator has gain $K = 1$ $(\text{ms})^{-1}$ and input

$$x(t) = x_0 e^{-t/\tau} u(t)$$

with $x_0 = 10$ V and $\tau = 20$ ms. Since we use this example to illustrate not only integration under CSMP but also good programming practice, we describe the simulation procedure in greater detail than would otherwise be warranted.

The first step in simulating a system using CSMP is to draw and fully label a block diagram for the system, identifying all signals and system parameters by their CSMP names. Such a block diagram for the integrator of this example is shown in Fig. 9-9. The second step is to check all names against the rules given above (six or fewer alphanumeric characters, letter for first character, and no reserved words).

The third step is to make a table of all parameters, giving symbols and values as specified in the problem and names and (standard SI) values to be used in the simulation. Table 9-1 is such a table for the present problem.

The fourth step is to choose reasonable values for duration of the simulation (FINTIM), integration step size (DELT), plotting step size (OUTDEL), and other quantities that control the simulation. In most practical problems selecting these quantities is a trial-and-error process, but here they can be selected almost by inspection because signals to be integrated and plotted can be determined easily. The input to the integrator is an exponential pulse having time constant $\tau = 20$ ms. The integration step size should be a small fraction of the time constant. A reasonable choice is $\tau/100 = 200$ μs; thus DELT $= 200.\text{E}-6$ (seconds). A reasonable value for the duration of the simulation is 5τ because most of the interesting behavior of the output occurs in the interval $0 < t < 5\tau$ as the output rises from its initial value to its final (steady-state) value; thus FINTIM $= 0.100$ (seconds). Finally, the plot should provide a good picture of the output using a reasonable number of pages. In this problem the output is smooth enough to be displayed using 50 (or even fewer) points; thus OUTDEL = FINTIM/50 = 0.002 (seconds).

The fifth (and, in this case, the last) step is to write the CSMP program. Use comments, blank lines, indentation, and spaces to make the program as readable as possible. In writing the initial section refer to the table of names and values to ensure that all parameters are correctly defined. When writing the dynamic section, refer to the labeled block diagram; first define the input(s); then describe components of the system, working from input to output in a systematic way. When writing the terminal section, be sure to include a timer statement containing correctly specified values for FINTIM and other quantities as needed; also be sure to include an output statement and (if needed) a method statement. Figure 9-10 shows a

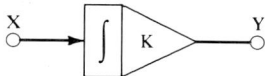

Figure 9-9 CSMP block diagram for the system of Example 9-2.

Table 9-1 Parameters used in the simulation of Example 9-2

Parameter	Value	CSMP symbol	CSMP value
x_0	10 V	X0	10 (V)
τ	20 ms	TAU	0.020 (s)
K	1 (ms)$^{-1}$	K	1000 (s^{-1})

588 INTRODUCTION TO SYSTEM ANALYSIS

```
TITLE   FIXED-STEP RUNGE-KUTTA INTEGRATION, EXPONENTIAL INPUT

INITIAL

*  X0 = PEAK AMPLITUDE OF INPUT (V)
*  TAU = TIME CONSTANT OF INPUT (S)
*  K = INTEGRATION SCALE FACTOR (1/S)

      PARAMETER X0 = 10., TAU = 0.02, K = 1000.

DYNAMIC

*  X = INPUT (V), Y = OUTPUT (V)

   X = X0 * EXP(- TIME/TAU)
   Y = INTGRL(0., K * X)

TERMINAL

   TIMER FINTIM = 0.1, DELT = 2.E-4, OUTDEL = 0.002
   PRTPLT Y
   METHOD RKSFX

END
STOP
```

Figure 9-10 CSMP program for the simulation of Example 9-2.

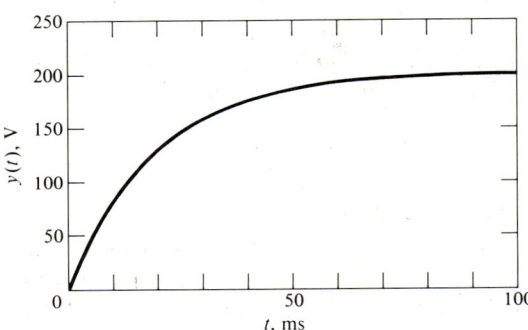

Figure 9-11 Output of the program of Fig. 9-10.

CSMP program for simulating the system of Fig. 9-9. Figure 9-11 shows a plot of the output $y(t)$ versus time produced by this program.

In simulating complicated systems the first run is often unsuccessful because of poor choices for FINTIM, DELT, or other quantities, and simulation is often a trial-and-error process. Further, when using simulation for design, additional runs may be required to home in on acceptable values for system parameters. Since these runs may be costly, a CSMP program must be checked for correctness. Methods for checking programs are described in Sec. 9-10.

9-1J Delay

The CSMP model for delay is invoked by the reserved word DELAY followed by three quantities enclosed in parentheses; e.g., a delay element is described in a CSMP simulation by a statement of the form

$$Y = \text{DELAY}(N, T0, X)$$

where X represents an input $x(t)$ and Y represents the corresponding output $y(t)$. The quantity T0 is delay time (in seconds) for the delay element being simulated. The quantity N is an integer (no decimal point) representing the number of points sampled in one delay time. The value of N must be supplied as a number (not a named variable) in the argument of the DELAY function. This quantity must be specified because CSMP simulates delay using N cascaded memory locations as a shift register, where a value entered in one end of the shift register appears at the other end N sample intervals later. If no integrators are present or if fixed-step integration is used, the number of points sampled in one delay time is $N = T0/\text{DELT}$. If variable-step integration is used, the number of points sampled in one delay time varies from one pass through the dynamic section to the next. Consequently, it is best to use only fixed-step integration in simulating a system containing both integration and delay. The time step DELT should be selected such that all delay times are integral multiples of DELT. The maximum permissible value for N is 16,384.

Example 9-3 This example illustrates use of the CSMP delay function in simulating the system of Fig. 9-12 for input

$$x(t) = x_0 r\left(\frac{t}{\tau}\right)$$

where $t_0 = 100$ ms, $K = 0.5$ V/(mAs), $x_0 = 50$ mA, and $\tau = 200$ ms. We wish to plot both the output $y(t)$ of the integration element and the output $z(t)$ of the delay element.

First, we draw and fully label a block diagram, as shown in Fig. 9-13. Second, we make a table of all system parameters showing names and values to be used in the simulation, as shown in Table 9-2 and we check names chosen for acceptability to CSMP. Third, we choose values for FINTIM, DELT, and other CSMP control parameters. Generally, the duration of a simulation should exceed the time required for all signals of interest to reach their steady-state values. In the present example the output of the integrator reaches steady state in 200 ms (the duration of the input), and the output of the delay element reaches steady state 100 ms later. The duration of the simulation should exceed $200 + 100 = 300$ ms; a reasonable value is FINTIM = 0.400 (400 ms). Since the signals to be plotted are reasonably well behaved, plots using 50 points should be satisfactory; thus

$$\text{OUTDEL} = \text{FINTIM}/50 = 0.008 \quad (8 \text{ ms})$$

Also, because the system contains a delay element, we use fixed-step integration with

$$\text{DELT} = \text{OUTDEL}/10 = 8.\text{E}-4 \quad (800 \; \mu\text{s})$$

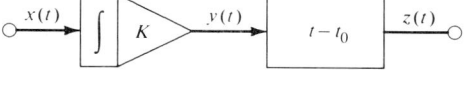

Figure 9-12 System of Example 9-3.

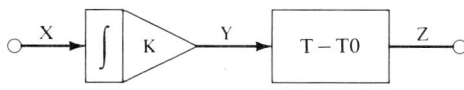

Figure 9-13 System of Example 9-3, labeled in preparation for simulation.

590 INTRODUCTION TO SYSTEM ANALYSIS

Table 9-2 Parameters for Example 9-3

Parameter	Value	Symbol	CSMP value
t_0	100 ms	T0	0.100 (s)
x_0	50 mA	X0	0.050 (A)
τ	200 ms	TAU	0.200 (s)
K	0.5 V/(mAs)	K	500 [V/(As)]

Finally, because fixed-step integration is used, the number of points sampled in one delay time is

$$N = T0/DELT = 125$$

Figure 9-14 shows a CSMP program for simulating the system of Fig. 9-12, and Fig. 9-15 shows the plots of $y(t)$ and $z(t)$ produced by this program. The program should

```
TITLE   R-K FIXED-STEP INTEGRATION, PULSE INPUT

INITIAL

* X0 = AMPLITUDE OF INPUT (A), TAU = DURATION OF INPUT (S)
* K = SCALE FACTOR FOR INTEGRATOR V/(A S)
* T0 = DELAY TIME (S)

        PARAMETER X0 = 0.05, TAU = 0.2, K = 500., T0 = 0.1

DYNAMIC

* X = INPUT, Y = OUTPUT OF INTEGRATOR, Z = OUTPUT OF DELAY

        X = X0 * (1. - STEP(TAU))
        Y = INTGRL(0., K * X)
        Z = DELAY(125, T0, Y)

TERMINAL

        TIMER FINTIM = 0.400, DELT = 8.E-4, OUTDEL = 0.008
        PRTPLT Y
        PRTPLT Z
        METHOD RKSFX

END
STOP
```

Figure 9-14 CSMP program for simulating the system of Fig. 9-12.

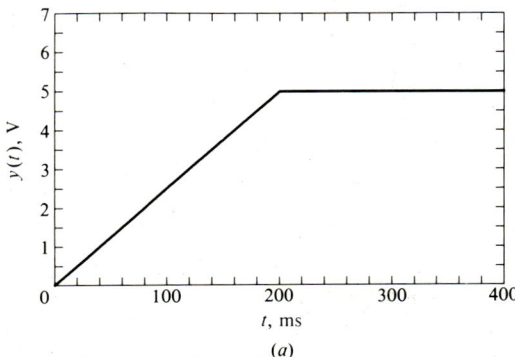

Figure 9-15 Outputs of the program of Fig. 9-14.

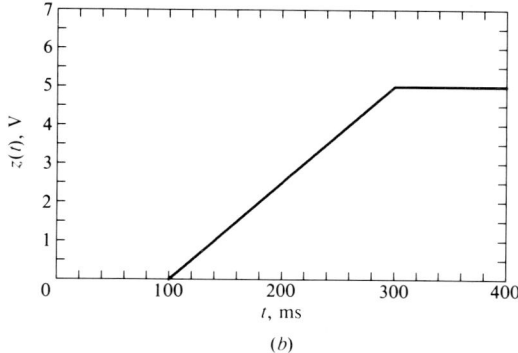

(b)

Figure 9-15 *Continued*

be self-explanatory. The plots extend from $t = 0$ to $t = 400$ ms and the plotting time step is 8 ms, as specified by FINTIM and OUTDEL, respectively.

9-1K Differentiation

A differentiation element having input $x(t)$ and output $y(t)$ is described by a transfer characteristic $y(t) = K\, dx(t)/dt$, where K is a constant having the unit of $y(t)t/x(t)$. This element is described in a CSMP simulation by the statement

$$Y = \text{DERIV}(Y0, K * X)$$

where X = input $x(t)$
 Y = corresponding output $y(t)$
 Y0 = initial value $y(0)$

This initial value must be specified because CSMP cannot calculate it. For $t = 0$ only one value of $x(t)$ is known, namely $x(0)$, whereas numerical differentiation of $x(t)$ requires at least two values of $x(t)$. When $y(0)$ cannot be determined, we must still provide a value for $y(0)$, remembering when we interpret results that the value provided may be incorrect. We cannot simply tell CSMP to leave $y(0)$ undefined; digital computers are notoriously intolerant of undefined quantities.

Example 9-4 To illustrate use of the CSMP differentiation model we simulate the system of Fig. 9-16 for input

$$x(t) = x_0 \cos(\omega_0 t + \theta_0)$$

where $x_0 = 5$ V, $f_0 = 10$ kHz, $\theta_0 = \pi/4$ rad, and $K = 10^{-4}$ mAs/V. The initial value of the output is

$$y(0) = K \left.\frac{dx(t)}{dt}\right|_{t=0} = \left.-Kx_0\omega_0 \sin(\omega_0 t + \theta_0)\right|_{t=0}$$

$$= (-10^{-7})(5)(2\pi \times 10^4) \sin \frac{\pi}{4} = -22.2 \text{ mA}$$

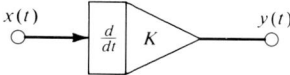

Figure 9-16 System of Example 9-4.

```
TITLE DIFFERENTIATOR WITH SINUSOIDAL INPUT

INITIAL

* X0 = PEAK AMPLITUDE OF INPUT (V)
* F0 = FREQUENCY OF INPUT (HZ)
* OMEGA = ANGULAR FREQUENCY OF INPUT (RAD/S)
* THETA = INITIAL PHASE OF INPUT (RAD)
* K = SCALE FACTOR FOR DIFFERENTIATOR (A S/V)
* Y0 = INITIAL VALUE OF DIFFERENTIATOR OUTPUT (A)

        PARAMETER X0 = 5., F0 = 1.E4, PI = 3.14, K = 1.E-7, ...
                  Y0 = - 0.0222

        OMEGA = 2. * PI * F0
        THETA = PI/4.

DYNAMIC

* X = INPUT (V), Y = OUTPUT (A)

        X = X0 * COS(OMEGA * TIME + THETA)
        Y = DERIV(Y0, K * X)

TERMINAL

        TIMER FINTIM = 2.E-4, OUTDEL = 4.E-6
        PRTPLT Y

END
STOP
```

Figure 9-17 CSMP program for simulating the system of Fig. 9-16.

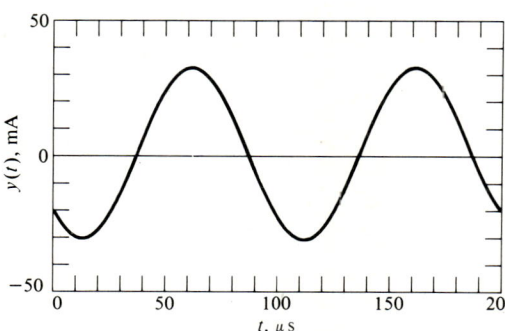

Figure 9-18 Output of the program of Fig. 9-17.

Figure 9-17 shows a CSMP program for simulating this system. Figure 9-18 shows the plot produced by the program.

9-1L Static Elements

Elementary static systems can be modeled in a CSMP simulation using FORTRAN statements and built-in CSMP functions. Example 9-1 illustrates use of a FORTRAN statement in modeling a full-wave rectifier. Below we illustrate use of CSMP functions in modeling comparators and limiters.

CSMP provides a comparator function invoked by a statement of the form

$$Y = \text{COMPAR}(X, \text{THOLD})$$

Figure 9-19 Transfer characteristic for the CSMP COMPAR function.

where X denotes an input, THOLD a threshold, and Y an output. The corresponding transfer characteristic is

$$Y = \begin{cases} 0 & X < \text{THOLD} \\ 1 & X \geq \text{THOLD} \end{cases}$$

as shown in Fig. 9-19. There are two differences between this transfer characteristic and the transfer characteristic for a comparator defined in Table 1-4. One is that an output of the CSMP function is nonzero if the input equals zero. This difference is insignificant in practical applications. The other is that a nonzero output of the CSMP function is unity whereas a nonzero output of a comparator as defined in Table 1-4 is a dimensioned quantity (denoted by y_0 in Table 1-4). Modeling a comparator as defined in Table 1-4 requires two statements in a CSMP program:

$$X1 = \text{COMPAR}(X, \text{THOLD})$$

$$Y = Y0 * X1$$

Example 9-5 Figure 9-20 shows the transfer characteristic for a comparator having input $x(t)$ and output $y(t)$. We wish to use CSMP to plot the output of this comparator for input

$$x(t) = 10 \cos \omega_0 t \quad \text{V}$$

where $f_0 = 4$ kHz. Figure 9-21 shows a CSMP program for this purpose. Figure 9-22 shows the plot of the output $y(t)$ versus time t produced by this program.

Figure 9-20 Transfer characteristic for the comparator of Example 9-5.

CSMP provides a limiting function invoked by a statement of the form

$$Y = \text{LIMIT}(XLO, XHI, X)$$

where X denotes an input and Y denotes an output. Figure 9-23 shows the transfer characteristic for this function. The transfer characteristic of Fig. 9-23 differs from that of Table 1-4 in two ways: first, the transfer characteristic of Fig. 9-23 is not necessarily

594 INTRODUCTION TO SYSTEM ANALYSIS

```
TITLE COMPARATOR WITH SINUSOIDAL INPUT

INITIAL

* X0 = PEAK AMPLITUDE OF INPUT (V)
* F0 = FREQUENCY OF INPUT (HZ)
* OMEGA = FREQUENCY OF INPUT (RAD/S)
* THOLD = COMPARATOR THRESHOLD (V)
* Y0 = COMPARATOR OUTPUT (V)

        PARAMETER X0 = 10., F0 = 4.E3, THOLD = 7., Y0 = 5., ...
                  PI = 3.14

        OMEGA = 2. * PI * F0

DYNAMIC

* X = INPUT (V), Y = OUTPUT (V)

        X = X0 * COS(OMEGA * TIME)
        X1 = COMPAR(X, THOLD)
        Y = Y0 * X1

TERMINAL

        TIMER FINTIM = 5.E-4, OUTDEL = 1.E-5
        PRTPLT Y

END
STOP
```

Figure 9-21 CSMP program for the system of Example 9-5.

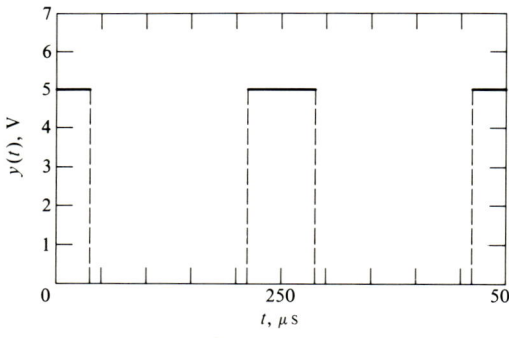

Figure 9-22 Output of the program of Fig. 9-21.

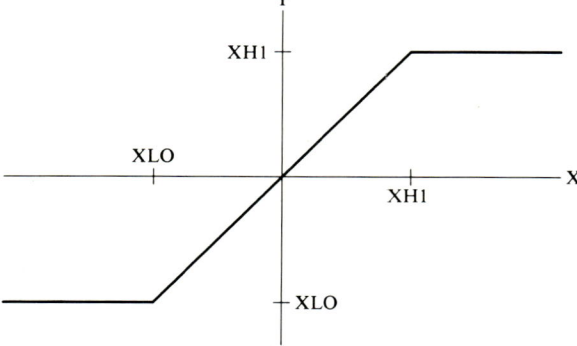

Figure 9-23 Transfer characteristic for the CSMP LIMIT function.

symmetric (i.e., XLO is not necessarily equal to −XHI); second, for the transfer characteristic of Fig. 9-23 input X and output Y have the same unit.

A limiter having the transfer characteristic given in Table 1-3 is modeled in a CSMP simulation by the statements

$$X1 = \text{LIMIT}(-X0, X0, X)$$
$$Y = Y0 * X1/X0$$

where $X0 = x_0$
$Y0 = y_0$
$X = \text{input } x(t)$
$Y = \text{output } y(t)$

Example 9-6 Figure 9-24 shows the transfer characteristic for a limiter having input $x(t)$ and output $y(t)$. Figure 9-25 shows a CSMP simulation of this limiter for input

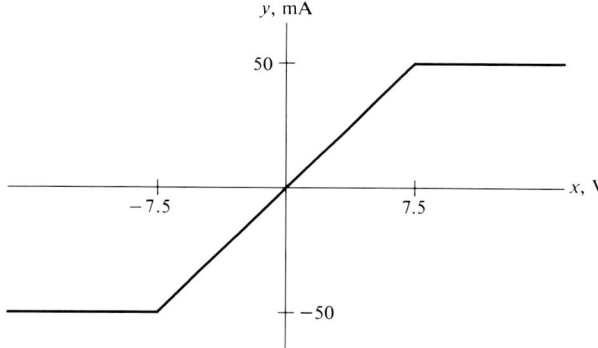

Figure 9-24 Transfer characteristic for the limiter of Example 9-6.

```
TITLE SOFT LIMITER WITH SINUSOIDAL INPUT
INITIAL
* X0 = PEAK AMPLITUDE OF INPUT (V)
* F0 = FREQUENCY OF INPUT (HZ)
* OMEGA = ANGULAR FREQUENCY OF INPUT (RAD/S)
* -XSAT,XSAT = INPUT AMPLITUDE FOR SATURATION (V)
* YSAT = SATURATED OUTPUT (A)
         PARAMETER X0 = 10., F0 = 1.E4, XSAT = 7.5, ...
                   YSAT = 0.05, PI = 3.14
         OMEGA = 2. * PI * F0
DYNAMIC
* X = INPUT (V), Y = OUTPUT (A)
         X = X0 * COS(OMEGA * TIME)
         X1 = LIMIT(-XSAT, XSAT, X)
         Y = YSAT * X1/XSAT
TERMINAL
         TIMER FINTIM = 2.E-4, OUTDEL = 4.E-6
         PRTPLT Y
END
STOP
```

Figure 9-25 CSMP program for the system of Example 9-6.

596 INTRODUCTION TO SYSTEM ANALYSIS

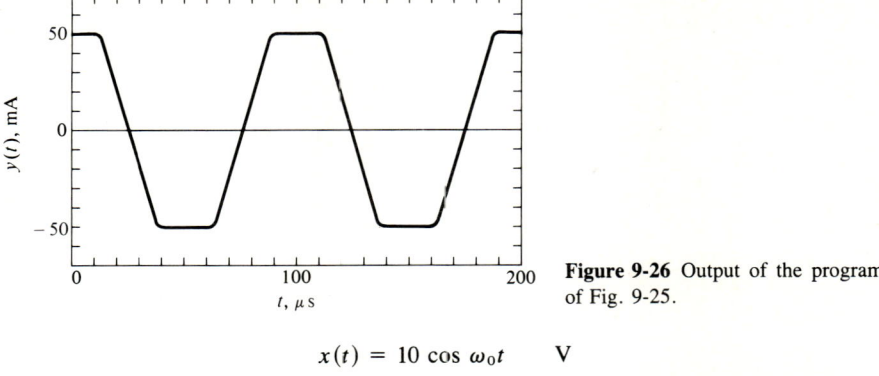

Figure 9-26 Output of the program of Fig. 9-25.

$$x(t) = 10 \cos \omega_0 t \quad \text{V}$$

where $f_0 = 10$ kHz. Figure 9-26 shows the output produced by this program.

9-1M Feedback Systems

One of the remarkable features of CSMP is that it automatically sorts statements (within a section) so that they are used in the proper order. We can even write the dynamic section of a simulation in reverse order (e.g., from output to input) and CSMP will put it in the right order before running the simulation. This feature is particularly helpful in simulating a feedback system, where a dynamic section can be written in various ways. Nonetheless, in simulating feedback systems it is helpful to follow a systematic procedure for writing the dynamic section of a simulation; otherwise it is surprisingly easy to omit an element or even an entire feedback path. The procedure we recommend is illustrated in Fig. 9-27. First write statements for all forward paths and then write statements for all feedback paths, working from innermost to outermost. For complicated systems it may be helpful to show and number all paths on a block diagram, as illustrated in Fig. 9-27.

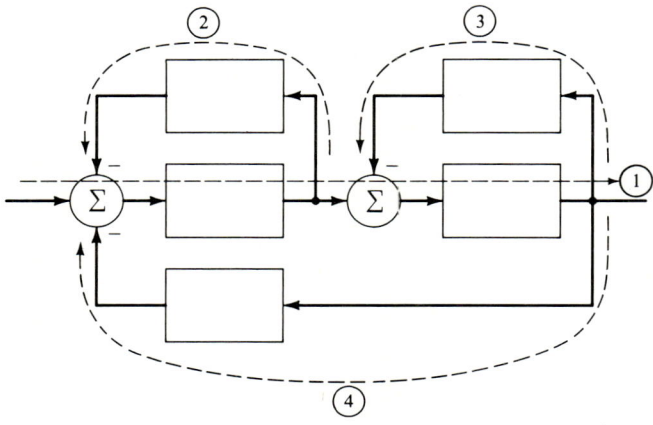

Figure 9-27 Systematic procedure for writing the dynamic section of a CSMP program for a multiloop feedback system.

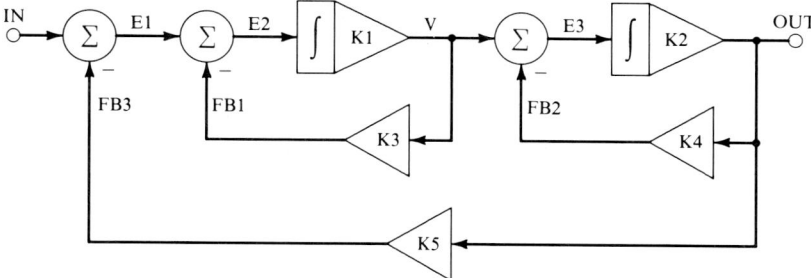

Figure 9-28 System of Example 9-7.

```
* FORWARD PATH

    E1  = IN - FB3
    E2  = E1 - FB1
    V   = INTGRL(V0, K1 * E2)
    E3  = V - FB2
    OUT = INTGRL(OUT0, K2 * E3)

* FIRST INNER FEEDBACK PATH

    FB1 = K3 * V

* SECOND INNER FEEDBACK PATH

    FB2 = K4 * OUT

* OUTER FEEDBACK PATH

    FB3 = K5 * OUT
```

Figure 9-29 Dynamic section of a CSMP program for the system of Fig. 9-28.

Example 9-7 Figure 9-28 shows a three-loop feedback system fully labeled in preparation for a CSMP simulation of the system. Figure 9-29 shows the dynamic section of a CSMP program for simulating the system. Comments in the program identify the paths followed in writing the program. The forward path is described first, then the two innermost feedback paths, and finally the outermost feedback path.

Example 9-8 In this example we simulate the system of Fig. 9-30 for input

$$v(t) = v_0 \cos \omega_0 t$$

with $v_0 = 100$ mV and $f_0 = 100$ Hz. We use variable-step integration.

First, we label the block diagram fully, using names acceptable to CSMP for all variables and parameters (already done in Fig. 9-30). Second, we express all parameters in standard SI units, as shown in Table 9-3.

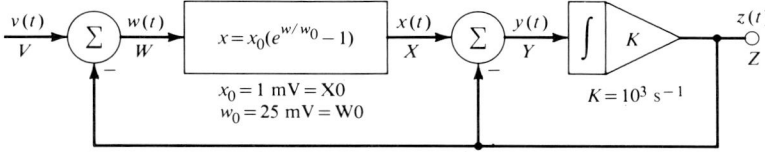

Figure 9-30 System of Example 9-8.

Table 9-3 Parameters for Example 9-8

Parameter	Value	Symbol	CSMP value
v_0	100 mV	V0	0.100 (V)
f_0	100 Hz	F0	100 (Hz)
x_0	1 mV	X0	0.001 (V)
w_0	25 mV	W0	0.025 (V)
K	1 (ms)$^{-1}$	K	1000 (s^{-1})

Third, we attempt to choose reasonable values for FINTIM, OUTDEL, and other CSMP control parameters. Here we must make some educated guesses because we cannot obtain a closed-form expression for the output $z(t)$. In some cases, several runs may be needed to home in on satisfactory values for these parameters. We choose FINTIM = 0.025 (25 ms), representing 2.5 periods of the input. We choose OUTDEL = 5.E−4 (500 μs), which gives 20 points plotted per period of the input. We use variable-step integration (the default method under CSMP). For nominal step size we use DELT = OUTDEL = 5.E−4 (500 μs), and for minimum step size we use DELMIN = OUTDEL/10 = 5.E−5. Figure 9-31 shows a CSMP program for simulating the system. Figure 9-32 shows the plot of output $z(t)$ versus time t produced by this program.

The plot of the output versus time in Fig. 9-32 indicates that our initial guess for the value of FINTIM was reasonable (because the output reaches steady state). To obtain a smoother plot we repeat the simulation using a smaller value for OUTDEL. This is left as an exercise.

```
TITLE NONLINEAR FEEDBACK SYSTEM

INITIAL

* V0 (V), F0 (HZ) = PEAK AMPLITUDE AND FREQUENCY OF THE INPUT
* X0 (V), W0 (V) = PARAMETERS OF THE STATIC ELEMENT
* K (1/S) = SCALE FACTOR FOR THE INTEGRATOR

    PARAMETER V0 = 0.10, F0 = 100., X0 = 1.E-3, W0 = 0.025, K = 1000.

* OMEGA = ANGULAR FREQUENCY (RAD/S) OF THE INPUT

    OMEGA = 6.28 * F0

DYNAMIC

    V = V0 * COS(OMEGA * TIME)
    W = V - Z
    X = X0 * (EXP(W/W0) - 1.)
    Y = X - Z
    Z = INTGRL(0., K * Y)

TERMINAL

    TIMER FINTIM = 0.025, DELT = 5.E-4, DELMIN = 5.E-5, OUTDEL = 0.0005
    PRTPLT Z

END
STOP
```

Figure 9-31 CSMP program for the system of Fig 9-30.

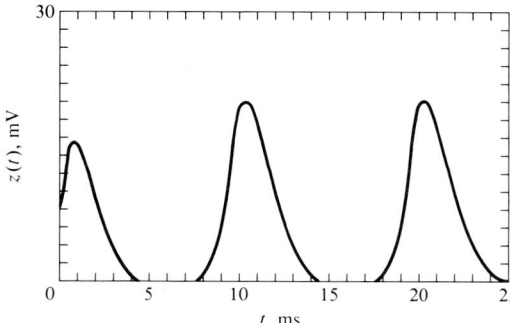

Figure 9-32 Output of the system of Example 9-8.

9-1N Algebraic Loops

Some feedback systems have *algebraic loops,* which require special handling under CSMP. To explain fully what constitutes an algebraic loop and how they are handled we would have to explain the inner workings of CSMP in more detail than is appropriate here; however, we can illustrate the basic principle involved. Figure 9-33 shows a feedback system consisting of only static elements. The dynamic section of a CSMP simulation for the system consists of the statements

$$E = X - Y$$
$$Y = Y0 * (E/E0) ** 2$$

Let us attempt to step through these statements in approximately the same way that CSMP would. First, we determine the variable (input) X. Next, we attempt to calculate E, for which calculation we need the value of Y. At this point, we begin chasing our tail, for in order to calculate E we must first calculate Y and in order to calculate Y we must first calculate E. This chicken-and-egg situation results because all variables in the loop are related algebraically; i.e., at any instant every variable in the loop depends on every other variable at the same instant. When CSMP encounters such a situation, it halts and prints the message "simulation involves an algebraic loop containing the following elements" followed by a list of variables encountered in the loop.

The presence of an algebraic loop often indicates an error, either in modeling or programming. In these cases, correcting the error eliminates the problem. But sometimes an algebraic loop results from a correct model; consequently, CSMP provides special facilities for handling algebraic loops. Since this topic is beyond the scope of this book, we avoid algebraic loops altogether, which requires that every loop contain an integration element or a delay element. Either element allows CSMP to update variables in the loop in a self-consistent manner, because an output of either element

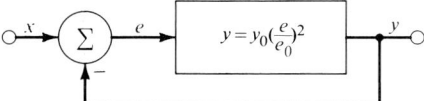

Figure 9-33 An algebraic loop.

is determined using only past inputs. A differentiation element, though dynamic, does not break an algebraic loop because CSMP uses the present value of an input to a differentiation element to calculate the corresponding value of the output.

9-10 Limit Checking

Establishing the correctness of a simulation is relatively easy for a system that can also be treated using only pencil and paper. It is much more difficult when simulation is a method of last resort, e.g., for a system that is mathematically intractable. Nonetheless, it must be attempted, for important decisions may be made on the basis of a simulation.

There are four ways to check the correctness of a simulation:

1. Closed-form solution
2. Independent simulation by another party
3. Breadboarding (building a prototype)
4. Limit checking, e.g., simulating degenerate cases

All these methods are useful in some cases, but closed-form solution is often impossible, independent simulation may leave us wondering which simulation (if either) is correct, and avoiding breadboarding may be the main reason for doing a simulation in the first place. Limit checking is generally the simplest, the least expensive, and the most widely useful of the methods listed above. Limit checking may include any or all of the following:

1. Modifying or removing troublesome elements and inputs
2. Checking initial and final values of an output
3. Using extreme values for selected quantities

Though limit checks are inconclusive, it is unlikely that an erroneous simulation can pass a number of well-designed limit checks.

Example 9-9 This example illustrates limit-checking methods for the system of Fig. 9-34a, where

$$w(t) = w_0 \left(\frac{t}{\tau}\right)^2 e^{-t/\tau}(\cos \omega_0 t) u(t)$$

This problem is complicated by three things: (1) feedback, (2) a nonlinear element (the limiter), and (3) the form of the input. Three possible limit checks are as follows:

1. Break the feedback path and determine (mathematically) the output for a simpler input (e.g., a step). In planning for this limit check, it is helpful to insert a dummy proportion element in the feedback path before writing the CSMP program, as shown in Fig. 34b, where we set $K_2 = 0$ for the limit check and $K_2 = 1$ for the actual simulation.
2. Replace the limiter by a proportion element having gain y_0/x_0; this should have no effect so long as the magnitude of $x(t)$ remains smaller than that of x_0.
3. Determine both mathematically and by simulation the steady-state output for a step input. In steady state the output for a step input should be constant (assuming that the

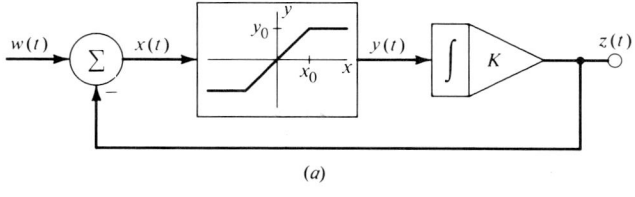

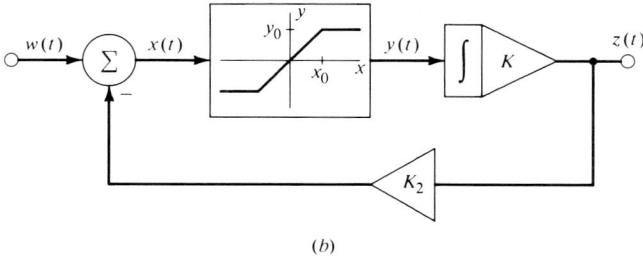

Figure 9-34 System of Example 9-9.

system is stable). This means that in steady state the input to the integration element should be zero, which in turn implies that in steady state the output should equal the input. If simulation gives other than a constant output, either the system is not stable or the simulation is erroneous.

9-1P Summary

To summarize this section we outline a procedure for carrying out a simulation using CSMP:

1. Decide in advance how the simulation will be tested and results checked. Plan all limit (and other) checks and (if necessary) add dummy elements to the block diagram to facilitate limit checking.
2. Check the block diagram for algebraic loops (loops containing neither a delay nor an integration element).
3. Fully label the block diagram for the system, assigning valid CSMP names to all variables and parameters. Use names that are mnemonic.
4. Make a table of all parameters and express all parameters in standard SI units without prefixes.
5. Determine trial values for FINTIM, DELT, OUTDEL, PRDEL, and (if variable-step integration is used) DELMIN. FINTIM must be an integral multiple of both OUTDEL and PRDEL. The larger of OUTDEL and PRDEL must be an integral multiple of the smaller. The smaller of OUTDEL and PRDEL must be an integral multiple of DELT. If delay elements are present, use fixed-step integration and make DELT an integral submultiple of every delay time.
6. Write the program. In writing the dynamic section describe all forward paths first. Then describe all feedback paths, working from innermost to outermost. It may be helpful to show and number all paths on the block diagram and to use comments

to identify corresponding statements in the program. Check to be sure that all parameters referred to in the dynamic section are defined in the initial section. In writing the terminal section, be sure to include a timer statement and at least one output statement. If the system contains both integration and delay elements, use the statement METHOD RKSFX to specify fixed-step integration. Terminate the program with END and STOP statements.
7. Run the test cases and check results. If fixed-step integration is used, halve DELT and run the program a second time. For the second run be sure to adjust (double) the first parameter (N) in the argument of each DELAY function and to maintain proper relationships between DELT, FINTIM, OUTDEL, and PRDEL. If both runs give the same result, the value selected for DELT is probably small enough. If these runs give different results, continue halving DELT and repeating the simulation until two consecutive runs agree. Adjust FINTIM, OUTDEL, and PRDEL if necessary before the final run.

9-2 LINEAR STATIONARY SYSTEMS

This section describes simulation and computer-aided analysis of a linear stationary system described either by an impulse response or by a differential equation. We first show that a transfer characteristic for a realizable, stable linear stationary system can be approximated using a tapped delay line and how such an approximation can be simulated using CSMP. We then show how a convolution integral can be computed numerically by a FORTRAN program and finally how CSMP can be used to simulate a linear stationary system described by a differential equation.

9-2A Simulation Using a Tapped Delay Line

From (2-3), the output $y(t)$ of a linear stationary system for input $x(t)$ is given by

$$y(t) = \int_{-\infty}^{\infty} x(t - \lambda) h(\lambda) \, d\lambda \qquad (9\text{-}1)$$

where $h(t)$ is the impulse response of the system. After approximating the integral by a sum having a finite number of terms, we show how a system described by (9-1) can be simulated using a tapped delay line.

In order to approximate the convolution integral of (9-1) by a sum having a finite number of terms we must replace the limits of integration by finite quantities. We do so by assuming that the system described by (9-1) is realizable and stable. For a realizable system,

$$h(t) = 0 \qquad t < 0 \qquad (9\text{-}2)$$

For a stable system,

$$\int_{-\infty}^{\infty} |h(t)| \, dt < \infty$$

This condition implies that the impulse response of a stable system is significantly

different from zero only over a finite interval, say, $0 < t < T$; thus
$$h(t) \approx 0 \quad t > T \quad (9\text{-}3)$$
Using (9-2) and (9-3), we can write (9-1) as
$$y(t) \approx \int_0^T x(t - \lambda)h(\lambda)\,d\lambda \quad (9\text{-}4)$$
The integral in (9-4) can be approximated by a sum
$$y(t) \approx \sum_{n=0}^{N-1} x(t - n\,\Delta\lambda)h(n\,\Delta\lambda)\,\Delta\lambda \quad (9\text{-}5)$$
where $N = T/\Delta\lambda$. This approximation improves as $\Delta\lambda$ is decreased. In particular applications $\Delta\lambda$ should be small compared with the duration of the narrowest significant feature of either the input $x(t)$ or the impulse response $h(t)$; for example, if the input is a sinusoid having period T and the impulse response is an exponential pulse having time constant τ, we require $\Delta\lambda$ to be small compared with the smaller of T and τ.

Figure 9-35 shows a block diagram for a system whose output $y(t)$ for input $x(t)$ is given by (9-5). A system described by a block diagram of this form is called a *tapped delay line*, where "delay line" refers to the series connection of delay elements and "tapped" refers to the fact that delayed values of an input are obtained by tapping the delay line (at delay times $0, \Delta\lambda, 2\Delta\lambda, \ldots$). The delay between adjacent taps is called *tap spacing*, the gains $h(0)\,\Delta\lambda$, $h(\Delta\lambda)\,\Delta\lambda$, ... are called *tap weights*, and the total delay $(N - 1)\,\Delta\lambda$ is called the *length* of the delay line. In most cases the number of taps N is large enough for the length of the delay line to be given to a good approximation by
$$(N - 1)\,\Delta\lambda \approx N\,\Delta\lambda$$
The length of the delay line must equal or exceed the effective duration of the impulse response of the system being simulated, and the tap spacing $\Delta\lambda$ must be small compared with the duration of the narrowest significant feature of either the impulse response $h(t)$ or an input $x(t)$.

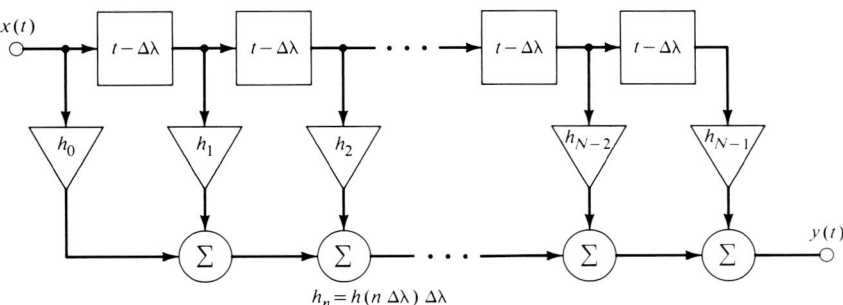

Figure 9-35 Block diagram for a tapped delay line.

Example 9-10 In this example we use a tapped delay line and CSMP to simulate a linear stationary system having impulse response

$$h(t) = \tau_1^{-1} e^{-t/\tau_1} u(t)$$

and input

$$x(t) = x_0 r\left(\frac{t}{\tau_2}\right)$$

where $\tau_1 = 50$ ms, $x_0 = 5$ V, and $\tau_2 = 100$ ms.

To specify the delay line we must specify the tap spacing $\Delta\lambda$ and the number of taps N (or the length of the line). Tap spacing should be small compared with the duration of the narrowest significant feature of either the input $x(t)$ or the impulse response $h(t)$. In this case the duration of the narrowest significant feature of the input is the pulse duration $\tau_2 = 100$ ms, and the duration of the narrowest significant feature of the impulse response is the time constant $\tau_1 = 50$ ms. Consequently, we require $\Delta\lambda < 50$ ms. To avoid using a large number of taps in this example we choose $\Delta\lambda = 10$ ms and take $T = 3\tau_1 = 150$ ms as the effective duration of the impulse response (the length of the delay line); thus the number of taps is $N = 150/10 = 15$. Figure 9-36 shows a block diagram for the tapped delay line of this example. The tap weights are given by

$$h(n\,\Delta\lambda)\,\Delta\lambda = \tau_1^{-1} e^{-n\,\Delta\lambda/\tau_1} \Delta\lambda\, u(n\,\Delta\lambda)$$

$$= \begin{cases} 0 & n = 0 \\ 0.2 e^{-0.2n} & n = 1, 2, \ldots, 15 \end{cases}$$

Figure 9-37 shows a CSMP program for simulating the tapped delay line of Fig. 9-36, and Fig. 9-38 shows the graph of the output $y(t)$ versus time t produced by the program.

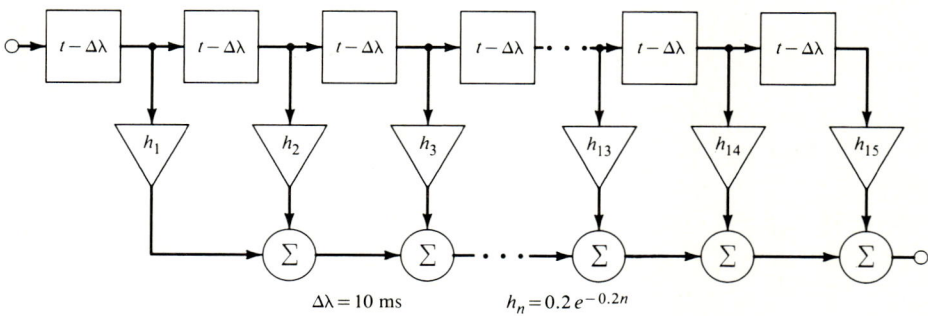

Figure 9-36 System of Example 9-10.

Simulating a system whose impulse response contains a delta function presents a problem. Since the duration of a delta function is zero, there is no tap spacing that is small compared with the duration of the narrowest significant feature of the impulse response. In order to simulate such a system we treat the delta function separately, as illustrated in the next example.

Example 9-11 We wish to simulate a system having impulse response

$$h(t) = K\delta(t) - K\tau_1^{-1} e^{-t/\tau_1} u(t)$$

```
TITLE TAPPED DELAY LINE
INITIAL
*   H1 - H15 ARE THE TAP WEIGHTS  X0 (V) AND TAU2 (S) ARE THE
*   AMPLITUDE AND DURATION OF THE INPUT. DT IS TAP SPACING.
    PARAMETER H1 = 0.1638, H2 = 0.1341, H3 = 0.1098,...
              H4 = 0.0899, H5 = 0.0736, H6 = 0.0602,...
              H7 = 0.0493, H8 = 0.0404, H9 = 0.0331,...
              H10 = 0.0271, H11 = 0.0222, H12 = 0.0181,...
              H13 = 0.0149, H14 = 0.0122, H15 = 0.0100,...
              X0 = 5., TAU2 = 0.100, DT = 0.010
DYNAMIC
*   INPUT X(T) = RECTANGULAR PULSE
    X = X0 * (1. - STEP(TAU2))
*   THE FOLLOWING STATEMENTS DESCRIBE THE DELAY LINE
    X1  = DELAY(10, DT, X)
    X2  = DELAY(10, DT, X1)
    X3  = DELAY(10, DT, X2)
    X4  = DELAY(10, DT, X3)
    X5  = DELAY(10, DT, X4)
    X6  = DELAY(10, DT, X5)
    X7  = DELAY(10, DT, X6)
    X8  = DELAY(10, DT, X7)
    X9  = DELAY(10, DT, X8)
    X10 = DELAY(10, DT, X9)
    X11 = DELAY(10, DT, X10)
    X12 = DELAY(10, DT, X11)
    X13 = DELAY(10, DT, X12)
    X14 = DELAY(10, DT, X13)
    X15 = DELAY(10, DT, X14)
*   THE NEXT STATEMENT SUMS THE WEIGHTED TAP OUTPUTS
*   TO OBTAIN THE OUTPUT Y(T)
    Y = H1 * X1 + H2 * X2 + H3 * X3 + H4 * X4 + H5 * X5 ...
      + H6 * X6 + H7 * X7 + H8 * X8 + H9 * X9 + H10 * X10 + ...
        H11 * X11 + H12 * X12 + H13 * X13 + H14 * X14 + H15 * X15
TERMINAL
    TIMER FINTIM = 0.300, OUTDEL = 0.006, DELT = 0.001
    PRTPLT Y
END
STOP
```

Figure 9-37 CSMP program for the system of Fig. 9-36.

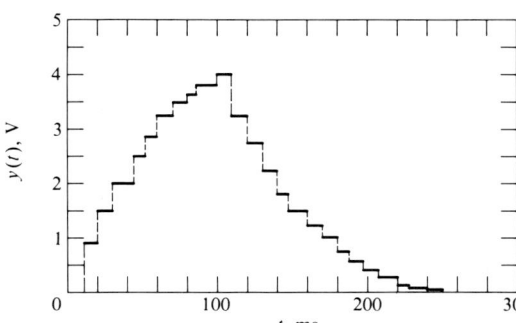

Figure 9-38 Output of the system of Example 9-10.

and input

$$x(t) = x_0 r\left(\frac{t}{\tau_2}\right)$$

with $K = 1$ V/V, $\tau_1 = 1$ ms, $x_0 = 5$ V, and $\tau_2 = 4$ ms. We cannot proceed exactly as in Example 9-10 because there is no tap spacing that is smaller than the duration of the narrowest significant feature of $h(t)$. We proceed as follows. The output for input $x(t)$ is given by

$$y(t) = \int_{-\infty}^{\infty} x(t-\lambda)h(\lambda)\,d\lambda = \int_{-\infty}^{\infty} x(t-\lambda)[K\delta(\lambda) - h_0(\lambda)]\,d\lambda$$

where $h_0(t) = K\tau_1^{-1}e^{-t/\tau_1}u(t)$. This reduces to

$$y(t) = Kx(t) - \int_{-\infty}^{\infty} x(t-\lambda)h_0(\lambda)\,d\lambda$$

Thus, we have separated the system into two subsystems, as shown by the block diagram of Fig. 9-39. To simulate this system we first design a tapped-delay-line simulation for the subsystem described by the impulse response

$$h_0(t) = K\tau_1^{-1}e^{-t/\tau_1}u(t)$$

This presents no special problems because $h_0(t)$ contains no delta function. We then subtract the output of the tapped delay line from the term $Kx(t)$ that arises from the delta function $K\delta(t)$ in the original impulse response $h(t)$.

The duration of the narrowest significant features of $h_0(t)$ and $x(t)$ are τ_1 and τ_2, respectively. Because $\tau_1 < \tau_2$, the tap spacing should be a fraction of τ_1. To keep the number of taps within reason (for this example), we use $\Delta\lambda = \tau_1/5 = 200$ μs. The duration of $h_0(t)$ is $T \approx 3\tau_1 = 3$ ms; thus the number of taps required is $N = T/\Delta\lambda = 15$. The tap weights are given by

$$h_{0n} = h_0(n\,\Delta\lambda)\,\Delta\lambda = K\tau_1^{-1}e^{-(n\Delta\lambda)/\tau_1}\Delta\lambda u(n\,\Delta\lambda)$$

$$= 0.2e^{-0.2n} \qquad n = 1, 2, \ldots, 15$$

Figure 9-40 shows a CSMP program for determining the output for input $x(t)$ given above, and Fig. 9-41 shows the graph of the output $y(t)$ versus time t produced by the program.

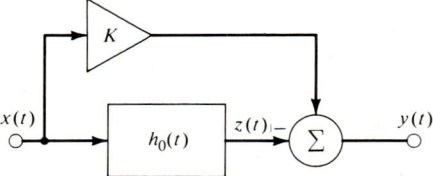

Figure 9-39 System of Example 9-11.

A tapped delay line is discussed above only as a means of simulating a linear stationary system. In some applications (notably in some radar systems), tapped delay lines also offer the best means of actually implementing certain operations.

9-2B Numerical Convolution

In practical cases accurate simulation (or realization) of a transfer characteristic using a tapped delay line may require hundreds or even thousands of taps. Simulation using

```
TITLE TAPPED DELAY LINE

INITIAL

*   H1 - H15 ARE THE TAP WEIGHTS  X0 (V) AND TAU2 (S) ARE THE
*   AMPLITUDE AND DURATION OF THE INPUT. DT IS TAP SPACING.
*   K IS THE COEFFICIENT OF THE DELTA FUNCTION IN H(T)

        PARAMETER K = 1.0, H1 = 0.1638, H2 = 0.1341, H3 = 0.1098,...
                  H4 = 0.0899, H5 = 0.0736, H6 = 0.0602,...
                  H7 = 0.0493, H8 = 0.0404, H9 = 0.0331,...
                  H10 = 0.0271, H11 = 0.0222, H12 = 0.0181,...
                  H13 = 0.0149, H14 = 0.0122, H15 = 0.0100,...
                  X0 = 5., TAU2 = 0.100, DT = 0.010

DYNAMIC

*   INPUT X(T) = RECTANGULAR PULSE

        X = X0 * (1. - STEP(TAU2))

*   THE FOLLOWING STATEMENTS DESCRIBE THE DELAY LINE

        X1  = DELAY(10, DT, X)
        X2  = DELAY(10, DT, X1)
        X3  = DELAY(10, DT, X2)
        X4  = DELAY(10, DT, X3)
        X5  = DELAY(10, DT, X4)
        X6  = DELAY(10, DT, X5)
        X7  = DELAY(10, DT, X6)
        X8  = DELAY(10, DT, X7)
        X9  = DELAY(10, DT, X8)
        X10 = DELAY(10, DT, X9)
        X11 = DELAY(10, DT, X10)
        X12 = DELAY(10, DT, X11)
        X13 = DELAY(10, DT, X12)
        X14 = DELAY(10, DT, X13)
        X15 = DELAY(10, DT, X14)

*   THE FOLLOWING STATEMENTS SUM THE WEIGHTED TAP OUTPUTS
*   TO OBTAIN THE SIGNAL Z(T) AND CALCULATE THE OUTPUT Y(T)

        Z = H1 * X1 + H2 * X2 + H3 * X3 + H4 * X4 + H5 * X5 ...
          + H6 * X6 + H7 * X7 + H8 * X8 + H9 * X9 + H10 * X10 + ...
            H11 * X11 + H12 * X12 + H13 * X13 + H14 * X14 + H15 * X15
        Y = K * X - Z

TERMINAL

        TIMER FINTIM = 0.300, OUTDEL = 0.006, DELT = 0.001
        PRTPLT Y

END
STOP
```

Figure 9-40 CSMP program for the system of Fig. 9-39.

CSMP is impractical when a large number of taps is required; it is easier (and considerably more economical) to calculate an output using FORTRAN (or any other language that is suitable for scientific computing). For such a calculation an input and the corresponding output must be represented by sequences of numbers because we can calculate the output only at certain predetermined times. A suitable formula for this purpose is obtained by taking $t = m\,\Delta\lambda$ in (9-5). This gives

608 INTRODUCTION TO SYSTEM ANALYSIS

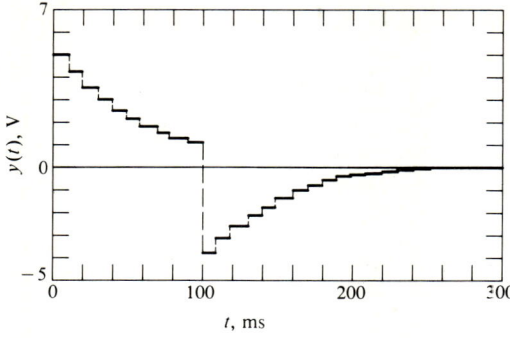

Figure 9-41 Output of the system of Example 9-11.

$$y(m\,\Delta\lambda) \approx \sum_{n=0}^{N-1} x(m\,\Delta\lambda - n\,\Delta\lambda)h(n\,\Delta\lambda)\,\Delta\lambda$$

or

$$y_m \approx \sum_{n=0}^{N-1} x_{m-n} h_n \qquad (9\text{-}6)$$

where $y_m = y(m\,\Delta\lambda)$, $x_{m-n} = x(m\,\Delta\lambda - n\,\Delta\lambda)$, and $h_n = h(n\,\Delta\lambda)\,\Delta\lambda$.

Example 9-12 In this example we use a FORTRAN program for numerical convolution to determine the output of a system having impulse response

$$h(t) = \omega_0 e^{-t/\tau}(\sin \omega_0 t)u(t)$$

and input

$$x(t) = x_0\left(1 - \frac{t}{\tau}\right)r\left(\frac{t}{\tau}\right)$$

where $f_0 = 1$ kHz, $\tau = 4$ ms, and $x_0 = 5$ V. Figure 9-42 shows graphs of the impulse response $h(t)$ and the input $x(t)$ versus time t. The duration of the narrowest significant feature of $h(t)$ is about $f_0^{-1} = 1$ ms, which is shorter than the duration of the narrowest significant feature of the input. The tap spacing $\Delta\lambda$ should be smaller than 1 ms. We use $\Delta\lambda = 50$ μs. The length T of the delay line should exceed the duration of the impulse response; we use $T = 4\tau = 16$ ms. The number of taps required is $N = T/\Delta\lambda = 320$. The output at times $t = 0, \Delta\lambda, 2\Delta\lambda, \ldots$ is calculated (approximately) using (9-6) with $N = 320$, $y_m = y(m\,\Delta\lambda)$,

$$x_{m-n} = x(m\,\Delta\lambda - n\,\Delta\lambda) = x_0\left[1 - (m-n)\frac{\Delta\lambda}{\tau}\right]r\left[(m-n)\frac{\Delta\lambda}{\tau}\right]$$

$$= 5[1 - 0.0125(m-n)]r[0.0125(m-n)] \quad \text{V}$$

and

$$h_n = h(n\,\Delta\lambda)\,\Delta\lambda = \omega_0 e^{-(n\,\Delta\lambda)/\tau} \sin(\omega_0 n\,\Delta\lambda) u(n\,\Delta\lambda)\,\Delta\lambda$$

$$= 0.1\pi e^{-0.0125n} \sin 0.1\pi n \qquad n = 1, 2, \ldots, 320$$

Figure 9-43 shows a FORTRAN program for calculating the sum above, and Fig. 9-44 shows a graph of the output $y(t)$ versus time t produced by the program.

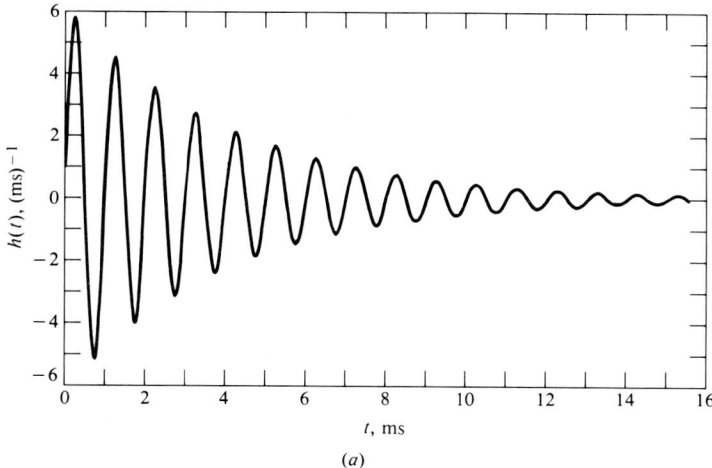

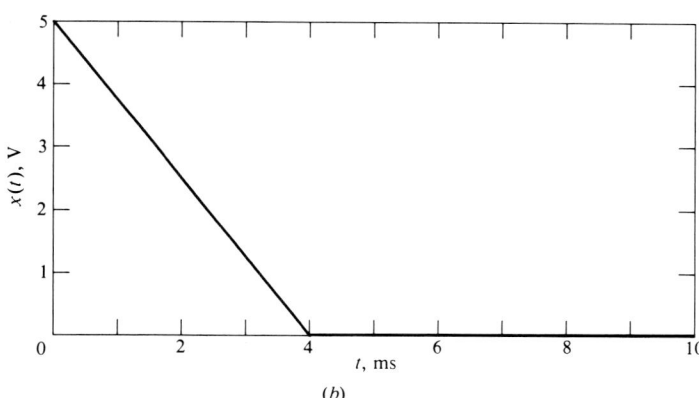

Figure 9-42 Impulse response $h(t)$ and input $x(t)$ for the system of Example 9-12.

9-2C Systems Described by Differential Equations

In this section we describe a method for simulating a system whose input $x(t)$ and output $y(t)$ are related by a differential equation of the form

$$\frac{d^n y}{dt^n} + a_{n-1}\frac{d^{n-1}y}{dt^{n-1}} + \cdots + a_0 y = b_n \frac{d^n x}{dt^n} + b_{n-1}\frac{d^{n-1}x}{dt^{n-1}} + \cdots + b_0 x \qquad (9\text{-}7)$$

We assume that the input $x(t)$ is zero for $t \leq 0$ and that the system described by (9-7) is realizable. It follows that the output $y(t)$, all derivatives of $y(t)$, and all derivatives of $x(t)$ are zero for $t < 0$ (see Sec. 2-4).

```
C       NUMERICAL CONVOLUTION
C
C       X = ARRAY CONTAINING SAMPLES OF THE INPUT
C       Y = ARRAY CONTAINING SAMPLES OF THE OUTPUT
C       H = ARRAY CONTAINING THE TAP WEIGHTS
C       T = ARRAY OF SAMPLING TIMES (FOR PLOTTING OUTPUT)
C
        REAL X(400), Y(400), H(320), T(400)
C
C       CALCULATE TAP WEIGHTS
C
        DO 10 N = 1, 320
        H(N) = 0.31416 * EXP(-0.0125 * N) * SIN(0.31416 * N)
10      CONTINUE
C
C       CALCULATE SAMPLES OF THE INPUT
C
        DO 20 N = 1, 400
        X(N) = 0.
        A = 0.0125 * N
        IF(A.GT.1.) GO TO 20
        X(N) = 5. * (1. - A)
20      CONTINUE
C
C       PERFORM CONVOLUTION NUMERICALLY
C
        DO 40 M = 1, 400
        Y(M) = 0.
        DO 30 N = 1, 320
        IF(M-N.LT.1) GO TO 40
        Y(M) = Y(M) + X(M-N) * H(N)
30      CONTINUE
40      CONTINUE
C
C       GENERATE ARRAY OF SAMPLING TIMES AND USE A
C       SUBROUTINE TO PLOT THE OUTPUT Y(T) VERSUS TIME T
C
        DO 50 N = 1, 400
        T(N) = N * 50.E-6
50      CONTINUE
        CALL SPLOT(T, Y, 400, 1, 'T_', 'Y(T)_')
        CALL PICSIZ(0., 0.)
        STOP
        END
```

Figure 9-43 Program for numerical convolution (see Example 9-12).

To simplify notation we use the operator D to denote differentiation with respect to time; thus

$$Df = \frac{df}{dt}$$

Using this notation and rearranging terms in (9-7) gives

$$D^n y = D^n(b_n x) + D^{n-1}(b_{n-1}x - a_{n-1}y) + D^{n-2}(b_{n-2}x - a_{n-2}y) + \cdots$$
$$+ D(b_1 x - a_1 y) + (b_0 x - a_0 y)$$

Integrating n times gives

$$y = b_n x + D^{-1}(b_{n-1}x - a_{n-1}y) + D^{-2}(b_{n-2}x - a_{n-2}y) + \cdots$$
$$+ D^{-(n-1)}(b_1 x - a_1 y) + D^{-n}(b_0 x - a_0 y)$$

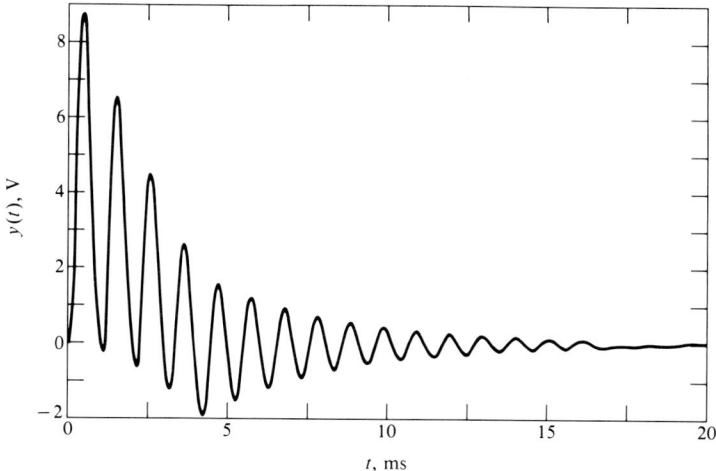

Figure 9-44 Output of the system of Example 9-12 (obtained by numerical convolution).

where
$$D^{-1}f = \int_{-\infty}^{t} f(\alpha)\,d\alpha$$

The expression above for $y(t)$ can be written

$$y = b_n x + D^{-1}w_1 = b_n x + \int_{-\infty}^{t} w_1(\alpha)\,d\alpha \tag{9-8}$$

where

$$w_1 = (b_{n-1}x - a_{n-1}y) + D^{-1}(b_{n-2}x - a_{n-2}y) + \cdots$$
$$+ D^{-(n-2)}(b_1 x - a_1 y) + D^{-(n-1)}(b_0 x - a_0 y) \tag{9-9}$$

Equation (9-8) is represented by the block diagram of Fig. 9-45. Because $x(t)$, $y(t)$, and all derivatives of $x(t)$ and $y(t)$ are zero for $t < 0$, we have $w_1(t) = 0$ for $t < 0$. Therefore the lower limit of integration in (9-8) can be changed to zero; thus

$$y(t) = b_n x(t) + \int_{0}^{t} w_1(\alpha)\,d\alpha$$

The corresponding CSMP statement is

$$Y = BN * X + INTGRL(0., W1)$$

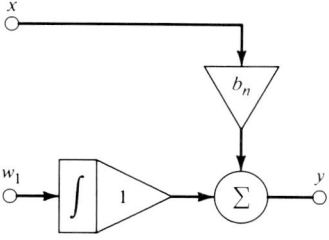

Figure 9-45 Block diagram for a CSMP statement of the form $Y = BN * X + INTGRL(0., W1)$.

612 INTRODUCTION TO SYSTEM ANALYSIS

Similarly, (9-9) can be written

$$w_1 = b_{n-1}x - a_{n-1}y + D^{-1}w_2 = b_{n-1}x - a_{n-1}y + \int_0^t w_2(\alpha)\,d\alpha \qquad (9\text{-}10)$$

where

$$w_2 = b_{n-2}x - a_{n-2}y + D^{-1}(b_{n-3}x - a_{n-3}y) + \cdots \\ + D^{-(n-3)}(b_1 x - a_1 y) + D^{-(n-2)}(b_0 x - a_0 y)$$

Equation (9-10) is represented by the block diagram of Fig. 9-46. The corresponding CSMP statement is

$$W1 = BN1 * X - AN1 * Y + INTGRL(0., W2)$$

Combining the block diagrams of Figs. 9-45 and 9-46 gives the block diagram of Fig. 9-47.

Continuing as indicated above, we obtain the block diagram of Fig. 9-48 and corresponding CSMP statements

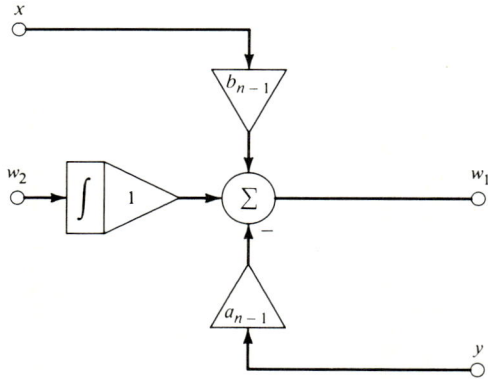

Figure 9-46 Block diagram for the CSMP statement of the form W1 = BN1 * X − AN1 * Y + INTGRL(0., W2).

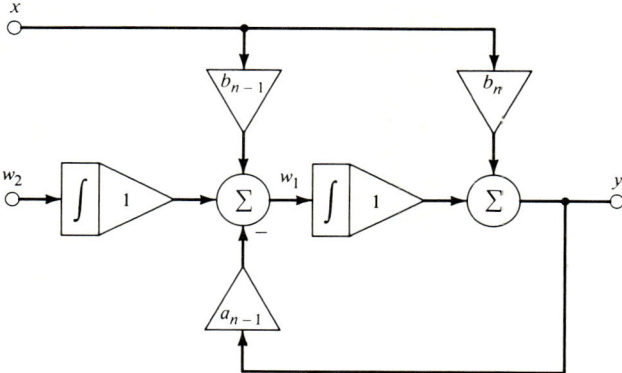

Figure 9-47 Block diagram for a segment of a linear stationary system.

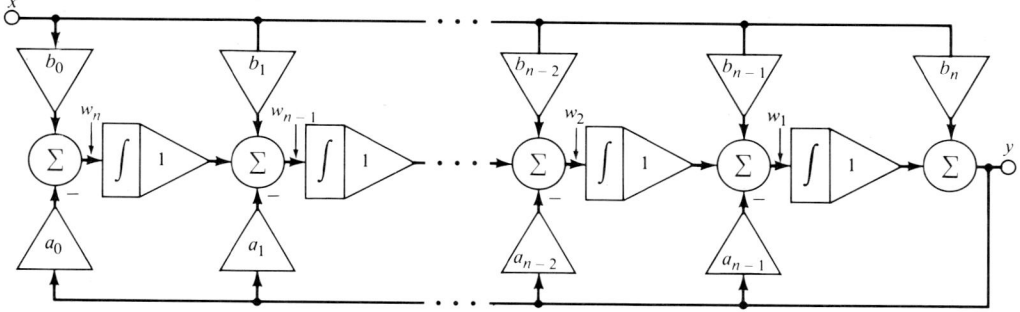

Figure 9-48 Block diagram for an nth-order linear stationary system.

$$WN = B0 * X - A0 * Y$$
$$WN1 = B1 * X - A1 * Y + INTGRL(0., WN)$$
$$\ldots$$
$$W2 = BN2 * X - AN2 * Y + INTGRL(0., W3)$$
$$W1 = BN1 * X - AN1 * Y + INTGRL(0., W2)$$
$$Y = BN * X + INTGRL(0., W1)$$

where $AN1 = a_{n-1}, \ldots, A0 = a_0, BN = b_n, BN1 = b_{n-1}, \ldots, WN = w_n$, $WN1 = w_{n-1}, \ldots, W1 = w_1$.

Example 9-13 A system having input $x(t)$ (voltage) and output $y(t)$ is described by the differential equation

$$\frac{d^2y}{dt^2} + a_1 \frac{dy}{dt} + a_0 y = b_1 \frac{dx}{dt} + b_0 x$$

with $a_1 = b_1 = 2.4$ (ms)$^{-1}$ and $a_0 = b_0 = 36$ (ms)$^{-2}$. We wish to use a CSMP simulation to determine the output of this system for input $x(t) = x_0 r(t/\tau)$, with $x_0 = 5$ V and $\tau = 3$ ms.

Table 9-4 gives the parameters for the simulation, Fig. 9-49 shows a block diagram for the simulation, Fig. 9-50 shows a CSMP program for the simulation, and Fig. 9-51 shows the graph of the output $y(t)$ versus time produced by this program. The duration of the simulation (FINTIM) is twice the duration of the input $x(t)$.

Table 9-4 Parameters for Example 9-13

Parameter	Value	CSMP symbol	CSMP value
a_0	36 (ms)$^{-2}$	A0	36×10^6 (s^{-2})
a_1	2.4 (ms)$^{-1}$	A1	2.4×10^3 (s^{-1})
b_0	36 (ms)$^{-2}$	B0	36×10^6 (s^{-2})
b_1	2.4 (ms)$^{-1}$	B1	2.4×10^3 (s^{-1})
x_0	5 V	X0	5 (V)
τ	3 ms	TAU	0.003 (s)

614 INTRODUCTION TO SYSTEM ANALYSIS

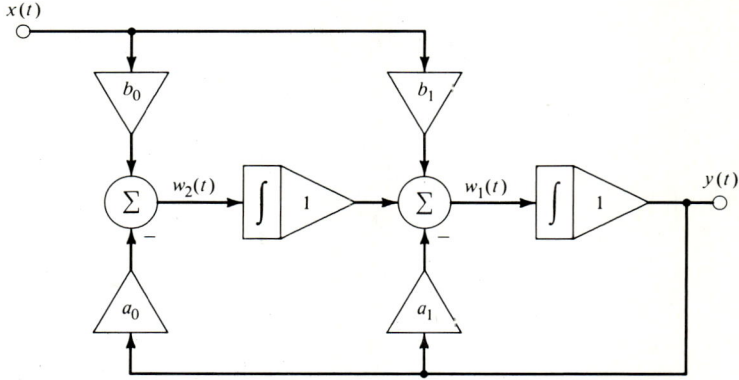

Figure 9-49 Block diagram for the system of Example 9-13.

```
TITLE SECOND-ORDER SYSTEM

* ILLUSTRATING SIMULATION OF A LINEAR STATIONARY
* SYSTEM DESCRIBED BY A DIFFERENTIAL EQUATION

INITIAL

* A1, A2, B1, B2 = COEFFICIENTS OF DIFF. EQN.
* UNIT OF A1, B1 IS 1/S; UNIT OF A0, B0 IS (1/S) ** 2
* X0 (V), TAU (S) = AMPLITUDE AND DURATION OF INPUT

      PARAMETER A1 = 2.4E3, B1 = 2.4E3, A0 = 36.E6,...
                B0 = 36.E6, X0 = 5., TAU = 0.003

DYNAMIC

* X (V) = INPUT; Y (V) = OUTPUT; W1, W2 = INTERMEDIATE
* SIGNALS (SEE FIG. 9-54).

      X = X0 * (1. - STEP(TAU))
      W2 = B0 * X - A0 * Y
      W1 = B1 * X - A1 * Y + INTGRL(0., W2)
      Y = INTGRL(0., W1)

TERMINAL

      TIMER FINTIM = 0.006
      PRTPLT Y

END
STOP
```

Figure 9-50 CSMP program for the system of Fig. 9-49.

The method described and illustrated above is one of several methods for simulating a linear stationary system described by (9-7). Another is obtained by separating a system described by (9-7) into feedforward and feedback subsystems, as described in Sec. 2-4 (see the discussion surrounding Figs. 2-5 and 2-6). This corresponds to writing (9-7) as two equations:

$$D^n w + a_{n-1} D^{n-1} w + \cdots + a_1 Dw + a_0 w = x \tag{9-11}$$

COMPUTER-AIDED ANALYSIS AND DESIGN **615**

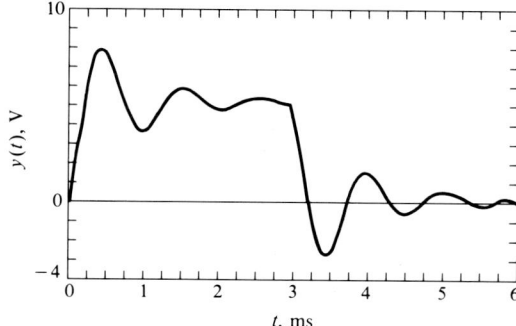

Figure 9-51 Output of the system of Example 9-13.

and
$$y = b_n D^n w + b_{n-1} D^{n-1} w + \cdots + b_1 Dw + b_0 w \tag{9-12}$$

Since (9-11) has the form of (9-7) with $b_0 = 1$ and $b_1 = b_2 = \cdots = b_n = 0$, the feedback subsystem described by (9-11) is represented by the block diagram of Fig. 9-52a. By inspection, the feedforward subsystem described by (9-12) is represented by the block diagram of Fig. 9-52b. A system described by (9-7) can be represented by cascading the subsystems of Fig. 9-52. In principle, the subsystems can be cascaded in either order, but for computational reasons it is generally best to place the feedforward section last (in order to provide smooth inputs to the differentiators).

The block diagram of Fig. 9-52 contains (in general) more elements than that of Fig. 9-48. This means that a simulation based on the block diagram of Fig. 9-52 will be more costly than one based on Fig. 9-48. Further, a simulation based on Fig. 9-52

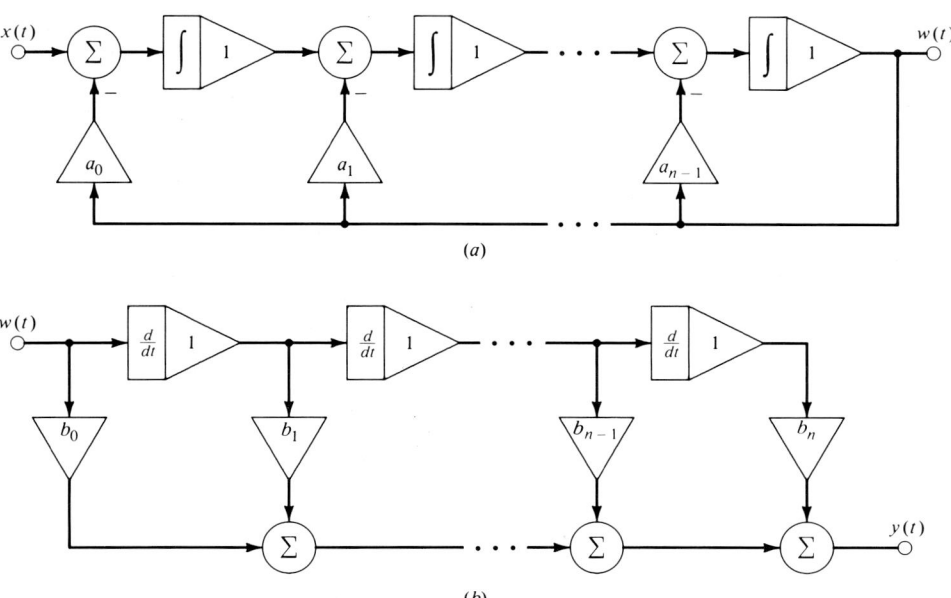

Figure 9-52 Separation of a linear stationary system into feedforward and feedback subsystems.

requires differentiation, whereas one based on Fig. 9-48 does not. In general, a simulation based on Fig. 9-48 will be more accurate than one based on Fig. 9-52, because accurate numerical differentiation is difficult to achieve. As a rule, the block diagram of Fig. 9-48 is a better basis for simulation than that of Fig. 9-52.

9-3 COMPUTER-AIDED FOURIER ANALYSIS

In this section we describe methods for calculating Fourier coefficients, Fourier series, Fourier transforms, and inverse Fourier transforms. These methods are useful in applications where closed-form expressions are difficult or impossible to obtain. Methods described here for Fourier analysis make use of the *fast-Fourier-transform* (FFT) algorithm, which greatly streamlines computer-aided Fourier analysis. We describe this algorithm first.

9-3A Fast Fourier Transformation

We show in Secs. 9-3B to 9-3E that numerical calculation of Fourier coefficients, numerical summation of Fourier series, numerical calculation of Fourier transforms, and numerical calculation of Fourier integrals (inverse Fourier transforms) all reduce to calculating sums of the form

$$\beta_k = \sum_{i=0}^{N-1} \alpha_i e^{\pm j2\pi ik/N} \qquad (9\text{-}13)$$

where α_i ($i = 0, 1, \ldots, N - 1$) and β_k ($k = 0, 1, \ldots, N - 1$) are (in general) complex. In this section, we describe an efficient algorithm for calculating such sums. We also describe two important properties of quantities β_k given by (9-13).

Direct calculation of any one of the quantities β_k ($k = 0, 1, \ldots, N - 1$) using (9-13) requires N complex multiplications and $N - 1$ complex additions because there are N products in the sum. We say that the calculation requires about N complex operations, where a complex operation is a complex multiplication followed by a complex addition (in most applications N is so large that $N - 1 \approx N$). Direct calculation of all N of these quantities requires about N^2 complex operations. In many applications (especially in real-time applications), the cost of such a calculation is prohibitive. Fortunately, in cases where N is a highly composite number (a number having many factors), the calculation can be streamlined greatly by using an algorithm called the fast Fourier transform (FFT). Usually things are arranged so that N is an integral power of 2; that is, $N = 2^m$, where m is an integer. In this case the number of complex operations required for an FFT is on the order of $N \log_2 N$, which in most applications is significantly smaller than the number ($\approx N^2$) required for a direct calculation. For example, for $N = 1024 = 2^{10}$ a direct calculation requires about $1024^2 \approx 10^6$ complex operations whereas an FFT requires only $1024 \log_2 2^{10} \approx 10^4$ complex operations. For $N = 1024$, using an FFT provides a hundredfold decrease in time required to calculate quantities β_k ($k = 0, 1, \ldots, N - 1$) given by (9-13); for larger values of N the savings are even greater.

Inner workings of FFT algorithms are irrelevant in applications considered in this book. We regard an FFT as a "black box" having N inputs α_i $(i = 0, 1, \ldots, N - 1)$ and N outputs β_k $(k = 0, 1, \ldots, N - 1)$ given by (9-13).

Subroutines implementing various FFT algorithms are available at virtually all installations for scientific computing. The FORTRAN FFT subroutine used for all examples in this book is given in Fig. 9-53. The quantity N [as in (9-13)] must be an integral power of 2, for example, $N = 1024 = 2^{10}$. The quantity JSIGN is either $+1$ or -1 depending on whether the argument of the exponential in (9-13) is $+j2\pi ik/N$ or $-j2\pi ik/N$, respectively. The array X (dimension $2N$) is used for both input and output. On input, odd-numbered elements $X(1), X(3), \ldots, X(N - 1)$ contain the real parts of α_i $(i = 0, 1, \ldots, N - 1)$ and even-numbered elements $X(2), X(4), \ldots, X(N - 2)$ contain the imaginary parts. On output, odd-numbered (even-numbered) elements contain the real (imaginary) parts of β_k $(k = 0, 1, \ldots, N - 1)$. The array X must be declared in the calling program; e.g., using

DIMENSION X(2048)

(for $N = 1024$). Use of this program is illustrated in Examples 9-14 to 9-17.

```
        SUBROUTINE FFT(X,N,JSIGN)
        DIMENSION X(1)
        N2=2*N
        J=1
        DO 5 I=1,N2,2
        IF(I-J) 1,2,2
1       TEMPR=X(J)
        TEMPI=X(J+1)
        X(J)=X(I)
        X(J+1)=X(I+1)
        X(I)=TEMPR
        X(I+1)=TEMPI
2       M=N2/2
3       IF(J-M) 5,5,4
4       J=J-M
        M=M/2
        IF(M-2) 5,3,3
5       J=J+M
        MMAX=2
6       IF(MMAX-N2) 7,9,9
7       ISTEP=2*MMAX
        DO 8 M=1,MMAX,2
        THETA=3.14159265*FLOAT(JSIGN*(M-1))/FLOAT(MMAX)
        WR=COS(THETA)
        WI=SIN(THETA)
        DO 8 I=M,N2,ISTEP
        J=I+MMAX
        TEMPR=WR*X(J)-WI*X(J+1)
        TEMPI=WR*X(J+1)+WI*X(J)
        X(J)=X(I)-TEMPR
        X(J+1)=X(I+1)-TEMPI
        X(I)=X(I)+TEMPR
8       X(I+1)=X(I+1)+TEMPI
        MMAX=ISTEP
        GO TO 6
9       RETURN
        END
```

Figure 9-53 Fast Fourier transform subprogram.

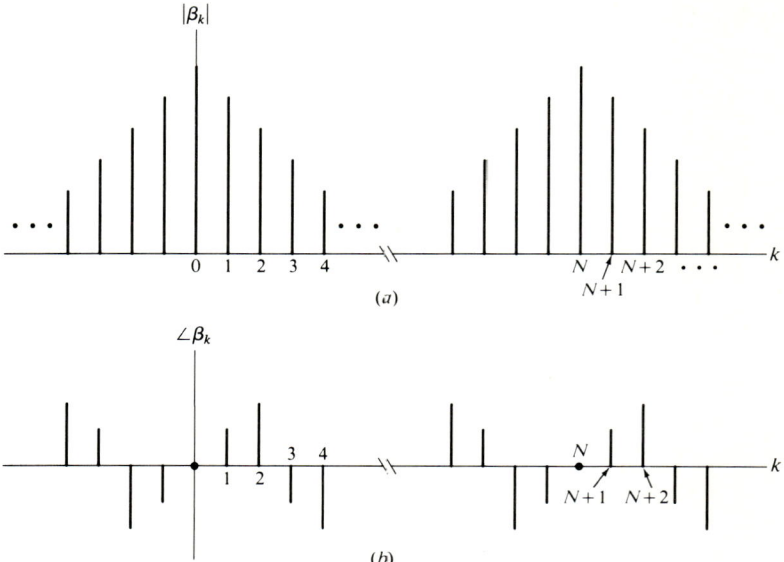

Figure 9-54 Periodicity and conjugate symmetry of quantities β_k given by (9-13).

The quantities β_k given by (9-13) have two properties that are important in subsequent developments: (1) the quantities β_k are periodic with period N; that is,

$$\beta_{k+N} = \beta_k \qquad (9\text{-}14)$$

and (2) if the quantities α_i are real, the quantities β_k have conjugate symmetry about $k = N/2$; that is, for α_i real,

$$\beta_{N-k} = \beta_k^* \qquad (9\text{-}15)$$

Figure 9-54 illustrates these properties.

To establish (9-14) let $k = k' + N$ in (9-13); this gives

$$\beta_{k'+N} = \sum_{i=0}^{N-1} \alpha_i e^{\pm j2\pi i k'/N} e^{\pm j2\pi i}$$

Equation (9-14) follows because $e^{\pm j2\pi i} = 1$. To establish (9-15) let $k = N - k'$ in (9-13); this gives

$$\beta_{N-k'} = \sum_{i=0}^{N-1} \alpha_i e^{j2\pi i k'/N} e^{\pm j2\pi i}$$

Equation (9-15) follows because $e^{\pm j2\pi i} = 1$ and (for α_i real) $\alpha_i e^{j2\pi i k'/N} = (\alpha_i e^{\pm j2\pi i k'/N})^*$.

9-3B Calculation of Fourier's Coefficients

Fourier's coefficients X_k ($k = 0, \pm 1, \pm 2, \ldots$) for a periodic signal $x(t)$ are given by

$$X_k = \frac{1}{T} \int_0^T x(t) e^{-jk\omega_0 t} \, dt \qquad (9\text{-}16)$$

where T is the period of $x(t)$ and $\omega_0 = 2\pi/T$. Using the rectangular rule to approximate the integral gives

$$\tilde{X}_k = \frac{1}{T} \Delta t \sum_{i=0}^{N-1} x(i\Delta t) e^{-jk\omega_0 i \Delta t}$$

where $\Delta t = T/N$ and $\tilde{X}_k$ denotes an approximation to X_k. Replacing Δt by T/N and ω_0 by $2\pi/T$ gives

$$\tilde{X}_k = \frac{1}{N} \sum_{i=0}^{N-1} x\left(\frac{iT}{N}\right) e^{-j2\pi ik/N} \qquad (9\text{-}17)$$

This has the form of (9-13), with

$$\beta_k = \tilde{X}_k \qquad \alpha_i = \frac{1}{N} x\left(\frac{iT}{N}\right) \qquad (9\text{-}18)$$

Therefore, if N is a composite number, an FFT can be used to calculate the quantities $\tilde{X}_k$ given by (9-17). From (9-14) and (9-15),

$$\tilde{X}_{k+N} = \tilde{X}_k \qquad \tilde{X}_{N-k} = \tilde{X}_k^* \qquad (9\text{-}19)$$

These relations show that approximate Fourier coefficients obtained from (9-17) have period N and have conjugate symmetry about $k = N/2$, regardless of the properties of the actual Fourier coefficients given by (9-16). This means that at best only X_0, $X_{\pm 1}, \ldots, X_{\pm N/2}$ can be determined using (9-17). The approximation $X_k \approx \tilde{X}_k$ worsens as k increases from zero to $N/2$ because the exponential $e^{-jk\omega_0 t}$ in (9-16) is described by N/k samples per cycle in (9-17). To ensure that $\tilde{X}_k$ given by (9-17) is a good approximation to X_k given by (9-16) we must choose N large such that the highest significant harmonic of $x(t)$ is sampled several times in one period. For simplicity we adopt the following rule of thumb: choose $N = 2^m$ such that $N \geq 10k_0$, where k_0 is the harmonic number of the highest significant harmonic of $x(t)$. Choosing $N = 2^m$ provides for using the FFT program of Fig. 8-53, and choosing $N \geq 10k_0$ ensures that all significant harmonics of $x(t)$ are sampled at least 10 times per cycle. If the harmonic number of the highest significant harmonic of $x(t)$ is unknown, we can use a cut-and-try approach to calculating the Fourier coefficients for $x(t)$, as follows. Let N_0 represent an initial guess for N. Calculate the Fourier coefficients using (9-17) with $N = N_0$, $N = 2N_0$, $N = 4N_0$, and so on, until two successive calculations yield essentially the same values for all Fourier coefficients of interest.

Example 9-14 The output $y(t)$ of a certain nonlinear amplifier for input $x(t)$ is given by

$$y = y_0 \tanh \frac{x}{x_0}$$

where $x_0 = 1$ mV and $y_0 = 5$ V. We wish to calculate third-harmonic distortion and fifth-harmonic distortion (both in percent) introduced by this amplifier for input

$$x(t) = A \cos \omega_0 t$$

with $A = 1.5$ mV and $f_0 = 5$ kHz. This requires that we calculate the peak amplitudes of the fundamental, third-harmonic, and fifth-harmonic components of the output $y(t)$ for input

$x(t)$ above. The complex amplitudes of the Fourier series for $y(t)$ are given by

$$Y_k = \frac{1}{T}\int_0^T y_0 \tanh\left(\frac{A}{x_0}\cos\omega_0 t\right) e^{-jk\omega_0 t}\, dt$$

where $T = 1/f_0 = 200\ \mu\text{s}$. Since the integration cannot be completed in closed form, we calculate the required amplitudes numerically, using the FFT subroutine given in Fig. 9-53. We do not know the harmonic number of the highest significant harmonic of $y(t)$; therefore we use the cut-and-try method described above to determine a suitable value for the number of samples per period of $y(t)$. Because the highest harmonic of interest is the fifth, we require the initial choice N_0 to equal or exceed $(10)(5) = 50$. Because we wish to use the FFT of Fig. 9-53, we take $N_0 = 2^6 = 64$ (an integral power of 2 no smaller than 50). Figure 9-55

```
            REAL Y(128), PEAK(5)
C
C           Y = ARRAY USED TO PASS SAMPLES OF Y(T) TO THE FFT
C           AND TO RETURN FOURIER COEFFICIENTS TO MAIN PROGRAM
C
C           PEAK = ARRAY OF PEAK AMPLITUDES FOR FIRST 5 AC TERMS OF
C           FOURIER'S SERIES FOR THE AMPLIFIER OUTPUT
C
            DATA X0/0.001/, Y0/5.0/, A/0.0015/, PI/3.1416/, F0/5000./
            N = 64
            W0 = 2. * PI * F0
            T = 1./F0
C
C           N = NUMBER OF SAMPLES PER PERIOD OF INPUT (OR OUTPUT)
C           X0, Y0 = PARAMETERS OF AMPLIFIER TRANSFER CHARACTERISTIC
C           A, F0 = PEAK AMPLITUDE AND FREQUENCY OF INPUT
C           W0, T = ANGULAR FREQUENCY AND PERIOD OF INPUT
C
C           CALCULATE 64 SAMPLES OF OUTPUT (IN ONE PERIOD).
C           PUT (REAL) SAMPLES DIVIDED BY N IN ODD-NUMBERED
C           ELEMENTS OF Y (SEE EQ. 9-18).
C           CLEAR EVEN-NUMBERED (IMAGINARY) ELEMENTS OF Y.
C
            DT = T/FLOAT(N)
            DO 10 I = 1, N
            Y(2 * I - 1) = Y0 * TANH(A * COS(W0 * (I-1) * DT)/X0)/FLOAT(N)
            Y(2 * I) = 0.
10          CONTINUE
C
C           USE FFT TO CALCULATE FOURIER COEFFICIENTS OF OUTPUT.
C
            CALL FFT(Y, N, - 1)
C
C           CALCULATE PEAK AMPLITUDES OF FIRST FIVE HARMONICS
C           (SKIP DC COMPONENT). SEE EQ. 3-4.
C
            DO 20 K = 2, 6
            PEAK(K - 1) = 2 * SQRT(Y(2 * K-1) ** 2 + Y(2 * K) ** 2)
20          CONTINUE
C
C           CALCULATE AND PRINT THIRD AND FIFTH HARMONIC DISTORTION
C           (SEE EXAMPLE 3-17).
C
            D3 = 100. * PEAK(3)/PEAK(1)
            D5 = 100. * PEAK(5)/PEAK(1)
            WRITE(3,30) D3, D5
30          FORMAT(' D3 = ', E10.3,' D5 = ', E10.3/)
            END
            STOP
```

Figure 9-55 FORTRAN program of Example 9-14.

shows a FORTRAN program for calculating third- and fifth-harmonic distortion. The comments explain the various segments of the program. The third- and fifth-harmonic distortion calculated by this program are $D_3 = 12.1$ percent and $D_5 = 1.89$ percent. These results can be checked by running the program again for $N = 128$ (left as an exercise).

In this example using (9-17) directly is actually more efficient than using the FFT because only three of the $N = 64$ components are needed; i.e., calculating Y_1, Y_3, and Y_5 directly using (9-17) requires only $3N = 192$ complex operations, whereas using the FFT to calculate all 64 coefficients requires $64 \log_2 64 = (64)(6) = 384$ complex operations. Nonetheless, since the computational savings are relatively small and writing and debugging a program for a direct calculation is costly, using the FFT is probably better than using a direct approach for this one-time calculation.

9-3C Summation of Fourier's Series

Virtually any realistic periodic signal $x(t)$ can be approximated as closely as we like using a finite Fourier's series

$$\tilde{x}(t) = \sum_{i=-i_0}^{i_0} X_i e^{ji\omega_0 t} \tag{9-20}$$

where $T = 2\pi/\omega_0$ is the period of $x(t)$ and X_i ($i = 0, \pm 1, \pm 2, \ldots$) are the Fourier coefficients for $x(t)$. We wish to show how this series can be summed using an FFT. We assume that $x(t)$ is to be determined at N equispaced instants over an interval $0 \le t < T$, where $N = 2^m$. Replacing t by kT/N and ω_0 by $2\pi/T$ in (9-20) gives

$$\tilde{x}\left(\frac{kT}{N}\right) = \sum_{i=-i_0}^{i_0} X_i e^{j2\pi i k/N} \tag{9-21}$$

To use an FFT to calculate $\tilde{x}(kT/N)$ we must write the sum as a one-sided sum [as in (9-13)]. Because $X_{-i} = X_i^*$, we have

$$\tilde{x}\left(\frac{kT}{N}\right) = \sum_{i=-i_0}^{-1} X_i e^{j2\pi i k/N} + X_0 + \sum_{i=1}^{i_0} X_i e^{j2\pi i k/N}$$

$$= \sum_{i=1}^{i_0} X_{-i} e^{-j2\pi i k/N} + X_0 + \sum_{i=1}^{i_0} X_i e^{j2\pi i k/N}$$

$$= \sum_{i=1}^{i_0} (X_i e^{j2\pi i k/N})^* + X_0 + \sum_{i=1}^{i_0} X_i e^{j2\pi i k/N}$$

It follows that

$$\tilde{x}\left(\frac{kT}{N}\right) = X_0 + 2 \operatorname{Re}\left(\sum_{i=1}^{i_0} X_i e^{j2\pi i k/N}\right)$$

which can be written

$$\tilde{x}\left(\frac{kT}{N}\right) = 2 \operatorname{Re} \beta_k \tag{9-22}$$

where

$$\beta_k = \sum_{i=0}^{N-1} \alpha_i e^{j2\pi i k/N} \qquad N = i_0 + 1 \tag{9-23}$$

622 INTRODUCTION TO SYSTEM ANALYSIS

and
$$\alpha_i = \begin{cases} \dfrac{X_0}{2} & i = 0 \\ X_i & i = 1, 2, \ldots, N-1 \end{cases} \quad (9\text{-}24)$$

Equation (9-23) has the form of (9-13). For $N = 2^m$ the sum can be calculated using the FFT of Fig. 9-53. Therefore, we have the following procedure for summing Fourier's series for a real signal $x(t)$:

1. Choose N such that $N - 1 \geq i_0$ and $N = 2^m$, where i_0 is the harmonic number of the highest significant harmonic and m is a positive integer.
2. Use the FFT of Fig. 9-53 to calculate the quantities β_k defined by (9-23), where the quantities α_i are as defined by (9-24).
3. Obtain the waveform of $x(t)$ by taking twice the real part of the quantities calculated in step 2.

Example 9-15 The rectangular pulse train of Fig. 9-56 is applied to the input of a linear stationary system having transfer function

$$H(j\omega) = \dfrac{0.5}{1 + j\omega/\omega_c}$$

where $\omega_c = 2$ krad/s. We wish to determine the waveform of the output $y(t)$.

From Appendix B, the Fourier coefficients for the input $x(t)$ are given by

$$X_i = \dfrac{x_0}{3} \operatorname{sa} \dfrac{i\pi}{3}$$

where $x_0 = 5$ V. The Fourier coefficients for the output are given by

$$Y_i = H(ji\omega_0)X_i = \dfrac{(x_0/6)\,\operatorname{sa}(i\pi/3)}{1 + ji\omega_0/\omega_c}$$

For $i > \omega_c/\omega_0 = 0.32$, the magnitude of Y_i decreases with i roughly as $1/i^2$, e.g., Y_{10i} is no more than 1 percent of Y_i. It seems reasonable to use only the first 10 harmonics of the Fourier series for $y(t)$ to approximate the waveform of $y(t)$; taking the smallest power of 2 that exceeds 10 gives $N = 2^4 = 16$ and

$$\tilde{y}(t) = \sum_{i=-15}^{15} Y_i e^{ji\omega_0 t}$$

where $\tilde{y}(t)$ denotes an approximation to $y(t)$ and

$$Y_i = \dfrac{(x_0/6)\,\operatorname{sa}(i\pi/3)}{1 + ji\pi}$$

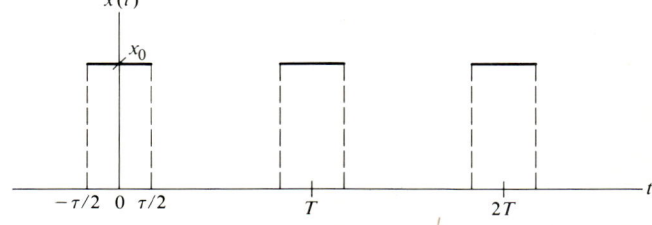

Figure 9-56 Input for the system of Example 9-15; $x_0 = 5$ V, $T = 1$ ms, and $\tau = T/3$.

We calculate $\tilde{y}(t)$ at N equispaced instants $t = kT/N$ ($k = 0, 1, \ldots, N - 1$). This gives

$$\tilde{y}\left(\frac{kT}{N}\right) = \sum_{i=-15}^{15} Y_i e^{j2\pi ik/N}$$

Following the procedure given above, we write the finite Fourier series above as

$$\tilde{y}\left(\frac{kT}{N}\right) = 2 \operatorname{Re} \beta_k$$

where $\beta_k = \sum_{i=0}^{15} \alpha_i e^{j2\pi ik/N}$ with $\alpha_0 = Y_0/2$
$\alpha_i = Y_i \quad i = 1, 2, \ldots, N - 1$

Figure 9-57 shows a FORTRAN program for calculating $\tilde{y}(kT/N)$ as described above, and

```
C       PROGRAM FOR SUMMING FOURIER SERIES
C
        COMPLEX BETA(16), ALPHA(16), J
        REAL YHAT(16), TIME(16)
        EQUIVALENCE (ALPHA(1), BETA(1))
        DATA J/(0.,1.)/, PI/3.1416/, X0/5./, BETA/16*(0.,0.)/
     1  ,T/0.001/, N/16/
C
C       ALPHA = QUANTITIES DEFINED BY (9-24)
C       BETA = QUANTITIES DEFINED BY (9-23)
C       J = SQRT(-1)
C       YHAT = WAVEFORM OF OUTPUT
C       TIME = SAMPLING INSTANTS
C       X0 = PEAK AMPLITUDE OF PULSE TRAIN
C       T = PERIOD OF PULSE TRAIN
C       N = NUMBER OF SAMPLES
C
C       CALCULATE ALPHA(I) (SEE EQ. 9-24)
C
        ALPHA(1) = X0/12.
        DO 10 IP1 = 2, 16
        I = IP1 - 1
        ALPHA(IP1) = (X0/6.) * SA(I * PI/3.)/(1. + J * I * PI)
  10    CONTINUE
C
C       USE FFT TO CALCULATE BETA(K); JSIGN = + 1
C
        CALL FFT(BETA, N, 1)
C
C       ON RETURN FROM FFT, 2 * RE(BETA(K)) = YHAT(K * T/N)
C       FOR K = 0, 1, ..., N-1
C
        DT = T/FLOAT(N)
        DO 30 KP1 = 1, N
        TIME(KP1) = FLOAT(KP1 - 1) * DT
        YHAT(KP1) = 2. * REAL(BETA(KP1))
  30    CONTINUE
C
C       PRINT VALUES OF OUTPUT
C
        WRITE(3,40) (TIME(I), YHAT(I), I = 1, N)
  40    FORMAT(' ', 2E15.3)
        STOP
        END
        FUNCTION SA(ALPHA)
        SA = 1.
        IF(ALPHA) 1, 2, 1
   1    SA = SIN(ALPHA)/ALPHA
   2    RETURN
        END
```

Figure 9-57 FORTRAN program of Example 9-15.

624 INTRODUCTION TO SYSTEM ANALYSIS

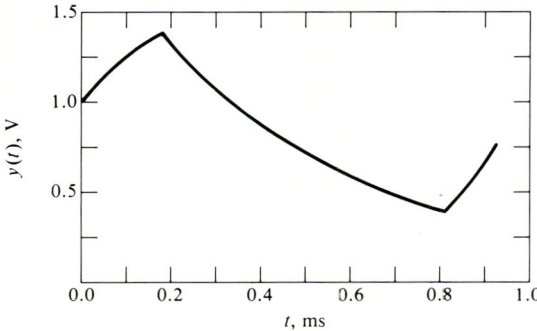

Figure 9-58 Output of the system of Example 9-15.

Fig. 9-58 shows a graph of the values of $\tilde{y}(t)$ obtained using this program. The accuracy of this result can be checked by repeating the calculation with $N = 32$ (left as an exercise).

9-3D Calculation of Fourier Transforms

The Fourier transform $X(j2\pi f)$ of a function $x(t)$ is given by

$$X(j2\pi f) = \int_{-\infty}^{\infty} x(t) e^{-j2\pi ft} \, dt$$

The transform exists (is finite for all f) only if

$$\int_{-\infty}^{\infty} |x(t)|^2 \, dt < \infty$$

which for all practical purposes implies that $x(t)$ has finite duration; i.e., for any function $x(t)$ having a Fourier transform we can find t_0 and T such that $x(t)$ is significantly different from zero only for $t_0 \leq t < t_0 + T$. It follows that in any practical application, (9-24) can be written

$$X(j2\pi f) \approx \int_{t_0}^{t_0+T} x(t) e^{-j2\pi ft} \, dt$$

In anticipation of subsequent developments, it is convenient to introduce the change of variable $\alpha = t - t_0$, so that the integration runs from zero to T. This gives

$$X(j2\pi f) \approx \int_0^T x(\alpha + t_0) e^{-j2\pi f(\alpha + t_0)} \, dt = e^{-j2\pi ft_0} \int_0^T x(\alpha + t_0) e^{-j2\pi f\alpha} \, d\alpha$$

$$= e^{-j2\pi ft_0} \int_0^T x(t + t_0) e^{-j2\pi ft} \, dt$$

Approximating the integral by a sum using the rectangular rule gives

$$\tilde{X}(j2\pi f) = e^{-j2\pi ft_0} \frac{T}{N} \sum_{i=0}^{N-1} x\left(\frac{iT}{N} + t_0\right) e^{-j2\pi fiT/N}$$

where $\tilde{X}(j2\pi f)$ denotes an approximation to $X(j2\pi f)$, N is the number of rectangles used, and T/N is the width of each rectangle. In order to use the FFT to calculate the sum we take $N = 2^m$, where m is an integer, and we calculate $\tilde{X}(j2\pi f)$ for N equispaced values $f = k\Delta f$ $(k = 0, 1, \ldots, N - 1)$ such that $k2\pi \Delta f iT/N = 2\pi ik/N$, [such that the argument of the second exponential above is the same as the argument of the exponential in (9-13)]; thus $\Delta f = 1/T$, and

$$\tilde{X}\left(\frac{j2\pi k}{T}\right) = e^{-j2\pi k t_0/T} \frac{T}{N} \sum_{i=0}^{N-1} x\left(\frac{iT}{N} + t_0\right) e^{-j2\pi ik/N}$$

or

$$\tilde{X}\left(\frac{j2\pi k}{T}\right) e^{j2\pi k t_0/T} = \frac{T}{N} \sum_{i=0}^{N-1} x\left(\frac{iT}{N} + t_0\right) e^{-j2\pi ik/N} \quad (9\text{-}25)$$

This relation has the form of (9-13) with

$$\alpha_i = \frac{T}{N} x\left(\frac{iT}{N} + t_0\right) \quad (9\text{-}26)$$

$$\beta_k = \tilde{X}\left(\frac{j2\pi k}{T}\right) e^{j2\pi k t_0/T} \quad (9\text{-}27)$$

From (9-27),

$$\left|\tilde{X}\left(\frac{j2\pi k}{T}\right)\right| = |\beta_k|$$

The quantities β_k $(k = 0, 1, \ldots, N - 1)$ satisfy (9-14) and (9-15). Therefore $\tilde{X}(j2\pi k/T)$ has period N and has conjugate symmetry about $k = N/2$. It follows that at most $N/2$ of the quantities $\tilde{X}(j2\pi k/T)$ approximate unique positive-frequency values of the spectral density (Fourier transform) $X(j2\pi f)$. It also follows that the highest frequency for which $X(j2\pi f)$ is calculated is $N/2T$. In using (9-25) we should choose N large enough to ensure that $N/2T$ exceeds the bandwidth of $x(t)$, and we should choose T large enough to ensure that the separation $1/T$ (in hertz) between adjacent values of $\tilde{X}(j2\pi k/T)$ is sufficiently small (to provide a good picture of the spectral density). A cut-and-try approach can be used if the shape of $X(j2\pi f)$ is completely unknown. Note that the approximation

$$X\left(\frac{j2\pi k}{T}\right) \approx \tilde{X}\left(\frac{j2\pi k}{T}\right)$$

worsens as k increases because the exponential function in (9-25) is represented by N/k samples per period. Consequently, it may be necessary to choose N such that $N/2T$ is considerably greater than the bandwidth of $x(t)$ if the spectral density must be determined accurately for frequencies near the bandwidth of $x(t)$.

The development above leads to the following procedure for calculating the Fourier transform (the spectral density) of a signal:

1. Determine t_0 and T in (9-25) such that all significant amplitudes of the signal are contained in the interval $t_0 < t < t_0 + T$. The quantity $1/T$ is the separation between calculated values of the spectral density. In order to obtain a good picture

of the spectral density it may be necessary to choose T larger than what ordinarily would be regarded as the duration of the signal.

2. Choose $N = 2^m$ such that $N/2T$ exceeds the highest frequency for which the spectral density is significantly different from zero. This frequency may be considerably greater than what ordinarily would be regarded as the bandwidth of the signal.

3. Use (9-25) to (9-27) and the FFT of Fig. 9-53 to calculate the spectral density. A cut-and-try approach may be needed to determine acceptable values for T and N.

Example 9-16 We use the procedure given above to calculate the Fourier transform of the signal

$$x(t) = x_0 \left(\frac{t}{\tau}\right) \ln\left(1 + \frac{t}{\tau}\right) e^{-t/\tau} u(t)$$

where $x_0 = 5$ V and $\tau = 2$ ms. Figure 9-59 shows a graph of this signal versus time.

First we choose the times t_0 and T to be used in (9-25). The signal is nonzero only for $t > 0$, and so we choose $t_0 = 0$. The amplitude of $x(t)$ decreases rapidly with time for $t > 4$ ms. For our first cut we use $T = 16$ ms $= 8\tau$. Next we choose the number N to be used in (9-25). We estimate the bandwidth of $x(t)$ as $F = 1/\tau$. For our first cut we use $N/2T > 5F$, which leads to $N/16\tau > 5/\tau$ or $N > 80$. Requiring N to be an integral power of 2 gives $N = 128$.

Figure 9-60 shows a FORTRAN program for calculating and printing the Fourier transform of $x(t)$ above. The comments explain the various segments of the program. Figure 9-61 shows a plot of $\tilde{X}(j2\pi f)$ (magnitude and angle). The result can be checked by repeating the calculation using, say, $T = 32\tau$ and $F = 10/\tau$. This calculation is left as an exercise.

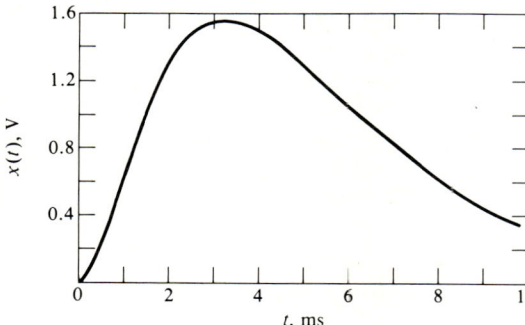

Figure 9-59 Signal of Example 9-16.

9-3E Calculation of Inverse Fourier Transforms

The inverse Fourier transform $x(t)$ of a function $X(j2\pi f)$ is given by

$$x(t) = \int_{-\infty}^{\infty} X(j2\pi f) e^{j2\pi ft} df$$

The inverse transform exists (is finite for all t) only if

$$\int_{-\infty}^{\infty} |X(j2\pi f)|^2 df < \infty$$

```
C         CALCULATION OF A FOURIER TRANSFORM USING AN FFT
C
          COMPLEX ALPHA(128), BETA(128), XHAT(128)
          REAL FREQ(64), XMAG(64), ANGLE(64)
          EQUIVALENCE (ALPHA(1), BETA(1), XHAT(1))
          DATA T/0.016/, N/128/, TAU/0.002/, ALPHA/128*(0.,0.)/
C
C         ALPHA, BETA, XHAT = QUANTITIES DEFINED IN (9-26), (9-27)
C         FREQ = FREQUENCY (HZ)
C         XMAG, ANGLE = MAGNITUDE AND ANGLE OF TRANSFORM
C         T= DURATION OF X(T), N = NUMBER OF SAMPLES OF SIGNAL
C
C         CALCULATE THE FFT INPUT ACCORDING TO (9-26)
C
          DO 10 IP1 = 2, N
          A = FLOAT(IP1 - 1) * T/(FLOAT(N) * TAU)
          ALPHA(IP1) = (T/FLOAT(N)) * 5. * A * ALOG(1. + A) * EXP(-A)
10        CONTINUE
          CALL FFT(ALPHA, N, - 1)
C
C         CALCULATE AND PRINT MAGNITUDE AND ANGLE OF TRANSFORM
C         NOTE1: IN THIS EXAMPLE THE EXPONENTIAL FACTOR IN (9-27) IS UNITY.
C         NOTE2: ONLY N/2 VALUES ARE UNIQUE (SEE EQ. 9-15)
C
          ND2 = N/2
          DO 20 KP1 = 1, ND2
          XR = REAL(BETA(KP1))
          XI = AIMAG(BETA(KP1))
          XMAG(KP1) = SQRT(XR * XR + XI * XI)
          ANGLE(KP1) = ATAN2(XI, XR)
          FREQ(KP1) = FLOAT(KP1 - 1)/T
20        CONTINUE
          WRITE(3,30) (FREQ(I), XMAG(I), ANGLE(I), I = 1, ND2)
30        FORMAT(' ', 3E15.3)
          STOP
          END
```

Figure 9-60 FORTRAN program of Example 9-16.

This implies (for all practical purposes) that $X(j2\pi f)$ has finite width, i.e., that $x(t)$ has finite bandwidth. Consequently, for any function $X(j2\pi f)$ having an inverse Fourier transform we can find a frequency F such that

$$x(t) \approx \int_{-F}^{F} X(j2\pi f)e^{j2\pi ft}\, df \qquad (9\text{-}28)$$

is a good approximation. We wish to approximate the integral using a one-sided sum like that of (9-13). The integrand has conjugate symmetry about $f = 0$; that is,

$$X(-j2\pi f)e^{-j2\pi ft} = [X(j2\pi f)e^{j2\pi ft}]^*$$

We take advantage of this symmetry in obtaining the desired one-sided sum. Figure 9-62 illustrates rectangular-rule approximation of the integral in (9-28), where A_n denotes the contribution of the nth rectangle to the integral; i.e.,

$$A_n = X(j2\pi f)e^{j2\pi ft}\, \Delta f$$

The integral can be approximated as

$$\tilde{x}(t) = A_0 + A_1 + A_{-1} + A_2 + A_{-2} + \cdots$$

where $\tilde{x}(t)$ denotes an approximation to $x(t)$. Because of the conjugate symmetry of the

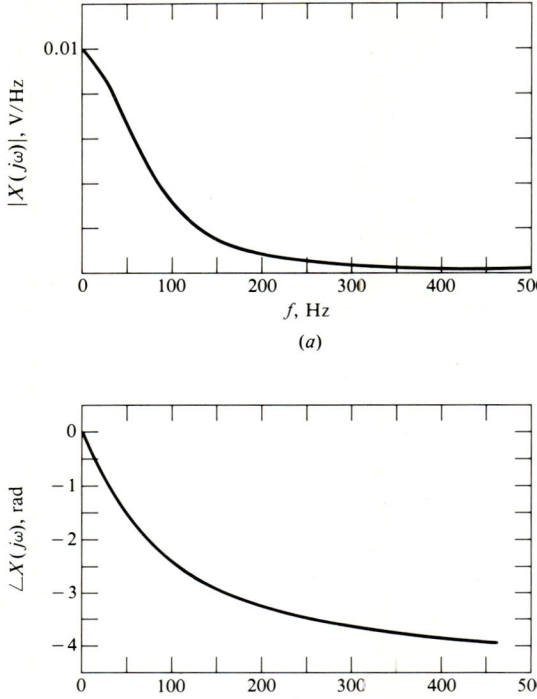

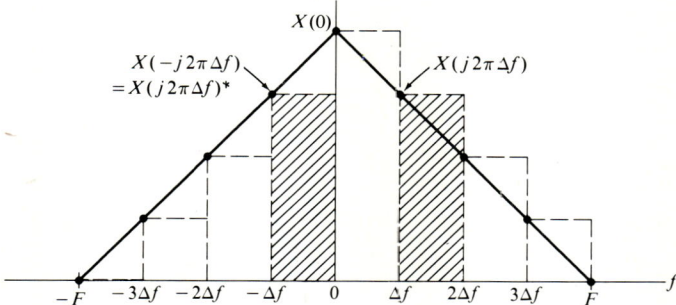

Figure 9-61 Fourier transform of the signal of Fig. 9-59.

Figure 9-62 Rectangular-rule approximation for a Fourier integral.

integrand, $A_{-n} = A_n^*$ and (note that A_0 is real)

$$\tilde{x}(t) = A_0 + 2 \, \text{Re}(A_1 + A_2 + \cdots)$$

$$= X(0)\,\Delta f + 2 \, \text{Re}\left[\sum_{i=1}^{N-1} X(j2\pi i \, \Delta f) e^{j2\pi i \Delta f t}\right] \Delta f$$

where N is the number of rectangles to the right of the origin and $\Delta f = F/N$. To put this in the form of (9-13) we calculate $x(t)$ at N equispaced instants

$0, \Delta t, 2\Delta t, \ldots, (N-1)\Delta t$ such that $j2\pi(i\,\Delta f)(k\,\Delta t) = j2\pi ik/N$; thus $\Delta t = 1/(N\,\Delta f) = 1/F$ and

$$\tilde{x}\left(\frac{k}{F}\right) = \frac{F}{N}X(0) + 2\,\text{Re}\left[\sum_{i=0}^{N-1}\frac{F}{N}X\left(\frac{j2\pi iF}{N}\right)e^{j2\pi ik/N}\right]$$

This relation can be written

$$\tilde{x}\left(\frac{k}{F}\right) = 2\,\text{Re}\,\beta_k \tag{9-29}$$

where

$$\beta_k = \sum_{i=0}^{N-1}\alpha_i e^{j2\pi ik/N} \tag{9-30}$$

with

$$\alpha_0 = \frac{1}{2}\frac{F}{N}X(0)$$

$$\alpha_i = \frac{F}{N}X\left(\frac{j2\pi iF}{N}\right) \qquad i = 1, 2, \ldots, N-1 \tag{9-31}$$

These relations can be used to calculate the inverse Fourier transform $x(t)$ of a function $X(j2\pi f)$. To interpret the result of such a calculation correctly we must recognize that the quantity $\tilde{x}(k/F)$ given by (9-29) is periodic with period N and that a calculated inverse transform $\tilde{x}(t)$ is (approximately) a periodic extension of the corresponding true inverse transform $x(t)$; that is,

$$\tilde{x}(t) \approx \sum_{m=-\infty}^{\infty} x(t + mT) \qquad \text{where } T = \frac{N}{F}$$

Example 9-17 We use the relations given above to calculate the inverse Fourier transform of the function

$$X(j2\pi f) = x_0\tau\,\text{sa}^2\,\pi f\tau$$

where $x_0 = 50$ mA and $\tau = 1$ ms.

Figure 9-63 shows a graph of the spectral density $X(j2\pi f)$ given above. The magnitude of $X(j2\pi f)$ decreases sharply with frequency. For a first cut at calculating the inverse transform we use $F = 5/\tau$ in (9-31).

We choose the number of samples N such that the narrowest significant feature of $X(j2\pi f)$ (the width of one lobe of the sampling function) is described by several samples.

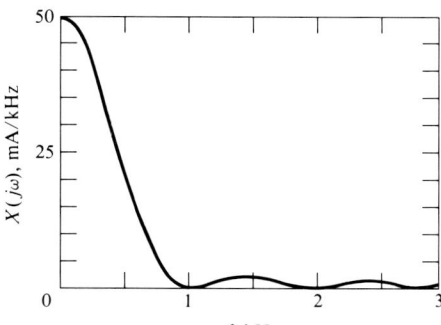

Figure 9-63 Fourier transform of Example 9-17.

630 INTRODUCTION TO SYSTEM ANALYSIS

The width of one lobe of the sampling function is $1/\tau$. For a first cut we take $\Delta f = 1/4\tau$ (four samples per lobe); thus

$$N = \frac{F}{\Delta f} = \frac{5/\tau}{1/4\tau} = 20$$

In order to use the FFT of Fig. 9-53 we use $N = 32$ (the smallest integral power of 2 that exceeds 20).

Figure 9-64 shows a FORTRAN program for calculating the inverse transform, and Fig. 9-65 shows a graph of the calculated inverse transform given by (9-28). To interpret this result correctly we must realize that the calculated inverse transform $x(t)$ of Fig. 9-65

```
C       FFT CALCULATION OF AN INVERSE FOURIER TRANSFORM
C
        COMPLEX ALPHA(32), BETA(32), J, C
        REAL XHAT(32), TIME(32)
        EQUIVALENCE (ALPHA(1), BETA(1), XHAT(1))
        DATA X0/0.05/, TAU/0.001/, PI/3.1416/, N/32/, J/(0.,1.)/
        BIGF = 10./TAU
C
C       ALPHA, BETA, XHAT = QUANTITIES DEFINED BY (9-29, 30, 31)
C       TIME = SAMPLING TIMES FOR XHAT
C       X0, TAU = PARAMETERS OF SPECTRAL DENSITY
C       BIGF = WIDTH OF SPECTRAL DENSITY
C       N = NUMBER OF SAMPLES
C       J = SQRT(-1)
C
C       CALCULATE ALPHA GIVEN BY (9-31)
C
        B = BIGF/FLOAT(N)
        C = B * X0 * TAU
        PITAU = PI * TAU
        ALPHA(1) = 0.5 * C
        DO 10 IP1 = 2, N
        I = IP1 - 1
        F = FLOAT(I) * B
        ALPHA(IP1) = C * SA(PITAU * F) ** 2
   10   CONTINUE
C
C       USE FFT TO CALCULATE BETA GIVEN BY (9-30)
C
        CALL FFT(ALPHA, N, 1)
C
C       CALCULATE XHAT USING (9-29)
C
        DO 20 KP1 = 1,N
        TIME(KP1) = FLOAT(KP1 - 1)/BIGF
        XHAT(KP1) = 2. * REAL(BETA(KP1))
   20   CONTINUE
C
C       PRINT RESULTS AND QUIT
C
        WRITE(3,300) (TIME(I), XHAT(I), I = 1, N)
  300   FORMAT(2E15.3)
        STOP
        END
        FUNCTION SA(A)
        SA = 1.
        IF(A.EQ.0.) RETURN
        SA = SIN(A)/A
        RETURN
        END
```

Figure 9-64 FORTRAN program of Example 9-17.

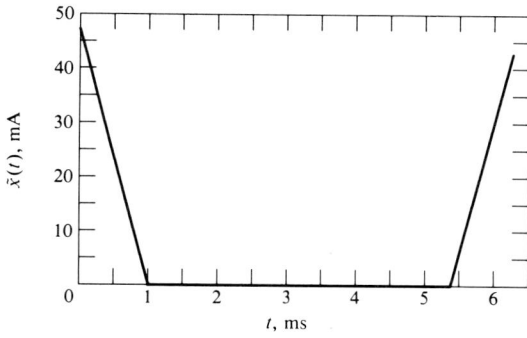

Figure 9-65 Inverse transform obtained using an FFT for the function of Fig. 9-63.

approximates one period (for $0 \leq t < N/F$) of a periodic repetition of the actual inverse transform $x(t)$; that is, the signal given by (9-28) is periodic with period N/F, as shown in Fig. 9-66. Consequently, the calculated inverse transform $\tilde{x}(t)$ is ambiguous; e.g., the periodic signal of Fig. 9-66 can be obtained by periodic repetition of any one of the waveforms shown in Fig. 9-67.

For the transform considered in this example, the ambiguity can be eliminated by noting that $X(j2\pi f)$ is real. This means that the inverse transform is symmetric about $t = 0$, which implies that the correct inverse transform is the waveform of Fig. 9-67a. In more difficult cases the ambiguity can be eliminated by calculating the inverse transform a second time after multiplying the transform by $e^{-j\pi fN/F}$. This translates (delays) the corresponding inverse transform by $t_0 = N/2F$ (by one-half the period of the calculated inverse transform). Figure 9-68 shows the result of this calculation for the transform considered in this example. The fact that the waveform shown in Fig. 9-68 must be a delayed replica of that shown in Fig. 9-65 implies that the correct inverse transform is the triangular pulse of Fig. 9-67a.

Example 9-18 This example illustrates numerical calculation of an output of a linear stationary system. The transfer function of the system considered in this example is

$$H(j2\pi f) = \frac{1}{[-(f/f_0)^2 + j(f/f_0) + 1][1 + j(f/f_0)]} \quad \text{V/V}$$

where $f_0 = 1$ kHz. The input is a rectangular pulse

$$x(t) = x_0 r\left(\frac{t}{\tau}\right)$$

with $x_0 = 5$ V and $\tau = 1$ ms. From Table C-1, the spectral density (Fourier transform) of the input is

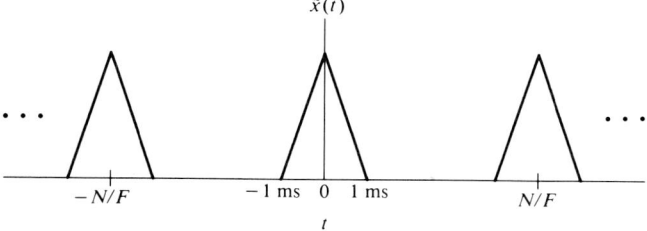

Figure 9-66 Periodicity of an inverse FFT.

632 INTRODUCTION TO SYSTEM ANALYSIS

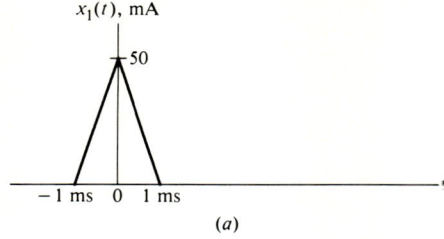

(a)

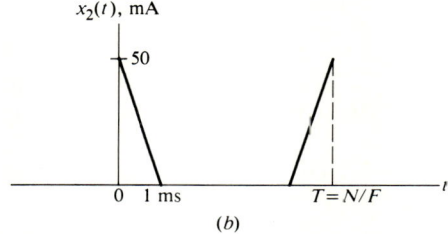

(b)

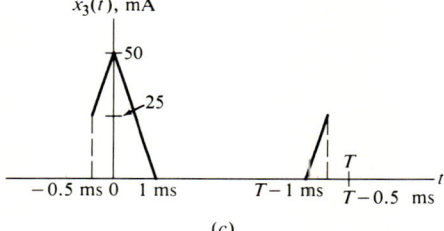

(c)

Figure 9-67 Ambiguity of an inverse FFT.

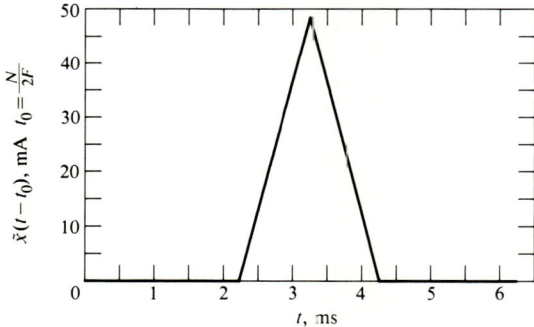

Figure 9-68 Inverse Fourier transform of the function of Fig. 9-63 delayed by $N/2F$.

$$X(j2\pi f) = x_0\tau(\text{sa } \pi f\tau)e^{-j\pi f\tau}$$

From (4-27), the output $y(t)$ is the inverse Fourier transform of

$$Y(j2\pi f) = H(j2\pi f)X(j2\pi f)$$

Figure 9-69 shows graphs of the input amplitude spectral density, the gain of the system, and the output spectral density.

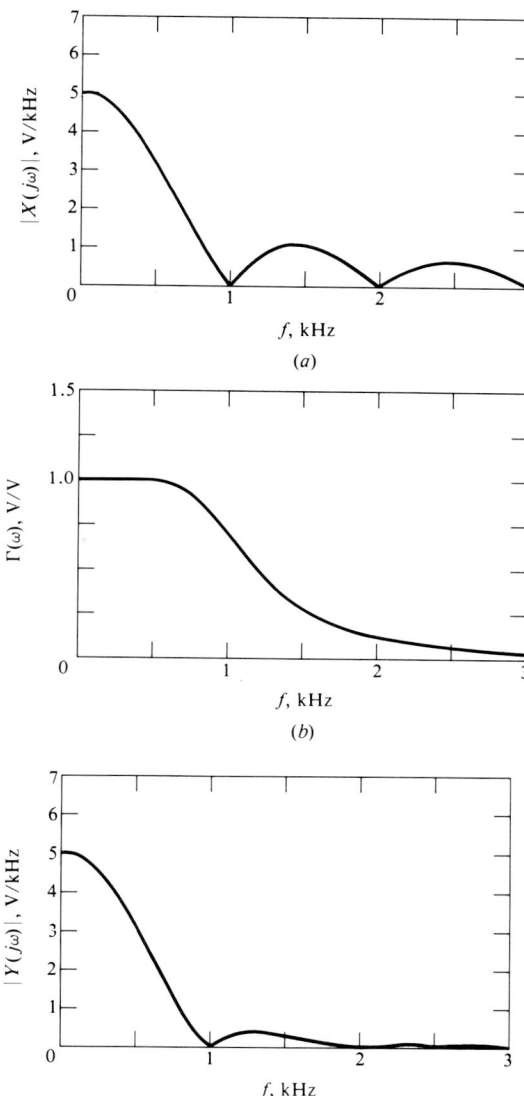

Figure 9-69 (a) Input spectral density, (b) gain, and (c) output spectral density for the system of Example 9-18.

We determine the number of samples N and the maximum frequency of interest F as follows. By inspection, the spectral density of $y(t)$ decreases sharply with frequency above $f = f_0$. A conservative choice for the maximum frequency of interest is $F = 10f_0 = 10$ kHz. Because the bandwidth of the system is on the order of f_0, we estimate the duration of the impulse response for the system as $1/f_0 = 1$ ms. Because the duration of the input is τ, we estimate the duration of the output as $\tau + 1/f_0 = 2$ ms. To be conservative we use $T = 4$ ms as the duration of the output. Requiring that the period of the calculated output exceed the duration of the output gives $N/F > T$ or $N > 40$. Choosing the smallest integral power of 2 exceeding 40 gives $N = 64$.

```
C     FFT CALCULATION OF AN OUTPUT OF A LINEAR STATIONARY SYSTEM
C
      COMPLEX H, X, HX, ALPHA(64), BETA(64), J
      REAL YHAT(64), TIME(64)
      EQUIVALENCE (ALPHA(1), BETA(1))
      DATA X0/5./, TAU/.001/, PI/3.1416/, N/64/, F0/1000./,
    1 BIGF/1.E4/, J/(0.,1.)/
C
C     H = TRANSFER FUNCTION
C     X = FOURIER TRANSFORM OF INPUT
C     HX = FOURIER TRANSFORM OF OUTPUT
C     ALPHA, BETA = QUANTITIES DEFINED IN (9-30, 31)
C     J = SQRT(- 1)
C     YHAT = WAVEFORM OF OUTPUT
C     TIME = SAMPLING TIMES
C     X0, TAU = AMPLITUDE AND DURATION OF INPUT
C     N = NUMBER OF SAMPLES
C     F0 = BANDWIDTH OF SYSTEM
C     BIGF = 10 * F0 = MAXIMUM FREQUENCY OF INTEREST
C
C     CALCULATE FOURIER TRANSFORM OF OUTPUT
C
      DF = BIGF/FLOAT(N)
      X0TAU = X0 * TAU
      PITAU = PI * TAU
      DO 10 IP1 = 1, N
      F = FLOAT(IP1 - 1) * DF
      X = X0TAU * SA(PITAU * F) * CEXP(- J * PITAU * F)
      H = 1./((- (F/F0) ** 2 + J * F/F0 + 1.) * (1. + J * F/F0))
      HX = H * X
      ALPHA(IP1) = BIGF * HX/FLOAT(N)
   10 CONTINUE
      ALPHA(1) = 0.5 * ALPHA(1)
C
C     CALCULATE INVERSE TRANSFORM
C     AND WAVEFORM OF OUTPUT
C
      CALL FFT(ALPHA, N, 1)
      DO 20 KP1 = 1, N
      YHAT(KP1) = 2. * REAL(BETA(KP1))
      TIME(KP1) = FLOAT(KP1 - 1)/BIGF
   20 CONTINUE
C
C     PRINT RESULTS AND QUIT
C
      WRITE(3,30) (TIME(I), YHAT(I), I = 1, N)
   30 FORMAT(' ', 2E20.7)
      STOP
      END
      FUNCTION SA(X)
      SA=1.
      IF(X.EQ.0.) RETURN
      SA=SIN(X)/X
      RETURN
      END
```

Figure 9-70 FORTRAN program of Example 9-18

Figure 9-70 shows a FORTRAN program for calculating the output $y(t)$, and Fig. 9-71 shows a graph of the output calculated by the program. It is possible that the period N/F of the calculated output is too small and that error is introduced by overlap of adjacent cycles. We can check overlap error by increasing N (note that this increases F) and repeating the calculation. This is left as an exercise.

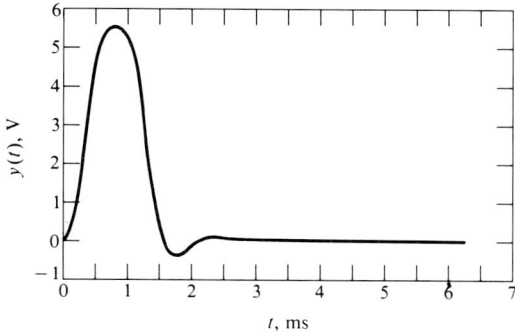

Figure 9-71 Calculated output of the system of Example 9-18.

9-4 INTERACTIVE COMPUTER-AIDED ANALYSIS

Many otherwise laborious operations entailed in analyzing and designing systems are made almost effortless by proper use of a digital computer. In this section we illustrate interactive computer-aided analysis using subprograms available (perhaps with minor variations) at almost all facilities for scientific computing. Illustrations given below include a program for finding roots (real and complex) of a polynomial (e.g., a characteristic polynomial) a program for finding a partial-fraction expansion for a rational function, and a program for designing a low-pass digital filter. These or similar programs can be used to find an inverse (Laplace or z) transform in a matter of minutes, whereas pencil-and-paper inversion of the same function may take hours (or even days). Even so, the programs given in this section are almost trivial compared with real-world software for computer-aided analysis and design.

9-4A Roots of a Polynomial

The problem of finding all roots of a polynomial arises frequently in system analysis and design, e.g., in obtaining a closed-form expression for step response from a differential equation, in examining a system for stability, and in designing a filter for a specified frequency response. A closed-form expression for the roots of a polynomial can be obtained (in general) only for a polynomial of degree 4 or less. Numerical methods are necessary for finding roots of polynomials of degree 5 or more. In practice, numerical methods generally are used for polynomials of degree 3 or more because closed-form expressions for the roots of third- and fourth-degree polynomials are cumbersome.

Subprogram POLRT, given in Fig. 9-72, implements a Newton-Raphson algorithm for finding all roots of a polynomial having real coefficients. Equivalent subprograms are available at almost all installations for scientific computing. An explanation of the Newton-Raphson algorithm is beyond the scope of this book. We regard the subprogram as a "black box," which accepts the coefficients of a polynomial as inputs and gives the roots of the polynomial as outputs. The subprogram is invoked by a statement of the form

$$\text{CALL POLRT(A, W, N, U, V, IER)}$$

```
      SUBROUTINE POLRT(XCOF,COF,M,ROOTR,ROOTI,IER)
      DIMENSION XCOF(1),COF(1),ROOTR(1),ROOTI(1)
      DOUBLE PRECISION XO,YO,X,Y,XPR,YPR,UX,UY,V,YT,XT,U,XT2,YT2,SUMSQ
     $DX,DY,TEMP,ALPHA,DABS
C          XCOF -VECTOR OF M+1 COEFFICIENTS OF THE POLYNOMIAL
C                ORDERED FROM SMALLEST TO LARGEST POWER
C          COF  -WORKING VECTOR OF LENGTH M-1
C          M    -ORDER OF POLYNOMIAL
C          ROOTR-RESULTANT VECTOR OF LENGTH M CONTAINING REAL ROOTS
C                OF THE POLYNOMIAL
C          ROOTI-RESULTANT VECTOR OF LENGTH M CONTAINING THE
C                CORRESPONDING IMAGINARY ROOTS OF THE POLYNOMIAL
C          IER  -ERROR CODE: WHERE IER=0 MEANS NORMAL RETURN, IER=1
C                MEANS DEGREE < 1, IER=2 MEANS DEGREE > 36, IER=3
C                MEANS ALGORITHM FAILED TO CONVERGE, AND IER=4
C                MEANS COEFF OF HIGHEST POWER = 0
      IFIT=0
      N=M
      IER=0
      IF(XCOF(N+1))10,25,10
   10 IF(N) 15,15,32
   15 IER=1
   20 RETURN
   25 IER=4
      GO TO 20
   30 IER=2
      GO TO 20
   32 IF(N-36) 35,35,30
   35 NX=N
      NXX=N+1
      N2=1
      KJ1 = N+1
      DO 40 L=1,KJ1
      MT=KJ1-L+1
   40 COF(MT)=XCOF(L)
   45 X0=.00500101
      Y0=0.01000101
      IN=0
   50 X=X0
      X0=-10.0*Y0
      Y0=-10.0*X
      X=X0
      Y=Y0
      IN=IN+1
      GO TO 59
   55 IFIT=1
      XPR=X
      YPR=Y
   59 ICT=0
   60 UX=0.0
      UY=0.0
      V =0.0
      YT=0.0
      XT=1.0
   55 IFIT=1
      XPR=X
      YPR=Y
   59 ICT=0
   60 UX=0.0
      UY=0.0
      V =0.0
      YT=0.0
      XT=1.0
      U=COF(N+1)
      IF(U) 65,130,65
   65 DO 70 I=1,N
```

Figure 9-72 FORTRAN subprogram for finding roots of a polynomial. *[Reprinted by permission from System/360 Scientific Subroutine Package (360A-CM-03x) Programmer's Manual (Version III). © 1968 by International Business Machines Corporation.]*

```
      L =N-I+1
      TEMP=COF(L)
      XT2=X*XT-Y*YT
      YT2=X*YT+Y*XT
      U=U+TEMP*XT2
      V=V+TEMP*YT2
      FI=I
      UX=UX+FI*XT*TEMP
      UY=UY-FI*YT*TEMP
      XT=XT2
   70 YT=YT2
      SUMSQ=UX*UX+UY*UY
      IF(SUMSQ)75,110,75
   75 DX=(V*UY-U*UX)/SUMSQ
      X=X+DX
      DY=-(U*UY+V*UX)/SUMSQ
      Y=Y+DY
   78 IF(DABS(DY)+DABS(DX)-1.0D-05) 100,80,80
C
C        STEP ITERATION COUNTER
C
   80 ICT=ICT+1
      IF(ICT-500) 60,85,85
   85 IF(IFIT) 100,90,100
   90 IF(IN-5) 50,95,95
C
C        SET ERROR CODE TO 3
C
   95 IER=3
      GO TO 20
  100 DO 105 L=1,NXX
      MT=KJ1-L+1
      TEMP=XCOF(MT)
      XCOF(MT)=COF(L)
  105 COF(L)=TEMP
      ITEMP=N
      N=NX
      NX=ITEMP
      IF(IFIT) 120,55,120
  110 IF(IFIT) 115,50,115
  115 X=XPR
      Y=YPR
  120 IFIT=0
  122 IF(DABS(Y)-1.D-4*DABS(X)) 135,125,125
  125 ALPHA=X+X
      SUMSQ=X*X+Y*Y
      N=N-2
      GO TO 140
  130 X=0.0
      NX=NX-1
      NXX=NXX-1
  135 Y=0.0
      SUMSQ=0.0
      ALPHA=X
      N=N-1
  140 COF(2)=COF(2)+ALPHA*COF(1)
      IF(N.GT.2)GOTO 145
      COF(3)=COF(3)+ALPHA*COF(2)-SUMSQ*COF(1)
      GO TO 155
  145 DO 150 L=2,N
  150 COF(L+1)=COF(L+1)+ALPHA*COF(L)-SUMSQ*COF(L-1)
  155 ROOTI(N2)=Y
      ROOTR(N2)=X
      N2=N2+1
      IF(N2.GT.M) GO TO 20
      IF(SUMSQ) 160,165,160
  160 Y=-Y
      SUMSQ=0.0
      GO TO 155
  165 IF(N) 20,20,45
      END
```

Figure 9-72 *Continued*

where N is the degree of a polynomial; A is an array containing the coefficients of the polynomial; W is a scratch array; U and V are arrays containing (after execution of the subprogram) the real and imaginary parts, respectively, of the roots of the polynomial; and IER is an error code. The quantities N and A are inputs. The quantities U and V are outputs. The degree of the polynomial is limited to 36 because the algorithm is generally unreliable for polynomials of higher degree (on machines using 32-bit words). The coefficients are given in order of increasing powers of the variable. For example, to obtain the roots of

$$c(s) = 3s^3 + 4s^2 + 2s + 5$$

we use the CALL statement above with N = 3, A(1) = 5, A(2) = 2, A(3) = 4, and A(4) = 3.

Figure 9-73 gives an interactive FORTRAN program for calculating roots of a polynomial using subprogram POLRT. The variable ICON is the logical unit number for the console (keyboard and screen). This unit number varies from one computer

```
              DIMENSION A(37),AWORK(37),U(36),V(36)
              DATA U/20*0./,V/20*0./
              ICON = 1
1             WRITE(ICON,5)
5             FORMAT(' ENTER DEGREE OF POLYNOMIAL (0<DEG<37) ')
              READ(ICON,10) N
10            FORMAT(I2)
              IF(N.LE.0) GOTO 200
              IF(N.GT.36) GOTO 210
              IR=1
              NCOEFS = N + 1
              DO 100 I=1,NCOEFS
              IM1 = I - 1
              WRITE(ICON,20) IM1
20            FORMAT(' ENTER COEF OF X**',I2,' ')
              READ(ICON,30) A(I)
30            FORMAT(F15.5)
100           CONTINUE
              CALL POLRT(A,AWORK,N,U,V,IER)
              GOTO(200,210,220,230),IER
              WRITE(ICON,105)
105           FORMAT(' ROOTS ARE:')
              WRITE(ICON,110)
110           FORMAT('0',9X,'  REAL PART',13X,'IMAGINARY PART')
              WRITE(ICON,4) (U(I),V(I),I=1,N)
4             FORMAT(' ',E20.3,2X,E20.3)
999           STOP
200           WRITE(ICON,205)
205           FORMAT('0**ERROR**DEGREE MUST EXCEED 0')
              GOTO 1
210           WRITE(ICON,215)
215           FORMAT('**ERROR**DEGREE MUST BE LESS THAN 37')
              GOTO 1
220           WRITE(ICON,225)
225           FORMAT('0**WARNING**ALGORITHM DID NOT CONVERGE TO SPECIFIED ACCU
             1RACY')
              GOTO 999
230           WRITE(ICON,235)
235           FORMAT('0**ERROR**COEFFICIENT OF HIGHEST POWER CANNOT = 0')
              GOTO 1
              END
```

Figure 9-73 Interactive program for finding roots of a polynomial.

COMPUTER-AIDED ANALYSIS AND DESIGN **639**

system to the next. Array A is the coefficient array, array AWORK is a scratch array used by POLRT, and arrays U and V contain (on return) the real and imaginary parts of the roots of the polynomial specified by N and A.

Example 9-19 In this example we use the interactive program of Fig. 9-73 to obtain the poles of the system function for the servomechanism treated in Sec. 5-3 (see Fig. 5-39 and the related discussion). A block diagram for the system is given in Fig. 9-74. We wish to determine the poles of the system for $K_1 = 5$, $K_1 = 10$, and $K_1 = 100$.

Using block-diagram reduction gives the overall transfer function

$$H(s) = \frac{a_0}{s^3 + a_2 s^2 + a_1 s + a_0}$$

where

$$a_0 = \frac{K_m K_e K_1}{JL} = 318.31 K_1 \text{ s}^{-3}$$

$$a_1 = \frac{FR + K_m^2}{JL} = 570 \text{ s}^{-2}$$

$$a_2 = \frac{R}{L} + \frac{F}{J} = 55 \text{ s}^{-1}$$

The poles of the system are the roots of the polynomial

$$c(s) = s^3 + a_2 s^2 + a_1 s + a_0$$

Figure 9-75 shows a console log (a printout of the terminal dialog) produced by the

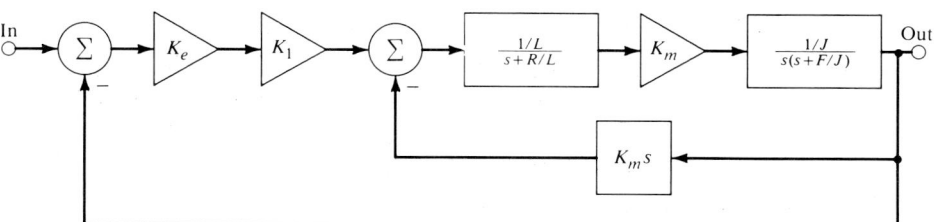

Figure 9-74 System of Example 9-19.

```
ENTER DEGREE OF POLYNOMIAL (0<DEG<37)   3

ENTER COEF OF X** 0   1591.5

ENTER COEF OF X** 1   570.

ENTER COEF OF X** 2   55.

ENTER COEF OF X** 3   1.

ROOTS ARE:
            REAL PART              IMAGINARY PART
          -.4913671E+01             .0000000E+01
          -.7628578E+01             .0000000E+01
          -.4245775E+02             .0000000E+01

                   (a)
```

Figure 9-75 Console log produced by the program of Fig. 9-73 used to find the characteristic roots (the poles) of the system of Fig. 9-74; (a) $k_1 = 5$.

640 INTRODUCTION TO SYSTEM ANALYSIS

```
ENTER DEGREE OF POLYNOMIAL (0<DEG<37)  3

ENTER COEF OF X** 0    3183.1

ENTER COEF OF X** 1    570.

ENTER COEF OF X** 2    55.

ENTER COEF OF X** 3    1.

ROOTS ARE:
            REAL PART              IMAGINARY PART
          -.4360138E+02             .0000000E+01
          -.5699308E+01            -.6365725E+01
          -.5699308E+01             .6365725E+01

                        (b)


ENTER DEGREE OF POLYNOMIAL (0<DEG<37)  3

ENTER COEF OF X** 0    31831.

ENTER COEF OF X** 1    570.

ENTER COEF OF X** 2    55.

ENTER COEF OF X** 3    1.

ROOTS ARE:
            REAL PART              IMAGINARY PART
          -.5513325E+02             .0000000E+01
           .6662649E-01            -.2402795E+02
           .6662649E-01             .2402795E+02

                        (c)
```

Figure 9-75 *Continued* (b) $k_1 = 10$, (c) $k_1 = 100$.

interactive program of Fig. 9-73 for this example (see also Fig. 5-46). For $K_1 = 5$, all three roots (poles) are real. The two poles nearest the origin are dominant, and the system is overdamped. For $K_1 = 10$, there is one real root and two complex roots. The complex roots are dominant, and the system is moderately damped. For $K_1 = 100$, the complex roots lie (barely) in the right half plane, and the system is unstable.

9-4B Systems of Linear Equations

The problem of solving a system of equations of the form

$$\begin{aligned} a_{11}x_1 + a_{12}x_2 + \cdots + a_{1n}x_n &= b_1 \\ a_{21}x_1 + a_{22}x_2 + \cdots + a_{2n}x_n &= b_2 \\ &\vdots \\ a_{n1}x_1 + a_{n2}x_2 + \cdots + a_{nn}x_n &= b_n \end{aligned} \quad (9\text{-}32)$$

arises often; e.g., in using the direct method for obtaining a partial-fraction expansion. Gaussian elimination is one of the simplest, fastest, and most widely used methods for solving such systems of equations. Essentially, gaussian elimination entails adding multiples of the equations to each other in an organized way to obtain a modified set of equations of the form

$$a'_{11}x_1 + a'_{12}x_2 + \cdots + a'_{1n}x_n = b'_1$$
$$a'_{22}x_2 + \cdots + a'_{2n}x_n = b'_2$$
$$\cdots\cdots\cdots\cdots\cdots\cdots\cdots\cdots\cdots$$
$$a'_{nn}x_n = b'_n$$

which can be solved directly (by backward substitution) for the unknowns. Algorithms for gaussian elimination vary in sophistication, reliability, and speed. Figure 9-76 shows a FORTRAN subprogram for solving a system of equations given by (9-32) where all quantities $(a_{i,j}, x_i, b_i)$ are real. The algorithm implemented by subprogram GAUSS is a moderately sophisticated algorithm that works in a large number of practical cases; however it may fail to give sufficiently accurate solutions if the number of unknowns is large (say, greater than 10) or if the coefficients $a_{11}, a_{12}, \ldots, a_{nn}$ range over several orders of magnitude, e.g., if $a_{11} = 1$ and $a_{12} = 10^5$. In practice, it is always wise to check solutions by substitution in the original equations, no matter what method is used.

Subprogram GAUSS of Fig. 9-76 is invoked by a statement of the form

$$\text{CALL GAUSS}(A, N, W1, W2, B, X)$$

Here N is the number of equations (and unknowns); A is an $N \times N$ array containing the coefficients, where $A(I, J) = a_{i,j}$; W1 and W2 are work arrays having dimension N; B is an array containing the quantities on the right side of (9-32), where $B(I) = b_i$; and X is an array containing (on return from GAUSS) values for the unknowns, where $X(I) = x_i$.

```
      SUBROUTINE GAUSS(A,IA,N,L,S,B,X)
      DIMENSION A(IA,N),L(N),S(N),B(N),X(N)
      DO 2 I=1,N
      L(I)=I
      S(I)=0.
      DO 2 J=1,N
2     S(I)=AMAX1(S(I),ABS(A(I,J)))
      DO 4 K=1,N-1
      RMAX=0.
      DO 3 I=K,N
      R=ABS(A(L(I),K))/S(L(I))
      IF(R.LE.RMAX) GO TO 3
      J=I
      RMAX=R
3     CONTINUE
      LK=L(J)
      L(J)=L(K)
      L(K)=LK
      DO 4 I=K+1,N
      XMULT=A(L(I),K)/A(LK,K)
      A(L(I),K)=XMULT
      DO 4 J=K+1,N
4     A(L(I),J)=A(L(I),J)-XMULT*A(LK,J)
      DO 20 J=1,N-1
      DO 20 I=J+1,N
20    B(L(I))=B(L(I))-A(L(I),J)*B(L(J))
      X(N)=B(L(N))/A(L(N),N)
      DO 40 I=1,N-1
      SUM=B(L(N-I))
      DO 30 J=N-I+1,N
30    SUM=SUM-A(L(N-I),J)*X(J)
40    X(N-I)=SUM/A(L(N-I),N-I)
      RETURN
      END
```

Figure 9-76 FORTRAN program for solving a system of N linear equations in N unknowns. *(Adapted from Numerical Mathematics and Computing. Copyright © 1980 by Wadsworth, Inc. Used by permission of Brooks/Cole Publishing Company.)*

```
            REAL A(100),B(10),X(10),L(10),S(10)
            ICON = 1
1           WRITE(ICON,2)
2           FORMAT(' ENTER NUMBER OF VARIABLES (1<NVAR<11) ')
            READ(ICON,10) NVAR
10          FORMAT(I2)
            IF(NVAR.LT.2.OR.NVAR.GT.10) GOTO 1
            WRITE(ICON,12)
12          FORMAT(' ENTER COEFFICIENT MATRIX BY ROWS')
            DO 50 J=1,NVAR
            DO 50 I=1,NVAR
            KA=J+(I-1)*NVAR
            WRITE(ICON,20) J,I
20          FORMAT('0ROW=',I2,' COL=',I2,' COEFFICIENT= ')
            READ(ICON,30) A(KA)
30          FORMAT(F15.5)
50          CONTINUE
            WRITE(ICON,52)
52          FORMAT(' ENTER B VECTOR')
            DO 55 I=1,NVAR
            WRITE(ICON,56) I
56          FORMAT('0B(',I2,')= ')
            READ(ICON,30) B(I)
55          CONTINUE
            CALL GAUSS(A,NVAR,NVAR,L,S,B,X)
            WRITE(ICON,58)
58          FORMAT(' SOLUTION')
            DO 70 I=1,NVAR
            WRITE(ICON,60) I,X(I)
60          FORMAT('0VARIABLE NO',I3,' = ',E20.5)
70          CONTINUE
            STOP
            END
```

Figure 9-77 Interactive program for solving a system of N linear equations in N variables.

Figure 9-77 shows an interactive program for solving a system of real linear equations. The next example illustrates use of this program for obtaining coefficients of a partial-fraction expansion.

Example 9-20 In this example we obtain a partial-fraction expansion for the system function of Example 9-19 (with $K_1 = 10$).
For $K_1 = 10$, the poles of the system function are

$$p_1 = -43.60 \text{ s}^{-1} \qquad p_2 = p_3^* = -5.70 + 6.37j \text{ s}^{-1}$$

It follows that the form of the partial-fraction expansion is

$$H(s) = \frac{c}{s^3 + as^2 + bs + c} = \frac{A}{s - p_1} + \frac{B}{s - p_2} + \frac{B^*}{s - p_2^*}$$

This can be written

$$\frac{c}{(s - p_1)(s^2 + ds + e)} = \frac{A}{s - p_1} + \frac{Cs + D}{s^2 + ds + e}$$

where $d = 2 \text{ Re } p_2 = -11.40 \text{ s}^{-1}$, $e = |p_2|^2 = 73.07 \text{ s}^{-2}$, and the real coefficients A, C, and D are to be determined. Clearing fractions gives

$$c = A(s^2 + ds + e) + (Cs + D)(s - p_1)$$
$$= (A + C)s^2 + (Ad + D - p_1 C)s + (Ae - p_1 D)$$

Equating coefficients of like powers of s gives

$$A + C = 0 \qquad Ad - p_1 C + D = 0 \qquad Ae - p_1 D = c$$

These are three equations in the three unknowns A, C, and D. In the notation of (9-32)

$$x_1 = A \qquad x_2 = C \qquad x_3 = D \qquad a_{11} = 1 \qquad a_{12} = 1$$
$$a_{13} = 0 \qquad a_{21} = d = -11.40 \text{ s}^{-1} \qquad a_{22} = -p_1 = -43.60 \text{ s}^{-1}$$
$$a_{23} = 1 \qquad a_{31} = e = 73.07 \text{ s}^{-2} \qquad a_{32} = 0 \qquad a_{33} = p_1 = -43.6 \text{ s}^{-1}$$
$$b_1 = b_2 = 0 \qquad b_3 = c = 3183.10 \text{ s}^{-3}$$

Figure 9-78 shows a console log produced by the program of Fig. 9-77 for this example. Note that a user must keep track of units, as a computer gives only numerical answers. The partial-fraction coefficients are

$$A = 2.16 \text{ s}^{-1} \qquad C = -2.16 \text{ s}^{-1} \qquad D = 69.40 \text{ s}^{-2}$$

where the units follow from the fact that the system function of this example is dimensionless.

In applications, quantities involved in (9-32) often are complex. Figure 9-79 shows a complex version (CGAUSS) of subprogram GAUSS given above. Figure 9-80 shows an interactive FORTRAN program for solving systems of linear equations involving complex quantities.

Example 9-21 In this example, we use interactive programs given above to obtain the step response of a system described by the differential equation

```
ENTER NUMBER OF VARIABLES (1<NVAR<11) 3
ENTER COEFFICIENT MATRIX BY ROWS
ROW= 1 COL= 1 COEFFICIENT=1.

ROW= 1 COL= 2 COEFFICIENT=1.

ROW= 1 COL= 3 COEFFICIENT=0.

ROW= 2 COL= 1 COEFFICIENT=-11.4

ROW= 2 COL= 2 COEFFICIENT=-43.60

ROW= 2 COL= 3 COEFFICIENT=1.

ROW= 3 COL= 1 COEFFICIENT=73.07

ROW= 3 COL= 2 COEFFICIENT=0.

ROW= 3 COL= 3 COEFFICIENT=-43.60
ENTER B VECTOR
B( 1)=0.

B( 2)=0.

B( 3)=3183.1
SOLUTION
VARIABLE NO  1 =          .21552E  1
VARIABLE NO  2 =         -.21552E  1
VARIABLE NO  3 =         -.69395E  2
```

Figure 9-78 Console log for the program of Fig. 9-77 applied to the system of equations of Example 9-20.

```
      SUBROUTINE CGAUSS(A,N,L,S,B,X)
      REAL L(N),S(N)
      COMPLEX A(N,N),B(N),X(N),SUM,XMULT
      DO 2 I=1,N
      L(I)=I
      S(I)=0.
      DO 2 J=1,N
2     S(I)=AMAX1(S(I),CABS(A(I,J)))
      DO 4 K=1,N-1
      RMAX=0.
      DO 3 I=K,N
      R=CABS(A(L(I),K))/S(L(I))
      IF(R.LE.RMAX) GO TO 3
      J=I
      RMAX=R
3     CONTINUE
      LK=L(J)
      L(J)=L(K)
      L(K)=LK
      DO 4 I=K+1,N
      XMULT=A(L(I),K)/A(LK,K)
      A(L(I),K)=XMULT
      DO 4 J=K+1,N
4     A(L(I),J)=A(L(I),J)-XMULT*A(LK,J)
      DO 20 J=1,N-1
      DO 20 I=J+1,N
20    B(L(I))=B(L(I))-A(L(I),J)*B(L(J))
      X(N)=B(L(N))/A(L(N),N)
      DO 40 I=1,N-1
      SUM=B(L(N-I))
      DO 30 J=N-I+1,N
30    SUM=SUM-A(L(N-I),J)*X(J)
40    X(N-I)=SUM/A(L(N-I),N-I)
      RETURN
      END
```

Figure 9-79 Program of Fig. 9-76 adapted for systems of equations involving complex quantities.

$$\frac{d^3y}{dy^3} + a_2\frac{d^2y}{dy^2} + a_1\frac{dy}{dt} + a_0 y = b_0 x$$

where $a_2 = 3 \text{ s}^{-1}$, $a_1 = 4 \text{ s}^{-2}$, and $a_0 = b_0 = 2 \text{ s}^{-3}$.

First, we use the program of Fig. 9-73 to obtain the characteristic roots for the system. The characteristic equation for the system is

$$s^3 + a_2 s^2 + a_1 s + a_0 = 0$$

Figure 9-81 shows a console log produced by the program of Fig. 9-73 for this example. The characteristic roots are $p_1 = -1 \text{ s}^{-1}$, $p_2 = p_3^* = -1 + j \text{ s}^{-1}$. The step response is given by

$$g(t) = b_0 a_0^{-1}[1 + g_c(t)]u(t) = [1 + g_c(t)]u(t)$$

where
$$g_c(t) = c_1 e^{p_1 t} + c_2 e^{p_2 t} + c_3 e^{p_3 t}$$

The coefficients must be determined such that the initial conditions

$$g_c(0^+) = -1 \qquad \left.\frac{dg_c(t)}{dt}\right|_{t=0^+} = 0 \qquad \left.\frac{d^2 g_c(t)}{dt}\right|_{t=0^+} = 0$$

are satisfied. This requires that

$$c_1 + c_2 + c_3 = -1 \qquad p_1 c_1 + p_2 c_2 + p_3 c_3 = 0 \qquad p_1^2 c_1 + p_2^2 c_2 + p_3^2 c_3 = 0$$

Next, we use the program of Fig. 9-80 to solve the system of equations above for the coefficients c_1, c_2, c_3. Figure 9-82 shows a console log produced by the program. The

COMPUTER-AIDED ANALYSIS AND DESIGN 645

```
              COMPLEX A(100),B(10),X(10),JJ
              REAL L(10),S(10)
              DATA JJ/(0.,1.)/
              ICON = 1
1             WRITE(ICON,2)
2             FORMAT(' NUMBER OF VARIABLES (1<NVAR<11) = ')
              READ(ICON,10) NVAR
10            FORMAT(I0)
              IF(NVAR.LT.2.OR.NVAR.GT.10) GOTO 1
              WRITE(ICON,12)
12            FORMAT(' ENTER COEFFICIENT MATRIX BY ROWS')
              DO 50 J=1,NVAR
              DO 50 I=1,NVAR
              KA=J+(I-1)*NVAR
              WRITE(ICON,20) J,I
20            FORMAT('0ROW=',I2,' COL=',I2,' RE COEFF = ')
              READ(ICON,30) AR
30            FORMAT(F15.5)
              WRITE(ICON,21) J,I
21            FORMAT('0ROW=',I2,' COL=',I2,' IM COEFF = ')
              READ(ICON,30) AI
              A(KA) = AR + JJ*AI
50            CONTINUE
              WRITE(ICON,52)
52            FORMAT(' ENTER B VECTOR')
              DO 55 I=1,NVAR
              WRITE(ICON,56) I
56            FORMAT('0RE B(',I2,')= ')
              READ(ICON,30) BR
              WRITE(ICON,57) I
57            FORMAT('0IM B(',I2,')= ')
              READ(ICON,30) BI
              B(I) = BR + JJ*BI
55            CONTINUE
              CALL CGAUSS(A,NVAR,NVAR,L,S,B,X)
              WRITE(ICON,58)
58            FORMAT(' SOLUTION')
              WRITE(ICON,59)
59            FORMAT('0',25X,'REAL PART',12X,'IMAG PART')
              DO 70 I=1,NVAR
              XR = REAL(X(I))
              XI = AIMAG(X(I))
              WRITE(ICON,60) I,XR,XI
60            FORMAT('0VARIABLE NO',I3,' = ',E20.5,E20.5)
70            CONTINUE
              STOP
              END
```

Figure 9-80 Interactive program for solving a system of N complex linear equations in N variables.

```
ENTER DEGREE OF POLYNOMIAL (0<DEG<37)   3

ENTER COEF OF X** 0   2.

ENTER COEF OF X** 1   4.

ENTER COEF OF X** 2   3.

ENTER COEF OF X** 3   1.

ROOTS ARE:

            REAL PART              IMAGINARY PART
         -.1000000E+01              .0000000E+00
         -.1000000E+01             -.1000000E+01
         -.1000000E+01              .1000000E+01
```

Figure 9-81 Console log for the program of Fig. 9-73 used to obtain the characteristic roots of the system of Example 9-21.

646 INTRODUCTION TO SYSTEM ANALYSIS

```
NUMBER OF VARIABLES (1<NVAR<11) = 3
ENTER COEFFICIENT MATRIX BY ROWS

ROW= 1 COL= 1 RE COEFF = 1.

ROW= 1 COL= 1 IM COEFF = 0.

ROW= 1 COL= 2 RE COEFF = 1.

ROW= 1 COL= 2 IM COEFF = 0.

ROW= 1 COL= 3 RE COEFF = 1.

ROW= 1 COL= 3 IM COEFF = 0.

ROW= 2 COL= 1 RE COEFF = -1.

ROW= 2 COL= 1 IM COEFF = 0.

ROW= 2 COL= 2 RE COEFF = -1.

ROW= 2 COL= 2 IM COEFF = -1.

ROW= 2 COL= 3 RE COEFF = -1.

ROW= 2 COL= 3 IM COEFF = 1.

ROW= 3 COL= 1 RE COEFF = 1.

ROW= 3 COL= 1 IM COEFF = 0.

ROW= 3 COL= 2 RE COEFF = 0.

ROW= 3 COL= 2 IM COEFF = 2.

ROW= 3 COL= 3 RE COEFF = 0.

ROW= 3 COL= 3 IM COEFF = -2.
ENTER B VECTOR

RE B( 1)= -1.

IM B( 1)= 0.

RE B( 2)= 0.

IM B( 2)= 0.

RE B( 3)= 0.

IM B( 3)= 0.
SOLUTION

                     REAL PART        IMAG PART

VARIABLE NO  1 =     -.20000E  1      .11921E -6

VARIABLE NO  2 =      .50000E  0     -.50000E  0

VARIABLE NO  3 =      .50000E  0      .50000E  0
```

Figure 9-82 Console log for the program of Fig. 9-80 used to obtain the step response of the system of Example 9-21.

coefficients are $c_1 = -2$ (the small imaginary part given by the program arises from roundoff error) and $c_2 = c_3^* = 0.5 - 0.5j$; thus

$$\begin{aligned} g_c(t) &= -2e^{p_1 t} + (0.5 - 0.5j)e^{p_2 t} + (0.5 + 0.5j)e^{p_3^* t} \\ &= -2e^{p_1 t} + 2\,\text{Re}[(0.5 - 0.5j)e^{p_2 t}] \\ &= -2e^{p_1 t} + 2e^{\sigma t}\cos(\omega t + \theta) \end{aligned}$$

where $p_1 = 1\text{ s}^{-1}$, $\sigma = \text{Re } p_2 = -1\text{ s}^{-1}$, $\omega = \text{Im } p_2 = -1\text{ s}^{-1}$, and $\theta = \angle p_2 = -\pi/4$ rad.

9-4C Partial-Fraction Expansion

Figure 9-83 shows an interactive program for expanding a rational function in partial fractions. The program implements a direct (not Heaviside) method of partial-fraction

```
          COMPLEX POLE(20),ZERO(20),COEFFS(21),J
          DATA J/(0.,1.)/
          ICON = 1
          WRITE(ICON,5)
5         FORMAT(' ENTER NUMBER OF POLES (1<NPOLES<21) ')
1         READ(ICON,10) NPOLES
10        FORMAT(I0)
          IF(NPOLES.LE.1.OR.NPOLES.GT.20) GOTO 1
          DO 30 I = 1,NPOLES
          WRITE(ICON,15) I
15        FORMAT(' RE PART POLE NO ',I3,' = ')
          READ(ICON,20) PRE
20        FORMAT(F15.5)
          WRITE(ICON,25) I
25        FORMAT(' IM PART POLE NO ',I3,' = ')
          READ(ICON,20) PIM
          POLE(I) = PRE + J*PIM
30        CONTINUE
          WRITE(ICON,31)
31        FORMAT(' ENTER NUMBER OF ZEROES (-1<NZEROS<21) ')
          READ(ICON,10) NZEROS
          IF(NZEROS.EQ.0) GOTO 55
          DO 50 I = 1,NZEROS
          WRITE(ICON,35) I
35        FORMAT(' RE PART ZERO NO ',I3,' = ')
          READ(ICON,20) ZRE
          WRITE(ICON,45) I
45        FORMAT(' IM PART ZERO NO ',I3,' = ')
          READ(ICON,20) ZIM
          ZERO(I) = ZRE + J*ZIM
50        CONTINUE
55        CALL PARF(POLE,NPOLES,ZERO,NZEROS,COEFFS,NCOEFS)
          WRITE(ICON,56)
56        FORMAT(' PARTIAL-FRACTION COEFFICIENTS ')
          WRITE(ICON,58)
58        FORMAT('0',7X,'REAL PART      IMAG PART')
          DO 70 I = 1,NCOEFS
          CRE = REAL(COEFFS(I))
          CIM = AIMAG(COEFFS(I))
          WRITE(ICON,60) CRE,CIM
60        FORMAT('0',F15.5,2X,F15.5)
70        CONTINUE
          STOP
          END
```

Figure 9-83 Interactive program for obtaining a partial-fraction expansion.

648 INTRODUCTION TO SYSTEM ANALYSIS

```
      SUBROUTINE PARF(POLES,NPOLES,ZEROES,NZEROS,COEFFS,NCOEFS)
      COMPLEX POLES(1),ZEROES(1),A(441),B(21),D(21),NUM(21)
      COMPLEX P(20),Z(20),COEFFS(21),SOLN(21)
      REAL SSG(21),LLG(21)
      INTEGER MUL(20)
C
C     SAVE INPUT
C
      NP=NPOLES
      NZ=NZEROS
      NEQ=NP
      IF(NZ.GE.NP) NEQ=NZ+1
      DO 30 I=1,NP
30    P(I)=POLES(I)
      IF(NZ.EQ.0) GO TO 50
      DO 40 I=1,NZ
40    Z(I)=ZEROES(I)
C
C     CLEAR WORK ARRAYS
C
50    DO 20 I=1,NEQ
      B(I)=(0.,0.)
      D(I)=(0.,0.)
      NUM(I)=(0.,0.)
      DO 20 J=1,NEQ
      K=I+(J-1)*NEQ
      A(K)=(0.,0.)
20    CONTINUE

C
C     COMPRESS POLE ARRAY
C
      CALL MULTI(P,NP,MUL,NPS)
C
C     FORM B-VECTOR IN EQUATION A*X=B TO BE SOLVED FOR
C     COEFFICIENTS OF PARTIAL FRACTION EXPANSION.
C
      NN=0
      NUM(1)=(1.,0.)
      IF(NZ.EQ.0) GO TO 200
      DO 100 I=1,NZ
100   CALL PROD(NUM,Z(I),NN)
200   NNP1=NN+1
      DO 250 I=1,NNP1
250   B(NEQ+1-I)=NUM(I)
      ICOL=0
C
      IF(NZ.LT.NP) GO TO 425
C
C     FORM COLUMNS OF A-MATRIX CORRESPONDING TO POWERS
C     OF S IN P.F.E.
C
      D(1)=(1.,0.)
      ND=0
      DO 325 I=1,NPS
      JJ=MUL(I)
      DO 300 J=1,JJ
300   CALL PROD(D,P(I),ND)
      III=ND+1
325   CONTINUE
      IMAX=NZ-NP+1
      NDP1=ND+1
      DO 400 I=1,IMAX
      DO 350 J=1,NDP1
      KA=J+I-1+(I-1)*NEQ
```

Figure 9-83 *Continued*

```
            KD=ND+2-J
            A(KA)=D(KD)
350         CONTINUE
400         ICOL=ICOL+1
C
C           FORM COLUMNS OF A-MATRIX CORRESPONDING TO COEFFICIENTS
C           OF 1/(S-P(I))**MUL(I) IN P.F.E.
C
425         DO 800 I=1,NPS
            ND=0
            D(1)=(1.,0.)
            DO 500 J=1,NPS
            IF(J.EQ.I) GO TO 500
            MJ=MUL(J)
            DO 450 K=1,MJ
450         CALL PROD(D,P(J),ND)
500         CONTINUE
            MI=MUL(I)
            DO 700 J=1,MI
            ICOL=ICOL+1
            NDP1=ND+1
            DO 600 K=1,NDP1
            IROW=NEQ-K+1
            KA=IROW+(ICOL-1)*NEQ
600         A(KA)=D(K)
            CALL PROD(D,P(I),ND)
700         CONTINUE
800         CONTINUE
            CALL CGAUSS(A,NEQ,LLG,SSG,B,SOLN)
            NCOEFS=NEQ
            DO 900 I=1,NEQ
900         COEFFS(I)=SOLN(I)
            RETURN
            END
C
            SUBROUTINE MULTI(P,NP,MUL,NPS)
            COMPLEX P(1)
            INTEGER MUL(1)
C
C           THIS SUBROUTINE COMPRESSES AN ARRAY OF COMPLEX
C           POLES; REPEATED POLES ARE ELIMINATED; THE
C           ORDERS OF THE DISTINCT POLES ARE RETURNED IN ARRAY MUL.
C
C           INPUT: P=ARRAY OF POLES
C                  NP=DIM(P)
C           OUTPUT: MUL=ARRAY OF MULTIPLICITIES
C                   NPS=NUMBER OF DISTINCT POLES
C                   P=COMPRESSED ARRAY
C
            DO 10 I=1,NP
            MUL(I)=1
10          CONTINUE
            NPS=NP
            DO 1 I=1,NP
            IF(MUL(I).EQ.0) GO TO 1
            IP1=I+1
            DO 1 J=IP1,NP
            IF(P(J).NE.P(I)) GO TO 1
            MUL(I)=MUL(I)+1
            MUL(J)=0
            NPS=NPS-1
1           CONTINUE
            IF(NPS.EQ.NP) RETURN
            NPM1=NP-1
            NM=NP-NPS
```

Figure 9-83 *Continued*

650 INTRODUCTION TO SYSTEM ANALYSIS

```
          DO 3 K=1,NM
          DO 3 I=1,NPM1
          IF(MUL(I).NE.0) GO TO 3
          DO 2 J=I,NPM1
          P(J)=P(J+1)
          MUL(J)=MUL(J+1)
2         CONTINUE
3         CONTINUE
          NPSP1=NPS+1
          DO 4 I=NPSP1,NP
4         P(I)=(0.,0.)
          RETURN
          END
C
          SUBROUTINE PROD(A,P,NA)
          COMPLEX A(1),P
C
C         THIS SUBROUTINE FORMS THE PRODUCT F(X)=(X-P)*F(X).
C         ON INPUT, F(X)=A(1)+A(2)*X+...+A(NA+1)*X**NA.
C         ON OUTPUT,NA=NA+1 AND A CONTAINS THE CCEFFICIENTS
C         FOR THE PRODUCT POLYNOMIAL.
C
          A(NA+2)=A(NA+1)
          NA=NA+1
          IF(NA.EQ.1) GO TO 2
          DO 1 I=2,NA
          J=NA+2-I
          A(J)=A(J-1)-P*A(J)
1         CONTINUE
2         A(1)=-P*A(1)
          RETURN
          END
```

Figure 9-83 *Continued*

expansion, where both sides of the expansion (containing undetermined coefficients) are cleared of fractions to obtain an equation equating two polynomials in the variable. A system of linear equations in the partial-fraction coefficients is obtained by equating coefficients of like powers of the variable, as illustrated in Example 5-15. Inputs requested by the program are the degree of the denominator of a function, the degree of the numerator of the function, the poles of the function, and the zeros of the function. The program gives a partial-fraction expansion for any rational function; it is not necessary that the degree of the denominator equal or exceed the degree of the numerator. If the degree of the numerator equals or exceeds the degree of the denominator, the program supplies coefficients for the leading polynomial in the variable.

Example 9-22 In this example, we use the program of Fig. 9-83 to obtain a partial-fraction expansion for the function

$$F(x) = \frac{(x-1)^2(x-2)^2(x-3)^2(x+j)(x-j)}{x^2(x+1)^2(x+2)(x^2+2x+2)}$$

Figure 9-84 shows a console log for this example. Since the degree of the numerator exceeds the degree of the denominator, the expansion has a leading polynomial part. The program prints this part first, in order of decreasing powers of the variable. The program then prints coefficients of terms $(s-p)^{-m}$, where p is a pole and m is the multiplicity of p, in the order in which the poles were entered and in order of decreasing values of m. The expansion is

```
ENTER NUMBER OF POLES (1<NPOLES<21) 7
RE PART POLE NO    1 =  0.
IM PART POLE NO    1 =  0.
RE PART POLE NO    2 =  0.
IM PART POLE NO    2 =  0.
RE PART POLE NO    3 = -1.
IM PART POLE NO    3 =  0.
RE PART POLE NO    4 = -1.
IM PART POLE NO    4 =  0.
RE PART POLE NO    5 = -2.
IM PART POLE NO    5 =  0.
RE PART POLE NO    6 = -1.
IM PART POLE NO    6 =  1.
RE PART POLE NO    7 = -1.
IM PART POLE NO    7 = -1.
ENTER NUMBER OF ZEROES (-1<NZEROS<21) 8
RE PART ZERO NO    1 =  1.
IM PART ZERO NO    1 =  0.
RE PART ZERO NO    2 =  1.
IM PART ZERO NO    2 =  0.
RE PART ZERO NO    3 =  2.
IM PART ZERO NO    3 =  0.
RE PART ZERO NO    4 =  2.
IM PART ZERO NO    4 =  0.
RE PART ZERO NO    5 =  3.
IM PART ZERO NO    5 =  0.
RE PART ZERO NO    6 =  3.
IM PART ZERO NO    6 =  0.
RE PART ZERO NO    7 =  0.
IM PART ZERO NO    7 =  1.
RE PART ZERO NO    8 =  0.
IM PART ZERO NO    8 = -1.
PARTIAL-FRACTION COEFFICIENTS
     REAL PART        IMAG PART

       1.00000          .00000

     -18.00029         -.00012

       9.00004         -.00002

     -64.50008          .00007

    1151.99851          .00000

   -2496.01173         -.00279

    2250.01907          .00512

     231.24790        -243.75648

     231.24599         243.75495
```

Figure 9-84 Console log for the program of Fig. 9-83 used to obtain a partial-fraction expansion for the function of Example 9-22.

$$F(x) = x - 18 + \frac{9}{x^2} - \frac{64}{x} + \frac{1152}{(x+1)^2} - \frac{2496}{x+1}$$
$$+ \frac{2250}{x+2} + \frac{231 - 243j}{x+1+j} + \frac{231 + 243j}{x+1-j}$$

9-4D Designing Recursive Digital Filters

Figure 9-85 shows a program for designing a Butterworth low-pass recursive digital filter using bilinear substitution. The program obtains specifications on a filter from the

```
              DATA PI/3.14159/
              ICON = 1
    1         WRITE(ICON,800)
    800       FORMAT('0SAMPLING FREQUENCY = ')
              READ(ICON,110) FS
              WRITE(ICON,100)
    100       FORMAT('0MINIMUM PASSBAND GAIN (NE 0 AND < 1) = ')
              READ(ICON,110) G1
              G11 = G1
              IF(G1.GE.1..OR.G1.EQ.0.) GOTO 1
              IF(G1.GT.0.) GOTO 107
              WRITE(ICON,105)
    105       FORMAT('0GAINS GIVEN IN DECIBELS')
              G1 = 10.**(G1/20.)
    107       GAM1 = (1./G1)**2 - 1.
    110       FORMAT(F15.5)
              WRITE(ICON,120)
    120       FORMAT('0MAXIMUM PASSBAND FREQUENCY = ')
              READ(ICON,110) F1
              F1 = FS*TAN(PI*F1/FS)/PI
              WRITE(ICON,122) F1
    122       FORMAT('0EXPANDED PASSBAND FREQ. = ',F15.5)
    2         WRITE(ICON,130)
    130       FORMAT('0MAXIMUM STOPBAND GAIN (NE 0 AND < PASSBAND GAIN) = ')
              READ(ICON,110) G2
              IF(G11/G2.GT.0.) GOTO 555
              WRITE(ICON,556)
    556       FORMAT('0PASSBAND AND STOPBAND GAINS MUST BE IN SAME UNIT.',/,
         1    REDO FROM START')
              GOTO 1
    555       IF(G2.GE.G1.OR.G2.EQ.0.) GOTO 2
              IF(G2.LT.0.) G2 = 10.**(G2/20.)
              GAM2 = (1./G2)**2 - 1.
    3         WRITE(ICON,140)
    140       FORMAT('0MINIMUM STOPBAND FREQUENCY (> PASSBAND FREQ) = ')
              READ(ICON,110) F2
              IF(F2.LE.F1) GOTO 3
              F2 = FS*TAN(PI*F2/FS)/PI
              WRITE(ICON,142) F2
    142       FORMAT('0EXPANDED STOPBAND FREQ. = ',F15.5)
              ORDER = 0.5*ALOG(GAM1/GAM2)/ALOG(F1/F2)
              N = ORDER
              IF(FLOAT(N).LT.ORDER) N = N + 1
              WRITE(ICON,150) N
    150       FORMAT('0ORDER OF FILTER = ',I3)
              FC1 = F1*EXP(-ALOG(GAM1)/(2.*FLOAT(N)))
              FC2 = F2*EXP(-ALOG(GAM2)/(2.*FLOAT(N)))
              FC = SQRT(FC1*FC2)
              WRITE(ICON,170) FC
    170       FORMAT('0EXPANDED HALF-POWER FREQ. = ',F15.5)
              FCA = FS*ATAN(PI*FC/FS)/PI
              WRITE(ICON,172) FCA
    172       FORMAT('0ACTUAL HALF-POWER FREQ. = ',F15.5)
              WRITE(ICON,175)
    175       FORMAT('0Z-PLANE POLES ')
              WRITE(ICON,176)
    176       FORMAT('0',5X,'RE PART',8X,'IM PART')
              WC = 2.*PI*FC
              CC = 4.*FS*FS + WC*WC
              DD = 4.*FS*FS - WC*WC
              DO 180 K = 1,N
              AA = FLOAT(2*K-1)*PI/FLOAT(2*N)
              SPR = -WC*SIN(AA)
              SPI = WC*COS(AA)
              PR = DD/(CC - 4.*FS*SPR)
              PIM = 4.*FS*SPI/(CC - 4.*FS*SPR)
              WRITE(ICON,190) PR,PIM
    180       CONTINUE
    190       FORMAT('0',F15.5,2X,F15.5)
    900       STOP
              END
```

Figure 9-85 Interactive program for designing digital low-pass Butterworth filters using bilinear substitution.

console, expands the specifications to account for compression produced by bilinear substitution, calculates the s-plane poles of the filter using (4-72b), and calculates the z-plane poles of the filter using (8-21).

Example 9-23 We use the program of Fig. 9-85 to design a low-pass digital filter meeting the specifications of Fig. 9-86. We then use the program of Fig. 9-83 to obtain the step response of the filter.

Figure 9-87 shows a console log produced by the program of Fig. 9-85 when the program is used to design a filter meeting the specifications of Fig. 9-86. Inputs requested by the program are sampling rate f_s, minimum passband gain Γ_1, maximum passband frequency f_1, maximum stopband gain Γ_2, and minimum stopband frequency f_2. Outputs given by the program are expanded maximum passband frequency, expanded minimum stopband frequency, expanded and actual half-power frequencies, and the poles of the digital filter. The system function for the filter is

$$H(z) = \frac{K(z+1)^3}{(z - 0.2843)(z^2 - 0.7382z + 0.4033)}$$

where the triple zero at $z = -1$ arises from the bilinear substitution and the constant K is to be determined such that the dc gain of the filter is unity. This gives

$$H(1) = \frac{8K}{(0.7157)(0.6651)} = 16.81K = 1$$

whence $K = 0.0595$.

To obtain the step response of the filter we first use the program of Fig. 9-83 to obtain a partial-fraction expansion for

$$\frac{H(z)}{K(z-1)} = \frac{(z+1)^3}{(z-1)(z-p_1)(z-p_2)(z-p_2^*)}$$

where $p_1 = 0.2843$ and $p_2 = 0.3691 + 0.5168j$. Figure 9-88 shows a console log produced by the program. The expansion is

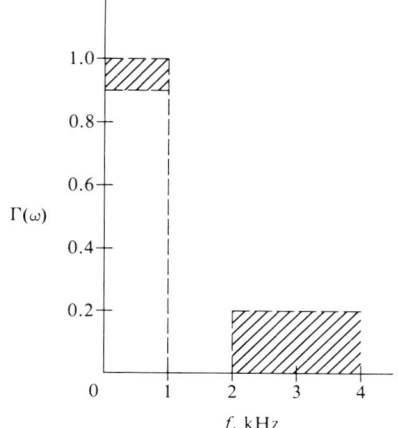

Figure 9-86 Specifications for the filter of Example 9-23; $f_1 = 1$ kHz, $\Gamma_1 = 0.9$; $f_2 = 2$ kHz, $\Gamma_2 = 0.2$; $f_s = 8$ kHz.

```
SAMPLING FREQUENCY = 8000.

MINIMUM PASSBAND GAIN (NE 0 AND < 1) = 0.9

MAXIMIUM PASSBAND FREQUENCY = 1000.

EXPANDED PASSBAND FREQ. =      1054.78601

MAXIMUM STOPBAND GAIN (NE 0 AND < PASSBAND GAIN) = 0.2

MINIMUM STOPBAND FREQUENCY () PASSBAND FREQ) = 2000.

EXPANDED STOPBAND FREQ. =      2546.47778

ORDER OF FILTER =    3

EXPANDED HALF-POWER FREQ. =    1419.09680

ACTUAL HALF-POWER FREQ. =      1294.66516

Z-PLANE POLES
     RE PART           IM PART
       .36911            .51677
       .28429            .00000
       .36911           -.51676
```

Figure 9-87 Console log for the program of Fig. 9-85 used to design a filter meeting the specifications of Fig. 9-86.

```
 ENTER NUMBER OF POLES (1<NPOLES<21) 4
RE PART POLE NO    1 = 1.0
IM PART POLE NO    1 = 0.
RE PART POLE NO    2 = 0.2843
IM PART POLE NO    2 = 0.
RE PART POLE NO    3 = 0.3691
IM PART POLE NO    3 = 0.5168
RE PART POLE NO    4 = 0.3691
IM PART POLE NO    4 = -0.5168
 ENTER NUMBER OF ZEROES (-1<NZEROS<21) 3
RE PART ZERO NO    1 = -1.
IM PART ZERO NO    1 = 0.
RE PART ZERO NO    2 = -1.
IM PART ZERO NO    2 = 0.
RE PART ZERO NO    3 = -1.
IM PART ZERO NO    3 = 0.
  PARTIAL-FRACTION COEFFICIENTS
     REAL PART         IMAG PART
       16.80578          -.00000
      -10.79153           .00000
       -2.50712          6.64135
       -2.50713         -6.64135
```

Figure 9-88 Console log for the program of Fig. 9-83 used to obtain a partial-fraction expansion for the system function of the filter of Example 9-23.

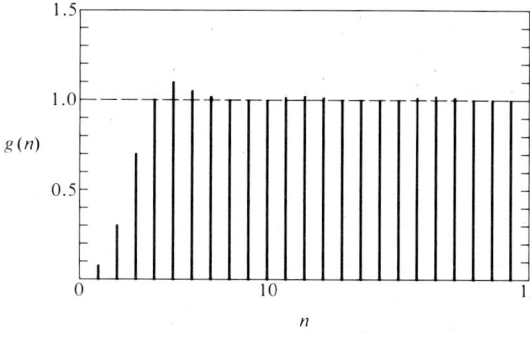

Figure 9-89 Step response of the filter of Example 9-23.

$$\frac{H(z)}{K(z-1)} = \frac{16.81}{z-1} - \frac{10.79}{z-0.2843}$$

$$+ \frac{-2.507 + 6.641j}{z - 0.3691 - 0.5168j} + \frac{-2.507 - 6.641j}{z - 0.3691 + 0.5168j}$$

The step response is

$$g(n) = \mathcal{L}^{-1}\left\{\frac{H(z)}{z-1}\right\} = [1.000 - (0.642)(0.284)^{n-1}$$

$$- (0.845)(0.635)^{n-1} \cos(0.951n - 1.210)]u(n)$$

Figure 9-89 shows a graph of step response $g(n)$ versus sample number n. The step response reaches steady state in about 10 sampling intervals, corresponding to time $t = 10/f_s = 1$ ms. This is in agreement with the time $t' = 1/f_1$, where $f_1 = 1$ kHz is the actual (not expanded) passband-edge frequency. The filter should provide high-fidelity transmission for a pulse whose bandwidth is much less than 1 kHz (for a pulse whose duration is much greater than 1 ms).

SUMMARY

Simulation is a powerful tool for analysis and design of otherwise intractable systems. It also provides a relatively fast and inexpensive means of verifying performance of systems that can (in time) be treated using conventional mathematical methods. One of the problems associated with simulation is obtaining algorithms for simulating analog operations, e.g., integration. Using one of several available simulation languages largely eliminates this problem, allowing a user to concentrate on behavior of a system under study. CSMP is a powerful and widely available simulation language that is particularly well suited for simulating analog systems of interest to electrical engineers. Sections 9-1 and 9-2 describe facilities of CSMP and illustrate use of CSMP for simulating systems.

Section 9-3 describes use of a fast Fourier transform (FFT) algorithm for calculating Fourier coefficients, Fourier series, Fourier transforms, and Fourier integrals. These calculations are necessary in many offline signal-processing applications; e.g., seismic signal analysis. FFT processors are also used in digital systems for real-time signal processing, notably in sonar systems.

Section 9-4 describes interactive computer-aided analysis, where a computer and special-purpose software are used to do tedious calculations necessary to system analysis and design. Programs described in Sec. 9-4 (or equivalent programs) are available at virtually all installations for scientific computing. Using these and other programs for system analysis requires only spending a little time at a terminal (or microcomputer). Time spent is repaid with interest in subsequent applications.

We wish to emphasize the positive aspects of simulation and other methods for computer-aided analysis and design because these methods are both powerful and useful. Nevertheless there are dangers that an intelligent user is wary of. The main danger is that a user will come to regard simulation as a substitute for mathematical and physical reasoning when in fact simulation is mainly an aid to such reasoning. Effective use of simulation or other methods for computer-aided analysis and design requires recognizing correct results, artifacts of simulation, and plain old errors. This in turn requires understanding of physical principles and facility with the classical methods of analysis treated in the first eight chapters of this book.

PROBLEMS

9-1 Use CSMP to obtain the output of the system of Fig. P9-1 for input

$$x(t) = A \cos(\omega_0 t)$$

with $\omega_0 = 2\pi$ krad/s and (a) $A = x_0/10$, (b) $A = x_0$, (c) $A = 10x_0$. Comment on the results. For what values of A/x_0 does the system act (approximately) as a proportion element? For what values of A/x_0 does the system act approximately as a hard limiter?

x(t) → [$y = y_0 \tan^{-1}(\frac{x}{x_0})$] → y(t)

Figure P9-1 $y_0 = x_0 = 1$ V.

9-2 Repeat Prob. 9-1 for the system of Fig. P9-2.

x(t) → [$y = y_0 \tanh(\frac{x}{x_0})$] → y(t)

Figure P9-2 $y_0 = 50$ mA, $x_0 = 20$ mV.

9-3 Use CSMP to obtain the output of the system of Fig. P9-3 for input $x(t) = 5(t/\tau)r(t/\tau)$ V, with $\tau = 2$ ms. Use the result and the output for another input to show that the system is conditionally stable.

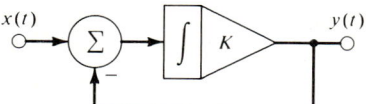

Figure P9-3 $y_0 = 1$ mA, $x_0 = 1$ V, and $K = 10^3$ s^{-1}.

9-4 Use CSMP to obtain the step response of the system of Fig. P9-4. Verify the result by obtaining a closed-form expression for the step response.

Figure P9-4 $K = 10^3$ s^{-1}.

9-5 Use CSMP to obtain the step response of the system of Fig. P9-5 for (a) $K = 1\ \text{s}^{-2}$, (b) $K = 2\ \text{s}^{-2}$, and (c) $K = -1\ \text{s}^{-2}$. Verify the results by obtaining a closed-form expression for the step response.

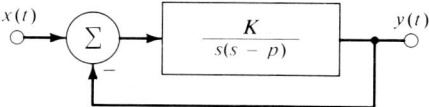

Figure P9-5 $p = -2\ \text{s}^{-1}$.

9-6 Use CSMP to model a voltage-controlled oscillator (VCO) whose output $y(t)$ for input $x(t)$ is given by

$$y(t) = y_0 \cos \phi(t) \qquad \text{where } \phi(t) = 2\pi\left[f_0 t + K \int_{-\infty}^{t} x(\alpha)\,d\alpha\right]$$

with $y_0 = 1$ V, $f_0 = 10$ kHz, and $K = 1$ kHz/V. Plot the output of the VCO for input $x(t) = 10(t/\tau)u(t)$ V, where $\tau = 500\ \mu\text{s}$.

9-7 Figure P9-7 shows a block diagram for a system called a phase detector. Simulate this system using CSMP and show that the dc component of the output for inputs

$$w(t) = A \cos \omega_0 t \qquad x(t) = B \cos(\omega_0 t + \theta)$$

is a linear function of the phase difference θ.

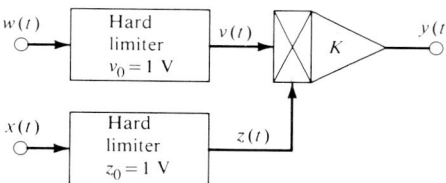

Figure P9-7 $K = 1\ \text{V}^{-1}$.

9-8 Determine the value of the gain K for which the servomechanism of Fig. P9-8 has damping ratio $\zeta = 0.7$. Use CSMP to simulate the system and verify your calculations. Determine the maximum torque, the maximum speed, and the maximum power required of the motor for input $\theta_i(t) = u(t)$ rad.

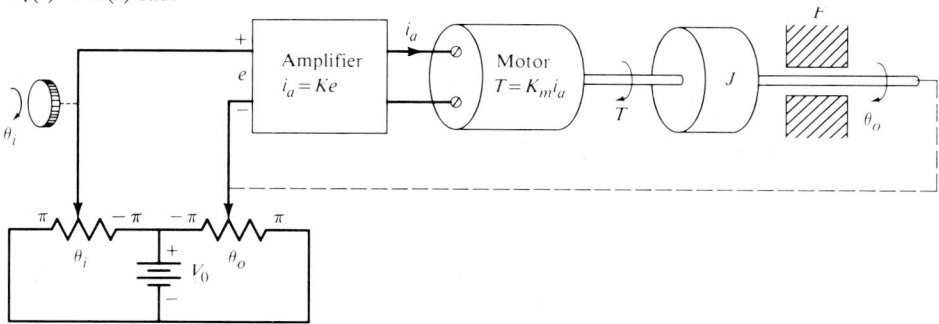

Figure P9-8 $K_m = 1$ Nm/A, $J = 0.2$ Nm/(rad/s^2), $V_0 = 5$ V and $F = 0.5$ Nm/(rad/s).

9-9 Using a dominant-pole approximation, obtain an approximate expression for the step response of the system of Fig. P9-9. Use CSMP to calculate the actual step response and comment on the accuracy of the approximation.

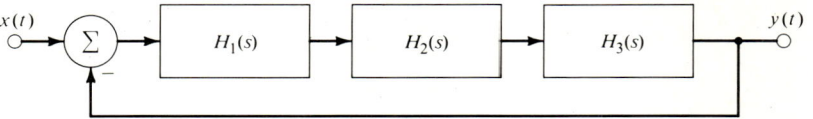

Figure P9-9 $H_1(s) = p_1/(s - p_1), H_2(s) = p_2/(s - p_2), H_3(s) = p_3/(s - p_3), p_1 = -1\ s^{-1}, p_2 = -2\ s^{-1}$, and $p_3 = -10\ s^{-1}$.

9-10 Use a CSMP simulation to calculate gain of the system of Fig. P9-10 for frequency $f = 100$ Hz; that is, use CSMP to obtain the output for input

$$x(t) = A \cos \omega_0 t$$

with $\omega_0 = 200\pi$ rad/s, and calculate the ratio of the peak amplitude of the output to the peak amplitude of the input after all transients have died out. What determines the duration of a simulation (FINTIM) necessary for this calculation?

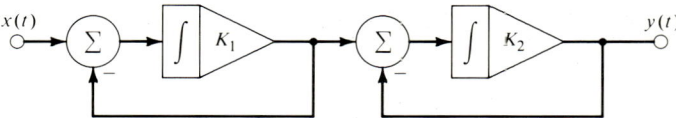

Figure P9-10 $K_1 = 10^2\ s^{-1}$, and $K_2 = 2 \times 10^2\ s^{-1}$.

9-11 Obtain the system function for a Butterworth filter meeting the specifications given in Fig. P9-11. Use CSMP to obtain the step response of the filter. From the result, estimate the duration of the transient response and determine whether the reciprocal of the filter bandwidth is a reasonable approximation for the duration of the transient. Comment on distortion introduced by the filter for input $x(t) = x_0 r(t/\tau)$ in terms of both frequency-domain (bandwidth) and time-domain relations. What features of the pulse response would probably be improved by delay equalization?

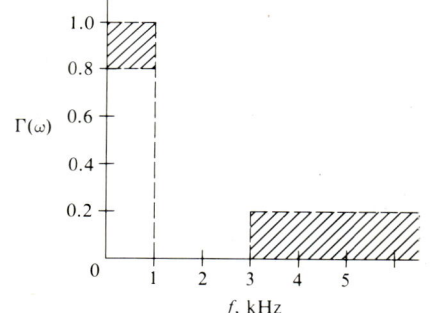

Figure P9-11

9-12 The block diagram of Fig. P9-12a represents a system called a phase-shift oscillator. For a particular value of K, the output of the (unforced) system is a sinusoid whose frequency is proportional to the value of the pole p. For larger values of K the system is unstable.

(a) Find the value K_0 of K for which the system produces a sinusoidal output and find the frequency of the output as a function of the pole p. Determine p such that the frequency of the output equals 10 Hz and use a CSMP simulation to check your calculations. *Note:* You may need to supply a pulse input to get the oscillation started. What would start the oscillation in a physical realization of the system?

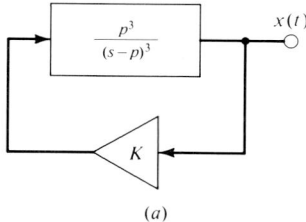

(a)

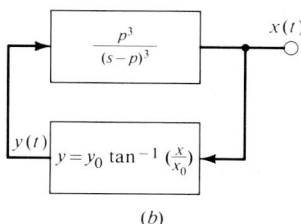

(b)

Figure P9-12

(b) Modify the system (and CSMP model) as shown in Fig. P9-12b, where $x_0 = 1$ V. Simulate the system for $y_0 = K_0 x_0$, $(K_0 - 1)x_0$, and $(K_0 + 1)x_0$. Comment on the results. In particular, explain why the system fails to oscillate for the smaller value of y_0 and why the output is approximately sinusoidal for the larger value of y_0 (in spite of the nonlinearity).

9-13 Refer to Prob. 1-44. Simulate the system of Fig. P1-44 using a delay element to model propagation delay in the feedback path. Find several values of delay time for which the system oscillates and plot frequency of oscillation versus delay time. Does moving the microphone closer to the speaker increase or decrease frequency of oscillation?

9-14 The block diagram of Fig. P9-14 represents a Butterworth low-pass filter. Obtain an expression for the sensitivity of the half-power bandwidth f_c to variations of the parameter K. Simulate the system using a sinusoidal input and verify that the half-power bandwidth changes as expected in response to changes in K.

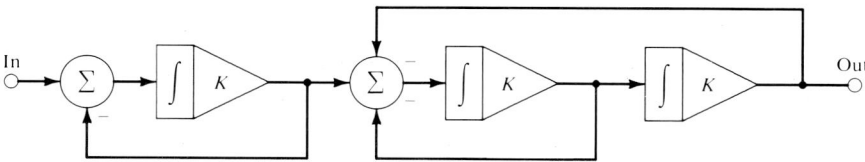

Figure P9-14

9-15 Use a tapped delay line and CSMP to simulate a system having impulse response $h(t) = \tau^{-1} r(t/\tau)$, where $\tau = 1$ ms. Plot the output of the system for input $x(t) = 5r(t/\tau_1)$ for:
  (a) $\tau_1 = 500$ μs   (b) $\tau_1 = 1$ ms   (c) $\tau_1 = 1.5$ ms

9-16 Use numerical convolution to obtain the output of a system having impulse response $h(t) = \tau^{-1} r(t/\tau)$ and input $x(t) = 5e^{-t/\tau} u(t)$ V, with $\tau = 5$ ms.

9-17 Use CSMP to obtain the outputs of systems described by the differential equations below for the specified inputs (x denotes an input and y denotes the corresponding output).

  (a) $D^2 y + a_1 Dy + a_0 y = b_1 Dx + b_0 x$ $x(t) = 5e^{-t/\tau} u(t)$ V for $\tau = 1$ s, $a_0 = b_0 = 2$ s^{-2}, and $a_1 = b_1 = 2$ s^{-1}.

(b) $D^2y + a_1 Dy + a_0 y = b_0 x$ $x(t) = 5\dfrac{t}{\tau} u(t)$ V

for $\tau = 1$ s, $a_1 = 2$ s^{-1}, and $a_0 = b_0 = 2$ s^{-2}.

(c) $D^3y + a_2 D^2y + a_1 Dy + a_0 y = b_0 x$ $x(t) = 5r\left(\dfrac{t}{\tau}\right)$ V

for $\tau = 10$ ms, $a_2 = 4.5 \times 10^3$ s^{-1}, $a_1 = 13.5 \times 10^6$ s^{-2}, and $a_0 = b_0 = 27 \times 10^9$ s^{-3}.

9-18 For each signal of Fig. P9-18 select a value for N such that an N-harmonic finite Fourier series provides a reasonable (say, ±10 percent) approximation to the waveform. Use an FFT subprogram to calculate the first N Fourier coefficients and then use an FFT subprogram to sum the associated finite Fourier series. Plot the finite Fourier series and compare the resulting waveform with that of the original signal.

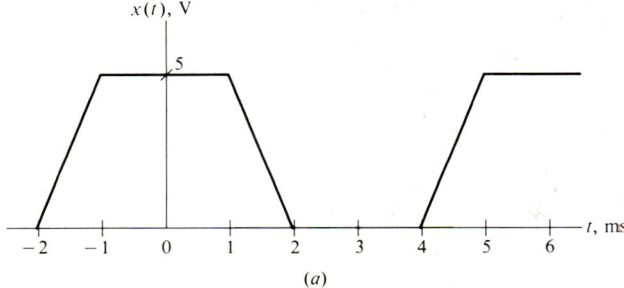

(a)

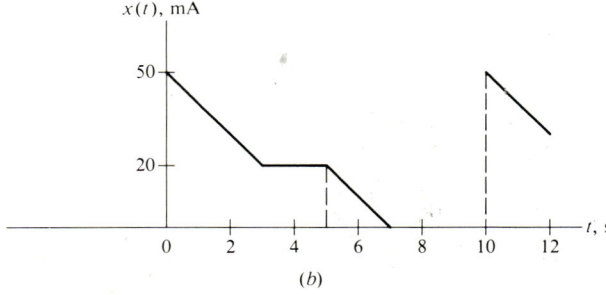

(b)

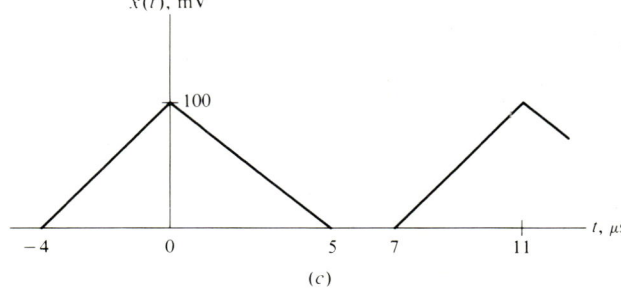

(c)

Figure P9-18

9-19 Use an FFT to calculate the Fourier transform of each signal of Fig. P9-19. Plot the magnitude and angle of the transform.

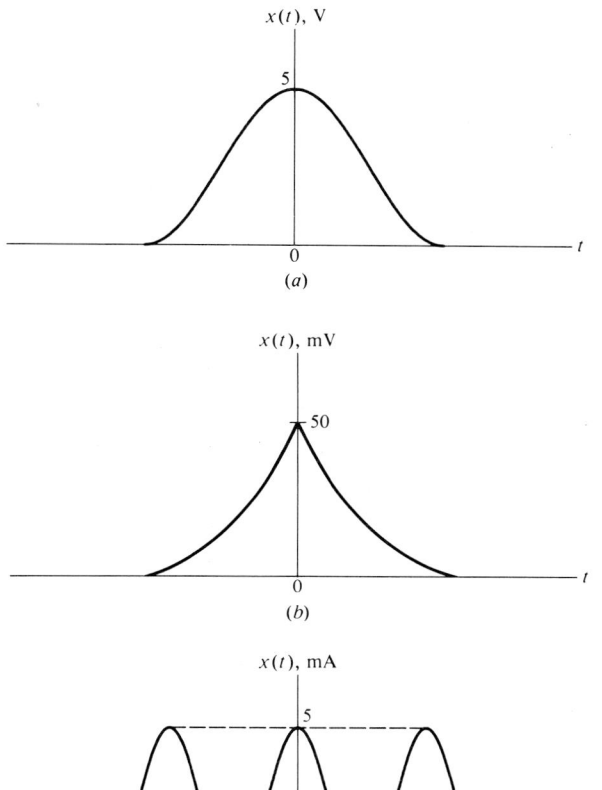

Figure P9-19 (a) $x(t) = 5e^{-(t/\tau)^2}$ V, $\tau = 1$ ms; (b) $x(t) = 50e^{-|t|/\tau}$ mV, $\tau = 100$ μs; (c) $x(t) = 5(\cos \omega_0 t) r[(t - t_0)/\tau]$, $f_0 = 1$ kHz, $\tau = 2.5$ ms, $t_0 = 1.25$ ms.

9-20 A sinusoidal signal $x(t) = 5 \cos \omega_0 t$ V is applied to the input of the system of Fig. P9-20. Obtain graphs of the amplitude spectrum and the waveform of the steady-state output $z(t)$. Calculate third- and fifth-harmonic distortion introduced by the system.

$x(t) \longrightarrow \boxed{y = y_0 \tanh\left(\frac{x}{x_0}\right)} \xrightarrow{y(t)} \boxed{\dfrac{\omega_0^2}{s^2 + 2\zeta\omega_0 s + \omega_0^2}} \longrightarrow z(t)$

Figure P9-20 $x_0 = 2.5$ V, $y_0 = 25$ V, $\omega_0 = 2\pi$ krad/s, $\zeta = 0.10$.

9-21 Use the interactive programs of Figs. 9-73 and 9-83 to obtain the inverse Laplace transforms of the following functions:

(a) $H(s) = \dfrac{s(s - z_1)^2}{(s^4 + a_3 s^3 + a_2 s^2 + a_1 s + a_0)}$

for $z_1 = -2 \text{ s}^{-1}$, $a_3 = 6 \text{ s}^{-1}$, $a_2 = 13 \text{ s}^{-2}$, $a_1 = 14 \text{ s}^{-3}$, and $a_0 = 6 \text{ s}^{-4}$.

(b) $H(s) = \dfrac{K}{s(s - p_1)(s^3 + a_2 s^2 + a_1 s + a_0)}$

for $p_1 = 500 \text{ s}^{-1}$, $a_2 = 1.80 \times 10^3 \text{ s}^{-1}$, $a_1 = 1.05 \times 10^6 \text{ s}^{-2}$, $a_0 = 2.50 \times 10^8 \text{ s}^{-3}$, and $K = 10 \text{ s}^{-5}$.

9-22 Use the interactive programs of Figs. 9-74 and 9-84 to obtain the inverse z transforms of the following functions:

(a) $H(z) = \dfrac{(z + 1)^3}{(z - 1)(z^3 + 1.60 z^2 + 1.28 z + 0.32)}$

(b) $H(z) = \dfrac{10(z + 1)}{z^4 - 1.40 z^3 - 0.10 z^2 + 0.90 z - 0.29}$

9-23 Use the interactive program of Fig. 9-85 to design a digital Butterworth filter meeting the specifications of Fig. P9-23. Then use the program of Fig. 9-83 to obtain a partial-fraction expansion for the system function of the filter and draw a block diagram representing a parallel realization of the filter.

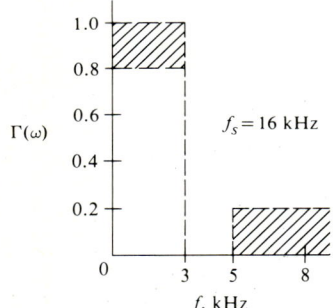

Figure P9-23

9-24 Use the program of Fig. 9-85 to design a digital Butterworth filter meeting the specifications of Fig. P9-24. Then use an FFT subprogram to obtain the output of the filter for input $x(t) = r(t/\tau)$ V, where $\tau = 1$ ms. Check the result using a cascade realization of the filter.

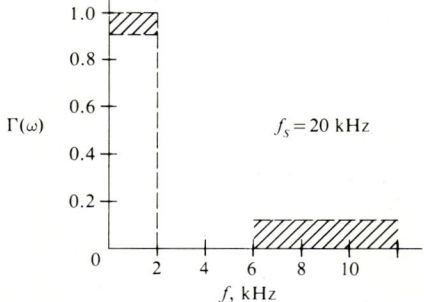

Figure P9-24

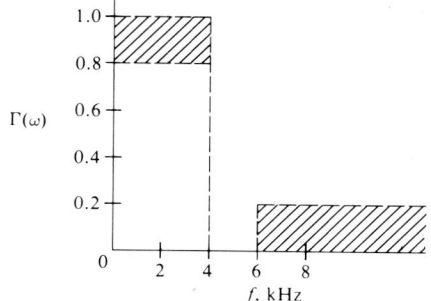

Figure P9-25

9-25 Write an interactive FORTRAN program for designing low-pass analog Butterworth filters. Use the program to design a filter meeting the specifications given in Fig. P9-25. Then use the interactive program of Fig. 9-85 to design a digital Butterworth filter meeting the same specifications with sampling rate $f_s = 16$ kHz. Explain why the order of the digital filter is smaller than that of the analog filter.

APPENDIX A

MATHEMATICAL FORMULAS

A-1 SPECIAL FUNCTIONS

α, β = real variables, n = integer variable.

$$u(\alpha) = \begin{cases} 0 & \alpha \leq 0 \\ 1 & \alpha > 0 \end{cases} \qquad \delta(\alpha) = \frac{du(\alpha)}{d\alpha}$$

$$r(\alpha) = \begin{cases} 1 & 0 < \alpha \leq 1 \\ 0 & \text{otherwise} \end{cases} \qquad \text{sa } \alpha = \begin{cases} 1 & \alpha = 0 \\ \dfrac{\sin \alpha}{\alpha} & \alpha \neq 0 \end{cases}$$

$$\text{sgn } \alpha = \begin{cases} -1 & \alpha \leq 0 \\ 1 & \alpha > 0 \end{cases} \qquad \text{int } \alpha = \text{largest integer less than } \alpha$$

$$u(n) = \begin{cases} 0 & n \leq 0 \\ 1 & n > 0 \end{cases} \qquad \Delta(n) = \begin{cases} 1 & n = 0 \\ 0 & n \neq 0 \end{cases}$$

$$\text{Tan}^{-1}(\beta, \alpha) = \begin{cases} \tan^{-1}(\beta/\alpha) & \alpha > 0 \\ 0 & \alpha = \beta = 0 \\ \pi + \tan^{-1}(\beta/\alpha) & \alpha < 0 \end{cases}$$

A-2 TRIGONOMETRIC IDENTITIES

$$\cos(-\alpha) = \cos \alpha$$

$$\sin(-\alpha) = -\sin \alpha$$

$$\cos\left(\alpha \pm \frac{\pi}{2}\right) = \mp \sin \alpha$$

$$\sin\left(\alpha \pm \frac{\pi}{2}\right) = \pm\cos\alpha$$

$$\cos(\alpha \pm \pi) = -\cos\alpha$$

$$\sin(\alpha \pm \pi) = -\sin\alpha$$

$$\cos^2\alpha + \sin^2\alpha = 1$$

$$\cos^2\alpha - \sin^2\alpha = \cos 2\alpha$$

$$\cos(\alpha \pm \beta) = \cos\alpha\cos\beta \mp \sin\alpha\sin\beta$$

$$\sin(\alpha \pm \beta) = \sin\alpha\cos\beta \pm \cos\alpha\sin\beta$$

$$\cos\alpha\cos\beta = \tfrac{1}{2}[\cos(\alpha - \beta) + \cos(\alpha + \beta)]$$

$$\sin\alpha\sin\beta = \tfrac{1}{2}[\cos(\alpha - \beta) - \cos(\alpha + \beta)]$$

$$\sin\alpha\cos\beta = \tfrac{1}{2}[\sin(\alpha - \beta) + \sin(\alpha + \beta)]$$

$$a\cos\alpha + b\sin\alpha = \sqrt{a^2 + b^2}\cos[\alpha - \mathrm{Tan}^{-1}(b, a)]$$

$$\cos^2\alpha = \tfrac{1}{2}(1 + \cos 2\alpha)$$

$$\cos^3\alpha = \tfrac{1}{4}(3\cos\alpha + \cos 3\alpha)$$

$$\cos^4\alpha = \tfrac{1}{8}(3 + 4\cos 2\alpha + \cos 4\alpha)$$

$$\cos^5\alpha = \tfrac{1}{16}(10\cos\alpha + 5\cos 3\alpha + \cos 5\alpha)$$

$$\cos^k\alpha = \sum_{n=0}^{k} \frac{2^{-k}k!\cos[(2n-k)\alpha]}{n!(k-n)!}$$

$$\sin^2\alpha = \tfrac{1}{2}(1 - \cos 2\alpha)$$

$$\sin^3\alpha = \tfrac{1}{4}(3\sin\alpha - \sin 3\alpha)$$

$$\sin^4\alpha = \tfrac{1}{8}(3 - 4\cos 2\alpha + \cos 4\alpha)$$

$$\sin^5\alpha = \tfrac{1}{16}(10\sin\alpha - 5\sin 3\alpha + \sin 5\alpha)$$

$$\sin^k\alpha = \begin{cases} \sum_{n=0}^{k} \dfrac{2^{-k}k!(-1)^{k-n+1}\cos[(2n-k)\alpha]}{n!(k-n)!} & k = 2, 6, 10, \ldots \\[1em] \sum_{n=0}^{k} \dfrac{2^{-k}k!(-1)^{k-n}\cos[(2n-k)\alpha]}{n!(k-n)!} & k = 4, 8, 12, \ldots \\[1em] \sum_{n=0}^{k} \dfrac{2^{-k}k!(-1)^{k-n}\sin[(2n-k)\alpha]}{n!(k-n)!} & k = 1, 5, 9, \ldots \\[1em] \sum_{n=0}^{k} \dfrac{2^{-k}k!(-1)^{k-n+1}\sin[(2n-k)\alpha]}{n!(k-n)!} & k = 3, 7, 11, \ldots \end{cases}$$

$$e^{j\alpha} = \cos\alpha + j\sin\alpha$$

$$\cosh\alpha = \cos j\alpha$$

$$\cos \alpha = \tfrac{1}{2}(e^{j\alpha} + e^{-j\alpha})$$

$$\sinh \alpha = -j \sin j\alpha$$

$$\sin \alpha = \frac{j}{2}(e^{-j\alpha} - e^{j\alpha})$$

$$\tanh \alpha = -j \tan j\alpha$$

A-3 DERIVATIVES

x and y denote functions of α; c denotes a constant.

$$\frac{dx^n}{d\alpha} = nx^{n-1}\frac{dx}{d\alpha} \qquad \frac{d(\cos x)}{d\alpha} = -\sin x \frac{dx}{d\alpha}$$

$$\frac{d(xy)}{d\alpha} = x\frac{dy}{d\alpha} + y\frac{dx}{d\alpha} \qquad \frac{d(\sin x)}{d\alpha} = \cos x \frac{dx}{d\alpha}$$

$$\frac{d(x/y)}{d\alpha} = \frac{1}{y^2}\left(y\frac{dx}{d\alpha} - x\frac{dy}{d\alpha}\right) \qquad \frac{d(\tan x)}{d\alpha} = \frac{1}{\cos^2 x}\frac{dx}{d\alpha}$$

$$\frac{d(\ln x)}{d\alpha} = \frac{1}{x}\frac{dx}{d\alpha} \qquad \frac{d(\cos^{-1} x)}{d\alpha} = -\frac{1}{\sqrt{1-x^2}}\frac{dx}{d\alpha}$$

$$\frac{de^x}{d\alpha} = e^x \frac{dx}{d\alpha} \qquad \frac{d(\sin^{-1} x)}{d\alpha} = \frac{1}{\sqrt{1-x^2}}\frac{dx}{d\alpha}$$

$$\frac{d(\tan^{-1} x)}{d\alpha} = \frac{1}{1+x^2}\frac{dx}{d\alpha} \qquad \frac{d(\cosh x)}{d\alpha} = \sinh x \frac{dx}{d\alpha}$$

$$\frac{d(\sinh x)}{d\alpha} = \cosh x \frac{dx}{d\alpha} \qquad \frac{d(\tanh x)}{d\alpha} = \frac{1}{\cosh^2 x}\frac{dx}{d\alpha}$$

$$\frac{du(\alpha)}{d\alpha} = \delta(\alpha) \qquad \frac{d[x(\alpha)u(\alpha)]}{d\alpha} = x(0)\delta(\alpha) + \frac{dx(\alpha)}{d\alpha}u(\alpha)$$

$$\frac{d}{d\alpha}\int_c^\alpha x(\lambda)\,d\lambda = x(\alpha) \qquad \frac{d}{d\alpha}\int_c^{y(\alpha)} x(\lambda)\,d\lambda = x[y(\alpha)]\frac{dy(\alpha)}{d\alpha}$$

A-4 ANTIDERIVATIVES

$$\int x(\alpha)\frac{dy(\alpha)}{d\alpha}\,d\alpha = x(\alpha)y(\alpha) - \int y(\alpha)\frac{dx(\alpha)}{d\alpha}\,d\alpha$$

$$\int \alpha^n\,d\alpha = \frac{\alpha^{n+1}}{n+1} \qquad n \neq -1 \qquad \int \frac{d\alpha}{\alpha} = \ln \alpha$$

$$\int e^{a\alpha}\,d\alpha = a^{-1}e^{a\alpha} \qquad \int \ln a\alpha\,d\alpha = \alpha(\ln a\alpha - 1)$$

$$\int \cos a\alpha \, d\alpha = a^{-1} \sin a\alpha \qquad \int \sin a\alpha \, d\alpha = -a^{-1} \cos a\alpha$$

$$\int \frac{d\alpha}{\alpha^2 + c^2} = c^{-1} \tan^{-1} \frac{\alpha}{c} \qquad \int \frac{d\alpha}{\sqrt{c^2 - \alpha^2}} = \sin^{-1} \frac{\alpha}{c}$$

$$\int \frac{d\alpha}{\sqrt{\alpha^2 \pm c^2}} = \ln(\alpha + \sqrt{\alpha^2 \pm c^2})$$

$$\int \frac{d\alpha}{(\alpha^2 + a^2)(\alpha^2 + b^2)} = \frac{a^{-1} \tan^{-1}(\alpha/a) - b^{-1} \tan^{-1}(\alpha/b)}{b^2 - a^2}$$

$$\int \sqrt{c^2 - \alpha^2} \, d\alpha = \frac{1}{2}\left(\alpha\sqrt{c^2 - \alpha^2} + c^2 \sin^{-1}\frac{\alpha}{c}\right)$$

$$\int \sqrt{\alpha^2 \pm c^2} \, d\alpha = \tfrac{1}{2}[\alpha\sqrt{\alpha^2 \pm c^2} \pm c^2 \ln(\alpha + \sqrt{\alpha^2 \pm c^2})]$$

$$\int \alpha \sin c\alpha \, d\alpha = c^{-2} \sin c\alpha - c^{-1}\alpha \cos c\alpha$$

$$\int \alpha \cos c\alpha \, d\alpha = c^{-2} \cos c\alpha + c^{-1}\alpha \sin c\alpha$$

$$\int \alpha e^{c\alpha} \, d\alpha = c^{-2}(c\alpha - 1)e^{c\alpha}$$

$$\int \alpha^2 e^{c\alpha} \, d\alpha = c^{-3}(c^2\alpha^2 - 2c\alpha + 2)e^{c\alpha}$$

$$\int e^{a\alpha} \cos b\alpha \, d\alpha = \frac{e^{a\alpha}}{a^2 + b^2}(a \cos b\alpha + b \sin b\alpha)$$

$$\int e^{a\alpha} \sin b\alpha \, d\alpha = \frac{e^{a\alpha}}{a^2 + b^2}(a \sin b\alpha - b \cos b\alpha)$$

A-5 INTEGRALS

$$\int_0^\infty \text{sa } \alpha \, d\alpha = \frac{\pi}{2} \qquad \int_0^\infty \text{sa}^2 \, \alpha \, d\alpha = \frac{\pi}{2}$$

$$\int_0^\infty e^{-a^2\alpha^2} \, d\alpha = \frac{\sqrt{\pi}}{2|a|}$$

$$\int_0^\infty \alpha^n e^{-a\alpha} \, d\alpha = \frac{n!}{a^{n+1}} \qquad \begin{array}{l} a > 0 \\ n = 0, 1, 2, \ldots \end{array}$$

$$\int_0^\infty \alpha^{2n} e^{-a\alpha^2} \, d\alpha = \frac{1 \cdot 3 \cdot 5 \cdots (2n - 1)}{2^{n+1} a^n} \sqrt{\frac{\pi}{a}} \qquad \begin{array}{l} a > 0 \\ n = 1, 2, 3, \ldots \end{array}$$

$$\int_a^b x(\alpha)\delta(\alpha - c)\,d\alpha = x(c)[u(b - c) - u(a - c)]$$

$$\int_{-\infty}^{\infty} x(\alpha)u(t - \alpha)\,d\alpha = \int_{-\infty}^{t} x(\alpha)\,d\alpha$$

$$\int_{-\infty}^{\infty} x(\alpha)u(\alpha)u(t - \alpha)\,d\alpha = u(t)\int_0^t x(\alpha)\,d\alpha$$

A-6 A SUMMATION FORMULA

$$\sum_{n=0}^{N-1} \alpha^n = \begin{cases} \dfrac{1 - \alpha^N}{1 - \alpha} & \alpha \neq 1 \\ N & \alpha = 1 \end{cases}$$

A-7 SERIES EXPANSIONS

$$e^\alpha = 1 + \alpha + \frac{\alpha^2}{2!} + \frac{\alpha^3}{3!} + \cdots$$

$$\ln(1 + \alpha) = \alpha - \frac{\alpha^2}{2} + \frac{\alpha^3}{3} - \frac{\alpha^4}{4} + \cdots \qquad |\alpha| < 1$$

$$\sin \alpha = \alpha - \frac{\alpha^3}{3!} + \frac{\alpha^5}{5!} - \frac{\alpha^7}{7!} + \cdots$$

$$\cos \alpha = 1 - \frac{\alpha^2}{2!} + \frac{\alpha^4}{4!} - \frac{\alpha^6}{6!} + \cdots$$

$$\tan \alpha = \alpha + \frac{\alpha^3}{3} + \frac{2\alpha^5}{15} + \frac{17\alpha^7}{315} + \cdots \qquad |\alpha| < \frac{\pi}{2}$$

$$\sin^{-1} \alpha = \alpha + \frac{\alpha^3}{6} + \frac{3}{2 \cdot 4}\frac{\alpha^5}{5} + \frac{3 \cdot 5}{2 \cdot 4 \cdot 6}\frac{\alpha^7}{7} + \cdots \qquad |\alpha| < 1$$

$$\cos^{-1} \alpha = \frac{\pi}{2} - \sin^{-1} \alpha$$

$$\tan^{-1} \alpha = \alpha - \frac{\alpha^3}{3} + \frac{\alpha^5}{5} - \frac{\alpha^7}{7} + \cdots \qquad |\alpha| < 1$$

$$\sinh \alpha = \alpha + \frac{\alpha^3}{3!} + \frac{\alpha^5}{5!} + \frac{\alpha^7}{7!} + \cdots$$

$$\cosh \alpha = 1 + \frac{\alpha^2}{2!} + \frac{\alpha^4}{4!} + \frac{\alpha^6}{6!} + \cdots$$

$$\tanh \alpha = \alpha - \frac{\alpha^3}{3} + \frac{2\alpha^5}{15} - \frac{17\alpha^7}{315} + \cdots \qquad |\alpha| < \frac{\pi}{2}$$

$$\sinh^{-1} \alpha = \alpha - \frac{\alpha^3}{6} + \frac{3}{2\cdot 4}\frac{\alpha^5}{5} - \frac{3\cdot 5}{2\cdot 4\cdot 6}\frac{\alpha^7}{7} + \cdots \qquad |\alpha| < 1$$

$$\tanh^{-1} \alpha = \alpha + \frac{\alpha^3}{3} + \frac{\alpha^5}{5} + \frac{\alpha^7}{7} + \cdots \qquad |\alpha| < 1$$

$$(1 + \alpha)^n = 1 + n\alpha + \frac{n(n-1)}{2!}\alpha^2 + \frac{n(n-1)(n-2)}{3!}\alpha^3 + \cdots \qquad |\alpha| < 1$$

APPENDIX B

FOURIER SERIES

In this appendix we give Fourier coefficients for the exponential form of the Fourier series expressed as

$$x(t) = \sum_{n=-\infty}^{\infty} X_n e^{jn\omega_0 t}$$

Thus $x(t)$ is a periodic signal having period T, $\omega_0 = 2\pi/T$ is the fundamental angular frequency of the Fourier series for $x(t)$, and X_n is the complex amplitude of the nth harmonic in the Fourier series for $x(t)$, given by

$$X_n = \frac{1}{T} \int_{t_0}^{t_0+T} x(t) e^{-jn\omega_0 t} \, dt$$

To express a Fourier series in amplitude-phase form as

$$x(t) = \sum_{n=0}^{\infty} A_n \cos(n\omega_0 t + \theta_n)$$

use the relations

$$A_0 = |X_0| \qquad \theta_0 = \angle X_0$$
$$A_n = 2|X_n| \qquad \theta_n = \angle X_n \qquad n \neq 0$$

To express a Fourier series in quadrature form as

$$x(t) = \sum_{n=0}^{\infty} (I_n \cos n\omega_0 t + Q_n \sin n\omega_0 t)$$

use the relations

$$I_0 = X_0 \qquad Q_0 = 0$$
$$I_n = 2 \operatorname{Re} X_n \qquad Q_n = -2 \operatorname{Im} X_n \qquad n > 0$$

APPENDIX B **671**

Rectangular pulse train See Fig. B-1:

$$X_n = \frac{a\tau}{T} \text{sa} \frac{n\pi\tau}{T}$$

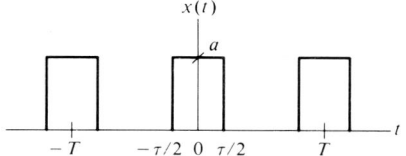

Figure B-1 Rectangular pulse train.

Triangular pulse train See Fig. B-2:

$$X_n = \frac{a\tau}{2T} \text{sa}^2 \frac{n\pi\tau}{2T}$$

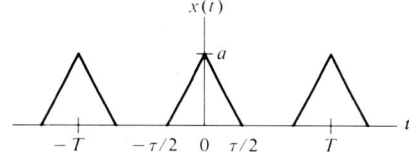

Figure B-2 Triangular pulse train.

Sawtooth pulse train See Fig. B-3:

$$X_0 = \frac{a\tau}{2T} \qquad X_n = \frac{a}{j2\pi n} e^{-jn\pi\tau/T}\left(\text{sa}\,\frac{n\pi\tau}{T} - e^{-jn\pi\tau/T}\right) \qquad n \neq 0$$

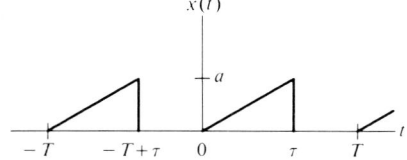

Figure B-3 Sawtooth pulse train.

Fractional sinusoid See Fig. B-4:

$$X_n = \frac{b\tau}{2T}\left[\text{sa}\,\frac{(n+1)\pi\tau}{T} + \text{sa}\,\frac{(n-1)\pi\tau}{T}\right] - \frac{(b-a)\tau}{T}\,\text{sa}\,\frac{n\pi\tau}{T}$$

where

$$b = \frac{a}{1 - \cos(\pi\tau/T)}$$

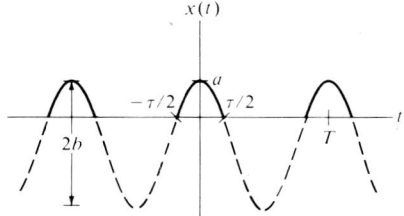

Figure B-4 Fractional sinusoid.

Impulse train See Fig. B-5:

$$X_n = \frac{a}{T}$$

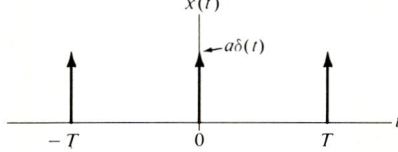

Figure B-5 Impulse train.

Half-wave-rectified sinusoid See Fig. B-6:

$$X_n = \frac{a}{4}\left[\operatorname{sa}\frac{(n+1)\pi}{2} + \operatorname{sa}\frac{(n-1)\pi}{2}\right]$$

Note that $X_n = 0$ for $n = \pm 3, \pm 5, \pm 7, \ldots$.

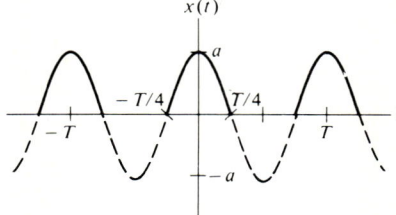

Figure B-6 Half-wave-rectified sinusoid.

Full-wave-rectified sinusoid See Fig. B-7:

$$X_n = \frac{a[1 + (-1)^n]}{4}\left[\operatorname{sa}\frac{(n+1)\pi}{2} + \operatorname{sa}\frac{(n-1)\pi}{2}\right]$$

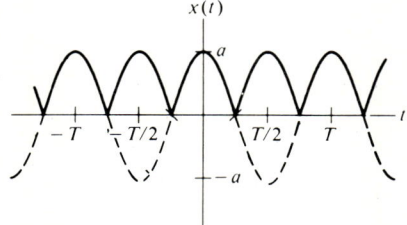

Figure B-7 Full-wave-rectified sinusoid.

Square wave See Fig. B-8:

$$X_0 = 0 \qquad X_n = \frac{a}{2}\operatorname{sa}\frac{n\pi}{2} \qquad n \neq 0$$

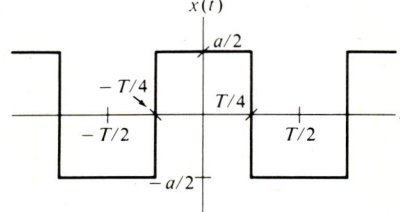

Figure B-8 Square wave.

Triangular wave See Fig. B-9:

$$X_0 = 0 \qquad X_n = \frac{a}{2} \operatorname{sa}^2 \frac{n\pi}{2} \qquad n \neq 0$$

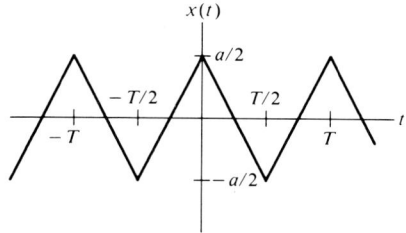

Figure B-9 Triangular wave.

Sawtooth wave See Fig. B-10:

$$X_0 = 0 \qquad X_n = \frac{ja(-1)^n}{2n\pi} \qquad n \neq 0$$

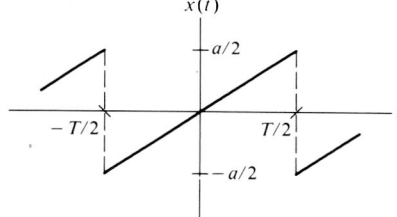

Figure B-10 Sawtooth wave.

Clipped sinusoid See Fig. B-11:

$$X_0 = 0 \qquad X_{\pm 1} = \frac{b}{2} - \frac{b\tau}{T}\left(\operatorname{sa}\frac{2\pi\tau}{T} + 1\right) + \frac{2a\tau}{T}\operatorname{sa}\frac{\pi\tau}{T}$$

$$X_n = [(-1)^n - 1]\left\{\frac{b\tau}{2T}\left[\operatorname{sa}\frac{(n+1)\pi\tau}{T} + \operatorname{sa}\frac{(n-1)\pi\tau}{T}\right] - \frac{a\tau}{T}\operatorname{sa}\frac{n\pi\tau}{T}\right\}$$

$$n = \pm 2, \pm 3, \ldots$$

where a = peak amplitude after clipping
b = peak amplitude before clipping
τ = duration of clipping = $(T/\pi)\cos^{-1}(a/b)$

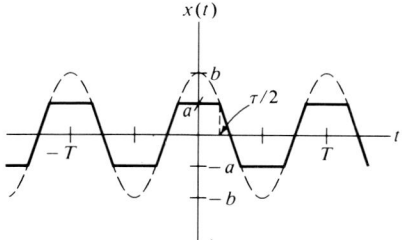

Figure B-11 Clipped sinusoid.

Exponential pulse train See Fig. B-12:

$$X_n = \frac{a(1 - e^{-T/\tau})}{T/\tau + j2\pi n}$$

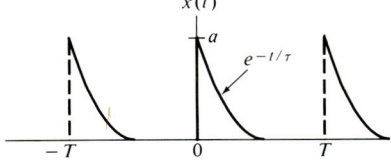

Figure B-12 Exponential pulse train.

APPENDIX C
FOURIER TRANSFORMATION

Table C-1 Transform pairs

$e(t) = \mathcal{F}^{-1}\{E(j\omega)\}$	$E(j\omega) = \mathcal{F}\{e(t)\}$
$\delta(t)$	1
$u(t)$	$\dfrac{1}{j\omega}$ (see Sec. 4-3H)
$e^{-t/\tau} u(t)$	$\dfrac{\tau}{1 + j\omega\tau}$
1	$2\pi\delta(\omega)$
$r\left(\dfrac{t}{\tau}\right)$	$\tau \operatorname{sa} \dfrac{\omega\tau}{2} e^{-j\omega\tau/2}$
$\dfrac{t}{\tau} e^{-t/\tau} u(t)$	$\dfrac{\tau}{(1 + j\omega\tau)^2}$
$\cos(\omega_0 t + \theta)$	$\pi e^{j\theta} \delta(\omega - \omega_0) + \pi e^{-j\theta} \delta(\omega + \omega_0)$
$(1 - e^{-t/\tau}) u(t)$	$\dfrac{1}{j\omega(1 + j\omega\tau)}$
$(e^{-t/\tau_1} - e^{-t/\tau_2}) u(t)$	$\dfrac{\tau_1 - \tau_2}{(1 + j\omega\tau_1)(1 + j\omega\tau_2)}$
$[1 + \beta^{-1} e^{-\zeta\omega_0 t} \sin(\beta\omega_0 t - \theta)] u(t)$ $\beta = \sqrt{1 - \zeta^2}$ $\theta = \operatorname{Tan}^{-1}(\beta, -\zeta)$	$\dfrac{1}{j\omega[(j\omega/\omega_0)^2 + 2\zeta j\omega/\omega_0 + 1]}$
$e^{-t/\tau} (\sin \omega_0 t) u(t)$	$\dfrac{\omega_0 \tau^2}{(1 + j\omega\tau)^2 + (\omega_0 \tau)^2}$
$e^{-|t|/\tau}$	$\dfrac{2\tau}{1 + (\omega\tau)^2}$
$e^{-t^2/2\tau^2}$	$\tau\sqrt{2\pi}\, e^{-(\omega\tau)^2/2}$

Table C-2 Operational properties

$e(t) = \mathscr{F}^{-1}\{E(j\omega)\}$	$E(j\omega) = \mathscr{F}\{e(t)\}$
$\displaystyle\int_{-\infty}^{\infty} e_1(\lambda)e_2(t-\lambda)\,d\lambda$	$E_1(j\omega)E_2(j\omega)$
$e(t)e^{j\omega_c t}$	$E(j\omega - j\omega_c)$
$e(t - t_0)$	$E(j\omega)e^{-j\omega t_0}$
$e(\eta t)$	$\dfrac{1}{|\eta|} E\left(\dfrac{j\omega}{\eta}\right)$
$\dfrac{d^n e(t)}{dt^n}$	$(j\omega)^n E(j\omega)$
$\displaystyle\int_{-\infty}^{t} e(\lambda)\,d\lambda$	$\dfrac{1}{j\omega} E(j\omega)$
$\displaystyle\int_{-\infty}^{\infty} |e(t)|^2\,dt$	$\dfrac{1}{2\pi} \displaystyle\int_{-\infty}^{\infty} |E(j\omega)|^2\,d\omega$
$e_1(t)e_2(t)$	$\dfrac{1}{2\pi} \displaystyle\int_{-\infty}^{\infty} E_1(j\lambda)E_2(j\omega - j\lambda)\,d\lambda$
$e(t) \displaystyle\sum_{n=-\infty}^{\infty} \delta(t - nT)$	$\dfrac{1}{T} \displaystyle\sum_{n=-\infty}^{\infty} E\left(j\omega - j\dfrac{2n\pi}{T}\right)$

APPENDIX D
LAPLACE TRANSFORMATION

Table D-1 Transform pairs

$e(t) = \mathcal{L}^{-1}\{E(s)\}$	$E(s) = \mathcal{L}\{e(t)\}$
$\delta(t)$	1
$u(t)$	$\dfrac{1}{s}$
$e^{pt}u(t)$	$\dfrac{1}{s-p}$
$r\left(\dfrac{t}{\tau}\right)$	$\dfrac{1 - e^{-s\tau}}{s}$
$\cos(\omega t - \theta)u(t)$	$\dfrac{s\cos\theta + \omega\sin\theta}{s^2 + \omega^2}$
$(e^{pt} - 1)u(t)$	$\dfrac{p}{s(s-p)}$
$t^n u(t)$	$\dfrac{n!}{s^{n+1}}$
$t^n e^{pt} u(t)$	$\dfrac{n!}{(s-p)^{n+1}}$
$e^{\sigma t}\cos(\omega t - \theta)u(t)$	$\dfrac{(s-\sigma)\cos\theta + \omega\sin\theta}{(s-\sigma)^2 + \omega^2}$

Table D-1 (continued)

$e(t) = \mathcal{L}^{-1}\{E(s)\}$	$E(s) = \mathcal{L}\{e(t)\}$
$At^{n-1}e^{\sigma t}\cos(\omega t - \theta)u(t)$ where $A = \dfrac{2\|B\|}{(n-1)!}$ $\quad \theta = \angle B$ $\sigma = \text{Re } p \quad \omega = \text{Im } p$	$\dfrac{B}{(s-p)^n} + \dfrac{B^*}{(s-p^*)^n}$ B, p complex
$ce^{-qt/2}\cos(\omega t - \theta)u(t)$ where $\omega = \sqrt{r - \left(\dfrac{q}{2}\right)^2}$ $c = \dfrac{\sqrt{rA^2 + B^2 - qAB}}{\omega}$ $\theta = \text{Tan}^{-1}\left(B - \dfrac{q}{2}A, A\omega\right)$	$\dfrac{As + B}{s^2 + qs + r}$
$\omega\beta^{-1}e^{-\zeta\omega t}(\sin\beta\omega t)u(t) \quad \beta = \sqrt{1-\zeta^2}$	$\dfrac{\omega^2}{s^2 + 2\zeta\omega s + \omega^2}$

Table D-2 Operational properties

$e(t) = \mathcal{L}^{-1}\{E(s)\}$	$E(s) = \mathcal{L}\{e(t)\}$
$e^{pt}e(t)$	$E(s-p)$
$e(t - t_0)$	$e^{-st_0}E(s)$
$\displaystyle\int_{-\infty}^{t} e(\lambda)\,d\lambda$	$\dfrac{1}{s}E(s)$
$\dfrac{d^n e(t)}{dt^n}$	$s^n E(s)$
$\displaystyle\int_{-\infty}^{\infty} e_1(\lambda)e_2(t-\lambda)\,d\lambda$	$E_1(s)E_2(s)$
$e(\eta t)$	$\dfrac{1}{\|\eta\|}E\left(\dfrac{s}{\eta}\right)$
$te(t)$	$\dfrac{-dE(s)}{ds}$

APPENDIX
E

z TRANSFORMATION

Table E-1 Transform pairs

Number	$x(n) = \mathscr{Z}^{-1}\{X(z)\}$	$X(z) = \mathscr{Z}\{x(n)\}$
1	$\Delta(n)$	1
2	$u(n)$	$\dfrac{1}{z-1}$
3	$c^n u(n)$	$\dfrac{c}{z-c}$
4	$nc^n u(n)$	$\dfrac{cz}{(z-c)^2}$
5	$n^2 c^n u(n)$	$\dfrac{cz(z+c)}{(z-c)^3}$
6	$n^3 c^n u(n)$	$\dfrac{cz(z^2 + 4cz + c^2)}{(z-c)^4}$
7	$c^n (\cos n\lambda) u(n)$	$\dfrac{c(\cos \lambda)z - c^2}{z^2 - 2c(\cos \lambda)z + c^2}$
8	$c^n (\sin n\lambda) u(n)$	$\dfrac{c(\sin \lambda)z}{z^2 - 2c(\cos \lambda)z + c^2}$
9	$c^n \cos(n\lambda + \theta) u(n)$	$\dfrac{c\cos(\lambda + \theta)z - c^2 \cos \theta}{z^2 - 2c(\cos \lambda)z + c^2}$
10	$c^{n-1} u(n)$	$\dfrac{1}{z-c}$

679

Table E-1 (continued)

Number	$x(n) = \mathscr{L}^{-1}\{X(z)\}$	$X(z) = \mathscr{L}\{x(n)\}$
11	$(n-1)c^{n-2}u(n-1)$	$\dfrac{1}{(z-c)^2}$
12	$\dfrac{(n-1)(n-2)c^{n-3}u(n-2)}{2}$	$\dfrac{1}{(z-c)^3}$
13	$\dfrac{(n-1)(n-2)(n-3)c^{n-4}u(n-3)}{6}$	$\dfrac{1}{(z-c)^4}$
14	$\dfrac{(n-1)!c^{n-m}u(n-m+1)}{(n-m)!(m-1)!}$	$\dfrac{1}{(z-c)^m}$
15	$2\|A\|\|c\|^{n-1}\cos(n\lambda+\theta)u(n)$ where $\lambda = \angle c$ $\quad \theta = \angle A - \angle c$	$\dfrac{A}{z-c} + \dfrac{A^*}{z-c^*}$

Table E-2 Operational properties

$x(n) = \mathscr{L}^{-1}\{X(z)\}$	$X(z) = \mathscr{L}\{x(n)\}$
$x(n-k)$	$z^{-k}X(z)$
$c^n x(n)$	$X(z/c)$
$\sum_{m=-\infty}^{\infty} x_1(m)x_2(n-m)$	$X_1(z)X_2(z)$

APPENDIX F

OUTLINE OF CSMP†

VARIABLE NAMES

Variable names consist of one to six letters and numbers. The first character must be a letter. All variable names denote real (not integer) variables by default. The words listed in Table F-1 have special meanings in CSMP and should be used only in their intended context. In addition, names of the form IZnnnn and ZZnnnn, where n is any integer, are reserved for system use.

Table F-1 Reserved names in CSMP

ABS	DELAY	FIND	INTRAN	OVERLAY	RST	
ABSERR	DELMAX	FINISH	IOR		SHIFT	
ADAMS	DELMIN	FINTIM	ISIGN	PARAMETER	SIGN	
AFGEN	DELT	FIXED		PAUSE	SINE	
ALPHA	DERIV	FLOAT	LABEL	PLOTR	SIMP	
AND	DIM	FORMAT	LEDLAG	PRDEL	SNGL	
	DIMENSION	FUNCTION	LIMIT	PREPARE	SORT	
BACKSPACE	DFLOAT			PRINT	SPLITR	
BOOLE	DO	GAUSS	MACRO	PROCEDURE	STEP	
BUILDER	DOUBLE	GO	MAINEX	PRTPLT	STOP	
	DSIGN	GOTO	MEMORY	PULSE	STORAGE	
CALL	DYNAMIC		METHOD		SUBROUTINE	
CENTRAL		HISTORY	MILNE	QNTZR		
CMPXPL	END	HSTRSS	MLEFT		TABLE	
COMMON	ENDDATA		MODINT	RAMP	TERMINAL	
COMPAR	ENDFILE	IABS	MRIGHT	RANGE	TIME	
COMPL	ENDJOB	IDIM		READ	TIMER	
CONSTANT	ENDMAC	IF	NALARM	REAL	TITLE	
CONTINUE	ENDPRO	IFIX	NAND	REALPL	TRAPZ	
CSTORE	EOR	IMPL	NLFGEN	RECT		
	EQUIV	IMPULS	NOR	RELERR	UND	
DABS	EQUIVALENCE	INCON	NOSORT	RENAME	UPDATE	
DATA	EXIT	INITIAL	NOT	RESET		
DBLE	EXTERNAL	INITLZ	NUMER	RETURN	WRITE	
DEADSP		INSW		REWIND		
DEBUG	F	INTGRL	OUTDEL	RKS	ZHOLD	
DECK	FCNSW	INTEGER	OUTSW	RNDGEN	ZOR	
DEFINE						

†Further details of CSMP can be found in Speckhart and Green.

F-2 SIGNAL AND SYSTEM MODELING STATEMENTS

Table F-2 Elementary systems

Element	Transfer characteristic	CSMP description
Proportion	$y(t) = Kx(t)$	Y = K * X
Delay	$y(t) = x(t - t_c)$	Y = DELAY(N, T0, X) T0 = delay time N = integer $\geq$ T0/DELT
Integration	$y(t) = K \int_{-\infty}^{t} x(\alpha)\, d\alpha$	Y = INTGRL(Y0, K * X) Y0 = $y(0)$
Differentiation	$y(t) = K \dfrac{dx(t)}{dt}$	Y = DERIV(Y0, K * X) Y0 = $y(0)$
Full-wave rectifier	$y(t) = y_0 \left\lvert \dfrac{x(t)}{x_0} \right\rvert$	Y = Y0 * ABS(X/X0)
Soft limiter	$y(t) = \begin{cases} x_{lo} & x \leq x_{lo} \\ x(t) & x_{lo} < x < x_{hi} \\ x_{hi} & x \geq x_{hi} \end{cases}$	Y = LIMIT(XLO, XHI, X)
Hard limiter	$y(t) = \begin{cases} -y_0 & x < 0 \\ y_0 & x \geq 0 \end{cases}$	Y = Y0 * SIGN(1., X)
Comparator	$y(t) = \begin{cases} 0 & x(t) < x_0 \\ y_0 & x \geq x_0 \end{cases}$	X1 = COMPAR(X, X0) Y = Y0 * X1
Half-wave rectifier	$y(t) = \begin{cases} 0 & x(t) < 0 \\ x(t) & x(t) \geq 0 \end{cases}$	Y1 = COMPAR(X, 0.) Y = X * Y1

Table F-3 Signal models

Signal	CSMP description	Parameters
$x_0 u(t - t_0)$	X0 * STEP(T0)	X0 = amplitude T0 = time origin
$x_0 \cos(\omega_0 t + \theta)$	X0 * COS(W0 * TIME + THETA)	X0 = amplitude W0 = frequency, rad/s THETA = initial phase, rad
$x_0 r\left(\dfrac{t}{\tau}\right)$	X0 * (1. - STEP(TAU))	X0 = amplitude TAU = duration
$x_0 e^{-t/\tau} u(t)$	X0 * EXP(-TIME/TAU)	X0 = amplitude TAU = time constant

Table F-4 System macros

System function	CSMP macro
$H(s) = \dfrac{1}{1 + s\tau}$	Y = REALPL(Y0, TAU, X) Y0 = y(0)
$H(s) = \dfrac{1 + s\tau_1}{1 + s\tau_2}$	Y = LEDLAG(TAU1, TAU2, X)
$H(s) = \dfrac{\omega_0^2}{s^2 + 2\zeta\omega_0 s + \omega_0^2}$	Y = CMPXPL(Y0, DY0, ZETA, W0, W02 * X) Y0 = y(0) ZETA = ζ W02 = ω_0^2 DY0 = $\dfrac{dy(t)}{dt}$ for $t = 0$

APPENDIX
G
REFERENCES

Campbell, G. A., and R. M. Foster: *Fourier Integrals for Practical Application,* Van Nostrand, New York, 1942.

Cheny, W., and D. Kinkaid: *Numerical Mathematics and Computing,* Brooks-Cole, Monterey, Calif., 1980.

Churchill, R. V.: *Fourier Series and Boundary Value Problems,* McGraw-Hill, New York, 1941.

D'Azzo, J. J., and C. H. Houpis: *Linear Control System Analysis and Design,* 2d ed., McGraw-Hill, New York, 1981.

Fitzgerald, A. E., C. Kingsley, Jr., and F. D. Uman: *Electric Machinery,* 4th ed., McGraw-Hill, New York, 1983.

Hayt, W. H., Jr., and J. E. Kemmerly: *Engineering Circuit Analysis,* 3d ed., McGraw-Hill, New York, 1983.

LePage, W. R.: *Complex Variables and the Laplace Transform for Engineers,* McGraw-Hill, New York, 1961.

Papoulis, A.: *The Fourier Integral and Its Applications,* McGraw-Hill, New York, 1962.

Speckhart, F. H., and W. L. Green: *A Guide to Using CSMP,* Prentice-Hall, Englewood Cliffs, N.J., 1976.

Stackgold, I.: *Green's Functions and Boundary Value Problems,* Wiley, New York, 1979.

System/360 Scientific Subroutine Package (360A-CM-03x) Version III Programmer's Manual, 4th ed., IBM, White Plains, N.Y., 1968.

Thomas, G. B., Jr., and R. L. Finney: *Calculus and Analytic Geometry,* 5th ed., Addison-Wesley, Reading, Mass., 1979.

Van Valkenburg, M. E.: *Network Analysis,* 3d ed., Prentice-Hall, Englewood Cliffs, N.J., 1974.

INDEX

Note: Problems are indicated by a *p* following the page numbers

ADC (*see* Analog-to-digital conversion)
Adder, 19, 421
Algebraic loops in CSMP, 599–600
Alias error, 411
   (*See also* Sampling)
All-pass biquad, 339–341
All-pass delay equalizer, 337–345
AM (amplitude modulation) radio, 153
Amplifier:
   audio, 2, 3, 128, 135
   nonlinear, 135, 187–188, 619–621
   nonstationary, 55–56
Amplitude:
   of an exponential pulse, 8
   of a rectangular pulse, 5
   of a sinusoidal signal, 8
   of a step signal, 5
Amplitude distortion, 226–234
Amplitude-phase form:
   of a dc signal, 115, 484
   of a Fourier integral, 196
   of a Fourier series, 177
   of a sinusoidal signal, 110, 479–480
Amplitude spectrum, 138–141
Analog signal processing, 392
Analog-to-digital conversion (ADC), 394–420
   electromechanical, 419–420
   error-free (ideal), 416–418
   error introduced by, 398–416
Angle operator, $110n$.

Angular frequency, 81, 110
   of a dc signal, 115, 159
Antenna, $1n$.
Anti-alias filter, 414
Aperture error, 415
Approximation:
   dominant-pole, 355–357
   in filter design, 244
   of high-order systems, 243
   of nonlinear transfer characteristics, 130–133
   of nonperiodic signals using Fourier series, 190–193
   of nonrealizable system functions, 326–327
   of signals using rectangular pulses, 58
Arbitrary constants, 81
Arithmetic element, 10
Armature-controlled dc motor, 353
Audio amplifier, 2, 3, 128, 135

Back emf (electromotive force), 353
Band occupied by a signal, 220
   (*See also* Bandwidth)
Bandpass filter (*see* Filter, bandpass)
Bandstop filter, 152
Bandwidth:
   and amplitude distortion, 230–231
   of a Butterworth filter, 248, 251
   definition of, 189–190, 220–221

Bandwidth (*Cont.*):
  of a double-sideband signal, 261
  and duration, 209–210
  effect of pole placement on, 333–334, 552–556
  of an exponential pulse, 221
  of a first-order digital filter, 494–495
  of a periodic signal, 189–190
  of a practical differentiator, 221
  of a rectangular pulse, 220
  and rise time, 235–238, 241
  of a single-sideband signal, 261
BIBO (bounded-input—bounded-output) condition for stability, 29–31, 432–433
Bilinear substitution, 557–562, 566–567
Billiard ball, 7
Biquad, all-pass, 339–341
Block-diagram reduction:
  in the frequency domain, 117–127, 486–489
  in the s domain, 308–311
  in the time domain, 67
  in the z domain, 531–533
Block diagrams, 2, 20, 459–460
  for elementary systems, 37, 124, 530
  in the frequency domain, 117
  for linear shift-invariant digital systems, 470
  for nth-order linear stationary systems, 224, 613, 615
  in the s domain, 309
  for systems described: by differential equations, 77–79, 613
    by impulse responses, 66
  in the time domain, 66–68, 421–422
  in the z domain, 530
Bounded-input—bounded-output (*see* BIBO)
Bounded signal, 29, 466–468
Boxcar filter, 405, 415
Boxcar reconstruction, 404–405
Butterworth, S., 246
Butterworth filter, 244–252, 562–567, 651–655
  block diagram for, 291p, 564–566

Butterworth filter (*Cont.*):
  characteristic roots of, 247–248
  conditions for high-fidelity transmission by, 251
  cutoff frequency of, 250
  delay introduced by, 251, 343–345
  digital, 562–567, 651–655
  equalization of, 341–345
  gain and phase shift of, 247–252
  order of, 249
  poles of, 247–248, 563
  realizability of, 290p
  specifications on, 249
  transfer function of, 247
Butterworth polynomial, 247
Butterworth power density spectrum, 413–415

Cancellation of poles, 322, 325, 546
Carrier, 154, 253
Cascade connection, 20
Causal output, 25, 431
Characteristic equation, 80, 474
Characteristic polynomial, 80, 474
Characteristic roots, 80, 475
  of a second-order system, 87
  and stability, 98
  (*See also* Poles)
Chebyshev filter, 291p
Circuit, 5–6, 84, 92, 364–365
Circuit analysis, analogy with system analysis, 1
Clipped sinusoid, 187–188
Clock, 392, 395
Coaxial cable, 45p
Comb filter, 281p
Commutativity, 45p
  of delay with stationary system, 54
  of digital systems, 459–460
  of feedforward and feedback sections of a system, 95, 615
  of linear stationary systems, 68–69
  of nonlinear and nonstationary systems, 68
  proof of, 68

INDEX **687**

COMPAR statement in CSMP, 592–593
Comparator, 12
  simulation of, 592–594
Complementary function, 80–81
Complete solution, homogeneous
    equation, 80–82, 474–477
Complex amplitude:
  dc signal, 115, 159, 484
  Fourier series, 178
  sinusoidal signal, 110, 480
  (*See also* Spectrum)
Complex frequency, 296
Complex spectrum, 141, 178
Compression, 18, 431–432
  impulse response of, 70–71
  made causal by adding delay, 27, 431–432
  nonrealizability of, 25, 26, 70–71, 431
  nonstationarity of, 54
  pulse response of, 18
  realized, using tape recorders, 46p
Compression ratio, 18
Computer-aided Fourier analysis, 616–635
Conditionally stable system, 30, 432
  (*See also* Stability)
Convergence:
  of a Fourier integral, 195
  of a Fourier series, 181–183
  of a geometric series, 73
  of a Laplace transform, 296–298
  of a Maclaurin series, 130
  of a z transform, 518–521
Convolution integral, 59–65
  derivation of, 59
  graphical interpretation of, 62–65
  implemented, using a tapped delay line, 602–606
  numerical calculation of, 606–609
  significance of, 62
  (*See also* Impulse response)
Convolution property:
  of the Fourier transformation, 206
  of the one-sided Laplace transformation, 389p

Convolution property (*Cont.*):
  significance of, 272–273
  of the two-sided Laplace transformation, 305–306
  of the z transformation, 527
Convolution sum, 455–457
  derivation of, 455
  graphical interpretation of, 457–459
  (*See also* Delta response)
Critically damped system, 90
CSMP (Continuous System Modeling Program):
  constants in, 577
  default values for DELT, OUTDEL, and PRDEL, 583
  elementary signal models, 583–584
  operators, 577–578
  procedure for using, 601–602
  program format, 578–579
  program verification, 600–601
  reserved words, 577
  statements, 577
  variable names, 577
Cue ball, 7
Cut-and-try method:
  for calculating: Fourier coefficients, 619
    Fourier transforms, 624–626
  for designing: a delay equalizer, 341–345
    a digital filter, 549–556

D operator, 610–611
DAC (*see* Digital-to-analog conversion)
Damping ratio, 108p, 238–243
  and classification of second-order systems, 351
  relationship of: to peak gain, 241
    to peak time, 242
    to percent overshoot, 242
    to rise time, 241
    to settling time, 242
Dc (direct current) gain, 115
  of a digital system, 484, 545
  of a first-order system, 84
  of a second-order system, 90

Dc (direct current) motor, 266
    armature-controlled, 293p
Dc (direct current) servomechanism:
    analysis of: using Fourier
        transformation, 264–271
      using one-sided Laplace
        transformation, 366–372
      using two-sided Laplace
        transformation, 352–359
Dc (direct current) signal:
    amplitude-phase representation of,
        115, 159, 484
    exponential representation of, 115,
        159, 484
    in input of integrator, 332
    in input of nonlinear system, 132–133
    (See also Dc gain)
Dead-band gain element, 40p, 283p
Deadband in a relay servomechanism,
    368
De Laplace, Pierre Simon, 295n.
Delay, 14, 421–422
    of analog and digital filters, 495–496
    in analog-to-digital conversion, 395
    of Butterworth filter, 251
    commutativity of, with stationary
        system, 54
    of first-order digital filter, 492–493
    in linear modulation-demodulation,
        260
    of nonrecursive digital filters, 498
    relation to phase shift, 226
    simulation of, 588–591
    used to achieve realizability, 26
Delay distortion, 226–234
Delay element, 13–14
Delay equalizer (see All-pass delay
    equalizer)
DELAY statement in CSMP, 588–589
DELMIN parameter in CSMP, 586
DELT parameter in CSMP, 582
Delta function, 6, 39p
    as a function of frequency, 198
    in impulse response: of a
        distortionless system, 73
      of a tapped delay line, 604–606

Delta function (Cont.):
    integral of, 61, 668
    in integrand of a convolution integral,
        61
    as limiting form: of an exponential
        pulse, 39p
      of a rectangular pulse, 39p, 58
      of a sampling function, 197, 218–
        219
    in spectral densities, 198, 219–220
    strength of, 6, 198
    in superposition integral, 58
    unit of, 6
Delta response, 454, 473–478
    of a digital filter, 515p
    and fidelity, 468–470
    realizability of, 465–466
    relationship of: to step response, 460–
        461
      to system function, 528
      to transfer function, 481
    and stability, 466–468
Demodulation (see Linear modulation)
De Moivre's theorem, 247
DERIV statement in CSMP, 591–592
Difference equation, 470–479
    and delta response, 473–477
    realizability of, 478
    recursive solution of, 425–426, 472–
        473
    relationship of: to system function,
        529
      to transfer function, 485–486
Differential equation, 76–99
    from block diagrams, 77–79
    homogeneous, 80n.
    and impulse response, 79–98
    for a linear stationary system, 76
      general block diagram for, 224,
        613
    and step response, 79–98
    relation to transfer function, 116,
        119, 123
Differentiator, 16–17
    bandwidth of, 221
    digital, 430–431, 502–504

Differentiator (*Cont.*):
  nonideal, 221
  pulse response of, 17
  simulation of, 591–592
Digital differentiator, 430–431, 502–504
Digital filter, 490–504, 651–655
  advantages and disadvantages of, 496, 504
  Butterworth, 562–567, 651–655
  delta response of, 515p
  design of: using bilinear substitution, 562–567, 651–655
    using cut-and-try pole selection, 553–556
    using Fourier series, 496–504
    using impulse invariance, 534–535
    using step invariance, 535–536
  direct, parallel, and series realizations of, 564–566
  first-order, compared with first-order analog filter, 494–496
  nonrecursive, 496–504
  recursive, 491–496
Digital frequency, 479–480
Digital integrator, 427–429
Digital signal, 420
Digital signal processing, 392–394
  (*See also* Digital filter)
Digital system, 420
  block diagram for, 470, 530
  delta response of, 454, 528
  difference equation for, 485–486, 529
  linear, 449–451
  realizability of, 478
  recursive and nonrecursive, 470
  shift-invariant, 451–452
  stability of, 478–479
  system function of, 527
  transfer function of, 481–483
Digital-to-analog conversion, 392–420
  electromechanical, 419–420
  error-free (ideal), 416–418
  and reconstruction error, 415–416
  (*See also* Analog-to-digital conversion)
Dimensions and units, 3–4

Diode, 11
Dirac, P. A. M., 6n.
Dirac delta, 6n.
Distortion, 31–33, 436–438
  amplitude, phase, and delay, 226–234
  in Butterworth filters, 251
  caused by inadequate bandwidth, 230
  caused by passband ripple, 230–233
  in digital systems, 437
  in a first-order low-pass system, 227–229
  harmonic, 133–134, 161, 619–621
  intermodulation, 136–138
  in linear modulators (double-sideband systems), 255, 260
  nonlinear, 133–134
  phase-intercept, 150–151, 260
  in single-sideband systems, 263
Distortionless transmission, 31–33, 436
  frequency-domain conditions for, 146–151, 225–226, 263
  in ideal filters, 152
  time-domain conditions for, 73–76, 436–438, 468–470
  (*See also* Distortion)
Dominant-pole approximation, 355–357
Dominant poles, 345–350
Double-sideband systems, 261
Doubly-symmetric function, 245–246
Duration-bandwidth relation, 209–210
Dynamic range of a signal, 403–404
Dynamic section of a CSMP program, 578–579
Dynamic system, 10, 13

Electric circuit, 5–6, 84, 92
Electromechanical AD and DA converters, 419–420
Electronic integrator, 15
Elementary signals, 4–9, 36, 420–422
  generation in CSMP, 583–584
Elementary systems, 9–20, 37, 421–424
  simulation of, in CSMP, 584–596
  symbols for, 12, 37, 124, 309, 422
END statement in CSMP, 578

Energy (*see* Normalized energy)
Energy density spectrum, 214–216
  of an exponential pulse, 216, 288*p*
  interpretation of, 215–216
  of a rectangular pulse, 215
  of a sinusoidal pulse, 287*p*
Energy signal, 217
Equalizer:
  all-pass delay, 337–345
  for a boxcar filter, 416
Error-detecting circuit, 264, 266, 353
Error-free AD and DA conversion, 416–418, 444*p*
  (*See also* Analog-to-digital conversion; Digital-to-analog conversion)
Euler, Leonhard, 89*n*.
Euler's identity, 89, 110, 480
Even harmonics, 131
Excitation, 2
Expansion, 18
  realizability of, 46*p*
  (*See also* Compression)
Exponential form:
  of a dc signal, 115, 480
  of a Fourier series, 177–178
  of a sinusoidal signal, 110, 479–480
  (*See also* Amplitude-phase form)
Exponential pulse, 8

Fast Fourier transformation, 616–618
  FORTRAN program for, 617
  periodicity property of, 618
  symmetry property of, 618
  used for calculating: Fourier coefficients, 618–621
    Fourier transforms, 624–626
    inverse Fourier transforms, 626–635
  used for summing Fourier series, 621–624
Feedback systems, 22
  compared with feedforward systems, 24
  recursive analysis of, 22, 425
  (*See also* Block-diagram reduction)
Feedforward and feedback sections of a system, 95, 615
Feedforward system, 24
Fidelity (*see* Distortion; Distortionless transmission)
Filter:
  all-pass, 337–345
  bandpass, 152, 236–238, 258–264, 553–556
  bandstop, 152
  Butterworth (*see* Butterworth filter)
  Chebyshev, 291*p*
  comb, 281*p*
  digital (*see* Digital filter)
  high-pass, 152, 236, 492–493
  ideal, 151–152
  low-pass, 152, 235–236, 491–496
Finite Fourier integral, 195
FINTIM parameter in CSMP, 578
First-order system, 83–86, 95–98, 491–493
Fixed-step integration, 584–585
Foldover error, 411
  (*See also* Sampling)
Forced response, 328–330, 543–545
  (*See also* Steady-state response; Transient response)
Forcing a system at a pole, 328, 544
FORTRAN programs (*see* Programs)
Four-quadrant inverse tangent function, 88*n*., 664
Fourier, Jean Baptiste Joseph, 175*n*.
Fourier coefficients, 177–178
  numerical methods for calculating, 618–621
  properties of, 276*p*, 283*p*
  table, 670–674
  as tap weights for nonrecursive digital filters, 499
Fourier integral, 190–200
  amplitude-phase form of, 196
  applications of, in system analysis, 199–200
  convergence of, 195
  finite, 195
  interpretation of, 196–199

INDEX **691**

Fourier integral (*Cont.*):
  as limiting form of Fourier series, 193–195
  numerical method for calculating, 626–635
  table of, 674–676
  (*See also* Fourier transform)
Fourier series, 175–192
  amplitude-phase form of, 177
  applications of, 183–189
  coefficients of, 177–178
  complex spectrum of, 178
  convergence of, 181–183
  exponential form of, 177
  finite, 182–183, 189–190
  interpretation of, 189–190
  minimum mean-squared-error property of, 283p
  of a nonperiodic signal, 190–193
  numerical methods for summing, 621–624
  operational properties of, 276p
  of a rectangular pulse train, 178–179
  table of, 670–674
Fourier-series design of nonrecursive digital filters, 499
Fourier transform:
  definition of, 201
  numerical methods for calculating, 624–626
  of a periodic signal, 213
  of a signal having a dc component, 209–211
  of a step function, 211–213
  table of, 675–676
Fourier transformation, 200–271
  analogy with logarithms, 200–201
  fast (*see* Fast Fourier transformation)
  linearity of, 203–204
  operator notation for, 202–203
  procedure for using, 204–205
  properties of, 206–211
    compression-expansion property, 208
    convolution property, 206–207
    delay property, 208

Fourier transformation, properties of (*Cont.*):
    differentiation property, 208–211
    duration-bandwidth relation, 209–210
    frequency-translation property, 207–208
    integration property, 208
    interpretation, 204–205
    linearity, 203–204
  relation to Laplace transformation, 299–302
  and spectral density, 204
  table of, 675–676
  and transfer function, 204
  using tables of, 205–206
Frequency counter, 39p, 45p
Frequency-division multiplexing, 264, 292p
Frequency domain, 143, 145, 161–162
Frequency-domain analysis, 138–159, 222–271
  flowchart for, 201
  of linear modulation, 152–159, 252–264
  of servomechanisms, 264–271
  (*See also* Fourier transformation)
Frequency doubler, 129–130
Frequency response (*see* Transfer function)
Frequency tripler, 281p
Full-wave rectifier, 12, 185–186
  simulation of, 580–581
Fundamental theorem of algebra, 80

Gain and phase shift, 113–115, 144
  of Butterworth filter, 248
  of differentiator, 16
  of digital system, 483–484
  of distortionless system, 146–147
  of first-order digital filter, 491–492
  of first-order system, 235–236, 494–496
  of linear shift-invariant system, 483–484
  of linear stationary system, 113–114, 144

Gain and phase shift (*Cont.*):
  of minimum- and nonminimum-phase systems, 335–336
  and pole-zero plots, 332–335
  of a second-order system, 236–240
  (*See also* Transfer function)
Gaussian elimination, 640–641
Gear train, 10
Geometric series, 73, 518–519
Gibbs, Josiah Willard, 182*n*.
Gibbs' oscillations:
  amplitude of, 181–182
  in finite Fourier integral, 195
  in finite Fourier series, 181–183
  in gain of a nonrecursive digital filter, 503–504
Graphical interpretation of convolution, 62–65, 457–459

Half-wave rectifier, 12
Hard limiter, 12
Harmonic distortion, 133–134, 161, 187–188
  compared with intermodulation distortion, 138
Heavily damped system, 90
Heaviside, Oliver, 314
Heaviside's formulas, 314–315
High-fidelity transmission, 31–33, 436–438
  frequency-domain conditions for, 149–151
  time-domain conditions for, 73–76, 468–470
  (*See also* Distortion; Distortionless transmission)
High-order systems, 97–98, 243–245, 345–350
High-pass filter (*see* Filter, high-pass)
Homogeneous difference equation, 474
Homogeneous differential equation, 80*n*.
Human hearing, 442*p*

Ideal ADC, 416–418
  (*See also* Analog-to-digital conversion)

Ideal DAC, 416–418
  (*See also* Digital-to-analog conversion)
Ideal filters, 151–152
Ideal integrator, 332
Impulse, 6
  (*See also* Delta function)
Impulse invariance, 534–535
Impulse response, 59
  and block diagrams, 66–68
  of a boxcar filter, 405
  and commutativity, 68
  of a compression element, 70
  of a distortionless system, 73–76
  of a first-order system, 83–86, 95–97
  interpretation of, 64–65, 69–76
  methods for finding, 79–98, 204, 306
  realizability of, 70–71
  relationship of: to step response, 65–66
    to system function, 306
    to transfer function, 111–112, 204
  of a second-order system, 87–95
  of a stable system, 71–73
  of a system described by a differential equation, 79–98
Impulse train, 404–409
Initial phase:
  of a dc signal, 115, 159
  of a sinusoidal signal, 110
Initial section of a CSMP program, 578–579
Initial values:
  in CSMP, 584, 591
  in one-sided Laplace transformation, 364
  of a step response, 81
Input, 2
Instantaneous distortion, 32, 437
Instantaneous error, 32
Int function, 249
Integrals, table of, 666–668
Integrated squared error, 32
Integrator, 14–15
  comparison of ideal and real, 167*p*, 332
  with dc input, 332

Integrator (*Cont.*):
  digital, 427–429
  with periodic input, 168*p*
  simulation of, 584–588
  step response of, 14
Interactive computer-aided analysis and design, 635–655
Intermodulation distortion, 136–138
  (*See also* Distortion)
INTGRL statement in CSMP, 584
Inverse Fourier transform, 201
  numerical methods for finding, 626–635
  table, 675–676
  (*See also* Fourier transform; Fourier transformation)
Inverse Laplace transform:
  one-sided, 360–362
  right-sided, 298–299
  table, 677–678
  two-sided, 296
  (*See also* Laplace transformation)
Inverse z transform, 517
  by long division, 525–526
  right-sided, 522
  table, 679–680
  using tables, 522–523
  (*See also* z transformation)
ISE (integrated squared error), 32

Jitter, 415
Job-control statements, 577*n*., 579

Kirchhoff's law, 16, 84, 92
Krauss, J. D., 3*n*.

Laplace transformation:
  one-sided, 359–372
    applied to initial-value problems, 364–372
    convergence of, 360
    inverse of, 360–362

Laplace transformation, one-sided (*Cont.*):
    linearity of, 360–361
    operational properties of, 362–364
    operator notation for, 359
    relation to two-sided Laplace transformation, 361–362
  two-sided, 295–359
    applied to analysis of linear stationary systems, 305–311
    convergence of, 296–298
    linearity of, 302–303
    operational properties of, 303–305
    operator notation for, 295–296
    relation to Fourier transformation, 299–302
    right-sided inverse, 298–299
    translation property, 305
    uniqueness of, 296–297
Left-sided function, 298
Length of a delay line, 603
L'Hospital's rule, 108
Lightly damped system, 90
Limit checking of a CSMP simulation, 600–601
LIMIT statement in CSMP, 593
Limiter, 12
  simulation of, 593–596
Linear distortion (*see* Distortion)
Linear modulation, 152–159, 252–264
  conditions for distortionless transmission, 255, 260
  delay introduced by, 260–261
  distortion in, 255–261
  frequency and phase errors in, 256–258, 173*p*
  frequency-domain analysis of, 152–159, 252–261
  using a subcarrier in, 258–260
Linear shift-invariant system, 453–454
Linear stationary system, 56
Linear system, 51, 449–450
Linearization of a nonlinear element, 134–135
Local oscillator, 256
Long division, 525
Low-pass filter (*see* Filter, low-pass)

Maclaurin series, 130–131
  table of, 668–669
Mean-squared amplitude, sum of
  sinusoids, 137n.
Mechanical damping, 85, 268–269
METHOD statement in CSMP, 586
Minimum-phase system, 335–337
Moderately damped system, 90
Modulated signal, 155, 253
Modulating signal, 155, 253
Modulation (*see* Linear modulation;
  Single-sideband system)
Momentum, 7
Motion picture, aliasing error in, 443p
Motor (*see* Dc motor)
Multiplier, 19, 421

Natural frequencies, 321–322
Natural modes, 321–322, 536–537,
  538–540
Negative-frequency components, 143
Newton-Raphson algorithm, 635
Newton's law, 7, 85, 266
Noncausal output, 25, 431
  (*See also* Realizability)
Nonlinear amplifier, 135, 187–188,
  619–621
Nonlinear distortion, 133–134
Nonlinear static systems:
  alias error in, 427
  linearization of, 134–135
  sinusoidal inputs, 127–138, 160–161
Non-minimum-phase system (*see*
  Minimum-phase system)
Nonrealizable system (*see* Realizability)
Nonrecursive digital system, 470, 473
  (*See also* Difference equation; Digital
  filter; Digital system)
Normalized energy, 213–221
  and bandwidth, 220–221
  of an exponential pulse, 214, 216
  interpretation of, 213, 215–216
  of a rectangular pulse, 215
  units of, 213
Normalized power, 216–220
  dc signal, 219
  interpretation of, 217

Normalized power (*Cont.*):
  periodic signal, 219
  rectangular pulse train, 219
  sinusoidal signal, 217
  units of, 217
Numerical encoding disk, 419
Numerical methods:
  for calculation: of Fourier
      coefficients, 618–621
    of Fourier transforms, 624–626
    of inverse Fourier transforms, 626–
      635
  convolution, 606–609
  inversion of z transforms, 525–526
  partial-fraction expansion, 647–651
  roots of a polynomial, 635–640
  for solution: of difference equations,
      472–473
    of systems of linear equations,
      640–647
  summation of Fourier series, 621–624

Odd harmonics, 131
Offline signal processing, 431
One-sided Laplace transformation (*see*
  Laplace transformation, one-sided)
One-sided spectral density (*see* Spectral
  density)
One-sided spectrum, 138–141
Operational amplifier, 15, 134, 168p
Order:
  of a difference equation, 470
  of a differential equation, 76
Oscilloscope, 14, 192
OUTDEL parameter in CSMP, 582
Output, 2
Overdamped system, 90
Overload error, 402

Paired echoes, 232
  (*See also* Distortion)
Paley-Wiener condition, 224–225, 289p
  applied to Butterworth filter, 290p
Parallel connection, 21
  (*See also* Block-diagram reduction)
PARAMETER statement in CSMP, 582

INDEX **695**

Parseval, Marc-Antoine, 219*n*.
Parseval's theorem, 219, 287*p*
Partial-fraction expansion, 212*n*., 311–320, 524
   FORTRAN program for, 647–651
Partial integration, 304*n*.
Passband edge, 249
Passband ripple, 230–232
   (*See also* Distortion; Paired echoes)
Peak amplitude (*see* Amplitude)
Peak gain, relation to damping ratio, 241
Peak time, 238–243
   relation to damping ratio, 242
Percent overshoot:
   relation to damping ratio, 242
   in step response of a servomechanism, 267
Phase detector, 657*p*
Phase distortion, 226–234
Phase-intercept distortion, 150–151, 260
   (*See also* Distortion)
Phase shift, 114, 144, 160
   of digital system, 483–484
   of first-order digital filter, 492
   of nonrecursive digital filter, 498
   relation to delay, 226
   (*See also* Gain and phase shift; Transfer function)
Phase-shift oscillator, 658*p*
Phase spectrum, 138–139, 161
   input-output relation for, 144
   (*See also* Complex spectrum; Spectral density; Spectrum)
Piecewise-stationary systems, linear, 365–372
Plancherel's theorem, 287*p*
Pole-zero diagram, 320–321, 536–537
   and frequency response, 332–335, 549–556
   (See also *s* plane; *z* plane)
Poles, 320–321
   cancellation of, 322, 325, 546
   dominant, 345–350
   forcing a system at, 328, 544
   relation to natural modes, 322–323
Polynomial transfer characteristics, 128–130, 160

Positive-frequency components, 143
Power (*see* Normalized power)
Power density spectrum, 217–218
   and alias error, 413
   of dc signal, 218–219
   of exponential pulse train, 220
   interpretation of, 218–220
   of rectangular pulse train, 288*p*
Power signal, 217
PRDEL parameter in CSMP, 582
Prefixes, 4
PRINT statement in CSMP, 579
Programs:
   CGAUSS (system of linear equations), 644
   for digital Butterworth filter design, 652
   FFT (fast Fourier transformation), 617
   GAUSS (system of linear equations), 641–642
   PARF (partial-fraction expansion), 647–650
   POLRT (roots of a polynomial), 636–638
Propagation delay, 14, 659*p*
Proportion element, 10, 421
   with time-varying gain, 55
PRTPLT statement in CSMP, 579
Public-address system, 29, 47*p*, 659*p*
Pull-in current, 390*p*
Pulse generator, 192
Pulse response, 66, 462–463
   of first-order system, 235–236
   of minimum- and nonminimum-phase systems, 337
   of tapped delay line, 605, 608

Quantizer, 400–404
   (*See also* Analog-to-digital conversion; Digital-to-analog conversion)
Quantizing error, 400–404

Radar, 14, 26, 46*p*, 47*p*
Ramp response, 358–359

Rayleigh, Lord (John William Strutt), 214n.
Rayleigh's theorem, 214, 287p
RC (resistance capacitance) circuit, 84–85
Realizability, 25–28, 431–432
　of compression, 25–28
　of a delta response, 465–466
　of a difference equation, 478
　in filter design, 245
　of an impulse response, 70–71
　in offline (non-real-time) signal processing, 431–432
　of a system function, 325–328, 542–543
　of a transfer function, 223–225
　using delay to achieve, 26–27, 326–327, 432
Reconstruction error, 415–416
　(See also Digital-to-analog conversion)
Rectangular function, 5
Rectangular pulse, 5, 421
　derivative of, 17
　as sum of step signals, 17
Rectifier, 12
　simulation of, 580–581
Recursion, 22–24, 472–473
Recursive digital filter, 491–496
Recursive digital system, 470, 473
Region of convergence:
　for one-sided Laplace transform, 360
　for two-sided Laplace transform, 296–298
　for z transform, 518–521
Relay, 367, 390p
Relay servomechanism, 366–372
Resonant frequency, 238–243
Response, 2
Right-sided function, 298
Rise time, 235, 434–436
　relationship of: to bandwidth, 235–238, 241
　　to damping ratio, 241
　　to loop gain, 435
RLC circuit, 92

RMS (root-mean-square) amplitude, sum of sinusoids, 137
Root-locus diagram, 350–352, 357, 541–542
Roots:
　of a Butterworth polynomial, 247
　characteristic (see Characteristic roots)
　of a polynomial, program for finding, 636–638
Runge-Kutta integration, 586

s plane, 296
Sag time, 236–238
Sample-and-hold, 394–398
Samples, 420
Sampling, 404–415
　error introduced by, 410–415
　　(See also Alias error; Aperture error; Jitter)
　ideal, 404–405
　interval, 406
　rate, 406
　　in digital filters, 499
　　in nonlinear digital systems, 427
　theorem, 406–409
　(See also Analog-to-digital conversion; Digital-to-analog conversion)
Sampling function, 179, 197
Scrambler, 292p
Second-order system:
　effect of bilinear substitution on, 560–562
　importance of, 98
　impulse response of, 87–95
　　step response and, table, 90
　root-locus diagram for, 351–352, 357, 541–542
　simulation of, 613–615
　step response of, 87–95, 236–238
　　impulse response and, table, 90
　transfer function of, 236, 238
　transient response of, 236–243
Semiconductor diode, 11

INDEX **697**

Sensitivity, 33–34, 438–440
Series connection, 20
  (*See also* Block-diagram reduction)
Servomechanism, 174*p*, 264–271, 352–359, 366–372
  tachometer feedback in, 269–271, 293*p*
  viscous damping in, 268–269
  (*See also* Dc servomechanism; Relay servomechanism)
Settling time, relation to damping ratio, 242
Shift element, 421, 487
Shift invariance, 451–453
  compared with time invariance, 463–465
Shift property of z transformation, 526
Shift register, 589
SI (Système Internationale d'Unités):
  prefixes, 4
  units, 3
Sideband filter, 261
Sidebands, 261
Sign function, 285*p*
Signal, 1, 28
  speech (*see* Speech signal)
  step (*see* Step signal)
Signal-to-alias-error ratio, 412–415
Signal-to-quantizing-error ratio, 403
Signal-transmission system, 152–153
  double-sideband, 252–261
  frequency-division multiplex, 264
  single-sideband, 261–263
Simulation:
  of feedback systems, 596–600
  using CSMP, 576–606
    of a comparator, 592–594
    of a differentiator, 591–592
    of feedback systems, 596–600
    of a full-wave rectifier, 580–581
    of an integrator, 586–588
    of a limiter, 593–596
    of linear stationary systems, 602–616
    of a second-order system, 613–615
    of static elements, 592–596

Simulation, using CSMP (*Cont.*):
    of systems described by differential equations, 609–616
    of a tapped delay line, 602–606
Single-sideband system, 173*p*, 261–263
  advantages of, 263
  distortion in, 263
  effects of frequency and phase errors in, 292*p*
  (*See also* Linear modulation)
Sinusoidal signal, 7, 110, 479–480
Sinusoidal step, 329
Soft limiter, 12
  simulation of, 593–596
Sonar, 14, 26, 46*p*, 442*p*
Spectral density, 193
  of a dc signal, 196–199
  delta functions in, 198
  interpretation of, 196–199
  of a modulated signal, 253
  numerical calculation of, 624–626
  of an output of a linear stationary system, 199
  of a rectangular pulse, 196–197
  of a sampled signal, 408–409
  of a sign function, 285*p*
  of a step signal, 211–213
  unit of, 198
  (*See also* Energy density spectrum; Power density spectrum; Spectrum)
Spectrum, 141, 161
  amplitude and phase, 138–141
  complex, 141, 178
  of a dc signal, 198–199
  input-output relation for, 144
  of a modulated signal, 155–158
  numerical calculation of, 618–621
  one-sided, 138–141
  of an output of a linear stationary system, 199
  of a periodic signal, 178
  of a rectangular pulse train, 180, 193–195
  translation of, 152–159
  of a triangular wave, 189

Spectrum (*Cont.*):
  two-sided, 141, 178
  (*See also* Energy density spectrum;
    Power density spectrum; Spectral
    density)
Speech signal, 403, 414
  insensitivity to phase distortion, 226
Speed of light, $45p$, $46p$, $47p$
Speed of sound, $46p$, $49p$
Square-law rectifier, 12
Stability, 29–31, 432–433
  of dominant-pole approximations, 357
  of filters, 245
  frequency-domain conditions for,
    222–223
  s-domain conditions for, 322–325
  time-domain conditions for, 71–73,
    98–99, 466–468, 478–479
  z-domain conditions for, 537–542
Stagecoach, $443p$
Static system, 9–13, 421
  simulation of, 592–596
  table of, 12
  transfer characteristic of, 11
Stationary system, 54
Steady-state component of a signal, 28
Steady-state response, 28–29, 330–332
  digital system, 433–436, 545–547
  relation to forced response, 331, 546–547
  (*See also* Forced response; Transient
    response; Unforced response)
Step function, unit, 4, 420
Step invariance, 535–536
Step response, 4, 65–98, 460–462
  continuity of, 81
  of a dc servomechanism, 267–271
  first-order system, 83–86, 95–97,
    235–236
  general form of, 80
  initial values of, 81
  relationship of: to delta response,
    460–461
    to frequency response, 235–243
    to impulse response, 65, 66, 79
  second-order system, 87–95, 236–243

Step response (*Cont.*):
  specification of, 97
  in superposition integral, $104p$
  of a system described by a differential
    equation, 79–97
Step signal, 4–5, 420–421
  in a CSMP simulation, 583
  Fourier transform of, 211–213
  integral of, 14
  (*See also* Unit step function)
Stepper motor, 419
STOP statement in CSMP, 578
Stopband edge, 249
Strength of an impulse, 6
  in the frequency domain, 198
Subcarrier, 258
Subroutine (*see* Programs)
Sum of squared errors, 437
Superposition, 51–53, 56, 449–451, 453
  sinusoidally excited system, 127,
    160, 489–490
Superposition integral, 58
Superposition sum, 455
Synthesis, 244
System, 2
  of linear equations, numerical method
    for solving, 640–647
System analysis, 1–2
  analogy with circuit analysis, 1
  importance of, 98
  objective of, 2
System function, 305–306, 527
  of elementary systems, 309, 530
  interpretation of, 320–359, 536–556
  poles and zeros of, 321–322, 536–537
  procedure for using, 306, 528
  realizability of, 325–328, 542–543
  relationship of: to delta response, 528
    to frequency response, 332–335,
      547–556
    to transfer function, 306, 332–335,
      547–556
    to transient and steady-state
      response, 330–332, 545–547
    to unforced and forced response,
      328–330

System function (*Cont.*):
  and stability, 322–325, 537–542
  of a system described by:
    a block diagram, 308–311, 530–533
    a difference equation, 529
    a differential equation, 307–308

Tables:
  of antiderivatives, 666–667
  of CSMP functions and reserved words, 681–683
  of derivatives, 666
  of elementary signals, 36, 421
  of elementary systems, 12, 37, 422
  of Fourier coefficients, 670–674
  of Fourier transforms, 675–676
  of integrals, 667–668
  of Laplace transforms, 677–678
  of nonlinear static elements, 12
  of references, 684
  of series expansions, 668–669
  of SI dimensions and units, 3
  of SI prefixes, 4
  of special functions, 664
  of step response and impulse response of a second-order system, 90
  of trigonometric identities, 664–666
  of z transforms, 679–680
Tachometer, 269–271, 293p
Tandem connection, 20
Tap spacing, 603
Tap weights, 497, 603
Tape recorder, 27, 46p
Tapped delay line, 602–606
Taps, 496
Taylor expansion, 131–133, 168p
  and realizability, 81
Terminal section of a CSMP program, 578–579
Third-harmonic distortion, 133–134, 619–621
  (*See also* Distortion)
Third-order system, 97
Time constant, 8

Time domain, 143, 145, 161–162
Time invariance, 54–56
  compared to shift invariance, 463–465
TIME variable in CSMP, 577
Time-varying gain, 55
TIMER statement in CSMP, 578, 582
TITLE statement in CSMP, 578
Torque, 85
Total differential, 33n.
Transfer characteristic, 9, 422
  of a distortionless system, 31
  of feedforward and feedback systems, 24
  and impulse response, 62
  of a nonlinear static system, 127–138
  of a realizable system, 25
  (*See also* Convolution integral; Convolution sum; System function; Transfer function)
Transfer function, 111, 159, 481–483
  and dc gain, 115
  of a distortionless system, 226
  effect of poles and zeros on, 332–335, 547–556
  of elementary systems, 124, 487
  existence of, 112, 119, 481
  of an ideal integrator, 211–213
  methods for finding, 159
  of a nonrecursive digital filter, 497
  procedure for using, 112, 482
  relationship of: to delta response, 481
    to gain and phase shift, 113–114, 160
    to impulse response, 111, 119
    to system function, 332–335, 547–556
  and stability, 119
  of a system described by: a block diagram, 117–127, 486–489
    a difference equation, 485–486
    a differential equation, 116, 119
  unit of, 111–112
  (*See also* System function)
Transformations, 200–201
  (*See also* Fourier transformation; Laplace transformation; z transformation)

Transient component of a signal, 28
Transient response, 28, 433–436
   of a dc servomechanism, 267–271
   effect of poles and zeros on, 321–322, 536–537
   first-order system, 83–86, 95–97, 235–236
   relationship of: to frequency response, 235–243
     to system function, 330–332, 545–547
     to unforced response, 331, 546–547
   of a second-order system, 87–95, 237–243
   (*See also* Impulse response; Pulse response; Step response; Unforced response)
Transition band, 249
Transmission line, 1n.
Trapezoidal rule, 584–586
Triangle inequality, 71n.
TV signal, 403, 442p, 443p
Two-sided function, 298
Two-sided spectrum, 141, 178

Undamped natural frequency, 108p, 238–243
Undamped system, 90
Underdamped system, 90, 238–243
   (*See also* Second-order system)
Undersampling, 411
Unforced response, 328–330, 543–544
   (*See also* Forced response; Transient response)

Unit circle, 537, 558
Unit delta, 420
Unit exponential pulse, 420
Unit-gain quantizer, 401–402
Unit impulse, 6n.
   (*See also* Delta function)
Unit rectangular pulse, 420
Unit sinusoid, 420
Unit step function, 4, 420
Units, 3–4
Unstable system (*see* Stability)

Variable-step integration, 585–588
Viscous damping, 85, 268–269
Voltage-controlled oscillator, 657p

z operator, 516
z plane, 517–518, 536–537
z transformation, 516–575
   applied to linear shift-invariant systems, 528–534
   linearity of, 523
   operational properties of, 526–527
   table of, 679–680
   uniqueness and convergence of, 518–521
   using tables of, 522–523
Zeros, 320–321, 536–537
   and dominant-pole approximations, 338–348
   and gain, 333–334
   and phase shift, 335–337
   (*See also* Poles)